Groundwater Assessment, Modeling, and Management

Groundwater Assessment, Modeling, and Management

Editors

M. Thangarajan
Vijay P. Singh

CRC Press
Taylor & Francis Group
Boca Raton   London   New York

CRC Press is an imprint of the
Taylor & Francis Group, an **informa** business

CRC Press
Taylor & Francis Group
6000 Broken Sound Parkway NW, Suite 300
Boca Raton, FL 33487-2742

First issued in paperback 2020

© 2016 by Taylor & Francis Group, LLC
CRC Press is an imprint of Taylor & Francis Group, an Informa business

No claim to original U.S. Government works

ISBN 13: 978-0-367-57469-7 (pbk)
ISBN 13: 978-1-4987-4284-9 (hbk)

This book is dedicated to the service rendered by all Global Groundwater Scientists (living and nonliving)

for the advancement of groundwater research and development in the past five decades.

Contents

Section IV   Transport Modeling

Section V   Pollution and Remediation

Section VI Management of Water Resources and the Impact of Climate Change on Groundwater

Foreword

Groundwater resources in arid and semiarid regions with limited renewable potential have to be managed judiciously in order to ensure adequate supplies of dependable-quality water now and in the future. Groundwater is a limited natural resource with economic, strategic, and environmental value, and is under stress due to climate change and anthropogenic influences. Therefore, management strategies ought to be aimed at the sustenance of this limited resource. In India, and also elsewhere in the world, major parts of the semiarid regions are characterized by hard rocks and it is of vital importance to understand the nature of the aquifer systems in these regions and their current stress conditions.

The ever-increasing demand for water resources has resulted in abstracting more water from the subsurface stratum than is being replenished, which, in turn, has forced the groundwater level to decline continuously. In the last four decades, both in India and elsewhere, there has been an uninterrupted decline of water level that has induced farmers to deepen their wells without knowing where the groundwater is and how much there is if it is there, and as a result they have fallen into a debt trap. Or they have been pumping groundwater indiscriminately, assuming that it will be there forever, regardless of pumping. The result has been that many areas in India are witnessing groundwater levels falling by as much as 20–50 m. This level of groundwater extraction is completely unacceptable and must be checked or else there is the potential risk of running out of groundwater and threatening the very existence of the people who solely depend on it. Therefore, policies for optimal management of water resources must be developed, which, in turn, would entail exploration, assessment, modeling, development, and management. Such policies, unfortunately, are not prevalent in India.

Exploration of groundwater by locating potential borehole sites in soft and hard rock regions is the first most important task, which can be tackled through electrical imaging techniques as well as through remote sensing and geographical information systems techniques. In order to understand the behavior of groundwater flow, the next step is to develop the knowledge of mechanisms as well as aquifer parameters that govern the movement of subsurface flow and its storage. In hard rock aquifers,

it is often noticed that the failure of boreholes is due to either the blocking of fracture through which the water moves or the fracture flow being interrupted by someone else. This can be rectified by applying hydraulic fracturing techniques to enhance fracture flow.

Recharge estimation through isotope techniques is very useful to quantify groundwater resources. As it is, groundwater is becoming increasingly limited, and its unchecked pollution is threatening its use. Once polluted, it is very difficult to clean groundwater or the cost of remediation and cleaning would be prohibitive. Therefore, appropriate mechanisms should be adopted either to prevent or mitigate pollution to start with or to clean the polluted aquifers.

The impact of climate change on the quantity and quality of water resources will be a constraint in evolving a water management program and that would impact crops and their patterns. Groundwater flow and transport models have become potential tools for simulating flow and contaminant movement in saturated and unsaturated media and hence for studying aquifer response for various input/output stresses. With the use of these models, one can evolve optimal management policies as well as predevelopment schemes.

The migration of people from one region to another has a great impact on the availability of water resources and the distribution thereof. The water manager must consider all of these issues and aspects in implementing management schemes. Further, one should integrate all of these issues for developing an effective understanding of groundwater systems and managing groundwater resources.

The book *Groundwater Assessment, Modeling, and Management*, edited by Dr. M. Thangarajan, retired scientist-G, NGRI, Hyderabad, India, and Professor Vijay P. Singh, Department of Biological and Agricultural Engineering and Zachry Department of Civil Engineering, Texas A&M University, USA, addresses many of the above aspects of groundwater. I am confident that this book will be very informative for those who are engaged in groundwater studies not only in India but also in other parts of the world. I wish to congratulate both the editors for bringing out this book.

M.S. Swaminathan

Preface

Water resources (surface and groundwater) are limited in arid and semi-arid regions and must be managed judiciously in order to ensure dependable supplies. Groundwater is a natural resource with economic, strategic, and environmental value, which is under stress due to changing climatic and anthropogenic factors. Therefore, management strategies need to be aimed at sustaining this limited resource. In the Indian subcontinent, and elsewhere in the world, major parts of the semi-arid regions are characterized by hard rocks and alluvial aquifers and it is of vital importance to understand the nature of the aquifer systems and their current stress conditions.

Extensive water level monitoring over the last few decades in many parts of the world has provided clear evidence of a long-term water-level decline, as a result of increased groundwater abstraction as well as reduced recharge. This has resulted in the deterioration of water quality and the widespread drying-up of wells. Deepening of wells does not usually provide a new source of water. Consequently, various research investigations have been carried out for prospecting, assessment, and management of groundwater resources and pollution. The development of electrical resistivity and electromagnetic devices, such as transient electromagnetic (TEM), very low frequency (VLF) magnetic techniques, to identify potential bore holes in hard rock regions as well as to delineate contaminant zones by making use of the electrical tomography is one of the advances in groundwater prospecting. At the same time, applications of remote sensing (RS) and geographical information systems (GIS) have started to play a vital role in the assessment of groundwater resources and water quality. Numerical simulation of groundwater flow and mass transport has become a very handy research tool for studying the aquifer response to various input/output stresses, which, in turn, is used to evaluate management options of groundwater resources and pollution. Pioneering research work is underway in the quantification of soil moisture movement and nutrient migration in the vadoze zone and sea water intrusion studies for coastal aquifer development. There is a renewed attempt to use nuclear magnetic resonance (NMR) magnetometers to directly identify groundwater zones.

The achievements of the past 50 years through scientific developments in exploration, assessment, modeling, and management are commendable; this can be considered a golden age of groundwater development and management. These achievements have prompted the preparation of this book containing invited papers from various international experts belonging to developed and developing countries.

Though a good number of books have been published in the last few decades dealing with groundwater resource evaluation, modeling, and management, they are focused either on theory or theory with some case studies of specific geographical areas. It was therefore decided to bring out a textbook with chapters from various scientists of the past decades and young dynamic academicians/scientists of the present from well-known academic and research institutions from all over the world.

We have taken advice from Professor M. S. Swaminathan, founder, Swaminathan Research Foundation, Chennai, India, Dr. Jim Mercer and Dr. L. F. Konikow, retired hydrologists, USGS, Reston, Virginia, and Professor Ghislain de Marsily, retired professor, School of Mines, Paris, France, while finalizing the various topics for the book. There are contributors from the United Kingdom, the United States, France, Sweden, Australia, Japan, South Korea, Brazil (South America), South Africa, and India. The book contains 32 chapters of various topics from groundwater scientists of developed and developing countries. There are about 30 case studies (real-life field problems) and 6 hypothetical studies from various countries on various themes (India—8, Africa—6, United States—4, Caribbean Islands—3, UK—1, South Korea—1, Egypt—1, Bangladesh—1, Japan—1, Indonesia—1, and Brazil—60 basins under three tectonic regions). The topics have been divided under six themes as given below:

Section I   *Groundwater Resources and Assessment*

Section II   *Groundwater Exploration*

Section III *Flow Modeling*

Section IV   *Transport Modeling*

Section V   *Pollution and Remediation*

Section VI   *Management of Water Resources and the Impact of Climate Change on Groundwater*

There are three chapters in Section I. The first chapter brings out the salient aspects of groundwater resources development and management in coastal aquifers in Korea, Egypt, and Bangladesh. The second chapter considers the deep Gangetic alluvial aquifer sequences and developments in India. In the third chapter, the current status of water resources development and management in 60 major basins, which fall in four major plate tectonic regions of various rock types that cut across Brazil (South America) have been brought out.

In Groundwater Exploration (Section II), there are three chapters of which the first two chapters consider the identification of potential bore holes in the hard rock regions through the electrical resistivity method and also delineation of fresh and salt water transition zone in a coastal aquifer using electrical imaging (tomography) technique. The third chapter describes how one can identify groundwater potential and contaminant zones by applying RS and GIS methods.

There are seven chapters in Section III (Groundwater Flow Modeling). The first two chapters provide methodologies to identify aquifer parameters, estimate stream conductance, and characterize the aquifer system. Chapter 10 provides details of data requirement and the procedure to conceptualize the groundwater flow system with a number of special case studies, providing caution on how to provide initial and boundary conditions effectively to model the groundwater flow system. Chapter 11 mainly focuses on the application of simulation optimization models for various groundwater management problems. A new approach for groundwater modeling based on connections (network theory) is presented in Chapter 12. It is an innovative technique without making use of meshes or nodes in solving groundwater management problems. A real-life simulation in Boro River Valley (Okavango Delta, Botswana, Southern Africa) to evolve optimal well field locations and pumping rates is included in Chapter 13. Parameter uncertainties in hard rock aquifers using the theory of regionalized variables (geostatistical modeling) are explored in Chapter 14, wherein both forward linear (kriging) and inverse modeling (pilot-points method) techniques have been applied in the Uppar-Awash Volcanic aquifer (hard rock) in Ethiopia (Eastern Rift). This study indicates how the parameter uncertainty can be minimized by applying geostatistical modeling.

In Transport Modeling (Section IV), there are five chapters that focus mainly on modeling aspects of various solution techniques of mass transport except in Chapter 18, which considers modeling of radionuclide migration. Transform techniques in solving mass transport equations are outlined in Chapter 15. Chapter 16 throws light on the methodology of precise hydrogeological facies modeling for vertical 2-D and 3-D groundwater flow, mass transport and heat transport (simulations in Vietnam and Japan). An extensive theory and practical approach for solving mass transport equations for practitioners are given in Chapter 17. Radionuclide migration studies through mass transport modeling are presented in Chapter 18, with a case study in India to identify a site for disposing the nuclear wastes.

Pollution and remediation aspects are covered in Section V, which contains six chapters. The first chapter describes the quantification of ionic exchange process in a sandstone aquifer in the UK, while Chapters 21 and 22 provide the theory of evolution of arsenic and fluoride contamination process and mobilization in the Gangetic plains of India and Bangladesh. Skimming of freshwater from a saline-fresh aquifer zone by using compound wells is given in Chapter 23 and the desalination of the contaminated water by the application of an electrochemical process is in Chapter 24. Chapter 25 describes fashionable techniques for the assessment of groundwater potential, pollution, prevention, and remedial measures.

Section VI of this book mainly deals with the management of water resources and the impact of climate change on groundwater potential, pollution, and agro-products. There are seven chapters in this group. The first one covers the management of groundwater resources in a complex environment: challenges and opportunities in Kenya. The second highlights water availability and food security—the impact on people's movement and migration in (Somalia and Ethopia) Sub-Saharan Africa (SSA). Utilization of groundwater for agriculture with respect to an Indian scenario is presented in Chapter 28. Chapter 29 describes the application of GIS tools for effective groundwater management. The estimation of recharge through natural and artificial isotopes is given in Chapter 30. Chapters 31 and 32 focus on the impact of climate change on groundwater resources with a case study in Cauvery River basin in Tamil Nadu, India.

The uniqueness of the book is the thoughtful selection of chapters covering all aspects of groundwater resource assessment, modeling, and management with modern theory and its practical application. Contributors have been selected based on their expertise, affiliation, and connectivity with the editors.

The editors believe that the present volume will cater to the needs of water engineers, geohydrologists, geoscientists, environmental scientists, and agricultural scientists to enhance their knowledge of both theory and application of the theory to real-life problems covering various types of rocks, namely, hard (crystalline, basaltic and lime stone) and soft (alluvial and sandstone) rocks and aquifers, such as confined, unconfined, multilayer with leaky condition, artesian, and coastal aquifers, and thereby evolve appropriate management policies through assessment and modeling.

Professor Ken Rushton, Retired Professor, Department of Civil Engineering (Birmingham and Cranfield Universities), United Kingdom, is thanked for his valuable inputs during the finalization of the themes for this book as well as providing valuable suggestions to improve the quality of the book.

We thank all contributors who helped in bringing out this book and also thank CRC Press (a unit of Taylor & Francis Group) for publishing this book.

M. Thangarajan
Vijay P. Singh

MATLAB® and Simulink® are registered trademarks of The MathWorks, Inc. For product information, please contact:

The MathWorks, Inc.
3 Apple Hill Drive
Natick, MA 01760-2098 USA
Tel: 508 647 7000
Fax: 508-647-7001
E-mail: info@mathworks.com
Web: www.mathworks.com

Editors

M. Thangarajan, formerly scientist-G and head of the groundwater modeling group at National Geophysical Research Institute (NGRI), Hyderabad, India, obtained his BS and MS in physics at Madras University, Tamil Nadu, India and PhD in applied geophysics at Indian School of Mines (ISM), Dhanbad. He has carried out extensive research work in the assessment and management of groundwater resources through electric analog and mathematical modeling. He has visited Germany, United Kingdom, Australia, United States, and Botswana. His research work and technical reports have been utilized for practical use in the assessment of groundwater resources, evolving optimal pumping schemes and quantification of pollutant migration in various projects in India and abroad. He has published over 140 research papers in national and international journals, and seminar proceeding volumes. He is the coeditor of five conferences proceeding volumes and authored two books on groundwater modeling. He has edited a book on groundwater, which was published jointly by Springer Ltd., and Capital Book Publishing Company, New Delhi. He has founded the Association of Global Groundwater Scientists (AGGS), Coimbatore and motivated in organizing six International Groundwater Conferences under the banner of IGWC. He is the recipient of Groundwater Scientist with Humanitarian Vision Award from Dindigul Chamber of Commerce for his dedicated effort to bring user community, decision makers, and scientists from India and abroad to a common platform in a rural town and discuss various issues regarding the identification of potential bore holes and its management aspect.

Vijay P. Singh is a distinguished professor and Caroline and William Leherer distinguished chair in water engineering in the Department of Biological and Agricultural Engineering as well as Zachry Department of civil engineering at Texas A & M University. He received his BS, MS, PhD, and DSc in engineering. He is a registered professional engineer, professional hydrologist, and an honorary diplomate in water resources engineering of American Academy of Water Resources Engineers, ASCE. He has published more than 790 journal articles; 24 textbooks; 55 edited reference books, including the massive *Encyclopedia of Snow, Ice and Glaciers*; 80 book chapters; 303 conference papers; and 70 technical reports in the areas of hydrology, groundwater, hydraulics, irrigation engineering, environmental engineering, and water resources. For his scientific contributions to the development and management of water resources, he has received more than 73 national and international awards and numerous honors, including the Arid Lands Hydraulic Engineering Award, Ven Te Chow Award, Richard R. Torrens Award, Norman Medal, and EWRI Lifetime Achievement Award, all given by American Society of Civil Engineers, 2010; Ray K. Linsley Award and Founder's Award, given by American Institute of Hydrology, 2006; Professor R.S. Garde Lifetime Achievement Award given by Indian Hydraulics Society, 2014; and Crystal Drop Award given by International Water Resources Association, 2015. He has received three honorary doctorates. He is a distinguished member of ASCE as well as a fellow of ASCE, EWRI, AWRA, IWRS, ISAE, IASWC, and IE and holds membership in 16 additional professional associations. He is a fellow and member of 10 international science and engineering academies. He has served as president and senior vice president of the American Institute of Hydrology (AIH).

Contributors

Shafique Ahamad
Department of Applied Mathematics
The Indian School of Mines (ISM) Dhanbad
Jharkhand, India

R. Annadurai
Department of Civil Engineering
SRM University
Tamil Nadu, India

Bhavna Arora
Earth and Environmental Sciences Area
Lawrence Berkeley National Laboratory
Berkeley, California

A. Arunachalam
Indian Council of Agricultural Research
New Delhi, India

S. Ayyappan
Indian Agricultural Research Institute (IARI)
New Delhi, India

Vivek S. Bedekar
S.S. Papadopulos & Associates, Inc.
Bethesda, Maryland

and

Department of Civil Engineering
Auburn University
Auburn, Alabama

K. Bhuvaneswari
Department of Agronomy
Agricultural College and Research Institute
Tamil Nadu Agricultural University
Tamil Nadu, India

Jens T. Birkholzer
Earth and Environmental Sciences Area
Lawrence Berkeley National Laboratory
Berkeley, California

Harriet Carlyle
Grontmij
Leeds, United Kingdom

D. K. Chadha
Global Groundwater Solutions
New Delhi, India

Manish Chopra
Radiation Safety Systems Division
Bhabha Atomic Research Centre
Maharashtra, India

V. M. Chowdary
Regional Remote Sensing Centre-East
NRSC
West Bengal, India

João Alberto O. Diniz
Department of Hydrology
CPRM—Geological Survey of Brazil
Pernambuco, Brazil

Dipankar Dwivedi
Earth and Environmental Sciences Area
Lawrence Berkeley National Laboratory
Berkeley, California

William Logan Dyer
Natural Resource Conservation Service
United States Department of Agriculture
Opelousas, Los Angeles

L. Elango
Department of Geology
Anna University
Tamil Nadu, India

T. I. Eldho
Department of Civil Engineering
Indian Institute of Technology (IIT) Bombay
Maharashtra, India

Hamdi El-Ghonemy
RSK Group
Manchester, United Kingdom

Vinit C. Erram
Indian Institute of Geomagnetism
Maharashtra, India

Fernando A. C. Feitosa
Department of Hydrology
CPRM—Geological Survey of Brazil
Pernambuco, Brazil

Edilton Carneiro Feitosa
LABHID—Hydrogeology Laboratory
and
UFPE—Federal University of Pernambuco
Pernambuco, Brazil

V. Geethalakshmi
Department of Agronomy
Agricultural College and Research Institute
Tamil Nadu Agricultural University
Tamil Nadu, India

Lennox A. Gladden
Department of Civil Engineering
Dong-A University
Busan, South Korea

R. Gowtham
Agro Climate Research Centre
Tamil Nadu Agricultural University
Tamil Nadu, India

Gautam Gupta
Indian Institute of Geomagnetism
Maharashtra, India

A. Shahul Hameed
Isotope Hydrology Division
Centre for Water Resources Development and
    Management
Kerala, India

X. Han
School of Civil and Environmental Engineering
The University of New South Wales
New South Wales, Australia

Omar Moalin Hassan
Environment Futures Research Institute
Griffith University
Brisbane, Australia

James E. Houseworth
Earth and Environmental Sciences Area
Lawrence Berkeley National Laboratory
Berkeley, California

A. Islam
Natural Resources Management Division (ICAR)
Krishi Anusandhan Bhawan-II
New Delhi, India

G. Jacks
Department of Sustainable Development
Environmental Science and Engineering
Stockholm, Sweden

Madan Kumar Jha
AgFE Department
Indian Institute of Technology Kharagpur
West Bengal, India

Nicholas Johnson
Garver Engineering LLC
Tulsa, Oklahoma

Deepak Kashyap
Department of Civil Engineering
Indian Institute of Technology Roorkee
Uttarakhand, India

Chang Hung Kiang
LEBAC—Basin of Studies Laboratory
and
UNESP—São Paulo State University
São Paulo, Brazil

Roberto Eduardo Kirchheim
Department of Hydrology
CPRM—Geological Survey of Brazil
São Paulo, Brazil

Alok Kumar
School of Environmental Sciences
Jawaharlal Nehru University (JNU)
New Delhi, India

Manoj Kumar
School of Environmental Sciences
Jawaharlal Nehru University (JNU)
New Delhi, India

Niraj Kumar
AgFE Department
Indian Institute of Technology Kharagpur
West Bengal, India

A. Lakshmanan
Department of Rice
Tamil Nadu Agricultural University
Tamil Nadu, India

Shrikant Daji Limaye
Water Institute
Maharashtra, India

Saumen Maiti
Department of Applied Geophysics
Indian School of Mines Dhanbad
Jharkhand, India

Partha Majumdar
Department of Civil Engineering
Indian Institute of Technology (IIT) Bombay
Maharashtra, India

Sonali Mcdermid
Department of Environmental Studies
New York University
New York, New York

Govinda Mishra
Department of Water Resources and Development
Indian Institute of Technology Roorkee
Uttarakhand, India

Sergi Molins
Earth and Environmental Sciences Area
Lawrence Berkeley National Laboratory
Berkeley, California

Sachikanta Nanda
Department of Civil Engineering
SRM University
Tamil Nadu, India

Mu. Naushad
Department of Chemistry
King Saud University
Riyadh, Saudi Arabia

J. M. Ndambuki
Department of Civil Engineering
Tshwane University of Technology
Pretoria, South Africa

Christopher J. Neville
S.S. Papadopulos & Associates, Inc.
Waterloo, Ontario, Canada

Namsik S. Park
Department of Civil Engineering
Dong-A University
Busan, South Korea

K. B. V. N. Phanindra
Department of Civil Engineering
Indian Institute of Technology Hyderabad (IITH)
Telangana, India

AL. Ramanathan
School of Environmental Sciences
Jawaharlal Nehru University (JNU)
New Delhi, India

A. P. Ramaraj
Department of Agronomy
Agricultural College and Research Institute
Tamil Nadu Agricultural University
Tamil Nadu, India

Moumtaz Razack
Department of Hydrogeology UMR CNRS 7285
University of Poitiers
Poitiers, France

Matthew T. Reagan
Earth and Environmental Sciences Area
Lawrence Berkeley National Laboratory
Berkeley, California

T. R. Resmi
Isotope Hydrology Division
Centre for Water Resources Development and
    Management
Kerala, India

K. R. Rushton
Department of Civil Engineering
Birmingham University
Birmingham, United Kingdom

and

Department of Civil Engineering
Cranfield University
Cranfield, United Kingdom

Sasmita Sahoo
AgFE Department
Indian Institute of Technology Kharagpur
West Bengal, India

K. Saravanan
School of Mechanical and Building Sciences
VIT University
Tamil Nadu, India

Nagothu Udaya Sekhar
Norwegian Institute of Bioeconomy Research
Akershus, Norway

K. Senthilraja
Agro Climate Research Centre
Tamil Nadu Agricultural University
Tamil Nadu, India

M. E. E. Shalabey (Deceased)
Irrigation Engineering and Hydraulics Department
Alexandria University
Alexandria, Egypt

Anupma Sharma
National Institute of Hydrology
Uttarakhand, India

Naoaki Shibasaki
Faculty of Symbiotic Systems Science
Fukushima University
Fukushima, Japan

A. K. Sikka
Natural Resources Management Division (ICAR)
Krishi Anusandhan Bhawan-II
New Delhi, India

Mritunjay Kumar Singh
Department of Applied Mathematics
The Indian School of Mines (ISM) Dhanbad
Jharkhand, India

Vijay P. Singh
Department of Biological and Agricultural
    Engineering & Zachry Department of Civil
    Engineering
Texas A&M University
College Station, Texas

D. C. Singhal
Department of Hydrology
Indian Institute of Technology Roorkee
Uttarakhand, India

B. Sivakumar
School of Civil and Environmental Engineering
The University of New South Wales
New South Wales, Australia

and

Department of Land, Air and Water Resources
University of California
Davis, California

Carl I. Steefel
Earth and Environmental Sciences Area
Lawrence Berkeley National Laboratory
Berkeley, California

William T. Stringfellow
Earth and Environmental Sciences Area
Lawrence Berkeley National Laboratory
Berkeley, California

and

Ecological Engineering Research Program
School of Engineering & Computer Science
University of the Pacific
Stockton, California

Faby Sunny
Radiation Safety Systems Division
Bhabha Atomic Research Centre
Maharashtra, India

Boddula Swathi
Department of Civil Engineering
Indian Institute of Technology (IIT) Bombay
Maharashtra, India

John Tellam
Earth and Environmental Sciences
University of Birmingham
Birmingham, United Kingdom

M. Thangarajan
Groundwater Modelling Group
CSIR—National Geophysical Research Institute (NGRI)
Telangana, India

M. Thirumurugan
Department of Geology
Anna University
Tamil Nadu, India

Ruth M. Tinnacher
Earth and Environmental Sciences Area
Lawrence Berkeley National Laboratory
Berkeley, California

Gurudeo Anand Tularam
Mathematics and Statistics, Griffith Sciences (ENV)
Environment Futures Research Institute
Griffith University
Brisbane, Australia

Avdhesh Tyagi
School of Civil and Environmental Engineering
Oklahoma State University
Stillwater, Oklahoma

Charuleka Varadharajan
Earth and Environmental Sciences Area
Lawrence Berkeley National Laboratory
Berkeley, California

S. Vasudevan
Electroinorganic Chemicals Division
CSIR-Central Electrochemical Research Institute
Tamil Nadu, India

F. M. Woldemeskel
School of Civil and Environmental Engineering
The University of New South Wales
New South Wales, Australia

Shailesh Kumar Yadav
School of Environmental Sciences
Jawaharlal Nehru University (JNU)
New Delhi, India

Section I

Groundwater Resources and Assessment

1

Coastal Groundwater Development: Challenges and Opportunities

Lennox A. Gladden and Namsik S. Park

CONTENTS

1.1 Introduction

Climate variability in the form of climate change is a phenomenon that is expected to make an impact on the hydrologic cycle through modified evapotranspiration, precipitation, and soil moisture patterns. The continuance of this event can significantly change both global and local climate characteristics, especially the temperature and the precipitation, whereby additional precipitation will be unevenly appropriated around the world. While some parts of the world will experience significantly more precipitation, other parts will experience significant reduction in precipitation, along with an altered season. Altered seasons mean a change in the timing of the wet and dry seasons. These alterations as documented by other studies will change the groundwater flow to coastal regions. Depletion of water levels will enhance the stress on the availability of groundwater and its usage in the coastal regions.

The coastal region areas are known to be home to a sizable and increasing proportion of the world's population. Many coastal zones, especially low-lying deltaic areas, accommodate a high density of populations. Two third of the world's population—4 billion people—are living within 400 km of the ocean shoreline; and just over half the world's population—around 3.2 billion people—occupy a coastal strip of 200 km. This was narrowed down even more, through the definition of a coastal zone being the area within 100 km from the coastline, when it was estimated that 41% of the global population inhabited this region. To emphasize further the population density in the coastal zone, highlights from various studies were that nine of the world's 10 most densely populated cities are located in the global coastal zone, seven of which are in developing countries.

The high population densities in the respective coastal zones have brought about many economic benefits, for example, income from tourism and food production and industrial and urban development. For ages, human race has been attracted to these areas because of the availability of an abundance of food (e.g., fisheries and agriculture) and the presence of economic activities (e.g., trade, harbors, ports, and infrastructure). However, this has its drawbacks because rapid population growth coupled with urban development threatens the coastal ecosystem. The most vulnerable is freshwater. A majority of these coastal regions rely on groundwater as their main source of freshwater for industrial, agricultural, and domestic purposes.

1.2 Challenges

1.2.1 Threats to Coastal Groundwater (Past, Present, and Future)

Several factors threaten coastal aquifers and in turn coastal groundwater resources. Some of these factors have existed for decades and with the expected continuous population growth due to economic opportunities in these coastal regions, the problems of these coastal aquifers will be further exacerbated. As shown in Figure 1.1, one can see how the various factors are affecting the coastal aquifer system (Essink, 2001), which would ultimately result in either salt-water intrusion, increased seepage, decreasing groundwater resources, salt-damaged crops, and ecosystem degradation, if left unchecked. Recently (not to say they did not garner attention before), two of these processes are of concern. These processes are the relative sea-level rise (SLR) and human activities.

1.2.2 Sea-Level Rise

The reference point in this case is the mean sea level (amsl), which is defined as the height of the sea with respect to a local benchmark, over a period of time, such as a month or a year, long enough for the fluctuations caused by waves and tides to be largely removed. While the relative sea-level change or rise is the combination of global SLR and local effects, which observe changes

in amsl as measured by coastal tide gauges, they can be caused either by the movement of the land on which the tide gauge is situated or by the changes in height of the adjacent sea surface. Therefore, SLR looks at an increased sea-level height at the coastal shoreline. It has been reported that since the late nineteenth century, the global sea level has risen by about 1.6 mm/year, whereas its rate did not exceed 0.6 mm/year during the two previous millennia.

Results from several studies have revealed that a large proportion of the population, especially in developing countries, will be displaced because of SLR within this century and accompanying economic and ecological damage will be severe if not devastating for many. They noted that at the country level results are extremely misinterpreted, considering that diverse coastal conditions are not taken into consideration during the assessment, and thus resulting in severe impacts being determined for a relatively limited number of countries. For some of these countries, for example, Egypt, Bangladesh, Vietnam, the Bahamas, and countries within the Caribbean region, however, the repercussions of SLR are probably disastrous.

1.2.2.1 Case Study 1: Egypt

The Nile Delta is considered one of the river deltas most vulnerable to SLR in the world. Like other assessments, they also concluded that SLR is expected to impact large agricultural areas through either inundation or higher levels of salinity of groundwater. It was discussed that SLR is one of the direct impacts of climate change and it can have a wide range of ramifications on health, tourism, agriculture, and other fields. It was furthermore emphasized that it may generally affect agriculture located in coastal areas in four different ways, namely increased frequency and magnitude of extreme weather events, higher salinity of groundwater, and higher levels of the groundwater table.

Another study attempted to estimate the economic value of potential impacts of higher levels of groundwater table due to salt-water intrusion associated with expected SLR on agricultural productivity in Damietta Governorate. Based on the projections within the study, agricultural activities within the study area—of which 76,266 acres are classified as agricultural lands—are expected to be considerably affected by higher levels of groundwater table induced by SLR. From the standpoint of economics, several billions of dollars are expected to be lost; this is a direct effect. The indirect effect deals with food security; a decline in crop yield would undoubtedly reduce the income generated by farmers, which would more than likely prompt them to diversify and plant crops that might not be vital for Egypt's overall food security.

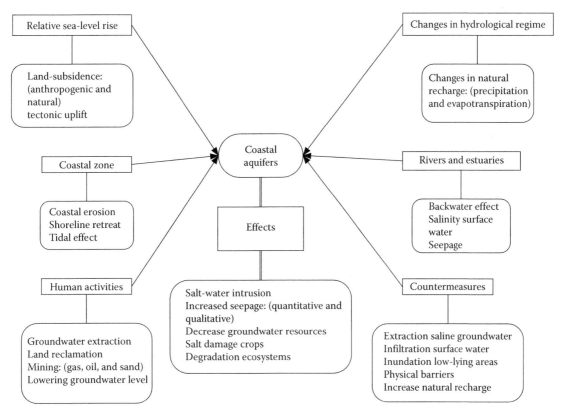

FIGURE 1.1
Features affecting the coastal aquifers.

1.2.2.2 Case Study 2: Bangladesh

Bangladesh's exposure to the growing hazard of SLR in the twenty-first century needs to be seen in the perspective of its current environmental hazards and its growing development needs. SLR has various impacts on Bangladesh, a coastal country facing a 710-km-long coast to the Bay of Bengal. Currently, SLR has already had a toll on Bangladesh by the loss of biodiversity, land erosion, and salinity intrusion. Climate change and other factors that are contributing to SLR are not subsiding and thus the threat to the coastal ecosystem of Bangladesh will increase as well. Like other regions, there are some similar resultant effects of SLR in Bangladesh, which include riverbank erosion, salinity intrusion of coastal aquifers, flood damage to infrastructures, crop failure, and, as previously mentioned, loss of biodiversity. In 2005, 28% of the country's population lived along the coast—a number that is currently obviously higher with their main economic activity being agriculture, shrimp farming, salt farming, and tourism.

It was mentioned that the main impacts of SLR on water resources are freshwater availability reduction by salinity intrusion, which would cause both water and soil salinity along the coast to increase because of destroying normal characteristics of coastal soil and water. Like the previous case, sea salinity intrusion due to SLR will decrease agricultural production by unavailability of freshwater and soil degradation, a case of which was investigated in a village in the Satkhira district in Bangladesh. The results showed that because of SLR for 6 years, there was a loss of 64% in rice production in 2003. The SLR affects coastal agriculture especially rice in two ways: salinity degrades the soil quality which decreases or inhibits rice production, and taking advantage of the saline intrusion and converting rice fields into ponds decreases the cropped area. The situation is applicable for other crops but with rice being the staple food of the Bangladeshi people, it is food security.

1.2.2.3 Case Study 3: Indonesia

A recent study looked at the coastal area of part of the Java province. This is a rapidly developing region of Indonesia with respect to economics, trading and industrial development, agricultural activities, and tourism and settlement areas. Coastal inundation, because of SLR, would impact it. However, this is not the only environmental problem, as there is also increased coastal storm flooding as well as increased salinity of estuaries and aquifers. Output from this assessment revealed that the water level in coastal areas could be disturbed by tidal behavior. Naturally, the flood tide will influence groundwater manner, particularly in the correlation of the salt-water component in the said flood tide. Furthermore, preliminary analysis evaluating the electroconductivity at the three wells all at varying distance from the baseline revealed that the well that was in the closest proximity to the coastline was already saline during the first testing phase and one well that had a distance of 4.16 km from the coastline was not. However, when the said well was tested after 5 years, it was determined to be saline. This result will inadvertently change the dynamics not only on the agricultural front but also on the traditional and developmental aspects.

1.2.2.4 Case Study 4: South Korea

Seawater intrusion-monitoring wells were established in Korea in 1999, results of which determined that groundwater levels were caused mainly by pumping for agricultural irrigation; at the same time, conductivity increases were detected in 45 of the 97 seawater-intrusive monitoring wells. Equally important from their findings was that groundwater levels at some of the monitoring wells were affected by sea-level variation. These findings were examined further by assessing the vulnerability of groundwater systems caused by SLR. Their assessment was based on the data collected from the monitoring wells, which were not influenced by anthropogenic activities and seven associated sea-level gauging stations. These monitoring wells were strategically located either on the mainland or at reclaimed land areas. Results illustrated that the vertical profiles or conductivity with depth show that the groundwater systems will reclaim land that is more susceptible than the wells on the mainland especially during the dry season.

The cases above are examples of the potential impacts of SLR on coastal zones and aquifers. From a regional view, areas that exhibit the greatest relative impact are East Asia and Middle East and North America; likewise, there are the regions that show minimal impacts due to SLR.

1.2.3 Overexploitation

Humans' actions or inactions may impede groundwater flow and in turn cause complicated changes in the salt water and freshwater relationship in coastal aquifers. Both effects, positive or negative, are caused by our direct or indirect impact on groundwater flow. Overexploitation of water from coastal aquifers that are adjacent or in contact with the sea may cause the decline of water quality of the coastal aquifer. However, this is not the only cause of a threat to the water quality of the aquifer; other risks may include contamination by urban use, industry, and agriculture. However, we first focus on the proximity of coastal aquifers to salt water, which creates a unique dilemma with respect to the sustainable development of groundwater in the coastal regions. Owing to increasing concentration of human settlements, agricultural development, and economic activities, shortages of fresh groundwater for municipal, agricultural, and industrial purposes become more striking in these zones.

In principle, we see that a delicate balance must be maintained between freshwater and salt water, to attain sustainability of fresh groundwater in coastal aquifers. However, it is not uncommon that the water table falls below the level of the sea during usage because development is planned. When continuous withdrawal occurs over a prolonged time period or at an excessive rate due to overexploitation, a permanent condition might be created as seawater propagates inland rendering the groundwater worthless for domestic or agricultural purposes. This is the first significant threat, and the second most-common significant threat to groundwater resources in the coastal zone is infiltration of pollutants from the landside, whereby groundwater of lesser quality is in the same vicinity as the aquifer, which is being exploited. If not properly managed and regulated, the lesser-quality water would readily invade the aquifer in question because of overexploitation. The battery of wells would then need to be abandoned if it cannot meet the desired standard for usage through treatment. Recurring use of lesser-quality water may damage agricultural soil, vegetation, animals, and biodiversity and may even cause human health problems. A condition is compounded during water-scarce conditions as it results in a steeper hydraulic gradient toward the wells.

1.2.4 Seawater Intrusion

Seawater intrusion is a natural process that occurs in virtually all coastal aquifers and is limited to coastal zones. It is also a slow process that occurs because of reversal of hydraulic gradients. The inflow pattern, which shows an imaginary line between the seawater and freshwater, is known as the interface. It is due to the density difference between freshwater ($1.0\,g/m^3$) and salt water ($1.025\,g/m^3$), which will show the seawater below the freshwater on the landside of the coast.

The first and the oldest formulations for salt and freshwater interface in a coastal unconfined aquifer under hydrostatic conditions were made and given by the following equation:

$$hs = \frac{df}{ds - df} hf \qquad (1.1)$$

where

hs = distance below amsl at the freshwater/salt-water interface

hf = distance from the groundwater table to amsl

ds = density of salt water (1.025 g/m^3)

df = density of freshwater (1.0 g/m^3)

When the groundwater table is lowered by 1 m, then the relationship between hs and hf is

$$hs = \frac{1}{1.025 - 1} hf = 40hf \qquad (1.2)$$

This relationship indicates that the salt-water interface will move upward to 40 m. It can be explained that for each unit of groundwater above sea level there are 40 units of freshwater below sea level. Therefore, the salt water and freshwater relationship is extremely sensitive to groundwater withdrawal.

1.2.4.1 Cases of Salt-Water Intrusion

1.2.4.1.1 Salt-Water Intrusion in the Caribbean (Trinidad and Tobago)

Owing to a high withdrawal of water at the rate of 30,000 m^3/d, the wells of El Soccoro located on the islands of Trinidad and Tobago were affected by salt-water intrusion. This occurred from the early 1960s until the late 1970s when the abstraction rate was reduced allowing the water level to recover (1979–1983). It was cited that on average, during the 1980s, the chloride concentration was above 600 mg/L, but after a reduction of the daily capacity, there has been a reduction in the chloride concentration as well. At the time of the report, the chloride concentration was recorded at 400 mg/L. This concentration value was still far from the original chloride concentration in 1959, which was 40 mg/L. This was not an isolated occurrence in Trinidad and Tobago, for salt-water intrusion has occurred at other locations on the islands (Valsayn aquifer and Mayaro Sandstone) due to the similar practice of overexploitation, which resulted in the abandonment of several wells.

1.2.4.1.2 Recommendations Made to Prevent Salt-Water Intrusion

"The pumping of aquifers within their safe yield values, drilling well further inland from the coastline and frequent monitoring of coastal observation wells for water level fluctuations and quality (chlorides) are measures in place to prevent salt water intrusion" (Water Resource Agency, Trinidad and Tobago, 2001, p. 26).

1.2.4.2 Salt-Water Intrusion in the Caribbean (Jamaica)

The freshwater resource in Jamaica comes from surface sources, for example, rivers and streams, underground sources, and from the harvesting of rainwater. In Jamaica, about 80% of the country's water demand, which represents 84% of the country's exploitable water, is from groundwater supplies. The aquifers that provide this freshwater are vulnerable like most aquifers to contamination. In Jamaica, there are some general considerations that increase the vulnerability of these coastal aquifers. Some of these considerations include an increase in population, agriculture and industry competing with domestic demands for groundwater, proximity of the aquifer to the sea, and the karstic nature of the limestone aquifers. All of these pose a threat to the groundwater in limestone and alluvial aquifers of Jamaica. According to a report by Jamaica's Water Resource Authority, one of the main sources of water pollution is saline intrusion caused from overpumping of the coastal karstic limestone aquifers.

As illustrated in Figure 1.2a, which shows a satellite image of Jamaica's exploited wells, a large number of the wells being exploited are concentrated in areas that are experiencing salt-water intrusion. This results in the abandonment of many of the production wells and sugarcane fields in the southern part of Jamaica. In particular, at the Rio Minho and Rio Cobre areas, an increase in salinity was observed at a distance more than 10 km from the coast. This occurred prior to the 1961 period when the control of licensing was introduced into the island. With respect to salt-water intrusion, the Water Resource Authority of Jamaica has calculated that the degradation of water quality has resulted in the loss of some 100 MCM annually, which is about 10% of all currently used groundwater, primarily because of overabstraction, which led to seawater intrusion in the areas observed (Figure 1.2b).

1.2.4.3 Salt-Water Intrusion in North America

Salt water has intruded into many coastal aquifers of the United States, Mexico, and Canada, but the extent of which the salt-water intrudes vary widely among the localities and hydrogeologic settings. According to the research, the extent of which salt water intrudes into the respective aquifers is dependent on several factors. These factors included the total rate of groundwater that is withdrawn from an aquifer compared to the total freshwater recharge to the aquifer, the distance between the groundwater discharge such as pumping from wells and drainage canals and the source or sources of salt water, the geology structure of the aquifer system, the hydraulic properties of an aquifer, and the presence of a confining unit that may prevent salt water from moving vertically or with the aquifer. Their research first looked at the United States where it was determined based on literature that was surveyed; the scale of salt-water intrusion varied greatly with several cases of intrusion being isolated in specific locations within the aquifer and had a negligible effect on the groundwater supplies.

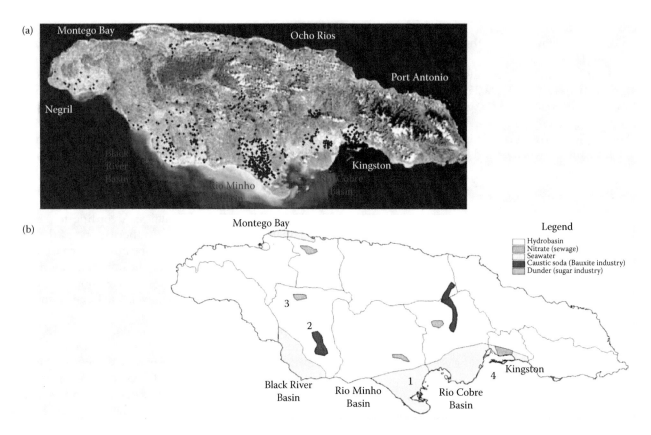

FIGURE 1.2
Well locations (a) and contaminated areas (b) on the island of Jamaica. (Adapted from Karanjac, J. 2005. Vulnerability of ground water in the Karst of Jamaica. *Water Resources and Environmental Problems in Karst*, Beograd and Kotor, Serbia and Montenegro, September 14–19; and Jamaica's Water Resource Authority.)

On the other hand, there were areas that were significantly affected by salt-water intrusion. Areas included Cape May County; New Jersey; southeastern Florida; and Monterrey, Ventura, Orange, and Los Angeles Counties in California, as shown in Figure 1.3.

In the case of Mexico, the most significant impact of salt-water intrusion was in the western states of Sonora, Baja California Norte, and Baja California Sur. Sonora region has been subjected to more salt-water intrusion due to overabstraction of groundwater for agricultural purposes. The last case that they observed was that of the coastal aquifers in Canada where only two cases (Prince Edward Island and Magdalen Islands) saw salt-water intrusion due to anthropogenic factors with other cases being due to natural intrusion.

1.2.4.3.1 Recommendations Made to Prevent Salt-Water Intrusion

"Sustainable management responses to salt-water intrusion will require multicomponent strategies that consider intrusion in the broader context of basin wide, integrated groundwater and surface water management. New engineering approaches will need to be considered (for example, utilizing physical barriers) and there will likely be increased use of recycled water for recharge and direct delivery to users. In addition, sophisticated decision-support systems will likely be developed that can link monitoring data with simulation and optimization models, and that provide for improved communication of simulation results to water-resource managers. Desalination is expected to become more widespread as the treatment methods become more energy efficient and the challenges of environmentally sound brine disposal are addressed. Finally, water managers will need to consider how saltwater intrusion may be affected by potential rises in sea level due to climate change" (Barlow and Reichard, 2010, p. 250).

1.3 Mitigating Approaches to Combat Salt-Water Intrusion

There are several guidelines to follow to prevent seawater intrusion. First, the annual abstraction from the

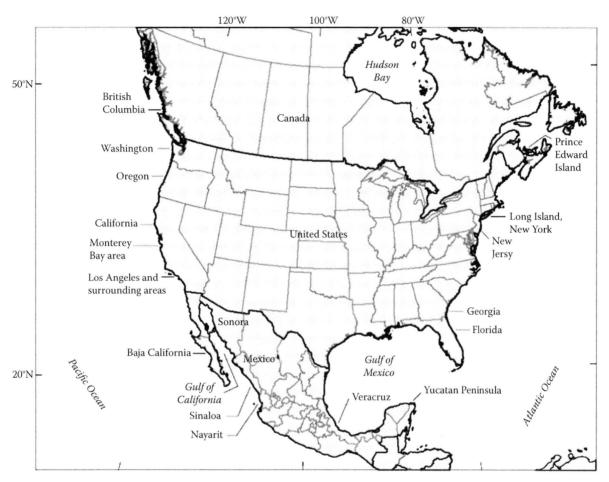

FIGURE 1.3
North American study areas for saltwater intrusion. (Adapted from Barlow, P. M. and E. G. Reichard. 2010. *Journal of Hydrogeology* 18:247–260.)

coastal aquifer should be less than the accumulated recharge and it is not recommended that a cluster of production wells be installed near the coastline, unless, of course, through testing it was determined that the combined abstraction rate would not result in seawater intrusion. Proper well placement is required especially in aquifers where the freshwater floats atop the saline water. The withdrawals from these wells must be properly gauged and the drawdown must be kept at a minimum. It is not only the well construction and site selection aspect that must be considered as a factor when trying to prevent seawater intrusion, but also the aquifer recharge sources need to be protected.

Although naturally there is a fluctuation in recharge, this phenomenon should not be magnified, as when coupled with the exploitation of a coastal aquifer, it can quite easily lead to salt-water intrusion. If, however, salt-water intrusion occurs within the aquifer, counter measures need to be adopted to rectify this condition. The rectification of seawater encroachment within a coastal aquifer is a tedious task that requires not only that the intruded water be removed and replaced with

freshwater but also that the salt residue within the aquifer medium be flushed out as well.

Seen in Figure 1.4 are several countermeasures that can be considered to prevent or retard the salinization process. The first seen are the freshwater injection hydraulic barriers, which, as was stated, are formed through the injection or (deep-well) infiltration of freshwater near the shoreline. Owing to the injection pressure, freshwater is forced to flow seaward; thus, seawater intrusion is impeded. The extraction of saline/brackish water commonly known as pumping hydraulic barriers can help to intercept the flow of salt water inland. The simplest method as seen in Figure 1.4 is coastal well screens in the lower part of the aquifer, which abstracts salt water and is then returned to the sea. Like the previous case with the injection hydraulic barriers, this strategy utilizes a battery of wells whose proximity to each other must be close enough as not to allow salt-water penetration to occur between them. The next countermeasure was previously mentioned in the Trinidad and Tobago case study whereby to counteract the effect of salt-water intrusion, they modified the pumping rate

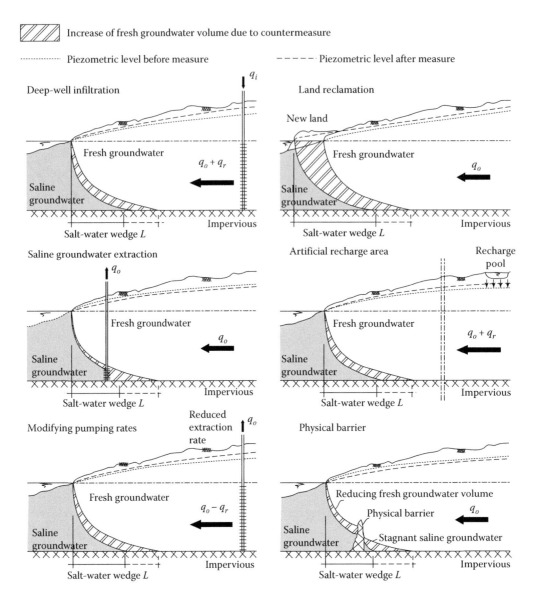

FIGURE 1.4
Counter measures to control saltwater intrusion. (Adapted from Essink, G. H. P. O. 2001. *Ocean and Coastal Management* 44:429–449.)

and suggested that extraction wells be relocated far-ther inland. The latter may be more easily accomplished from a sectoral point of view, as it is difficult to justify reduction in groundwater abstraction. Therefore, reduc-tion in groundwater abstraction should be comple-mented with a new water source. The next procedure that can be used is land reclamation, which creates a foreland where a freshwater body may develop, which could delay the inflow of saline groundwater. Aquifer recharge or artificial recharge is also a viable counter-measure to control salt-water intrusion. The concept of which describe the use of water of a given quality that is introduced at a point (upland areas) that is intended to allow water to flow into well field production zones (to enlarge outflow of fresh groundwater through the

coastal aquifer) resulting in a reduced length of the salt-water wedge. The last-seen examples are physical barri-ers (clay trenches, sheet piles, and chemical injections), which are only applicable for shallow aquifers.

1.3.1 There Is However Another Concept Known as the Salinity Barrier

"The salinity barrier concept may also be applied to address the potential upconing problem in source aquifers. Upconing is the upward (vertical) movement of saltwater underneath the production well that is withdrawing groundwater. Because the aquifer head is lowered in the production zone, decreased freshwa-ter head allows the salt water to rise from the lower

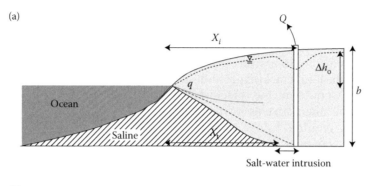

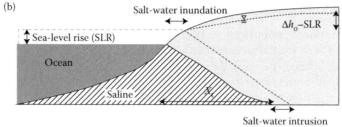

FIGURE 1.5
Coastal aquifers are affected both by (a) groundwater extraction and (b) sea level rise. (Adapted from Ferguson, G. and T. Gleeson. 2012. *Nature Climate Change* 2:342–345.)

zones. The concept of aquifer reclamation applied to upconing includes the installation of injection wells at depths below that of the production wells; this would cause the injection of higher quality water to displace poorer-quality water upconing into the production zone" (Bloestcher et al., 2004, p. 104).

To conclude this section on challenges facing coastal groundwater water development, the conceptual model used in the previous study is seen in Figure 1.5, which best summarizes the two main factors to affect coastal aquifers, namely, SLR and overexploitation, both of which would result in salt-water intrusion.

1.4 Opportunities

Various ways to protect and exploit coastal freshwater have been explained but there is an opportunity to use brackish or saline groundwater. Brackish groundwater is defined as distastefully salty and it differs from seawater in that seawater contains a higher salt content of about 35,000 ppm and brackish water has a range of 1000–10,000 ppm in total dissolved solids (TDSs).

There are two ways in which to use brackish groundwater; it either can undergo direct use or treated use. For direct usage, brackish groundwater is utilized in the oil and gas industry for tasks such as drilling, enhancing recovery and hydraulic fracturing, and cooling water for power generation and aquaculture. For treated use,

it is used to combat a decreasing supply of fresh groundwater and surface water, increased cost of water rights, and increased competition for surface water resources. Brackish water can be treated through reverse osmosis or other desalination processes to lower dissolved solids content. Low dissolved solids content is important especially for brackish water being treated for drinking water purposes.

1.4.1 Cooling Water for Power Generation

There are several cases that can be observed and assessed but here two cases are looked at—one in which brackish water was used for a power plant-cooling tower operation and the second case where the brackish water is recommended to be used as a source of cooling water for a nuclear power plant in Israel. The motivation behind the first case was an attempt to deal with both current and forecasted constraints on a new power plant to be put up for short- and long-term operation. Therefore, it was recommended that if the electric industry wanted to boost its abilities, new technology would be required to expand the industry's water resources. It was acknowledged that due to an ever-increasing competing demand, water resources must be prioritized. To expand the current sources of water used for power plant cooling, degraded and non-traditional water resources were used, one of which was brackish water.

The second case where brackish water was recommended for usage was in Israel's Negev Desert as an alternative source of cooling water for a nuclear power plant. Since the Negev Desert does not have a source of surface water, the cooling water options for any future nuclear power plant project would be either water piped in from the Mediterranean or the local brackish groundwater. Preliminary results were that the brackish groundwater would be a viable alternative as it would be able to supply the capacity required by the nuclear power plant without any large changes to the regional water table.

1.4.2 Treated Use of Brackish Groundwater

To achieve drinking water standards, salts need to be removed from the freshwater. There are two commonly used methods to accomplish this. The distillation method is replicated artificially but often seen in nature whereby the sun causes water to evaporate from a body of water and when it eventually comes into contact with cooler air it condenses. The second method is known as the reverse osmosis approach, which utilizes a semipermeability membrane and pressure to separate the salt from the water.

1.4.3 Desalination

Desalination (also called desalinization) is the process of removing dissolved salts from water, thus producing freshwater from either brackish or seawater. Of the many applications for which it can be used, the most common use is to produce potable water for domestic use by removing salt from saline or brackish water. International Development Association (IDA) reported that as of 2013, there were over 17,000 desalination plants worldwide spread across 150 countries with a global capacity of 80 million m^3/d, which is equivalent to 66.5 million m^3/d. This desalinated water is relied on by 300 million people for some if not all of their daily needs. Out of approximately 17,346 contracted and online desalination plants across the globe, the Gulf Cooperation Council (GCC), which comprises Qatar, The Kingdom of Saudi Arabia, Bahrain, Oman, Kuwait, and the United Arab Emirates, holds 7499, which is 43% the total number of online desalination plants. The GCC also accounts for 62,340,000 m^3/d of the 94,500,000 online capacity.

Although the majority of the desalination plants are concentrated in one region, the technology is becoming more prominent in other parts of the world as its importance grows as a mechanism of supplying the population especially those who reside in areas where this resource is scarce. As improvements to the technology are made, popularity will increase among the various sectors (domestic, agricultural, and industrial) as an alternative source of water supply.

1.4.4 Brackish Water for Agricultural Production

Water scarcity severely affects the agricultural sector of most countries in the Middle East and North America and many other areas around the world. There is agricultural water management practices that could enhance water use efficiency and prevent wastage; for example, irrigation practices that would mitigate the problem of water scarcity agriculturists are encountering. However, these strategies to combat water scarcity are slow to implement and in some areas not applicable, given the fact that nontraditional water sources, for example, marginal-quality groundwater aquifers, desalination, and rainwater harvesting, are being exploited to satisfy the demand.

Irrigation has always been an important factor in agricultural development and irrigation with brackish water from coastal aquifers has been growing in popularity especially in India and Middle Eastern countries. This method has its shortcomings, however, the first of which is the level of salinity. A high level of salinity would result in an osmotic imbalance. This osmotic imbalance would reduce the water uptake and transpiration by the crop, which would lower the crop yield. Second, there is only a narrow selection of crops that are salt tolerant and care must still be taken as these crops have their respective salinity threshold. Finally, there is the accumulation near the roots of the plant. This can be corrected but an unsustainable volume of irrigation water would be needed to leach the soil. It is vital when applying saline/brackish irrigation water to crops that the salinity of the soil be constantly monitored because although the salinity of the irrigation water will remain constant, the soil salinity will vary depending on drainage characteristics of the soil, the type and management of crops, prevailing weather conditions, and the method and frequency of irrigation.

Therefore, it is recommended that irrigation be done with desalinated water, as it is more water efficient and economical than brackish water irrigation. However, the desalination process needs to use nanofiltration, as it is more cost-effective due to its lower energy requirement. In addition, unlike desalination by reverse osmosis, it is able to retain a majority of the ions required for plant growth. Reverse osmosis and distillation not only separate the undesirable from the water, but also remove ions that are essential for plant growth. In addition, desalinized water typically replaces irrigated water that previously provided basic nutrients such as calcium, magnesium, and sulfate at levels sufficient to preclude additional fertilization. This was the case with the Ashkelon plant in Israel, in which the water does not have any magnesium. The farmers needed to apply fertilizer to their crops after irrigation with the desalinated water because their crops started to exhibit magnesium deficiency.

1.4.5 Aquaculture Using Brackish or Saline Water

The management of groundwater salinity problems often results in large amounts of unwanted saline or brackish water, which is generally free of other pollution and which fulfills the basic requirement for most aquaculture systems. Where qualities of seawater and saline groundwater are somewhat the same, the usage of the saline groundwater for aquaculture production would only require matching the system with the saline water as well as the specie. The specie need to be matched based on the known tolerance range. The optimum conditions as well as the prevailing conditions of the specie's original or designed environment are required to attain the desired production levels. As in the case of desert aquaculture in Egypt, it was mentioned that brackish water and brine would play an important role in the sustainable development of the aquaculture industry.

The use of saline groundwater for aquaculture can also be expanded to the agricultural sector, as seen in from a previous study conducted by researchers in China. In their experiment, they irrigated Jerusalem artichoke and sunflower using saline aquaculture wastewater mixed with brackish groundwater at various ratios. Their results demonstrated that when saline aquaculture wastewater is applied to the artichoke at the proper ratio, the crop could grow properly without any adverse effects.

1.4.6 Subsurface Water Storage in Coastal Aquifers

Coastal aquifers may also be used as a reservoir in which water can be stored. The manner in which the water reaches the target zone is usually done using the technology known as aquifer storage and recovery (ASR) or aquifer storage transfer and recovery (ASTR). To compare and contrast ASR and ASTR, the explicit definition found in literature must be utilized where ASR is the recharge of an aquifer via a well for subsequent recovery from the same well and ASTR is the recharge of an aquifer through injection into a well and then recovery from another well. It has several applications: seasonal storage, long-term storage, emergency storage and supply, reclaimed water storage for reuse, prevention of salt-water intrusion, etc., all of which can be conducted in a coastal aquifer.

1.5 Developing Coastal Groundwater

It has been established that groundwater aquifers are an integral part of the coastal ecosystem and are currently facing numerous threats, both natural and manufactured. For the optimal exploitation of water (freshwater/brackish/saline) from a coastal aquifer to occur, efficient management strategies are needed. Whereby the development of any worthwhile management strategy would require the development of management models that can not only simulate the respective processes—for example, salt-water intrusion and SLR—but also be able to possibly optimize. One of the best methods of determining the sustainable management strategy for a groundwater system may be the combined use of simulation/optimization (S/O) models. While simulation models provide solutions for the governing equations of groundwater flow, optimization models identify an optimal management and planning strategy from a set of feasible alternative strategies. Through the utilization of linear and nonlinear formulations, it helps to find the answer that yields the best results for minimizing or maximizing an objective function subjected to various constraints.

The conventional development of a coastal aquifer focuses on freshwater development subjected to salt-water intrusion. However, within this chapter, the opportunities were discussed which observed that brackish or saline water can also be regarded as

TABLE 1.1

Types of Coastal Groundwater Management Problem

Type of Well	Application	Example Objective	Penalty	Constraints
Pumping FW	FW development	Maximize FW pumping rate	SW intrusion at FW pumping wells	Maximum FW pumping rates
Pumping SW	SW intrusion control	Minimize SW pumping rate	FW and SW intrusion at SW and FW pumping wells, respectively	Minimum SW pumping rates
	SW fish farm	Maximize SW pumping rate	FW intrusion at SW pumping well	Target SW concentration
Pumping BW	R/O desalination	Minimize SW contents	SW concentration at intake	Minimum BW pumping rates
	BW fish farm	Maximize BW pumping rate	FW intrusion at SW pumping well	Target BW concentration
Injecting FW	SW intrusion control	Minimize FW injection rate	SW contents at FW pumping well	Minimum FW injection rate
	FW augmentation	Objectives and penalties differ widely	Varied	Varied

beneficial. Table 1.1 reviews and summarizes the types of coastal groundwater problems. The summary table indicates that there are diverse applications in coastal groundwater apart from freshwater development, all of which have the potential to yield an optimal management strategy when treated separately as documented in previous studies, where the optimization approach has been used for freshwater development, salt-water intrusion control, operation of desalination plants, and freshwater augmentation.

1.6 Summary

The present and future need for water to support an ever-increasing coastal population along with their economic endeavors shows no indication of subsiding. This situation will prompt both water resource managers and decision makers to be more cautious and aware when developing coastal groundwater. As discussed and illustrated from the case studies there are challenges and opportunities with regard to coastal groundwater. Great strides have been taken to better understand the complex system that is the coastal aquifer. Through this understanding, it was realized that what was once considered a problem—like in the case of brackish groundwater—could also be an opportunity.

In an age where there is and will be an increase in competition for this resource, it is only fitting that optimal management strategies be considered, which clearly identify a comprehensive groundwater management solution.

Bibliography

Abdrabo, M. A. and M. A. Hassaan. 2014. Economic valuation of sea level rise impacts on agricultural sector: Damietta Governorate, Egypt. *Journal of Environmental Protection* 5:87–95.

Alauddin, M. and M. A. R. Sarker. 2014. Climate change and farm-level adaptation decisions and strategies in drought-prone and groundwater-depleted areas of Bangladesh: An empirical investigation. *Ecological Economics* 106:204–213.

Aoustin, E., F. Marechal, F. Vince, and P. Bréant. 2008. Multi-objective optimization of RO desalination plants. *Desalination* 222:96–118.

Ayvaz, M. T. 2010. A linked simulation–optimization model for solving the unknown groundwater pollution source identification problems. *Journal of Contaminant Hydrology* 117 (1–4):46–59.

Barlow, P. M. and E. G. Reichard. 2010. Saltwater intrusion in coastal regions of North America. *Journal of Hydrogeology* 18:247–260.

Bloestscher, F., A. Muniz, and M. W. Gerhardt. 2004. *Groundwater Injection Modelling, Risks and Regulations.* United States: McGraw-Hill.

Brammer, H. 2014. Bangladesh's dynamic coastal regions and sea-level rise. *Climatic Risk Management* 1:51–62.

Cashman, A., L. Nurse, and C. John. 2010. Climate change in the Caribbean: The water management implications. *The Journal of Environment and Development* 19:42–67.

Cheng, A. H.-D., D. Halhal, A. Naji, and D. Ouazar. 2000. Pumping optimization in saltwater-intruded coastal aquifers. *Water Resources Research* 36(8):2155–2165.

Church, J. A., N. J. White, L. F. Konikow, C. M. Domingues, J. G. Cogley, E. Rignot, J. M. Gregory, M. R. van den Broeke, A. J. Monaghan, and I. Velicogna. 2011. Revisiting the Earth's sea-level and energy budgets from 1961 to 2008. *Geophysical Science Letters* 38: L18601.

Council, G. W. and C. J. Richards. 2008. A saltwater upconing model to evaluate well field feasibility. *20th Saltwater Intrusion Meeting*, Naples, FL.

Cozannet, G., M. Le, M. Garcina, D. I. Yates, and B. Meyssignac. 2014. Approaches to evaluate the recent impacts of sea-level rise on shoreline changes. *Earth-Science Reviews* 138:47–60.

Das, A. and B. Datta. 2001. Simulation of seawater intrusion in coastal aquifers: Some typical responses. *Sadhana* 26(4):317–352.

Dasgupta, S., B. Laplante, C. Meisner, D. Wheeler, and J. Yan. 2007. *The Impact of Sea Level Rise on Developing Countries: A Comparative Analysis.* Washington: World Bank.

El-Guindy, S. 2006. The use of brackish water in agriculture and aquaculture. *Panel Project on Water Management Workshop on Brackish Water Use in Agriculture and Aquaculture*, Cairo, Egypt, December 2–5.

Essink, G. H. P. O. 2001. Improving fresh groundwater supply—Problems and solutions. *Ocean and Coastal Management* 44:429–449.

Eusuff, M. M. and K. E. Lansey. 2004. Optimal operation of artificial groundwater recharge systems considering water quality transformations. *Water Resource Management* 18:379–405.

Ferguson, G. and T. Gleeson. 2012. Vulnerability of coastal aquifers to groundwater use and climate change. *Nature Climate Change* 2:342–345.

Gandure, S., S. Walker, and J. J. Botha. 2013. Farmers' perceptions of adaptation to climate change and water stress in a South African rural community. *Environmental Development* 5:39–53.

Gengmao, Z., S. K. Mehtac, and L. Zhaopu. 2010. Use of saline aquaculture wastewater to irrigate salt-tolerant Jerusalem artichoke and sunflower in semiarid coastal zones of China. *Agricultural Water Management* 97(6):1987–1993.

Ghermandi, A. and R. Messalem. 2009. The advantages of NF desalination of brackish water for sustainable irrigation: The case. *Desalination and Water Treatment* 10:101–107.

Ghyben, W. B. 1888. *Nota in Verband met de Voorgenomen putboring nabij* Amsterdam. Tijdschr. Kon. Inst. Ing. 9:8–22.

Hashemi, R. A., S. Zarreen, A. Al Raisi, F. A. Al Marzooqi, and S. W. Hasan. 2014. A review of desalination trends in the Gulf Cooperation Council countries. *International Interdisciplinary Journal of Scientific Research* 1(2):72–96.

Hernandez, E. A., V. Uddameri, and M. A. Arreola. 2014. A multi-period optimization model for conjunctive surface water–ground water use via aquifer storage and recovery in Corpus Christi, Texas. *Environmental Earth Science* 71:2589–2604.

Herzberg, A. 1901. Die Wasserversorgung einiger Nordseebader. *Gasbeleucht Wasserversorg* 44:478–487.

Hinkel, J. and R. J. T. Klein. 2009. Integrating knowledge to assess coastal vulnerability to sea-level rise: The development of the DIVA tool. *Global Environmental Change* 19(3):384–385.

Hinrichsen, D. 2010. Ocean planet in decline. peopleandplanet. net. Accessed March 11. http://www.peopleandtheplanet. com/index.html@lid=26188&section=35&topic=44.html.

Hong, S. 2004. *Optimization Model for Development and Management of Groundwater in Coastal Areas.* PhD thesis, Civil Engineering, Dong-A University.

Ibáñez, C., J. W. Day, and E. Reyes. 2014. The response of deltas to sea-level rise: Natural mechanisms and management options to adapt to high end scenarios. *Ecological Engineering* 65:122–130.

IPCC. 2001. *Climate Change 2001: IPCC Third Assessment Report— The Scientific Basis.* Edited by The IPCC Secretariat. Switzerland: World Meteorological Organization.

IPCC. 2007. *Climate Change 2007: Impacts, Adaptation, and Vulnerability.* Edited by Contribution of Working Group II to the fourth assessment report of the Intergovernmental Panel on Climate Change. New York: IPCC.

Karanjac, J. 2005. Vulnerability of ground water in the Karst of Jamaica. *Water Resources and Environmental Problems in Karst,* Beograd and Kotor, Serbia and Montenegro, September 14–19.

Kemp, A. C., B. P. Horton, J. P. Donnelly, M. E. Mann, M. Vermeer, and S. Rahmstorf. 2011. Climate related sea-level variations over the past two millennia. *Proceedings of the National Academy of Sciences of the United States of America* 108(27):11017–11022.

Kløve, B., A. Allan, G. Bertrand, E. Druzynska, A. Ertürk, N. Goldscheider, S. Henry et al. 2011. Groundwater dependent ecosystems. Part II. Ecosystem services and management in Europe under risk of climate change and land use intensification. *Environmental Science and Policy* 14(7):782–793.

Knowling, M. J., A. D. Werner, and D. Herckenrath. 2015. Quantifying climate and pumping contributions to aquifer depletion using a highly parameterised groundwater model: Uley South Basin (South Australia). *Journal of Hydrology* 523:515–530.

Kurylyk, B. L. and K. T. B. MacQuarrie. 2013. The uncertainty associated with estimating future groundwater recharge: A summary of recent research and an example from a small unconfined aquifer in a northern humid-continental climate. *Journal of Hydrology* 492:244–253.

Lee, J. Y., M. J. Yi, S. H. Song, and G. S. Lee. 2008. Evaluation of seawater intrusion on the groundwater data obtained from the monitoring network in Korea. *Water* 33:127–146.

Ludwig, F., E. van Slobbe, and W. Cofino. 2014. Climate change adaptation and integrated water resource management in the water sector. *Journal of Hydrology.* 518:235–242.

Mahapatra, M., R. Ratheesh, A. S. Rajawat, S. Bhattacharya, and Ajai. 2013. Impact of predicted sea level rise on land use and land cover of Dahej Coast, Bharuch District, Gujarat, India. *International Journal of Geology, Earth and Environmental Sciences* 3 (2):21–27.

Mantoglou, A. and M. Papantoniou. 2008. Optimal design of pumping networks in coastal aquifers using sharp interface models. *Journal of Hydrology* 361 (1–2):52–63. doi: org/10.1016/j.jhydrol.2008.07.022.

Mantoglou, A., M. Papantoniou, and P. Giannoulopoulos. 2004. Management of coastal aquifers based on nonlinear optimization and evolutionary algorithms. *Journal of Hydrololgy* 297(1–4):209–228. doi: org/10.1016/j.jhydrol.2004.04.011.

Marfai, M. A. 2011. Impact of coastal inundation on ecology and agricultural land use case study in Central Java, Indonesia. *Quaestiones Geographicae* 30 (3):19–32.

Maurer, A. 2012. *Combined Source Infrastructure Assessment Model.* Master's dissertation, Department of Civil and Environmental Engineering, Colorado State University.

Nicholls, R. J. and C. Small. 2002. Improved estimates of coastal population exposure to hazards released. *EOS Transactions* 83:303–305.

NSW Public Works. 2011. *Brackish Groundwater: A Viable Community Water Supply Option?* Canberra: National Water Commission.

Palanisamy, H., A. Cazenave, B. Meyssignac, L. Soudarin, G. Wöppelmann, and M. Becker. 2014. Regional sea level variability, total relative sea level rise and its impacts on islands and coastal zones of Indian Ocean over the last sixty years. *Global and Planetary Change* 116:54–67.

Park, N., L. Cui, and L. Shi. 2009. Analytical design curves to maximize pumping or minimize injection in coastal aquifers. *Ground Water* 47(6):797–805.

Pendleton, E. A., E. R. Thieler, and S. J. Williams. 2004. Coastal vulnerability assessment of Cape Hatteras national seashore (CAHA) to sea level rise. In *Open-File Report 2004-1064,* edited by U.S. Department of Interior, U.S. Geological Survey, Reston, Virgina.

Pyne, R. D. G. 1995. *Groundwater Recharge and Wells, a Guide to Aquifer Storage Recovery.* Boca Raton, FL: Lewis Publishers.

Pyne, R. D. G. 2005. *Aquifer Storage Recovery: A Guide to Groundwater Recharge through Wells.* Gainesville, FL: ASR Systems.

Ranjan, S. P., S. Kazama, and M. Sawamoto. 2006. Effects of climate and land use changes on groundwater resources in coastal aquifers. *Journal of Environmental Management* 80(1):25–35.

Sarwar, M. G. M. 2005. *Impacts of Sea Level Rise on the Coastal Zone of Bangladesh.* Master's dissertation, Environmental Science, Lund University.

Shearer, T. R., S. J. Wagstaff, R. Calow, J. F. Muir, G. S. Haylor, and A. C. Brooks. 1997. The potential for aquaculture using saline groundwater. *BGS Technical Report.* Nottingham, UK, British Geologic Survey. (WC/97/058).

Sherif, M. M. and K. I. Hamza. 2001. Mitigation of seawater intrusion by pumping brackish water. *Transport in Porous Media* 43:29–44.

Shi, L. 2010. *Decision-Support Model for Sustainable Development of Groundwater Subject to Seawater Intrusion.* PhD thesis, Civil Engineering, Dong-A University.

Singh, A. 2014a. Groundwater resources management through the applications of simulation modeling: A review. *Science of the Total Environment* 499:414–423.

Singh, A. 2014b. Optimization modelling for seawater intrusion management. *Journal of Hydrology* 508:43–52.

Singh, A., S. Pathirana, and H. Shi. 2005. *Assessing Coastal Vulnerability: Developing a Global Index for Measuring Risk.* Nairobi: Division of Early Warning and Assessment.

Small, C. and R. J. Nicholls. 2003. A global analysis of human settlement in coastal zones. *Journal of Coastal Research* 19:584–599.

Song, S. H. and G. Zemansky. 2012. Vulnerability of groundwater systems with sea level rise in coastal aquifers, South Korea. *Environmental Earth Science* 65:1865–1876.

Soni, A. K. and P. R. Pujari. 2012. Sea-water intrusion studies for coastal aquifers: Some points to ponder. *The Open Hydrology Journal* 6:24–30.

Sreekanth, J. and B. Datta. 2010. Multi-objective management of saltwater intrusion in coastal aquifers using genetic programming and modular neural network based surrogate models. *Journal of Hydrology* 393:245–256.

Stokke, K. B. 2013. Adaptation to sea level rise in spatial planning—Experiences from coastal towns in Norway. *Ocean and Coastal Management.* 9:66–73.

Vorosmarty, C. J., P. Green, J. Salisbury, and R. B. Lammers. 2000. Global water resources: Vulnerability from climate change and population growth. *Science* 289:284–288.

WRA. 2001. *Integrating the Water Management of Watersheds and Coastal Areas in Trinidad and Tobago.* Port of Spain: The Water Resources Agency.

Yermiyahu, U., A. Tal, A. Ben-Gal, A. Bar-Tal, and J. Tarchitzky. 2007. Rethinking desalinated water quality and agriculture. *Science* 318:920–9211.

2

Gangetic Alluvial Plains: Uniqueness of the Aquifer System for Food Security and for Carbon Dioxide Sequestration

D. K. Chadha

CONTENTS

2.1 Introduction

The Ganga River Basin of India is an active foreland basin having an east–west elongated shape. The basin has been formed in response to the uplift of the Himalayas after the collision of Indian and Asian plates. The Ganga Basin forms one of the largest underground reservoirs in the world and came into existence as the result of sedimentation in the foredeep in front of the Himalayas. The sedimentation period belongs to the upper tertiary period and it continued all through the Pleistocene up to the present time. The various geophysical and exploratory-drilling investigations have indicated that the depth of the alluvial is variable from about 1000 m to over 2000 m and encompasses a number of aquifer zones, both fresh and saline. On the basis of various studies, depth to basement rock is shown in Figure 2.1.

2.2 Food Security

Intergovernmental Panel on Climate Change (IPCC) has indicated the adverse impact of climate change due to rising temperatures and extreme weather events such as tropical cyclones, associated storm surges, and extreme rainfall events on food production. Low-lying regions, including small islands, will face the highest exposure to rising sea levels, which will increase the risk of floods bringing more cultivable areas under submergence and degradation. The vulnerability of India in the event of climate change is more pronounced due to its ever-increasing dependency on agriculture, excessive pressure on natural resources, and poor coping mechanisms. The likely change in temperature and frequency and intensity of rainfall will impact yield decline as follows:

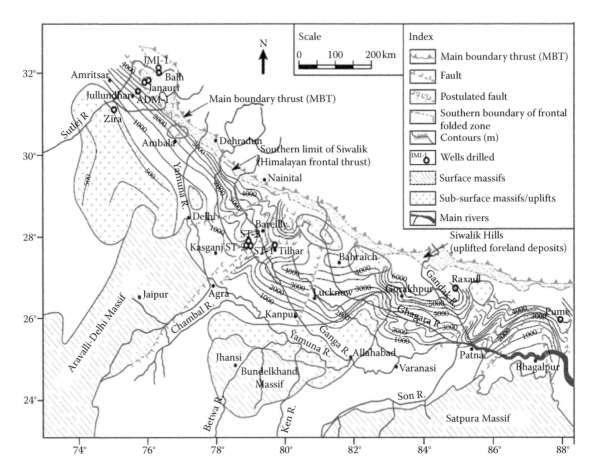

FIGURE 2.1
Basement depth contour map of Gangetic alluvial plains.

- A 1°C rise in mean temperature is likely to affect wheat yield in the heartland of green revolution. There is already an evidence of negative impacts on the yield of wheat and paddy in parts of India due to increased temperatures, increased water stress, and reduction in the number of rainy days.

- Crop-specific simulation studies, though not conclusive due to inherent limitations, project a significant decrease in cereal production by the end of this century.

- Irrigation requirements in arid and semiarid regions are estimated to increase by 10% for every 10°C rise in temperature. Rise in sea level is likely to have adverse effects on the livelihoods of fisher and coastal communities.

In India, the population is projected to increase to about 1.6 billion thereby increasing the food demand from the current production of 225 million ton (Mt) to 377 Mt by 2050. The increase in food production will evidently require more surface storages and development of deep aquifers to meet the stipulated irrigation water demand to the extent of 807 billion cubic meters (BCM) (minimum) in 2050; surface water is stipulated to contribute about 463 BCM and 344 BCM of groundwater. The Ganga River Basin constitutes the main dependable area for food security and therefore, dependence of developing groundwater resources will enhance in case the surface water resources are not available to increase the intensity of irrigation.

2.3 Groundwater Resource Potential

The estimation of groundwater resources is prepared separately as (1) dynamic resource (annual replenishment) potential and (2) in-storage (referred to as static) potential, that is, below the level of groundwater fluctuation.

2.3.1 Dynamic Groundwater Potential

The dynamic groundwater resource potential of Uttar Pradesh, which covers almost the entire Ganga Basin, shows that the annual recharge potential from various components of recharge to groundwater is estimated as 71.66 BCM and the groundwater draft for agriculture, domestic, and industrial uses is about 52.78 BCM. This indicates that the stage of groundwater development has already reached 74% of the annual recharge. The groundwater development being a continuous process, the groundwater draft will increase further and the groundwater development will exceed 100% of the recharge. Currently, the estimation of a dynamic resource potential and the groundwater draft, thereof for all the 75 districts of Uttar Pradesh, indicates that 11 of the districts have already fully utilized the dynamic recharge potential and are using in-storage potential by constructing deep tubewells.

2.3.2 In-Storage Groundwater Resources

In addition to the estimation of a dynamic resource potential, the in-storage potential below the dynamic resource potential has been estimated up to the depth of 400 m based on the cumulative thickness of aquifer zones as revealed by exploratory drilling. The in-storage potential of the alluvial aquifers of the Ganga Basin is estimated as 7769.1 BCM, out of which the in-storage potential of the alluvial aquifers of Uttar Pradesh has 3470 BCM. The estimated potential of the hard-rock aquifer of the Ganga Basin as a whole is estimated as 65 BCM and for Uttar Pradesh, part of the hard-rock aquifers of Ganga Basin has been computed as 30 BCM.

The in-storage potential of Ganga Basin and of Uttar Pradesh, both for the hard rock and the alluvial aquifers is shown in Table 2.1.

2.4 Disposition of Deep-Aquifer System (Freshwater)

The Ganga River Basin has been extensively explored mostly up to the depth of 400 m to identify the aquifer zones and their disposition. A number of exploratory boreholes have been drilled to understand the configuration of the different aquifer zones and the basement rock, wherever it is encountered, up to a depth of 400 m. The exploratory-drilling data were also supplemented with the drilling data of exploration undertaken by the oil sector to have a comprehensive understanding of the lateral and vertical extension of the different litho units.

The subsurface lithological section between Kadmaha-Ghur Khauli, indicating different aquifer zones, is shown in Figure 2.2.

The above section covers a total distance of about 275 km, and it shows that the basement rock (i.e., Vindhayan) is encountered at a very shallow depth toward the south. At the northern end, the Vindhyan basement rock is encountered at a depth of 4679-m bgl (Oil and Natural Gas Corporation [ONGC]).

The northern and southern parts of the section are represented by arenaceous sediments, while the central part is mostly represented by clayey (argillaceous) sediments. There are a number of granular zones recorded in the central part, which are formed by the river channels and their shifting in time and space. However, the granular zones are not extensive. The granular zones are mostly dominant in the northern part where four to five such zones can be demarcated up to 750 m.

The subsurface cross section along Haldwani–Pachomi is shown in Figure 2.3.

Another subsurface lithological section has been prepared to understand the change in the disposition of the aquifer system. The subsurface lithological section from Haldwani to Pachomi in the Bareilly district is over a distance of about 130 km across the Bhabhar and the Terai belts. It is observed that there are thick granular horizons in the north comprising coarse-grained sands, gravels, and boulders within the Bhabhar zone. The thickness diminishes toward the south and there is an increase in the content of argillaceous (clay) sediments.

The above two sections indicate that the deep aquifer zones are very thick and remain more or less unexploited. The future water requirement to enhance the food production and intensity of irrigation can be ensured if only 5% (i.e., 175 BCM) of the in-storage potential is exploited.

TABLE 2.1

The In-Storage Groundwater Resources of Uttar Pradesh and Ganga Basin

States	Static Fresh Groundwater Resource		
	Alluvium/ Unconsolidated Rocks (BCM)	Hard Rocks (BCM)	Total (BCM)
Uttar Pradesh	3470	30	3500
Ganga Basin	7769.1	65	7834.1

2.5 Climate Change

In India, CO_2 emissions have increased from 560 Mt in 1990 to 668 Mt in 1994 to 1069 Mt in 2000. Emission projections of CO_2 only due to fossil fuel burning indicate that the emission is going to increase to 1670 Mt in 2010 and 2934 Mt in 2020. India's initial national communication to the United Nations Framework Convention on

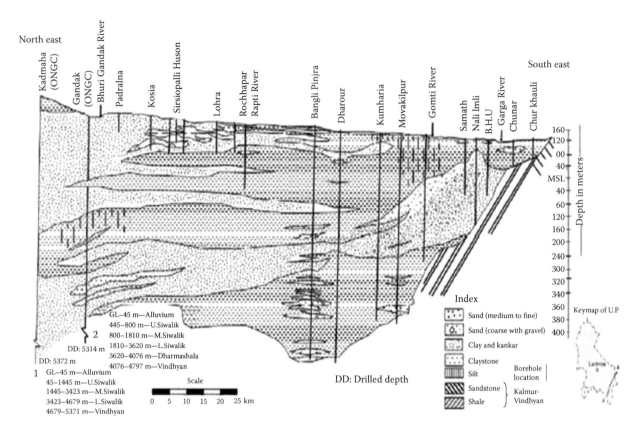

FIGURE 2.2
Subsurface cross section along Kadmaha–Ghur Khauli, Uttar Pradesh.

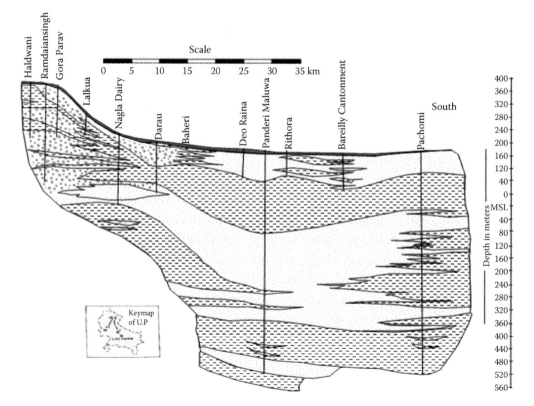

FIGURE 2.3
Subsurface cross section along Haldwani–Pachomi, Uttar Pradesh.

Climate Change (Ministry of Environment and Forest, 2004) has shown that the total CO_2 emitted in 1994 from all the sectors was 817,023 gigagrams (Gg) and removal by sinks was 23,533 Gg, resulting in a net emission of 793,490 Gg of CO_2, which was mainly contributed by activities in the energy sector, industrial processes, land use change, and forestry.

2.6 Methods of Mitigation

It is important to conduct scientific studies to identify the possible geological sinks near the emitting sources to capture and sequestrate them.

Capture of CO_2 from the major stationary sources and its storage into deep geological formations is considered a potential mitigation option to combat global warming. Geological storage of CO_2 can be undertaken in a variety of geological settings in sedimentary basins. Geological storage options for CO_2 are

- Depleted oil and gas reservoirs
- Deep unmineable coal seams/enhanced coal bed methane recovery
- Oceans
- Other geological media (basalt, shale, and cavities)
- Deep unused saline water-saturated formations

The deep-saline aquifers are present in the different sedimentary basins near the power plants, which makes it easy to capture and sequestrate in the saline aquifers that is technically feasible and very cost-effective.

Moreover, the saline aquifers provide a significant advantage because of the following reasons:

- They are deep seated and not exploited for any surface water use by irrigation, industry, or domestic purposes.
- Owing to high salinity, it is not economical to exploit them but they have now been recognized for their usefulness in injection of wastewater, radioactive waste, and carbon dioxide sequestration.
- These aquifers being present inland can help in achieving near-zero emissions for the existing power plant and industrial units located above the deep-saline aquifers.
- Negative impact and unintended damages are limited being deep seated.
- India is one of the few countries having totally unutilized saline aquifers of sufficient thickness and are regionally extensive.

TABLE 2.2

Storage Capacity for Geological Storage Options

Geological Storage Option	Global Capacity	
	Reservoir-Type Lower Estimate of Storage Capacity of CO_2 (in Gt)	Upper Estimate of Storage Capacity of CO_2 (in Gt)
Depleted oil and gas fields	675	900
Unmineable coal seams	3–15	200
Deep-saline reservoirs	1000	Uncertain, but possibly 10^4

- The options to capture and sequestrate the carbon dioxide lie in the location of different geological sinks such as abundant coalmines, oil exploratory wells, and offshore or deep-saline aquifers as discussed. The recent studies that have estimated the storage capacity for different geological storage options is shown in Table 2.2.

2.7 Emission of Greenhouse Gases

The state of Uttar Pradesh has a number of coal-based power plants that emit huge quantities of greenhouse gases. The distribution of power plants is shown in Figure 2.4.

The capacity of power plants and the estimated emission thereof is shown in Table 2.3. It is estimated that the emission from the power plant is about 71.89 Mt. It is apparent that the reduction in the emission of carbon dioxide is of great importance under the different international agreements to mitigate the emission of carbon dioxide to prevent further global warming. In this context, it may be noted that the geological sinks such as an ocean, deep coalmines, and depleted oil and coal reservoirs are far off from the emitting sources. It is thus required to map the deep-saline aquifers, in part of the Ganga Basin, below 700-m depth to meet part of the screening criteria established for carbon dioxide sequestration in deep-saline aquifers.

2.8 Saline Aquifers of Ganga River Basin

2.8.1 Disposition of Shallow-Saline Aquifers

A number of studies have been conducted through geophysical and hydrogeological survey identifying the disposition of saline aquifers in unconfined aquifers. The areal extent of brackish (saline) groundwater in Uttar Pradesh is shown in Figure 2.5.

FIGURE 2.4
Location of thermal power plants, Uttar Pradesh.

TABLE 2.3

Power Station in Uttar Pradesh with Installed Capacity and CO_2 Emission

Power Station	Installed Capacity (mw)	Power Generation in 2009 (mw)	CO_2 Emissions in 2009 (Mt)	Storage Capacity of Ganga Basin (Mt)
Obra Tps	1322	511.18	5.49	
Anpara Tps	1630	1252.62	10.58	
Panki Tps	210	122.37	1.54	
Parichha Tps	640	371.11	4.18	
Harduaganj Tps	220	141.98	1.58	
Singrauli Super Tps	2000	1723.51	14.75	48.3
Rihand Tpp	2000	1785.73	14.94	
Nctpp	1820	824.77	7.07	
Feroz Gandhi Unchahar Tpp	1050	940.18	8.09	
Tanda Tpp	440	359.81	3.67	
TOTAL	11,332	8033.26	71.89	

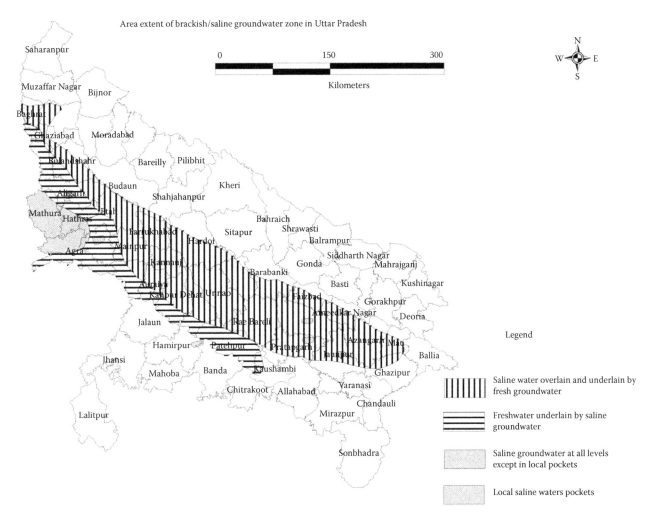

FIGURE 2.5
Areal extent of brackish/saline groundwater zone in Uttar Pradesh, Ganga Basin.

The recent mapping of the shallow-saline aquifers shows that the saline aquifers occurred at varying depths all along the area from Meerut to Ghazipur. The distribution of saline aquifers has been divided into three groups (Trivedi, 2012).

- The groundwater salinity commencing from a very shallow level and extending up to the bedrock with the localized pockets of fresh groundwater, namely:
- Salinity occurring at a depth overlain and underlain by freshwater aquifers and
- Salinity occurring at two depth zones—the shallow and deeper one separated by fresh water aquifers.

The distribution of saline aquifers based on geophysical survey and monitoring of the water quality is shown in Figure 2.6.

2.8.2 Disposition of Deep-Saline Aquifers

The understanding of the disposition of deep-saline aquifers is of great significance under a climate change scenario as in one of the potential geological sinks, a deep vertical electrical sounding (VES) survey was carried out at 12 selected locations where the shallow aquifers are also saline in nature. The depth of the saline aquifers was probed up to a depth of 1100 m to satisfy the screening criteria to identify the deep-saline aquifers more than 750-m bgl. As the carbon dioxide increases in density with depth and becomes a super critical fluid below 750-m bgl therefore it diffuses better than either gases or liquids.

It was thus important to map the deep-saline aquifers in the part of the Ganga Basin where some primary information was available of the possible presence of deep-saline aquifers. To identify the deep-saline aquifers, a deep-resistivity survey was carried out at 12 locations with depth ranging from 850 to 1100 m. The basic

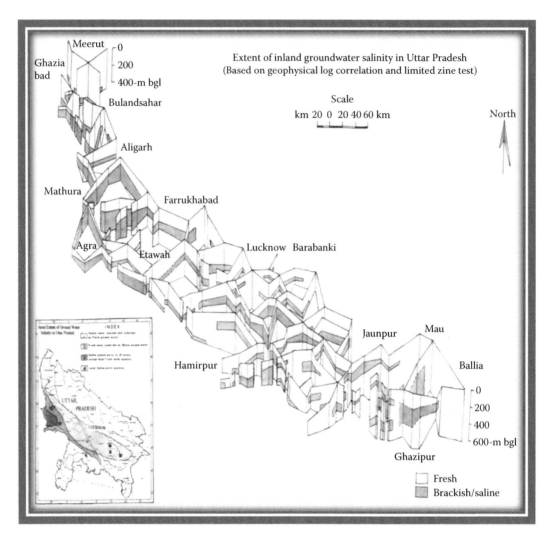

FIGURE 2.6
Extent of inland groundwater salinity in Ganga Basin (UP).

data and the interpreted result for VES 1 of the Ganga Basin are shown in Figure 2.7 as a representation of the 12 such curves with interpreted results. The summarized results are shown in Table 2.4.

Subsurface lithological sections and the saline fresh interface have been prepared using different software. As shown in Figures 2.8 and 2.9, it revealed the definite presence of thick-saline aquifers below the freshwater aquifers but at varying depths. The saline aquifers overlie the Shivalik rocks and other rock formations but the continuity of the deep-saline aquifers in the basement rocks has not been deciphered. The exploratory drilling by ONGC has also indicated the presence of deep-saline aquifers overlying the bedrock as shown in Figure 2.10.

Figure 2.8 shows the fresh–saline groundwater interface based on deep-resistivity measurement, in the Budaun and Bareilly districts, Uttar Pradesh.

In Figure 2.9, a subsurface lithological cross section shows groundwater-quality variation with depth based on Schlumberger deep-resistivity-sounding results, in the Bareilly and Budaun districts, Uttar Pradesh.

2.9 First Approximation of Carbon Dioxide Storage Potential

Koide et al. (1992) studied the carbon dioxide capacity estimated for deep-saline aquifers and made several assumptions in the calculation of storage capacity based on the idea that the carbon dioxide will be stored by being dissolved in water. On the underground disposal of carbon dioxide, the calculation of storage capacity assumed that 3% of the aquifer volume would be trapped.

The methodology adopted for the purpose of a first approximation of the storage capacity involves the consideration of the water held in the formation and that

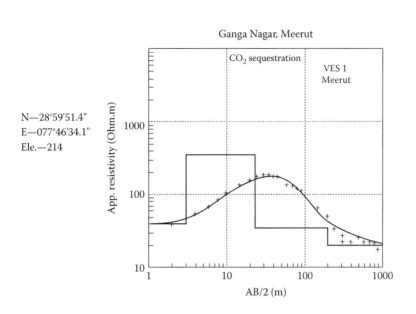

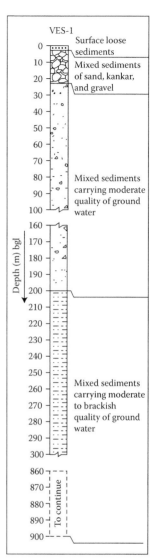

			Resistivity layer interpretation			
No.	Resistivity (ohm.m)	Depth (m)	Thickness (m)	From (m)	To (m)	Tentative lithology
1	40	3	3	0	3	Surface loose soil
2	350	23	20	3	23	Mixed sediments of sand, kankar, and gravel
3	35	200	177	23	200	Mixed sediments carrying moderate quality of groundwater
4	20	Indeterminate		Below 200		Mixed sediments carrying moderate to brackish quality of groundwater

FIGURE 2.7
Composite VES data processing for the recommended site VES 01, Ganga Basin.

the water can be displaced to incorporate the carbon dioxide in the aquifer. Equation 2.1 used for computing the first approximation of the storage potential in a part of the Ganga Basin is given below:

$$Ff = VSA \times \% \text{ fluid} \qquad (2.1)$$

where
Ff = fluid fraction
VSA = volume of the saline aquifer
% fluid = percentage of the fluid fraction

On the basis of the present mapping of the deep-saline aquifers, the potential area in the part of the area of the Ganga Basin investigated for deep-saline aquifers is estimated as 75 km² and the thickness of saline aquifers is about 200 m, that is, the total saturated volume is 15 km³.

Therefore, the carbon dioxide storage capacity with fluid fraction is shown in Table 2.5.

2.10 Conclusions

The dynamic groundwater resources will be fully exploited to meet the immediate water demand but the deep aquifers that have an area fairly well established having an in-storage potential of 3500 BCM can be exploited to the extent of 5%, that is, 175 BCM for food security.

The deep-saline aquifers are now recognized worldwide as the best option for CO_2 sequestration as they

TABLE 2.4

Summary Statement of Results of Ganga Basin, Budaun District, Uttar Pradesh

						Global Hydrogeological Solutions				
						Project: Development of Screening Criteria for Saline Aquifers and Other Geological Sinks in Ganges Basin, Adjoining Rajasthan and Vindhyan Basins for CO_2 Sequestration				
						Resistivity Layer Interpretation				
Ves No.	Location	Latt/Long and Surface Elevation (m)	No.	Resistivity (ohm.m)	Depth (m)	Thickness (m)	Ele. (m)	From (m)	To (m)	Tentative Lithology
1	Ganganagar, Meerut	N—28°59′51.4″ E—077°46′34.1″ Ele—214	1	40	3	3	211	0	3	Surface loose soil
			2	350	23	20	191	3	23	Mixed sediments of sand, kankar, and gravel
			3	35	200	177	14	23	200	Mixed sediments carrying moderate quality of groundwater
			4	20	Indeterminate			Below 200		Mixed sediments carrying moderate to brackish quality groundwater
2	Abdullapur, Meerut	N—28°59′18.5″ E—077°45′08.7″ Ele—203	1	90	8	8	195	0	8	Surface loose sediments
			2	800	21	13	182	8	21	Partially dry sediments consisting of sand, boulders, and kankars
			3	15	45	24	158	21	45	Mixed sediments of sand and clay carrying brackish quality of groundwater
			4	35	Indeterminate			Below 45		Mixed sediments of sand, clay, and kankar carrying moderate quality of groundwater
3	Abdullaganj Ujhani, Budaun	N—28°00′56.0″ E—079°02′57.0″ Ele—149	1	80	3	3	146	0	3	Surface loose sediments
			2	300	25	22	124	3	25	Mixed sediments consisting of sand, boulders, and kankars
			3	50	145	120	4	25	145	Mixed sediments of sand, kankar, and clay carrying fresh quality of groundwater
			4	15	Indeterminate			Below 145		Mixed sediments of sand and clay carrying brackish quality of groundwater
4	Anola, Bareilly	N—28°00′57.2″ E—077°02′59.2″ Ele—148	1	105	4	4	144	0	4	Surface loose sediments
			2	10	11	7	137	4	11	Clay predominant mixed sediments
			3	70	250	239	−102	11	250	Mixed sediments of sand, kankar, and clay carrying fresh quality of groundwater
			4	10	Indeterminate			Below 250		Mixed sediments of sand and clay carrying brackish quality of groundwater
5	Nagla Basela, Budaun	N—28°03′05.1″ E—079°25′22.9″ Ele—128	1	22	6	6	122	0	6	Surface loose sediments
			2	32	720	714	−592	6	720	Mixed sediments of sand, kankar, and clay carrying a moderate quality of groundwater
			3	6	Indeterminate			Below 720		Mixed sediments of sand and clay carrying brackish quality of groundwater

(Continued)

TABLE 2.4 (*Continued***)**

Summary Statement of Results of Ganga Basin, Budaun District, Uttar Pradesh

Global Hydrogeological Solutions

Project: Development of Screening Criteria for Saline Aquifers and Other Geological Sinks in Ganges Basin, Adjoining Rajasthan and Vindhyan Basins for CO_2 Sequestration

Ves No.	Location	Latt/Long and Surface Elevation (m)	No.	Resistivity (ohm.m)	Depth (m)	Thickness (m)	Ele. (m)	From (m)	To (m)	Tentative Lithology
						Resistivity Layer Interpretation				
6	Sobrangpur, Budaun	N—28°02′35.1″ E—079°10′17.2″ Ele—150	1	180	7	7	143	0	7	Surface loose sediments
			2	700	31	24	119	7	31	Partially dry sediments consisting of sand, boulders, and kankars
			3	25	732	701	−582	31	732	Mixed sediments of sand, kankar, and clay carrying saline-quality groundwater
			4	7	Indeterminate			Below 732		Mixed sediments of sand and clay carrying brackish quality of groundwater
7	Bamiana, Bareilly	N—28°15′10.7″ E—079°20′13.4″ Ele—144	1	70	10	10	134	0	10	Loose mixed sediments
			2	42	270	260	−126	10	270	Mixed sediments carrying a moderate quality of groundwater
			3	20	Indeterminate			Below 270		Mixed sediments carrying moderate-to-brackish quality of groundwater
8	Digoi, Budaun	N—28°12′30.8″ E—079°08′54.2″ Ele—153	1	65	1.5	1.5	151.5	0	1.5	Surface loose sediments
			2	2440	6	4.5	147	1.5	6	Dry sediment consisting of pebble, gravel, and sand
			3	70	190	184	−37	6	190	Mixed sediments of sand, kankar, and clay carrying fresh quality of groundwater
			4	20	Indeterminate			Below 190		Mixed sediments of sand and clay carrying brackish quality of groundwater
9	Berojpur, Budaun	N—28°15′47.3″ E—078°53′51.2″ Ele—158	1	250	5	5	153	0	5	Surface loose sediments
			2	1150	11	6	147	5	11	Dry sediment consisting of pebble, gravel, and sand
			3	46	140	129	18	11	140	Mixed sediments of sand, kankar, and clay carrying fresh quality of groundwater
			4	23	Indeterminate			Below 140		Mixed sediments of sand and clay carrying moderate-to-brackish quality of groundwater
10	Karia Mai, Budaun	N—28°12′44.0″ E—078°49′30.2″ Ele—133	1	120	4	4	129	0	4	Surface loose sediments
			2	1200	10	6	123	4	10	Dry sediment consisting of pebble, gravel, and sand
			3	60	100	90	33	10	100	Mixed sediments of sand, kankar, and clay carrying fresh quality of groundwater
			4	20	Indeterminate			Below 100		Mixed sediments of sand and clay carrying brackish quality of groundwater

(*Continued*)

TABLE 2.4 (*Continued*)

Summary Statement of Results of Ganga Basin, Budaun District, Uttar Pradesh

					Global Hydrogeological Solutions					
					Project: Development of Screening Criteria for Saline Aquifers and Other Geological Sinks in Ganges Basin, Adjoining Rajasthan and Vindhyan Basins for CO$_2$ Sequestration					
					Resistivity Layer Interpretation					
Ves No.	Location	Latt/Long and Surface Elevation (m)	No.	Resistivity (ohm.m)	Depth (m)	Thickness (m)	Ele. (m)	From (m)	To (m)	Tentative Lithology
11	Sadullagang, Budaun	N—28°08′02.8″ E—079°25′18.5″ Ele—110	1	25	8	8	102	0	8	Surface loose sediments
			2	45	345	337	−235	8	345	Mixed sediments of sand and clay carrying fresh groundwater
			3	4	550	205	−440	345	550	Mixed sediments of sand and clay carrying brackish quality of groundwater
			4	2650	Indeterminate			Below 550		Existence of hard rocks (Siwalik?)
12	Ghatpuri, Budaun	N—28°08′42.8″ E—079°12′10.2″ Ele—141	1	80	6	6	135	0	6	Surface loose sediments
			2	400	33	27	108	6	33	Mixed sediments consisting of sand, boulders, and kankars
			3	34	245	212	−104	33	245	Mixed sediments of sand and clay carrying moderate quality of groundwater
			4	25	Indeterminate			Below 245		Mixed sediments of sand and clay carrying moderate-to-brackish quality of groundwater

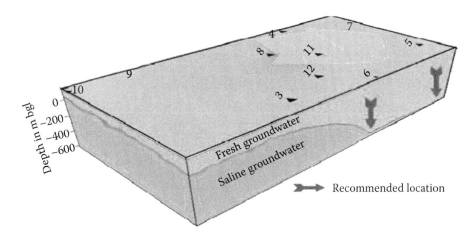

FIGURE 2.8
Sub-surface-layered feature showing a variation in depth to a saline groundwater zone.

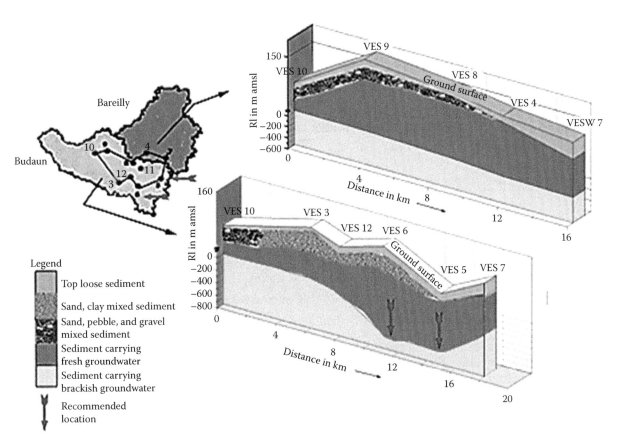

FIGURE 2.9
Panel diagram showing a variation in depth to a saline groundwater zone.

have a great promise for being regionally extensive and not utilized for any purpose. The present geological and geophysical mapping of the deep-saline aquifers in parts of the Ganga Basin has revealed that it has deep-saline aquifers in the quaternary alluvial formation.

The total storage capacity for carbon dioxide sequestration is about 48.3 Mt; therefore, more areas are required to be investigated for deep-saline aquifers so that the present emission from the power plants can be mitigated in saline aquifers without requiring the need to transfer the gases to other geological sinks.

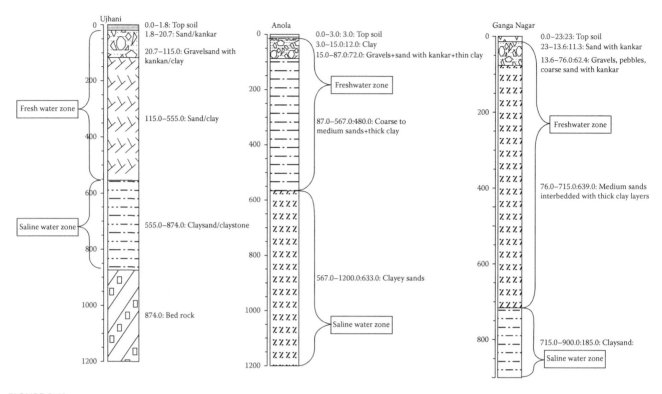

FIGURE 2.10

Deep-saline aquifers. (Adapted from Oil and Natural Gas Corporation Limited (ONGC). 1994–1998. Unpublished drilling report.)

TABLE 2.5

Carbon Dioxide Storage Capacity with Fluid Fraction

Sl. No.	Name of the Basin	Volume of the Saline Aquifer (cu. km)	Reservoir Porosity	Storage Efficiency	Density (km/cu. m)	Mt CO$_2$
1	Uttar Pradesh—a part of Ganga Basin	15	0.23	0.02	700	48.3

Notes: I. CO$_2$ storage efficiency is estimated as 2% and all the aquifers are assumed closed.

II. CO$_2$ density in the offshore sites is assumed to be 700 kg/m^3.

Acknowledgment

This research program has been supported by the Department of Science and Technology (IS-STAC Division), Government of India. Thanks to Dr. B.B. Trivedi and to Dr. A.N. Bhowmick for interpreting the field geophysical data. The technical assistance provided by Nidhi Jha is duly acknowledged.

Bibliography

Central Ground Water Board (CGWB). 2002. Hydrogeology and deep ground water exploration in Ganga Basin, unpublished report.

Central Ground Water Board (CGWB). 2011. Dynamic ground water resources of India, unpublished report.

Central Ground Water Board (CGWB). 2012. Uttar Pradesh state geophysical report, unpublished report.

Chadha, D. K. and S. P. Sinha Ray. 2001. Inland ground water salinity in India, Central Ground Water Board, Ministry of Water Recourses, Government of India, unpublished report.

Chadwick, A., R. Arts, C. Benstone, F. May, S. Thibeau, and P. Zweigel. 2008. Best practices for the storage of CO$_2$ in saline aquifer—Observations and guidelines from the SACS and CO$_2$ Store Projects. Nottingham, UK, British Geological Survey, 267pp. (British Geological Survey Occasional Publication, 14).

Chandra, P. C., B. B. Trivedi, and M. Adil. 1996. Geophysical log study of groundwater salinity in Ganga Alluvial Plains of Uttar Pradesh (presented at *Recent Researches in Sedimentary Basins*, BHU, Varanasi).

Cooperative Research Centre. 2008. *Storage Capacity Estimation, Site Selection and Characterization for CO$_2$ Storage Projects*, Cooperative Research Centre for Greenhouse Gas Technologies, Canberra, CO$_2$ CRC Report No: RPT08-1001.

Flett, M., A. Gurton, and I. J. Taggart. 2004. Heterogeneous saline formations: Long-term benefits for geo-sequestration of greenhouse gases. *Seventh International Conference on Greenhouse Gas Technologies*, September 2004, Vancouver, Canada.

Garcia, J. E. 2003. Fluid dynamics of carbon dioxide disposal into saline aquifers. A dissertation submitted in partial satisfaction of requirements for the degree of Doctor of Philosophy in Engineering—Civil and environmental engineering in the Graduate Division of the University of California, Berkeley.

Global Hydrogeological Report. 2013. Development of screening criteria for saline aquifers and other geological sinks in Ganges Basin adjoining Rajasthan and Vindhyan Basin for CO_2 sequestration, unpublished report.

Global Hydrogeological Solutions. 2007. Study on identification of deep underground aquifers and their suitability for carbon dioxide sequestration, unpublished report.

Koide, H., Y. Tazaki, Y. Noguchi, S. Nakayama, M. Iijima, K. Ito, and Y. Shindo. 1992. Subterranean containment and long term storage of carbon dioxide in unused aquifers and depleted natural gas reservoirs. Energy conversion management. *Proceedings of the First International Conference on Carbon Dioxide Removal*, Amsterdam.

Ministry of Environment and Forest. 2004. India's Initial National Communication to the United Nations Framework Convention on Climate Change.

Oil and Natural Gas Corporation Limited (ONGC). 1994–1998. Unpublished drilling report.

Parkash, B. and S. Kumar. 1991. The Indogangetic Basin. In: Tandon, S.K., Pant, C.C., and Casshyap, S.M. (eds.), *Sedimentary Basins of India: Tectonic Context*. Gyanodaya Prakashan, Nainital, India, pp. 147–170.

Pruess, K. 2006. Carbon sequestration in saline aquifers. *Presented at the Forum on Carbon Sequestration*, Lawrence Berkeley National Laboratory, Yale University.

Trivedi, B. B. 2012. *Inland Groundwater Salinity in Uttar Pradesh*. Central Groundwater Board (CGWB) Lucknow.

Umar, R. 2006. Hydrogeological environment and groundwater occurrences of the alluvial aquifers in parts of the Central Ganga Plain, Uttar Pradesh, India, *Hydrogeology Journal*, 14: 969–978.

3

Assessment of Groundwater Resources in Brazil: Current Status of Knowledge

Fernando A. C. Feitosa, João Alberto O. Diniz, Roberto Eduardo Kirchheim,
Chang Hung Kiang, and Edilton Carneiro Feitosa

CONTENTS

3.1 Geological Framework and Tectonic Scenario

The South American continent hosts three large tectonic domains: the Andes, the Patagonic platform, and the South American platform. Exception should be made to a small part of Venezuela that belongs to the Caribbean plate. The South American plate, where Brazilian territory is situated, corresponds to the continental portion of the homonymous plate, which has remained stable as a foreland against the Andean and Caribbean mobile belt and continental drifting during the Meso-Cenozoic period. It has undergone multiple tectonic cycles between the Paleoarchean up to the Ordovician resulting in a complex framework. Phanerozoic covers have developed since then, setting the beginning of its stabilization phase (CPRM, 2003). The Brazilian orogenic cycle activities lasted up to the Upper Ordovician/Lower Silurian whereby the actual tectonic framework of the Brazilian territory has been built. Microcontinents and continental blocks were transformed giving rise to the actual cratonic areas (Amazônico, São Francisco, São Luis, and Paraná) allowing ocean development (Borborema, São Francisco, Goiano, and Adamastor), where a whole set of sedimentary rocks, insular and juvenile continental arc portion shad has undergone metamorphism, deformation, and emplacement of granitic intrusions across multiple events. During its stabilization period, large synecleses have been developed, such as Amazonas (500,000 km^2), Solimões (600,000 km^2), Parnaíba (700,000 km^2), and Chaco-Paraná (1,700,000 km^2). Besides the large basins, many other small ones were originated (Parecis/Alto Xingu, Alto Tapajós, Tacutu, Recôncavo/Tucano/Jatobá, Araripe, Iguatu, Rio do Peixe, and Bacia Sanfranciscana). Across its continental margin, a great number of mesozoic basins have been developed (Pelotas, Santos, Campos, Espírito Santo/Mucuri, Cumuruxatiba, Jequitinhonha/Camumu/Almada/Jacuípe, Sergipe/Alagoas, Pernambuco/Paraíba, Potiguar, Ceará, Barreirinhas, Pará/Maranhão, Foz do Amazonas, Cassiporé, Marajó, Bragança/São Luís, Barra de São João, and Taubaté). Widespread Cenozoic deposits with heterogeneous thickness cover large portions of the territory. The main units are formation Solimões, Içá, Boa Vista, Pantanal, Araguaia, and Barreiras. The continental area occupied by sedimentary basins is 4,898,050 km^2, from which 4,513,450 km^2 (70%) are intracratonic and the remaining 384,600 km^2 (30%) are lying on the continental margin. Figure 3.1 presents the main basins and sedimentary cover within Brazil. The cratonic areas are composed of plutonic rocks, gneisses, migmatites TTG, and greenstone belts sequences while in the orogenic belts there is a predominance of metasedimentary sequences and intrusive bodies. Some neo-proterozoic and cambri-ordovician basins such as AltoParaguai, Bambuí, Chapada Diamantina, Paranoá, Santo Onofre, Estancia, Rio Pardo, and Jaibaras, among others contain sedimentary sequences bearing primary structures and low metamorphic grades behaving mostly as fractured aquifers.

3.2 The Hydrogeological Map of Brazil

The hydrogeological map of Brazil, launched by the Brazilian Geological Survey—CPRM/SGB at the end of 2014, represents a synthesis of the hydrogeological information data sets available in the country. It aims at offering an overview of the water well location, aquifer use, and groundwater potential nationwide. It constitutes five main thematic layers.

3.2.1 Planialtimetry Base Map

The planialtimetry base map was obtained from the vector base—1:1.000.000 BCIM/IBGE (Brazilian Institute for Geography and Statistics 2010)—generated throughout the integration of the World International Chart (CIM) with the following information categories: hydrography, landform, political boundaries, transport system, economical structure, energy, communication, reference points, and vegetation.

3.2.2 Geological Database

The geological database was obtained from the GIS Brazil from the CPRM (2003) and based on a simplification of unit attributes and conversion into hydrogeology characteristics, such as groundwater transmissivity and storage.

3.2.3 Tubular Well Database

Data on tubular wells were taken from the SIAGAS—Groundwater Information System, operated and kept by the CPRM/SGB, which is equipped with query modules and report preparation modules. Regarding the hydrogeological map development, a set of 241,692 tubular wells was available for analysis.

3.2.4 Water-Level Database

Water level contour lines based on groundwater level data were developed for the following regional aquifers: Boa

Bacias e Coberturas Sedimentares Brasileiras

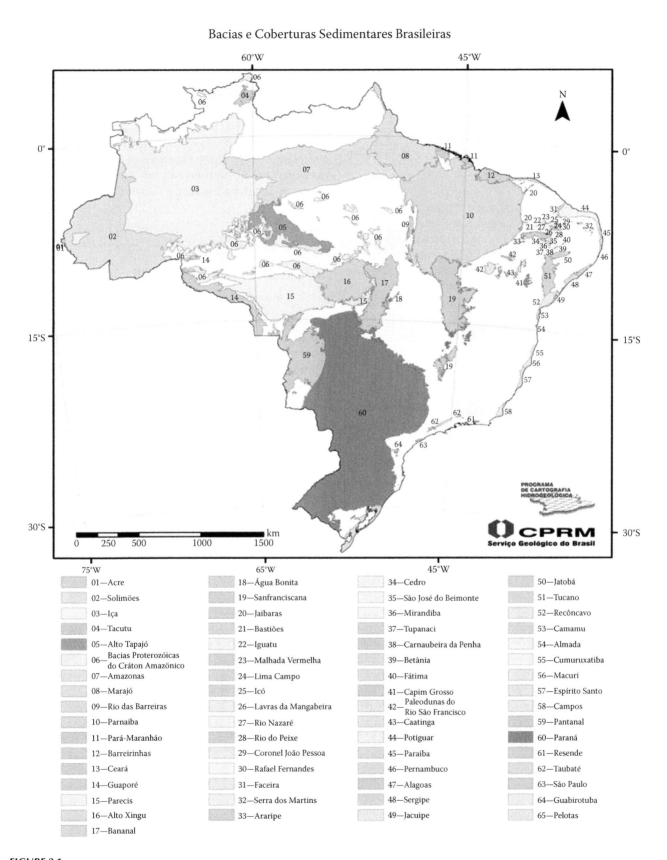

FIGURE 3.1
Basins and sedimentary covers of Brazil. (Modified from Diniz, J. A. O. et al. 2014. de, Mapa hidrogeológico do Brasil ao milionésimo: Nota técnica. Recife: CPRM—Serviço Geológico do Brasil, Recife.)

Vista, Itapecuru, Parecis, Guarani, Cabeças, and Urucuia. These outputs were transformed into potentiometric surfaces using topographical data gathered in the field with GPS. When such data were missing, groundwater level was determined after a digital terrain model (DTM).

3.3 Hydrological Database

The concept of "hydrographic regions" defined by the Water Resources National Council—CNRH has been adopted. The country was divided into 12 main hydrographic regions: Amazônica, Tocantins-Araguaia, Atlântico Nordeste Ocidental, Parnaíba, Atlântico Nordeste Oriental, São Francisco, Atlântico Leste, Atlântico Sudeste, Paraná, Paraguai, Uruguai, and Atlântico Sul.

The hydrogeological potential of each one of the mapped units has been classified according to the contribution of Struckmeier and Margat (1995), who defined six classes (Table 3.1):

1. Very high
2. High
3. Moderate
4. Generally low but locally moderate
5. Generally low but locally very low
6. Nonproductive or nonaquifer

According to this methodology, there are 18 hydrostratigraphic mapping units that have similar storage and transmissibility properties with production classes

at the same magnitude order, whose attributes need to be described. The map has also used the International Legend for Hydrogeological Maps developed by UNESCO (1970). Rebouças et al. (1969) and SADC (2009) were also used as important references in this chapter.

The hydrogeological map of Brazil (Figure 3.2) presents an innovative feature dealing with a simplification of the geological background information together with complementary representations of outcropping and nonoutcropping aquifers within its thematic layout. It has been developed within a GIS under a 1:1000.00 scale.

3.4 Paleozoic Sedimentary Basins

3.4.1 Amazonas Sedimentary Basin (1, 2, 3, 7)

The Amazonas Province sedimentary basins (Acre, Solimões, and Amazonas) start at the Andean region where they spread over into the Atlantic coast assuming a fan layout. Their greatest occurrence area corresponds to the flood plains of the Solimões River.

The Acre Basin is situated at the Brazilian territory of the Marañon–Ucayali–Acre Basin, whose total area comprises 905,000 km^2 (CPRM, 2003), being the most distal part of the sedimentary edge dated between the Cretaceous and the Pliocene. The Iquitos arc acts as its eastern border toward the Solimões Basin. It presents thicknesses up to 6000 m distributed into four supersequences: (i) Permian–Carboniferous, (ii) Jurassic, (iii) Cretaceous, and (iv) Tertiary.

The Permian–Carboniferous supersequence comprises the Apuí Formation, made up of a clastic edge of

TABLE 3.1

Productivity Classes

Q/s (m³/h/m)	T (m²/s)	K (m/s)	Q (m³/h)	Productivity	Classes
≥4.0	≥10^{-2}	≥10^{-4}	≥100	Very high—Regional relevance (supply source for urban and irrigation demands). Aquifers with national importance.	(1)
2.0 ≤ Q/s < 4.0	10^{-3} ≤ T < 10^{-2}	10^{-5} ≤ K < 10^{-4}	50 ≤ Q < 100	High—Same relevance of class 1 in terms of supply demands, but less-productive aquifers.	(2)
1.0 ≤ Q/s < 2.0	10^{-4} ≤ T < 10^{-3}	10^{-6} ≤ K < 10^{-5}	25 ≤ Q < 50	Moderate—Source of water supply for small communities, factories, and small irrigation scheme demands.	(3)
0.4 ≤ Q/s < 1.0	10^{-5} ≤ T < 10^{-4}	10^{-7} ≤ K < 10^{-6}	10 ≤ Q < 25	Generally low, but locally moderate—Source of water supply for local private demands.	(4)
0.04 ≤ Q/s < 0.4	10^{-6} ≤ T < 10^{-5}	10^{-8} ≤ K < 10^{-7}	1 ≤ Q < 10	Generally low but locally very low—Source of intermittent water supply for local private demands.	(5)
<0.04	<10^{-6}	<10^{-8}	<1	Nonproductive or nonaquifer—Insignificant water supply. Extraction restricted to manual devices.	(6)

Source: Modified from Struckmeier, W. F. and J. Margat. 1995. *Hydrogeological Maps: A Guide and a Standard Legend.* International Association of Hydrogeologists—Hannover (International Contributions to Hydrogeology; Vol. 17).

Q/s = specific yield; Q = flow rate; T = transmissivity; K = hydraulic conductivity.

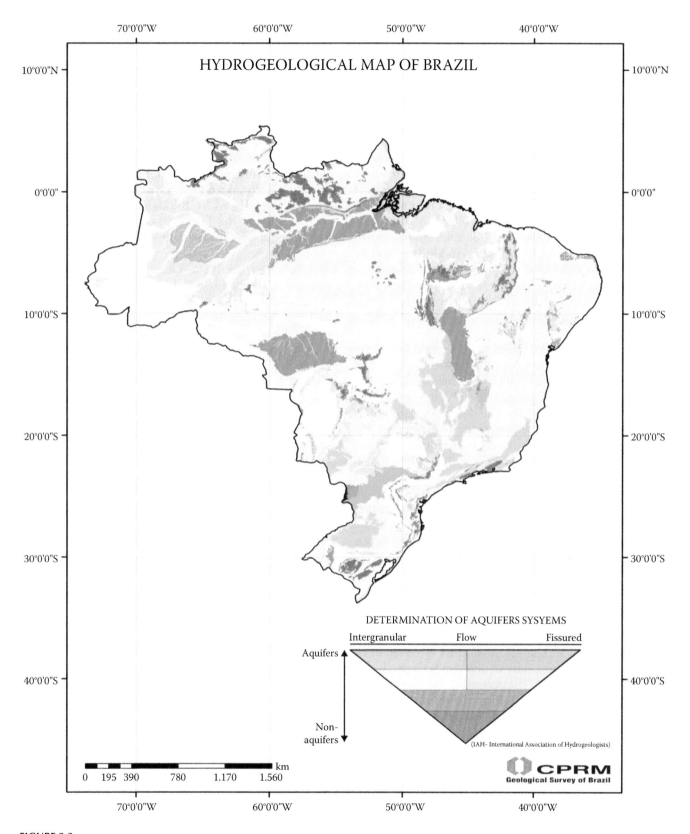

FIGURE 3.2

Hydrogeologic map of Brazil. (Modified from CPRM—Serviço Geológico do Brasil. 2014. Mapa Hidrogeológico do Brasil ao Milionésimo. Organizadores: Diniz, J. A. O., Monteiro, A. B., De Paula, T. L. F., and Silva, R. C. Recife.)

conglomerates, Cruzeiro do Sul, containing carbonates, evaporites, and sandstones (Rio do Moura). The Jurassic supersequence is entirely made up of the JuruáMirim formation bearing sandstones and red beds intercalated with evaporites and basalt flows, deposited at continental environment. Several formations belong to the Cretaceous supersequence: Moa, Rio Azul, Divisor, and Ramón, which are constituted by sandstones, shales, and calcarenites from fluviatile–lacustrine environments. The Tertiary supersequence is represented by the Solimões formation with onlap deposition against the basement. Together, they both (the Cretaceous and Tertiary sequences) sum up to 3000 m of thickness (Milani and Zalán, 1998). The Pliocene sandy–clayish sediments of the Solimões formation and the Pleistocene deposits of the Içá formation cover the entire basin.

The Solimões Basin has an area of about 500,000 km^2 and a total sediment fill of 3800 m, divided by clear marked discordances building up six supersequences (Eiras et al., 1994).

The ordovician and silurian–devonian supersequences comprising, respectively, Benjamim Constant formation (neritic clastic) and Jutaí formation (clastic and neritic limestone) are restricted to the Jandiatuba subbasin (Eiras et al., 1994a). The devonian–carboniferous supersequence encompasses the marine sedimentary rocks and glacial–marine rocks from the Marimari group (Uerê and Jandiatuba formation), which outreach the Caruari arc, extending up to the Juruá subbasin.

The carboniferous–permian supersequence is made up of clastic sediments, limestones, marine evaporites, and continental evaporites from the Tefé group (Juruá, Carauari, and Fonte Boa formations). The Cretaceous sequence corresponds to the fluviatile deposits of the Alter do Chão formation, which are preserved due to the subsidence effects related to the Andean orogeny. Finally, the pelites and the Pliocene sandstones from the Solimões formation constitute the Tertiary supersequence, while the Içá formation is a Pleistocene sedimentation product. The Içá formation is covered by eolic deposits that originate in the Araçá, Anauá, and Catrimâni dune fields. The sedimentary rocks of the Amazon Basin are in onlap form disposition covering basement rocks from the Guianas and the Brasil Central shields, limited by the Solimõesbasin (Purus arc) on the western side and by the Marajómesozoic rift through the Gurupá arc on the eastern side. Total rock thickness reaches 5000 m. Sedimentation begins at the rift phase, with the cambrian–ordovician rocks of the Prosperança formation, basically on analluvial–fluviatile fan environments. The syneclese phase started with the deposition of marine clastic sediments from the Autás-Mirim, Nhamundá, Pitinga, and Manacapuru, arranged in the Trombetas group, belonging to the ordovician–devonian supersequence. The devonian–carboniferous

supersequence is composed of the Maecuru, Ererê, Curiri, Oriximiná, and Faro formations, which represent the fluviatile–deltaic and neritic sediments from the Urupadi and Curuá groups. This last one has been followed by a glacial sedimentation period and a posterior depositional gap.

The Tapajós group, constituted by the Monte Alegre, Itaituba, Nova Olinda, and Andirá formations, has a wide variety of sedimentation environments such as clastic, continental, and marine, building up the Permian–Carboniferous supersequence. This supersequence is followed by the Sanrafaélica orogeny (*ca.* 260 Ma.) and by the Juruá diastrophism. At the very beginning of the Jurassic, an expressive basalt-type magmatism occurred placing Penatecaua dikes and flows between the Nova Olinda and Alter do Chão formations. The sedimentation of the Amazonas Basin ceased after the deposition of the continental sequences, one from the upper Cretaceous (Alter do Chão formation) and another Cenozoic (Solimões and Içá formations), generated by a fluviatile and fluviatile–lacustrine systems. The groundwater research in this region is still incipient and deals mainly with the Alter do Chão formation aquifer. There is overall information about the Solimões and Içá formations as well (Figure 3.3).

The geologic framework described before and the assessment of the water well logs and oil soundings suggest that the Alter do Chão formation is the main regional aquifer functioning under an unconfined regime. Based on existing data, the Alter do Chão aquifer covers an area of about 410,000 km^2. Considering a mean thickness of 400 m and an effective porosity of 20%, the saturation volume reaches more than 30,000 km^3. According to Souza et al. (2013), even though there are not sufficient data for the estimation of the pressure component, it is clear that this pressure volume is by far much less than saturation volumes, since the parameter S, in confined aquifers, does have magnitudes less than 10^{-4}. Therefore, in terms of a regional estimation, the saturation volume may be a reasonable magnitude for the aquifer permanent reserve.

Regarding hydrodynamic parameters, the available data that have been taken as references were estimated by Tancredi (1996) for the region of Santarém, at Pará State. According to the author, T ranges from a minimum value of 1.5×10^{-3} m^2/s and a maximum value of 9.1×10^{-3} m^2/s. The storage coefficient, S, has values varying from 4.1×10^{-4} to 3.3×10^{-4} and finally the K values fall in between 2.1×10^{-4} and 5.0×10^{-5} m/s. The hydrodynamic parameters for the Içá-Solimões aquifers are T = 3×10^{-3} m^2/s, S = 5×10^{-4}, and K = 1×10^{-4} m/s, whose magnitudes are similar to the minimum values that were determined for the Alter do Chãosystem, in Santarém. Regarding the Içá-Solimões system, covering an area of 948,600 km^2, the estimated reserves reach

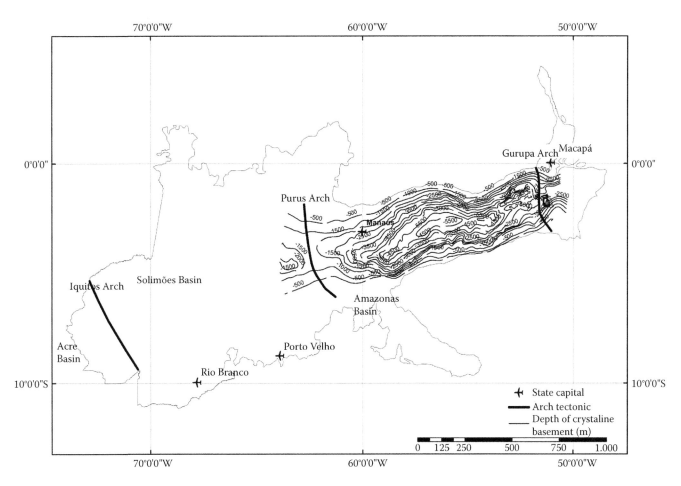

FIGURE 3.3
Main aquifers in the Amazonic Basin.

7200 km³, less expressive and 22% than the ones found for the Alter do Chãoaquifer system. The water quality in almost all aquifers from the Amazonic Basin show generally low contents of cations and anions, with sodium-bicarbonate waters, bearing values for Na⁺ and HCO₃⁻ lower than 7 and 30 mg/L and expressive for K⁺ (maximum concentration reaching 5.5 mg/L). The lower ionic concentrations determine lower values for the electric conductivity, which ranges between 1212 and 100 µS/cm. The groundwater is generally acid and has pH values between 4 and 5.

3.4.2 Parecis Sedimentary Basin (15, 16)

The Parecis sedimentary basin is one of the largest intracratonic basins from Brazil, which is situated at the southwest border of the Amazon craton, assuming an elongated W–E form with 1250-km width. It occupies an area of about 500,000 km² between the latitudes 10° and 15° S and longitudes of 64° and 54° W covering the states of Rondônia and Mato Grosso with almost 6000 m of siliclastic paleozoic, mesozoic, and cenozoic sediments

(Figure 3.4). The paleozoic sequence is constituted by the Cacoal, Furnas, Ponta Grossa, Pimenta Bueno, and Fazenda da Casa Branca formation, outcropping on the west, southwestern, and southeastern border of the basin. The mesozoic sequence, on the other hand, formed by the Anari/Tapirapuã and Rio Ávila units and the Parecis group (Salto das Nuvens and Utiariti formations) occurs in the central and western portion of the basin. Finally, the Cenozoic sequence, represented by the detrital–laterite covers belonging to the Ronuro formation and by the quaternary sediments from the Guaporé river, is concentrated in the Alto-Xingú region. The Furnas aquifer constituted by sandstones, conglomerates, and siltstones show productivity classes between 3 and 4 (according to Table 3.1), with low-to-medium productivity, showing specific yields between 0.4 and 2.0 m³/h/m and mean discharge of about 10 and 50 m³/h. The Ponta Grossa formation, composed mainly by pelites (shales, fine sandstones, siltstones, and claystones) belong to class 6 (less productive or nonaquifer). The Pimenta Bueno formation (sandstones, conglomerate, shales, and siltstones), Fazenda Casa Branca formation (conglomerate, arcosean

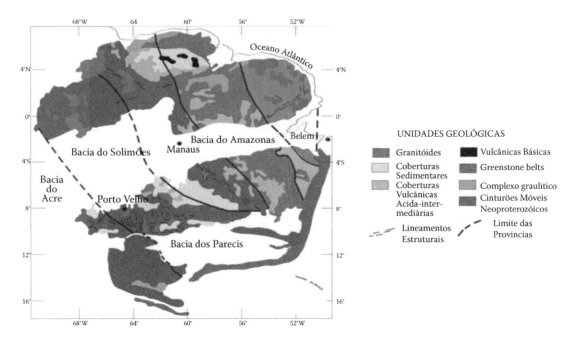

FIGURE 3.4
Geotectonical framework of the Amazon and Parecis Basins within the Amazonic Craton. (Modified from Bahia, R. B. C. Evolução Tectonossedimentar d Bacia dos Parecis—Amazônia. 2007. 115 f. Tese (Doutoramento em Ciências Naturais)—Escola de Minas, Universidade Federal de Ouro Preto, Ouro Preto. 2007.)

sandstones, and shales), and Anari/Tapirapuã formation (basalts and diabases) vary according to the productivity classification (Table 3.1) from classes 4 and 6. The Parecis group (sandstones, siltstones, and conglomerate) is considered to be the most important aquifer of the Parecis basin. It is classified as class 1 (very high productivity) showing high values for specific yield, reaching more than $4 \, m^3/h/m$ and discharges higher than $100 \, m^3/h$. Finally, the Ronuro formation and the undifferentiated quaternary deposits constituted by sand, clay, and gravel were classified as class 4; nevertheless, due to the fact that they are easily tapped, a great part of the population tends to use them.

3.4.3 Parnaíba Sedimentary Basin

The Parnaíba sedimentary basin occupies an area of about 600,000 km², embracing almost the entire area of the states of Piaui and Maranhão and expressive areas of the Pará and Tocantins States. The São Vicente Ferrer-Urbano Santos-Guamá Arc acts as its northern border, whereas the Tauá fault zone, the Senador Pompeu fault zone, the Tocantins-Araguaia fault zone, and the Tocantins arc are their borders on the eastern, southeastern, western, and northwestern portions. According to Goés and Feijó (1994) the basin hosts four depositional sites: Parnaíba basin, Alpercata basin, Grajaú basin, and Espigão Mestre basin (Figure 3.5). The depositional site called Parnaíba basin covers

approximately half of the total area of the entire basin and is situated mainly in the center and southern area (Figure 3.5). It comprises the Silurian supersequences (Serra Grande group), devonian (Canindé group), and triassic–carboniferous (Balsas group). The Serra Grande group is composed of the Ipu, Tianguá, and Jaicós formation whereas the Canindé group is composed of the Itaim, Pimenteiras, Cabeças, Longá, and Poti formations. The Piauí, Pedra-de-Fogo, Motuca, and Sambaíba formations constitute the Balsas group (Figure 3.5). The Alpercatas basin covers 70,000 km² (Figure 3.5) and is composed of the Jurassic supersequence (Mearim group), which is constituted of the Pastos Bons and Corda formations sealed, respectively, at the bottom and the top, by the igneous formations Mosquito (Jurassic) and Sardinha (lower Cretaceous). The Grajaú basin is situated at the northern side of the Alpercatas basin and gets isolated from the São Luis basin by the Ferrer-Urbano Santos arc, which does not exert any influence on the sedimentation continuity between both basins. It is filled by the Grajaú, Codó formations and the Itapecuru Group belonging to the Cretaceous supersequence (Figure 3.5). The Espigão Mestre basin is covered by eolic sandstones and lies discordantly above the Parnaíba basin. It corresponds to the northern part of the Urucuia basin, which is the setentrional part of the Sanfranciscana basin.

The Parnaíba sedimentary basin has the largest groundwater potential in the northeast region of Brazil.

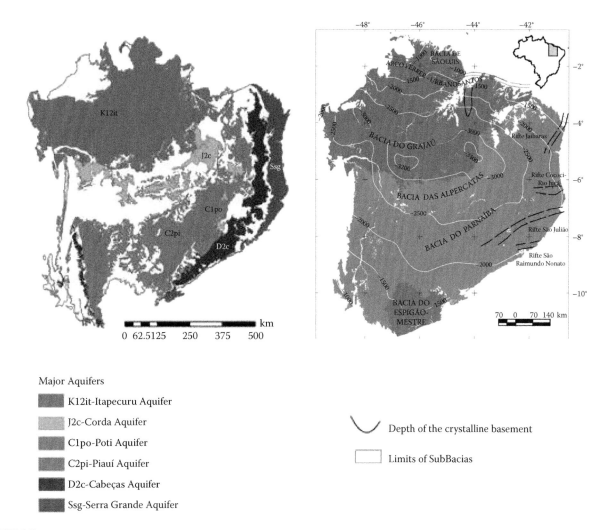

FIGURE 3.5
Parnaíba sedimentary basin. (Adapted from Góes, A. M. and F. J. Feijó. 1994. Bacia do Parnaíba. *Boletim de Geociências da PETROBRAS. Rio de Janeiro*, 8 (1), 57–67.)

The multilayered permeable geological strata gives rise to regional aquifer systems in heterogeneous hydraulic regimes, varying from unconfined to confined and sometimes artesian conditions (Figure 3.6). The most important aquifer units are the Serra Grande group followed by the Cabeças and Sambaíba formations and the Poti/Piauí system. The Corda and Itapecuru aquifers show slightly lower potential but a wide geographical distribution within the basin. On the upper part of the sedimentary sequence, one may find the Grajaú aquifer, the Barreira/Pirabas group, and the quaternary cover with low potential for groundwater production. The Serra Grande aquifer represents an extensive and important aquifer unity, which lies discordantly over the crystalline basement. It is constituted by an essentially clastic sequence, with conglomerates and consolidated kaolinite conglomerate sandstones (Ipu formation) followed by arcosean, fine to middle size grained sandstones (Tianguá formation) with conglomerate layers. The sequence ends with clastic pelites, which are situated predominantly in the southern part of the basin. This aquifer extends over 38,000 km² in the eastern, southeastern, and southern border of the Parnaíba basin exhibiting lower potential at its recharge area, a narrow 2–15-km-width fringe. The region under confined conditions shows excellent hydrogeological properties with expressive artesianism regime in some areas. Its thickness varies from 400 to 1000 m. According to the classification adopted by the hydrogeological map of Brazil, the aquifer is classified as class 1, even though its outcropping areas have lower productivity, being classified as classes 5 and 6. The mean hydrodynamic coefficients are $T = 3.0 \times 10^{-3}$ m²/s; $K = 1.0 \times 10^{-5}$ m/s, and $S = 4.3 \times 10^{-4}$.

The Cabeças aquifer is constituted by sandstones with clay material, outcropping over 42,000 km² of the middle

Period	Geological formation		Lithology predominant	Aquifer potential
Quaternary		Dunes, Alluvion	Sands/Sandstones	(4)
Tertiary		Group barreiras/Pirabas	Sandstones/Silex	(4)
Cretaceous		Itapecuru/Urucuia	Sandstones	(3)
		Codó	Clay	(6)
		Grajaú	Sandstones	(4)
	Group mearim	Sardinha	Intrusive rocks	(6)
Jurassic		Corda	Sandstones	(3)
		Pastos bons	Clay/Sands	(5)
		Mosquito	Intrusive rocks	(6)
Triassic	Group balsas	Sambaíba	Sandstones	(1)
Permian		Motuca	Shales	(5)
		Pedra de fogo	Sandstones	(5)
Carboniferous	Group canindé	Piauí	Sandstones	(2)
		Poti		
		Longá	Shales	(6)
Devonian		Cabeças	Sandstones	(1)
		Pimenteiras	Shales	(6)
		Itaim		
Silurian	G. Serra grande	Serra grande indiviso	Sandstones	(1)
Ordovician				

FIGURE 3.6
Hydrogeological potential for the geological formations of the Parnaíba basin, according to the classification of Table 3.1.

part of the basin reaching a mean thickness of 300 m. It is classified as class 1, similar to the Serra Grande aquifer. Due to the topographical context of their outcropping areas, productivity in those areas is situated between classes 3 and 5; hydrodynamic parameters are $T = 1.3 \times 10^{-2}$ m^2/s, K = 5.4×10^{-5} m/s, and S = 3.7×10^{-4} (Feitosa and Demetrio, 2009). The Sambaiba aquifer occurs at the southeastern parts of the Maranhão and northeastern part of Tocantins states, both as an unconfined aquifer and as a confined aquifer as well. It is composed mainly by well-sorted sandstones bearing high permeability and therefore high to very high potential (classes 1 and 2). Its inflow is fed by direct infiltration from rainfall in recharge areas at plain areas covered by unconsolidated sands and by the drainage network. Its principal outlets are the natural drainage and river beds which keep basin discharges over the year, evapotranspiration, when clay-rich sequences hinder vertical infiltration, vertical bottom drainage, and artificial discharge as an effect of the well operation. It shows yields of more than 200 m^3/h in some cases. The Poti and Piauí aquifer units constitute an important aquifer system

covering an area of about 92,250 km^2. In the largest part of its extension, mainly close to the Parnaíba River, it behaves as an unconfined aquifer whereas toward the middle of the basin it changes to a confined condition. It presents a lithological constitution based on massive sandstones with few intercalations of shale at the inferior part of the sequence. Their recharge originates directly from the rain vertical infiltration, drainage throughout-confining units, and the superficial drainage network. The main aquifer outlets are the drainage system and evapotranspiration at some aquifer portions richer in clay content.

3.4.4 Paraná Sedimentary Basin

The Paraná sedimentary basin is an intracratônica phanerozoic basin established over the Archean and Proterozoic continental crust that is situated in the southern part of the South American platform (Almeida et al., 2000). The stratigraphic register of the basin comprehends a succession of approximately 7000-m thickness of sedimentary and volcanic rocks developed during the neo-Ordovician and the neo-Cretaceous, under marine to continental sedimentation environments (Milani, 2004; Milani et al., 2007). The Paraná basin occupies an area of about 1.1 million km² in Brazil, distributed in eight Brazilian states, complemented by more than 400,000 km² in Argentina, Paraguay, and Uruguay. Within its geological framework, the Paraná basin contains a diversity of sedimentary aquifers with intergranular porosity that has been generated in different depositional environments, such as fluviatile, marine, glacial, and desert. The predominant sedimentary processes and the postdiagenetical modifications ended up defining distinct hydraulic characteristics, resulting in different groundwater potential. Among these aquifers, the most important ones are the Tubarão aquifer (SAT), the Guarani aquifer (SAG), and the Bauru aquifer (SAB), whose storage and transmission conditions allow their wide exploitation for fulfilling of domestic, industrial, and agricultural demands. Emphasis should be given to the fractured aquifer developed by the volcanic rocks called the Serra Geral aquifer (SASG). Besides the main aquifers, there are also some aquifers with low permeability, such as the Passa Dois aquiclude, which is composed by a thicker permian elite sequence disrupting the hydraulic continuity between the SAT and SAG aquifers (Figure 3.7).

3.4.4.1 Tubarão Aquifer System (SAT)

The Tubarão aquifer system has its outcropping areas of about 99,000 km² in a narrow fringe close to the northwestern, eastern, and southwestern borders of the basin. At the subsurface, it spreads over almost the entire basin reaching 750,000 km². It is considered a granular porous aquifer constituted by the Tatuí, Palermo, Rio Bonito, Aquidauana, and Itararé stratigraphic units. Its lithological composition varies a lot, from diacmetites, siltstones, pelites, shales, ritmites, sandstones, and conglomerate sandstones deposited by marine, glacier, coastal, and fluviatile processes. It may reach 800 m of thickness in the outcropping areas (DAEE, 2005) and sometimes more than 1000 m in the remaining areas as shown by well drilling logs. The hydraulic conductivities range from 2.31×10^{-8} to 8.10×10^{-6} m/s (Diogo et al., 1981) whereas transmissivities vary between 3.5×10^{-6} and 4.63×10^{-4} m²/s. Locally, transmissivity values may reach 150 m²/day (DAEE, 1981, 1982). These values allow this aquifer to be classified as classes 3 and 4. The porosities are generally low in clay sandstones, but may reach up to 30% in sandstones with lower clay content (França and Potter, 1989). The porosity and the permeability of this reservoir are controlled mainly by grain size, grain selection and, secondarily, by the presence of carbonate cementation (Vidal, 2002). The high

subsidence rates of this basin during deposition of the SAT unities has also affected the permo-porosity characteristics of the aquifer due to the chemical compaction effect, followed by an increase of the pressure and temperature conditions (Bocardi et al., 2008). Frequent intercalations between coarser and fine sediments, with distinct thicknesses set up a very heterogeneous framework that affects the groundwater storage and flow within aquifer media. (DAEE, 1981; Diogo et al., 1981). The pelite lithologies interlayered to the sandstones hinder the groundwater flow downward increasing its heterogeneity where vertical permeability is lower than horizontal permeability (DAEE, 1981; Diogo et al., 1981). The same happens with the frequent diabase sills, with variable thickness, which may affect badly the regional or local flow continuity (DAEE, 1981). At small depths, the SAT behaves generally as an unconfined aquifer (DAEE, 1981). At the outcropping areas, the permeable sediments receive direct recharge from rainfall and they do discharge expressive amounts of water into the fluvial network. Besides its expressive thickness, the SAT is being exploited by tubular wells not deeper than 300 m, extracting moderate yields between 10 and 20 m³/h (Diogo et al., 1981). The groundwater tends to be slightly saline, with total dissolved solids content between 100 and 200 mg/L and being classified as sodium bicarbonate or calcium bicarbonate (DAEE, 1984). Under extreme confined conditions, in depths more than 400 m, the groundwater may present elevated saline concentrations, above potable thresholds. This is why it is not being intensively exploited thus far.

3.4.4.2 Guarani Aquifer System (SAG)

The SAG is the most important hydroestratigraphic unit from the southern part of the South American continent and is considered to be one of the world's largest transboundary aquifers, extending across wide territories of Brazil, Argentina, Paraguay, and Uruguay. The largest part of the aquifer is situated in Brazilian territory comprising 736,000 km². The outcropping areas reach only 88,000 km² and are situated along the basin border as a narrow belt whereas the confined areas, covered by volcanic sequences, sum up 648,000 km² (OEA, 2009). The SAG is constituted by a sequence of mesozoic continental clastic rocks within the Paraná basin, being delimited by a regional permo-triassic discordancy (250 million) at the base and by volcanic flows from the Serra Geral formation (145–130 million) at the top. In almost all compartments of the basin, the stratigraphic units that constitute the SAG are, exclusively, the Piarambóia and Botucatu formations. Nevertheless at the southern parts of the basin, the SAG is also locally represented by the Santa Maria, Caturrita, and Guará formations (Machado and Faccini, 2004). The SAG framework comprehends

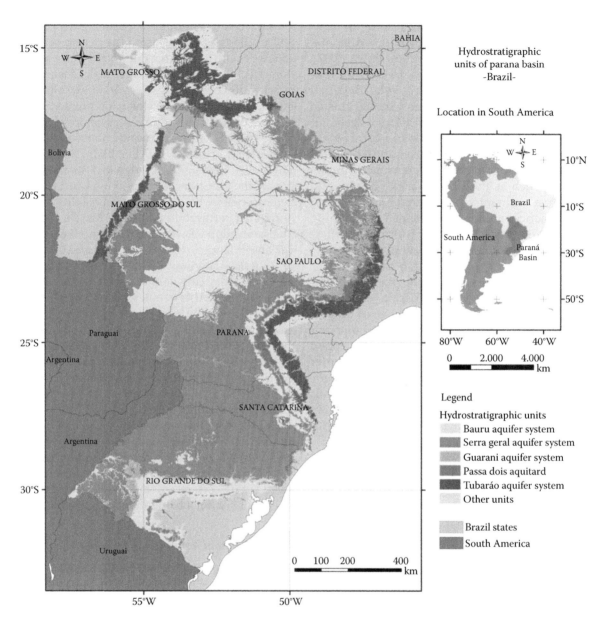

FIGURE 3.7
Main hydrostratigraphic units of Paraná basin.

predominant eolic continental deposits represented by fine-to-medium-size sandstones exhibiting large-size cross-stratification and secondarily fluviatile lacustrine sandstones and sandy pelites (Caetano-Chang, 1997). At the southern compartment of the basin, there is a basal succession composed of sandstones and pelites deposited by a fluviatile lacustrine system (Machado, 2005; Soares et al., 2008). In almost the entire basin extension, the SAG sequences are layered on top of thick permian units of low permeability, which integrate the Passa Dois aquiclude. At the western part of the basin, the SAG covers carbo-permian sediments of the Aquidauana formation (LEBAC, 2008). The thickness of the SAG increases

gradually from outcropping areas, where they are only partially preserved, to the main axis of the basin. At the northern region, there is an elongated depocenter parallel to the main basin axis that accumulates a sediment thickness of 600 m (LEBAC, 2008). Close to the internal tectonic arcs (Ponta Grossa arc and Rio Grande arc) and to the Torres sinclinal, the thicknesses get drastically smaller until they are less than 100 m (LEBAC, 2008). At the eastern compartment of the basin, where there is an intensive groundwater exploitation, SAG thickness varies from 100 m at outcropping areas to 400 m toward the basin major axis (DAEE, 2005). Generally, the mean thickness of the SAG ranges between 200 and 250 m. At its largest part, SAG is covered by about 1700 m of volcanic rocks (LEBAC, 2008). The SAG eolic sandstones have mean porosities of 20% up to 30%, but fluvial ones may show lower values (OEA, 2009). The conductivity of the aquifer has been estimated at 2.6 m/day for the confined areas and 3.0 m/day at the unconfined areas (DAEE, 2005); transmissivity has been estimated at 3×10^{-3} m²/s for the outcropping areas and more than 1.4×10^{-2} m²/s for the confined areas (DAEE, 2005). The storage coefficient varies between 10^{-3} and 10^{-5} (DAEE, 1974). Due to the faciological features of the main hydroestratigraphic units of the SAG (Botucatu and Pirambóia formation) and the shallow burying history, the diagenetical modifications were not efficient enough to change original permo-porosity of these rocks (Gesicki, 2007). On the other hand, diluted water inflows, acting in depths up to 250 m within SAG layers, have removed carbonate cement and have leached feldspatic grains giving rise to a secondary porosity (França et al., 2003). The unconfined aquifer potentiometry reveals local and regional flows, being ruled by the topography within the hydrographic basin in the first case and from outcropping areas dipping toward the interior parts of the basin under confined flow regime. From there on, the aquifer remains mostly confined where in regional terms, flow tends to be from north to south, following the main basin axis (LEBAC, 2008). In some specific areas, water levels go far beyond surface levels building artesianism. The SAG presents excellent potentials, turning it into a strategic reservoir for satisfying water demands at small- and medium-sized cities. At the outcropping areas, well discharges are about 80–100 m³/h whereas in the confined areas, they yield more than 200 m³/h with specific capacity varying from 2 to 15 m³/h. The recharge rates of the SAG, from 1 to 3 km³/annual, are very small considering its extension (OEA, 2009) and extraction volumes for a variety of uses. In both situations, it is considered to be class 1 in terms of productivity. The hydrochemistry of the SAG shows different patterns depending on the aquifer flow regime. At the outcropping areas with unconfined regime, the water tends to be calcium bicarbonate with low electric conductivity. At the confined areas, in the other side, waters are sodium bicarbonate with higher mineralization degree. At the main basin axis, the groundwater tends to be sulfate, sodium chlorinated highly mineralized, however, and presenting great possibility of mixture with water originated in underlying formations.

3.4.4.3 Serra Geral Aquifer System (SASG)

The Serra Geral Aquifer System spreads over an area of about 735,000 km² within the Paraná basin, from which 409,000 km² constitutes outcropping areas of the Serra Geral formation. These volcanic sequences are partially covered by the sediments of the Bauru group and they may reach almost 1700 m of thickness at the main basin axis. Due to their wide spatial distribution, this system is considered to be an important groundwater reservoir with capacity to fulfill small-to-medium-size demands and work as a complementary water source. The storage and flow of the water occur through physical discontinuities such as fractures, faults, and interflow surfaces, which constitute a heterogeneous, anisotropic, and discontinuous media (Rebouças, 1978). The fracture systems are related to tectonic stresses and also to cooling processes generating subvertical and subhorizontal fractures, respectively (Campos, 2004; Lastoria et al., 2006). Water extraction is done through wells with 100–200-m depth, which allow yields varying from 100 m³/h (when intercepting productive fracture systems) to null, a situation that may happen very often. The relationship between water yield and lineament density proved to be weak according to studies carried out by DAEE (2005). The explanation given is that subhorizontal surfaces such as lava spill contacts do have an important influence controlling water flow, but are not detectable by remote-sensing techniques. The water from the SASG is mainly calcium bicarbonate and secondarily calcium–magnesium bicarbonate and sodium bicarbonate with saline contents less than 250 mg/L (Campos, 2004). According to the classification scheme adopted, these aquifers fall between class 2, in clearly confined scenarios, and 6 due to their topographic setting.

3.4.4.4 Bauru Aquifer System (SAB)

The Bauru Aquifer System comprises a succession of Cretaceous sedimentary rocks that were deposited over 370,000 km² of the center-setentrional part of the Paraná basin covering the basalt floods of the Serra Geral formation. Its mean thickness is 100 m, but it may reach 300 m in certain sectors of the basin. Due to the fact that it is a superficial aquifer, well drilling gets easier and exploitation costs are low. On the other hand, it shows high vulnerability toward inorganic and organic contaminant leakages (DAEE, 1976, 1979). The SAB is a multilayered hydroestratigraphic system composed of the Marília, Adamantina, Birigui, Santo Anastácio, and Caiuá aquifers and the aquitards Araçatuba and Pirapozinho (Paula and Silva, 2003, 2005). The sedimentation environments of these aquifers are mainly fluviatile with eolic interactions. The aquitards relate to pelites developed at lacustrine environment (Paula and Silva, 2005). Their hydrodynamic behavior is heterogeneous according to their lithological framework in which sediments with different porosities and permeabilities share lateral and vertical contact relationships. Consequently, the registered yields in these aquitards are variable (Paula and Silva, 2003). The SAB is considered to have moderate permeability according to its clay and silt contents and to the presence of less permeable and impervious layers along the profile (DAEE, 1976). The conductivities in the SAB range from 2.31×10^{-8} to 4.24×10^{-5} m/s whereas the transmissivities vary from 1.62×10^{-6} to 3.8×10^{-3} m²/s. Values lower than 50 m²/day are very often the case (DAEE, 2005). In areas where sedimentation is predominantly eolic, clearly sandy, the transmissivities reach far beyond 200 m²/day (Iritani et al., 2000). The effective porosities are about 5% in clayish sandstones and between 10% and 20% in sandstones with less clay (DAEE, 1979). These hydraulic characteristics set up exploitable yields that start at 10 m³/h and reach up to 120 m³/h. Discharges more than 80 m³/h, however, are not recommended (DAEE, 2005). Multilayered aquifer systems or single confined aquifer units, such as the SAB, may exhibit more than one potentiometric surface, which reflects the equilibrium among different aquifer unit hydraulic charges. So, at one single place, one can recognize an unconfined potentiomentric surface, at shallow depths, and a deeper confined one (Paula and Silva, 2005). The aquifer unconfined potentiometric surface is ruled by the groundwater flow within the watershed. The water from the SAB is calcium to calcium–magnesium bicarbonate and more rarely sodium bicarbonates (Coelho, 1996; Celligoi and Duarte, 2002; Barison, 2003). Stradioto (2007a,b) has also found the presence of calcium chloride–sulfate and sodium chloride–sulfate water. Generally, the SAB presents lower saline concentration with dry residue showing values rarely higher than 300 mg/L (DAEE, 2005).

3.5 Mesozoic and Meso-Cenozoic Sedimentary Basin

In respect to their strategic importance for the semiarid region in Brazil, among all the Mesozoic and Meso-Cenozoic basins, only the ones situated in the northeast area of the country are going to be emphasized (Figure 3.1).

3.5.1 Potiguar Basin

This basin is located in the north coast of the state of Rio Grande do Norte and southeast of Ceará state. Its entire extension comprises an area that can vary between 41,000 and 60,000 km², including its outcropping and subsurface portions. The main aquifers are represented by the Jandaíra and Açu formations. The aquifer Açu, whose thickness varies between 400 and 700 m, corresponds to the inferior portion of the Açu formation and is constituted by sandstones and conglomerate at the lower portion of the sequence evolving gradually to fine sandstones at the upper part of the sequence. It is perceived as the most important groundwater storage system within the Potiguar Basin.

The outcropping areas are situated along a marginal belt whose widths vary from 5 km at the eastern side to 20 km at the western corner. The first deep well drilled in this aquifer was in 1967 and has revealed artesianism conditions, discharging about 80 m³/h of excellent water quality. This favorable scenario was followed by an intense economic development of the region and increase of the water demand due to agro-industrial plants based on irrigation schemes. The Açu aquifer exploitation has been accelerated since the 1970s, reaching overall discharge rates of about 42 hm³/year generating expressive drawdowns of more than 160 m in the most critical areas. Despite the fact that there is a vertical drainage from the limestone above, studies are still not conclusive thus far. The incontestable fact is that the discharge increase has triggered the deepening of the groundwater levels depleting the storage capacity of the aquifer. Meanwhile admitting that the CAERN (Water and Sewage Company of the State of Rio Grande do Norte) has slowed down the use of this aquifer for domestic purposes and that irrigation plants have started to use water from the Jandaíra Aquifer, the potentiometric depression tends to recover. However, the Açu aquifer will always play an important and strategic role in providing low cost solutions. It normally shows high magnitudes of yield allowing it to be classified as productivity class 1. Mean hydrodynamic parameters are $T = 2.3 \times 10^{-4}$ m²/s, $K = 7.5 \times 10^{-6}$ m/s, and $S = 1 \times 10^{-4}$. The Jandaíra aquifer must be addressed in the limestone.

3.5.2 Araripe Basin

The Araripe basin is situated in the states of Ceará, Pernambuco, and Piauí and covers an area of about 11,000 km². It can be divided into two different sectors: Araripe Highlands and Cariri Valley. Almost the entire groundwater exploitation takes place inside the valleys with few water wells on the highlands. The most important aquifers are the Mauriti and the Batateira/Abaiara/MissãoVelha system. The Mauriti aquifer is constituted by a uniform sequence of coarse-grained sandstones, generally silicified, contributing to significant losses of primary porosity. In this case, groundwater flow is controlled by the secondary porosity (fractures and faults). In general, they show only a moderate potential with thickness about 100 m.

Discharge from wells is low (<5 m^3/h), excepting fault zones, where yields tend to be much higher. The Rio da Batateira/Abaiara/MissãoVelha system is constituted by course to fine size sandstones with siltstones, claystones, and shales, at the intermediate to upper part of the sequence reaching 500 m of thickness. Actually, it is the most used aquifer in the region with wells yielding up to 300 m^3/h. Recent studies carried out by the CPRM and the Federal University of Ceará had proposed for the Cariri valley (including the both aquifers) the following estimates: 360 million/m^3 of renewable resources, 14 billion/m^3 of permanent resources, 450 million/m^3/year of exploitable resources, and availability of 54 million/m^3/year.

3.5.3 Interior Basins

3.5.3.1 Iguatu/Malhada Vermelha/ Lima/Campos Icó Basins

At the southeast of the Ceará state, there is a group of small basins situated between the Iguatu and Icó cities, occupying an area of approximately 1000 km^2. The sedimentation within these basins is mainly clastic with pelite intercalations composing the following aquifer unities: Icó, Malhada Vermelha, and Lima Campos. On top of them, there are some unconsolidated clastic formations that may exhibit groundwater storing capacity. Well drilling is done intercepting all these aquifers at depths lower than 100 m. However, the exploitation in this region is still small due to the large hydric availability imposed by the Orós Lake. There is no well deeper than 100 m and underground information has been generated by geophysical assessments. The aquifers show a small hydrogeological potential delivering yields of about 3 m^3/h. The greatest potential remains in the banks of the Jaguaribe River where reservoirs may have 25 m of thickness and 500 m of width. The high conductivities shown by these alluvial bars allow expressive groundwater extraction fulfilling water demands of Iguatu City.

3.5.3.2 Lavras da Mangabeira Basin

This represents a group of small basins situated in the southeast region of the Ceará state covering an area of about 60 km^2. The Serrote, Limoeira, and Iborepi formations show groundwater potential. Assessments carried out by the CPRM and the Federal University of Ceará (CPRM/UFC, 2008b) indicates 4.6 million/m^3/year of potential and an installed availability of 1 million/m^3/year. The greatest part of this volume is used by the state-owned water company CAGECE for public supply.

3.5.3.3 Coronel João Pessoa/Marrecas and Pau dos Ferros Basins

These small basins with total area of 16, 27, and 65 km^2, respectively, are situated in the west side of the state of Rio Grande do Norte. They are constituted by fine-to-coarse-grained sandstones, siltstones, and claystones from the Antenor Navarro formation. Despite the inexistence of data, through an analogy with other similar sequences, one may estimate that they bear reasonable groundwater potential at their sand-rich zones.

3.5.3.4 Rio do Peixe Basin

This basin is located at the far northwest side of the Paraíba State covering an area of 1300 km^2. Sedimentary filling comprises the coarse-to-fine sandstones of the Antenor formation, the siltstones, shale, and calciferous sandstones of the Souza formation, and the fine sandstones and conglomerates of the Rio Piranhas formation. This stratigraphic profile conditions the existence of two major aquifers, namely, the Rio Piranhas and the Antenor Navarro, separated by the Souza aquitard. Recent studies developed by the Brazilian Geological Survey in partnership with the Federal University of Campina Grande (CPRM/UFCG, 2008) led to significant advances in the understanding of the groundwater flow network within the basin. The reserves estimates had not been calculated because there are still some incongruences concerning aquifer geometry. Both aquifers show a small potential with yields of about 10 m^3/h.

3.5.3.5 Cedro Basin

This basin is situated at the northwest corner of the state of Pernambuco and has an area of 168 km^2. The most important aquifer is represented by the Mauriti formation, whose hydrogeological behavior was already described. Detailed information is still missing but expectations converge to moderate groundwater potential.

3.5.3.6 São José do Belmonte Basin

It is situated at the center–north of the Pernambuco state and has an area of 755 km^2. The predominant aquifer is the Tacaratu formation composed of heterogeneous medium-size-to-coarse sandstones with kaolinite levels and strong diagenesis. It shows a very heterogeneous hydrodynamic behavior, where secondary porosity prevails over the primary one. As a consequence of that, a wide discharge magnitude variation occurs (starting at 1 m^3/h to more than 50 m^3/h). Besides the existence of more than 1000 wells registered by the CPRM, the knowledge on the aquifer is still very incipient.

3.5.3.7 Mirandiba/Carnaubeira/Betânia and Fátima Basins

These basins are located in the center portion of the state of Pernambuco and present the following dimensions: 143, 136, 280, and 270 km^2, respectively. Their hydrogeological potential is given by the Tacaratu formation, which is constituted by medium-to-coarse heterogeneous sandstones with kaolinite levels and strong diagenesis as well. In the outcropping areas, groundwater behavior is similar to the Mauriti aquifer. The knowledge is still very incipient, but potential is expected to be moderate to low. Tubular wells completed by the Brazilian Geological Survey in the Fátima basin with depths starting at 300–420 m delivered yields of 30 and 100 m^3/h, respectively.

3.5.3.8 Recôncavo/Tucano/Jatobá Basins

The sedimentary basins of the Recôncavo and Tucano cover an area of about 50,000 km^2 across the coastal areas of the Bahia State and Pernambuco. In these two basins there are three major aquifer systems: (i) the upper aquifer represented by the Marizal and São Sebastião formations; (ii) the intermediate aquifer represented by the Ilhas Group and Candeias formation; and (iii) the lower aquifer represented by the Sergi and Aliança formation. The upper aquifer system is the most exploited one and the São Sebastião formation is the most productive unit with wells reaching up to 100 m^3/h and thickness of 3000 m. This aquifer is responsible for the water supply of the Camaçari petrochemical plant, where a strict water-quality monitoring control is taking place. There is no consistent data on stored volumes, mainly in the intermediate and lower aquifers. In general, one can assume a moderate to high hydrogeological potential with wells having a specific capacity of 3 $m^3/h/m$. Until 800 m the water is considered to be of good quality. The Jatobá basin is situated in the central region of the Pernambuco and northeast of the Alagoas state, covering 5941 km^2. It shows an excellent hydrogeological potential represented by the Inajá/Tacaratu aquifer system. This system is constituted by a sequence of coarse-grained sandstones with pelite intercalations at the base (Tacaratu formation) and fine, ferruginous sandstones with siltstones intercalations at the upper part (Inajá formation). Thickness estimates reach about 500 m for the entire sedimentary sequence, whereas 350 m refers to the Tacaratu formation and 150 m to the Inajá formation. Studies carried out by the Brazilian Geological Survey together with the Pernambuco Federal University revealed reserves of about 6.192 hm^3 (only for the areas under unconfined behavior regime), renewable resources in order of 3.1 $hm^3/year$, potential about 12.4 $hm^3/year$, installed availability of 0.7 $hm^3/$ year, and exploitable resources of about 9.3 $hm^3/year$ for the next 50 years (CPRM/UFPE, 2008). The groundwater resources are being used for the supply of the surrounding cities (Sertânia and Arcoverde in Pernambuco state).

3.5.3.9 Sanfranciscana Basin/Urucuia Aquifer

The Urucuia Aquifer, in the context of the Sanfranciscana basin, covers territories of six different states of Brazil (Bahia, Tocantins, Minas Gerais, Piauí, Maranhão, and Goiás) and occupies an area estimated as 120,000 km^2. The greatest area, 90,000 km^2 occurs in the western side of the Bahia State. For a long time, due to the lack of information, the Urucuia Aquifer has been considered a low hydrogeological potential unit. However, recent studies have shown that wells with 250–300-m depths delivering up to 500 m^3/h and bearing specific capacity higher than 10 $m^3/h/m$ are frequent. From a lithological point of view, they are represented by a succession of friable fine-to-coarse-size kaolinite sandstones with conglomerate levels reaching thickness of 600 m. It acts as the watershed boundaries between the São Francisco river at the east, the Tocantins river at the west, and the head of the Parnaíba river at the north. Under such conditions, it is expected that the aquifer exerts an important role keeping basal flow in those rivers, a scenario where the integrated water resources management concepts are crucial. In the last few years, the aquifer exploitation has risen vertiginously following the accentuated expansion of irrigated agriculture. Aquifer knowledge is still insufficient and restricted to pilot areas after studies carried out by the Brazilian Geological Survey, the Water Resources Secretary of the Bahia State, and universities. The unconfined characteristics of the Urucuia Aquifer make it the largest groundwater reservoir in the Bahia State and one of the largest within the entire country. Gaspar (2006) has estimated permanent and renewable reserves in 3×10^{12} m^3 and 3×10^{10} $m^3/year$, respectively. The exploitable reserves were estimated to be 4×10^{11} m^3.

3.6 Limestone

Karstic limestone formations are always or nearly always present in all Brazilian sedimentary basins with varying degrees of economic interest both as groundwater reservoirs and as raw material for the cement industry. The most extensive water-bearing limestone formations occur, nevertheless, in Neo-Proterozoic terrains, in the states of Bahia and Minas Gerais, within the drainage basin of the São Francisco River. Overall, four major hydrogeologic karstic provinces may be recognized in

Brazil, at the current stage of knowledge. These are (1) Jandaira Aquifer, in the Cretaceous Potiguar Basin, state of Rio Grande do Norte, northeast Brazil; (2) Pirabas Aquifer of Tertiary age, in the sedimentary coast basin of the state of Pará, north Brazil; (3) the Una Group of Neoproterozoic age, in the north of the state of Bahia; and (4) the Bambuí Group of Neoproterozoic age, in the west of the states of Bahia and Minas Gerais.

3.6.1 Jandaira Aquifer

The Jandaira Formation is a sedimentary marine deposit made mostly of carbonate rocks whose thickness may attain 600 m in some places of the Potiguar Basin, such as the valley of the Mossoró River. The formation traces back the widespread marine transgression which closed the Cretaceous sedimentary history of the Potiguar Basin in the north coast of the state of Rio Grande do Norte (Figure 3.8). Karst structures such as solution channels, caves, and sinkholes developed in the upper 80 m of the formation, giving place to the so-called Jandaira Aquifer. Although occurring all over the Potiguar Basin, this aquifer shows its uppermost expression in the region west of the Apodi River known geologically as Platform of Aracati. There, since the early decade of 1990, the Jandaira Aquifer has been giving extensive support to fruit crops such as melon, pineapple, papaya, and others, mainly for exportation to Europe and the United States. Recent studies carried out by the Brazilian government counted about 2000 wells in the Platform of Aracati, with depths in

the range of 60–120 m, and discharges commonly vary from 10 to 70 m³/h. Discharges from 70 to 250 m³/h also occur although less frequently. The groundwater storage, namely, the water table, is very sensitive to the recurring droughts that affect the region, which may cause serious water crises, when a great number of wells can go dry. In the year 2002, the fruit crops, which are the basis of the economy of the region, were impacted strongly by such a crisis. Nevertheless, the typical karstic landscape with sinkholes densely scattered all over the surface area greatly improves the response of the water table to infiltration in years when precipitation is above the annual mean value of 700 mm. On occasion, one or two very generous rainy seasons may provide the replenishment of the reservoir to what its reserves were prior to the draught period. In this way, the water table of the Jandaira Aquifer, as well as its reserves, seems to undergo a long-term fluctuation whose behavior is to be better understood for the sake of groundwater management in the region. In the year 2010, 244 hm³/year were being extracted, which represented 41% of the renewable resources, estimated as 591 hm³/year with a 50% chance (Oliveira et al., 2012).

3.6.2 Pirabas Aquifer

Although present all over the coastal region of the Pará State, the main area of occurrence of the Pirabas Formation is the region west of Belém City. This area measures about 24,000 km² and is known geologically

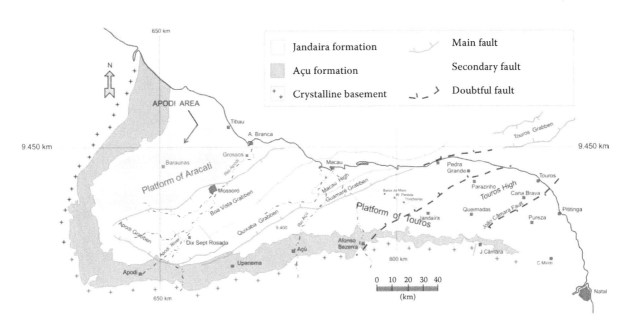

FIGURE 3.8
The Potiguar Basin and the Platform of Aracati. (Modified from Feitosa, E. C. and J. G. Melo. 1998. Estudos de Base—Caracterização Hidrogeológica dos Aquíferos do Rio Grande do Norte. In: Hidroservice, Plano estadual de recursoshídricos do Rio Grande do Norte.)

as the Bragantina Platform (Figure 3.9). The Pirabas Formation, of Tertiary age, shows two distinct sections. The upper section is made up of limestone and marls with intercalations of black and greenish-gray shale and carbonate sandstones. Light-gray sandstones dominate in the lower section.

The Upper Pirabas Aquifer developed in the upper section of Pirabas Formation, in depths between 70 and 180 m. Wells can yield up to 200 m³/h being, nevertheless, very expensive, which makes drilling accessible only to government or big industries (Matta, 2002). Quite preliminary studies suggest groundwater storage of about 400,000 hm³. The discharge being recovered is something around 100 hm³/year corresponding to 0.03% of the groundwater storage. No data are available yet on renewable resources. The Lower Pirabas Aquifer developed in the lower section of Pirabas Formation, in depths within the range of 180–280 m. Wells may produce as much as 600 m³/h of excellent potable water (Matta, 2002). Due to excessive costs of drilling, though, this aquifer is little exploited.

3.6.3 Salitre Aquifer

The Salitre Formation, of the Neo-Proterozoic age, is the most important formation in the Una Group, in the state of Bahia. It spreads itself over an area of about 38,000 km²

forming four separate bodies. The most important of them is the so-called Basin of Irecê. The dominant lithology is black-to-gray limestone exhibiting a certain degree of metamorphism. Tectonic style goes from near-horizontal layers, in the region along the rims of the basin, to folded layers striking E–W and dipping near vertically in the central regions. The Una Group is physically separated from the Bambuí Group by high ridges sculptured in Proterozoic quartzites. The Salitre Aquifer corresponds to karst structures which are widely developed in the Salitre Formation to depths up to 80 m. About 20,000 wells this deep are reported to exist in the Irecê Basin giving support both to agriculture and public supply. As with the Jandaira Aquifer, climate hazards affect seriously on occasion the economy of the region. Nowadays the Brazilian Water Agency is undertaking hydrogeologic studies in the karst provinces of the São Francisco Basin, aiming at the knowledge of the amount of water being withdrawn from the aquifer and being recharged to it. The main goal of the studies is to establish a groundwater budget in order to assess the sustainability of groundwater exploitation in the near future.

3.6.4 Bambuí Aquifer

The Bambuí Group, of Proterozoic age, is composed of five geological formations (Três Marias Formation,

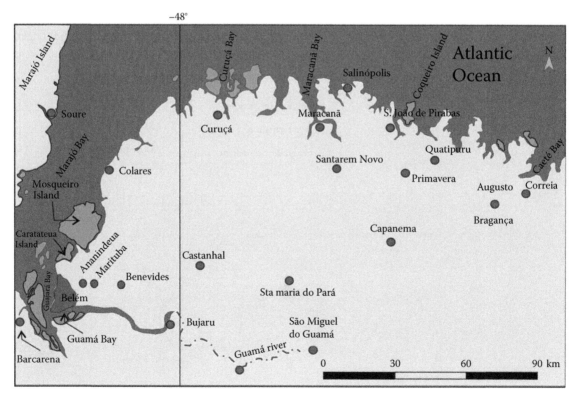

FIGURE 3.9
Main area of occurrence of the Pirabas Formation.

Serra da Saudade Formation, Lagoa do Jacaré Formation, Santa Helena Formation, and Sete Lagoas Formation), which occur in the states of Bahia, Minas Gerais, Goiás, and Tocantins, spreading itself over an area of about 120,000,00 km² (Figure 3.10). Excepting the uppermost Três Marias Formation, all the other formations of the Bambuí Group include carbonate rocks in greater or lesser amounts. The term Bambuí Aquifer, therefore, applies to the water-bearing karst structures developed in the various formations of the Bambuí Group. The formations Sete Lagoas and Lagoa do Jacaré, particularly, are the ones mostly made up of limestone. In this way, these formations are the most susceptible to development of karst structures. Due to the variation in the carbonate content of the formations of the Bambuí Group, groundwater storage and availability in the Bambuí Aquifer varies widely throughout its domain of occurrence. The Bambuí Aquifer has

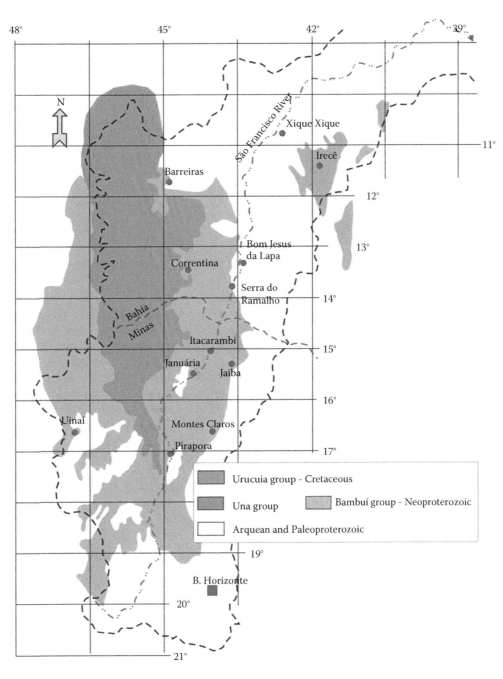

FIGURE 3.10
Una and Bambuí Groups in the basin of the São Francisco River.

become of utmost importance in providing support to irrigation and public supply in the states of Bahia and Minas Gerais, mainly in times of droughts. Prolonged droughts, however, such as the one the region is undergoing, impact the economy and public supply in exactly the same way as with the Jandaíra Aquifer. The lack of knowledge concerning withdrawals and recharge brings about incertitude as to sustainability of groundwater exploitation in the future. The Brazilian Water Agency is carrying out extensive studies to provide better knowledge on the matter.

3.7 Pre-Cambrian Crystalline Basement

In the crystalline region of the Brazilian semiarid, where there is practically no weathering cover, the groundwater flows through interconnected rock fracture and discontinuity systems, building up reservoirs with limited extension. Considering a certain control volume of rock, which is representative of the whole rock mass of the basement region, there are "n" discontinuity systems, independent among themselves, but with the ability to store and transmit water. Manoel Filho (1996) introduced the concept of hydraulic conductor (HC), in order to define the interconnected fracture systems that are associated with a certain well and that represent the water storage and production at crystalline rocks. Therefore, the fissured aquifer would be the sum of all existing HCs within an area, being represented as

$$\sum_{i=1}^{n} CH_i(X,Y,Z)$$

where X and Y are location coordinates and Z is the depth of the well.

At crystalline rock terrains, the water prospection approaches still miss deeper technical affirmation. A great number of unsuccessful drillings or wells bearing saline water are still taking place. There are no conceptual models strong enough to fully sustain well location and exploitation and the variables conditioning groundwater quality and quantity. The utilization of these water sources is always associated with risk components to the extent that the groundwater-sustainable yields and overall reserves cannot be safely estimated still. Despite this, since the early 20th century, in the entire northeast region, there is a great number of water wells discharging uninterruptedly. In many cases, unconfined aquifer characteristics and the high

hydraulic conductivities associated with fracture systems allow a direct and prompt recharge that keeps permanent exploitation conditions which, in turn, are only disturbed under long periods of drought. The major restrictive factor, for instance, for the use of groundwater resources within this region, is the quality. In general, waters are sodium chlorinated and show total dissolved solids above potable limits. The issue regarding the high heterogeneity and anisotropy of the fractured media depends directly on the assessment scale. At a punctual scale, practically, every single well may represent a single aquifer, which may differ from other ones. The differences in quantity and quality between neighboring wells, which intercept distinct HCs, are surprising. Regionalization approaches, dealing with fractured aquifer data sets, therefore, are not consistent. However, for smaller scales ($\geq$ 1:1,000,000) it may be possible to establish some zones showing similar tendencies regarding determined variables. Figure 3.11 shows 18,600 determinations of electrical conductivity of the groundwater found in fractured aquifers in the states of Ceará, Rio Grande do Norte, Paraíba, and Pernambuco. Each source is classified according to conductivity values, chosen for expressing water quality in terms of salinity: freshwater (CE $\leq$ 500 μS/cm), brackish water (1000 μS/cm < CE $\leq$ 2500 μS/cm), and salt water (CE > 2500 μS/cm).

It can be clearly seen that there are zones with different water qualities.

Water classified as brackish appears in the form of contour belts between fresh and salt water. A generic assessment suggests that there are four large zones: Zone 1—predominantly freshwater (southeast coastal area); Zone 2—predominantly salt water (a northeast–southwest range); Zone 3—predominantly freshwater (midwest); and Zone 4—predominantly salt water (north–northwest). Regarding quantity issues, every attempt at reserve evaluation would be mere speculation. However, it is believed that the quantities that can be extracted from fractured aquifers are enough to supply great parts of the diffuse-located population within the semiarid region of Brazil. The occurrence area of the basement rocks in the northeast region is about 750,000 km^2 and alone in the region called the drought polygon, the area would be 600,000 km^2. Considering the hypothesis of the existence of one single working tubular well in each 5-km^2 cell, there would be a total number of 120,000 wells exploiting groundwater resources within this region. The average depths of the wells are 60 m and the mean yields are situated between 1 and 3 m^3/h. Statistically, yield distribution assumes a lognormal model, with median oscillating around 1 and 2 m^3/h. Möbus et al. (1998) found the value of 1.7 m^3/h for the yield of

FIGURE 3.11
Groundwater quality in the domain of crystalline rocks in the northeast region. (From Feitosa, F. A. C. 2008. Compartimentação qualitativa das águas subterrâneas das rochas cristalinas do Nordeste oriental. UFPE, Proposta de Tese de Doutoramento, Feitosa, F. A. C., Manoel Filho, J., Feitosa, E. C., and J. G. A. Demetrio. Hidrogeologia: conceitos e aplicações. 3ª Edição Revisada e Ampliada. Rio de Janeiro: CPRM, 2008.)

tubular wells at crystalline regions of the state of Ceará. Adopting the lower value found for the median, that is, 1 m³/h, and admitting a pumping regime of 6/24 h (considered low), the daily quantity of water delivered would be 720 million L/day, supplying 3.6-million inhabitants at a daily consumption rate of 200 L/inhabitant/day. However, according to the assessment done by the Brazilian Geological Service, the percentage of freshwater within this region is about 20%–30%, reducing the production of freshwater drastically. The most important factor that hinders groundwater use within this region is given by quality constraints. At the southeast area, the existing water boreholes tend to have depths between 100 and 150 m due to the occurrence of thick-weathering covers. The yields are higher than in the northeast, averaging about 5–10 m³/h and delivering water of good physical and chemical quality (Feitosa and Feitosa, 2011).

3.8 Use of Groundwater in Brazil

The urban and rural population growth together with an expressive expansion of agriculture and industrial activities experienced by the nation have led to a remarkable increase in the use of groundwater. There is a close spatial connection between population growth and quality and quantity availability. Groundwater potability can be analyzed after electrical conductivities values because that directly reproduces the dissolved salt content of samples for the whole country, as it was represented in the hydrogeological map of Brazil (Diniz et al., 2014). A great part of the Brazilian territory shows water of excellent quality (electrical conductivity <150 μS/cm), mainly at the north and midwest. Similar to this one, other regions, such as the south and the southeast and the northeast states of Maranhão and

Piauí, do have good water quality (electrical conductivities between 150 and 500 μS/cm) suitable for all kinds of uses. These low saline content zones coincide with the major Paleozoic basins (Amazonas, Parecis, Paraná, and Parnaíba). Besides them, low electrical conductivity values are also found within the Urucuia Aquifer at the borders with the states of Bahia and Goiás, Tocantins and Maranhão, and Piauí. Areas with high saline contents are found at crystalline areas in the northeast region of the country. Waters bearing intermediate quality are found at the Recôncavo—Tucano-Jatoba Basin and Potiguar and Araripe Basin as well. Figure 3.12 illustrates the spatial distribution of the electrical conductivity values for the entire country. Analyzing Table 3.2 it is clear that in many areas of the northern and midwestern regions, despite the good groundwater quality, well density is very small due to the low urban density development and enormous superficial water availability. The northeast region of semiarid climate conditions, which encompasses large regional capitals such as Recife, Fortaleza, Natal, Joao Pessoa, and Maceió, some of them partially supplied with groundwater, present the greatest well density of the country—more than 50% of registered wells consuming 41% of the total volume of extracted groundwater in the entire country.

Although these waters are frequently saline, they still represent the only available water source. From a general overview, the largest groundwater exploitation takes place at the coastal areas, increasing toward the east and south, coherent with the main urbanization and industrialization axis of the country. The overall volumes according to each region in the country are found in Table 3.2.

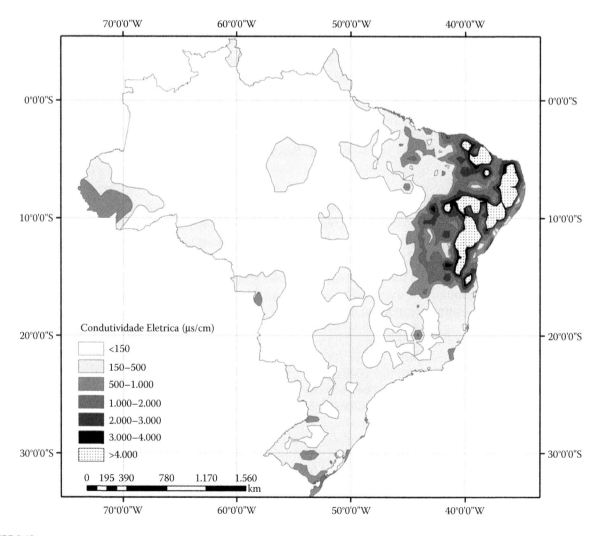

FIGURE 3.12

Distribution of the electrical conductivity in Brazil. (Modified from Diniz, J. A. O. et al. 2014. de, Mapa hidrogeológico do Brasil ao milionésimo: Nota técnica. Recife: CPRM—Serviço Geológico do Brasil, Recife.)

TABLE 3.2

The Use of the Groundwater in Brazil

Geographic Region	State	Área (km²)	Number of Wells	Wells/10 0 km²	Brazil (%)	Annual Volume Exploited (m³)	Brazil (%)
Norte	Acre	164.123,04	647	0.39	0.27	15.742.165	0.17
	Amazonas	1.559.159,15	7134	0.46	2.96	617.083.709	6.59
	Amapá	142.828,52	105	0.07	0.04	11.400.410	0.12
	Pará	1.247.954,67	6809	0.55	2.82	456.689.795	4.88
	Rondonia	237.590,55	1794	0.76	0.74	96.057.123	1.03
	Roraima	224.300,51	906	0.4	0.38	55.572.734	0.59
	Tocantins	277.720,52	1211	0.44	0.5	68.539.787	0.73
Total Norte		3.853.676,96	18606	0.48	7.72	1.321.085.723	14.11
Ne	Alagoas	27.778,51	1211	4.36	0.5	33.149.230	0.35
	Bahia	564.733,18	21943	3.89	9.1	833.999.175	8.91
	Ceará	148.920,47	21098	14.17	8.75	345.643.058	3.69
	Maranhão	331.937,45	11332	3.41	4.7	658.216.578	7.03
	Paraíba	56.469,78	17781	31.49	7.37	207.294.650	2.21
	Pernambuco	98.148,32	25416	25.9	10.54	536.515.417	5.73
	Piauí	251.577,74	27721	11.02	11.5	776.444.593	8.3
	R.G. Do Norte	52.811,05	9557	18.1	3.96	385.816.506	4.12
	Sergipe	21.915,12	4956	22.61	2.06	67.658.849	0.72
Total Nordeste		1.554.291,62	141015	9.07	58.43	3.844.738.056	41.08
Centro Oeste	Distrito Federal	5.780,00	198	3.43	0.08	8.219.712	0.09
	Goiás	340.111,78	3181	0.94	1.32	100.760.244	1.08
	Mato Grosso	903.366,19	3535	0.39	1.47	183.807.837	1.96
	Mato G. Do Sul	357.145,53	1377	0.39	0.57	185.152.320	1.98
Total C. Oeste		1.606.403,50	8291	0.52	3.44	477.940.113	5.11
Sudeste	Espírito Santo	46.095,58	1010	2.19	0.42	23.703.519	0.25
	Minas Gerais	586.522,12	19316	3.29	8.01	712.075.045	7.61
	Rio De Janeiro	43.780,17	488	1.11	0.2	13.054.442	0.14
	São Paulo	248.222,80	18607	7.5	7.72	1.498.161.956	16.01
Total Sudeste		924.620,67	39421	4.26	16.35	2.246.994.962	24.01
Sul	Paraná	199.307,92	12429	6.24	5.15	683.350.782	7.3
	R.G. Do Sul	281.731,44	14670	5.21	6.08	581.358.666	6.21
	Santa Catarina	95.737,00	7260	7.58	3.01	204.217.799	2.18
Total Sul		576.776,36	34359	5.96	14.24	1.468.927.247	15.69
Total Brazil		8.515.767,05	241.692	2.84	—	9.359.686.101	

Source: Based on SIAGAS—Groundwater Information System—www.cprm.gov.br.

References

Almeida, F. F. M., B. B. Brito Neves, and Carneiro, C. D. R. 2000. The origin and evolution of the South American platform. *Earth Science Reviews*, 50:77–111.

Bahia, R. B. C. Evolução Tectonossedimentar da Bacia dos Parecis—Amazônia. 2007. 115 f. Tese (Doutoramento em Ciências Naturais)—Escola de Minas, Universidade Federal de Ouro Preto, Ouro Preto. 2007.

Barison, M. R. 2003. Estudo Hidroquímico da Porção Meridional do Sistema Aquífero Bauru no Estado de São Paulo. 2003, 153p. Tese (Doutorado em Geociências e Meio Ambiente)—Instituto de Geociência e Ciências Exatas, UNESP, Rio Claro—SP.

Bocardi, L. B., Rostirolla, S. P., Deguchi, M. G. F., and F. Mancini. 2008. História de soterramento e diagênese em arenitos do Grupo Itararé—implicações na qualidade de reservatórios. *Rev. Bras. Geoc.*, 38:207–216.

Caetano-Chang, M. R. 1997. A Formação Pirambóia no centro-leste do estado de São Paulo. 1997. Tese (Livre Docência em Geologia) Instituto de Geociências e Ciências Exatas—Rio Claro, Universidade Estadual Paulista (UNESP), Rio Claro.

Campos, H. C. S. 2004. Águas subterrâneas na Bacia do Paraná. *Geosul, Florianópolis*, 19(37): 47–65.

Celligoi, A. and U. Duarte. 2002. Hidrogeoquímica do Aquífero Caiuá no Estado do Paraná. Boletim Paranaense de Geociências, n. 51, pp. 19–32, 2002. Editora UFPR.

Coelho, R. O. 1996. Estudo hidroquímico e isotópico do aquífero Bauru, Sudoeste do Estado de São Paulo. Dissertação (Mestrado em Recursos Minerais e Hidrogeologia)—Instituto de Geociências, Universidade de São Paulo, São Paulo.

CPRM—Serviço Geológico do Brasil. 2003. Geologia, tectônica e recursos minerais do Brasil: Texto, mapas & SIG. Organizadores: Bizzi, L. A., Schobbenhaus, C., Vidotti, R. M., and Gonçalves, J. H. Brasília.

CPRM—Serviço Geológico do Brasil. 2014. Mapa Hidrogeológico do Brasil ao Milionésimo. Organizadores: Diniz, J. A. O., Monteiro, A. B., De Paula, T. L. F., and Silva, R. C. Recife.

CPRM—Serviço Geológico do Brasil; Universidade Federal do Ceará—UFC. 2008a. Projeto Comportamento das Bacias Sedimentares da Região Semiárida do Nordeste Brasileiro/Hidrogeologia da Porção Oriental da Bacia Sedimentar do Araripe-Ceará. Fortaleza.

CPRM—Serviço Geológico do Brasil; Universidade Federal do Ceará—UFC. 2008b. Projeto Comportamento das Bacias Sedimentares da Região Semiárida do Nordeste Brasileiro/Hidrogeologia da Bacia Sedimentar de Lavras da Mangabeira-Ceará. Fortaleza.

CPRM—Serviço Geológico do Brasil; Universidade Federal de Campina Grande—UFCG. 2008. Projeto Comportamento das Bacias Sedimentares da Região Semiárida do Nordeste Brasileiro/Hidrogeologia da Bacia Sedimentar do Rio do Peixe-Paraíba. Fortaleza.

CPRM—Serviço Geológico do Brasil; Universidade Federal de Pernambuco—UFPE. 2008. Projeto Comportamento das Bacias Sedimentares da Região Semiárida do Nordeste Brasileiro/Hidrogeologia da Bacia de Jatobá—Sistema Aquífero Tacaratu/Inajá. Fortaleza.

DAEE—Departamento de Águas e Energia Elétrica do Estado de São Paulo. 1974. Estudo de águas subterrâneas—Região Administrativa 6—Ribeirão Preto. São Paulo: DAEE. 4 v.

DAEE—Departamento de Águas e Energia Elétrica do Estado de São Paulo. 1976. Estudo de Águas Subterrâneas—Regiões Administrativas 7, 8, 9—Bauru, São José do Rio Preto, Araçatuba. São Paulo: DAEE. v. 1 e 2.

DAEE—Departamento de Águas e Energia Elétrica do Estado de São Paulo. 1979. Estudo de Águas Subterrâneas—Regiões Administrativas 10 e 11—Presidente Prudente e Marília. São Paulo: DAEE, v.1 e 2.

DAEE—Departamento de Águas e Energia Elétrica do Estado de São Paulo. 1981. Estudo de Águas Subterrâneas—Região Administrativa 5—Campinas. São Paulo: DAEE. 2 v.

DAEE—Departamento de Águas e Energia Elétrica do Estado de São Paulo. 1982. Estudo de Águas Subterrâneas—Região Administrativa 4—Sorocaba. São Paulo: DAEE. 2 v.

DAEE—Departamento de Águas e Energia Elétrica do Estado de São Paulo. 1984. Caracterização dos recursos hídricos no Estado de São Paulo. DAEE, São Paulo.

DAEE—Departamento de Águas e Energia Elétrica: IG-Instituto Geológico: IPT-Instituto de Pesquisas Tecnológicas do Estado de São Paulo: CPRM-Serviço Geológico do Brasil, 2005. Mapa de águas subterrâneas do Estado de São Paulo. Escala 1:100.000. Coordenação Geral Gerôncio Rocha. São Paulo.

Diniz, J. A. O., Monteiro, A. B., Silva, R. de C. T. L. F., and Paula. 2014. de, Mapa hidrogeológico do Brasil ao milionésimo: Nota técnica. Recife: CPRM—Serviço Geológico do Brasil, Recife.

Diogo, A., Bertachini, A. C., Campos, H. C. N. S., and R. B. G. S. Rosa. 1981. Estudo preliminar das características hidráulicas e hidroquímicas do Grupo Tubarão no estado de São Paulo. In: SBG, Simp. Reg. Geol., 3, Curitiba, Atas, 1:359–368.

Eiras, J. F., Becker, C. R., Souza, E. M., Gonzaga, J. E. F., Silva, L. M., Daniel, L. M. F., Matsuda, N. S., and F. J. Feijó. 1994. Bacia do Solimões. *Boletim de Geociências da PETROBRAS, Rio de Janeiro*, 8(1): 17–45.

Eiras, J. F., Kinoshita, E. M., and F. J. Feijó. 1994a. Bacia do Tacutu. Boletim de Geociências da PETROBRAS, *Rio de Janeiro*, 8(1): 83–89.

Feitosa, E. C. and J. G. A. Demetrio. 2009. Os Aquíferos Cabeças e Serra Grande no vale do Gurguéia: síntese dos conhecimentos e perspectivas de explotação. LABHID/UFPE. Recife: Relatório Inédito.

Feitosa, E. C. and J. G. Melo. 1998. Estudos de Base—Caracterização Hidrogeológica dos Aquíferos do Rio Grande do Norte. In: Hidroservice, Plano estadual de recursoshídricos do Rio Grande do Norte.

Feitosa, F. A. C. and E. C. Feitosa. 2011. Realidade e perspectivas de uso racional de águas subterrâneas na região semiárida do Brasil. In: Medeiros, S. S., Gheyi, H. R., Galvão, C. O. and V. P. S. Paz. Recursos Hídricos em Regiões Áridas e Semiáridas, Campina Grande, INSA, pp. 269–305.

Feitosa, F. A. C. 2008. Compartimentação qualitativa das águas subterrâneas das rochas cristalinas do Nordeste oriental. UFPE, Proposta de Tese de Doutoramento, Feitosa, F. A. C., Manoel Filho, J., Feitosa, E. C., and J. G. A. Demetrio. Hidrogeologia: conceitos e aplicações. 3ª Edição Revisada e Ampliada. Rio de Janeiro: CPRM, 2008.

França, A. B., Araujo, L. M., Maynard, J. B., and P. E. Potter. 2003. Secondary porosity formed by deep meteoric leaching: Botucatu eolanite, southern South America. *AAPG Bulletin*, 87(7): 1073–1082.

França, A. B. and P. E. Potter. 1989. Estratigrafia, ambiente deposicional do Grupo Itararé (Permocarbonífero), Bacia do Paraná (Parte 2). Boletim Geociências Petrobrás, 3:17–28.

Gaspar, M. T. P. 2006. Sistema aquífero Urucuia: caracterização regional e propostas de gestão. Brasília: UNB. Tese de doutorado.

Gesicki, A. 2007. Evolução diagenética das Formações Pirambóia e Botucatu (Sistema Aquífero Guarani) no Estado de São Paulo. Tese de Doutorado. Instituto de Geociências—USP. São Paulo.

Góes, A. M. and F. J. Feijó. 1994. Bacia do Parnaíba. *Boletim de Geociências da PETROBRAS. Rio de Janeiro*, 8(1): 57–67.

IBGE—Instituto Brasileiro de Geografia e Estatística, 2010. Base cartográfica vetorial contínua do Brasil ao milionésimo—BCIM. Ver. 3. Rio de Janeiro.

Iritani, M. A., Oda, G. H., Kakazu, M. C., Campos, J. E., Ferreira, L. M. R., Silveira, E. L., and A. A. B. Azevedo. 2000. Zoneamento das características hidrodinâmicas (transmissividade e capacidade específica) do Sistema Aquífero Bauru no Estado de São Paulo—Brasil. In:

Congresso Mundial Integrado de Águas Subterrâneas, 1 e (ABAS, 11 e ALHSUD, 5), Fortaleza—CE, Brasil, Boletim de Resumos. p. 147.

Lastoria, G., Sinelli, O., Chang, H. K., Hutcheon, I., Paranhos Filho, A. C., and D. Gastmans. 2006. Hidrogeologia da Formação Serra Geral no estado de Mato Grosso do Sul. *Águas Subterrâneas*, 20(1): 139–150.

LEBAC—Laboratório de Estudo de Bacias. 2008. Informe Final de Hidrogeologia Regional do SAG. In: Gastmans, D. and H. K. Chang (Coord.). Informe Técnico do Consórcio Guarani. Rio Claro.

Machado, J. L. F. 2005. Compartimentação Espacial e Arcabouço Hidroestratigráfico do Sistema Aquífero Guarani no Rio Grande do Sul. 2005, 238 p. Tese (Doutorado em Geologia).—Programa de Pós-Graduação em Geologia Sedimentar, Universidade do Vale do Rio dos Sinos (UNISINOS). São Leopoldo (RS).

Machado, J. L. F. and U. F. Faccini. 2004. Influência dos falhamentos regionais na estruturação do Sistema Aquífero Guarani no Estado do Rio Grande do Sul. In: ABAS, Congresso Brasileiro de Águas Subterrâneas, 13, Cuiabá, Anais CD-ROM, pp. 1–14.

Manoel, F. J. 1996. Modelo de dimensão fractal para avaliação de parâmetros hidráulicos em meio fissural. São Paulo: USP, Tese de Doutorado.

Matta, M. A. S. 2002. Fundamentos hidrogeológicos para a gestão integrada dos recursos hídricos da região de Belém/Ananindeua, Pará/Brasil. Tese de Doutoramento apresentada à Universidade Federal do Pará, Belém.

Milani, E. J. 2004. Comentários sobre a origem e a evolução tectônica da Bacia do Paraná. In: Montesso-Neto, V., Bartorelli, A., Carneiro, C. D. R., and B. B. Brito Neves. Geologia do Continente Sul-Americano—Evolução da obra de Fernando Flávio Marques de Almeida. Ed. Becca, pp. 265–279.

Milani, E. J., Melo, J. H. G., Souza, P. A., Fernandes, L. A., and A. B. França. 2007. Bacia do Paraná. *Boletim de Geociências da Petrobras*, 15(2): 265–287.

Milani, E. J. and P. V. Zalán. 1998. Brazilian Geology Part 1: the Geology of Paleozoic Cratonic Basins and Mesozoic Interior Rifts of Brazil. In: AAPG, International Conference and Exhibition, Rio de Janeiro, Brazil. Short Course Notes.

Möbus, G., Silva, C. M. S. V., and F. A. C. Feitosa. 1998. Perfil estatístico de poços no cristalino cearense. III Simpósio de Hidrogeologia do Nordeste. Recife, Anais.

OEA—Organização dos Estados Americanos. 2009. Aquífero Guarani: programa estratégico de ação—PEA. OEA: Brasil; Argentina; Paraguai; Uruguai, 2009.

Oliveira, F. R., Cardoso, F. B. F., Manoel Filho, J., Kirchheim, R., Feitosa, E. C., Teixeira, H. R., Varella Neto, P. L., Gonçalves, M. V. C., and F. S. Nascimento. 2012. Gestão Compartilhada de Águas Subterrâneas na Chapada do Apodi, entre os Estados do Ceará e Rio Grande do Norte. XVII Congresso Brasileiro de Águas Subterrâneas, Bonito/MS.

Paula e Silva, F. 2003. Geologia de subsuperfície e hidroestratigrafia do Grupo Bauru no Estado de São Paulo. Tese de Doutorado, Instituto de Geociências e Ciências Exatas, Universidade Estadual Paulista (UNESP), Rio Claro.

Paula e Silva, F., Kiang, C. H., and M. R. Caetano-Chang. 2005. Hidroestratigrafia do Grupo Bauru (K) no Estado de São Paulo. *Águas Subterrâneas*, 19(2): 19–36.

Rebouças, A. C. 1978. Potencialidades hidrogeológicas dos basaltos da Bacia do Paraná no Brasil. In: CONGRESSO BRASILEIRO DE GEOLOGIA, 30, Recife, Anais, v. 6, Recife: SBG. pp. 2963–2976.

Rebouças, A. da C., Manoel Filho, J., and B. B. Brito Neves. 1969. de Inventário hidrogeológico básico do Nordeste— Programa e Normas Técnicas. Recife: SUDENE— Superintendência de Desenvolvimento do Nordeste. Divisão de Documentação.

SADC—Southern African Development Community. 2009. Folheto Explicativo do Mapa e Atlas Hidrogeológico da Comunidade para o Desenvolvimento da África Austral (SADC). [Lusaka, Zambia]: SADC, March 2009. (Projecto de Elaboração do Mapa Hidrogeológico da SADC).

Soares, A. P., Soares, P. C., and M. Holz. 2008. Heterogeneidades hidroestratigráficas no sistema Aquífero Guarani. *Revista Brasileira de Geociências*, 38(4): 600–619.

Souza de E. L., Galvão, P. H. F., and C. S. do Pinheiro. 2013. Síntese da hidrogeologia nas bacias sedimentares do Amazonas e do Solimões: Sistemas Aquíferos Içá-Solimões e Alter do Chão. *Geological USP, Sér. cient.* 13(1), São Paulo mar.

Stradioto, M. R. 2007a. Hidroquímica e aspectos diagenéticos dos Sistema Aquífero Bauru na região sudoeste do estado de São Paulo. 103 f. Dissertação (Mestrado)— Curso de Geociências e Meio Ambiente, Universidade Estadual Paulista Júlio de Mesquita Filho, Rio Claro.

Stradioto, M. R. 2007b. Hidroquímica e aspectos diagenéticos do Sistema Aquífero Bauru na região sudeste do estado de São Paulo. Dissertação (Mestrado em Geociências e Meio Ambiente)—Instituto de Geociências e Ciências Exatas, Universidade Estadual Paulista. Rio Claro.

Struckmeier, W. F. and J. Margat. 1995. *Hydrogeological Maps: A Guide and a Standard Legend*. International Association of Hydrogeologists—Hannover (International) Contributions to Hydrogeology; Vol. 17).

Tancredi, A. C. F. N. S. 1996. Recursos hídricos subterrâneos de Santarém: fundamentos para uso e proteção. 154 p. Tese (Doutorado)—Curso de Pós-Graduação em Geologia e Geoquímica, Centro de Geociências, Universidade Federal do Pará, Belém.

UNESCO—United Nations Educational, Scientific and Cultural Organization, 1970. International Legend for Hydrogeological Maps. Paris, published by Cook, Hammond & Kell Ltd, England, 101 pp.

Vidal, A. C. 2002. Estudo Hidrogeológico do Aquífero Tubarão na Área de Afloramento da Porção Central do Estado de São Paulo. IGCE/UNESP, Rio Claro SP. 109 p. (Tese de Doutorado).

Section II

Exploration

4

An Overview of Six Decades of Research and Evolution in Resistivity Exploration for Groundwater in Hard-Rock Terrain of India

Shrikant Daji Limaye

CONTENTS

4.1 Introduction (the Preresistivity Era in Groundwater Exploration)

Well digging to obtain water for drinking purposes and for irrigation has been practiced in India for thousands of years. Exploration for locating suitable sites to get groundwater is also an old technique from the fifth century starting from Varahmihir's "Drukggargal Shastra" in which anthills, certain species of trees, and certain types of strata have been mentioned as indicators for groundwater. These indications are still widely believed and "water diviners" in Indian villages still use them. Even in tribal areas, the local diviners know where to look for water in the summer season when streams go dry. Historically, although the kings were mainly responsible for sponsoring dug wells and providing drinking water supply to the population, their administration was not officially involved in the exploration of groundwater. This was the domain of private diviners. However, around the 1920s, the geologists, agriculturists, and civil engineers became involved in groundwater exploration, based on the study of strata met within existing wells, topography, rock exposures in streams or rivers, and their own experience.

D.G. Limaye, the father of the author and one of the pioneering hydrogeologists in India, started working in 1933 in Pune, as a private consultant for farmers for selecting locations for dug wells for irrigation, mainly in the Deccan traps or basaltic terrain in western India and the pre-Cambrians in south India. He had a bachelor's degree in agriculture and basic knowledge of geology. As the words "hydrogeologist" and "geohydrologist" were born much later, in those days, he was called an "underground water geologist." Farmers even mocked him as an "educated water-diviner." However, in a letter dated September 30, 1941, written to the political agent for the erstwhile Kathiawar States at Rajkot, in the preindependence British-ruled India (now Saurashtra region of Gujarat State), he mentioned the "science of hydro-geology" as a useful tool for groundwater exploration in the basaltic region of Kathiawar. This letter is kept in the National Archives of Government of India in New Delhi. His first technical paper on groundwater exploration appeared in 1940 in the *Journal of Geological, Mining and Metallurgical Society of India* (Limaye, 1940).

The farmers expected him to select a good spot for well digging in their farms and advise them on the depth and expected yield. They also sought his guidance on revitalization of existing wells by deepening or by drilling horizontal or vertical bores in the well bottom by using long crowbars. The farmers paid him 5 rupees (about 8 cents in the current rate of exchange of the rupee to U.S. dollar) as a consultation fee. Some farmers even paid in kind, such as a few kilos of wheat, millet, or rice. During the 4 months of Monsoon rains, the exploration work for farmers almost came to a standstill, unless the year was a drought year. However, he used the Monsoon time to work on data collection, making notes on the occurrence of intertrappean beds in the Deccan trap or basalts and carrying out laboratory experiments on the porosity and specific yields. He had observed that intertrappean beds or the red boles in the basaltic terrain were important for the success of a well or a bore because the upper and lower junctions of these beds with adjoining lava flows were often permeable. Apart from publishing technical papers on the geological exploration of groundwater in the basaltic terrain, he started writing articles in newspapers. Later on, his clientele expanded to industries and departments of the erstwhile government of Bombay. The industries were in favor of drilling bores rather than digging large-diameter wells where gunpowder had to be used for blasting the hard rock. In those days, drilling a bore of 100 mm or 150 mm diameter up to about 25–30-m depth in basalts used to take anywhere between 1 and 2 months with hammer percussion rig or calyx rotary rig, depending on the hardness of strata and thickness of overburden. "Piston and cylinder"-type reciprocating pumps were used for pumping water from such bores. Water from dug wells was pumped by diesel engine pump sets as in those days rural electrification was minimal. The mode of transportation for a hydrogeologist for exploration in rural areas was bullock cart, bicycle, or simply on foot.

Government departments related to rural development in the old-Bombay Province, which were thus far taking advice from "water diviners" invited from England by the erstwhile British Government of India, took note of this young agriculture graduate working as an "underground water geologist." In their rural drinking water programs, D.G. Limaye was given a contract every year for surveying the area around a village and locating a suitable spot for the construction of a "drinking water dug-well." A typical well in this program was 2.5–3.0 m in diameter and about 8 m in depth. With such contracts of surveying 60–80 villages per year for about 10 years, my father had an opportunity to move all over the basaltic terrain of western India and also in some adjoining portions of the basement complex and alluvial terrains.

After about 23 years of work in groundwater exploration, D.G. Limaye gradually realized the limitations of a purely hydrogeological method of groundwater exploration comprising the study of topography, stratigraphy, weathering pattern, occurrence of hard rock or sand in stream or river beds, stream curvatures and shifting of streams with a possibility of occurrence of old stream beds, role of dikes, if any, and the inventory of existing wells. The book *Groundwater* by C.F. Tolman (1937) was the Bible for all hydrogeologists at that time. Additional equipment used in those days was a "water-finding machine," which had a box with east–west coils and a long "specialized" magnetic needle. Government departments had imported these "patented" machines in the 1930s from Mansfield and Schmidt Companies in England. However, the success of these machines depended on the degree of accuracy of the data obtained using purely geological investigations.

4.2 Early Resistivity Work in India

Around 1955–1956, news came to India about the use of geophysical equipment for exploration for groundwater. In 1956, the first symposium on groundwater was organized in New Delhi by the Ministry of Natural Resources and Scientific Research and the volume of proceedings was published by Central Board of Geophysics. During this symposium, experts discussed some new small-sized equipment from UK and German manufacturers for groundwater exploration similar to the ones used in the western world for hydrocarbon or oil exploration. The geophysical instruments for seismic and resistivity surveys used in oil exploration were bulky but had a good depth of investigation. The depth of investigation for groundwater with the new small-sized instrument was only up to 100 m or so but it was adequate for most applications. This new-resistivity equipment was lightweight and provided an additional tool for shallow groundwater exploration in the hands of hydrogeologists and engineers.

The first available literature was from a German company named "Gesellschaft fur Angewandt Geophysik" (Company for Applied Geophysics) but the price of their resistivity meter was far beyond the means of any private consultant from India. Information brochures for DC resistivity meters were later available from companies such as H. Tinsley from the United Kingdom, but again the price was high. However, for learning the theory of resistivity exploration and being acquainted with Wenner and Schlumberger systems of resistivity survey using four electrodes, imported books such as those written by Nettleton, Dobrin, Heiland, or Jakowsky were available in the Indian market. The first resistivity meter in India was purchased around 1956–1957 by the Central Water and Power Research Station (CWPRS) at Khadakwasla near Pune. This was a DC resistivity meter from H. Tinsley of the United Kingdom, employing dry batteries to introduce electric current (measured in mA, i.e., milli amperes) into the ground between two outer electrodes C1 and C2. The DC potential difference (measured in mV, i.e., milli volts) caused by this electric current between the two "nonpolarizing $Cu–CuSO_4$ electrodes" P1 and P2 was balanced on a standard potentiometer. The apparent resistivity (AR) for Wenner configuration with an electrode separation of "a" meters was then calculated using the formula (AR) = 2 π a mV/mA. This instrument had another potentiometer to balance the stray potential difference between electrodes P1 and P2, when no current was introduced in the ground. The potentiometers were calibrated using an "in-built, standard galvanic" cell.

After calculation of the resistivity in Schlumberger or Wenner configuration, graphs were drawn with electrode separation on the X-axis and AR on the Y-axis. Local electronic workshops in India were capable of repairing or even copying the H. Tinsley resistivity meter. D.G. Limaye purchased one DC resistivity meter for groundwater exploration. This was probably the first resistivity meter purchased in the private sector in India in April 1958. Dr. S. L. Banerjee, the chief geophysicist with CWPRS in 1958, was very happy that the resistivity method would now also be used in the private sector. The first fieldwork carried out by the author and his father was in May 1958, in metamorphic basement complex rocks for The Sandur Manganese and Iron Ore Company in Karnataka State. A paper on this new method was presented in a symposium in the city of Baroda, India—now known as the city of Vadodara (Limaye, 1959).

The DC instrument worked satisfactorily in rural areas. But in urban areas, the stray (spontaneous) potential difference between P1 and P2 kept changing and it was difficult to balance it on the potentiometer. These stray potentials were often much larger than the signal potential difference generated between P1 and P2, when the current was switched on. To get a satisfactory reading, the average of two measurements had to be taken—one when the current was going from C1 to C2 and another when it was going from C2 to C1. Sometimes, it was impossible to get reliable, consistent readings especially in Mumbai, where DC traction was employed for local trains. Getting durable porous ceramic pots for nonpolarizing electrodes was also a problem. After each survey, the remaining concentrated $CuSO_4$ solution from P1 and P2 electrodes had to be stored in a plastic bottle for reuse.

The only way to get rid of the menace of stray potentials was to use a commutator, such as a "dual split ring—hand-driven" commutator. One split ring with four carbon brushes converted the DC from batteries into low-frequency, square-wave AC to be fed to C1–C2 electrodes. The other split ring mounted on the same shaft converted the square-wave potential difference received from electrodes P1–P2 back into DC for measuring on the potentiometer. The use of a commutator considerably improved the performance of the DC resistivity meter. An electronic vibrator with four pairs of contacts was also tried in those days, but somehow, it was not as successful as the hand-driven commutator.

With the advent of transistor circuitry, AC resistivity meters were introduced in the market. The use of DC instruments virtually stopped. In the drought year of 1972, "Terrameter" from the Swedish company ABEM was probably used for the first time in the states of Maharashtra, Andhra Pradesh, and Karnataka in India, mainly by the Christian mission personnel engaged in drought relief work and rural development. Terrameter had two boxes; one G-box for generating low-frequency, square-wave AC to be fed to C1–C2 electrodes and the other V-box for measuring potential differences between P1 and P2 electrodes. In the same year (1972), Maharashtra State Government's Ground Water Survey and Development Agency (GSDA) was established. However, its activity in geophysical exploration using AC resistivity meters started much later.

4.3 Data Acquisition and Interpretation

The introduction of integrated circuits, microprocessors, band-pass and band-stop filters, operational amplifiers for signal boosting, etc., improved the performance of AC resistivity meters to a great extent. The Schlumberger system became more popular for sounding while the Wenner system was popular for both sounding and profiling. Indian companies from Mumbai and Hyderabad started manufacturing AC resistivity meters so that their prices became affordable. However, one side effect of the low prices was that many "spurious experts" without much knowledge of geology or geophysics purchased resistivity meters and started giving advice to farmers, using any convenient electrode arrays of their own design. Their failures earned a bad name for the resistivity method in rural areas.

Data acquisition became easier but interpretation with the help of standard two-layer and three-layer master curves was too theoretical especially in hard-rock terrain. It took some years for the fact to sink in that the interpretation of results must be consistent with field geology and local observations. In the field, a clayey bed in riverside alluvium would indicate low resistivity and may be interpreted as a water-bearing horizon while a saturated coarse sand bed would be neglected because of its higher resistivity. Later on, programs for computerized interpretation of resistivity curves also became available but their correlation with the field conditions was still lacking. Even today, interpretation of field data and recommendations for good sites for obtaining groundwater are often made without testing the "ground-truth" by drilling at least one bore at a recommended site. This is probably because many groundwater research projects are being carried out by universities that do not have funds for drilling.

The practical question is, "Does the assumption of a horizontally layered, isotropic, semi-infinite ground hold good in practice, especially in hard-rock terrain?" Wenner electrode separation of about 150–200 m between C1 and C2 for groundwater exploration often includes several variations of surface resistivity along the line on which the electrodes move for sounding, even if one tries to avoid any hard-rock exposures on such a line. Erratic variation of near-surface resistivity near P1 or P2 gives erratic readings. Furthermore, in a typical three-layer case comprising a dry soil–subsoil and weathered rock at the surface (thickness h1), underlain by a saturated weathered rock (thickness h2), followed by hard rock of very large thickness at the bottom, the three-layer H-type curve obtained by vertical electrical sounding (VES) would only give a fair idea of occurrence of shallow water only above the top surface of the hard rock. But would a water-bearing flow junction of an intertrappean bed in basalt or a water-bearing fracture in granite, which is just a few centimeters thick and is located at about 60-m depth within the hard rock, give its signature in resistivity measurements of a VES? Probably not.

Instead of using the two-layer and three-layer standard master curves, a better method of interpretation of VES data was given by Sankar Narayan and Ramanujachary (1967). The inverse slope vertical electrical resistivity (VES) method as suggested by these two authors works very well in engineering geology and also in groundwater exploration for a 3–5-layered ground. This method assumes considerable importance in the field of groundwater exploration because of its ease of operation, low cost, and its capability to distinguish between the saline and freshwater zone. This method is being widely used in India in different geological situations. It was found to give good results correlating well with borehole data. This method is simple and gives resistivities and depths directly from the plot of the field data on a linear graph (Sankar Narayan and Ramanujachary, 1967). The author tested it in India and in Canadian glacial till with good results.

Wenner profiling with electrode separations of 10, 20, and 30 m is also helpful in groundwater exploration, especially in hard-rock terrain. Here, it is possible to distinguish between a hard-rock area devoid of groundwater and a weathered zone in which groundwater would occur and would be available in a well. Assuming that even the deeper flow junctions or fractures would be recharged from phreatic groundwater, bore wells could also be located in such weathered zones confirmed by profiling carried out 2 or 3 times on the same profiling line with different electrode spacings. In field practice, once the profile line is selected, it is advisable to dig pits at specific distances and water these pits before starting profiling from one end. Even if the electrode separation is 20 or 30 m, the "jump" after each reading could be restricted to 10 m if the pits are spaced at 10-m distances along the line. If the site has a productive bore and a failed bore, it is a good idea to do profiling along the line joining these two bores and repeat the profile with different electrode spacings to see the difference between the resistivity values obtained near the two bores. The difference in resistivity near a good bore and a failed bore in an area is a good guide for interpretation of profiling data in the area.

A typical "resistivity survey party" in the private sector comprises one expert to take the readings of the resistivity meter, two to three assistants to ensure correct placing of the electrodes, and 10 helpers (preferably 2–4 women among these helpers). Two women hold umbrellas; one umbrella over the instrument and the other over the expert's head. Two women are necessary to bring water from a nearby well/bore in buckets for watering the electrodes. The cost of a resistivity survey is affordable only to industries, large commercial complexes, residential developments such as satellite townships, and cooperative factories of farmers. The scope of consultation work is thus limited and that too only for about 8–9 months of the summer and winter seasons. The private sector, therefore, has not grown appreciably over the last few decades. Some nongovernmental organizations (NGOs) are active in this field. Staff members of government organizations, institutions, and university departments conduct resistivity surveys in their projects and research work. For them, the cost and time is not much of a problem and they do a lot of survey work over a small-project area. These days, funds to purchase resistivity meters from Indian manufacturers and conduct fieldwork are available to various university departments. However, their services are either not available to farmers or are very expensive. Recently, some private consultants have purchased resistivity meters but they take just two or three readings with short spreads of their own design to cut the cost. This quasi-scientific procedure is often unsuccessful, earning a bad name for the resistivity method. Therefore, most of the farmers still depend on local diviners or hydrogeologists using only the hydrogeological method comprising geological observations, topography, well inventory, and sometimes bioindicators.

Recently, the GSDA has started using ABEM's electromagnetic instrument "Wadi" using very-low-frequency (VLF) signals in the range of 15–20 kHz for groundwater exploration. However, the cost of Wadi is prohibitive for private hydrogeologists, who still prefer to depend on AC resistivity meters. Even the GSDA of Maharashtra has the Wadi instruments but uses them sparingly.

4.4 Conclusions

In the past 56 years or so, resistivity methods for groundwater exploration have come a long way, especially in instrumentation and interpretation techniques. Data collection has been easier and more reliable. However, while interpreting the data and modeling, especially in hard-rock terrain, it is necessary to keep in mind the limitations because the ground itself does not comply with all the assumptions made while deriving the theoretical formulae. The interpretation has to be consistent with the local field conditions. Although based on sound geophysical principles, the resistivity method has not achieved more than 90% success in exploration for shallow phreatic groundwater and is still out of reach for a common farmer due its high cost. Drilling trial bores based on a low-cost hydrogeological survey or on advice from water diviners is still the method of choice by farmers because taking trial bores is cheap. With the advent of modern drilling technology, farmers are now drilling trial bores up to 150-m depth in hard rock in search of groundwater in fracture networks or lava flow junctions—which are not likely to register their signatures in VES graphs obtained using the resistivity method.

Such deep bores, if successful, give a good amount of water (sometimes as high as 20 m^3/h). Farmers pump them continuously, as they are the absolute owners of groundwater. The rise in exploration and exploitation of deeper groundwater resources in the past 30 years has resulted in an appreciable depletion of the water table and drying of many old dug wells (Limaye, 2012). The depletion in water level is resulting in higher consumption of electricity for pumping water. It is the duty of geoscientists to motivate the farmers for rainwater harvesting and watershed management with forestation of degraded watersheds for augmenting recharge to groundwater and for ensuring sustainability of the supply. Such geoethical activities need to be promoted by NGOs at the village level through decisions taken in "Village Meetings" under the guidance of elected members of the "Village Council."

References

Limaye D.G. 1940. Laboratory tests for water bearing capacity of trap and associated recent rocks in the Deccan and Konkan areas. *Quarterly Journal of Geological, Mining and Metallurgical Society of India, Calcutta*, 12 (2), 31–50.

Limaye S.D. 1959. Ground water investigation by electrical resistivity method in Dharwar rocks. *Proceedings of the Symposium on Geophysical Exploration*, Baroda, India.

Limaye S.D. 2012. Observing geoethics in mining and in ground-water development: An Indian experience. *Annals of Geophysics*, 55 (3), 379–381; doi: 10.4401/ag-5573.

Sanker Narayan P.V. and K.R. Ramanujachary. 1967. Inverse slope method for determining absolute resistivities. *Geophysics*, 32, 1036. Society of Exploration Geophysicists.

Tolman C.F. 1937. *Groundwater*, McGraw-Hill, New York.

5

Application of Electrical Resistivity Tomography in Delineation of Saltwater and Freshwater Transition Zone: A Case Study in West Coast of Maharashtra, India

Gautam Gupta, Vinit C. Erram, and Saumen Maiti

CONTENTS

5.1 Introduction

The coastal zones are subject to rapid development with growing and conflicting demands on natural resources, and often they are subject to irreversible degradation. The critical phenomena that mainly affect these areas are coastal erosion, flooding due to river floods, tidal waves, or rising sea level, and contamination of the aquifers (e.g., salt-wedge intrusion of seawater). Along the coast, there are many sites of community interest (SCI) subjected to a strong incidence of human activities mainly linked to agriculture and tourism. According to the Recommendation on Integrated Coastal Zone Management (ICZM) of the European Commission, coastal areas are of great environmental, economic, social, and cultural relevance. Therefore, the implementation of suitable monitoring and protection actions is fundamental for their preservation and for assuring the future use of this resource. Such actions have to be based on an ecosystem perspective for preserving coastal environment integrity and functioning and for planning sustainable resource management of both the marine and terrestrial components for the promotion of economic and social welfare of coastal zones.

Of all nonintrusive surface geophysical techniques, the electrical resistivity profiling and vertical electrical sounding (VES) methods have been applied most widely to detect and monitor saline water/freshwater transitions in different coastal areas in the last several years with proven efficiency (Barker, 1996; Sherif et al., 2006; Song et al., 2007; Omosuyi et al., 2008; Adeoti et al., 2010; Hermans et al., 2012; Maiti et al., 2012; Mondal et al., 2013; Gupta et al., 2014). Mondal et al. (2009) have envisaged the fact that contaminants play a vital role in both inland and coastal aquifers. These authors further suggest that the electrical resistivity of the aquifer is reduced due to contaminants, which reflect as an anomaly. Sankaran et al. (2012) systematically studied the hydrogeological, geophysical, and hydrochemical properties in SIPCOT area in southern India to distinguish between groundwater pollution and saline water intrusion through the Uppanar River. Integrated geophysical investigation and geochemical analysis were employed to assess the subsurface geologic formations, aquifer geometry, and seawater intrusion in Godavari Delta Basin (Gurunadha Rao et al., 2011). These authors suggested that the lowering of resistivity was due to the encroachment of seawater into the freshwater zones and also infiltration of seawater from the high tides. These studies also indicated thick marine clays, which possess the palaeosalinity due to recession of sea level. Systematic collection and analysis of hydrological, geophysical, and hydrochemical data from vulnerable parts of Andrott Island, Lakshadweep (Singh et al., 2009) divulge that fresh groundwater is only available between 2.5 and 5.0-m depths and provide an indication of deterioration in groundwater quality in the peripheral parts of eastern and western coasts of this island. Barker et al. (2002) have successfully used the resistivity-imaging technique for borehole sitting in hard-rock regions of India.

Groundwater chemistry plays a vital role for the study of its quality in coastal aquifers (Mondal et al., 2008; Maiti et al., 2013) and thus assessing seawater incursion through an aquifer in coastal belts is a periodic analysis of groundwater chemistry (Sukhija et al., 1996; Saxena et al., 2003). Several researchers have used hydrochemical parameters to evaluate the seawater incursion process, which can be helpful to control the water quality in coastal areas. Sukhija et al. (1996) studied the coastal groundwater of Karaikal and Tanjavur, Tamil Nadu and differentiated the palaeomarine and modern salinities using inorganic chemistry, organic biomarker fingerprints, and tritium and radiocarbon measurements. Mondal et al. (2010a) successfully tested the relationship between total dissolved solids (TDS) with chloride (Cl^-), sodium, magnesium, and sulfate concentrations of groundwater for pre- and postmonsoon seasons from a watershed along the southeastern coast of India.

Mondal et al. (2010b) examined trace element content in the groundwater of a coastal island (Pesarlanka) of the eastern coast of India. These authors concluded that the concentration of most trace elements in the groundwater exceeded the WHO (1984) limits. The main source of most elements has been attributed to marine sediments. Studies have been carried out in the Krishna delta to observe the influence of strontium and boron on the groundwater of coastal regions as well as their correlation with the chemical parameters, which are commonly used for seawater ingression studies (Saxena et al., 2004).

Hydrochemical characteristics of coastal aquifers in Tuticorin, Tamil Nadu (Mondal et al., 2011) seem to be influenced by various processes together with saline water mixing, anthropogenic contamination, and water–rock interaction, which is reflected by very wide ranges and high standard deviations of most hydrochemical parameters, such as TDS, Cl^-, SO_4^{2-}, Mg^{2+}, and Na^+ exceeding the limit of drinking water standards (WHO, 1984). Nonetheless, there are some serious limitations in such investigations as they fall short to differentiate between formations of similar resistivities such as saline clay and saline sand, and the causes of low resistivity due to water quality (fresh or saline).

Water resources in coastal regions of Maharashtra, India assume a special significance because of rapid strides in developmental activities thereby depleting the available groundwater. Not much information is available on the role of lineaments in the hydrogeological setup as well as their role in the occurrence and movement of groundwater in the coastal region of Maharashtra. The ingress of saline water through inland drains due to tidal influence also makes the potable water unfit for consumption (Gurunadha Rao et al., 2011). In such areas, exploration and differentiation of freshwater aquifers from saline water aquifers become the primary objectives (Bear et al., 1999). An attempt is made here to carry out geoelectrical studies along the coastal tract of western Maharashtra to analyze the effects of seawater intrusion and in locating fresh groundwater pockets to meet the water demands of society.

Little is understood of the dynamics and complex mixing relationships between fresh surface waters and saline groundwater in the western coastal region of Maharashtra. Conventional methods of characterization utilizing point measurements offer limited information about processes occurring at depth. The high levels of salinity here have provided a unique opportunity with which to utilize geoelectrical characterization methods such as electrical resistivity imaging (ERI). Contrasts in resistivity values between saline groundwater and fresh surface waters have allowed current researchers to view images of the deep subsurface; thus, distinct zones of saline groundwater migrating from depth can be tracked in the resistivity images collected and subsurface processes serving to control salinity can be inferred. Additionally, ERI offers the potential for a greater understanding of this system through its use as a temporal monitoring tool; preliminary results have revealed that continued efforts in this manner will aid in developing a better understanding of the various processes occurring within the saline wetlands, which serve to make each site unique.

In this chapter, we show the first results of a study based on ERI for the characterization of a coastal area of Maharashtra in southwest India. This region is in contact with the Arabian Sea and groundwater level fluctuates in response to tidal variations (Song et al., 2007). We are confronted with two major problems in this region. The first one is contamination of fresh groundwater by seawater occurring in locations where saline water displaces or mixes with freshwater (Todd, 1979), which leads to the infiltration of saline fluids into the fresh aquifer thereby changing the near-surface distribution pattern of electrical properties. Second, the Sindhudurg district of Maharashtra is covered by Deccan volcanic rocks and most of the soils are derived from lateritic rocks. Groundwater flows preferentially through a network of voids, conduits, joints, and fractures (CGWB, 2009). Hence, monitoring the shallow distribution of the true resistivity patterns in the area is vital for mapping faults, fractures, joints, preferential groundwater conduits, and lineaments affecting groundwater circulation patterns. Modeling and interpretation of resistivity imaging in this region is therefore of special interest to help understand inhomogeneous infiltrations of fluids through pores and geologically weak zones as well as fluid percolation patterns at the subsurface.

5.2 Theoretical Background

The resistivity method is one of the oldest geophysical survey techniques to determine the subsurface resistivity distribution by making measurements on the ground surface. From these measurements, the true resistivity of the subsurface can be estimated. The ground resistivity is related to various geological parameters such as the mineral and fluid content, porosity, and degree of water saturation in the rock. Electrical resistivity surveys have been used for many decades in hydrogeological, mining, geotechnical, environmental, and even hydrocarbon exploration (Keller and Frischknecht, 1966; Bhattacharya and Patra, 1968; Loke, 2011). The fundamental physical law used in resistivity surveys is Ohm's Law, which governs the flow of current in the ground. Here, a known current (I) is passed into the ground with two current electrodes (C1 and C2) and the potential difference (ΔV, i.e., potential drop) between two potential electrodes (P1 and P2) is measured (Figure 5.1) (Koefoed, 1979). The ratio (ΔV/I) gives the resistance (R), which is multiplied by the geometrical factor (K) of electrode separation to get apparent resistivity of the ground

$$\rho_a = k\Delta V / I$$

Figure 5.2 shows the common arrays used in resistivity surveys together with their geometric factors. In a

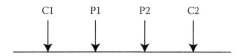

FIGURE 5.1
General electrode array used in resistivity surveys. C1 and C2 are the two current electrodes while, P1 and P2 are the two potential electrodes.

later section, we will discuss the advantages and disadvantages of some of these arrays. Resistivity meters normally give a resistance value, R=V/I; so, in practice, the apparent resistivity value is calculated by

$$\rho_a = kR$$

The calculated resistivity value is not the true resistivity of the subsurface, but an "apparent" value, which is the resistivity of a homogeneous ground that will give the same resistance value for the same electrode arrangement. The relationship between the "apparent" resistivity and the "true" resistivity is a complex relationship. To determine the true subsurface resistivity, an inversion of the measured apparent resistivity values using a computer program must be carried out. However, in the field, such conditions are rarely observed as in most cases the ground is heterogeneous, inhomogeneous, and anisotropic; hence,

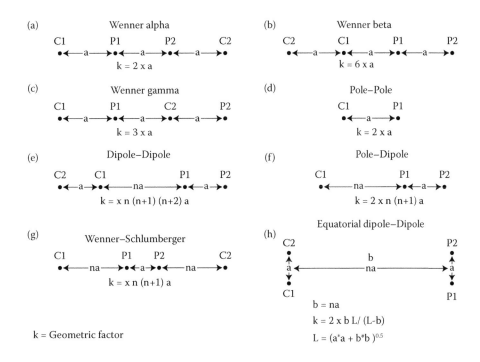

FIGURE 5.2
Common arrays used in resistivity surveys and their geometric factors. Note that the dipole–dipole, pole–dipole, and Wenner–Schlumberger arrays have two parameters, the dipole length "a" and the dipole separation factor "n." While the "n" factor is commonly an integer value, noninteger values can also be used.

the resistivity values vary with the relative position of the electrodes on the ground. Resistivity obtained under such conditions is termed apparent resistivity. In homogeneous ground, the depth of current penetration increases as the separation of the current electrodes is increased.

The measured apparent resistivity values are normally plotted on double-log graph paper. To interpret the data from such a survey, it is normally assumed that the subsurface consists of horizontal layers. In this case, the subsurface resistivity changes only with depth, but does not change in the horizontal direction. Despite this limitation, the method has given useful results for geological situations, such as water table, where the one-dimensional (1-D) model is approximately true. Another classical survey technique is the profiling method. In this case, the spacing between the electrodes remains fixed, but the entire array is moved along a straight line. This gives some information about lateral changes in the subsurface resistivity, but it cannot detect vertical changes in the resistivity. Interpretation of data from profiling surveys is mainly qualitative.

The most severe drawback of the resistivity-sounding method is that horizontal (or lateral) changes in the subsurface resistivity are commonly found. Lateral changes in the subsurface resistivity will cause changes in the apparent resistivity values, which might be misinterpreted as changes with depth in the subsurface resistivity. In many engineering and environmental studies, the subsurface geology is very complex where the resistivity can change rapidly over short distances. The resistivity-sounding method might not be sufficiently accurate for such circumstances.

5.2.1 Two-Dimensional ERI Technique

ERI has proven to be a useful methodology in groundwater exploration, civil engineering, and environmental and mining applications because of the wide range of variations in electrical resistivity. Electrical resistivity is the inverse of electrical conductivity (EC), and thus, it is a measure of how much an earth material resists the flow of electricity. When earth materials or fluids are highly conductive, their resulting electrical resistivity values are low. Values of resistivity for most earth materials are well established. As with most geophysical methods, a degree of nonuniqueness exists for all earth materials, and thus, most materials have a range of values that tend to overlap. Ranges for a single material can result from minor differences in the composition of a given rock type; however, ranges are largely attributed to differences in the size and/or availability of pore space as the interstitial pore fluid strongly influences the resistivity signature of a given material (Telford et al., 1976; Loke, 2011).

There are several advantages of using a multielectrode ERI system over the conventional vertical electrical technique (Dahlin, 1996). This is because the multielectrode scheme is a fast computer-aided data-acquisition system and simultaneously studies both lateral and vertical changes of resistivity below the entire profile length. There are, therefore, an increasing number of users for the ERI technique in India and abroad for mapping the accurate location of subsurface geological formations and structures such as faults, fractures, joints for delineation of water-bearing zones, geothermal, etc. (Griffiths and Barker, 1993; Loke and Barker, 1996; Singh et al., 2006; Francese et al., 2009; Kumar et al., 2010; Zarroca et al., 2012). This technique has also been found to be a powerful tool to delineate a sub-surface-contaminated zone over the Kodaganar Basin in Dindigul District, Tamil Nadu, when there is sufficient resistivity contrast (Barker et al., 2001).

5.3 Instrumentation and Survey Design

ERI surveys are usually carried out using a large number of electrodes, 25 or more, connected to a multicore cable (Loke, 2011). A laptop microcomputer together with an electronic switching unit is used to automatically select the relevant four electrodes for each measurement (Figure 5.3). Currently, field techniques and equipment to carry out two-dimensional (2-D) resistivity surveys are fairly well developed. The necessary field equipment is commercially available from a number of international companies.

Figure 5.3 shows the typical setup for a 2-D survey with a number of electrodes along a straight line attached to a multicore cable. Normally, a constant spacing between adjacent electrodes is used. The multicore cable is attached to an electronic switching unit, which is connected to a laptop computer. The sequence of measurements to take, the type of array to use, and other survey parameters (such as the current to use) are normally entered into a text file that can be read by a computer program in a laptop computer. After reading the control file, the computer program then automatically selects the appropriate electrodes for each measurement. In a typical survey, most of the fieldwork is in laying out the cable and electrodes. After that, the measurements are taken automatically and stored in the computer. Most of the survey time is spent waiting for the resistivity meter to complete the set of measurements.

To obtain a good 2-D picture of the subsurface, the coverage of the measurements must be 2-D as well. As an example, Figure 5.3 shows a possible sequence of measurements for the Wenner electrode array for a

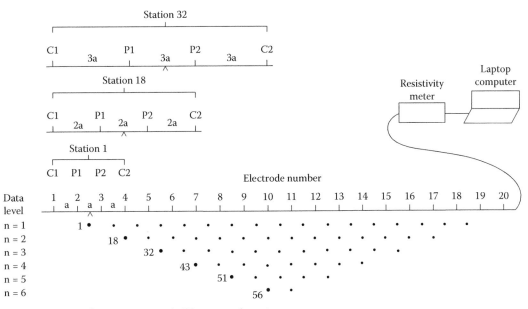

Sequence of measurements to build up a pseudosection

FIGURE 5.3
Schematic illustrating how a pseudosection is developed from acquisition of data from a large number of electrodes. Varying the unit spacing "a" of the current and potential electrodes allows for a greater depth of influence but also decreases the number of measurements that can be made. (Adapted from Loke, M.H. 2000. Topographic modelling in resistivity imaging inversion, *62nd EAGE Conference and Technical Exhibition Extended Abstracts, D-2.*)

system with 20 electrodes (Loke, 2011). Here, the spacing between adjacent electrodes is "a." The first step is to make all the possible measurements with the Wenner array with electrode spacing of "1a." For the first measurement, electrodes number 1, 2, 3, and 4 are used. It may be noted that electrode 1 is used as the first current electrode C1, electrode 2 as the first potential electrode P1, electrode 3 as the second potential electrode P2, and electrode 4 as the second current electrode C2. For the second measurement, electrodes number 2, 3, 4, and 5 are used for C1, P1, P2, and C2, respectively. This is repeated down the line of electrodes until electrodes 17, 18, 19, and 20 are used for the last measurement with "1a" spacing. For a system with 20 electrodes, there are 17 (20−3) possible measurements with "1a" spacing for the Wenner array. After completing the sequence of measurements with "1a" spacing, the next sequence of measurements with "2a" electrode spacing is made. First, electrodes 1, 3, 5, and 7 are used for the first measurement. The electrodes are chosen so that the spacing between adjacent electrodes is "2a." For the second measurement, electrodes 2, 4, 6, and 8 are used. This process is repeated until electrodes 14, 16, 18, and 20 are used for the last measurement with spacing "2a." For a system with 20 electrodes, note that there are 14 (20−2×3) possible measurements with "2a" spacing.

The same process is repeated for measurements with "3a," "4a," "5a," and "6a" spacings. To get the best results, the measurements in a field survey should be carried out in a systematic manner so that, as far as possible, all the possible measurements are made. This will affect the quality of the interpretation model obtained from the inversion of the apparent resistivity measurements (Dahlin and Loke, 1998).

As the electrode spacing increases, the number of measurements decreases. The number of measurements that can be obtained for each electrode spacing for a given number of electrodes along the survey line depends on the type of array used. The Wenner array gives the smallest number of possible measurements compared to the other common arrays that are used in 2-D surveys. The survey procedure with the pole–pole array is similar to that used for the Wenner array. For a system with 20 electrodes, first 19 measurements with a spacing of "1a" are made, followed by 18 measurements with "2a" spacing, followed by 17 measurements with "3a" spacing, and so on. For the dipole–dipole, Wenner–Schlumberger, and pole–dipole arrays, the survey procedure is slightly different. As an example, for the dipole–dipole array, the measurement usually starts with a spacing of "1a" between the C1 and C2 (and the P1–P2) electrodes. The first sequence of measurements is made with a value of 1 for the "n" factor (which is the ratio of the distance between the C1 and P1 electrodes to the C1–C2 dipole spacing), followed by "n" equals 2 while keeping the C1–C2 dipole pair spacing fixed at "1a." When "n" is equal to 2, the distance of the C1 electrode from the P1 electrode is twice the C1–C2 dipole pair spacing. For subsequent measurements, the "n"-spacing factor is usually increased to a maximum value of about 6, after which accurate measurements of the potential are difficult due to very low potential values. To increase the depth of investigation, the spacing between the C1 and C2 dipole pair is increased to "2a," and another series of measurements with different values of "n" is made. If necessary, this can be repeated with larger values of the spacing of the C1–C2 (and P1–P2) dipole pairs. A similar survey technique can be used for the Wenner–Schlumberger and pole–dipole arrays where different combinations of the "a" spacing and "n" factor can be used.

One technique used to horizontally extend the area covered by the survey, particularly for a system with a limited number of electrodes, is the roll-along method (Loke, 2011). After completing the sequence of measurements, the cable is moved past one end of the line by several unit electrode spacings. All the measurements that involve the electrodes on a part of the cable that do not overlap the original end of the survey line are repeated (Figure 5.4).

A number of different array types can be utilized so that acquisition of the most optimal vertical and/or lateral distributions can be achieved. Each array type differs in the geometric arrangement of electrode stake pairs and also in the coverage it is best able to provide. For example, a Wenner array is set up such that the spacing between each current and potential electrode is constant for the entire survey. The Wenner array gives an excellent depth resolution, but is not the optimal array type for acquiring high-resolution data in the near subsurface (Loke et al., 2010). Conversely, the dipole–dipole array type, in which the current-to-current electrodes and potential-to-potential electrodes maintain the same

spacing for a survey but the distance between the two pairs may change, boasts good lateral resolution but is not the optimal array type for achieving high-resolution data at depth.

In addition to the variety of array types available for collecting the most optimal data for a given field site, advances in technology have also allowed for the development of systems capable of collecting data at relatively unlimited horizontal distances. Survey lines are no longer limited in the length by the available cable-stake setup due to the advent of roll-along surveys. Roll-along surveys begin like a regular survey in that stakes are inserted at constant spacing and connected to cables, which send and receive information to and from an earth resistivity meter. However, after the first survey is completed, a portion of the cable setup can be advanced in front of the survey line and data collection can continue. The configuration can be advanced as many times as a given field site will allow such that data can be collected along a continuous line with relatively few limits to the distance achieved. Figure 5.4 is a schematic of a roll-along survey illustrating the general procedure employed during field data collection. Once data are collected along the first line, then some portion of the configuration is advanced to the end of the line and data collection continues. Although this procedure has made collection of electrical resistivity data more streamlined and less labor intensive, it is not without limitations. While great distances can be achieved with a roll-along survey, limitations in the depth of penetration associated with spacing are such that a gap will remain between each of the respective roll-along sections. For the most optimal data resolution, the smallest section of cable should be advanced during a single roll along so that the subsurface gap will occupy the

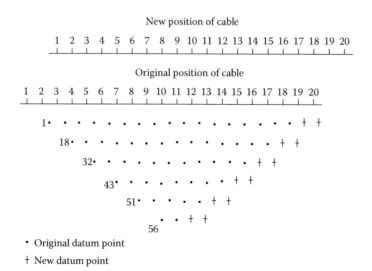

FIGURE 5.4
The use of the roll-along method to extend the area covered by a survey.

smallest possible area. While this configuration allows excellent lateral coverage, greater depth of penetration cannot be achieved with this method, as the depth achieved through a given survey is always determined in a single deployment (AGI, 2006).

5.4 Data Processing and Inversion

Data inversion is the procedure whereby the apparent resistivity measurements obtained by an earth resistivity meter are processed into a cross section that provides a model of the true spatial distribution of resistivity beneath an ERI array. The set of data initially acquired from a resistivity meter produces apparent resistivity data, which can be plotted as an apparent resistivity pseudosection, representing averages of all of the material encountered on a given electrical field path before it arrives at the potential electrodes where the potential difference (in volts) is measured. In an inversion process, a model domain is fragmented into a grid (i.e., rows and columns) where each column in the grid is striped of the weighted average of that column, and each square of the grid is assigned its own resistivity value (a "true" or "model" resistivity) (Loke, 2011). The inversion process results in a model cross section of true resistivity that most closely approximates the subsurface distribution required to generate the apparent resistivity values obtained with the resistivity meter at the surface. A number of different software packages, including freeware that can be downloaded online, are available for processing of electrical resistivity data (Loke and Barker, 1996; Binley, 2003). Each software package contains a number of different tools and algorithms for processing data so that each user can choose the most appropriate method based on his or her knowledge of the field site with which he or she is working. A process of least-squares inversion is one of the most common algorithms used in resistivity data processing. Least-squares inversion works to replicate the measured values in the model by reducing the square of the difference between measured and calculated apparent resistivity values (Loke, 2011).

A radically different approach is a boundary-based inversion method (Loke, 2011). This method subdivides the subsurface into different regions. The resistivity is assumed homogenous within each region. The resistivity is allowed to change in an arbitrary manner across the boundaries, and thus, it is useful in areas with abrupt transitions in the geology. The resistivity of each region and the depths to the boundaries are changed by the least-squares optimization method so that the calculated apparent resistivity values match the observed

values. While this method works well for synthetic data from numerical models, for many field data sets, it can lead to unstable results with highly oscillating boundaries (Olayinka and Yaramanci, 2000). Its greatest limitation is probably the assumption of a constant resistivity within each region. In particular, lateral changes in the resistivity near the surface have a very large effect on the measured apparent resistivity values. Since this model does not take into account such lateral changes, they are often mistakenly modeled as changes in the depths of the boundaries.

The choice of the best array for a field survey depends on the type of structure to be mapped, the sensitivity of the resistivity meter, and the background noise level. In practice, the arrays that are most commonly used for 2-D-imaging surveys are the (a) Wenner, (b) dipole–dipole, (c) Wenner–Schlumberger, (d) pole–pole, and (d) pole–dipole. Among the characteristics of an array that should be considered are the (i) depth of investigation, (ii) sensitivity of the array to vertical and horizontal changes in the subsurface resistivity, (iii) horizontal data coverage, and (iv) signal strength.

In the present case, data acquisition was performed using a Wenner–Schlumberger configuration, a hybrid between the Wenner and Schlumberger arrays (Pazdirek and Blaha, 1996) with a constant interelectrode spacing of 5 m. This array is moderately sensitive to both horizontal and vertical geological structures. The average investigation depth is greater than the Wenner array and the intensity of the signal is weaker than that of the Wenner array but greater than that of the dipole–dipole array and twice that of the pole–dipole array, resulting in a higher signal-to-noise ratio (Dahlin and Zhou, 2004). The horizontal data coverage is somewhat wider than the Wenner arrangement, but narrower than that achieved using the dipole–dipole array (Loke, 2011). An ERI survey was carried out at 12 stations using SYSCAL R1plus switch 48 system (Iris Instruments, Orléans, France) with 5-m interelectrode separation. The maximum length of the profile was 240 m, which resulted in a depth of investigation of about 50 m. The length of the profiles surveyed depends on the availability of free-stretch land. The contact resistance was checked before data acquisition and was kept below 2 kΩ (Zarroca et al., 2014).

The acquired apparent resistivity datasets were tomographically inverted to obtain the true electrical resistivity distribution of the study area using the "RES2DINV" finite-difference software, based on the smoothness-constrained least-squares inversion by a quasi-Newton optimization method (Loke and Barker, 1996). An initial 2-D electrical resistivity model is generated, from which a response is calculated and compared to the measured apparent resistivity values of the field data. The optimization method then attunes the resistivity value of the model block iteratively until the calculated apparent resistivity values of the model are in close agreement with the measured values of the field data. The absolute error provides a measure of the differences between the model response and the measured data, which is an indication of the quality of the model obtained. Using this scheme, 2-D-inverted models of true resistivity variation of subsurface geological formations for all the 12 sites have been computed.

The RES2DINV software offers two inversion options—robust inversion (Loke et al., 2003) and smoothness-constrained least-squares inversion (Loke and Dahlin, 2002). It has been reported by Dahlin and Zhou (2004) that the robust inversion is better than the smoothness-constrained least-squares inversion. In situations where the subsurface geology comprises a number of almost-homogeneous regions but with sharp boundaries between different regions, the robust inversion scheme attempts to find a model that minimizes absolute changes in the model resistivity values (also known as L1 norm or blocky inversion method) thereby giving appreciably superior results. The smoothness-constrained optimization method (also known as L2 norm) on the other hand tries to minimize the squares of the spatial changes (or roughness) of the model resistivity values and tends to construct a model with a smooth variation of resistivity values. This approach is used only if the subsurface resistivity varies in a smooth or gradational manner.

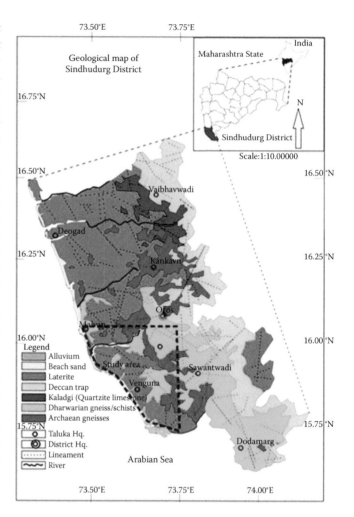

FIGURE 5.5
Geological map of the study area.

5.5 Case Study from Coastal Maharashtra, India

The study area lies between latitude 15.7°–16.2°N and longitude 73.4°–73.8°E in the Sindhudurg district, western Maharashtra, India (Figure 5.5). Kudal, Malvan, and Vengurla are some of the important townships in the district. Geologically, the study area exposes rocks ranging in age from Archaean to Quaternary period. The Archaeans are represented by granite gneiss and is seen in the southern part of the Sindhudurg district near Vengurla and Sawantwadi. The Palaeo- to Meso-Proterozoic represented by Dharwar supergroup overlie the Archaeans and occupy a major part of the area comprising psammatic metasediments consisting of metagabbro, quartz chromites, amphibolites schist, and ferruginous phyllite. The general geologic succession encountered in the Sindhudurg district is given in Table 5.1.

The Archaean granite and gneisses are medium-to-coarse grained and consist of quartz, microcline, orthoclase biotite, and hornblende (CGWB, 2009; Maiti

et al., 2013). Dharwarian metasediments (Archaean), Kaladgi formation (pre-Cambrian), Deccan Trap lava flows (upper Cretaceous to lower Eocene), laterites (Pleistocene), and alluvial deposits (recent to subrecent) are the water-bearing formation observed in Sindhudurg district. However, the Kaladgi formation occurs only

TABLE 5.1

General Geological Succession in Sindhudurg District, Western Maharashtra, India

Geological Time	Formation
Recent to subrecent	Alluvium beach sand
Pleistocene	Laterite and lateritic spread
Miocene	Shale with peat and pyrite nodules
Cretaceous to Eocene	Deccan Trap basalt lava flows
Upper pre-Cambrian	Kaladgi series, quartzite, sandstone, shale, and associated limestone
Dharwar super group	Phyllite, conglomerate, and quartzite

Source: CGWB. 2009. Groundwater information, Sindhudurg district, Maharashtra. Technical Report, 1625/DB/2009.

in very-limited patches and does not form a potential aquifer in the district. The alluviums also have a limited areal extent found mainly along the coast (CGWB, 2009; Maiti et al., 2013). Dharwarian gneiss/schists are devoid of primary porosity and permeability. Kaladgi rocks are mainly represented by orthoquartzite, limestone, sandstone, and shales. They are jointed in diverse directions and the weathered portion actually controls their water-bearing properties (CGWB, 2009; Maiti et al., 2013). Since primary porosity is negligent in the Deccan Trap basalts, secondary porosity due to jointing and fracturing plays an important role in groundwater circulation. Laterite has more porosity than the Deccan Trap basalt, thereby forming many potential aquifers in the area. Groundwater level in the study area varies from 2 to 20 m (Maiti et al., 2013).

Geophysical information was obtained earlier (Maiti et al., 2013) by inverting 85 of the VES data from the study area (latitude 15.7°–16.2°N and longitude 73.4°–73.8°E) (Figure 5.6). Geochemical information was also obtained by analyzing 36 water samples collected from representative dug wells and bore wells distributed throughout the area. The location of the geochemical sample collection points has been marked in Figure 5.6.

True resistivity and layer thickness of each geologic layer is deduced by the inversion of VES apparent resistivity data. Inversion results suggest that there are three to four geologic layers in the study area (Maiti et al., 2013). The true resistivity contour map reflects the distribution pattern of true resistivity at a depth of 20 m. The contour map of ρ at a depth of 20 m (Figure 5.7) suggests

that the true resistivity found at Kelus is the minimum (ρ~1 Ωm). The true resistivity found at Shiroda and Malvan is of the order of 0.95 and 2 Ωm, respectively. The low resistivity of these places is attributed to the saline water intrusion from the Arabian Sea. The coastal parts are conductive possibly due to the effect of saline water intrusion whereas the northeast and southeast parts of the study area are highly resistive. The relatively low resistivity at Nerur may be attributed to leaching of normal salt.

Geochemical analysis of groundwater in the study area (Figure 5.8a–e) (Maiti et al., 2013) reported that the EC ranged between 70 and 4450 μS/cm (mean = 421.47 μS/cm). The EC values recorded at Shiroda and Kelus fall beyond the acceptable level for drinking prescribed by World Health Organization (WHO, 1984), and were attributed to the intrusion of saline water from the Arabian Sea. The total dissolved salts (TDS) value at Kelus (2845 mg/L) also exceeds the acceptable limit prescribed by WHO (1984). Nutrient enrichment due to fertilizers and saline water intrusion could enhance TDS and, in turn, increase the EC in the study area, a fact that can be observed in other parts of the study area also reflecting high S values (VES points 7–11, 19) (Gupta et al., 2014). The ranges of Na^+ and Cl^- ions were from 3.78 to 276.1 mg/L (mean = 33.90 mg/L), and 13.19 to 832 mg/L (mean = 54.50 mg/L), respectively. Cl^- concentration recorded at Kelus and Shiroda falls beyond the permissible limit (200 mg/L). The high concentration of Cl^- at these places is primarily due to the saline water intrusion and secondarily due to the influence of discharged agricultural, industrial, and domestic wastewaters.

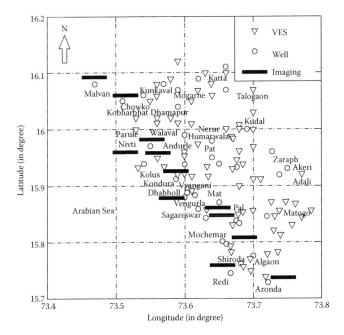

FIGURE 5.6
Location map of the geochemical sampling point (well), VES point, and the ERI profile.

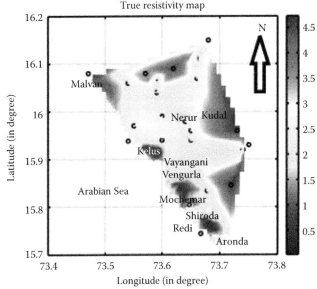

FIGURE 5.7
Concentration map of true resistivity. Units are expressed in log 10.

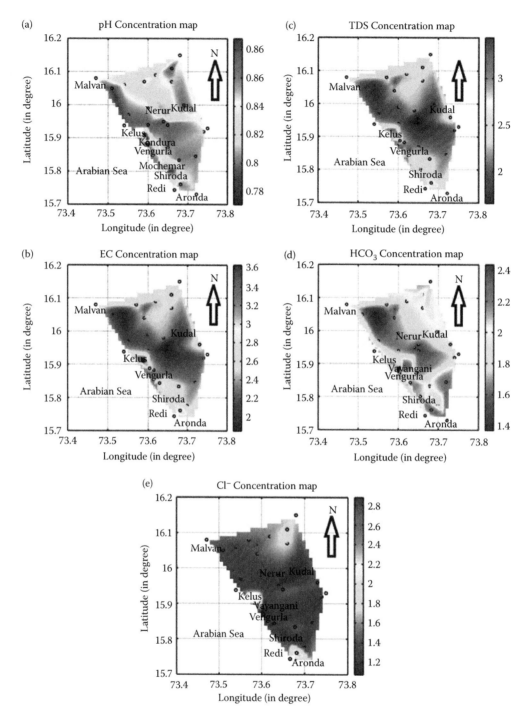

FIGURE 5.8
Concentration map of (a) pH, (b) EC, (c) TDS, (d) bicarbonate (HCO_3^-), and (e) Cl^-. Units are expressed in log 10.

Note that the mismatches between EC and ρ maps are likely because the EC map is based on the direct groundwater sample analysis whereas the true resistivity map is derived from gross resistivity values of a geologic layer saturated with or without water.

Motivated by the prior field results suggesting the possible movement of saline water and thus contaminating the freshwater aquifer zones, it was decided to carry our ERI over selected profiles for coastal characterization, which has the potential to provide an insight into the complexity of a shallow coastal aquifer system.

In this chapter, the 2-D inversion of the field data along the 12 Wenner–Schlumberger profiles was carried out using the robust (L1 norm) inversion approach.

As the survey area was anticipated to be infested with urban noise, the robust inversion scheme was applied to the model resistivity values as well. The noisy data at a few sites were automatically filtered by removing the resistivity records having negative resistivity values or with a standard variation coefficient over 2%. The convergence between the measured and calculated data was achieved after 5–10 iterations. The absolute error in the inverted models was below 5% at all the stations, except at Shiroda where the absolute error was 5.2%. To reduce the distortion caused by the large-resistivity variations near the ground surface and to obtain significantly better results, an inversion model with a cell width of half the unit electrode spacing was used for all the 12 imaging profiles (Loke, 2011). All the stations are oriented perpendicular to the coast. The resistivity profile locations are shown in Figure 5.6.

The interpretation of the 2-D resistivity models of 12 ERI profiles has been carried out to ascertain groundwater potential zones and the extent of saline water intrusion in the Konkan coastal region in view of the hydrogeological scenario. A generalized resistivity range for different litho units vis-à-vis water-bearing zones in the Deccan basalts (after Rai et al., 2013) is given in Table 5.2.

The inverted resistivity models obtained are discussed from the southernmost station (Aronda) to the northernmost station (Malvan) below.

An inverted resistivity model at Aronda (Figure 5.9a) suggests that the western part is highly conductive having resistivities of the order of 1–3 Ωm up to lateral distances of 50 m. Further east, the top layer is 3–10-m thick consisting of alluvium/weathered formation saturated with water having a resistivity of about 3–40 Ωm. On the eastern part between lateral distances of 160–240 m, a high resistive (>70 Ωm) shallow layer is revealed indicating hard rock. Beneath 10 m, the entire profile exhibits very low resistivity (<1 Ωm). The inverse resistivity section converged after seven iterations with an absolute error of 1.51%. It has been reported by Maiti et al. (2013) that the spatial distribution of Ca^{2+} at Aronda is 80.4 mg/L and falls beyond the permissible limit

TABLE 5.2

Resistivity Values for Different Litho Units in Deccan Traps

Litho Units	Resistivity Range (Ωm)
Alluvial, black cotton soil, and bole beds	5–10
Weathered/fractured vesicular basalt saturated with water	20–40
Moderately weathered/fractured vesicular basalt saturated with water	40–70
Massive basalt	>70

(75 mg/L). It may be noted that the crystalline limestone and prolonged agricultural activities could influence directly or indirectly to augment the mineral dissolution in groundwater, thus enhancing the conductivity of groundwater. The low resistivity observed here is possibly due to the high calcium value and thus the water may not be fit for domestic use. This station is also near the Arabian Sea and thus it is presumed that the low resistivity is also influenced by the saline water ingress from the coastal side.

The resistivity model at Shiroda (Figure 5.9b) indicated that the entire profile is infested with saline water intrusion, wherein the resistivity values range from 0.1 to 7 Ωm. The inverse resistivity section converged after five iterations with absolute error as high as 5.2%. The true resistivity model obtained at Shiroda (Maiti et al., 2013) indicates a resistivity of 0.95 Ωm at a depth of 20 m. The imaging results reveal that the western part of the profile up to lateral distances of 100 m is more conductive up to depths of 41 m than the eastern part, where only the top 5–10 m is affected by saline intrusion. The EC value at Shiroda is reported to be 1240.00 μS/cm (Maiti et al., 2013), which falls beyond the acceptable level for drinking prescribed by WHO (1984). The high EC value at Shiroda is due to the intrusion of saline aerosol from the Arabian Sea. These authors also reported a high TDS value at Shiroda (TDS value 792.00 mg/L), which exceeds the acceptable limit prescribed by WHO (1984). Probably nutrient enrichment due to fertilizers and saline water intrusion could enhance TDS and, in turn, increase the EC in the study area. Cl⁻ is the dominant ion of seawater (Song et al., 2007). Cl⁻ concentration recorded at Shiroda (Cl⁻ value 255.2 mg/L) falls beyond the permissible limit (200 mg/L). The high concentration of Cl⁻ at these places is primarily due to the saline water intrusion and secondarily due to the influence of discharged agricultural, industrial, and domestic wastewaters (Maiti et al., 2013). Both calcium and sodium values are 104.2 and 136.8 mg/L, respectively, and fall beyond permissible limits. Thus, it is seen that this station is very vulnerable to saline water intrusion and thus the water here is not fit for domestic use.

The inverted resistivity model at Mochemar (Figure 5.9c) indicates that the top 10–15 m is highly conductive (resistivity values <1 Ωm). Underlying this layer, a relatively resistive layer with resistivity values ranging from 1 to 10 Ωm is revealed up to depths of about 25 m. The last layer in the model having resistivity values >10 Ωm is delineated up to depths of investigation. The low resistive top layer is presumably due to the effect of saltwater intrusion. The second layer in the model with resistivities of the order of 1–10 Ωm is perhaps the contact between the saline water–freshwater zone. The inverse resistivity section converged

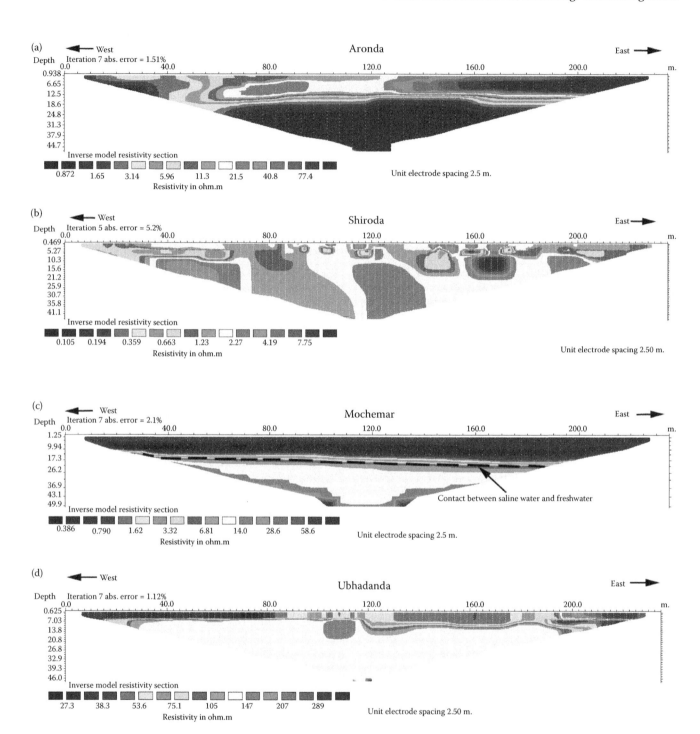

FIGURE 5.9
(a) Showing Wenner–Schlumberger 2-D-inverted resistivity section at Aronda. (b) Showing Wenner–Schlumberger 2-D-inverted resistivity section at Shiroda. (c) Showing Wenner–Schlumberger 2-D-inverted resistivity section at Mochemar. The zone of contact between saline and freshwater is also marked. (d) Showing Wenner–Schlumberger 2-D-inverted resistivity section at Ubhadanda.

after seven iterations with an absolute error of 2.1%. Geochemical studies (Maiti et al., 2013) observed that the hydrogen ion concentration (pH) value at Mochemar is 5.92 indicating the fact that the water is acidic in nature. In such a scenario, the acidic nature of groundwater can also hasten the corrosion rate of metallic

substances in water, which in turn causes the rise of metallic substances in groundwater, thus contaminating it. These authors also revealed that the calcium content at Mochemar is 96.19 mg/L (which is higher than the permissible limit of 75 mg/L), signifying that crystalline limestone and extended agricultural activities could impact directly or indirectly to enhance the mineral dissolution in groundwater.

The resistivity model for the profile Ubhadanda is depicted in Figure 5.9d. Here, the inverse resistivity section converged after seven iterations with an absolute error of 1.12%. The top 5–10 m is conductive with resistivity varying from 27 to 70 Ωm up to a lateral distance of 190 m from the west. At 200-m distance and beyond, the top layer is resistive (>200 Ωm). This is a lateritic terrain and hence the high resistivity is due to top layer lateritic formation. Beneath this layer, the entire profile exhibits high resistivity (about 147 Ωm) representing hard-rock formation. However, potential aquifer zones are not revealed over the entire profile.

The station Vengurla is at a higher altitude, although it is near the coast. However, both stations Ubhadanda and Vengurla exhibit similar subsurface resistivity features. The inverted resistivity section converged after seven iterations with an absolute error of 1.13%. Figure 5.10a suggests that the top 7 m is conductive (resistivity values of 30–50 Ωm). Beneath this layer, a very thin layer is delineated with resistivity ranging from 80 to 377 Ωm. Underlying this layer, the entire profile exhibits high resistivity greater than 377 Ωm. The top layer is possibly due to reasonably weathered/fractured basalt saturated with water. The second and third layers are characterized by massive basalts. It can be noted that the top part of the profile can be explored for shallow wells at least up to 15 m.

The inverted resistivity section at station Dhaboli (Figure 5.10b) reveals a potential aquifer zone at 160-m lateral distance, which is characterized by resistivities of the order of about 50 Ωm extending from shallow levels up to the depth of investigation. The low-resistive feature shows downward extension of resistivity decreasing with depth, which appears to be linked with a fault zone extended to deeper levels beyond the depth of probing. It has been reported by Zhu et al. (2009) that if a low-resistivity zone extends to a near-surface terrain from the deep, only then can it be interpreted that the low-resistivity zone can be an indication of the location of a fault zone. This is a potential zone for groundwater exploration. Beneath a lateral distance of 120 m, an anticlinal-shaped high-resistivity (579 Ωm) feature is delineated. The rest of the profile exhibits resistivities of the order of about 100 Ωm. It is pertinent to mention here that the inverted model converged after eight iterations with an absolute error of 1.11%.

Figure 5.10c depicts the inverted resistivity model of Kelus, which converged after 10 iterations giving an absolute error of 1.05%. The top 10–13 m on the western part up to a lateral distance of 120 m is highly conductive (resistivity of 1–10 Ωm). The eastern part beyond 120-m distance is relatively resistive (28–82 Ωm). A thin high-resistive (about 200 Ωm) layer is delineated at depths of

13 m. Underlying this layer, the entire profile reveals very-high-resistivity values greater than 234 Ωm. The geophysical and geochemical studies carried out here earlier by Maiti et al. (2013) suggest that the EC, TDS, and Cl⁻ values obtained at Kelus are very high and are beyond permissible limits. In addition, the calcium, magnesium, and sodium concentrations are high and beyond permissible limits. The true resistivity at a depth of 20 m is found to be about 1 Ωm at Kelus (Maiti et al., 2013). It can be thus surmised that station Kelus is severely affected by intrusion of saline aerosols from the Arabian Sea. The groundwater quality is also found to be unsuitable at Kelus. Thus, it can be advocated from the inverted resistivity model that the top 10–13 m are intruded by saline water thus giving rise to such low-resistivity values.

The station at Nivti is barely 500 m away from the coast. The inverted resistivity model of this profile is shown in Figure 5.10d. The model converged after nine iterations with an absolute error of 4.3%. The image suggests that the entire stretch from shallow levels to deeper levels is infested by saline water intrusion characterized by very low resistivities (<1 Ωm). At a lateral distance of 80–120, a 20-Ωm subhorizontal patch is seen. This region might be the zone of contact between the saline and freshwater. Maiti et al. (2013) reported that the calcium and bicarbonate concentration at Nivti is high and is beyond permissible limits. These high values could be due to prolonged agricultural activity in the region. In addition, the primary source of bicarbonate ions in groundwater is attributed to the dissolution of carbonate minerals (e.g., calcite). The secondary source of bicarbonate is due to the reaction of water with the carbon dioxide gas.

Another profile was surveyed in the Nivti region, which was about 4 km away from the coast. Figure 5.11a shows the inverted resistivity model that converged after nine iterations giving an absolute error of 1.39%. The top 7 m toward the west is conductive (resistivities of about 50 Ωm) up to a lateral distance of 140 m. Further east at distances of 160 m, the top layer is highly resistive (>200 Ωm) indicating hard rock/lateritic formation at the top. A high-resistive block (>200 Ωm) is revealed at depths of 7 m and beneath at lateral distances of 40–100 m. This high-resistive feature is juxtaposed by a relatively low-resistive (about 100 Ωm) zone toward the east continuing up to the depth of investigation. Beneath a lateral distance of 110–120 m, a small aquifer zone is seen up to a depth of 20 m. This feature is less than 50 Ωm and is a potential groundwater precinct. There is no indication of saline water invasion at this station, thus suggesting that the extent of saline water ingress is less than 4 km in this region. There were no open and suitable sites in between the coastal and inland station at Nivti for further work.

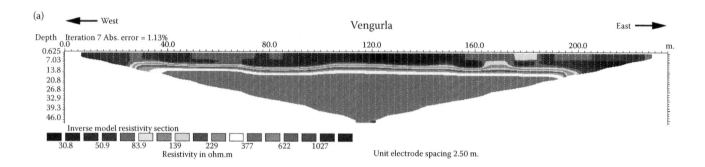

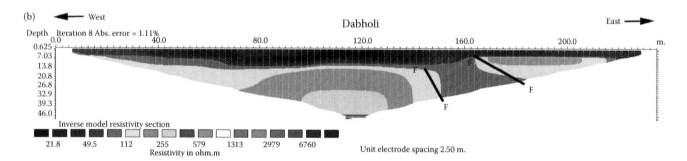

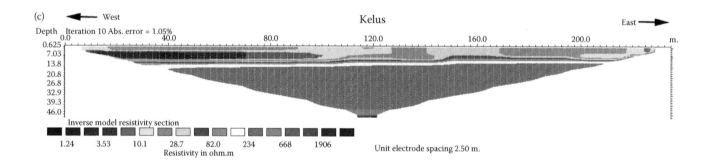

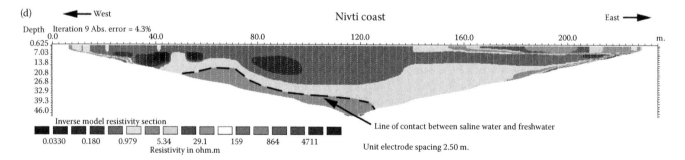

FIGURE 5.10

(a) Showing Wenner–Schlumberger 2-D-inverted resistivity section at Vengurla. (b) Showing Wenner–Schlumberger 2-D-inverted resistivity section at Dabholi. Also shown are probable faults (F–F). (c) Showing Wenner–Schlumberger 2-D-inverted resistivity section at Kelus. (d) Showing Wenner–Schlumberger 2-D-inverted resistivity section at Nivti coast. Also marked is the line of contact between saline and freshwater.

The inverted resistivity model of station Pat is shown in Figure 5.11b. The model converged after seven iterations with an absolute error of 1.06%. The top 5 m reflects a resistivity ranging from 50 to 80 Ωm throughout the profile length. This zone is the weathered basalt saturated with water. Underlying this layer, a very-high-resistive zone (>400 Ωm) is revealed in the western part continuing up to the depth of study. A similar structure is also seen toward the eastern part; however, the resistivity values are lower (about 200 Ωm)

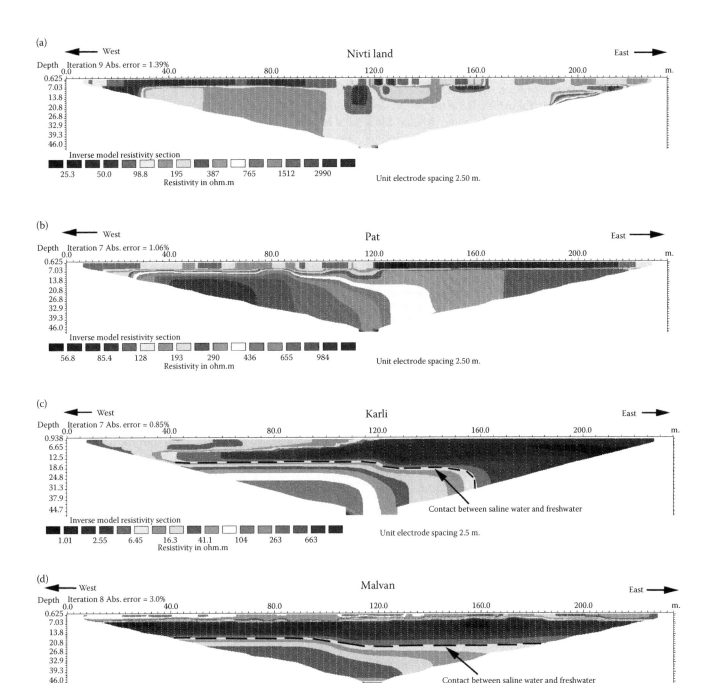

FIGURE 5.11

(a) Showing Wenner–Schlumberger 2-D-inverted resistivity section at Nivti land. (b) Showing Wenner–Schlumberger 2-D-inverted resistivity section at Pat. (c) Showing Wenner–Schlumberger 2-D-inverted resistivity section at Karli. The contact between saline water and freshwater is marked. (d) Showing Wenner–Schlumberger 2-D-inverted resistivity section at Malvan. Also marked is the line of contact between saline and freshwater.

compared to its western counterpart. This entire zone is the signature of compact basalt underneath the weathered basalt.

Figure 5.11c depicts the inverted resistivity model of station Karli. This profile is situated near the Karli River, which is essentially the backwater of the Arabian Sea. The model converged after seven iterations with an absolute error of 0.85%. The top layer up to 110 m from the west is composed of weathered and fractured basalt with resistivities around 40 Ωm having a thickness of about 7 m. Further east, at lateral distances of 120–240 m, the top layer is conductive (resistivity values of 1–6 Ωm). This low-resistive zone continues up to a lateral distance of 40 m to the west and up to the depth of investigation at lateral distances of 160–240 m. This is due to saline water incursion, presumably from the saline backwater of the Arabian Sea. Beneath lateral distances of 50–120 m, a high-resistive zone (40–260 Ωm) is revealed at depths of about 15 m presumably due to weathered/compact basalt. However, at lateral distances of 120–150 m, a low-resistivity (15–40 Ωm) zone is delineated, which represents a freshwater body. A clear demarcation of saline and freshwater is observed as shown in Figure 5.11c.

The inverted resistivity model of the northernmost station Malvan is shown in Figure 5.11d. The model converged after eight iterations with an absolute error of 3%. The top couple of meters beneath 0–40 and 110–135-m lateral distance reflect high-resistive (about 70 Ωm) formation probably due to laterites. A low resistive (2–8 Ωm) is underlying the top layer throughout the spread of the profile up to depths of about 25 m. Such low-resistivity values could be attributed to saline water intrusion from the Arabian Sea. Maiti et al. (2013) reported a value of 2 Ωm at 20-m depth from VES studies. These authors further reported that sulfate and calcium concentrations are beyond permissible limits. Thus, the plausible cause of this low-resistivity zone is due to both saline water intrusion and anthropogenic activities. The last layer in this section is in the resistivity range of 17–35 Ωm and continues up to the depth of investigation. Again, a conspicuous demarcation of saline and freshwater boundary is revealed (Figure 5.11d).

5.6 Conclusion

The resistivity models obtained after inversion of measured apparent resistivity data at 12 profiles suggests that the subsurface structure is fragmented into multiple units due to weathering, fracturing, and faulting. The massive basaltic units are covered by a thin veneer of alluvium and weathered and jointed rocks formed by erosion and subsequent deposition, which form unconfined aquifer zones that are the main sources of groundwater to the dug wells in Deccan volcanic province (DVP). The VES and water sample analysis results available in the study region reveal that the top layer is composed of alluvium/laterites, weathered/fractured basalts, and compact basalts as bedrock. The low-resistive feature shows downward extension of resistivity decreasing with depth, which appears to be linked with a fault zone extended to deeper levels beyond 47 m. At imaging the profiles of Aronda, Shiroda, Mochemar, Kelus, Nivti coast,

Karli, and Malvan, widespread saline water intrusion is evident. The present results are in good agreement with the VES and geochemical studies carried out earlier here.

The study demonstrates the efficacy of using electrical resistivity in imaging the subsurface from which the underlying structures and extent of saline water incursion, fractures, and faults that influence the occurrence of groundwater in basaltic rocks can be evaluated, thus enhancing the accuracy of interpretation with minimum error. Resistivity models produced by inverse modeling of measured apparent resistivity data indicate prospective groundwater zones at several sites in the top layer, which can be explored for groundwater. Likewise, resistivity models have further deciphered saline water-affected zones within and below traps. In general, L1-based resistivity inversion results are stable and well correlated with the available geological information. Moreover, 2-D resistivity models based on robust inversion appear to be appropriate to infer sharp lateral resistivity variation caused by multiple episodes of lava flows and genesis of hard-rock terrain of DVP. It is worthwhile to note that an L1-based inversion scheme is more robust than that of an L2-based inversion scheme to take care of the uncontrolled error/outliers in the data, and hence provide some confidence to apply the algorithm for modeling the resistivity data. The reliability of the resistivity models of the subsurface formations is also validated by the litho log of the available dug wells in the study area. In addition to sharply mapping the detail of the geological features such as faults, lineaments, fractures, etc., in the hard-rock terrain, the present analysis also defines the extent of saline water ingress and potential groundwater-prospecting zones, which is of considerable significance for groundwater exploration. Further, these results are useful to gain better insights of the hydrogeological system of the study area.

Acknowledgments

The authors are indebted to Dr. D.S. Ramesh, director, IIG for according permission to publish this chapter. The authors are thankful to Shri B.D. Kadam, Shri M. Laxminarayana, and Shri S.H. Mahajan for helping in data acquisition. The authors also express their gratitude to Professor N.J. Pawar for many fruitful discussions. Thanks are also due to Dr. M. Thangarajan for many useful suggestions. Our third author, Dr. Saumen Maiti, is thankful to the director, Indian School of Mines (ISM), Dhanbad for encouragement and motivation.

References

Adeoti, L., Alile, O.M., and Uchegbulam, Q. 2010. Geophysical investigation of saline water intrusion into freshwater aquifers: A case study of Oniru, Lagos State, *Scientific Research and Essays*, 5, 248–259.

Advanced Geosciences, Inc (AGI). 2006. Instruction manual for the SuperSting™ with Swift™ automatic resistivity and IP system, http:///www.agiusa.com.

Barker, R.D. 1996. The application of electrical tomography in groundwater contamination studies, *EAGE 58th Conference and Technical Exhibition Extended Abstracts*, Amsterdam, The Netherlands, June 3–7, P082.

Barker, R.D., Venkateswara Rao, T., and Thangarajan, M. 2001. Delineation of contaminant zone through electrical imaging technique, *Current Science*, 81(3), 277–283.

Barker, R.D., Venkateshwara Rao, T. and Thangarajan, M. 2002. Application of electrical imaging for borehole siting in hard rock regions on India, *Journal of the Geological Society of India*, 61(2), 147–158.

Bear, J., Ouazar, D., Sorek, S., Cheng, A., and Herrera, I. 1999. *Seawater Intrusion in Coastal Aquifers*, Springer, Berlin.

Bhattacharya, P.K., and Patra, H.P. 1968. *Direct Current Geoelectric*, Elsevier, Amsterdam.

Binley, A. 2003. Free resistivity software; download available through Lancaster University: http://www.es.lancs.ac.uk/people/amb/Freeware/freeware.htm.

CGWB. 2009. Groundwater information, Sindhudurg district, Maharashtra. Technical Report, 1625/DB/2009.

Dahlin, T. 1996. 2D resistivity surveying for environmental and engineering applications, *First Break*, 14, 275–283.

Dahlin, T., and Loke, M.H. 1998. Resolution of 2D Wenner resistivity imaging as assessed by numerical modelling, *Journal of Applied Geophysics*, 38, 237–249.

Dahlin, T., and Zhou, B. 2004. A numerical comparison of 2D resistivity imaging with ten electrode arrays, *Geophysical Prospecting*, 52, 379–398.

Francese, R., Mazzarini, F., Bistacchi, A.L.P., Morelli, G., Pasquarè, G., Praticelli, N., Robain, H., Wardell, N., and Zaja, A. 2009. A structural and geophysical approach to the study of fractured aquifers in the Scansano–Magliano in Toscanaridge, southern Tuscany, Italy, *Hydrogeology Journal*, 17, 1233–1246.

Griffiths, D.H., and Barker, R.D. 1993. Two dimensional resistivity imaging and modeling in areas of complex geology, *Journal of Applied Geophysics*, 29, 211–226.

Gupta, G., Maiti, S., and Erram, V.C. 2014. Analysis of electrical resistivity data in resolving the saline and fresh water aquifers in west coast Maharashtra, *Journal of the Geological Society of India*, 84, 555–568.

Gurunadha Rao, V.V.S., Tamma Rao, G., Surinaidu, L., Rajesh, R., and Mahesh, J. 2011. Geophysical and geochemical approach for seawater intrusion assessment in the Godavari Delta Basin, AP, India, *Water, Air and Soil Pollution*, 217, 503–514, DOI 10.1007/s11270-010-0604-9.

Hermans, T., Vandenbohede, A., Lebbe, L., Martin, R., Kemna, A., Beaujean, J., and Nguyen, F. 2012. Imaging artificial salt water infiltration using electrical resistivity tomography constrained by geostatistical data, *Journal of Hydrology*, 438–439, 168–180.

Keller, G.V., and Frischknecht, F.C. 1966. *Electrical Methods in Geophysical Prospecting*, Pergamon, Oxford.

Koefoed, O. 1979. *Geosounding Principles 1: Resistivity Sounding Measurements*, Elsevier Science Publishing Company, Amsterdam.

Kumar, D., Rao, V.A., Nagaiah, E., Raju, P.K., Mallesh, D., Ahmeduddin, M., and Ahmed, S. 2010. Integrated geophysical study to decipher potential groundwater and zeolite-bearing zones in Deccan Traps, *Current Science*, 98(6), 803–814.

Loke, M.H. 2000. Topographic modelling in resistivity imaging inversion, *62nd EAGE Conference and Technical Exhibition Extended Abstracts, D-2*, Glasgow, Scotland, May 29–June 2.

Loke, M.H. 2011. *Electrical Imaging Surveys for Environmental and Engineering Studies—A Practical Guide to 2-D and 3-D Surveys*, Penang, Malaysia, http://www.geoelectrical.com/coursenotes.zip.

Loke, M.H., Acworth, I., and Dahlin, T. 2003. A comparison of smooth and blocky inversion methods in 2D electrical imaging surveys, *Exploration Geophysics*, 34, 182–187.

Loke, M.H., and Barker, R.D. 1996. Rapid least-squares inversion of apparent resistivity pseudosections by a quasi-Newton method, *Geophysical Prospecting*, 44, 131–135.

Loke, M.H., and Dahlin, T. 2002. A comparison of the Gauss–Newton and quasi-Newton methods in resistivity imaging inversion, *Journal of Applied Geophysics*, 49, 149–162.

Loke, M.H., Wilkinson, P.B., and Chambers, J.E. 2010. Fast computation of optimized electrode arrays for 2D resistivity surveys, *Journal of Computational Geosciences*, 36(11), 1414–1426.

Maiti, S., Erram, V.C., Gupta, G., Tiwari, R.K., Kulkarni, U.D., and Sangpal, R.R. 2013. Assessment of groundwater quality: A fusion of geochemical and geophysical information via Bayesian neural networks, *Environmental Monitoring and Assessment*, 185, 3445–3465, http://dx.doi.org/10.1007/s10661-012-2802-y.

Maiti, S., Erram, V.C., Gupta, G., and Tiwari, R.K. 2012. ANN based inversion of DC resistivity data for groundwater exploration in hard rock terrain of western Maharashtra (India), *Journal of Hydrology*, 464–465, 281–293, http://dx.doi.org/10.1016/j.jhydrol.2012.07.020.

Mondal, N.C., Singh, V.P., and Ahmed, S. 2013. Delineating shallow saline groundwater zones 18 from Southern India using geophysical indicators, *Environmental Monitoring and Assessment*, 185, 4869–4886.

Mondal, N.C., Singh, V.P., Singh, S., and Singh, V.S. 2011. Hydrochemical characteristic of coastal aquifer from Tuticorin, Tamil Nadu, India, *Environmental Monitoring and Assessment*, 175(1–4), 531–550.

Mondal, N.C., Singh, V.P., Singh, V.S., and Saxena, V.K. 2010a. Determining the interaction between groundwater and saline water through groundwater major ions chemistry, *Journal of Hydrology*, 388, 100–111.

Mondal, N.C., Singh, V.S., Puranik, S.C., and Singh, V.P. 2010b. Trace element concentration in groundwater of

Pesarlanka Island, Krishna Delta, India, *Environmental Monitoring and Assessment*, 163(1–4), 215–227.

Mondal, N.C., Singh, V.S., Sarwade, D.V., and Nandakumar, M.V. 2009. Appraisal of groundwater resources in an island condition, *Journal of Earth System Science*, 183(3), 217–229.

Mondal, N.C., Singh, V.S., Saxena, V.K., and Prasad, R.K. 2008. Improvement of groundwater quality due to fresh water ingress in Potharlanka Island, Krishna delta, India, *Environmental Geology*, 55, 595–603, DOI 10.1007/s00254-007-1010-5.

Olayinka, A.I., and Yaramanci, U. 2000. Use of block inversion in the 2-D interpretation of apparent resistivity data and its comparison with smooth inversion, *Journal of Applied Geophysics*, 45, 63–82.

Omosuyi, G.O., Ojo, J.S., and Olorunfemi, M.O. 2008. Geoelectric sounding to delineate shallow aquifers in the coastal plain sands of Okitipupa area, Southwestern Nigeria, *The Pacific Journal of Science and Technology*, 9(2), 562–577.

Pazdirek, O., and Blaha, V. 1996. Examples of resistivity imaging using ME 100 resistivity field acquisition system, *EAGE 58th Conference and Technical Exhibition Extended Abstracts*, Amsterdam.

Rai, S.N., Thiagarajan, S., Ratnakumari, Y., Anand Rao, V., and Manglik, A. 2013. Delineation of aquifers in basaltic hard rock terrain using vertical electrical soundings data, *Journal of Earth System Science*, 122(1), 29–41.

Sankaran, S., Sonkamble, S., Krishnakumar, K., and Mondal, N.C. 2012. Integrated approach for demarcating subsurface pollution and saline water intrusion zones in SIPCOT area: A case study from Cuddalore in Southern India, *Environmental Monitoring and Assessment*, 184, 5121–5138, DOI 10.1007/s10661-011-2327-9.

Saxena, V.K., Singh, V.S., Mondal, N.C., and Jain, S.C. 2003. Use of chemical parameters to delineation fresh groundwater resources in Potharlanka Island, India, *Environmental Geology*, 44(5), 516–521.

Saxena, V.K., Mondal, N.C., and Singh, V.S. 2004. Identification of sea-water ingress using strontium and boron in Krishna Delta, India, *Current Science*, 86, 586–590.

Sherif, M., El Mahmoudi, A., Garamoon, H., Kacimov, A., Akram, S., Ebraheem, A., and Shetty, A. 2006. Geoeletrical and hydrogeochemical studies for delineating seawater intrusion in the outlet of Wadi Ham, UAE, *Environmental Geology*, 49, 536–551.

Singh, V. S., Sarwade, D.V., Mondal, N.C., Nanadakumar, M.V., and Singh, B. 2009. Evaluation of groundwater resources in a tiny Andrott Island, Union Territory of Lakshadweep, India, *Environmental Monitoring and Assessment*, 158, 145–154, DOI 10.1007/s10661-008-0569-y.

Singh, K.K.K., Singh, A.K.S., Singh, K.B., and Sinha, A. 2006. 2D resistivity imaging survey for sitting water-supply tube well in metamorphic terrains: A case study of CMRI campus, Dhanbad, India, *The Leading Edge*, 25, 1458–1460.

Song, S.H., Lee, J.Y., and Park, N. 2007. Use of vertical electrical soundings to delineate seawater intrusion in a coastal area of Byunsan, Korea, *Environmental Geology*, 52, 1207–1219.

Sukhija, B.S., Varma, V.N., Nagabhushanam, P., and Reddy, D.V. 1996. Differentiation of paleomarine and modern seawater intruded salinities in coastal groundwater (of Karaikal and Tanjavur, India) based on inorganic chemistry, organic biomarker fingerprints and radiocarbon dating, *Journal of Hydrology*, 174, 173–201.

Telford, N.W., Geldart, L.P., Sheriff, R.S., and Keys, D.A. 1976. *Applied Geophysics*, Cambridge University Press, London.

Todd, D.K. 1979. *Groundwater Hydrology*, Wiley, New York, pp. 277–294.

World Health Organization (WHO). 1984. *Guideline of Drinking Quality*, World Health Organization, Washington, pp. 333–335.

Zarroca, M., Linares, R., Bach, J., Roqué, C., Moreno, V., Font, L., and Baixeras, C. 2012. Integrated geophysics and soil gas profiles as a tool to characterize active faults: The Amer fault example (Pyrenees, NE Spain), *Environmental Earth Sciences*, 67, 889–910.

Zarroca, M., Linares, R., Roqué, C., Rosell, J., and Gutiérrez, F. 2014. Integrated geophysical and morphostratigraphic approach to investigate a coseismic (?) translational slide responsible for the destruction of the Montclús village (Spanish Pyrenees), *Landslides*, 11, 655–671, DOI 10.1007/s10346-013-0427-z.

Zhu, T., Feng, R., Hao, J., Zhou, J., Wang, H., and Wang, S. 2009. The application of electrical resistivity tomography to detecting a buried fault: A case study, *Journal of Environmental and Engineering Geophysics*, 14(3), 145–151.

6

GIS-Based Probabilistic Models as Spatial Prediction Tools for Mapping Regional Groundwater Potential

Madan Kumar Jha, Sasmita Sahoo, V. M. Chowdary, and Niraj Kumar

CONTENTS

6.1 Introduction

The stark reality in the beginning of the twenty-first century is that freshwater has already become a limiting resource for increasing human population, sustaining healthy ecosystems, and poverty alleviation in many parts of the globe (WWAP, 2009). Bearing in mind the mounting water scarcity and accelerating environmental degradation in several regions of the world, the twenty-first century is actually facing four grand socio-environmental challenges, namely, *water security*, *food security*, *energy security*, and *environmental security*, which are becoming increasingly daunting due to looming climate change (Jha, 2010). Sustainable management of freshwater resources is the key to effectively addressing these great and interrelated challenges. Groundwater is a treasured earth's resource and today, it constitutes a major and more dependable source of water supply across the world. It is the largest available source of freshwater on the earth, which supports human health and hygiene, socioeconomic development, and sustains ecosystems and biodiversity. However, the growing overuse and pollution of groundwater are threatening the existence of present generations as well as the life of subsequent generations. How to maintain a sustainable water supply from existing aquifers for meeting both human and ecosystem needs is one of the most important environmental concerns now and in coming decades (Vörösmarty et al., 2000; Mooney et al., 2005; Biswas et al., 2009; Grayman et al., 2012; Jha, 2013). Therefore, there is a pressing need in both developing and developed nations for the efficient management of our dwindling freshwater resources to ensure their long-term sustainability. The complexities of the processes governing the occurrence and movement of groundwater make the problem of groundwater assessment somewhat difficult because not only enormous field data are to be obtained at adequate spatial and temporal resolutions, but also a multidisciplinary scientific approach is to be adopted. Certainly, the appraisal of groundwater resources is often vital but no single comprehensive technique is yet identified that is capable of estimating accurate groundwater resources in a basin (Lowry et al., 2007; Vrba and Lipponen, 2007). This calls for widespread use of a set of emerging tools/techniques for the estimation of hidden groundwater resources at a basin/subbasin scale. In this context, mapping of groundwater potential using geospatial techniques can play a vital role in the identification of zones depicting spatial variation of probable groundwater occurrence in a catchment or basin (Jha and Peiffer, 2006; Jha et al., 2007).

The application of geospatial techniques such as remote sensing (RS) and geographic information system (GIS) to the identification of groundwater potential zones has been reported by numerous researchers from different parts of the world, mostly from developing countries (e.g., Krishnamurthy et al., 1996; Sander et al., 1996; Saraf and Choudhury, 1998; Jaiswal et al., 2003; Sener et al., 2005; Solomon and Quiel, 2006; Srivastava and Bhattacharya, 2006; Tweed et al., 2007; Madrucci et al., 2008; Chowdhury et al., 2009; Jha et al., 2010; Machiwal et al., 2011; Jasrotia et al., 2013). However, the studies on groundwater potential mapping using GIS and probabilistic models are highly limited. Although the frequency ratio (FR) and weight-of-evidence (WOE) probabilistic models have been applied for landslide susceptibility mapping (e.g., Lee and Choi, 2004; Oh et al., 2009) and ground subsidence hazard mapping (e.g., Kim et al., 2006), only a couple of studies are reported to date that deal with probabilistic modeling for mapping groundwater potential (Oh et al., 2011; Ozdemir, 2011; Lee et al., 2012).

This chapter demonstrates the efficacy of two GIS-based probabilistic models—*FR* and *WOE*—for the spatial prediction of groundwater potential in a river basin.

6.2 Applications of RS and GIS in Groundwater Studies: An Overview

In the past, remotely sensed data have been more frequently used for the assessment of surface resources such as land use/cover and vegetation structure/density mapping (e.g., Franklin, 1986; Engman and Gurney, 1991; Kite et al., 1997; Chen et al., 2004) or for general studies on water resources management (e.g., Sharma and Anjaneyulu, 1994). The use of satellite data in the field of groundwater hydrology started with the availability of Landsat MSS and then TM data. These data were later complemented by SPOT multispectral (XS), IRS (LISS sensors), panchromatic and stereo pairs, and then by the merging of radar and visible data. Engman and Gurney (1991) suggested that the satellite imagery can be most effectively used for regional groundwater exploration, and emphasized that the analysis of satellite imagery is a rapid and inexpensive means of obtaining reconnaissance groundwater information. Initially, visual interpretation was the main tool for the evaluation of groundwater potential zones for over two decades. Aerial photographs have been used to eliminate the areas of potential low-water-bearing state (Engman and Gurney, 1991), to assess abandoned flowing artesian wells, and to estimate artesian well flow rates (Jordan and Shih, 1988, 1991; Shih and Jordan, 1990). Subsequently, digital enhancement techniques such as linear stretching, band combination, filtering, and edge enhancement techniques have been employed for deriving geological, structural, and geomorphological details from RS data (e.g., Colwell, 1983; Lillesand and Kiefer, 2000). The current trend is toward a better distribution of technical tasks and a better adaptation of available tools to find appropriate solutions to real-world problems.

Compared with the information acquired by traditional methods, RS data in general offer a number of advantages: (i) less need for field work and slower, more expensive exploration methods; (ii) identification of promising areas for a more detailed study and ground exploration; (iii) new or better geologic and hydrologic information; (iv) information can be acquired for the same area at a high rate of repetition (2–3 times a month or even faster) thus permitting selection of the most appropriate seasonal data; (v) satellite imagery is recorded in various wavelengths (visible and non-visible), which provide accurate information on ground conditions; (vi) special capability of synthetic aperture radar (SAR) systems to acquire information even in the presence of clouds is a great advantage in many countries; (vii) the data can be obtained for any part of the world without encountering administrative restrictions; (viii) the perspective of large areal coverage available from satellite imagery, which may be unavailable from other means of exploration; and (ix) the availability of data archives with historical data, which in most cases is the only means to conduct long-term historical studies and analysis of the evolution of natural resources.

Satellite data provide quick and useful baseline information about the factors controlling the occurrence and movement of groundwater such as geology, geomorphology, soil types, land use/land cover, topography, drainage pattern, lineaments, etc. (e.g., Waters et al., 1990; Engman and Gurney, 1991; Meijerink, 2000). Satellite imagery enables hydrologists/hydrogeologists to infer aquifer location from surface features, and to find regions with a high potential for containing well sites (Jordan and Shih, 1991). However, all the controlling factors have rarely been studied together because of the unavailability of data, integrating tools, and/or modeling techniques. Structural features such as faults, fracture traces, and other such linear or curvilinear features can indicate the possible presence of groundwater (Todd, 1980; Engman and Gurney, 1991). Similarly, other features such as sedimentary strata or certain rock outcrops may indicate potential aquifers. The presence of oxbow lakes and old river channels are good indicators of alluvial deposits. Geomorphic units have been found to provide direct links to potential near-surface groundwater both in terms of feature morphology and sediment content in Botswana and Australia (Shaw and de Vries, 1988; McFarlane et al., 1994; Ringrose et al., 1998). Shallow groundwater could also be inferred by soil moisture measurements and by changes in vegetation types and pattern. Cultural features such as farming practices can also be used to infer aquifers. In arid regions, vegetation characteristics may indicate groundwater depth and quality. The recharge and discharge areas in drainage basins can be detected from soils, vegetation, and shallow/perched groundwater (Meijerink, 2000). By measuring surface temperature, differences evolved through RS have also been used to identify alluvial deposits, shallow groundwater, and springs or seeps (van de Griend et al., 1985). Thus, a variety of earth's surface features can be derived from satellite imagery and/or aerial photographs (Table 6.1) and they can be used for evaluating groundwater conditions (e.g., occurrence, depth, flow patterns, quantity, or quality) under different hydrogeologic settings.

Moreover, airborne radar can be used to show surface features even under dense-vegetation canopies, and is particularly valuable for revealing topographic relief and roughness. For example, side-looking airborne radar (SLAR) mounted on aircraft or satellites has been successfully used to map and identify structural features in regions of the world which because of either perpetual cloud cover or thick vegetation had not been adequately mapped earlier (Engman and Gurney, 1991). Also, the SAR, because of its penetrating capability and the capability to provide the best soil moisture information, has a great potential for groundwater exploration, especially in arid and hyperarid regions (Jackson, 2002). In the recent past, the mapping of land subsidence due

to excessive groundwater pumping using interferometric synthetic aperture radar (InSAR) has also been reported by some researchers (e.g., Galloway et al., 1998; Hoffmann et al., 2001). Ground-penetrating radar (GPR) can be used to detect water table depth and characterize the nature of phreatic surface in coarse sediments. Landsat MSS data have also been used in conjunction with aerial photographs to assess perched water table. It is worth mentioning that the RS data gain additional value for groundwater hydrology or hydrogeology when physical models are used to extrapolate/translate them into the subsurface (Meijerink, 2000; Hoffmann, 2005). Excellent reviews on RS applications in groundwater hydrology are presented in Waters et al. (1990), Engman and Gurney (1991), Meijerink (2000), and Jha et al. (2007). These reviews indicate that RS has been widely used as a tool, mostly to complement standard geophysical techniques. Meijerink (2000) recognizes the value of RS in recharge-based groundwater studies wherein it can aid conventional assessment or modeling techniques for groundwater recharge.

As the use of RS technology involves a large amount of spatial data management, it requires an efficient system to handle such data. GIS provides a suitable platform for the efficient management of large and complex databases. Various thematic layers generated using RS data such as geology, geomorphology (landforms), land use/land cover, lineaments, etc., can be integrated with slope, drainage density, and other collateral data in a GIS framework and they can be analyzed using a GIS-based spatial model developed with logical conditions to identify potential groundwater zones, recharge/discharge areas, and suitable sites for artificial recharge and rainwater harvesting (e.g., Hinton, 1996; Jha and Peiffer, 2006; Jha et al., 2007, 2010, 2014; Madrucci et al., 2008; Chowdhury et al., 2009, 2010; Machiwal et al., 2011).

The geographic location and attribute data of individual wells have also been successfully stored in a GIS database for the monitoring and management of abandoned wells. The monitoring of abandoned wells using RS, GIS, and global positioning system (GPS) is a faster alternative to the conventional methods. Furthermore, the integrated use of RS and GIS is a valuable tool for the analysis of voluminous hydrogeologic data from different sources and for the simulation modeling of complex subsurface flow and transport processes under saturated and unsaturated conditions (e.g., Watkins et al., 1996; Loague and Corwin, 1998; Gogu et al., 2001; Gossel et al., 2004). The current status of RS and GIS applications in groundwater hydrology and their future prospects are discussed in Jha and Peiffer (2006) and Jha et al. (2007), whereas the challenges of using RS and GIS in developing nations are discussed in Jha and Chowdary (2007). Jha and Peiffer (2006) and Jha et al. (2007) have reported that the current

TABLE 6.1

Relevant Physical Features of the Landscape Obtained from RS Technique for Evaluating Groundwater Condition

Landscape Feature	Interpretation for Groundwater Condition
1. Topography	The local and regional relief setting gives an idea about the general direction of groundwater flow and its influence on groundwater recharge and discharge.
Low slope (0–5°)	Presence of high groundwater potential.
Medium slope (5–20°)	Presence of moderate-to-low groundwater potential.
High slope (>20°)	Presence of poor groundwater potential.
2. Natural vegetation	Dense vegetation indicates the availability of adequate water where groundwater may be close to the land surface.
Phreatophytes	Shallow groundwater under unconfined conditions.
Xerophytes	Appreciably deep groundwater under confined or unconfined conditions.
Halophytes	Shallow brackish or saline groundwater under unconfined conditions.
3. Geologic landform	
Modern alluvial terraces, alluvial plains, floodplains, and glacial moraines	Favorable sites for groundwater storage.
Sand dunes	Gives an idea about the presence of underlying sandy glacio-fluvial sediments, which indicate the presence of groundwater.
Rock outcrops	Presence of a potential aquifer.
Thick-weathered rocks	Moderate groundwater potential.
Rocks with fractures/fissures	Very good or excellent potential of groundwater.
Rocks without fractures/fissures	Unfavorable sites for groundwater occurrence.
Hillocks, mounds, and residual hills	Unfavorable sites for groundwater existence.
4. Lakes and streams	
Oxbow lakes and old river channels	Favorable sites for groundwater extraction.
Perennial rivers and small perennial rivers	High-to-moderate potential of groundwater.
Drainage density	High-drainage density indicates an unfavorable site for groundwater existence, moderate-drainage density indicates moderate groundwater potential, and less/no-drainage density indicates high groundwater potential.
Drainage pattern	Gives an idea about the joints and faults in the bedrock, which in turn indicates the presence or absence of groundwater.
5. Spring types (tentatively inferred from RS data)	
Depression springs, contact springs, and artesian springs	Presence of a potential aquifer.
Moist depressions, seeps, and marshy environments	Presence of shallow groundwater under unconfined conditions.
6. Lineaments (applicable only to rocky terrains)	Gives an idea about the underground faults and fractures, and thereby indicates the occurrence of groundwater.

Source: Jha, M.K. et al., 2007. *Water Resources Management*, 21(2), 427–467.

applications of RS and GIS in groundwater hydrology could be grouped into six major areas: (i) exploration and assessment of groundwater resources, (ii) selection of sites for artificial recharge and water harvesting, (iii) GIS-based subsurface flow and transport modeling, (iv) groundwater-pollution hazard assessment and protection planning, (v) estimation of natural recharge distribution, and (vi) hydrogeologic data analysis and process monitoring.

Thus, the ability of RS technology for obtaining systematic, synoptic, quick (within a short time), and repetitive coverage in different windows of the electromagnetic spectrum, and covering large and inaccessible areas from space, together with the expanding capability of GIS have made these two geospatial techniques unique and powerful in the era of information technology. Consequently, the domain of their applications in groundwater evaluation and management is gradually expanding. In recent years, GRACE satellite data have been used to estimate the extent of groundwater depletion in some parts of the world, including India. In addition, in the recent past, the RS technology has proved to be a major source for an increasing number of hydroclimatic data, which have enhanced our capabilities to reliably predict the variations in the global energy and water cycle (Su et al., 2014). The availability of high-resolution satellite imagery and emerging hyperspectral satellite imagery, and an increasing number of earth observation (EO) missions in the near future will provide unprecedented opportunities as well as open up new and exciting avenues for hydrological sciences, including groundwater hydrology.

6.3 Probabilistic Models for Spatial Prediction: An Overview

6.3.1 FR Model

The FR is the probability of occurrence of a certain attribute (Bonham-Carter, 1994). If we create an event E and certain factors attributed to F, the probability–FR of F is the ratio of the conditional probability. It is mathematically expressed as follows:

$$P(E/F) = \frac{P(E \cap F)}{P(F)} \qquad (6.1)$$

A FR model can provide a simple geospatial assessment tool to calculate the probabilistic relationship between dependent and independent variables, including multiclassified maps (Oh et al., 2011). The idea behind FR is that the relationship between groundwater occurring in an area and groundwater-related factors can be distinguished from the relationship between groundwater not occurring in an area and groundwater-related factors. It can be expressed as an FR that represents the quantitative relationship between groundwater occurrences and different causative parameters. Thus, FR for a class of the groundwater-affecting factors can be expressed as

$$FR = \frac{A/B}{C/D} = \frac{E}{F} \qquad (6.2)$$

where
A = area of a class for the factor
B = total area of the factor
C = number of pixels in the class area of the factor
D = number of total pixels in the study area
E = percentage for area with respect to a class for the factor
F = percentage for the entire domain
FR = frequency ratio of a class for the factor

To obtain a groundwater potential index (GWPI), the ratings of the factors are summed as

$$GWPI = \sum (FR)_i \quad (i = 1, 2, 3 \ldots N) \qquad (6.3)$$

where
GWPI = groundwater occurrence potential index
FR = frequency ratio of a factor
N = total number of input factors

6.3.2 WOE Model

The WOE model calculates the weight for the presence or absence of each groundwater predictive factor's class (F or F^*) based on the presence or absence of the groundwater (S or S^*) within the area under study. Accordingly, the WOE is calculated using the following equations:

$$W^+ = \ln \frac{P(F/S)}{P(F/S^*)} \qquad (6.4)$$

$$W^- = \ln \frac{P(F^*/S)}{P(F^*/S^*)} \qquad (6.5)$$

where
P = probability
F = presence of a dichotomous pattern
F^* = absence of a dichotomous pattern
S = presence of event occurrence
S^* = absence of event occurrence
W^+ = WOE when a factor is present (i.e., relevant)
W^- = WOE when a factor is absent (i.e., not relevant)

The weight contrast (C) is given as

$$C = W^+ - W^- \qquad (6.6)$$

The weight contrast (C) reflects the overall spatial association between a predictable variable and groundwater occurrence. A contrast value equal to zero indicates that the considered class of casual factors is not significant for the analysis, while a positive contrast indicates a positive spatial correlation, and vice versa for a negative contrast (Corsini et al., 2009). The value of posterior probability [$P(S)$] is given as (Barbieri and Cambuli, 2009)

$$\ln P(S) = \left\{ \sum Wi^* \times \ln P_{P(S)} \right\} \qquad (6.7)$$

$$P = \exp \left\{ \sum W^+ + \ln P_{P(S)} \right\} \qquad (6.8)$$

where $P_{P(S)}$ is prior probability that is given as

$$P_{P(S)} = \frac{(\text{groundwater cell number})}{(\text{study area cell number})} \qquad (6.9)$$

6.4 Application of Probabilistic Models: A Case Study

6.4.1 Study Area

The Kushabhadra–Bhargavi groundwater basin of Odisha is the study area for this case study because of the availability of necessary field data. It is situated in the Mahanadi Delta Stage-II region of Odisha, eastern India (Figure 6.1). It lies between 19° 49′ 04″ N and 20° 18′ 45″ N latitude and 85° 54′ 47″ E to 86° 03′ 26″ E longitude.

It has a geographical area of 620 km² and is formed by the branching of the River Kuakhai into Kushabhadra River and Bhargavi River, which form the eastern and western boundaries of the study area, respectively. The topography of the area is almost flat with the elevation

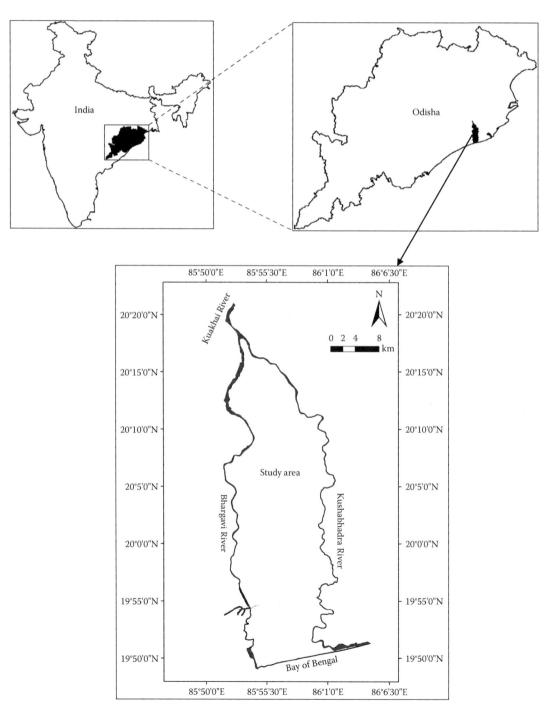

FIGURE 6.1
Location map of the study area.

varying from 0 to 26-m MSL. The climate of the study area is characterized as a tropical monsoon. The average annual rainfall in the study area is about 1416 mm, with the majority of rainfall occurring during mid-June to end of October. The mean monthly maximum and minimum temperatures in the area are 42°C in the month of May and 17°C in the month of December. The mean monthly humidity varies from 41% in the month of December to 86% during the months of July and August. The study area is underlain by laterite and alluvium types of geologic formations, which offer important sources of freshwater in the study area.

6.4.2 Acquisition of Spatial Data

The soil map, land use/land cover map, and geology map of the study area at a scale of 1:250,000 were collected from Odisha Space Applications Centre (ORSAC), Department of Science and Technology (DST), Government of Odisha. The DEM (digital elevation model) of the study area was extracted from Shuttle Radar Topography Mission (SRTM), which was used for generating slope and drainage density thematic layers of the study area.

6.4.3 Preparation of Thematic Layers

Five significant thematic layers, namely, slope, geology, land use/land cover, soil, and drainage density, were selected for this study. These layers were classified into their features in the GIS environment, which helped interpret their importance from the standpoint of groundwater occurrence or storage. The important features of these five thematic layers (i.e., thematic maps) are succinctly described in the subsequent subsections.

6.4.3.1 Slope Map

The prevailing slope in the study area varies from 0 to 4%. The slope statistics of the study area reveal that a major portion of the study area (nearly 79%) falls in the 0–1% slope category. This slope class can be considered "very good" from the groundwater occurrence viewpoint due to a nearly flat terrain, and hence relatively high infiltration potential. The area having 1–2% slope can be considered "good" for groundwater due to a slightly undulating topography with some runoff. On the other hand, the area having a slope of 2–4% is likely to produce a relatively high runoff and low infiltration, and hence it can be categorized as "moderate."

6.4.3.2 Geology Map

The geology/lithology map of the study area indicates two types of geology features, namely, alluvium and

laterite. A major portion of the study area (434 km^2) is covered by the laterite formation, which encompasses about 61% of the total area, and laterite is prevalent in the middle and southern portions of the study area. Laterite is a somewhat porous subsurface formation, which can form potential aquifers along topographic lows and moderate groundwater potential can be expected in this area. On the other hand, about 39% of the study area (186 km^2) is occupied by the alluvium formation. Alluvium constitutes a very good water-bearing formation and is often considered an important source of groundwater in the deltaic regions.

6.4.3.3 Land Use/Land Cover Map

Land use/land cover plays an important role in deciding the extent of infiltration rate, which in turn gives an idea about recharge potential. The study area comprises six major land use/land cover categories, namely, agricultural land, dense forest, degraded forest, wasteland, settlements, and rivers and other water bodies.

6.4.3.4 Soil Map

The soil classes found in the study area are silty loam, clayey loam, coarse sand, very fine sand, and sandy loam. According to their hydraulic characteristics, these soil classes can be considered "very good," "good," "moderate," and "poor" depending on their contribution to groundwater recharge.

6.4.3.5 Drainage Density Map

The density of surface drainage network has a relation with surface runoff and soil permeability, and therefore it is considered one of the important hydrological indicators of groundwater occurrence in the study area. Based on the drainage density of microwatersheds in the study area, the drainage density was grouped into five classes. A major portion of the study area (nearly 88%) falls under 0–0.50 km/km^2 drainage density category, which can be considered "very good" for groundwater occurrence. Similarly, the remaining drainage density classes such as 0.50–0.75 km/km^2 can be considered "good," 0.75–1 km/km^2 "moderate," and >1 km/km^2 can be considered "poor" from the viewpoint of groundwater storage.

6.4.4 Preparation of Groundwater Potential Map

As mentioned earlier, GIS-based probabilistic models namely *FR* and *WOE* were used to predict groundwater potential zones over the study area. These models require additional data on the location of pumping wells over the study area, together with the thematic layers having an influence on groundwater occurrence or storage. A well-location map of the study area was also generated depicting all 77 existing pumping wells, of which 55 pumping wells were used for training and 22 pumping wells for testing the FR and WOE probabilistic models. Thereafter, the modeling results were implemented over the entire study area. The testing wells were used solely for the verification of FR and WOE modeling results. The selected thematic layers (factors) were overlaid with the well-location map. On the basis of these intersections, the FRs and WOE probability (P) values were calculated for each of the five factors (thematic layers). The procedures for the application of these probabilistic models to the spatial prediction of groundwater potential are briefly described next.

6.4.4.1 GIS-Based FR Modeling

For computing the FR of groundwater potential, the *"area ratio"* and *"well occurrence ratio"* were calculated

for different classes of each factor. Then, FR for different classes of each factor was calculated by dividing the *"well occurrence ratio"* with the *"area ratio."* The FR values thus obtained were used for generating a groundwater potential map of the study area by using the overlay function of GIS.

6.4.4.2 GIS-Based WOE Modeling

In this case, the selected groundwater-related thematic layers (factors) were overlaid with the training well map using the overlay function of GIS. Based on these intersections, the weight and WOE probability values were calculated for the individual classes of different factors. Thereafter, a groundwater potential map of the study area was prepared based on the range of WOE (P) probability values over the study area.

6.4.5 Validation of Groundwater Potential Map

Finally, a comparative evaluation of the probabilistic models used for the spatial prediction of groundwater potential was performed in terms of prediction accuracy. The prediction accuracy was calculated in the GIS environment by overlaying the well-yield map on

TABLE 6.2

Number of Pixels and Number of Pumping Wells in the Features of the Factors

Factor	Features of the Factor	Number of Pixels (30 × 30 m)	Number of Wells
1. Geology	Alluvium	203,858	23
	Laterite	481,971	32
	Total	685,829	55
2. Soil	Coarse sand	50,881	2
	Sandy loam	1361	0
	Very fine sand	26,634	0
	Silty loam	540,580	49
	Clayey loam	66,373	4
	Total	685,829	55
3. Land use/land cover	Rivers and water bodies	9490	0
	Agricultural land	559,421	40
	Dense forest	19,699	0
	Degraded forest	759	0
	Wasteland	12,173	0
	Settlements	84,287	15
	Total	685,829	55
4. Drainage density (km/km²)	0–0.25	168,801	8
	0.25–0.50	438,470	34
	0.50–0.75	52,164	6
	0.75–1	11,958	6
	>1	14,436	1
	Total	685,829	55
5. Slope (%)	0–1	540,707	39
	1–2	142,240	16
	2–4	2882	0
	Total	685,829	55

TABLE 6.3

Values of FR for the Features of the Five Factors

Factor	Features of the Factor	Area (%)	Well (%)	FR
1. Geology	Alluvium	29.7	0.418	1.407
	Laterite	70.3	0.582	0.827
2. Soil	Coarse sand	7.40	3.63	0.490
	Sandy loam	0.30	0	0
	Very fine sand	3.80	0	0
	Silty loam	78.80	89.10	1.130
	Clayey loam	9.70	7.27	0.749
3. Land use/land cover	Rivers and water bodies	1.38	0	0
	Agricultural land	81.57	72.73	0.89
	Dense forest	2.87	0	0
	Degraded forest	0.12	0	0
	Wasteland	1.77	0	0
	Settlements	12.29	27.27	2.21
4. Drainage density (km/km²)	0–0.25	24.5	14.54	0.593
	0.25–0.50	63.7	61.82	0.970
	0.50–0.75	7.6	10.91	1.435
	0.75–1	2.1	10.91	5.455
	>1	2.1	1.82	0.866
5. Slope (%)	0–1	78.84	70.91	0.899
	1–2	20.74	29.09	1.403
	2–4	0.42	0	0

the groundwater potential maps predicted by the FR and WOE models. Thus, the values of prediction accuracy were obtained for the two predicted groundwater potential maps.

6.4.6 Spatial Prediction of Groundwater Potential

The results of groundwater potential prediction by FR and WOE models are discussed in this section. The number of pumping wells and the number of pixels present in individual features of the five factors, as obtained from GIS analysis/modeling, are summarized in Table 6.2.

6.4.6.1 Groundwater Potential Prediction by FR Model

The values of FRs for each feature of the five factors are presented in Table 6.3. The GWPI obtained from the FR

values provided a basis for identifying groundwater potential zones in the study area. A groundwater potential map of the study area was thus generated, which indicated four groundwater potential zones in the study area, namely, "poor," "moderate," "good," and "very good" as shown in Figure 6.2. A summary of groundwater potential statistics of the study area based on FR modeling is provided in Table 6.4.

It is apparent from Figure 6.2 that the groundwater potential zones "very good" and "good" exist mainly in the northern part and in some patches scattered over the study area. The area under "very good" and "good" groundwater potential zones is about 236 km^2 (38%), while the "moderate" groundwater potential zone covers an area of 322 km^2, which is 52% of the total study area (Table 6.4). The southern part and scattered small patches in the central portion of the study area fall in the "poor" groundwater potential zone that covers an

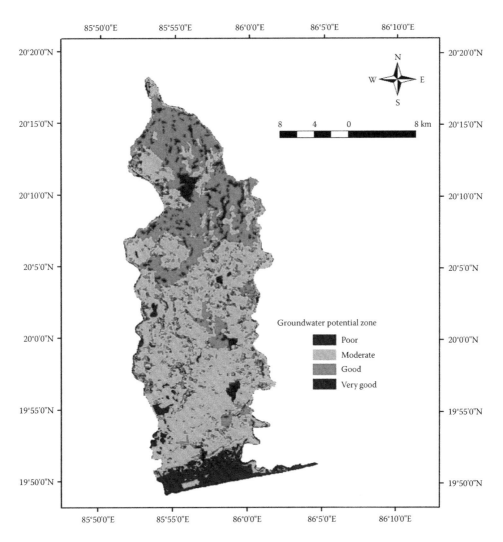

FIGURE 6.2
Groundwater potential map of the study area obtained by FR modeling. (From Sahoo, S. et al., 2015. 74(3), 2223–2246. *Environmental Earth Sciences*, DOI 10.1007/s12665-015-4213-1.)

TABLE 6.4

Groundwater Potential Statistics of the Study Area Based on FR Modeling

Groundwater Potential Zone	Zone Area (km²)	Percentage Coverage
Poor	63	10.2
Moderate	322	52
Good	192	31
Very good	44	7

area of 63 km² (10.2%). Thus, the results of FR modeling revealed that the "moderate" groundwater potential zone is predominant in the study area.

6.4.6.2 Groundwater Potential Prediction by WOE Model

The values of WOE (W^+) and WOE probability (P) for each feature of the five factors are presented in Table 6.5. Based on the WOE probability values, a groundwater potential map of the study area was generated using GIS and four distinct zones of groundwater potential were identified: "poor," "moderate," "good," and "very good" (Figure 6.3). Figure 6.3 depicts that the "good" groundwater potential zone occurs in the northern portion and some parts of the central portion of the study area, but the "very good" groundwater potential zone occurs in a

few patches in the northern, central, and southern parts of the study area.

The area under "very good" and "good" groundwater potential zones is about 44.4% (Table 6.6). The central portion, some parts of the southern portion, and some patches in the northern portion of the study area have "moderate" groundwater potential. The lower southern part and a few small patches/strips in the central and northern parts of the study area have "poor" groundwater potential, which encompasses an area of 7.6% (Table 6.6).

6.4.7 Evaluating Probabilistic Models

The verification of probabilistic models was performed using measured well yields of 22 testing pumping wells in the study area. The verification of the groundwater potential map predicted by the FR model revealed that five out of eight "high-discharge" pumping wells exist in the "good" zone, two in the "moderate" zone, and one in the "poor" zone. However, two out of 11 "medium-discharge" wells exist in the "good" zone, eight in the "moderate" zone, and one in the "poor" zone. Based on these findings, the prediction accuracy of the FR model is computed to be 68.18%.

On the other hand, the verification of the groundwater potential map predicted by the WOE model indicated that six out of eight "high-discharge" wells exist in the "good" zone and two exist in the "moderate" zone. Further, two

TABLE 6.5

Values of WOE (W^+) and WOE Probability (P) for the Features of the Five Factors

Factor	Features of the Factor	Area (%)	Well (%)	W^+	P
1. Geology	Alluvium	29.7	0.418	0.3414	0.0001128
	Laterite	70.3	0.582	−1.8882	0.0000121
2. Soil	Coarse sand	7.40	3.63	−0.7131	0.0000393
	Sandy loam	0.30	0	0.0000	0.0000802
	Very fine sand	3.80	0	0.0000	0.0000802
	Silty loam	78.80	89.10	0.1226	0.0000906
	Clayey loam	9.70	7.27	−0.1909	0.0000662
3. Land use/land cover	Rivers and water bodies	1.38	0	0.0000	0.0000802
	Agricultural land	81.57	72.73	−0.1146	0.0000715
	Dense forest	2.87	0	0.0000	0.0000802
	Degraded forest	0.12	0	0.0000	0.0000802
	Wasteland	1.77	0	0.0000	0.0000802
	Settlements	12.29	27.27	0.7978	0.0001780
4. Drainage density (km/km²)	0–0.25	24.5	14.54	−0.5286	0.0000472
	0.25–0.50	63.7	61.82	−0.0336	0.0000775
	0.50–0.75	7.6	10.91	0.361	0.0001152
	0.75–1	2.1	10.91	1.8357	0.0005027
	>1	2.1	1.82	−0.1466	0.0000692
5. Slope (%)	0–1	78.84	70.91	−0.1104	0.0000718
	1–2	20.74	29.09	0.3549	0.0001143
	2–4	0.42	0	0.0000	0.0000802

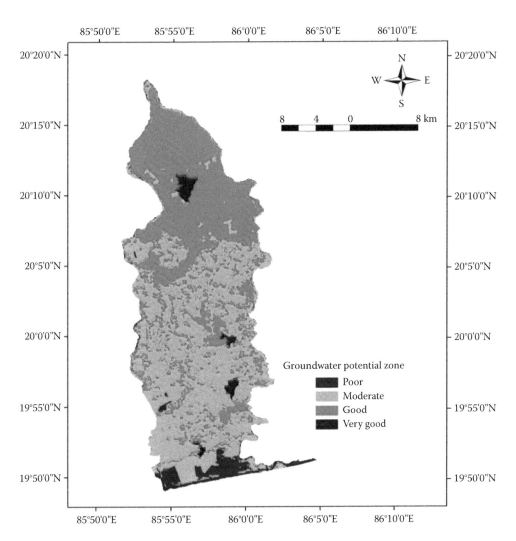

FIGURE 6.3
Groundwater potential map of the study area obtained by WOE modeling. (From Sahoo, S. et al., 2015. *Environmental Earth Sciences*, 74(3), 2223–2246. DOI 10.1007/s12665-015-4213-1.)

out of 11 "medium-discharge" wells exist in the "good" zone, eight in the "moderate" zone, and one in the "poor" zone. Based on these findings, the prediction accuracy of the WOE model is computed to be 72.72%.

It is obvious from the above verification results that although the results of FR and WOE probabilistic models are satisfactory, the prediction accuracy of the WOE model is higher than that of the FR model. Thus, the

WOE model is recommended for the spatial prediction of groundwater potential in the study area as well as in the areas/regions having more or less-similar hydrogeologic settings.

TABLE 6.6

Groundwater Potential Statistics of the Study Area Based on WOE Modeling

Groundwater Potential Zone	Zone Area (km²)	Percentage Coverage
Poor	47	7.6
Moderate	297	48
Good	254	41
Very good	21	3.4

6.5 Conclusions

The case study discussed in this chapter employed two GIS-based probabilistic models, namely, *FR* and *WOE* for the spatial prediction of regional groundwater potential. The groundwater potential map predicted by the WOE model revealed that about 3% and 41% of the study area fall in "very good" and "good" groundwater potential zones, respectively, while 48% falls in the "moderate" zone and 8% falls in the "poor" groundwater potential zone. In contrast, the groundwater potential map predicted by the FR model revealed that about 7% of the study area falls in the "very good" groundwater potential zone, 31% in the "good" zone, 52% in the "moderate" zone, and 10% in the "poor" zone. The validation results of the two models indicated that the prediction accuracy of the WOE model is about 73% and that of the FR model is about 68%. That is, the performance of the WOE model is better than the FR model in the spatial prediction of groundwater potential in a river basin. However, a major drawback of these models is that their application requires a large number of well records that may not be available in many basins or catchments. Also, if a large number of pumping wells or springs are already available in a basin or subbasin, the necessity of qualitative spatial prediction of groundwater potential by FR and WOE models is significantly undermined. Thus, the practical application of these models is greatly limited.

References

Barbieri, G. and P. Cambuli. 2009. The weight of evidence: Statistical method in landslide susceptibility mapping of the Rio Pardu Valley. Sardinia, Italy. In: *18th World IMACS/ MODSIM Congress*, Cairns, Australia, July 13–17, 2009.

Biswas, A.K., Tortajada, C., and R. Izquierdo. 2009. *Water Management in 2020 and beyond*. Springer, Berlin, Germany.

Bonham-Carter, G.F. 1994. *Geographic Information Systems for Geoscientists: Modeling with GIS*. Pergamon Press, Ottawa.

Chen, Y., Takara, K., Cluckie, I.D., and F.H.D. Smedt. 2004. *GIS and Remote Sensing in Hydrology, Water Resources and Environment*. IAHS Press, Wallingford. IAHS Publication No. 289

Chowdhury, A., Jha, M.K., and V.M. Chowdary. 2010. Delineation of groundwater recharge zones and identification of artificial recharge sites in West Medinipur district, West Bengal using RS, GIS and MCDM techniques. *Environmental Earth Sciences*, 59(6), 1209–1222.

Chowdhury, A., Jha, M.K., Chowdary, V.M., and B.C. Mal. 2009. Integrated remote sensing and GIS-based approach for assessing groundwater potential in West Medinipur district, West Bengal, India. *International Journal of Remote Sensing*, 30(1), 231–250.

Colwell, R.N. 1983. *Manual of Remote Sensing. Vols. I and II*. American Society of Photogrammetry, Falls Church, VA.

Corsini, A., Cervi, F., and F. Ronchetti. 2009. Weight of evidence and artificial neural networks for potential groundwater spring mapping: An application to the Mt. Modino area Northern Apennines, Italy. *Geomorphology*, 111, 79–87.

Engman, E.T. and R.J. Gurney. 1991. *Remote Sensing in Hydrology*. Chapman and Hall, London.

Franklin, J. 1986. Thematic mapper analysis of coniferous forest structure and composition. *International Journal of Remote Sensing*, 7, 1287–1301.

Galloway, D.L., Hudnut, K.W., and S.E. Ingebritsen et al. 1998. Detection of aquifer-system compaction and land subsidence using interferometric synthetic aperture radar, Antelope Valley, Mojave Desert, California. *Water Resources Research*, 34, 2573–2585.

Gogu, R.C., Carabin, G., Hallet, V., Peters, V., and A. Dassargues. 2001. GIS-based hydrogeological databases and groundwater modeling. *Hydrogeology Journal*, 9, 555–569.

Gossel, W., Ebraheem, A.M., and P. Wycisk. 2004. A very large scale GIS-based groundwater flow model for the Nubian sandstone aquifer in Eastern Sahara (Egypt, northern Sudan and eastern Libya). *Hydrogeology Journal*, 12(6), 698–713.

Grayman, W.M., Loucks, D.P., and L. Saito (editors). 2012. *Towards a Sustainable Water Future: Visions for 2050*. American Society of Civil Engineers (ASCE), Reston, VA.

Hinton, J.C. 1996. GIS and remote sensing integration for environmental applications. *International Journal of Geographical Information Systems*, 10(7), 877–890.

Hoffmann, J. 2005. The future of satellite remote sensing in hydrogeology. *Hydrogeology Journal*, 13(1), 247–250.

Hoffmann, J., Galloway, D.L., Zebker, H.A., and F. Amelung. 2001. Seasonal subsidence and rebound in Las Vegas Valley, Nevada; observed by synthetic aperture radar interferometry. *Water Resources Research*, 37, 1551–1566.

Jackson, T.J. 2002. Remote sensing of soil moisture: Implications for groundwater recharge. *Hydrogeology Journal*, 10, 40–51.

Jaiswal, R.K., Mukherjee, S., Krishnamurthy, J., and R. Saxena. 2003. Role of remote sensing and GIS techniques for generation of groundwater prospect zones towards rural development: An approach. *International Journal of Remote Sensing*, 24(5), 993–1008.

Jasrotia, A.S., Bhagat, B.D., Kumar, A., and R. Kumar. 2013. Remote sensing and GIS approach for delineation of groundwater potential and groundwater quality zones of Western Doon Valley, Uttarakhand, India. *Journal of Indian Society of Remote Sensing*, 41(2), 365–377.

Jha, M.K. 2010. Sustainable management of disasters: Challenges and prospects. In: M.K. Jha (editor), *Natural and Anthropogenic Disasters: Vulnerability, Preparedness and Mitigation*, Springer, Berlin, Germany, Chapter 26, pp. 598–609.

Jha, M.K. 2013. Sustainable management of groundwater resources in developing countries: Constraints and challenges. In: Mu. Ramkumar (editor), *On a Sustainable Future of the Earth's Natural Resources*, Springer, Berlin, Germany, Chapter 18, pp. 325–348.

Jha, M.K. and V.M. Chowdary. 2007. Challenges of using remote sensing and GIS in developing nations. *Hydrogeology Journal*, 15(1), 197–200.

Jha, M.K. and S. Peiffer. 2006. *Applications of Remote Sensing and GIS Technologies in Groundwater Hydrology: Past, Present and Future*. BayCEER, Bayreuth, Germany.

Jha, M.K., Chowdary, V.M., and A. Chowdhury. 2010. Groundwater assessment in Salboni Block, West Bengal (India) using remote sensing, geographical information system and multi-criteria decision analysis techniques. *Hydrogeology Journal*, 18, 1713–1728.

Jha, M.K., Chowdary, V.M., Kulkarni, Y., and B.C. Mal. 2014. Rainwater harvesting planning using geospatial techniques and multicriteria decision analysis. *Resources, Conservation and Recycling*, 83, 96–111.

Jha, M.K., Chowdhury, A., Chowdary, V.M., and S. Peiffer. 2007. Groundwater management and development by integrated remote sensing and geographic information systems: Prospects and constraints. *Water Resources Management*, 21(2), 427–467.

Jordan, J.D. and S.F. Shih. 1988. Use of remote sensing in abandoned well assessment. *Transactions of the American Society of Agricultural Engineers,* 31(5), 1416–1422.

Jordan, J.D. and S.F. Shih. 1991. Satellite and aerial photographic techniques for use in artesian well assessment. *Proceedings of the International Conference on Computer Application in Water Resources*, Taipei, Taiwan, 2, 991–998.

Kim, K.D., Lee, S., Oh, H.J., Choi, J.K., and J.S. Won. 2006. Assessment of ground subsidence hazard near an abandoned underground coal mine using GIS. *Environmental Geology*, 50, 1183–1191.

Kite, G.W., Pietroniro, A., and T. Pultz. 1997. Application of remote sensing in hydrology. *Third International Workshop, Goddard Space Flight Center*, Washington, DC, October 16–18, 1996, NHRI Symposium Series No. 17.

Krishnamurthy, J., Kumar, N.V., Jayaraman, V., and M. Manivel. 1996. An approach to demarcate groundwater potential zones through remote sensing and a geographic information system. *International Journal of Remote Sensing*, 17(10), 1867–1884.

Lee, S. and J. Choi. 2004. Landslide susceptibility mapping using GIS and the weight of evidence model. *International Journal of Geographical Information Science*, 18(8), 789–814.

Lee, S., Kim, Y.S., and H.J. Oh. 2012. Application of a weights-of-evidence method and GIS to regional groundwater productivity potential mapping. *Journal of Environmental Management*, 96, 91–105.

Lillesand, T.M. and R.W. Kiefer. 2000. *Remote Sensing and Image Interpretation*. 4th edition, John Wiley & Sons, New York.

Loague, K. and D.L. Corwin. 1998. Regional-scale assessment of non-point source groundwater contamination. *Hydrological Processes*, 12(6), 957–966.

Lowry, C.S., Walker, J.F., Hunt, R.J., and M.P. Anderson. 2007. Identifying spatial variability of groundwater discharge in a wetland stream using a distributed temperature sensor. *Water Resources Research*, 43, W10408, doi: 10.1029/2007WR006145.

Machiwal, D., Jha, M.K., and B.C. Mal. 2011. Assessment of groundwater potential in a semi-arid region of India using remote sensing, GIS and MCDM techniques. *Water Resources Management*, 25(5), 1359–1386.

Madrucci, V., Taioli, F., and C.A. Carlos. 2008. Groundwater favorability map using GIS multi-criteria data analysis on crystalline terrain, São Paulo State, Brazil. *Journal of Hydrology*, 357, 153–173.

McFarlane, M.J., Ringrose, S., Guisti, L., and P.A. Shaw. 1994. The origin and age of karstic depressions in the Darwin–Koolpinyah area, Northern Territory, Australia. In: A.G. Brown (editor), *Geomorphology and Groundwater*, Wiley, Chichester, pp. 93–120.

Meijerink, A.M.J. 2000. Groundwater. In: G.A. Schultz and E.T. Engman (editors), *Remote Sensing in Hydrology and Water Management*, Springer, Berlin.

Mooney, H., Cropper, A., and W. Reid. 2005. Confronting the human dilemma. *Nature*, 434(7033), 561–562.

Oh, H.J., Kim, Y.S., Choi, J.K., and S. Lee. 2011. GIS mapping of regional probabilistic groundwater potential in the area of Pohang City, Korea. *Journal of Hydrology*, 399, 158–172.

Oh, H.J., Lee, S., Chotikasathien, W., Kim, C.H., and J.H. Kwon. 2009. Predictive landslide susceptibility mapping using spatial information in the Pechabun area of Thailand. *Environmental Geology*, 57, 641–651.

Ozdemir, A. 2011. GIS-based groundwater spring potential mapping in the Sultan Mountains (Konya, Turkey) using frequency ratio, weights of evidence and logistic regression methods and their comparison. *Journal of Hydrology*, 411, 290–308.

Ringrose, S., Vanderpost, C., and W. Matheson. 1998. Evaluation of vegetative criteria for near-surface groundwater detection using multispectral mapping and GIS techniques in semi-arid Botswana. *Applied Geography*, 18(4), 331–354.

Sahoo, S., Jha, M.K., Kumar, N., and V.M. Chowdary. 2015. Evaluation of GIS-based multicriteria decision analysis and probabilistic modeling for exploring groundwater prospect. *Environmental Earth Sciences*, 74(3), 2223–2246. DOI 10.1007/s12665-015-4213-1.

Sander, P., Chesley, M.M., and T.B. Minor. 1996. Groundwater assessment using remote sensing and GIS in a rural groundwater project in Ghana: Lessons learned. *Hydrogeology Journal*, 4(3), 40–49.

Saraf, A.K. and P.R. Choudhury. 1998. Integrated remote sensing and GIS for groundwater exploration and identification of artificial recharge sites. *International Journal of Remote Sensing*, 19(10), 1825–1841.

Sener, E., Davraz, A., and M. Ozcelik. 2005. An integration of GIS and remote sensing in groundwater investigations: A case study in Burdur, Turkey. *Hydrogeology Journal*, 13(5–6), 826–834.

Sharma, P. and A. Anjaneyulu. 1994. Application of remote sensing and GIS in water resource management. *International Journal of Remote Sensing*, 14(17), 3209–3220.

Shaw, P.A. and J.J. de Vries. 1988. Duricrust, groundwater and valley development in the Kalahari of Southeast Botswana. *Journal of Arid Environments*, 14(7), 245–254.

Shih, S.F. and J.D. Jordan. 1990. Remote-sensing application to well monitoring. *Journal of Irrigation and Drainage Engineering*, ASCE, 116(4), 497–507.

Solomon, S. and F. Quiel. 2006. Groundwater study using remote sensing and geographic information system (GIS) in the central highlands of Eritrea. *Hydrogeology Journal*, 14(5), 729–741.

Srivastava, P.K. and A.K. Bhattacharya. 2006. Groundwater assessment through an integrated approach using remote sensing, GIS and resistivity techniques: A case study from a hard rock terrain. *International Journal of Remote Sensing*, 27(20), 4599–4620.

Su, Z., Fernàndez-Prieto, D., and J. Timmermans et al. 2014. First results of the earth observation water cycle multi-mission observation strategy (WACMOS). *International Journal of Applied Earth Observation and Geoinformation*, 26, 270–285.

Todd, D.K. 1980. *Groundwater Hydrology*. 2nd edition, John Wiley & Sons, New York, NY, pp. 111–163.

Tweed, S.O., Leblanc, M., Webb, J.A., and M.W. Lubczynski. 2007. Remote sensing and GIS for mapping groundwater recharge and discharge areas in salinity prone catchments, southeastern Australia. *Hydrogeology Journal*, 15(1), 75–96.

Van de Griend, A.A., Camillo, P.J., and R.J. Gurney. 1985. Discrimination of soil physical parameters, thermal inertia and soil moisture from diurnal surface temperature fluctuations. *Water Resources Research*, 21, 997–1009.

Vörösmarty, C.J., Green, P., Salisbury, J., and R.B. Lammers. 2000. Global water resources: Vulnerability from climate change and population growth. *Science*, 289(5477), 284–288.

Vrba, J. and A. Lipponen. 2007. *Groundwater Resources Sustainability Indicators*. UNESCO, Paris, France.

Waters, P., Greenbaum, P., Smart, L., and H. Osmaston. 1990. Applications of remote sensing to groundwater hydrology. *Remote Sensing Reviews*, 4(2), 223–264.

Watkins, D.W., McKinney, D.C., Maidment, D.R., and M. D. Lin. 1996. Use of geographic information systems in ground-water flow modeling. *Journal of Water Resources Planning and Management, ASCE*, 122(2), 88–96.

WWAP. 2009. *The United Nations World Water Development Report 3: Water in a Changing World*. UNESCO, Paris, France, and Earthscan, London, UK.

7

Hydraulic Fracturing from the Groundwater Perspective

Ruth M. Tinnacher, Dipankar Dwivedi, James E. Houseworth, Matthew T. Reagan,
William T. Stringfellow, Charuleka Varadharajan, and Jens T. Birkholzer

CONTENTS

7.1 Hydraulic Fracturing for Hydrocarbon Recovery

The extraction of shale gas from formations with lower permeabilities compared to other rock formations has became economical with the emergence of relatively newer technologies such as hydraulic fracturing (commonly known as "fracking") and precision drilling of wells (Arthur et al., 2008). The extraction of shale gas using hydraulic fracturing has increased estimates of natural gas resources enormously in many countries (Table 7.1), with an overall worldwide increase from 18 to 118 trillion cubic meters (Tcm) (Howarth et al., 2011).

Currently, there are only four countries—the United States (U.S.), Canada, China, and Argentina—that produce shale gas and shale oil commercially (U.S. Energy Information Administration: http://www.eia.gov/todayinenergy/detail.cfm?id=19991). China is estimated to possess the world's largest shale gas reserves followed by the U.S.; however, the U.S. is the largest producer of both shale gas and shale oil (Howarth et al., 2011). Owing to the high abundance of U.S. shale hydrocarbon resources, combined with its dominance in the production of these resources, the U.S. will be the primary focus in this chapter.

The main classes of reservoirs where hydraulic fracturing has been used intensively in the U.S. include very-low-permeability unconventional shale reservoirs and tight-gas sand reservoirs, accounting for over 73% of the hydraulic fracturing activity (Beckwith, 2010). Most of

TABLE 7.1

Estimates of Proven and Technically Recoverable Shale Gas
Resources in Trillion Cubic Meters (Tcm)

Countries	Proven Gas Reserves (Tcm)	Technically Recoverable Shale Gas Resources (Tcm)	Increase (Times)
Mexico	0.3	19	63.33
France	0.006	5	833.33
Argentina	0.4	22	55.0
China	3	36	12.0
Venezuela	5	0.3	0.06
Canada	1.8	11	6.1
USA	7.7	24.4	3.17

Source: Compiled from Howarth, R.W., A. Ingraffea, and T. Engelder.
2011. *Nature*, 477(7364), 271–275, doi:10.1038/477271a.

the unconventional shale reservoirs contain natural gas, with the exceptions of the Eagle Ford, which produces oil in the shallower portion of the formation, and the Bakken and Niobrara plays, which mainly contain oil. Shale gas resources are found in a variety of basins including the Barnett Shale (Fort Worth Basin, Texas), Haynesville Shale (East Texas and Louisiana), Antrim Shale (Michigan), Fayetteville Shale (Arkansas), New Albany Shale (Illinois Basin), Bakken Shale (North Dakota), and Marcellus Shale (Pennsylvania). The Marcellus Shale probably represents the most expansive shale gas play in the U.S., with recoverable reserves as large as 13.8 Tcm (Kargbo et al., 2010). According to a report submitted to the U.S. Energy Information Administration (EIA), technically recoverable shale oil resources amount to 23.9 billion barrels in the lower 48 states, with California providing the largest reserve amounting to 15.4 billion barrels (~64%) (U.S. Energy Information Administration: http://www.eia.gov/analysis/studies/usshalegas/pdf/usshaleplays.pdf). Currently, hydraulic fracturing is used to produce a significant amount (~20%) of oil and gas in California with most of the production occurring in the San Joaquin Valley. It is important to note that each shale formation has unique geologic, chemical, mineralogical, and physical properties (Arthur et al., 2008; Kargbo et al., 2010) that need to be considered for hydraulic fracturing operations.

7.2 Hydraulic Fracturing for Enhancing Groundwater Yields

Although hydraulic fracturing is primarily used for the extraction of hydrocarbons, this technique can also be applied to enhance groundwater yields from

sedimentary or crystalline-rock aquifers (Banks et al., 1996). Crystalline-rock aquifers, in this context, refer to fractured igneous and metamorphic rocks with negligible porosity and permeability. In these aquifers, groundwater flows through fractured crystalline rocks. Typically, one or two high-yielding fractures that are interconnected through a wider fracture network supply the vast majority of water from a successful borehole (Gustafson and Krásný, 1994). In sedimentary aquifers, water supply wells are frequently clogged due to chemical (e.g., mineral precipitation), physical (e.g., suspended solids), mechanical (e.g., gas entrapment), or biological (e.g., growth of algae) processes (Martin, 2013). Hydraulic fracturing artificially enhances fractures or cleans out non-water-producing veins that are clogged, thereby providing a clear transport pathway for groundwater flow into the well. Additionally, hydraulic fracturing is also used for effectively increasing groundwater recharge (Martin, 2013).

For these groundwater-related applications, hydraulic fracturing is a two-step process that involves the setting of packers, followed by high-pressure pumping (Figure 7.1a). First, the inflation of packers is used to seal the borehole hydraulically. Then, a high volume of water (~5000 gal) is pumped into the borewell at high pressure (~3000 psi) through a pipe connecting the packers. This process induces new fractures, or opens and clears previously obstructed fracture paths (Figure 7.1b) (Banjoko, 2014).

For the remainder of this chapter, we will focus on hydraulic fracturing in the context of enhanced hydrocarbon recovery, and its implications on groundwater

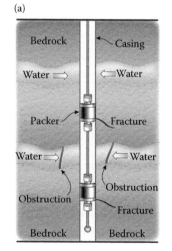 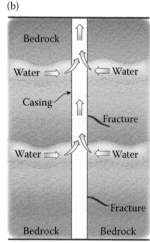

FIGURE 7.1

Hydraulic fracturing process for enhancing groundwater yields: (a) packers and packer piping are used to open and clear clogged fracture paths and (b) hydraulic fracturing results in enhanced groundwater flow into the bore well. (Adapted from Banjoko, B., *Handbook of Engineering Hydrology: Environmental Hydrology and Water Management*, CRC Press, Boca Raton, FL, 2014.)

resources. However, we would like to emphasize that hydraulic fracturing has tremendous potential to increase groundwater yields from drinking water wells. For this purpose, hydraulic fracturing has been used in many countries, including the U.S., Australia, India, and South Africa, given its low cost compared to drilling additional boreholes (Less and Andersen, 1994).

7.3 Hydraulic Fracturing Operations for Hydrocarbon Recovery

Hydrocarbon reservoirs that have insufficient natural permeability, which limits fluid flow for economic hydrocarbon recovery, are called *unconventional* reservoirs (California Council on Science and Technology [CCST] and Lawrence Berkeley National Laboratory [LBNL], 2015a). The general permeability levels used to distinguish between high- and low-permeability reservoirs cannot be precisely defined, but are generally in the range of about 10^{-15} square meters (m^2; about 1 millidarcy, md) (e.g., King, 2012). Low-permeability reservoirs are candidates for reservoir stimulation techniques such as hydraulic fracturing, which opens permeable flow paths between the undisturbed reservoir rock and the well such that it can provide economic rates of hydrocarbon production. Typically, unconventional shale reservoirs are relatively simple geologic systems with nearly horizontal deposition and layer boundaries, and much longer dimensions along the horizontal directions compared with the (usually) vertical dimension perpendicular to the bedding. Horizontal drilling combined with some form of hydraulic fracturing allows a well to access the reservoir over a longer distance than could be achieved with a traditional vertical well (McDaniel and Rispler, 2009).

There are two hydraulic fracturing techniques: hydraulic fracturing using proppant (traditional hydraulic fracturing) or acid (also known as acid fracturing) (Economides and Nolte, 2000). Traditional hydraulic fracturing is a stimulation technique that uses high-pressure fluid injection to create fractures in the rock and then fill the fractures with a granular material called proppant to retain the fracture openings after the fluid pressure is relieved. The large-permeability fractures then act as pathways for a hydrocarbon to flow through to the well. Acid fracturing is similar in that the fluid is injected under pressure to create fractures, but then acid is used to etch channels into the fracture walls to retain fracture permeability instead of injecting proppant.

Traditional hydraulic fracturing induces fractures by injecting fluid into the well until the pressure exceeds the threshold for fracturing. The induced fractures emanate from the well into the reservoir and provide a high-permeability pathway from the formation to the well. One of the goals of the fracturing operation is to only fracture rock within the target reservoir; if the hydraulic fracturing strays out of the low-permeability target zone, there will be a "short-circuiting" effect, as more permeable units will contribute to production fluids. During portions of a hydraulic fracture treatment, "proppant" (natural sand or manufactured ceramic grains) is generally pumped in the frac fluid to prop the fracture(s) open, to maintain fracture conductivity after the treatment is completed, and the well is put on production. The effective stress imposed on a fracture plane and the proppant within the fracture is the total stress perpendicular to the fracture plane *minus* the pore pressure within the fracture. The use of proppant becomes particularly important for maintaining fracture permeability as formation fluids, a load-supporting element of formation strength, are removed by production. The creation of a highly permeable fracture network allows for the effective drainage of a much larger volume of low-permeability rock, and thus increases the hydrocarbon flow rates and total recovery.

Another variation of hydraulic fracturing is called acid fracturing, where acid is injected instead of proppant, typically in strongly reactive carbonate reservoirs characterized by a high number of natural fractures and relatively high permeability (Economides et al., 2013). The acid etches channels into the fracture surfaces, which then prevent the natural overburden stress from closing the fractures and allows fluid-flow pathways to remain along the fractures even after the injection pressure is removed. Industry comparisons of stability of propped fractures to that of acid fractures indicate that propped fractures are usually more stable over time, especially in sandstones and soft carbonates (Abass et al., 2006). Because these methods do not reduce viscosity, they are primarily targeted at rock formations containing gas or lower-viscosity oil, although they may be used with thermal stimulation for heavy oil.

7.3.1 Typical Phases in Hydraulic Fracturing Operations

The typical hydraulic fracture operation involves four process steps to produce the fractures (Arthur et al., 2008; CCST and LBNL, 2015a). The long production intervals present in most horizontal wells lead to a staged approach to hydraulic fracturing. In this staged approach, a portion of the well is hydraulically isolated in order to concentrate the injected fracture fluid pressure on an isolated interval, which is called a "stage." After isolating the stage, the first phase of the fracturing process is the "pad," in which fracture fluid is injected without proppant to initiate and propagate the fracture from the well. The second phase adds proppant to the injection fluid; the proppant is needed to keep the fractures open after the fluid pressure dissipates. This phase is also used to further open the hydraulic fractures. The third phase, termed the "flush," entails injection of fluid without proppant to push the remaining proppant in the well into the fractures. The fourth phase is the "flowback," in which the hydraulic fracture fluids are removed from the formation, and fluid pressure dissipates. The term "flowback water/fluids" operationally refers to fluids recovered at the surface after well stimulation is completed, and before a well is placed into production. The duration of flowback periods can vary regionally as well as between operators within a region ranging anywhere from 2 days to a few weeks (e.g., Hayes, 2009; Stepan et al., 2010; Barbot et al., 2013; Warner et al., 2013). Once the well is placed into production, the water recovered from the operations is operationally defined as "produced water."

7.3.2 Hydraulic Fracture Geomechanics and Fracture Geometry

Fractures are formed when the pressure of the injection fluid exceeds the existing minimum rock compressive stress by an amount greater than the tensile strength of the rock (Thiercelin and Roegiers, 2000). The orientation of the hydraulic fractures cannot be controlled by the operator, but is rather determined by stress conditions in the rock. The hydraulic fracture will preferentially push open against the least-compressive stress for a rock with the same strength in all directions (Economides et al., 2013). Therefore, the fracture plane develops in the direction perpendicular to the minimum compressive stress. Stress field orientation, however, can and does vary with time in producing oilfields as a result of fluid injections and withdrawals (Minner et al., 2002).

Natural fractures are generally present to some degree in natural rock and affect the formation of hydraulic fractures, for instance resulting in fracture lengths that are greater than fracture heights (Weijers et al., 2005; Fisher and Warpinski, 2012). Natural fracture features of the rock are often the flow pathways from the reservoir to the hydraulic fractures (Weijermars, 2011). The contact area developed by opening natural fractures is considerably larger than what can be achieved by planar fractures. However, the filling of natural fractures with secondary calcite or quartz may make them more or less susceptible to reopening during fracturing operations (Gale and Holder, 2010).

Typically, most rocks at depth experience greater vertical than horizontal stress, which favors the formation of hydraulic fractures that are vertical. Consequently, the question of the vertical fracture height growth is important when considering the potential migration

of fracture fluid or other reservoir fluids out of the typically very-low-permeability target oil reservoir (see Section 7.6 for a detailed discussion). Overall, fracture height is limited by a number of mechanisms, including variability of _in situ_ stress, material property contrasts across layered interfaces, weak interfaces between layers, leakoff of fracturing fluid into formations, and the volume of fracture fluid required to generate extremely large fracture heights (Fisher and Warpinski, 2012). Finally, at shallow depths (305–610 m or 1000–2000 ft), the minimum stress is typically in the vertical and not the horizontal direction, like at greater depths. The latter favors a horizontal fracture orientation, and generally prevents vertical fracture growth from deeper into shallower depths (Fisher and Warpinski, 2012), but exceptions are possible (Wright et al., 1997).

In addition, fracture height growth and hydraulic fracture development may also be affected by the interaction of hydraulic fracture fluids with faults (Flewelling et al., 2013; Rutqvist et al., 2013), the presence of neighboring wells (King, 2010), the magnitude of the stimulation pressure, and the fracturing fluid viscosity, which are discussed in further detail below.

7.4 Composition and Potential Hazards of Injection Fluids

Understanding the composition, or formulations, of hydraulic fracturing fluids is an important step in defining the upper limits of potential direct environmental impacts from hydraulic fracturing and other well stimulation technologies. The amounts of chemicals added to injection fluids define the maximum possible mass and concentrations of chemical additives that can be released to contaminate groundwater or otherwise impact the environment. Chemical additives in hydraulic fracturing fluids may also influence the release of metals, salts, and other materials from rocks and sediments found naturally in oil- and gas-bearing geological formations.

Recent studies, discussed next, have increased our understanding of the types and numbers of chemical additives used in hydraulic fracturing and are informing the development of groundwater monitoring plans and other activities directed at protecting underground sources of drinking water (e.g., Esser et al., 2015; U.S. EPA, 2015a). However, extensive knowledge gaps on the effects of chemical mixtures and the environmental exposures associated with hydraulic fracturing operations still exist. In the past, chemical disclosures were largely voluntary and, due to the economic value of individual injection fluid formulations and the competition between companies, operators and service companies were often reticent about releasing detailed information concerning the types and amounts of chemicals used in specific formulations. This lack of transparency has heightened uncertainty and concerns not only about the chemicals used in fracturing fluids, but also for the assessment of hazards and risks to groundwater resources.

A primary source of public data regarding the composition of hydraulic fracturing fluids has been the FracFocus Chemical Disclosure Registry (http://fracfocus.org/), which is based on voluntary disclosures made by the oil and gas industry. This database contains records of individual hydraulic fracturing operations, typically listing all chemicals used in the treatment. A complete treatment record includes information about the volume of water used as a "base fluid" and the concentration of each chemical used in percent of total treatment fluid mass. However, since these disclosure records are not required to go through any form of independent review or confirmation process, the released information may often be incomplete (e.g., Konschnik et al., 2013). For instance, in a recent review of hydraulic fracturing operations in California, it was concluded that the information in voluntary disclosures that describe the purpose of a specific additive in the hydraulic fracturing fluid was very frequently inaccurate or misleading (Chapter 2 in CCST and LBNL, 2015b). Furthermore, while a total of 5000–7000 hydraulic fracturing treatments has been estimated for the state of California for the time frame between 2011 and 2014, voluntary disclosure records covered only one-fifth to one-third of these treatments (Chapter 2 in CCST and LBNL, 2015b). Nevertheless, chemical additives reported in voluntary disclosures were considered consistent with additives described in industry literature, patents, scientific publications, and other sources, such as government reports (e.g., Gadberry et al., 1999; U.S. EPA, 2004; Baker Hughes Inc., 2011, 2013; Stringfellow et al., 2014), and hence representative for the chemical use in hydraulic fracturing operations in California (Chapter 2 in CCST and LBNL, 2015b).

In many regions of the U.S., new laws are now requiring mandatory reporting of the composition of the injection fluids. The objectives are to gain a better understanding of the risks associated with hydraulic fracturing and other well stimulation technologies, and to provide a scientific basis for the development of regulatory frameworks for hydraulic fracturing operations on the state level. For instance, driven by regulations under the new California Senate Bill 4 (SB 4), operators are required to apply for well stimulation permits (via "Well Stimulation Notices") and report on the completion of well stimulations (in "Well Stimulation Treatment Disclosure Reports") (DOGGR, 2014a,b,c).

Overall, the potential risks associated with any individual chemical are a function of the hazardous properties of the material, how the material is released into the environment, how much material is released, the persistence of the compound in the environment, and many other properties and variables that allow a pathway to human or environmental receptors. Therefore, a range of information on hazards, toxicology, and other physical, chemical, and biological properties of these chemicals is needed in order to understand the potential environmental and health risks associated with hydraulic fracturing operations in general, and for the handling of injection fluids specifically. As described in detail next, the specific characteristics of geologic reservoirs, local geology, and rock types govern the technologies involved in hydraulic fracturing operations, and in turn, the selection of effective chemical additives in injection fluids. Therefore, broad generalizing statements regarding the composition, use, and associated hazards of hydraulic fracturing fluids across the U.S. should be avoided.

7.4.1 Technical Considerations for the Selection of Hydraulic Fracturing Fluids

The design of a hydraulic fracture requires a specification of the hydraulic fracturing fluid composition (CCST and LBNL, 2015a). While there are many additives used in hydraulic fracture fluids, most of these are used to mitigate adverse chemical and biological processes, and are the same as those used in drilling. The viscosity* of the fluid, however, is one of its most important characteristics influencing (1) the mechanics of fracture generation, and the resulting fracture length and fracture-network complexity, and (2) the ability of the fluid to emplace proppant efficiently.

The lowest-viscosity fracturing fluid is slickwater, which contains a friction-reducing additive (commonly polyacrylamide) and has a viscosity on the order of 4 cP (about 4 times that of pure water) (Kostenuk and Browne, 2010). Gelled fracture fluids generally use guar gum or cellulose polymers to increase the viscosity (King, 2012). Further increases in viscosity in a guar-gelled fluid can be achieved by adding a cross-linking agent to the gel that is typically a metal ion, such as in boric acid or zirconium chelates (Lei and Clark, 2004). The cross-linking binds the gel's polymer molecules into larger molecules, causing an increase in the solution viscosity. Linear gels have viscosities about 10 times that of slickwater, and cross-linked gels have viscosities that are on the order of 100–1000 times larger (Montgomery, 2013). Furthermore, fracturing fluids can be "energized" with nitrogen and surfactant to create foams, which leads to increased

---

* Viscosity is a fluid property that quantifies resistance to fluid flow. It takes little effort to stir a cup of water (viscosity ~1 centipoise [cP]), noticeably more effort to stir a cup of olive oil (viscosity ~100 cP), and significantly more effort to stir a cup of honey (viscosity ~10,000 cP).

viscosities of the energized fluids compared to the original linear or cross-linked gels (Harris and Heath, 1996; Ribeiro and Sharma, 2012).

Both laboratory and field data indicate that low-viscosity fracture fluids tend to create complex fractures with a large fracture–matrix area and narrow fracture apertures—as compared to higher-viscosity fracture fluids, which tend to create simpler planar-style fractures with a low fracture–matrix area and wide fracture apertures, as illustrated in Figure 7.2 (Rickman et al., 2008; Cipolla et al., 2010). Fracture lengths increase with the volume of injected fracture fluid, and are roughly proportional to fracture heights with proportionality factors ranging from 0.5 to 1 (Flewelling et al., 2013), while stratigraphic limitations on fracture height growth may also play a role (Weijers et al., 2005; Fisher and Warpinski, 2012). Fracture apertures (or widths) are typically on the order of a few tenths of an inch (~8 mm) (Barree et al., 2005; Shapiro et al., 2006; Bazan et al., 2012) and tend to increase with viscosity, rate, and volume of the fluid injected (Economides et al., 2013).

With regard to transporting proppant, cross-linked gels are generally more effective than slickwater (Lebas et al., 2013), while the effective viscosity may also be influenced by the proppant concentration itself (Montgomery, 2013).

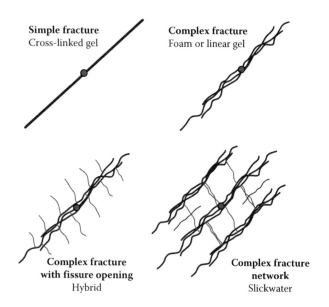

Simple fracture
Cross-linked gel

Complex fracture
Foam or linear gel

Complex fracture with fissure opening
Hybrid

Complex fracture network
Slickwater

FIGURE 7.2
Effects of fracture fluid viscosity on fracture complexity. (Modified from Warpinski et al., 2009. *Journal of Canadian Petroleum Technology.* 48(10): 39–51; California Council on Science and Technology (CCST), and Lawrence Berkeley National Laboratory (LBNL). 2015a. *An Independent Scientific Assessment of Well Stimulation in California: Volume I, Well Stimulation Technologies and Their Past, Present, and Potential Future Use in California.* Sacramento, CA. Available from: http://ccst.us/SB4.)

The selection of fluid composition also depends on the properties of the reservoir rock, specifically the rock permeability and brittleness (Rickman et al., 2008; Cipolla et al., 2010). Formations with relatively higher intrinsic permeability are generally stimulated using a higher-viscosity fracture fluid to create a simpler and wider fracture (Cipolla et al., 2010). The rationale for this selection is that the fracture is needed both to increase the contact area with the formation and to provide a high conductivity flow path toward the wellbore. As reservoir permeability decreases, the resistance to fluid movement through the unfractured portion of the formation increases. Therefore, a denser fracture pattern, with narrower spacing between the fractures, is needed to minimize the distance that reservoir fluids must flow in the rock matrix to enter the hydraulically induced fractures (Economides et al., 2013). This leads to the application of lower-viscosity fracturing fluids to create more dense (and complex) fracture networks. With respect to rock brittleness, wider fracture apertures are needed as rock brittleness decreases (or as ductility increases) because of the greater difficulty maintaining fracture permeability after pressure is withdrawn (Rickman et al., 2008).

The general trends in fracture fluid types, fluid volumes used, and fracture complexity as a function of rock properties are summarized in Figure 7.3. Overall, a ductile, relatively higher permeability reservoir rock with a low natural fracture density tends to receive a hydraulic fracture treatment using low volumes of viscous cross-linked gels, but with a large concentration and total mass of proppant. This reservoir type responds by producing a simple single fracture from the well into the rock that has a relatively large aperture filled with proppant. As rock brittleness and degree of natural fracturing increase, and as permeability decreases, hydraulic fracturing treatments tend to use a higher-volume, lower-viscosity fracture fluid that carries less proppant. This results in more complex fracture networks, with fractures of narrower apertures and a more asymmetric cross section in the vertical direction as a result of proppant settling. In short, ductile and more permeable rocks usually receive gel fracture treatments, while more brittle, lower-permeability rocks with existing fractures are more amenable to slickwater fracturing.

In acid-fracturing treatments, HCl, formic acid, acetic acid, or blends thereof are typically used to etch the faces of the fracture surfaces (Kalfayan, 2008). The presence of the etched channels allows fractures to remain permeable even after the fracture-fluid pressure is removed and compressive rock stress causes the fractures to close (Economides et al., 2013). However, acid fracturing generally results in relatively short fractures as compared with fractures secured with proppant (Economides et al., 2013).

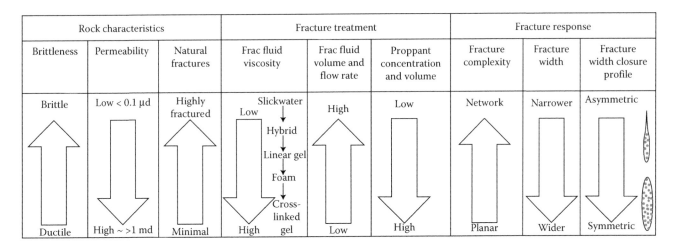

FIGURE 7.3
General trends in rock characteristics, hydraulic fracture treatment applied, and hydraulic fracture response. (Modified from Rickman, R., M. Mullen, E. Petre, B. Grieser, and D. Kundert. 2008. A practical use of shale petrophysics for stimulation design optimization: All shale plays are not clones of the Barnett Shale. In: *SPE 115258 Annual Technical Conference and Exposition*, pp. 1–11, Society of Petroleum Engineers, Denver, CO; California Council on Science and Technology (CCST), and Lawrence Berkeley National Laboratory (LBNL). 2015a. *An Independent Scientific Assessment of Well Stimulation in California: Volume I, Well Stimulation Technologies and Their Past, Present, and Potential Future Use in California*. Sacramento, CA. Available from: http://ccst.us/SB4.)

7.4.2 Composition of Injection Fluids

Previous studies have evaluated and characterized chemical additives to well stimulation fluids that are in common use in the U.S. nationally (U.S. House of Representatives Committee on Energy and Commerce, 2011; King, 2012; U.S. EPA, 2012, 2015b; Stringfellow et al., 2014), or specifically in the state of California (Chapter 2 in CCST and LBNL, 2015b). As a whole, fracturing fluids can contain hundreds of chemicals, which can be initially bewildering, even to experts. Hence, it is helpful to understand the significance of individual chemicals or chemicals in mixtures in the context of their purpose and frequency of use, the amounts used, and their hazardous properties, such as toxicity.

An overview of additives in fracturing fluids, categorized based on their various functions during the fracturing process, is provided in Table 7.2. As described above, sand is typically used as a "proppant" that ensures that the newly created fractures remain open. Other compounds are added to facilitate the development of the desired fracture geometry, to provide an efficient delivery of the proppant to the areas of interest underground, to prevent the growth of bacteria, and to minimize the scaling of the well. Classes of relevant chemicals include gelling and foaming agents, friction reducers, cross-linkers, breakers, pH adjusters/buffers, biocides, corrosion inhibitors, scale inhibitors, iron control chemicals, clay stabilizers, and surfactants (NYSDEC, 2011; King, 2012; Stringfellow et al., 2014). Over 80% of the treatments use an identified biocide and many formulations also include chemicals such as clay control additives.

While in sum a large number of different chemicals may be used in hydraulic fracturing operations across the U.S., the actual number of additives used for each single treatment may be much lower. For instance, over 300 unique chemicals were identified as being used in hydraulic fracturing fluids in California; however, in an individual treatment, a median of only 23 individual components—including base fluids (such as water), proppants, and chemical additives—were applied (Chapter 2 in CCST and LBNL, 2015b). Furthermore, nationwide, only between 10 and 20 chemical additives were used per hydraulic fracturing treatment, not including proppants and base fluids (U.S. EPA, 2015b).

7.4.3 Potential Hazards of Injection Fluids

Potential hazards of fracturing fluid constituents are based on their characteristics in terms of corrosivity, ignitability, and chemical reactivity, as well as on their potential toxic effects on humans and the environment. With regard to potential adverse effects on humans, mammalian toxicity values are often considered a first proxy in evaluations. Furthermore, given the hydraulic link between groundwater and surface water, potential toxic effects on aquatic species also need to be considered.

In two studies (Stringfellow et al., 2014; Chapter 2 in CCST and LBNL, 2015b), described in further detail next, chemicals were ranked and classified with regard to their toxic impacts on mammals and aquatic organisms in the context of the U.N. Globally Harmonized System (GHS). In the GHS system, lower numbers

TABLE 7.2

Additives to Aqueous Fracturing Fluids

Additive Type	Description of Purpose	Examples of Chemicals
Proppant	"Props" open fractures and improves gas/fluid flow to the well bore.	Sand (sintered bauxite; zirconium oxide; and ceramic beads)
Acid	Removes cement and drilling mud from casing perforations prior to fracturing fluid injection.	Hydrochloric acid (3%–28%) or muriatic acid
Breaker	Reduces fluid viscosity to release proppant into fractures and enhance fracturing fluid recovery.	Peroxydisulfates
Bactericide/biocide/ antibacterial agent	Inhibits growth of organisms that could produce gases (e.g., hydrogen sulfide) and contaminate methane gas. Prevents bacterial growth that can negatively affect proppant delivery into fractures.	Gluteraldehyde; 2,2-dibromo-3-nitrilopropionamide
Buffer/pH adjusting agent	Adjusts and controls fluid pH to maximize the effectiveness of other additives, such as cross-linkers.	Sodium or potassium carbonate; acetic acid
Clay stabilizer/control/KCl	Minimizes swelling and mobility of formation clays, which could reduce permeability by blocking pore spaces.	Salts (e.g., tetramethyl ammonium chloride, potassium chloride [KCl])
Corrosion inhibitor (including oxygen scavengers)	Reduces rust formation on steel tubing, well casings, tools, and tanks (used only in fluids containing acid).	Methanol; ammonium bisulfate for oxygen scavengers
Cross-linker	Increases fluid viscosity using phosphate esters combined with metals (referred to as cross-linking agents) to carry more proppant into fractures.	Potassium hydroxide; borate salts
Friction reducer	Allows fracture fluids to be injected at optimum rates and pressures by minimizing friction.	Sodium acrylate–acrylamide copolymer; polyacrylamide (PAM); and petroleum distillates
Gelling agent	Increases fracturing fluid viscosity, to emplace proppant efficiently.	Guar gum; petroleum distillates
Iron control	Controls the precipitation of metal oxides, which could plug off the formation.	Citric acid
Scale inhibitor	Prevents the precipitation of carbonates and sulfates (calcium carbonate, calcium/barium sulfate), which could decrease permeability.	Ammonium chloride; ethylene glycol
Solvent	Additive, soluble in oil, water-, and acid-based fluids, that allow to control the wettability of contact surfaces, or to prevent or break emulsions.	Various aromatic hydrocarbons
Surfactant	Reduces fracturing fluid surface tension and improves fluid recovery.	Methanol; isopropanol; and ethoxylated alcohol

Source: Adapted from NYSDEC, 2011. Revised Draft Supplemental Generic Environmental Impact Statement on the Oil, Gas and Solution Mining Regulatory Program—Well Permit Issuance for Horizontal Drilling and High-Volume Hydraulic Fracturing to Develop the Marcellus Shale and Other Low-Permeability Gas Reservoirs. Albany, NY. Retrieved from http://www.dec.ny.gov/data/dmn/rdsgeisfull0911.pdf.

indicate greater toxicity, with "1" representing the most toxic compounds.

For U.S.-wide applied constituents in fracturing fluids, a recent, extensive review by Stringfellow et al. (2014) provides a summary of mammalian toxicity values (LD_{50} values) for additives predominantly used in the U.S. A list of 81 chemicals was compiled after analyzing the FracFocus Chemical Disclosure Registry website (http://fracfocus.org), the SkyTruth database (http://frack.skytruth.org), a relevant U.S. EPA report (U.S. EPA, 2004), and textbook (Economides and Nolte, 2000), as well as published literature from industry sources (Nishihara et al., 2000; Gartiser and Urich, 2003; Miller, 2007; Arthur et al., 2008; Halliburton Energy Services, 2009). After evaluating inhalation and oral toxicity data from the literature, the authors concluded that most of the listed fracturing fluid constituents were of low toxicity; only four chemicals were classified as Category 2 *oral* toxins by U.N. standards. However, for 34 additives, specific toxicity information could not be located (Stringfellow et al., 2014).

Furthermore, a recent report, with a specific focus on hydraulic fracturing operations in the state of California (Chapter 2 in CCST and LBNL, 2015b), provides a similar overview for acute aquatic toxicity data (Figure 7.4). Thirty-three chemicals with reported use in hydraulic fracturing treatments in California have a GHS ranking of 1 or 2 for at least one aquatic species. This indicates that these compounds are hazardous to aquatic species and could potentially present a risk to the environment if released. However, significant data gaps exist for these test species with 65%, 76%, and 79% of chemical toxicity

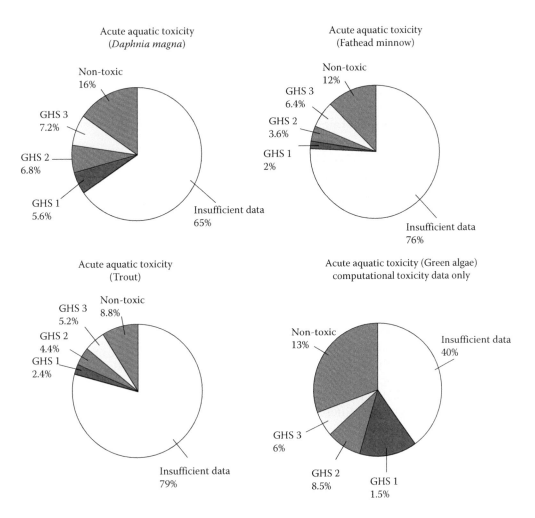

FIGURE 7.4
Aquatic toxicity data for hydraulic fracturing and acid treatment chemicals in California (Chapter 2 in CCST and LBNL, 2015b), categorized according to the U.N. GHS of classification and labeling of chemicals with a scale from 1 to 3 (1 representing the highest toxicity).

data missing for *Daphnia magna*, fathead minnows, and trout, respectively. Furthermore, there is a large lack of data for algal toxicity (60%) and even for simulated data from toxicity models developed by the U.S. EPA (Estimation Program Interface [EPI] Suite: http://www.epa.gov/opptintr/exposure/pubs/episuite.htm.), which are simulation-based predictions of toxic effects.

While the studies described above were focused on acute lethality to aquatic organisms or mammals, sublethal impacts from acute or chronic exposures—such as impacts on reproduction and development, physiological status, disease or debilitation, and avoidance or migratory behavior—are also important to population viability (U.S. EPA, 1998, 2003). For instance, hazardous chemicals may include carcinogens (substances that can cause cancer), endocrine disrupting compounds (chemicals that may interfere with the body's endocrine system and produce adverse developmental, reproductive, neurological, and immune effects), and bioaccumable materials (chemicals that

increase in concentration in a biological organism over time compared to their environmental concentrations). At this point, there is a significant lack of data on chronic and sublethal impacts of chemicals used in hydraulic fracturing treatments. However, experimental results from a recent study focused on unconventional gas operations in Colorado suggest that natural gas drilling chemicals can have harmful effects on reproductive organ development in fetuses exposed *in utero* as well as their offspring (Kassotis et al., 2013). Similarly, little is known about potential bioaccumulation effects of hydraulic fracturing chemicals due to a lack of basic physical and chemical characterization data and results from standardized biodegradation tests (Chapter 2 in CCST and LBNL, 2015b). Some additives, however, such as surfactants, related compounds known as quaternary ammonium compounds (QACs), and halogenated biocides, are expected to be fairly persistent in the environment (Chapter 2 in CCST and LBNL, 2015b). Additional relevant data gaps regarding the potential toxicological impacts of hydraulic fracturing chemicals include (1) the response of organisms to the mixture of chemicals in the fluid in contrast to the response to a single chemical (which may be different due to chemical interactions between compounds), and (2) potential changes in fluid compositions due to dilution effects and interactions with rock formations in the subsurface.

7.5 Composition of Flowback and Produced Waters

Flowback fluids consist of (1) injection fluids pumped into the well previously, which include water, proppants, and additives as described above, (2) new compounds that may have formed due to chemical reactions between additives, (3) dissolved substances from waters naturally present in the shale formation, (4) substances that have become mobilized from the shale formation due to the well stimulation treatment, and (5) oil and/or gas (Stepan et al., 2010; NYSDEC, 2011). Thus, the water chemistry of flowback waters is typically different from that of the injection fluids.

Furthermore, the composition of flowback fluids often changes over the course of the flowback time period, and is expected to gradually evolve from being more similar to the injection fluids to approaching the chemical characteristics of the formation waters. For example, changes in fluid composition were observed in studies conducted in the Marcellus Shale (Hayes, 2009; Barbot et al., 2013) and the Bakken (Stepan et al., 2010), indicating concentration increases for constituents such as total dissolved solids (TDS), chloride, and some cations/metals over time. In the Marcellus study, water hardness and radioactivity levels were found to increase during the flowback period, but sulfate and alkalinity levels decreased with time.

Once the well is put on production, the waters recovered from the operations are "operationally-defined" as "produced waters." A limited amount of data on compositions of produced water from hydraulic fracturing operations is available in the literature (Table 7.3) and from the USGS-Produced Water Database (Blondes et al., 2014). Some constituents present in produced waters from hydraulically fractured wells are similar to those present in produced water from conventional oil and gas wells (e.g., Benko and Drewes, 2008; Alley et al., 2011). The most concentrated constituents measured in produced water from both conventional and unconventional wells are typically salts, that is, sodium and chloride (Blauch et al., 2009; Warner et al., 2012; Barbot et al., 2013; Haluszczak et al., 2013; CCST and LBNL, 2015b).

TABLE 7.3

Comparison of Constituent Concentrations between Produced Waters from Hydraulic Fracturing and Conventional Oil Operations

Parameter	Marcellus[a]	Bakken[b]	Conventional Oil[c]
pH	5.1–8.4	5.5–6.5	5.2–8.9
Conductivity (mS/cm)		205–221	
Alkalinity (mg/L as $CaCO_3$)	8–580		300–380
TSS (mg/L)	4–7600		
TDS (mg/L)	680–350,000	150,000–219,000	
Chloride (mg/L)	64–200,000	90,000–130,000	36–240,000
Sulfate (mg/L)	0–2000	300–1000	
Bicarbonate (mg/L)	0–760	300–1000	8–14,000
Bromide (mg/L)		300–1000	1–2
Nitrate (mg/L)	5–800		0–92
Oil and grease– HEM (mg/L)	195–37,000		
COD (mg/L)	1–1500		
TOC (mg/L)			15–3500
Aluminum (mg/L)		ND	0.0–0.1
Arsenic (mg/L)			0.2–0.9
Barium (mg/L)	0–14,000	0–25	0.1–7.4
Boron (mg/L)		40–190	
Calcium (mg/L)	38–41,000	7500–14,000	4–53,000
Cadmium (mg/L)			0.0–0.2
Chromium (mg/L)			0.1–1.0
Copper (mg/L)		ND	0.3–2.7
Iron (mg/L)	3–320	ND	0.1–0.5
Potassium (mg/L)		0–5800	2–43
Magnesium (mg/L)	17–2600	630–1800	2–5100
Manganese (mg/L)		4–10	1–8
Sodium (mg/L)	69–120,000	47,000–75,000	400–127,000
Nickel (mg/L)			3–10
Strontium (mg/L)	1–8500	520–1000	0–2
Zinc (mg/L)		2–11	6–17
Ra-226 (pCi/L)	3–9300		0–10
Ra-228 (pCi/L)	0–1400		
U-235 (pCi/L)	0–20		
U-238 (pCi/L)	0–500		
Gross alpha (pCi/L)	37–9600		
Gross beta (pCi/L)	75–600,000		

Source: Modified after California Council on Science and Technology (CCST), Lawrence Berkeley National Laboratory (LBNL), and Pacific Institute. 2014. *Advanced Well Stimulation Technologies in California: An Independent Review of Scientific and Technical Information.* Sacramento, CA. Available from: http://ccst.us/BLMreport.

Note: Numbers have been rounded to two significant digits.

[a] Barbot et al. (2013).
[b] Stepan et al. (2010).
[c] Alley et al. (2011).

Magnesium and calcium can also be present at high levels and can contribute to increased water hardness.

Formation brines can also contain high concentrations of trace elements, such as boron, barium, strontium, and heavy metals, which may be brought up to the surface in the produced water. For example, several studies report measuring high levels of trace elements (e.g., barium, strontium, iron, arsenic, and selenium) in waters recovered from fracturing operations in the Marcellus Shale (e.g., Hayes et al., 2009; Balaba and Smart, 2012; Barbot et al., 2013; Haluszczak et al., 2013). Produced waters from oil and gas operations also contain many organic substances, for example, organic acids, polycyclic aromatic hydrocarbons (PAHs), phenols, benzene, toluene, ethylbenzene, xylenes (BTEXs), and naphthalene (e.g., Fisher and Boles, 1990; Higashi and Jones, 1997; Veil et al., 2004). Wastewaters from some shale formations have been found to contain high levels of naturally occurring radioactive materials (NORMs), which were several hundred times above U.S. drinking water standards (NYSDEC, 2009; Rowan et al., 2011; Barbot et al., 2013; Haluszczak et al., 2013). Concentration ranges for these constituents during hydraulic fracturing operations in the Marcellus and Bakken Shales are listed in Table 7.3.

7.6 Potential Pathways for Groundwater Contamination

Well stimulation and associated activities may result in the release of contaminants into the environment, including surface and groundwater resources. Releases can occur during chemical transport, storage, mixing, well stimulation, well operation and production, and wastewater storage, treatment, and disposal (Figure 7.5). A *physical connection* or "transport pathway," either natural or induced, between the release location and the impacted surface or groundwater body is required for releases to occur, along with a *driving force* (e.g., pressure difference, buoyancy) for contaminant migration. The probability of contaminant migration is regulated by a number of factors, including reservoir depth, the physical and hydrological properties of the formation, reservoir production strategies, drilling and casing practices, plus the unique geologies of each oil and gas-producing region.

Surface releases are typically easier to identify and associate with a particular activity, while subsurface releases are generally more difficult to detect, associate with a particular release mechanism, and mitigate. Reservoir and stimulation fluids can also migrate through the subsurface if surface releases percolate into groundwater, produced water is directly injected into

protected groundwater, or transport pathways have been created through stimulation operations, either through direct fracturing into overlying aquifers or connection to a preexisting pathway (e.g., a fault or some other permeable feature). While transport through preexisting or induced subsurface pathways has been documented in conventional oil and gas operations, it is not known whether stimulation is likely to exacerbate these concerns. Data concerning such release mechanisms are currently limited, and ongoing assessments by the U.S. Environmental Protection Agency (U.S. EPA) have not been conclusive (U.S. EPA, 2012, 2015a).

In 2015, the U.S. EPA released an external draft assessment of the impact of hydraulic fracturing operations on drinking water resources (U.S. EPA, 2015a). The scope of this investigation included all aspects of hydraulic fracturing operations, such as water use, off-site chemical handling, the injection process itself, the handling of flowback and produced waters, and wastewater treatment and disposal, but was limited to impacts on drinking water resources only. According to the U.S. EPA, hydraulic fracturing has occurred in at least 25 states, with Texas, Colorado, Pennsylvania, and North Dakota experiencing the highest activities. The study noted that—between 2011 and 2014—25,000–30,000 wells had been drilled in the U.S. annually. A total of 9.5 million people and 6800 drinking water sources, serving a population of 8.6 million people, were located within one mile of a fractured well.

According to this study (U.S. EPA, 2015a), both above- and below-ground mechanisms exist, associated with hydraulic fracturing operations, that have the potential to impact drinking water resources. These mechanisms include subsurface pathways (fracturing directly into aquifers, subsurface migration), surface spills, inadequate treatment of wastewater, and increased use of groundwater for fracturing operations. The importance of deteriorating wells, improper surface casings, and poor well construction were all implicated, in agreement with the later, more detailed discussion included in this chapter.

However, this U.S. EPA study did not provide evidence of widespread or systemic impacts of hydraulic fracturing operations on drinking water resources. While specific incidents were found to have affected drinking water wells, they had only limited or local impacts. For instance, on a local scale, below-ground movement of fluids, and cases of fractured fluids being injected directly into water resources have occurred. Furthermore, only a limited number of cases resulting in impacts could be identified relative to the overall number of hydraulically fractured wells. The latter could be due to the unlikeliness of such impacts, but it could also be an indication for insufficient data, the lack of long-term monitoring studies, the presence of

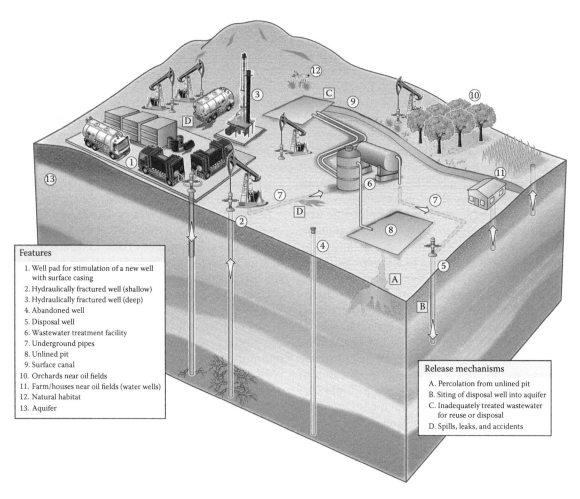

FIGURE 7.5

Potential surface and near-surface contaminant release mechanisms related to stimulation, production, and wastewater management and disposal activities (Chapter 2 in CCST and LBNL, 2015b).

preexisting contamination, and a general inaccessibility of data on hydraulic fracturing operations.

In this section, we discuss the set of plausible processes that might create transport pathways and the related driving forces. The mechanisms include processes related to drilling, well completion, hydraulic fracturing, and production: (1) leakage through hydraulic fractures, (2) leakage through failed inactive wells, (3) failure of active wells, and (4) transport through subsurface pathways. Mechanisms also include scenarios related to the disposal of fluids, such as (5) injection of produced water into protected groundwater, and (6) use of unlined pits for produced water disposal. Other accidental discharges could be related to (7) spills and leaks or (8) operator error, or secondary processes including (9) treatment or reuse of produced water.

7.6.1 Leakage through Hydraulic Fractures

The possibility that the hydraulic fracturing process itself may create both a permeable pathway and a driving

force for fluid migration is the concern most commonly raised in discussions of the hazards of hydraulic fracturing. By definition, a hydrocarbon reservoir is likely to be capped or bounded by low-permeability layers, and thus understanding the process of pathway formation is a key for understanding the potential hazard. The degree to which induced fractures may extend beyond the target formation needs to be understood, and whether connections to overlying groundwater, or other natural or manufactured pathways, are possible.

It is generally thought that fractures created through stimulation have limited vertical extent. Basic work on understanding induced fractures spans decades (Perkins and Kern, 1961; Hubbert and Willis, 1972; Nordgren, 1972), but the latest published work directly addresses concerns about possible leakage of gas and fracturing fluids. These studies find that geophysical processes favor containment of gas and fluids within the reservoir (Flewelling et al., 2013; Flewelling and Sharma, 2014). Flewelling and Sharma (2014) also capped potential vertical fracture propagation at 600 m (2000 ft) or less and observed that shallow formations are more likely to fracture horizontally rather than vertically (see also Section 7.3). Fisher and Warpinski (2012) compared microseismic data on fracture extent for gas shales, finding that deep hydraulic fracturing operations should not bring the fractures in close contact with shallow aquifers. This work also indicated that fractures in shallower formations (<1200 m or 3900 ft) have a greater horizontal component, although earlier work found the fracture directionality was inconclusive, with orientations dependent on the unique stress profiles and rock fabric of a given location (Walker et al., 2002), in addition to depth of the reservoir. Likewise, Davies et al. (2012) found that the majority of induced fractures (with data focused on the Barnett Shale) range from less than 100 m (330 ft) to about 600 m (2000 ft) in height, with approximately a 1% and 50% probability of a fracture exceeding 350 m (1100 ft) or 130 m (427 ft), respectively. This leads to a suggested minimum separation of 600 m (2000 ft) between shale reservoirs and overlying groundwater resources for high-volume fracturing operations (King, 2012), although local geology must always be evaluated.

Coupled hydro-geomechanical modeling (Kim and Moridis, 2012) found inherent physical limitations to the extent of fracture propagation—for example, the presence of overlying confining formations may slow or stop fracture growth, thus containing fractures within the shale reservoir (Kim et al., 2014). An industry study (Cardno ENTRIX, 2012) evaluated the effects of ten years of hydraulic fracturing and gas production from a Los Angeles Basin oil and gas field. Microseismic monitoring indicates that fractures were contained within the reservoir zone, extending to within no more than 2350 m (7700 ft) of the base of the freshwater zone. However, some proposed fracturing operations, particularly in California, may occur at much shallower depths (CCST and LBNL, 2015b), possibly as shallow as 200 m, and as such site-specific evaluations may be needed to understand hazards.

Fault activation resulting in the formation of fluid pathways is an additional concern when stimulation operations occur in faulted geologies. Fault activation is a remote possibility for faults that can admit stimulation fluids during injection (Rutqvist et al., 2013), possibly increasing the permeability of previously sealed faults or creating new subsurface pathways analogous to induced fractures (possibly on a larger scale). Fault activation could also give rise to (small) microseismic events, but fault movement is limited to centimeter scales across fault lengths of 10–100 m. Chilingar and Endres (2005) document a California incident, where the migration of gas via permeable faults (among other pathways) created a gas pocket below a populated area in Los Angeles that resulted in an explosion. While the incident was not related to stimulation operations,

it shows how naturally faulted geologies can provide pathways for the migration of gas and fluids.

7.6.2 Leakage through Failed Inactive Wells

Reservoir fluids may reach the surface through degraded or improperly sealed wells. Regions with a history of oil and gas production are likely to have a large number of inactive (abandoned, buried, idle, or orphaned) wells. Many of these wells are undocumented, unknown, and either degraded, improperly abandoned, or substandard in construction. Fractures created during hydraulic fracturing could create connectivity to such degraded inactive wells, particular in high-density fields. However, the inactive wells have to fail (e.g., creating permeable pathways due to the degradation of cement or casings), and sufficient driving forces must be present for a leakage of gas or formation fluids to occur through inactive wells.

Old and inactive wells are a problem in states with a long history of petroleum production. For example, in Pennsylvania there are thousands of abandoned wells, with 200,000 dating from a time before formal record-keeping began and perhaps 100,000 that are essentially unknown (Vidic et al., 2013). Programs to locate, assess, and cap previously abandoned wells have been initiated in Ohio and Texas (Kell, 2011). In California, there are more inactive than active wells. Of a total of about 221,000 wells on file with the state, only 116,000 wells have been plugged and abandoned according to current standards. Nearly 1800 wells are "buried," that is, are older wells that have not been abandoned to standards and whose location is approximate. The status of 388 wells is unknown (CCST and LBNL, 2015b).

Chilingar and Endres (2005) document an incident where well corrosion at shallow depths led to casing failure of a producing well and the subsequent migration of gas via a combination of abandoned wells and fault pathways, as well as multiple cases of gas leakage from active oil fields and natural gas storage fields in the Los Angeles Basin and elsewhere. The most common pathway was identified as gas migration through faulted and fractured rocks, which were penetrated by abandoned and leaking wellbores. While stimulation technologies are not implicated in these events, they illustrate the real possibility of degraded abandoned wells as pathways.

The hazards of degraded abandoned wells are not just limited to their proximity to stimulated wells, but are also relevant to the issue of disposal of wastewater from stimulated wells by injection into Class II wells. A 1989 U.S. Government Accountability Office (GAO) study of Class II wells across the U.S. (U.S. GAO, 1989) found that one-third of contamination incidents were caused by injection near an improperly plugged abandoned oil and gas well. Current permitting requirements require a search for abandoned wells within a quarter mile of a new injection wellbore, and the plugging and remediation of any suspect wellbores (40 CFR 144.31, 146.24). However, Class II wells operating prior to 1976 are exempt from this requirement. About 70% of the disposal wells reviewed in this study were permitted prior to this date, grandfathered into the program, and allowed to operate without investigating nearby abandoned wells (U.S. GAO, 1989).

7.6.3 Failure of Active (Production and Class II) Wells

Operating wells (whether used for production or injection/disposal) can serve as transport pathways for subsurface migration in a similar fashion to deteriorated abandoned wells. Failures in well design and construction may allow migration of gas and fluids from the reservoir, or from shallower gas and fluid-bearing formations intersected by the wellbore. Pathways can be formed due to inadequate design, imposed stresses unique to stimulation operations, shear failure due to subsidence, or other forms of human/operator error. Class II deep injection wells with casing or cement inadequacies would also have similar potential for contamination as a failed production well or a well that fails due to stimulation pressures. Examples of potential subsurface releases through wells are illustrated in Figure 7.6.

Wells can thus serve as pathways for gas migration to overlying aquifers or even to the surface (Brufatto et al., 2003; Watson and Bachu, 2009). Multiple factors over the operating life of a well may lead to failure (Bonett and Pafitis, 1996; Dusseault et al., 2000; Brufatto et al., 2003; Chilingar and Endres, 2005; Watson and Bachu, 2009; Carey et al., 2013); however, the most important

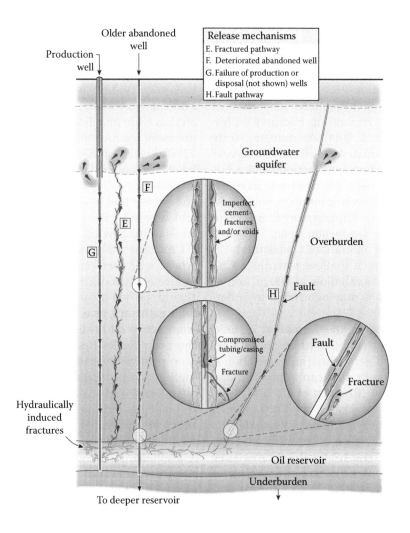

FIGURE 7.6
Potential subsurface mechanisms and related transport pathways: Illustrated in the figure are well failure and consequent leakage through an abandoned well, migration through intercepted fractures, and fault activation. Well failure can occur in production, abandoned, or disposal wells (Chapter 2 in CCST and LBNL, 2015b).

mechanism leading to gas and fluid migration is poor well construction or an exposed or uncemented casing (Watson and Bachu, 2009). A surface casing, constructed to shield groundwater resources, may not adequately protect those resources if the casing does not extend to a sufficient depth below the aquifer (Harrison, 1983, 1985).

Watson and Bachu (2009, p. 121) also noted that deviated wellbores, defined as "any well with total depth greater than true vertical depth," show a higher occurrence of gas migration than vertical wells, likely due to the challenges of deviated well construction increasing the likelihood of gaps, bonding problems, or thin regions in the cement that could create connectivity to other formations. This is important as economic production from hydraulically fractured shales typically requires long horizontal wellbores to maximize the extent of the fracture network. In a review of the regulatory record, Vidic et al. (2013) noted a 3.4% rate of cement and casing problems in Pennsylvania shale-gas wells (that all had some degree of deviation) based on filed notices of violation. Pennsylvania inspection records, however, show a large number of wells with indications of cement/casing impairments for which violations were never noted suggesting that the actual rate of occurrence could be even higher.

Stimulated wells may be subject to greater stresses than nonstimulated wells, due to the high-pressure stimulation process and the special drilling practices used to create deviated (often horizontal) wells (Ingraffea et al., 2014). During hydraulic fracturing operations, multiple stages of high-pressure injection may result in the cyclic expansion and contraction of the steel casing (Carey et al., 2013). This could lead to radial fracturing and/or shear failure at the steel–concrete or concrete–rock interfaces, or even separation between the casing and the cement. These gaps or channels could serve as pathways between the reservoir and overlying aquifers. Current practice does not typically use the innermost casing as the direct carrier of stimulation fluids (or produced fluids and gases)—additional tubing (injection or production tubing) is run down the innermost well casing without being cemented into place and thus carries the stresses associated with injection. However, less complex stimulation treatments may not require such additional steps, and some fracturing operations use the innermost casing to convey the fracturing fluids and the pressures associated with the fracturing operation.

The bulk of the peer-reviewed work on contaminant migration associated with stimulation focuses on the Marcellus Shale, and contains robust debate on the role of deteriorated or poorly constructed wells. A sampling study by Osborn et al. (2011a) and Jackson et al. (2013) noted that methane concentrations in wells increased with increasing proximity to gas wells, and that the sampled gas was similar in composition to gas from nearby production wells. Follow-up work by Davies (2011) and

Schon (2011) found that leakage through well casings was a better explanation than other fracturing-related processes (also see Vidic et al., 2013). Most recently, other sampling studies (Molofsky et al., 2013; Darrah et al., 2014) found gas compositions that do not match the Marcellus, suggesting that intermediate formations are providing the source for the methane, and thus migration through poorly constructed wells is a more likely scenario. The Darrah et al. (2014) study in particular identifies eight locations in the Marcellus (and also for one additional case in the Barnett shale in Texas) where annular migration through/around poorly constructed wells is considered the most plausible mechanism for measured methane contamination of groundwater.

7.6.4 Transport through Induced or Natural Subsurface Pathways

If a permeable pathway has been created via some combination of the processes listed above, fluids may move through the subsurface and possibly reach groundwater. Several modeling studies have attempted to elucidate these transport mechanisms through numerical simulation. A well-publicized study by Myers (2012) found potential transport between fractured reservoirs and an overlying aquifer, but did so using a highly simplified flow model regarded as unrepresentative of a shale reservoir system (Vidic, 2013). Modeling work by Kissinger et al. (2013) suggests that transport of liquids, fracturing fluids, or gas is not an inevitable outcome of fracturing into connected pathways, at least for the specific reservoir modeled. Modeling work by Gassiat et al. (2013) found that migration of fluids from a fractured formation is possible for high-permeability fractures and faults, and for permeable bounding formations, but on a 1000-year timeframe. Flewelling and Sharma (2014) conclude that upward migration through permeable bounding formations, if possible at all, is likely an even slower process operating at much longer timescales (in their estimate, ~10^6 years). Three-dimensional modeling of permeable fractures and degraded wells connecting gas shales to overlying groundwater (Reagan et al., 2015) showed that transport can be rapid through very permeable pathways, but that the duration and magnitude of release is limited by the fracture volume, shale permeability, and the capillary forces associated with gas shales. Additional modeling studies are underway to clarify these results (U.S. EPA, 2012, 2015a).

Sampling and field studies have also sought evidence of migration. A key conclusion is that pathways and mechanisms are difficult to characterize and the role of fracturing or transport through fractures has not been clearly established. Methane concentrations in water wells increase with proximity to gas wells, and the gas is similar in composition to gas produced nearby (Osborn et al., 2011a; Jackson et al., 2013), but evidence of contamination from brines or stimulation fluids was not found (Jackson et al., 2011, 2013; Osborn et al., 2011b). The most recent sampling studies (Molofsky et al., 2013; Darrah et al., 2014) conclude that migration through poorly constructed wells is a more likely scenario than fracture-related pathways. Work on the properties of gas shales (Engelder, 2014) proposed that a "capillary seal" would restrict the ability of fluids to migrate out of gas shales (see also Reagan et al., 2015), but some reservoirs may contain more mobile water than others.

7.6.5 Injection of Produced Water into Protected Groundwater

Subsurface injection was the second most common disposal method for produced water from stimulated wells. Studies show that with proper siting, construction, and maintenance, subsurface injection is less likely to result in groundwater contamination than disposal in unlined surface impoundments (Kell, 2011).

7.6.6 Use of Unlined Pits for Produced Water Disposal

There is evidence of groundwater contamination associated with unlined pits used for the storage and disposal of fluids in the context of drilling and production. Kell (2011) reviewed incidents of groundwater contamination caused by oil field activities in Texas, finding that 27% were associated with unlined pits, which were banned in Texas in 1969 and closed no later than 1984. For Ohio, a similar analysis found that 5% of incidents were associated with unlined pits (Kell, 2011). Such pits are no longer permitted in either state, and no incidents have been reported since the mid-1980s.

While these studies and others linking unlined pits to groundwater contamination are not specific to well stimulation fluids, they illustrate the implications of this disposal method. A case in Pavillion, WY, raises additional concerns. According to the U.S. EPA draft report, released in 2011, high concentrations of hydraulic fracturing chemicals found in shallow monitoring wells near surface pits "indicate that pits represent a source of shallow ground water contamination in the area of investigation" (Digiulio et al., 2011, p. 33). At least 33 unlined pits were used to store/dispose of drilling muds, flowback, and produced water in the area. These findings were not contested by the company responsible for the natural gas wells, or the other stakeholders (Folger et al., 2012). There was, however, considerable controversy about U.S. EPA's other findings, that is, the presence of hydraulic fracturing chemicals in deep water wells and thermogenic methane in monitoring and domestic wells.

7.6.7 Spills and Leaks

Oil and gas production involves some risk of surface or groundwater contamination from spills and leaks. Well stimulation, however, raises additional concerns, due to the use of additional chemicals during the stimulation process, the generation of wastewaters that contain these chemical additives, as well as formation brines with potentially different compositions from conventional produced water, and the increased transportation requirements to haul these materials to the well and disposal sites. Surface spills and leaks can occur at any time in the stimulation or production process—during chemical or fluid transport, prestimulation mixing, or as the stimulation process is taking place. In addition, storage containers used for chemicals and well stimulation fluids can leak (NYSDEC, 2011). For instance, in September 2009, two pipe failures and a hose rupture in Pennsylvania released 8000 gallons of a liquid gel mixture during the hydraulic fracturing process, polluting a local creek and wetland (Pennsylvania Department of Environmental Protection, 2010).

7.6.8 Treatment or Reuse of Produced Water

Produced water is commonly reused for beneficial purposes, including reservoir steam flooding and industrial cooling. The produced water may be treated prior to reuse, or simply blended with fresh water to lower the levels of salts and other constituents. There is growing interest in expanding the beneficial reuse of produced water for agriculture, particularly for irrigation, due to the frequent colocation of oil, gas, and agricultural operations and regional water scarcity concerns. The use of produced water from unconventional production raises specific or unique concerns, because of the variety of chemicals used during hydraulic fracturing that may end up mingled with produced water.

The treatment of produced water has been the subject of intensive investigation and standard treatment practices have evolved for the reuse of produced water (e.g., Federal Remediation Technologies Roundtable, 2007). Treatment of constituents commonly found in produced water (e.g., oil and grease, dissolved solids, suspended particles, bacteria, etc.) is generally well documented (Arthur et al., 2005; Drewes, 2009; Fakhru'l-Razi et al., 2009; Igunnu and Chen, 2012; M-I SWACO, 2012), but little information is available on how commonly used produced water treatment systems may handle hydraulic fracturing chemicals.

Sewer systems are not typically equipped to handle produced water. In Pennsylvania, for example, the high salt content of oil and gas wastewater discharged to sewage treatment plants resulted in increased salt loading to Pennsylvania rivers (Kargbo et al., 2010; Wilson and VanBriesen, 2012; Vidic et al., 2013; Brantley et al., 2014). Warner et al. (2013) studied the effluent from a brine treatment facility in Pennsylvania and found an increase of salts downstream, despite significant reduction in concentrations due to the treatment process and dilution from the river. Moreover, radium activities in the stream sediments near the point of discharge were 200 times higher than in upstream and background sediments, and were above radioactive waste disposal thresholds. Much of the research on disposal has focused on produced water constituents and not specifically on stimulation chemicals commingled with produced water.

7.6.9 Operator Error

Human error during the well completion, stimulation, or production processes could also lead to contamination of groundwater. Operator error could create connectivity to other formations that could serve as transport pathways. For example, poor monitoring or control of the fracturing operation could increase the extent of fractures beyond desired limits. Such errors could lead to an unexpected migration of fluids, or connection between wells that impacts production activities themselves. Fracturing beyond the reservoir bounds due to operator error may also be of particular concern in the case of shallower fracturing operations.

An example of operator error during stimulation is a 2011 incident in Alberta, Canada (ERCB, 2012), where a misjudged fracturing depth led to fracturing fluids being injected into a water-bearing strata below an aquifer. A hydraulic connection between the fractured interval and the overlying aquifer was not observed, but groundwater samples contained elevated levels of chloride, BTEX, petroleum hydrocarbons, and other chemicals.

7.6.10 Conclusions

Contamination of water supplies due to spills, leaks from surface facilities, disposal of inadequately treated water, leakage from wells, and illegal discharges have occurred. Several plausible release mechanisms and transport pathways exist for surface and groundwater contamination. However, the issue of contamination of groundwater via processes specific to hydraulic fracturing has not been proven to have occurred. It is clear that the hazard exists, and practices that mitigate the potential for contamination should be in place at each step of the process. Additional research on fracture propagation, the hydrological processes in and around unconventional reservoirs, and additional sampling studies to investigate potential transport scenarios are all required to understand the risks associated with the pathways discussed herein.

Acknowledgments

This chapter was supported by the California Natural Resources Agency and the U.S. Bureau of Land Management, under a Work for Others Agreement with the U.S. Department of Energy (DOE) at LBNL, under contract number DE-AC02-05CH11231. We appreciate the cooperation and leadership of Jane Long and Laura Feinstein of the CCST in the execution of this chapter. We also acknowledge partial support from the Scientific Focus Area at LBNL funded by the U.S. DOE, Office of Science, Office of Biological and Environmental Research under Award Number DE-AC02-05CH11231.

References

Abass, H.H., A.A. Al-Mulheim, M.S. Alqem, and K.R. Mirajuddin. 2006. Acid fracturing or proppant fracturing in carbonate formation? A rock mechanic's view. In: *SPE 102590, SPE Annual Technical Conference and Exhibition*, San Antonio, TX.

Alley, B., A. Beebe, J. Rodgers, and J.W. Castle. 2011. Chemical and physical characterization of produced waters from conventional and unconventional fossil fuel resources, *Chemosphere*, 85(1), 74–82. Retrieved from http://www.ncbi.nlm.nih.gov/pubmed/21680012.

Arthur, D.J., B.G. Langhus, and C. Patel. 2005. *Technical Summary of Oil and Gas Produced Water Treatment Technologies*, ALL Consulting, LLC, Tulsa, OK.

Arthur, J.D., B. Bohm, B.J. Coughlin, and M. Layne. 2008. Evaluating the environmental implications of hydraulic fracturing in shale gas reservoirs. In: *SPE 121038-MS, SPE Americas E and P Environmental and Safety Conference*, pp. 1–21, ALL Consulting, San Antonio, TX.

Arthur, J.D., B. Bohm, and M. Layne. 2008. Hydraulic fracturing considerations for natural gas wells of the Marcellus Shale. *The Ground Water Protection Council 2008 Annual Forum*, September 21–24, Cincinnati, OH.

Baker Hughes Inc. 2011. *Material Safety Data Sheet: Clay Master-5C*. Baker Hughes, Sugar Land, TX, https://oilandgas.ohiodnr.gov/portals/oilgas/_MSDS/baker-hughes/ClayMaster5C_US.pdf.

Baker Hughes Inc. 2013. *Master Fracturing Chemical List—Arkansas*. Trican, http://aogc2.state.ar.us/B-19/1242_ChemConst.pdf.

Balaba, R.S., and R.B. Smart. 2012. Total arsenic and selenium analysis in Marcellus Shale, high-salinity water, and hydrofracture flowback wastewater, *Chemosphere*, 89(11), 1437–1442, doi: http://dx.doi.org/10.1016/j.chemosphere.2012.06.014.

Banjoko, B. 2014. *Handbook of Engineering Hydrology: Environmental Hydrology and Water Management*, S. Eslamian (Ed.), CRC Press, Boca Raton, FL.

Banks, D., N.E. Odling, H. Skarphagen, and E. Rohr-Torp. 1996. Permeability and stress in crystalline rocks, *Terra*

Nova, 8(3), 223–235. Retrieved August 19, 2015 http://doi.wiley.com/10.1111/j.1365-3121.1996.tb00751.x.

Barbot, E., N.S. Vidic, K.B. Gregory, and R.D. Vidic. 2013. Spatial and temporal correlation of water quality parameters of produced waters from devonian-age shale following hydraulic fracturing, *Environmental Science and Technology*, 47(6), 2562–2569, doi:10.1021/es304638h.

Barree, B., H. Fitzpatrick, J. Manrique, M. Mullen, S. Schubarth, M. Smith, and N. Stegent. 2005. Propping-up production. *Conductivity Endurance*, 4–11. Supplement to E&P Journal.

Bazan, L.W., M.G. Lattibeaudiere, and T.T. Palisch. 2012. Hydraulic fracture design and well production results in the Eagle Ford Shale: One operator's perspective. In: *SPE 155779, Americas Unconventional Resources Conference*, vol. 2012, pp. 1–14, Society of Petroleum Engineers, Pittsburgh, PA.

Beckwith, R. 2010. Hydraulic fracturing: The fuss, the facts, the future, *Journal of Petroleum Technology*, 62(12), 34–41.

Benko, K., and J. Drewes. 2008. Produced water in the Western United States: Geographical distribution, occurrence, and composition, *Environmental Engineering Science*, 25(2), 239–246.

Blauch, M.E., R.R. Myers, T.R. Moore, and B.A. Lipinski. 2009. Marcellus Shale post-frac flowback waters—Where is all the salt coming from and what are the implications? In: *SPE 125740, SPE Regional Eastern Meeting*, pp. 1–20, Society of Petroleum Engineers, Charleston, WV. Retrieved from http://www.onepetro.org/mslib/servlet/onepetropreview?id=SPE-125740-MS.

Blondes, M.S., K.D. Gans, J.J. Thordsen, M.E. Reidy, B. Thomas, M.A. Engle, Y.K. Kharaka, and E.L. Rowan. 2014. U.S. Geological Survey National Produced Waters Geochemical Database v2.1 (Provisional).

Bonett, A., and D. Pafitis. 1996. Getting to the root of gas migration, *Oilfield Review*, 8(1) (Spring 1996), 36–49.

Brantley, S.L., D. Yoxtheimer, S. Arjmand et al. 2014. Water resource impacts during unconventional shale gas development: The Pennsylvania experience, *International Journal of Coal Geology*, 126 (June 1, 2014), 140–156. doi:10.1016/j.coal.2013.12.017.

Brufatto, C., J. Cochran, L. Conn et al. 2003. From mud to cement—Building gas wells, *Oilfield Review*, 15(3) (Autumn 2003), 62–76.

California Council on Science and Technology (CCST), and Lawrence Berkeley National Laboratory (LBNL). 2015a. *An Independent Scientific Assessment of Well Stimulation in California: Volume I, Well Stimulation Technologies and Their Past, Present, and Potential Future Use in California*. Sacramento, CA. Available from: http://ccst.us/SB4.

California Council on Science and Technology (CCST), and Lawrence Berkeley National Laboratory (LBNL). 2015b. *An Independent Scientific Assessment of Well Stimulation in California: Volume II, Potential Environmental Impacts of Hydraulic Fracturing and Acid Stimulations*. Sacramento, CA. Available from: http://ccst.us/SB4.

Cardno ENTRIX. 2012. Hydraulic Fracturing Study—PXP Inglewood Oil Field, http://www.scribd.com/doc/109624423/Hydraulic-Fracturing-Study-Inglewood-Field10102012.

Carey, J.W., K. Lewis, S. Kelkar, and G.A. Zyvoloski. 2013. Geomechanical behavior of wells in geologic sequestration, *Energy Procedia*, 37, 5642–5652, doi:10.1016/j.egypro.2013.06.486.

Chilingar, G.V., and B. Endres. 2005. Environmental hazards posed by the Los Angeles Basin urban oilfields: An historical perspective of lessons learned, *Environmental Geology*, 47(2), 302–317, doi:10.1007/s00254-004-1159-0.

Cipolla, C.L., N.R. Warpinski, M.J. Mayerhofer, E.P. Lolon, and M.C. Vincent. 2010. The relationship between fracture complexity, reservoir properties, and fracture-treatment design, *SPE Production and Operations*, 25(4), 438–452.

Darrah, T.H., A. Vengosh, R.B. Jackson, N.R. Warner, and R.J. Poreda. 2014. Noble gases identify the mechanisms of fugitive gas contamination in drinking-water wells overlying the Marcellus and Barnett Shales, *Proceedings of the National Academy of Sciences*, 111(39), 14076–14081, doi:10.1073/pnas.1322107111.

Davies, R.J. 2011. Methane contamination of drinking water caused by hydraulic fracturing remains unproven, *Proceedings of the National Academy of Sciences*, 108(43), E871, doi:10.1073/pnas.1113299108.

Davies, R.J., S.A. Mathias, J. Moss, S. Hustoft, and L. Newport. 2012. Hydraulic fractures: How far can they go? *Marine and Petroleum Geology*, 37(1), 1–6, doi:10.1016/j.marpetgeo.2012.04.001.

Digiulio, D.C., R.T. Wilkin, C. Miller, and G. Oberly. 2011. *DRAFT: Investigation of Ground Water Contamination Near Pavillion*, U.S. Environmental Protection Agency, Wyoming.

DOGGR (Division of Oil, Gas and Geothermal Resources). 2014a. Well Stimulation Treatment Disclosure Reporting Form Instructions. Retrieved from http://www.conservation.ca.gov/dog/Documents/FINAL_Reporting%20Forms%20and%20Instructions_04-17-2014.pdf#search=well%20stimulation%20treatment%20disclosure%20reporting%20form%20instructions.

DOGGR (Division of Oil, Gas and Geothermal Resources). 2014b. Interim Well Stimulation Treatment Notices Index. Retrieved from http://maps.conservation.ca.gov/doggr/iwst_index.html.

DOGGR (Division of Oil, Gas and Geothermal Resources). 2014c. Well Completion Reports.

Drewes, J.E. 2009. *An Integrated Framework for Treatment and Management of Produced Water: Technical Assessment of Produced Water Treatment Technologies*, Colorado School of Mines, Golden, CO.

Dusseault, M.B., M.N. Gray, and P.A. Nawrocki. 2000. Why oilwells leak: Cement behavior and long-term consequences. In: *SPE International Oil and Gas Conference and Exhibition, SPE 64733*, p. 8, Society of Petroleum Engineers, Beijing, China.

Economides, M.J., A.D. Hill, C. Ehlig-Economides, and D. Zhu. 2013. *Petroleum Production Systems*, 2nd ed. Prentice-Hall, Upper Saddle River, NJ.

Economides, M.J., and K.G. Nolte. 2000. *Reservoir Stimulation*, 3rd ed. John Wiley & Sons, West Sussex, England.

Engelder, T., L.M. Cathles, and L.T. Bryndzia. 2014. The fate of residual treatment water in gas shale, *Journal of Unconventional Oil and Gas Resources*, 7, 33–48.

ERCB (Energy Resources Conservation Board). 2012. Caltex Energy Inc. Hydraulic Fracturing Incident 16-27-068-10W6M September 22, 2011, ERCB Investigation Report (Released: December 20, 2012).

Esser, B.K., H.R. Beller, S.A. Carroll et al. 2015. *Recommendations on Model Criteria for Groundwater Sampling, Testing, and Monitoring of Oil and Gas Development in California*, (July) Lawrence Livermore National Laboratory LLNL-TR-669645, Livermore, CA.

Fakhru'l-Razi, A., A. Pendashteh, L.C. Abdullah, D.R.A. Biak, S.S. Madaeni, and Z.Z. Abidin. 2009. Review of technologies for O&G produced water treatment, *Journal of Hazardous Materials*, 170, 530–551.

Federal Remediation Technologies Roundtable. 2007. The Remediation Technologies Screening Matrix. Retrieved from: https://frtr.gov/scrntools.htm.

Fisher, J.B., and J.R. Boles. 1990. Water–rock interaction in tertiary sandstones, San Joaquin Basin, California, USA: Diagenetic controls on water composition, *Chemical Geology*, 82(0), 83–101, doi: http://dx.doi.org/10.1016/0009-2541(90)90076-J.

Fisher, K., and N. Warpinski. 2012. Hydraulic-fracture-height growth: Real data, *Society of Petroleum Engineers Production and Operations*, 27(1), 8–19.

Flewelling, S., and M. Sharma. 2014. Constraints on upward migration of hydraulic fracturing fluid and brine, *Ground Water*, 52(1), 9–19.

Flewelling, S.A., M.P. Tymchak, and N. Warpinski. 2013. Hydraulic fracture height limits and fault interactions in tight oil and gas formations, *Geophysical Research Letters*, 40(14), 3602–3606, doi:10.1002/grl.50707.

Folger, P., M. Tiemann, and D.M. Bearden. 2012. The EPA Draft Report of Groundwater Contamination Near Pavillion, Wyoming: Main Findings and Stakeholder Responses. Retrieved from http://wyofile.com/wp-content/uploads/2012/01/R42327-2.pdf.

FracFocus, What chemicals are used? Ground Water Protection Council and Interstate Oil and Gas Compact Commission. Retrieved from http://fracfocus.org/chemical-use/what-chemicals-are-used.

Gadberry, J.F., M.D. Hoey, R. Franklin, G. del Carmen Vale, and F. Mozayeni. 1999. Surfactants for hydraulic fractoring compositions, Akzo Nobel NV, U.S. Patent 5,979,555.

Gale, J.F.W., and J. Holder. 2010. Natural fractures in some U.S. shales and their importance for gas production. In: *Petroleum Geology: From Mature Basins to New Frontiers: Proceedings of the 7th Petroleum Geology Conference*, B.A. Vining, and S.C. Pickering, Eds., pp. 1131–1140, Geological Society, London, UK.

Gartiser, S., and E. Urich. 2003. Elimination of cooling water biocides in batch tests at different inoculum concentrations. In: *Society of Environmental Toxicology and Chemistry Europe 13th Annual Meeting*, Hamburg, Germany, April 28–May 1, 2003.

Gassiat, C., T. Gleeson, R. Lefebvre, and J. McKenzie. 2013. Hydraulic fracturing in faulted sedimentary basins: Numerical simulation of potential contamination of shallow aquifers over long time scales, *Water Resources Research*, 49, 8310–8327.

Gustafson, G., and J. Krásný. 1994. Crystalline rock aquifers: Their occurrence, use and importance, *Hydrogeology Journal*, 2(2), 64–75. Retrieved August 19, 2015, http://link.springer.com/10.1007/s100400050051.

Halliburton Energy Services. 2009. GBW-30 Breaker MSDS, Duncan, OK.

Haluszczak, L.O., A.W. Rose, and L.R. Kump. 2013. Geochemical evaluation of flowback brine from Marcellus gas wells in Pennsylvania, USA, *Applied Geochemistry*, 28, 55–61. Retrieved from http://linkinghub.elsevier.com/retrieve/pii/S0883292712002752.

Harris, P., and S. Heath. 1996. Rheology of crosslinked foams, *SPE Production and Facilities*, 11(2), 113–116.

Harrison, S.S. 1983. Evaluating system for ground-water contamination hazards; due to gas-well drilling on the glaciated Appalachian Plateau, *Ground Water*, 21(6), 689–700.

Harrison, S.S. 1985. Contamination of aquifers by overpressuring the annulus of oil and gas wells, *Ground Water*, 23(3), 317–324, doi: 10.1111/j.1745-6584.1985.tb00775.

Hayes, T. 2009. Sampling and analysis of water streams associated with the development of Marcellus Shale gas. Retrieved from http://energyindepth.org/wp-content/uploads/marcellus/2012/11/MSCommission-Report.pdf.

Higashi, R.M., and A.D. Jones. 1997. Identification of bioactive compounds from produced water discharge/characterization of organic constituent patterns at a produced water discharge site: Final technical summary, Final Technical Report. MMS 97-0023, U.S. Department of the Interior, Minerals Management Service, Pacific OCS Region, http://www.coastalresearchcenter.ucsb.edu/scei/Files/97-0023.pdf.

Howarth, R.W., A. Ingraffea, and T. Engelder. 2011. Natural gas: Should fracking stop? *Nature*, 477(7364), 271–275, doi:10.1038/477271a.

http://www.eia.gov/analysis/studies/usshalegas/pdf/usshaleplays.pdf. (accessed on August 21, 2015)

http://www.eia.gov/todayinenergy/detail.cfm?id=19991. (accessed on July 29, 2015).

Hubbert, M.K., and D.G. Willis. 1972. Mechanics of hydraulic fracturing, *American Association of Petroleum Geologists*, 18, 239–257.

Igunnu, E.T., and G.Z. Chen. 2012. Produced water treatment technologies, *International Journal of Low-Carbon Technologies*, 1–21. Published online, doi:10.1093/ijlct/cts049.

Ingraffea, A.R., M.T. Wells, R.L. Santoro, and S.B.C. Shonkoff . 2014. The integrity of oil and gas wells, *Proceedings of the National Academy of Sciences, USA*, 111(30), 10902–10903.

Jackson, R.B., S.G. Osborn, A. Vengosh, and N.R. Warner. 2011. Reply to Davies: Hydraulic fracturing remains a possible mechanism for observed methane contamination of drinking water, *Proceedings of the National Academy of Sciences*, 108(43), E872, doi:10.1073/pnas.1113768108.

Jackson, R.B., A. Vengosh, T.H. Darrah et al. 2013. Increased stray gas abundance in a subset of drinking water wells near Marcellus shale gas extraction, *Proceedings of the National Academy of Sciences, USA*, 110(28), 11250–11255, doi:10.1073/pnas.1221635110.

Kalfayan, L. 2008. *Production Enhancement with Acid Stimulation*, M. Patterson, Ed. p. 252. PennWell Corporation, Tulsa, Oklahoma.

Kargbo, D.M., R.G. Wilhelm, and D.J. Campbell. 2010. Natural gas plays in the Marcellus Shale: Challenges and potential opportunities, *Environmental Science and Technology*, 44(15), 5679–5684, doi:10.1021/es903811p.

Kassotis, C.D., D.E. Tillit, J.W. Davis, A.M. Hormann, and S.C. Nagel. 2013. Estrogen and androgen receptor activities of hydraulic fracturing chemicals and surface and ground water in a drilling-dense region—en.2013-1697. *Endocrinology*, 1–11. Retrieved from http://press.endocrine.org/doi/pdf/10.1210/en.2013-1697.

Kell, S. 2011. *State Oil and Gas Agency Groundwater Investigations and Their Role in Advancing Regulatory Reforms, a Two-State Review*, Groundwater Protection Council, Ohio and Texas, http://www.gwpc.org/sites/default/files/State%20 Oil%20%26%20Gas%20Agency%20Groundwater%20 Investigations.pdf.

Kim, J., and G.J. Moridis. 2012. Gas flow tightly coupled to elastoplastic geomechanics for tight and shale gas reservoirs: Material failure and enhanced permeability. In: *Proceedings of Americas Unconventional Resources Conference*, Pittsburgh, PA.

Kim, J., E.S. Um, and G.J. Moridis. 2014. SPE 168578 fracture propagation, fluid flow, and geomechanics of water-based hydraulic fracturing in shale gas systems and electromagnetic geophysical monitoring of fluid migration. In: *Proceedings of SPE Hydraulic Fracturing Technology Conference*, Woodlands, TX.

King, G.E. 2010. Thirty years of gas shale fracturing: What have we learned? In: *SPE 133456, SPE Annual Technical Conference and Exhibition*, pp. 1–50, Society of Petroleum Engineers, Florence, Italy.

King, G.E. 2012. Hydraulic fracturing 101: What every representative, environmentalist, regulator, reporter, investor, university researcher, neighbor and engineer should know about estimating frac risk and improving frac performance in unconventional gas and oil wells. In *SPE 152596, SPE Hydraulic Fracturing Technology Conference*, pp. 1–80, Society of Petroleum Engineers, Woodlands, TX. Retrieved from http://fracfocus.org/sites/default/files/publications/hydraulic_fracturing_101.pdf.

Kissinger, A., R. Helmig, A. Ebigbo, H. Class, T. Lange, M. Sauter, M. Heitfeld, J. Klünker, and W. Jahnke. 2013. Hydraulic fracturing in unconventional gas reservoirs: Risks in the geological system, Part 2, *Environmental Earth Sciences*, 70(8), 3855–3873, doi: 10.1007/s12665-013-2578-6.

Kostenuk, N., and D.J. Browne. 2010. Improved proppant transport system for slickwater shale fracturing. In: *CSUG/SPE 137818, Canadian Unconventional Resources and International Petroleum Conference*, pp. 19–21, Society of Petroleum Engineers, Calgary, Alberta.

Lebas, R., P. Lord, D. Luna, and T. Shahan. 2013. Development and use of high-TDS recycled produced water for cross-linked-gel-based hydraulic fracturing. In: *SPE 163824, SPE Hydraulic Fracturing Technology Conference*, pp. 1–9, Society of Petroleum Engineers, The Woodlands, TX.

Lei, C., and P.E. Clark. 2004. Crosslinking of guar and guar derivatives. In: *SPE 90840, SPE Annual Technical Conference and Exhibition*, p. 12, Society of Petroleum Engineers, Houston, TX.

Less, C., and N. Andersen. 1994. Hydrofracture: State of the art in South Africa, *Hydrogeology Journal*, 2(2):59–63.

Retrieved August 19, 2015. http://link.springer.com/10.1007/s100400050050.

Martin, R. ed. 2013. *Clogging Issues Associated with Managed Aquifer Recharge Methods*, IAH Commission on Managing Aquifer Recharge, Australia.

McDaniel, B., and K. Rispler. 2009. Horizontal wells with multistage fracs prove to be best economic completion for many low-permeability reservoirs. In: *SPE 125903, Proceedings of SPE Eastern Regional Meeting*, pp. 1–15, Society of Petroleum Engineers, Charleston, WV.

Miller, J. 2007. Biodegradable surfactants aid the development of environmentally acceptable drilling fluid additives. In: *International Symposium on Oilfield Chemistry*, Society of Petroleum Engineers, Houston, TX, February 28–March 2, 2007.

Minner, W.A., C.A. Wright, G.R. Stanley et al. 2002. Waterflood and production-induced stress changes dramatically affect hydraulic fracture behavior in Lost Hills infill wells. In: *SPE 77536, SPE Annual Technical Conference and Exhibition*, pp. 1–13, Society of Petroleum Engineers, San Antonio, TX.

M-I SWACO. 2012. *Fracturing Fluid Flowback Reuse Project: Decision Tree and Guidance Manual*, Petroleum Technology Alliance of Canada, Science and Community Environmental Knowledge.

Molofsky, L.J., J.A. Connor, A.S. Wylie, T. Wagner, and S.K. Farhat. 2013. Evaluation of methane sources in groundwater in Northeastern Pennsylvania, *Groundwater*, 51(3), 333–349, doi:10.1111/gwat.12056.

Montgomery, C. 2013. Fracturing fluids. In: *Effective and Sustainable Hydraulic Fracturing*, A.P. Bunger, J. McLennan, and R. Jeffrey, Eds., pp. 3–24, InTech. ISBN: 978-953-51-1137-5, DOI: 10.5772/56192. Available from: http://www.intechopen.com/books/effective-and-sustainable-hydraulic-fracturing/fracturing-fluids.

Myers, T. 2012. Potential contaminant pathways from hydraulically fractured shale to aquifers, *Ground Water*, 50(6), 872–882.

NYSDEC (New York State Department of Environmental Conservation) 2009. Draft Supplemental Generic Environmental Impact Statement on the Oil, Gas and Solution Mining Regulatory Program—Well Permit Issuance for Horizontal Drilling and High-Volume Hydraulic Fracturing to Develop the Marcellus Shale and Other Low-Permeability Gas Reservoirs. Retrieved from: ftp://ftp.dec.ny.gov/dmn/download/OGdSGEISFull.pdf.

NYSDEC (New York State Department of Environmental Conservation) 2011. Revised Draft Supplemental Generic Environmental Impact Statement on the Oil, Gas and Solution Mining Regulatory Program—Well Permit Issuance for Horizontal Drilling and High-Volume Hydraulic Fracturing to Develop the Marcellus Shale and Other Low-Permeability Gas Reservoirs. Albany, NY. Retrieved from http://www.dec.ny.gov/data/dmn/rdsgeisfull0911.pdf.

Nishihara, T., T. Okamot, and N. Nishiyama. 2000. Biodegradation of didecyldimethylammonium chloride by *Pseudomonas fluorescens* TN4 isolated from activated sludge, *Journal of Applied Microbiology*, 88(4), 641–647. Retrieved from http://www.ncbi.nlm.nih.gov/pubmed/10792522.

Nordgren, R.P. 1972. Propagation of a vertical hydraulic fracture, *Society of Petroleum Engineers Journal*, 12(4), 306–314.

Osborn, S.G., A. Vengosh, N.R. Warner, and R.B. Jackson. 2011a. Methane contamination of drinking water accompanying gas-well drilling and hydraulic fracturing, *Proceedings of the National Academy of Sciences USA*, 108(20), 8172–8176, doi:10.1073/pnas.1100682108.

Osborn, S.G., A. Vengosh, N.R. Warner, and R.B. Jackson. 2011b. Reply to Saba and Orzechowski and Schon: Methane contamination of drinking water accompanying gas-well drilling and hydraulic fracturing, *Proceedings of the National Academy of Sciences USA*, 108(37), E665–E666, doi:10.1073/pnas.1109270108.

Pennsylvania Department of Environmental Protection. 2010. DEP Fines Atlas Resources for Drilling Wastewater Spill in Washington County, Press Release (August 17). Retrieved from http://www.portal.state.pa.us/portal/server.pt/community/newsroom/14287?id=13595andtypeid=1.

Perkins, T.K., and L.R. Kern. 1961. Widths of hydraulic fractures, *Journal of Petroleum Technology*, 13(9), 937–949. doi:10.2118/89-PA.

Reagan, M.T., G.J. Moridis, N.D. Keen, and J.N. Johnson. 2015. Numerical simulation of the environmental impact of hydraulic fracturing of tight/shale gas reservoirs on near-surface groundwater: Background, base cases, shallow reservoirs, short-term gas and water transport, *Water Resources Research*, 51(4), 2543–2573, doi: 10.1002/2014WR016086.

Ribeiro, L.H., and M.M. Sharma. 2012. Multiphase fluid-loss properties and return permeability of energized fracturing fluids, *SPE Production and Operations*, 27(3), 265–277.

Rickman, R., M. Mullen, E. Petre, B. Grieser, and D. Kundert. 2008. A practical use of shale petrophysics for stimulation design optimization: All shale plays are not clones of the Barnett Shale. In: *SPE 115258 Annual Technical Conference and Exposition*, pp. 1–11, Society of Petroleum Engineers, Denver, CO.

Rowan, E.L., M.A. Engle, C.S. Kirby, and T.F. Kraemer. 2011. Radium content of oil- and gas-field produced waters in the northern Appalachian Basin (USA)—Summary and discussion of data: U.S. Geological Survey Scientific Investigations Report 2011–5135.

Rutqvist, J., A.P. Rinaldi, F. Cappa, and G.J. Moridis. 2013. Modeling of fault reactivation and induced seismicity during hydraulic fracturing of shale-gas reservoirs, *Journal of Petroleum Science and Engineering*, 107, 31–44, doi:10.1016/j.petrol.2013.04.023.

Schon, S.C. 2011. Hydraulic Fracturing is not responsible for methane migration, *Letters to Proceedings of National Academy of Science*, 108(37), E664, doi:10.1073/pnas.1107960108.

Shapiro, S.A., C. Dinske, and E. Rothert. 2006. Hydraulic-fracturing controlled dynamics of microseismic clouds, *Geophysical Research Letters*, 33, L14312.

Skytruth. 2013. Fracking Chemical Database. Retrieved from http://frack.skytruth.org/fracking-chemical-database.

Stepan, D.J., R.E. Shockey, B.A. Kurz, N.S. Kalenze, R.M. Cowan, J.J. Ziman, and J.A. Harju. 2010. Bakken Water Opportunities Assessment—Phase I, Technical Report Number 2010-EERC-04-03, North Dakota Industrial Commission, Bismarck, ND. http://www.nd.gov/ndic/ogrp/info/g-018-036-fi.pdf.

Stringfellow, W.T., J.K. Domen, M.K. Camarillo, W.L. Sandelin, and S. Borglin. 2014. Physical, chemical, and biological characteristics of compounds used in hydraulic fracturing, *Journal of Hazardous Material*, 275, 37–54.

Thiercelin, M.C., and J.-C. Roegiers. 2000. Formation characterization: Rock mechanics. In: *Reservoir Stimulation*, M.J. Economides, and K.G. Nolte, Eds. p. 856, John Wiley & Sons, West Sussex, England.

U.S. EPA (Environmental Protection Agency). 1998. Guidelines for Ecological Risk Assessment, EPA/630/R-95/002F, Washington, DC, http://www2.epa.gov/sites/production/files/2014-11/documents/eco_risk_assessment1998.pdf.

U.S. EPA (Environmental Protection Agency). 2003. Generic Ecological Assessment Endpoints (GEAEs) for Ecological Risk Assessment, EPA/630/P-02/004F, Washington, DC, http://www.epa.gov/osainter/raf/publications/pdfs/GENERIC_ENDPOINTS_2004.PDF.

U.S. EPA (Environmental Protection Agency). 2004. Chapter 4: Hydraulic fracturing fluids. In: *Evaluation of Impacts to Underground Sources of Drinking Water by Hydraulic Fracturing of Coalbed Methane Reservoirs*, EPA 816-R-04-003, Washington, DC.

U.S. EPA (Environmental Protection Agency). 2012. Study of the Potential Impacts of Hydraulic Fracturing on Drinking Water Resources: Progress Report, EPA 601/R-12/011, Office of Research and Development, Washington, DC, http://www2.epa.gov/sites/production/files/documents/hf-report20121214.pdf.

U.S. EPA (Environmental Protection Agency). 2015a. Assessment of the Potential Impacts of Hydraulic Fracturing for Oil and Gas on Drinking Water Resources (External Review Draft). EPA/600/R-15/047. U.S. Environmental Protection Agency, Washington, DC.

U.S. EPA (Environmental Protection Agency). 2015b. Analysis of Hydraulic Fracturing Fluid Data from the FracFocus Chemical Disclosure Registry 1.0: Data Management and Quality Assessment Report, EPA/601/R-14/006, U.S. Environmental Protection Agency, Washington, DC.

U.S. GAO (Government Accountability Office). 1989. Drinking water, safeguards are not preventing contamination from injected oil and gas wastes, GAO/RCED-89-97. Available from: http://www.gao.gov/assets/150/147952.pdf.

U.S. House of Representatives Committee on Energy and Commerce. 2011. *Chemicals Used in Hydraulic Fracturing*, Washington, DC, http://democrats.energycommerce.house.gov/sites/default/files/documents/Hydraulic-Fracturing-Chemicals-2011-4-18.pdf.

Veil, J., M. Puder, D. Elcock, and R. Redweik. 2004. A White Paper Describing Produced Water from Production of Crude Oil, Natural Gas, and Coal Bed Methane, Argonne National Laboratory, Prepared for: U.S.

Department of Energy, National Energy Technology Laboratory, Contract W-31-109-Eng-38. Retrieved from http://ww.ipd.anl.gov/anlpubs/2004/02/49109.pdf.

Vidic, R.D., S.L. Brantley, J.M. Vandenbossche, D. Yoxtheimer, and J.D. Abad. 2013. Impact of shale gas development on regional water quality, *Science*, 340(6134), 1235009, doi:10.1126/science.1235009.

Walker, T., S. Kerns, D. Scott, P. White, J. Harkrider, C. Miller, and T. Singh. 2002. Fracture stimulation optimization in the redevelopment of a mature waterflood, Elk Hills Field, California. *SPE Western Regional/AAPG Pacific Section Joint Meeting*, Society of Petroleum Engineers, Anchorage, AK, May 20–22.

Warner, N.R., T.M. Kresse, P.D. Hays, A. Down, J.D. Karr, R.B. Jackson, and A. Vengosh. 2013. Geochemical and isotopic variations in shallow groundwater in areas of the Fayetteville Shale development, north-central Arkansas, *Applied Geochemistry*, 35, 207–220. Retrieved from http://linkinghub.elsevier.com/retrieve/pii/S0883292713001133.

Warpinski, N.R., M.J. Mayerhofer, M.C. Vincent, C.L. Cipolla, and E.P. Lolon. 2009. Simulating unconventional reservoirs: Maximizing network growth while optimizing fracture conductivity. *Journal of Canadian Petroleum Technology*. 48(10): 39–51.SPE-114173-PA. doi: 10.2118/114173-PA.

Watson, T., and S. Bachu. 2009. Evaluation of the potential for gas and CO_2 leakage along wellbores, *SPE Drilling & Completion*, 24(1), 115–126.

Weijermars, R. 2011. Analytical stress functions applied to hydraulic fracturing: Scaling the interaction of tectonic stress and unbalanced borehole pressures. In: *45th U.S. Rock Mechanics/Geomechanics Symposium, ARMA 11-598*, pp. 1–13, American Rock Mechanics Association, San Francisco, CA.

Weijers, L., C. Wright, M. Mayerhofer, and C. Cipolla. 2005. Developing calibrated fracture growth models for various formations and regions across the United States. In: *SPE 96080, SPE Annual Technical Conference and Exhibition*, pp. 1–9, Society of Petroleum Engineers, Dallas, TX.

Wilson, J.M., and J.M. VanBriesen. 2012. Oil and gas produced water management and surface drinking water sources in Pennsylvania, *Environmental Practice*, 14(04), 288–300.

Wright, C.A., E.J. Davis, L. Weijers, W.A. Minner, C.M. Hennigan, and G.M. Golich. 1997. Horizontal hydraulic fractures: Oddball occurrences or practical engineering concern? In: *SPE 38324, SPE Western Regional Meeting*, pp. 1–14, Society of Petroleum Engineers, Long Beach, CA.

Section III

Flow Modeling

8

Significance of Dar Zarrouk Parameters in Groundwater Management and Aquifer Protection

D. C. Singhal

CONTENTS

8.1 Introduction

Groundwater is an important resource for sustaining agriculture in India and many other parts of the world. The country, with its population crossing the 1.25 billion mark, is also dependent on this resource for meeting its drinking requirements. The availability of noncontaminated groundwater requires the use of efficient methods for locating potential aquifers. Owing to such reasons, the use of geophysical methods has been playing an increasingly important role in groundwater exploration. Surface electrical resistivity methods of groundwater investigation are one of the most powerful techniques for locating aquifers up to depths of a few hundred meters. The judicial use of the resistivity methods enables selection and pinpointing of borehole sites for locating productive aquifers containing quality groundwater. An optimum combination of drilling and geophysical resistivity investigations can provide reliable data for delineation of viable aquifers on a regional basis.

The evaluation of hydraulic characteristics of aquifers often necessitates conducting pumping tests of prolonged duration on the drilled boreholes for pumping and observation purposes, which entail a huge requirement of funds. Such a large expenditure can be reduced considerably by employing the alternative approaches of surface resistivity investigations. It was demonstrated by Ungemach et al. (1969), Kelly (1977), and Sri Niwas

and Singhal (1981, 1985) that surface electrical resistivity techniques that cover large volumes of earth materials can enable the estimation of hydraulic parameters of aquifers at the aquifer scale. The subsequent works by Kosinski and Kelley (1981), Kelly and Reiter (1984), Mazac et al. (1985), Shakeel et al. (1998), Singhal et al. (1998), and Sinha et al. (2009) helped in extending these relationships to include variation in groundwater quality as well as anisotropy of aquifers.

In this chapter, the application of the concept of transverse resistance has been reviewed in the evaluation of hydraulic parameters of aquifers having groundwater with varying quality and anisotropy using field data of Ganga–Yamuna Interfluve in North India. An attempt has also been made for assessing the protective capacity of the unconfined alluvial aquifer in Saharanpur town from the "total longitudinal conductance" of the unsaturated overburden.

8.2 Theoretical Background

Geoelectrical methods can be used for the estimation of aquifer properties, from field measurements of resistivity of aquifers. The geoelectrical methods are based on the analogy between Darcy's law and Ohm's law relating hydraulic conductivity with electrical conductivity.

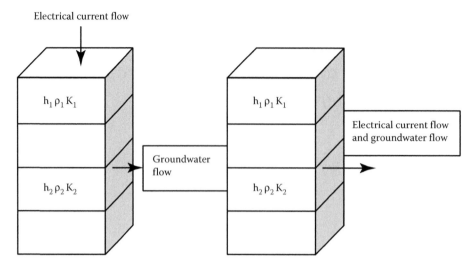

FIGURE 8.1
Layered model showing transverse and longitudinal current flow.

In a clay-free porous formation, fully saturated with water, the formation resistivity factor (F) is given as

$$F = \rho_o / \rho_w \qquad (8.1)$$

where ρ_o is the resistivity of the water-saturated formation, and ρ_w is the resistivity of groundwater in the aquifer.

For a two-layer formation, having thicknesses of h_1 and h_2 and the corresponding electrical resistivities ρ_1 and ρ_2, the average longitudinal resistivity (ρ_l) and transverse resistivity (ρ_t) are given by the following equations:

$$(h_1 + h_2)\rho_t = h_1\rho_1 + h_2\rho_2 \qquad (8.2)$$

and

$$(h_1 + h_2)/\rho_l = (h_1/\rho_1) + (h_2/\rho_2) \qquad (8.3)$$

Based on the above approach, Maillet (1947) introduced the concept of transverse unit resistance (R) and longitudinal unit conductance (S), which can be expressed as

$$R = \sum_{i=1}^{n} \rho_i h_t \qquad (8.4)$$

$$S = \sum_{i=1}^{n} h_i / \rho_i \qquad (8.5)$$

R and S are also known as Dar Zarrouk variable and Dar Zarrouk function, respectively. These parameters play a significant role in the interpretation of resistivity-sounding data.

Owing to electrical anisotropy $(\lambda = \sqrt{\rho_t/\rho_l} > 1)$ in electrical prospecting, a layer of thickness, h, average conductivity, σ, and anisotropy, λ, will be found to be exactly equivalent in its outside effects to an isotropic layer of thickness, λh, and conductivity, σ_l (Maillet, 1947).

Figure 8.1 shows the layered model, which assumes horizontal groundwater flow with vertical electrical flow in the transverse case and horizontal electrical flow in the longitudinal case.

After generalization, Equations 8.2 and 8.3 can be rewritten as

$$\text{Transverse resistivity } (\rho_t) = \frac{\sum_{i=1}^{n} h_i\rho_i}{\sum_{i=1}^{n} h_i} \qquad (8.6)$$

$$\text{Longitudinal resistivity } (\rho_l) \frac{\sum_{i=1}^{n} h_i}{\sum_{i=1}^{n} h_i/\rho_i} \qquad (8.7)$$

8.3 Overview of Available Relations

Sri Niwas and Singhal (1981) proposed that Equations 8.4 and 8.5 offer the possibilities of estimating transmissivity and hydraulic conductivity from the values of transverse resistance and longitudinal conductance, once the nature of variation of products (K σ) and K/σ is known. Three types of aquifer materials, namely, gravel, coarse sand, and sand with clay, were considered for discussion. It could be established that if the hydraulic conductivity of these aquifer materials decreased, their electrical conductivity also increases in the same order. This implied that an assumption could be made that the product "$K\sigma$" remained unchanged in areas with a similar geologic setting and water quality. Under these conditions, the following equations were found to be useful for calculating aquifer transmissivity (T) from surface resistivity data (Sri Niwas and Singhal, 1981):

$$T = \alpha R \qquad (8.8)$$

where $\alpha = K\sigma$ (a constant)

For incorporating the effects of variation in the quality of groundwater, Equation 8.10 was modified by Singhal and Sri Niwas (1983) as

$$K = \alpha' \rho' \qquad (8.9)$$

where

$$\alpha' = (K\sigma') = \text{constant for an aquifer} \qquad (8.10)$$

and

$$\sigma' = \frac{\sigma \rho_w}{\overline{\rho_w}}, \quad \text{or} \quad \rho' = \rho\left[\frac{\overline{\rho_w}}{\rho_w}\right] \qquad (8.11)$$

where $\overline{\rho_w}$ is the average groundwater resistivity, σ' is the modified aquifer conductivity, and ρ' is the modified aquifer resistivity. Arising from the above, the relation between transmissivity (T) and hydraulic conductivity (K) for varying groundwater quality will be

$$T = (K\sigma')R' \qquad (8.12)$$

where $R' = $ modified transverse resistance $= R\left[\overline{\rho_w}/\rho_w\right]$.

The value of $K\sigma'$ is area specific depending on the rock type and water quality. It was estimated to be 0.5–1.15 for alluvial formations of the districts of Saharanpur, Banda, and Varanasi in North India (Sri Niwas and Singhal, 1985).

8.4 Effect of Anisotropy of Aquifers

The above approach was used for 23 sites in parts of the Ganga–Yamuna Interfluve for anisotropic aquifers in Saharanpur–Roorkee areas of North India (Figure 8.2). Initially, using the average value of aquifer water resistivity (measured from collected groundwater samples), an average aquifer water resistivity (17.1 Ω-m) and

"modified transverse" resistance (R') have been calculated. Figure 8.3 shows a scatter plot of transmissivity (T) and modified transverse resistance (R). The following linear relationship is obtained with a correlation coefficient of 0.92 (RMS error; 238.57):

$$T = 0.1653 R' + 209.08 \qquad (8.13)$$

However, when the transmissivity values are sorted and plotted separately on the basis of hydraulic units 1 and 2 (of the above interfluve), which are composed of relatively more clayey aquifer in the first unit (having Yamuna deposition) as compared to the relatively clay-free second unit (having Ganga deposition), the plot (Figure 8.3) showed two parallel lines with lesser scatter. Yet, it is worth noticing here that the slopes of the lines are approximately the same with a lateral shift that increases with anisotropy. The linear relationship for the more clayey hydraulic unit-1 (Yamuna deposition) took the following form (Sinha et al., 2009):

$$T = 0.159 R' + 402.05 \qquad (8.14)$$

And for the clay-free hydraulic unit-2 (Ganga deposition), the relationship was found as

$$T = 0.1682 R' - 18.716 \qquad (8.15)$$

Equations 8.13 through 8.15 were used to compute transmissivity and were compared with the observed data in Table 8.1. It was revealed that the values computed from Equations 8.14 and 8.15 for separate hydraulic units (1 and 2) were generally closer to measured values in comparison to those computed from Equation 8.13 for the two units taken together. This reflects the effect of increased anisotropy of the sediments on the transmissivities.

In the case of anisotropic aquifers, if the aquifer is clayey and is underlain by a conducting matrix, the dominant current flow is vertical and there is a component of electrical current flow in the lateral direction as well (Sri Niwas and Lima, 2003). However, if in such an aquifer some dispersed clay is present along with the sand, the horizontal component of current may be significant due to the presence of the conducting clay. Kelly and Reiter (1984) proposed a three-parameter model of the following form for aquifers having constant anisotropy:

$$K = A \cdot B^n \rho_l^{\,m} \qquad (8.16)$$

where $B\ (= K_h/K_v)$ is the hydraulic anisotropy with K_v and K_h being the vertical and horizontal hydraulic conductivities, respectively. The exponent n varies with

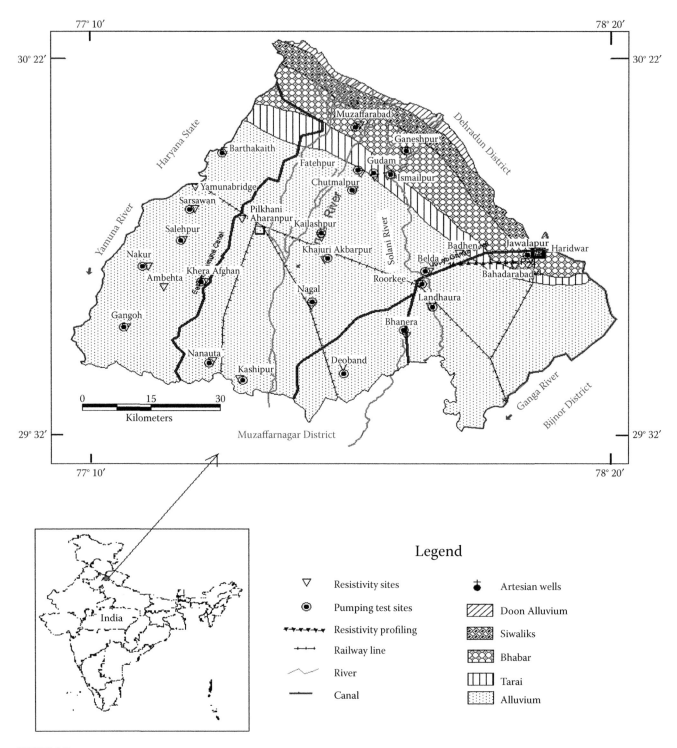

FIGURE 8.2
Location map of Ganga–Yamuna Interfluve.

anisotropy in Equation 8.16 whereas coefficient A and exponent m can be empirically derived from the following relationship (Mazac and Landa, 1979) between formation factor (FF) and hydraulic conductivity (K):

$$K = A.FF^m \qquad (8.17)$$

The values of the coefficient (A) and the exponent (m) in Equation 8.17 were found to be 4.81 and 0.84, respectively, for the Ganga–Yamuna Interfluve. Thus, Equation 8.17 can be written as

$$K = 4.81(FF)^{0.84} \qquad (8.18)$$

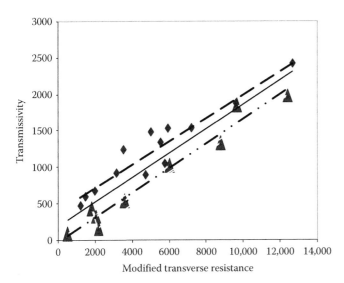

FIGURE 8.3

Transmissivity (m²/day) plotted against modified transverse resistance (ohm-m²) for different hydraulic units. The solid line in the middle represents the linear relationship when both hydraulic units are combined to one unit. Dashed lines above and below the solid line represent the linear relationships for hydraulic unit-1 and unit-2, respectively. (Adapted from Sinha, R., M. Israil, and D.C. Singhal. 2009. *Hydrogeology Journal*, 17: 495–503.)

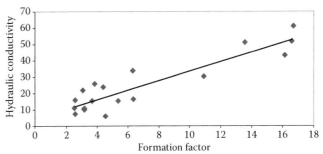

FIGURE 8.4

Hydraulic conductivity (m/day) plotted against FF for Ganga–Yamuna Interfluve. (Adapted from Sinha, R., M. Israil, and D.C. Singhal. 2009. *Hydrogeology Journal*, 17: 495–503.)

The plot of the empirical relationship (Equation 8.18) is shown in Figure 8.4 for the field data from the Ganga–Yamuna Interfluve. The analysis is restricted to unconsolidated sediments where aquifers are anisotropic due to layering of fine and coarse sediments with dispersed clay.

The relationship is likely to be further influenced by the longitudinal resistivity if the aquifer is underlain

TABLE 8.1

Observed and Computed Transmissivity Values for Ganga–Yamuna Interfluve Using Modified Transverse Resistance in Different Equations

Name of the Site	Modified Transverse Resistance (ohm-m²)	Observed Transmissivity (m²/day)	Calculated Transmissivity Equation 8.13 (m²/day)	Calculated Transmissivity Equation 8.14[a]	Calculated Transmissivity Equation 8.15[b]	Hydraulic Unit
Gangoh	1463.39	595	450.92	635.75	–	1
Nagal	1969.83	674.16	534.63	716.63	–	1
Nanauta	1197.99	472.09	407.05	593.37	–	1
Nakur	5917.77	1532.29	1187.23	1347.12	–	1
Khajuri Akbarpur	4712.93	895	988.07	1154.70	–	1
Fatehpur	5519.85	1340.41	1121.45	1283.57	–	1
Kashipur	7190.18	1536.72	1397.56	1550.32	–	1
Barthakaith	3510.59	1240	789.32	962.69	–	1
Khera Afghan	12,674.07	2420	2304.04	2426.10	–	1
Salehpur	5749.90	1050	1159.48	1320.31	–	1
Sarsawan	4999.03	1481.49	1035.36	1200.40	–	1
Kailashpur	9600.62	1874.81	1796.00	1935.27	–	1
Gudam	3137.04	920	727.57	903.04	–	1
Muzafrabad	502.61	95.95	292.10	–	65.82	2
Bhanera	6016.96	1028.56	1203.62	–	993.33	2
Belda	3604.42	541.58	804.83	–	587.54	2
Landhaura	8787.97	1335	1661.67	–	1459.42	2
Chutmalpur	2158.76	167.58	565.86	–	344.38	2
Deoband	1803.62	439.88	507.16	–	284.65	2
Ismailpur	2036.95	338.4	545.73	–	323.89	2
Ganeshpur	3594.87	542.6	803.25	–	585.94	2
Jawalapur	12,432.61	1985	2264.13	–	2072.44	2
Roorkee	9678.26	1850	1808.84	–	1609.16	2

[a] Refers to Yamuna deposition.
[b] Refers to Ganga deposition.

by an infinitely resistive basement and by transverse resistivity if it is underlain by a conductive bottom layer (Sri Niwas and Lima, 2003).

In view of the above logic, the model (Equation 8.18) was modified by introducing the longitudinal (ρ_1') and transverse resistivity (ρ_t') component separately (Sinha et al., 2009)

$$K_h = A.B^n (\rho_t')^m (\rho_1')^\lambda \qquad (8.19)$$

where A and m are the empirical constants derived from Equation 8.18. The value of n, which varies with anisotropy (λ), will depend on the percentage of dispersed clay present in the aquifer.

By fitting the model to the observed field data, the values of the unknown parameters were obtained, and after substituting these values, Equation 8.19 was rewritten as

$$K_h = 4.81.B^{-0.43} (\rho_t')^{-0.85} (\rho_1')^{-0.17} \qquad (8.20)$$

It was observed that the values estimated from the proposed model (Equation 8.20) are quite close to the observed ones (Table 8.2). Thus, the modified relationship (Equation 8.20) shows a better fit with the values obtained from field observations in comparison to the values estimated by the existing models.

TABLE 8.2

Observed and Computed Hydraulic Conductivity (K_h) Values for Saharanpur–Roorkee Area Using the Proposed Model (Equation 8.20)

Name of Site	Observed Hydraulic Conductivity (m/day)	Calculated Hydraulic Conductivity (m/day)
Gangoh	25.87	24.23
Nagal	33.71	30.90
Nanauta	21.76	20.43
Nakur	61.29	55.78
Khajuri Akbarpur	10.17	9.68
Fatehpur	15.23	14.03
Kashipur	51.22	46.44
Barthakaith	15.7	14.65
Khera Afghan	23.63	21.80
Salehpur	10.71	9.83
Muzaffarabad	10.49	9.61
Bhanera	51.43	46.63
Belda	16.41	16.29
Landhaura	43.06	38.50
Deoband	11.00	10.04

Source: Adapted from Sinha, R., M. Israil, and D.C. Singhal. 2009. *Hydrogeology Journal,* 17: 495–503.

8.5 Groundwater Protection

Arising from Equation 8.5, the magnitude of "total longitudinal conductance" of the unsaturated zone overlying an unconfined aquifer can be used for estimating the degree of its protection from the pollutants infiltrating and percolating into an aquifer. The ability of the overburden to retard and filter percolating waste effluents is a measure of its protective capacity (Belmonte et al., 2004). The longitudinal conductance (S) gives a measure of the permeability of the confining clay/shale layers. Such layers have low hydraulic conductivity and low resistivity. Protective capacity (P_c) of the overburden layers is proportional to its longitudinal conductance (S) (Braza et al., 2006). Therefore

$$P_c = S = \sum_{i=1}^{n} h_i / \rho_i \qquad (8.21)$$

Besides, for aquifers underlain by a resistive basement rock (represented by H-type field resistivity curves), the depth to the hard rock can be readily estimated by evaluating the total longitudinal conductance of the overburden by the asymptotic method (Bhattacharya and Patra, 1968).

8.6 Evaluation of Protective Capacity of the Vadose Zone: Field Application

Field data of 33 numbers of vertical electrical resistivity and induced polarization soundings recorded in the Saharanpur town of Western Uttar Pradesh, India (Figure 8.5, after Sharma, 2007) have been utilized for evaluating the thickness of the overburden overlying the unconfined aquifer of the area. It was observed from the computations that the total longitudinal conductance of the clayey sediments overlying the unconfined aquifer varied between 0.03 and 0.74 mho.

Figure 8.6 showing contours of total longitudinal conductance of the overburden for the Saharanpur town indicates that areas toward the northwest, west, and southeast parts broadly aligned in WNW–ESE direction possess high longitudinal conductance (characterized by a lighter shade) near VES locations 9, 22, and 23. This indicates that these parts of the town offer greater protection to the underlying unconfined aquifer from the pollutants percolating from the surface through the clays, which tend to absorb the effluents. On the other hand, the localities toward north–northeast, south, and

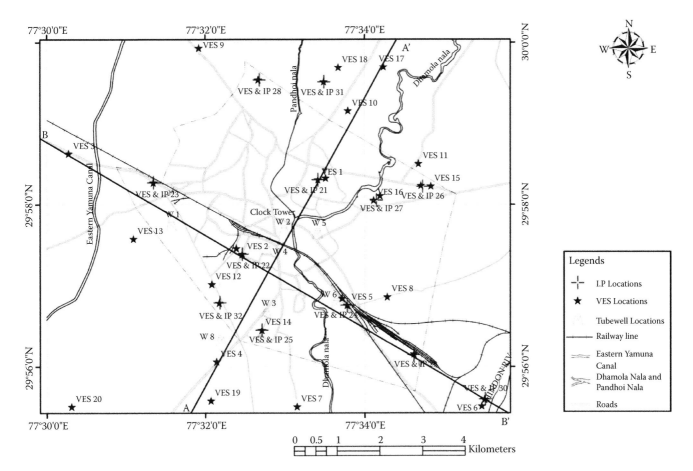

FIGURE 8.5
Location of VES/IP soundings in Saharanpur city.

southwest (shown by dark shades) around VES sites 1, 4, 7, 8, 10, 16, 17, 20, 27, 28, 30, and 31 offer a relatively lesser degree of protection to the unconfined aquifer from the infiltrating effluents. This conclusion appears to correlate fairly well with the groundwater vulnerability map prepared using the DRASTIC approach (Singhal et al., 2008).

Such differentiation of the area on the basis of the total electrical conductance distribution in an alluvial area can prove to be of immense value for planning protection of groundwater resources.

8.7 Conclusions

A review of research developments in the area of evaluation of hydraulic characteristics of aquifers from surface geophysical resistivity data indicates that transverse resistance of an aquifer can be successfully used to evaluate its transmissivity and hydraulic conductivity especially in the alluvial areas even in cases of varying quality of groundwater and when anisotropy of the aquifer is significant due to the presence of varying amounts of clay. This has been demonstrated by comparing the transverse unit resistance data of two hydraulic units (having varying clay contents) located in Ganga–Yamuna Interfluve of Saharanpur–Roorkee areas of North India (Sinha et al., 2009). Such estimations can substantially contribute to reducing capital costs incurred in indiscriminate drilling of tubewells for conducting pumping tests. Further, the total longitudinal conductance of the overburden occurring above an aquifer can give a fair idea of its protective capacity of the clayey sediments against waste effluents percolating from the surface. The validity of the concept has been demonstrated, in this chapter, from the analysis of resistivity data for Saharanpur town of Uttar Pradesh (India). Thus, the estimation of the Dar Zarrouk parameters can be meaningfully used in effective groundwater management and aquifer protection planning.

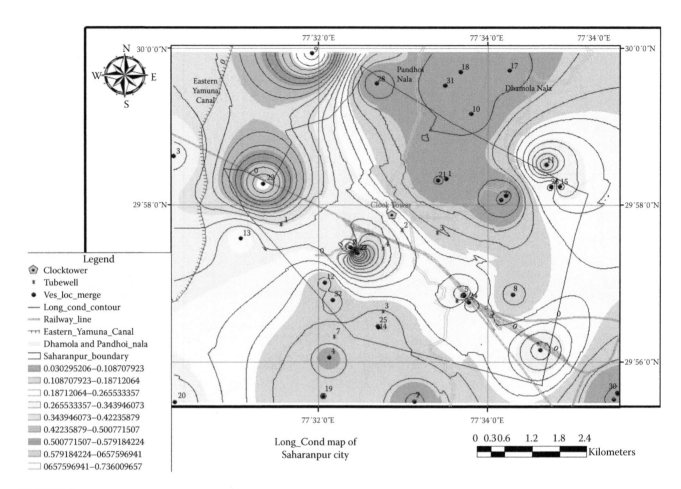

FIGURE 8.6
Map showing the total longitudinal conductance of overburden above the unconfined aquifer in Saharanpur city.

Acknowledgments

The assistance rendered by Manish Tungaria and Iqbal Sheikh, MTech, hydrology students (IITR), in preparing Figure 8.6 is gratefully acknowledged.

References

Belmonte, S.J., J.O. Enrigueze, and M.A. Zamora. 2004. Vulnerability to contamination of Zaachila Aquifer, Oaxaca, Mexico. *Geophysical International*, 44(3): 283–300.

Bhattacharya, S.K. and P.K. Patra. 1968. *Direct Current Geoelectrical Sounding*, Elsevier, Amsterdam, 135p.

Braza, A.C., M.F. Walter, W.M. Filho, and J.C. Dourado. 2006. Resistivity (DC) method applied to Aquifer Protection Studies. *Revisita Journal of Geophysics*, 24(4): 573–581.

Kelly, W.E. 1977. Geoelectric sounding for estimating aquifer hydraulic conductivity. *Groundwater*, 15(6): 420–425.

Kelly, W.E. and P.F. Reiter. 1984. Influence of anisotropy on relations between electrical and hydraulic properties. *Journal of Hydrology*, 74:311–321.

Kosinski, W.K. and W.E. Kelly. 1981. Geoelectric sounding for predicting aquifer properties. *Groundwater*, 19:163–171.

Maillet, R. 1947. The fundamental equations of electrical prospecting. *Geophysics*, 12(4): 529–556.

Mazac, O. and I. Landa. 1979. On the determination of hydraulic conductivity and transmissivity of granular aquifer by vertical electrical sounding. *The Journal of Geological Science*, 16:123–139.

Mazac, O., W.E. Kelly, and I. Landa. 1985. A hydrogeophysical model for relations between electrical and hydraulic properties of aquifers. *Journal of Hydrology*, 79:1–19.

Shakeel, A., G. De Marsily, and A. Talbot. 1998. Combined use of hydraulic and electrical properties of an aquifer in a geostatistical estimation of transmissivity. *Groundwater*, 26(1): 78–86.

Sharma, V.K. 2007. Groundwater occurrence and quality in Saharanpur Town, India: Impact of urbanization. PhD thesis (unpublished). Indian Institute of Technology, Roorkee, India.

Singhal, D.C. and Sri Niwas. 1983. Estimation of aquifer transmissivity from surface geoelectrical measurements. *Proceedings of UNESCO Symposium on Methods and Instrumentation of Investigating Groundwater Systems, Noordwijkerhout*, The Netherlands, May 1983: 405–414.

Singhal, D.C., Sri Niwas, M. Shakeel, and E.M. Adam. 1998. Estimation of hydraulic characteristics of alluvial aquifers from electrical resistivity data. *Journal of the Geological Society of India*, 51:461–470.

Singhal, D.C., H. Joshi, A.K. Seth, and V.K. Sharma. 2008. Groundwater quality in alluvial aquifers of North India: Strategy for protection. *Proceedings of the 36th IAH Congress,* Toyama, Japan, October 2008.

Sinha, R., M. Israil, and D.C. Singhal. 2009. A hydrogeophysical model of the relationship between geoelectric and hydraulic parameters of anisotropic aquifers. *Hydrogeology Journal*, 17: 495–503.

Sri Niwas and D.C. Singhal. 1981. Estimation of aquifer transmissivity from Dar-Zarrouk parameters in porous media. *Journal of Hydrology*, 49: 393–399.

Sri Niwas and D.C. Singhal. 1985. Aquifer transmissivity of porous media from resistivity data. *Journal of Hydrology*, 82:143–153.

Sri Niwas and O.A.L. de Lima. 2003. Aquifer parameter estimation from surface resistivity data. *Groundwater*, 41(1): 94–99.

Ungemach, P., F. Mosthaghini, and A. Duport. 1969. Essai de determination du coefficient d'emmagasinementennapprelibre: Application a la nappe alluviale du Rhin. *Bulletin of International Association of Scientific Hydrology*, 14(3): 160–190.

9

Estimation of Stream Conductance

Govinda Mishra

CONTENTS

9.1 Introduction

Streams are important components of earth's ecosystem. They are the natural drains of watersheds. A part of the precipitation falling within the boundary of a watershed is drained by a stream. Depending on the hydrogeology and meteorological conditions prevailing in a region, a stream may be an ephemeral or a perennial one. An ephemeral stream flows intermittently for a short period after a major rainfall event, whereas a perennial stream flows throughout a water year. An aquifer adjacent to a stream gets recharged during a rise in the stream stage above the groundwater table in the aquifer. A part of the recharged water returns back to the stream after the stream stage declines. The effluent seepage contributes to the lean flow in a perennial stream. Stream aquifer interaction has been analyzed by several hydrologists.

In a steady or quasi-steady-state flow condition, the exchange of flow between a hydraulically connected stream and an aquifer, in which the water table lies at a shallow depth, is often assumed to be linearly proportional to the boundary potential difference (Ernst 1962, Aravin and Numerov 1965, Herbert 1970, Morel-Seytoux and Daly 1975, Besbes et al. 1978, Flug et al. 1980). Hydraulic connection implies that when there is a rise in water level in the aquifer below the stream bed, the seepage from the stream gets reduced and vice versa. Bouwer (1969) has reported that the seepage from a canal to an aquifer is directly proportional to the difference in the water levels in the canal and in the aquifer in the vicinity of the canal, where the water table lies at a shallow depth. Theoretically, in the case of steady and confined flow, the relation between seepage

and boundary potential difference causing flow is linear. The constant of proportionality is known as reach transmissivity (Morel-Seytoux and Daly 1975), which is related to the hydraulic conductivity of the aquifer medium, the stream cross section, and the distance of the observation well location from the stream boundary at which the head difference is measured. The reach transmissivity constant is specific for the piezometer at which the potential difference is measured (Morel-Seytoux et al. 1975). The seepage from the stream under steady as well as quasi-steady-state flow is estimated by multiplying the difference in hydraulic heads at the stream boundary and at a piezometer or at an observation well in the vicinity of the stream with the corresponding reach transmissivity constant. Several investigators (Aravin and Numerov 1965, Bouwer 1969) have analyzed steady seepage from a canal relating to various canal geometry and boundary conditions. These solutions for steady-state flow yield reach transmissivity constants, which are applicable to streams with similar sections.

The exchange of flow between a stream and an aquifer is also linearly related to the difference in heads at the stream boundary and at a point on the stream line in the aquifer under the stream axis. The constant of proportionality is known as the stream conductance. However, construction of a piezometer under a stream is not practicable. Alternatively, the piezometric head at any location can be predicted by numerical modeling of groundwater flow in a stream aquifer system. For simulation of a piezometric surface, the flow domain is discretized by a grid pattern, and at each grid node a mass balance equation is written. The set of simultaneous linear equations in terms of unknown hydraulic heads are solved in a discrete time domain satisfying the initial and boundary conditions. The stream is treated as a head-dependent-type hydrologic boundary. In the mass balance equation near the stream nodes, the exchange between the stream reach and the aquifer is incorporated assuming the exchange rate to be linearly proportional to the difference in stream stage and the unknown piezometric head at the node under the stream axis. The constant of proportionality is known as stream conductance or river constant.

9.2 Herbert's Stream Conductance for Confined Flow Condition

Using Darcy's law for radial flow from a partially penetrating stream with a semicircular cross section, Herbert (1970) derived an expression relating influent seepage to potential difference between the stream and at half of the aquifer thickness below the stream bed. Herbert's stream constant for the idealized stream and confined aquifer system is presented in Figure 9.1.

Let the maximum depth of water in the stream be d_w and the thickness of the aquifer below the stream bed be D_1. Let a datum be chosen at the bottom of the stream bed. Hence, the hydraulic head along the stream boundary, that is, at $r = r_r$ is d_w. Let the hydraulic head in the aquifer at $r = (D_1 + r_r)/2$ be h_a as shown in Figure 9.1. The hydraulic head h_a is unknown unless it is measured in a piezometer, or predicted by a groundwater flow model.

In a steady-state flow condition, the radial flow q_r passing through a semicircular flow area at a radial distance r according to Darcy's law is given by

$$q_r = -\pi r k \frac{dh(r)}{dr} \tag{9.1}$$

$$dh(r) = -\left(\frac{q_r}{k\pi}\right)\frac{dr}{r} \tag{9.2}$$

Integrating

$$h(r) = -\left(\frac{q_r}{k\pi}\right)\ln r + A \tag{9.3}$$

At $r = r_r, h(r_r) = d_w$, and at $r = (r_r + D_1)/2, h\{(r_r + D_1)/2\} = h_a$. Applying these boundary conditions

$$q_r = \frac{k\pi}{\ln\{0.5(D_1 + r_r)/r_r\}}(d_w - h_a) \tag{9.4}$$

Hence, Herbert's stream conductance Γ_s, is

$$\Gamma_s = \frac{k\pi}{\ln\{0.5(1 + D_1/r_r)\}} \tag{9.5}$$

For a natural stream, whose wetted perimeter is w_p, the radius of the equivalent semicircular section of the

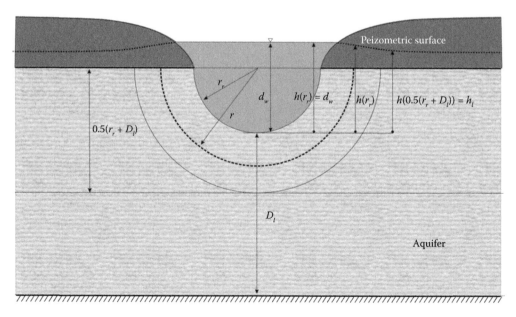

FIGURE 9.1
A partially penetrating stream in a confined aquifer with a semicircular section. (Adapted from Herbert, R., *Groundwater*, 8, 29–36, 1970.)

stream is equal to w_p/π. The stream constant for a natural stream is expressed as

$$\Gamma_s = \frac{k\pi}{\ln\{0.5(1+\pi D_1/w_p)\}} \quad (9.6)$$

A stream section is rarely semicircular in nature. When a section of a natural stream is idealized as semicircular, the depth of water in the idealized section would be very much more than the actual depth of water leading to reduction in the actual flow path under the idealized stream and resulting in a higher stream conductance. For $\pi D_1/w_p < 1$, $\ln\{0.5(1 + \pi D_1/w_p)\}$ is negative. Thus, Herbert's formula is applicable for $\pi D_1/w_p > 1$. Herbert (1970) has made an assumption that the partially penetrating stream is small implying a small width compared to the thickness of the aquifer. This assumption takes care of the above limitation.

Herbert's derivation also yields reach transmissivity constant. Let a piezometer be located at a distance l from the stream bank and let the piezometric head recorded at the piezometer be h_l. Applying the head boundary conditions, from Equation 9.3, we obtain

$$q_r = \frac{k\pi}{\ln(1+l/r_r)}(d_w - h_l) = \frac{k\pi}{\ln(1+l/r_r)}\Delta h = \Gamma_r \Delta h \quad (9.7)$$

The reach transmissivity constant is given by

$$\Gamma_r = \frac{k\pi}{\ln(1+l/r_r)} \quad (9.8)$$

The Herbert's stream conductance and the reach transmissivity constant are applicable for confined flow if the stream section conforms approximately to a semicircular section.

9.3 Modified Herbert's Stream Conductance for Confined Flow Condition

Herbert's stream constant for the partially penetrating stream in a confined aquifer system is presented in Figure 9.2. Let b_0 be the bed width of the partially penetrating stream in the confined aquifer and the depth of penetration of the stream into the aquifer be d_0. Let the slope of the stream banks be equal to m. Unlike in Herbert's derivation, we conserve b_0 and d_0 while we derive the modified Herbert's stream conductance. The width B_0 of the stream at the level of the upper impervious boundary of the confined aquifer, that is, at a height

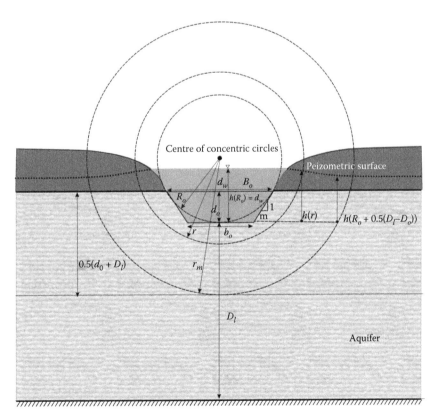

FIGURE 9.2
A partially penetrating stream in a confined aquifer section conforming to the segment of a circle.

d_0 from the stream bed level is given by $B_0 = b_0 + 2md_0$. Let R_0 be the radius of the circle whose chord is B_0. The radius R_0 is given by

$$R_0 = \frac{d_0}{2} + \frac{B_0^2}{8d_0} \tag{9.9}$$

We represent arcs of concentric circles of radii $r \geq R_0$ within the flow domain as equipotential lines and radials of the concentric circles as flow lines. But for the upper and lower impervious boundaries of the aquifer, the flow lines and the equipotential lines are orthogonal.

At any radial distance r, the flow area, $A(r)$, is given by

$$A(r) = r\left\{\pi - 2\sin^{-1}\frac{R_0 - d_0}{r}\right\} \tag{9.10}$$

Applying Darcy's law, the radial flow at any radius r is given by

$$q_r = -kr\left\{\pi - 2\sin^{-1}\frac{R_0 - d_0}{r}\right\}\frac{dh(r)}{dr} \tag{9.11}$$

Separating the variables and integrating

$$\int_{h(R_0)}^{h(r)} dh(r) = \{h(r) - h(R_0)\} = -\frac{q_r}{k}\int_{R_0}^{r}\frac{dr}{r\{\pi - 2\sin^{-1}((R_0 - d_0)/r)\}} \tag{9.12}$$

The radial distance of the concentric circle up to the middle of the aquifer is given by

$$r_m = R_0 - d_0 + 0.5(d_0 + D_1) = R_0 + 0.5(D_1 - d_0) \tag{9.13}$$

Let the hydraulic head at the middle of the aquifer under the stream bed be equal to $h(r_m)$.

$$q_r = k\frac{\{h(R_0) - h(r_m)\}}{\displaystyle\int_{R_0}^{R_0 + 0.5(D_1 - d_0)}\frac{dr}{r\{\pi - 2\sin^{-1}((R_0 - d_0)/r)\}}}$$

$$= \Gamma_s\{h(R_0) - h(r_m)\} \tag{9.14}$$

The modified Herbert's stream conductance from Equation 9.14 is given by

$$\Gamma_s = \frac{k}{I_1}; \quad I_1 = \int_{R_0}^{R_0 + 0.5(D_1 - d_0)}\frac{dr}{r\{\pi - 2\sin^{-1}((R_0 - d_0)/r)\}} \tag{9.15}$$

Substituting $r = R_0 + 0.25(D_1 - d_0)(1 + u)$; $dr = 0.25(D_1 - d_0)\,du$, the integral reduces to

$$I_1 = \frac{(D_1 - d_0)}{4}\int_{-1}^{1}\frac{du}{\begin{bmatrix}\{R_0 + 0.25(D_1 - d_0)(1+u)\}[\pi - 2\sin^{-1} \\ \{(R_0 - d_0)/(R_0 + 0.25(D_1 - d_0)(1+u))\}]\end{bmatrix}} \tag{9.16}$$

The integral I_1 can be evaluated numerically applying Gauss integration.

In general, a natural stream does not have a regular section. Let d_0 be the maximum depth of penetration of the stream into the confined aquifer and let w_p be the perimeter of the stream corresponding to the stream section within the confined flow domain. The equivalent width B_0 that preserves d_0 and w_p is given by

$$w_p = R_0\left\{\pi - 2\sin^{-1}\frac{R_0 - d_0}{R_0}\right\}$$

$$= \left\{\frac{B_0^2}{8d_0} + 0.5d_0\right\}\left\{\pi - 2\sin^{-1}\left(1 - \frac{d_0}{(B_0^2/8d_0) + 0.5d_0}\right)\right\} \tag{9.17}$$

The equivalent B_0 can be computed from Equation 9.17 following an iteration procedure.

9.4 Results

We make a comparison between Herbert's stream conductance and the revised Herbert's stream conductance in Table 9.1.

The modified stream conductance derived preserving the depth of penetration and stream width or the one derived preserving the wetted perimeter of the stream and depth of penetration should be used in stream aquifer interaction studies. As seen from Table 9.1, Herbert's stream conductance is applicable to a stream with width about one-tenth of the aquifer thickness. Stream conductance increases with the decreasing value of $d_0/(d_0 + D_1)$.

9.5 Conclusion

Herbert's stream conductance is not applicable to streams with a large width as the depth of penetration of the stream is not maintained in the idealized semi-circular stream section while preserving the wetted

TABLE 9.1

Herbert's Stream Conductance and Modified Herbert's Stream Conductance, $m = 1$

		Dimensionless Stream Conductance Γ_s/K		
$B_0/(d_0 + D_1)$	$d_0/(d_0 + D_1)$	Stream Perimeter Only Conserved	Stream Width B_0 and Penetration Depth d_0 Conserved	Stream Perimeter and Penetration Depth d_0 Conserved
2.0	0.01	12.950	16.5079	16.5749
2.0	0.005	12.732	22.2458	22.2913
1.0	0.01	4.3995	8.3077	8.3732
1.0	0.005	4.3573	11.1583	11.2034
0.5	0.01	2.4744	4.2574	4.3193
0.5	0.005	2.4525	5.6483	5.6921
0.25	0.01	1.6755	2.3106	2.3649
0.25	0.005	1.6582	2.9524	2.9933
0.1	0.01	1.1641	1.2558	1.2931
0.1	0.005	1.1460	1.4446	1.4773

perimeter of the stream. Herbert's stream conductance is applicable to streams with width about one-tenth of the aquifer thickness. The stream conductance derived maintaining the depth of penetration is the appropriate parameter for stream aquifer interaction study.

Acknowledgment

The author acknowledges Mr Vikrant Vijay Singh, Research Fellow, NIH, Roorkee, and Mr Kailash Bishnoi, research scholar, WRD, IIT Roorkee for their kind help provided in preparing the chapter.

References

Aravin, V.I. and S.N. Numerov. 1965. *Theory of Fluid Flow in Underformable Porous Media*. Israel Programme for Scientific Translation, Jerusalem.

Besbes, M., J.P. Delhomme, and G. De Marsily. 1978. Estimating recharge from ephemeral streams in arid regions. A case study at Kaironan, Tunisia. *Water Resources Research*, 14(2), 281–290.

Bouwer, H. 1969. Theory of seepage from open channels. *Advances in Hydro Science*, V.T. Chow, ed., Vol. 5, Academic Press, New York, pp. 121–172.

Ernst, L.F. 1962. *Groundwater stromingen in de Verzadigde Zone en hun berckening bij de aanwezigheid van horizontale even-wijddige open leidingen*. Verslag, Landbouwk, Onderzoek, 67, 15.

Flug, M., G.V. Abi-Ghanem, and L. Duckstein. 1980. An event based model of recharge from an ephemeral stream. *Water Resources Research*, 16(4), 685–690.

Herbert, R. 1970. Modelling partially penetrating rivers on aquifers model. *Groundwater*, 8, 29–36.

Morel-Seytoux, H.J. and C.J. Daly. 1975. A discrete kernel generator for stream aquifer studies. *Water Resources Research*, 2(2), 253–260.

10

Conceptual Understanding and Computational Models of Groundwater Flow in the Indian Subcontinent

K. R. Rushton

CONTENTS

10.1 Introduction

Although groundwater has been exploited in the Indian subcontinent for several millennia, significant changes in exploitation occurred in the last few decades of the twentieth century. Important factors included the ability to construct deeper wells and the wide availability of powered pumps, many of them powered by electricity. Before these advances, abstraction and recharge tended to be in balance because it was only possible to abstract groundwater that was readily available. However, the construction of deeper wells and the ability to pump sufficient water to irrigate dry season crops means that questions have arisen about the long-term viability of this increased abstraction. A report by the World Bank (2010) has the title Deep Wells and Prudence: Towards Pragmatic Action for Addressing Groundwater OverExploitation in India; the report explains that there are more than 20 million wells in India. In 2004, a nation-wide assessment found that 29% of groundwater blocks were in the semicritical or overexploited categories with the situation deteriorating rapidly. The situation in Sri Lanka is considered by Villholth and Rajasooriyar (2010); they state that "uncontrolled groundwater use and contamination or natural poor quality, are leading to access limitations and health concerns." They explain that "despite an emerging awareness, groundwater management is in its infancy, with the attitude of groundwater development still not converted into an approach of active management" (p. 1489).

Five important features of groundwater are considered in this chapter; flow process within hard rock and alluvial aquifers, water losses in irrigation schemes, the estimation of the recharge entering the aquifer systems, artificial or enhanced recharge, and finally the exploitation of aquifers of limited saturated thickness. For each issue, fieldwork is the first essential step followed by the development of a conceptual understanding; this can lead to computational modeling.

Two broad types of hydrogeological settings are considered in detail: shallow, low-storage hard-rock aquifers and deep, high-storage alluvial aquifers. In hard-rock aquifers, it was widely considered that the construction of bore wells into the fractured zone would tap a new source of water. However, in locations where bore wells were constructed and the water was used for irrigation (e.g., for grape gardens), farmers, who had for many years used dug wells to irrigate their crops (often only growing crops on a fraction of their land during the dry season) found it necessary to deepen their wells due to a steady decline in the water table. Problems also occurred in alluvial aquifers. The construction of bore wells that are 100 m deep or more, and the availability of subsidized electricity for their pumps, resulted in a dramatic increase in crop production. Early studies suggested that these deep bore wells tapped a separate aquifer system with water originating from a distant "common recharge zone." However, following the construction of many deep bore wells, water tables began to decline. Also, pumping levels in the bore wells fell with the yield of some of the earlier bore wells decreasing substantially. During the 1980s, several investigations were carried out to understand the flow processes in hard rock and alluvial aquifers. These investigations were based on careful fieldwork and the formulation of conceptual models, followed by the construction of

computational models to confirm and quantify the conceptual understanding.

A second important issue that has been of concern for many years is the low efficiency of many irrigation schemes. In 1883, soon after the construction of canals in Punjab, measurements were made that indicated significant losses (Ahmad 1974). Waterlogging in the vicinity of canals was a clear evidence of losses. It is generally assumed that the losses are proportional to the wetted perimeter of the canal, but is this a valid assumption? This assumption was shown to be incorrect by Wachyan and Rushton (1987) during field and modeling studies. As the awareness increased that canal losses can cause serious waterlogging problems, interceptor drains, which run parallel to the canal, were installed to collect the water lost from the canal and pump it back into the canal. During a visit to an interceptor drain installed alongside a major canal in Pakistan, the engineers expressed serious concerns that, although the system had been designed using computer modeling, the pumps had insufficient capacity to discharge all the water collected by the drain. The reason was found to be that, in their conceptual understanding, the designers had assumed that the interceptor drains would only collect water lost from the canal and failed to recognize that the drains would also collect water from the overlying water table. This is a typical example of the failure of investigators to develop a valid conceptual understanding. Irrigated rice fields provide a further example of serious losses arising from irrigation. When irrigation water is provided on a rotational basis, perhaps weekly, the depth of water in a rice field following the irrigation is typically 150 mm. A rapid initial fall in the water level in the rice field frequently occurs, a fall far larger than the evapotranspiration of the crop plus the loss through the puddled bed. In a simple but imaginative experiment in which the bunds of a rice field were covered with plastic sheeting, the rate of fall of the water level in the rice field was greatly reduced indicating significant water losses through the bunds. From the development of conceptual and computational models of flooded rice fields, it was confirmed that unexpectedly large water losses do occur through the bunds of rice fields. Each of these sources of loss from irrigation schemes can provide recharge to aquifers.

Realistic estimates of the volume of water (recharge) actually entering an aquifer system are essential if overexploitation is to be avoided. Many approaches to the estimation of recharge have been developed; the methodology introduced in this chapter is based on a daily soil moisture balance. This is a direct method using the actual physical components of rainfall, crop evapotranspiration, and runoff (RO). The calculation takes account of the soil type and cropping patterns; it also represents reduced evapotranspiration when the soil moisture content is low. Irrigation can also be included in the balance. Many of the concepts and parameter values of the FAO Irrigation and Drainage Paper 56, Crop Evapotranspiration (FAO 1998) are used in the calculations. This soil moisture balance approach provides an understanding of the processes involved in the soil zone leading to a determination of how much water passes through the soil zone to become recharge. Nevertheless, for overall water resources planning, it may be impractical to carry out detailed water balances for every location, soil type, and crop type. The Central Ground Water Board (CGWB), India, has developed an approach for recharge estimation that is effective in assessing the degree of exploitation of groundwater resources. The water table fluctuation method is used to estimate the recharge. This is an indirect method because the rise in the water table is a result of recharge and other processes. It is also necessary to estimate the specific yield of the aquifer, a parameter that is difficult to determine accurately. Adjustments are also made for additional contribution to recharge such as return flow from irrigation. Despite its limitations, the CGWB methodology is a pragmatic approach to recharge estimation.

For overexploited aquifers, artificial recharge is a remedial measure that can, in part, alleviate the problems. Artificial recharge pilot projects were carried out in Gujarat in the 1980s using excess water available during the monsoon season. The detailed fieldwork and monitoring during the recharge experiments demonstrated that there are many difficulties in achieving satisfactory artificial recharge over extended periods. Artificial or enhanced recharge has become a priority in India during the past 20 years with the implementation of many schemes. However, there has been only limited detailed monitoring to assess the success of the schemes. A comprehensive investigation of three schemes (Gale et al. 2006) indicates that the actual increase in recharge may be an order of magnitude less than the value estimated during the planning stages.

Certain aquifers are considered to have a restricted potential for water supply; these include aquifers of limited saturated thickness. Examples of the successful exploitation of these aquifers are discussed. Large-diameter dug wells are suitable when water is required for small-scale irrigation with power for pumps available for only a limited time period. The potential of horizontal wells in water table aquifers of limited saturated thickness is explored. For aquifers of limited saturated thickness at greater depths, an imaginative approach using wells with an inverted screen has proved to be highly successful in a location in Bangladesh where international experts considered that only small domestic needs could be met from the aquifer system.

10.2 Groundwater Resources of Hard Rock and Alluvial Aquifers

There is a contrast between two of the commonly occurring aquifers in the Indian subcontinent. Hard-rock aquifers are relatively shallow, often with a weathered zone underlain by a fractured zone; the volume of water stored in these aquifers is limited. Alluvial aquifers often extend to considerable depths and large volumes of water are stored. When drilling equipment and powerful pumps became available, groundwater exploitation increased rapidly. This resulted in declining water tables and even greater falls in the pumping water levels, especially in alluvial aquifers. Field studies, supported by conceptual and computational modeling, showed that for both hard rock and alluvial aquifers, vertical components of flow are of major importance.

10.2.1 Hard-Rock Aquifers

10.2.1.1 Identifying the Flow Processes in Hard-Rock Aquifers

For centuries, water has been collected from the weathered zone of hard-rock aquifers by means of large-diameter dug wells. Owing to increasing demands for water or due to falling water tables, bores were sometimes constructed through the bottom of the dug well penetrating into the fractured hard rock. Later, as equipment became available for constructing wells into the fractured hard rock, numerous bore wells were drilled into what was considered to be a separate aquifer. Indeed the existing farmers were told that it was safe for bore wells to be constructed in the vicinity of their dug wells because a new groundwater source would be tapped.

However, as more wells were drilled into the fractured zone, the farmers with dug wells experienced a fall in the water table in the shallow aquifer; deepening of their dug wells became essential. Initially, the bore wells in the fractured zone provided a reliable supply of water and in some locations the water was used to irrigate grape gardens. However, within a few years, the declining yield from bore wells meant that most of the grape gardens were abandoned.

The reasons for the falling groundwater yield in both the shallow weathered and the deeper-fractured aquifers were identified from a study of a pumping test in a weathered–fractured granitic aquifer near Hyderabad, India (Rushton and Weller 1985). A bore well that penetrated 37 m, cased through the weathered zone into the fractured zone of the aquifer (Figure 10.1a). Shallow observation wells were constructed in the weathered zone to monitor the water table. Pumping, using an air-lifting technique, continued for 7.5 hours; the

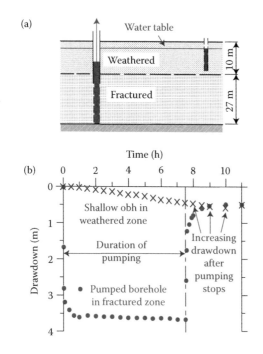

FIGURE 10.1

Pumping from a bore well in a fractured zone and response in a piezometer in a weathered zone; obh signifies an observation borehole.

flow processes as illustrated in Figure 10.2; note that this is an idealized diagram because flows in the fractured zone occur through fissures, joints, and fractures rather than the smooth flowlines of the diagram.

The next step in the investigation was to select a computational model that could represent these flow processes. A two-layer radial flow mathematical model was chosen (Rathod and Rushton 1991; Rushton 2003, Section 7.4). With suitable aquifer parameters (weathered zone, radial and vertical hydraulic conductivity $K = 0.5$ and 0.15 m/d, and specific yield $S_Y = 0.01$: fractured zone, radial and vertical $K = 4.0$ and 0.6 m/d, and confined storage coefficient $S_C = 0.004$), this model adequately reproduced the pumping and recovery phases in the bore well and in the shallow observation well. From the mathematical model results, it is possible to deduce flows from the weathered to the fractured zone. At the end of the pumping phase, the transfer from the weathered to the fractured zone was 45% of the rate of abstraction (Section 7.4.6 of Rushton 2003).

The hydrogeologists, who recommended that bore wells should be drilled in the vicinity of dug wells, failed to recognize that, although water is transmitted through the fractured zone toward the bore well, the actual source of the water is the water table in the overlying weathered zone. After a few years, these vertical flows result in a substantial fall in the water table to below the base of many dug wells. Not only do dug wells dry up but also the quantity of water pumped from the fractured zone far exceeds the recharge to the weathered zone, with the result that most bore wells fail within a few years. This analysis demonstrates that most of the water pumped from bore wells, penetrating not more than 50 m into the fractured zone, originates from the overlying weathered zone. As with alluvial aquifers (considered in Section 10.2.2), hydrogeologists found it difficult to accept that bore wells into the fractured zone did not collect water from a separate aquifer system.

final discharge rate was 288 m³/d. Drawdowns for the pumped bore well (solid circles) and one of the shallow observation wells (crosses) are plotted in Figure 10.1b.

In the pumped well, water levels in the fractured zone recovered rapidly once pumping ceased. However, in the shallow piezometers that monitor the water table elevation, water levels continued to fall for a further 2.5 hours before any recovery occurred. Even though the continuing fall in the water table is relatively small, it does indicate that there is a downward flow of water from the weathered zone to the fractured zone even when pumping stops. This finding leads to the conceptual understanding of the

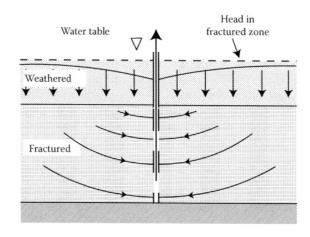

FIGURE 10.2

Schematic, idealized diagram of flows to a bore well pumping from the fractured zone and drawing water from the weathered zone.

10.2.1.2 Current Situation

The fact that in most situations the weathered and fractured zones in hard-rock aquifers act as a single system is now widely accepted. Exploitation of hard-rock aquifers has increased substantially with the result that the fall in the water table in the weathered zone causes wells to become dry before the onset of monsoon rainfall (Foster 2012). The challenges of sharing the limited groundwater available in hard-rock aquifers is addressed by Limaye (2010) who considers that, although state governments have passed groundwater protection acts controlling well digging or drilling in overexploited watersheds, wells continue to be dug or deepened.

An example of community-based groundwater management in Madhya Pradesh, India, is provided by a "well commune" of seven farmers who pump not more than two wells at any one time with the water shared between the participating farmers. The low-production wells are pumped first followed by the more-productive wells (Kulkarni et al. 2004).

An informative report by Garduno et al. (2009) highlights the serious challenges faced in hard-rock areas of Andhra Pradesh where significant declines in the water table elevation have been observed. Recharge rates are estimated to be 70–100 mm/a, but groundwater abstraction rates have grown to 120–180 mm/a. Premonsoon water tables have fallen 10–15 m between 1995 and 2005 with almost all the dug wells (which penetrate just below the bottom of the weathered zone) drying up early in the postmonsoon season. In addition, yield reduction/failure of bore wells commonly occurs. Improving irrigation efficiency through the introduction of high-technology precision irrigation techniques does not solve the problem because most of the current water losses due to seepage return to groundwater.

Community-based groundwater management has been introduced in a number of districts (Garduno et al. 2009). The objective is to equip farmers with the necessary knowledge, data, and skills to manage the groundwater resources available to them in a sustainable manner, mainly through controlling demand. For instance, at the end of the main monsoon, comparisons are made between the available groundwater reserves and the proposed abstraction of the farmers for the forthcoming dry season crops. This comparison helps the farmers to refine their cropping patterns. Farmers are free to make crop-planting decisions and extract groundwater; the project relies on the impact of groundwater education, which is offered to all the farmers.

10.2.2 Alluvial Aquifers

10.2.2.1 Identifying the Flow Processes in Alluvial Aquifers

By the 1970s, equipment became available to drill deep wells in alluvial aquifers and provide powerful pumps; this led to a dramatic increase in the availability of groundwater for irrigation. Often, the assumption was that a new source of groundwater had been found that originated from a "distant recharge area." Initially there were few bore wells pumping water from the deeper aquifer; this resulted in only small pumped drawdowns (Figure 10.3a). However, as more bore wells were drilled, the area irrigated by groundwater in Gujarat doubled between the 1970s and 1990s (Panda et al. 2012).

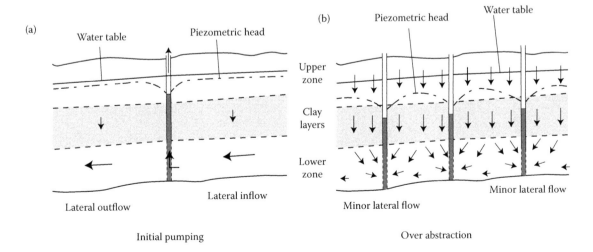

FIGURE 10.3
Schematic diagrams illustrating: (a) when exploratory bore wells are first drilled into deeper aquifers, the discharge is achieved with only a small pumped drawdown by intercepting water moving laterally through deeper aquifers; (b) when many bore wells are pumping, most water is drawn from the overlying water table. (Adapted from Rushton, K.R., *Groundwater Hydrology, Conceptual and Computational Models,* Wiley, Chichester, 2003.)

Pumping water levels declined with the result that even deeper bore wells were drilled to maintain the yields. Consequently, there was a fall in water tables in the uppermost aquifers that resulted in many dug wells becoming dry. However, the view prevailed that a new source of groundwater had been tapped by the deep bore wells and the declines in pumped water levels (pwls) and water tables were not a serious concern.

Nevertheless, a team of hydrogeologists and groundwater engineers, working on an artificial recharge pilot project in Gujarat, questioned the current understanding of flow processes in alluvial aquifers. From the interpretation of field information and the development of conceptual models, which were subsequently confirmed by computational modeling, the concept of a distant source of groundwater was shown to be incorrect.

Important insights that led to the revised understanding include

a. A careful examination of the groundwater head hydrographs in two observation wells (Figure 10.4) provided crucial information. One observation well was constructed in the shallow aquifer and the second penetrated about 200 m into the strata from which the deep bore wells were pumping. A superficial examination indicates that both hydrographs responded to the

monsoon rainfall. The shallow piezometer certainly recovers during the monsoon. However, the deeper piezometer exhibits a recovery in piezometric head in February, which is in the middle of the dry season. When the period of major pumping for a postmonsoon crop is considered (indicated by the shaded areas in the lower half of Figure 10.4) it becomes clear that the fluctuations in the deeper aquifer are primarily due to pumping from the deep bore wells. The difference between the shallow and deep piezometric heads of almost 30 m is also significant. In the vertical distance of 200 m between these observation wells, there are layers of sand, sandy clay, and clay. Assuming that the effective vertical permeability of these strata is 0.01 m/d, the vertical velocity from Darcy's law is $(30/200) \times 0.01 = 1.5$ mm/d or 550 mm in a year. This is consistent with the crop water requirement for a postmonsoon crop of wheat. This calculation is not more than a rough assessment but it does indicate that vertical flows through the overlying strata are the main source of water, not a distant recharge area. This is supported by modeling studies described in Rushton and Phadtare (1989), Rushton and Tiwari (1989), and Kavalanekar et al. (1992).

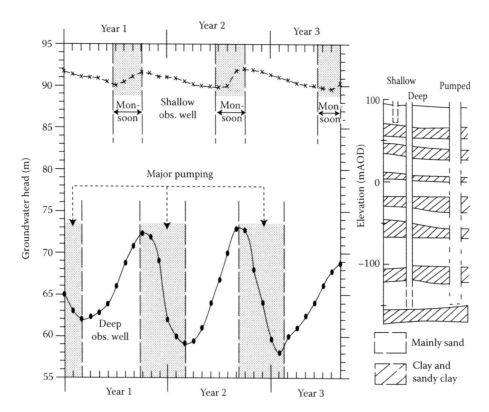

FIGURE 10.4
Fluctuation of groundwater head in shallow and deep piezometers.

b. Falls occur in piezometric heads in nonpumping observation wells in the deeper aquifers. Table 10.2 of Rushton (2003) indicates that the decline in piezometric head in an alluvial aquifer in Gujarat over a 11-year period is equivalent to 3.6 m/yr. In addition, the fall in the water table in the uppermost permeable stratum was 1–2 m/yr.

c. Decreasing specific capacities were identified in many bore wells. Table 10.2 of Rushton (2003) lists a reduction from 310 to 198 $m^3/d/m$ for one deep bore well. Further confirmation of the significance of vertical flows from the water table was gained from an investigation of one of the first bore wells to be drilled. After further deeper bore wells were constructed in its vicinity, operation of the pump in the original bore well resulted in a very small discharge. All of the bore well pumps used a mains electrical supply. In an attempt to understand the reason for the negligible yield, a generator was used to supply power so that this "problem" well could be pumped when none of the other pumps were operating. The discharge returned almost to the rate when the well was constructed. This indicates that bore wells compete for water with the deeper bore wells achieving higher yields.

These insights led to a conceptual understanding of the flow processes as illustrated schematically in Figure 10.3b. Most of the groundwater entering the deep bore wells originates from the overlying water table. Furthermore, substantial falls in the piezometric heads in the deeper aquifers are required to draw the water through the overlying lower permeability layers. Finally, these bore wells compete for water so that they can only draw water from a limited area of the overlying water table.

An important question is whether the falls both in pumping levels in the deeper aquifer and the decline in water table elevation could have been identified when bore wells were first constructed. Figure 10.3a illustrates the situation when the first deep wells were drilled. There was a small lateral flow in the deeper aquifer originating from an up-gradient area where the deeper strata outcropped. When the first bore wells were drilled, they intercepted these lateral flows and only a small pumped drawdown was required to divert some of this lateral flow into the wells. However, as many more bore wells were drilled, the lateral flows were insufficient to meet the demands; consequently, lower groundwater heads developed in the deeper aquifers to draw water from the overlying water table (Figure 10.3b). This outcome demonstrates that an initial test pumping is rarely sufficient to identify the long-term aquifer response.

Despite the above evidence it remained difficult to persuade hydrogeologists that they were not tapping a separate aquifer with a distant source of recharge. Farmers also believed that they had tapped a new source of water. However, by the early 1990s, there was an increasing recognition that deep bore wells in alluvial aquifers drew most of their water from the overlying water table.

10.2.2.2 Current Situation

Since the 1980s when the above investigations were carried out, there has been a substantial growth in the number of bore wells. For example, in north Gujarat where the study was carried out, the Central Ground Water Board (2011a) estimated that, in 2009, groundwater abstraction in districts was between 100% and 165% of the annual replenishable groundwater resource. For the basis of the CGWB methodology for these estimations, see Section 10.4.4 in this chapter and also Chatterjee and Purohit (2009).

How can abstraction be managed in deep alluvial aquifers? Unlike hard rock aquifers with limited storage, farmers do not have an unambiguous indication that overexploitation is occurring. The World Bank study (World Bank 2010) suggests that little progress is likely to be made in resolving the problems of overexploitation unless management strategies are identified for promoting sustainable groundwater use in India, within a systematic, economically sound, and politically feasible framework.

Panda et al. (2012) indicate that average declines in the water table in north Gujarat are greater than 1.7 m/yr; the fall in the pumping water levels is more substantial. Another consequence of the increasing abstraction of groundwater from the alluvial aquifer is the occurrence of high fluoride content in the groundwater in some areas of the north Gujarat aquifer. Panda et al. (2012) also consider that, with these large falls in pumping levels, only wealthy farmers can afford the increased drilling and pumping costs. Shah et al. (2008) indicated that by the year 2000, most of the talukas in the alluvial aquifers in north Gujarat were withdrawing more groundwater than the long-term average recharge. At that time, Gujarat had more than 1-million irrigation wells.

Of the many potential policies for groundwater regulation, one approach adopted in Gujarat was to separate electricity feeders for agriculture from the feeders for nonagricultural purposes. The nonagricultural users have a 24-hour supply while power is only available for agricultural users for 8 hours a day on a preannounced schedule. This scheme was implemented during 2004–2006. A preliminary, limited survey (Shah et al. 2008) indicates that the electricity consumption for agriculture is about 63% of that before the introduction of the scheme. Another positive outcome is that bore well owners and their customers know, in advance, precisely when power will be available. Previously they had to wait anxiously, not knowing when electricity would be available so that they could start irrigation.

Several other approaches for responding to the economic and social consequences of the excessive withdrawal of groundwater for irrigation in the alluvial aquifers of north Gujarat are proposed in Groundwater Management in North Gujarat: Sustainable Agriculture Improving Rural Livelihoods and Hydrological Integrity of Groundwater (http://www.cgiar.org/pdf/53_cn.pdf). Since pumping levels in bore wells are so depressed, pumping costs are very high. However, demand side options are available to reduce groundwater use in agriculture without comprising the farm economy. Examples of successful approaches include the introduction of new low water-intensive fruit crops and the use of water-saving technologies including microirrigation techniques such as drip irrigation. It is claimed, "The north Gujarat sustainable groundwater initiative is the first project in Asia that attempts to rescue a regional aquifer that supports millions of rural households through interventions that are purely farmer-initiated, and not supported by any subsidies" (p. 4). The document explores the possibility of mitigating the imbalance in groundwater use through water demand management in agriculture. The standard approach to address groundwater problems in India during the past couple of decades has been water harvesting and artificial recharge. "These interventions, carried out in arid and semiarid regions on a large-scale, involve huge public investments with little impact on achieving the desired outcomes" (p. 4).

10.3 Losses from Irrigation Schemes

Substantial water losses occur from most irrigation schemes. Not only are these losses significant in understanding the poor efficiency of irrigation schemes but also they increase recharge to aquifers and possibly lead to water logging. This section considers how these losses can be identified and quantified. Once the processes associated with the losses are conceptualized they can then be represented in mathematical models. The following section includes losses from canals and the use of interceptor drains to collect water lost from canals; losses from irrigated rice fields are also considered.

10.3.1 Canals

10.3.1.1 Magnitude and Nature of Losses from Irrigation Canals

Losses from irrigation canals have been identified both from reduced irrigation efficiencies (Bos and Nugteren 1990) and from the water logging, which often occurs in the vicinity of leaking canals. Attempts have been made to quantify these losses. More than 100 years ago, losses from canals per unit area of wetted perimeter were estimated to be 0.25 m/d from main canals and 0.2 m/d from water courses (see Wachyan and Rushton 1987). More recently, detailed experiments in canals near Madurai, India, indicated losses from unlined and lined main canals to be 0.37 and 0.11 m/d; for small distributaries, the corresponding figures were 0.06 and 0.045 m/d (Wachyan and Rushton 1987).

Losses from canals are usually assumed to be a function of the wetted perimeter but Bouwer (1969), using electrical analog models, showed that losses also depend on the hydraulic conductivity of the underlying aquifer and the hydraulic conditions within the aquifer. Subsequently, Wachyan and Rushton (1987) examined a range of hydraulic conditions within aquifers, the dimensions of the canal, and the effect of lining.

Selecting appropriate boundary conditions to represent realistic conditions in the canal and aquifer system is vitally important (Waychan and Rushton 1987). The conventional approach for boundary conditions is to assume that, at a sufficient lateral distance on either side of the canal, there is a specified fixed head; this approach is still commonly used. This assumption is illustrated in Figure 10.5a; due to symmetry, only one half of the canal and aquifer is illustrated. This fixed head boundary is, in effect, a column of water (or channel) extending for the full depth of the aquifer; the only flows in the system are from the canal to this artificial column of water. It might appear that this is an extreme case which tends to overestimate the loss, but it usually leads to an underestimate of canal losses.

A situation more in accord with realistic field conditions is shown in Figure 10.5b. At some distance below the ground surface there is a low permeability layer. Beneath this low permeability layer there is another aquifer that may be pumped or may drain to springs or a river. Computational model results indicate that, even if the vertical hydraulic conductivity of the low hydraulic conductivity layer is 1/500 of the overlying aquifer, the loss from the canal is 2.05 times the flow to the fictitious fully penetrating body of water (Figure 10.5a). With the lateral fixed head boundary (Figure 10.5a), the groundwater flow is through a relatively small depth of water. However, for the vertical outflow through the low permeability layer, the discharge is spread over the far larger plan area. Furthermore, some of the flow paths between the canal and the underlying layer are much shorter. For more details of this analysis, see Rushton (2003, Section 4.2) and Rushton (2007).

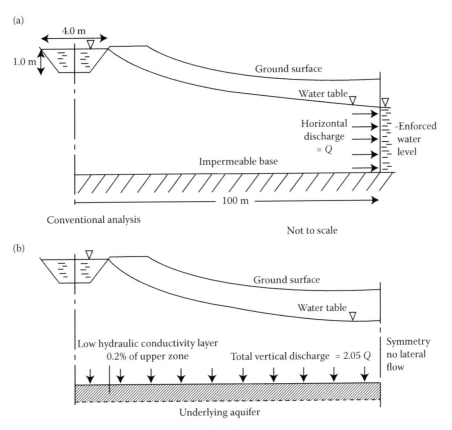

FIGURE 10.5
Canal losses: (a) flow to an imaginary lateral boundary and (b) flow through an underlying low-permeability layer.

10.3.1.2 Dimensions and Lining of Canals

Two further important issues concern the impact of the dimensions and lining of canals on canal losses. For the arrangement shown in Figure 10.5b, a series of mathematical model solutions has been obtained with different canal dimensions and with different forms of lining. The results are summarized in Table 10.1.

The effect of the size of the canal (usually described in terms of the wetted perimeter) on the magnitude of the losses can be identified from rows 1 to 3 in Table 10.1. The fifth column of the table records the wetted perimeters; the final column lists the canal loss per unit length of canal determined from the mathematical model. Increasing the wetted perimeter by a factor of 2.5 (rows 1–3) only leads to an increase in the seepage loss Q/K from 1.80 to 1.95. The reason for the small differences is that the flow paths from the canal to the outlet (the underlying low permeability layer in this instance) primarily determine the outflow; the actual dimensions of the canal have only a minor effect unless it is a very large canal. The losses per unit wetted perimeter for these three examples equal 0.47, 0.38, and 0.20 m/d

TABLE 10.1

Canal Losses for Different Dimensions of the Canal and for Lining of the Canal

	Width of Water Surface (m)	Width of Canal Base (m)	Depth of Water (m)	Wetted Perimeter (m)	Hydrology Condition of Lining ÷ K of Aquifer	Canal Loss Q/K per Unit Length (m)
1	3.0	1.0	1.0	3.83	1.0	1.80
2	4.0	2.0	1.0	4.83	1.0	1.86
3	8.0	4.0	2.0	9.66	1.0	1.95
4	4.0	2.0	1.0	4.83	0.30	1.83
5	4.0	2.0	1.0	4.83	0.20	1.81
6	4.0	2.0	1.0	4.83	0.10	1.74
7	4.0	2.0	1.0	4.83	0.05	1.49

(column 7 divided by column 5); therefore, it is incorrect to estimate losses as being proportional to the wetted perimeter. Consequently, losses from smaller tertiary and distributary canals can be significant.

The second issue is concerned with the effect of the lining of canals. Field investigations indicate that unless the lining is almost perfect, it has only a small effect in reducing losses. Rushton (2003) examined the losses when there are small gaps in the lining and showed that this does not lead to a substantial reduction in the losses. Rows 4–7 of Table 10.1 refer to a canal having a lining of 10 cm thick with a reduced hydraulic conductivity relative to the aquifer. When the hydraulic conductivity of the lining is 30% of that for the aquifer (row 4), the canal loss of 1.83 is 98.6% of the value with no lining (row 2). Even if the hydraulic conductivity of the lining is 5% of the unlined case, this is equivalent to cracks occurring between individual panels, the loss is 80% of the unlined. For lower lining hydraulic conductivities, unsaturated conditions occur under the canal (Section 4.2.5 of Rushton 2003).

10.3.2 Interceptor Drains

Water lost from irrigation canals can lead to water logging. There are two common approaches to counter water logging—abstraction of groundwater using vertical wells—for recent studies in Pakistan see Qureshi et al. (2004), Latif (2007), and Basharat (2012)—or alternatively horizontal pipe interceptor drains are installed.

An interceptor drain designed to prevent water logging resulting from losses from the Chesma Right Bank canal (http://www.wapda.gov.pk/htmls/chashrightcanal082011.htm.) is used as an example. The interceptor drain is located almost 80 m from the bank of the canal, which was constructed by excavating a channel and forming banks on either side of the channel; the canal water level is about 2 m above ground level (Rushton 2003). The interceptor drain is located about 6 m below ground level; a series of sumps are situated along the interceptor drain with pumps provided to return water to the canal.

The conceptual model, used by the consultants to design the interceptor drain and determine the pump capacity, is sketched in Figure 10.6a.

A finite difference computational model was used to represent the conceptual model; from the computational model results the pumping capacity was estimated. Note that the only inflow of water to the aquifer is from the canal; this water is either collected by the interceptor drain or flows to a "distant" fixed head boundary. Furthermore, the water table extends from the water surface in the canal to the interceptor drain and then

(a)

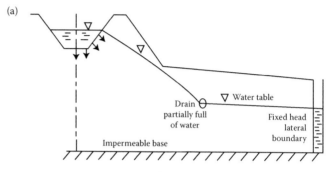

Unsatisfactory conceptual model

(b)

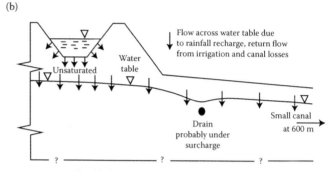

Conceptual model including flows from water table

FIGURE 10.6
Conceptual models of flow to an interceptor drain: (a) the model ignores recharge with a fixed head lateral boundary and (b) model with a perched canal and inflows from the canal and water table.

to the top of the fictitious fixed head boundary. At each sump two pumps were provided, the second intended only as a standby. However, after a year of continuous operation of both the main pumps and the standby pumps, the water table had not reached equilibrium with the water table some distance above the top of the interceptor drain.

During operation of the interceptor drain, field information about water table elevations was collected from a series of monitoring piezometers including one that penetrated through the bed of the canal. Figure 10.6b shows the location of the water table; there are substantial differences from the assumptions of Figure 10.6a. From this information a new conceptual model was developed as illustrated in Figure 10.6b.

a. The drain does not only collect water from the canal; it also collects water from the overlying water table due to rainfall recharge and return flow from irrigation. Therefore, the canal is not the only source of groundwater reaching the interceptor drain.

b. There is an unsaturated zone beneath the canal.

c. The water table is likely to be above the drain; only rarely does a water table intersect a drain (Khan and Rushton 1996).

d. The water in the drain may be above atmospheric pressure resulting in a surcharge.

e. Flow into and through the drain may be influenced by the material surrounding the drain, the perimeter of the drain, and the carrying capacity of the drain.

f. There is unlikely to be an impermeable layer beneath the drain; instead, there may be a low permeability layer perhaps with groundwater abstraction from the deeper aquifer.

g. A realistic lateral boundary must be selected; for this example there is a small canal 600 m from the main canal.

With this conceptual understanding, it is possible to identify why the design capacity of the interceptor drain pumps is insufficient. The initial assumption, that the interceptor drain only collects water from the canal, is a serious misconception and leads to a substantial underestimate of the quantity of water that needs to be pumped from the interceptor drain.

10.3.3 Flooded Rice Fields

Frequently there are problems in maintaining the water depth in a flooded rice field, especially if the rice field only receives water every 7th or 10th day. Field experiments were conducted by Walker to investigate water losses (Walker and Rushton 1986). A typical flooded rice field (Figure 10.7a) is separated from the adjacent field by bunds; a plough layer is developed in the field and puddled, a hard pan usually forms below the plough layer. By constructing small lysimeters within the field, one with a solid base and the second without a base but penetrating through the plough layer and hard pan, it is possible to examine the pathways by which water is lost from the rice field. Experimental results, for a 7-day period, of the fall in water level in each lysimeter and the fall in level in the field, are plotted in Figure 10.7b.

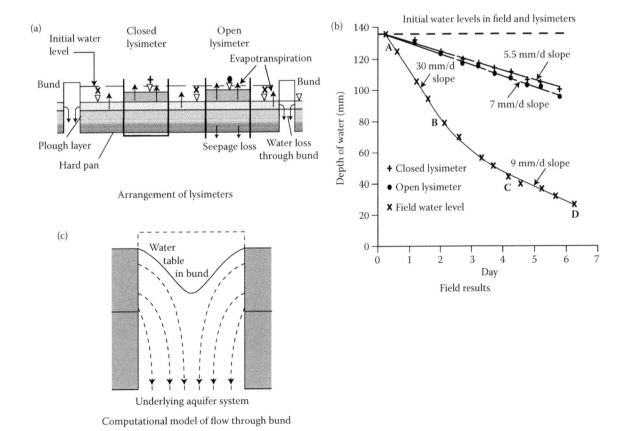

FIGURE 10.7
Water losses from a flooded rice field: (a) provision of lysimeters indicating water levels in the lysimeters and field, (b) field results for the fall in water levels in each of the lysimeters and in the field, and (c) diagram of flows through the bund. (Adapted from Rushton, K.R., *Groundwater Hydrology, Conceptual and Computational Models*, Wiley, Chichester, 2003.)

For the closed lysimeter, the fall of 5.5 mm/d is entirely due to evapotranspiration. For the open lysimeter, the loss is 7 mm/d; this is a combination of 5.5 mm/d evapotranspiration plus a loss of 1.5 mm/d through the plough layer and hard pan. Losses from the whole field are far larger starting at 30 mm/d (fitted line **A-B**) and reducing at 6 days to about 9 mm/d (**C-D**). The only possible route for this large loss is through the bunds which are not puddled. Water flow through the bunds is indicated in Figure 10.7a. Evidence for the significant flows through the bunds was obtained by covering the bunds with plastic sheeting that was tucked into the puddled layer on either side of the bund; this resulted in a substantial reduction in the rate of fall of the water surface in the field. Losses through the bunds were also confirmed by preparing a vertical section computational model of flow through the bund to the underlying aquifer. Figure 10.7c illustrates the flow lines through the bund as deduced from the computational model. Note that there is a dip in the water table in the bund between the water surfaces. Careful examination of the water table shape within bunds shows that this is the actual profile of the water table.

A further important conclusion that can be deduced from the field result is that the water loss through the bunds reduces with lower water levels in the rice field. With the water level in the field at 135 mm, the loss through the bund is 30 – 7 = 23 mm/d. However, after 6 days when the depth of water in the field falls to about 30 mm, the loss through the bund reduces to 9 – 7 = 2 mm/d. This behavior was reproduced in the computational model. Certain irrigation engineers recognized that high losses occur when a rice field is supplied with water on a rotational basis (perhaps weekly) with an initial depth of water in the field of up to 160 mm, which is necessary to store water. On the other hand, when irrigation occurs every day (e.g., using groundwater) a water depth of 30–40 mm is sufficient and the losses are small. Further information about the flow of water from rice fields through bunds into aquifers can be found in Walker and Rushton (1984) and Rushton (2003, Section 4.5).

The above discussion refers to rice fields in a relatively level area with the regional groundwater table beneath the puddled plough layer and hard pan. Different conditions occur in sloping areas with terraced rice fields. In a preliminary study by Lowe (Lowe and Rushton 1990) on a terrace in Sri Lanka, information was collected about water levels in the fields and also the underlying groundwater heads. Figure 10.8 refers to two fields toward the top of the terrace and two fields toward the bottom; only the parts of the rice fields adjacent to the bunds are included in the diagram. For the higher fields, bunds **A** and **B**, the regional water table is below the fields; this situation is similar to Figure 10.7.

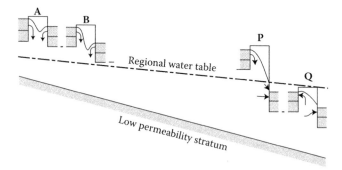

FIGURE 10.8
Schematic diagram of a terraced rice field system with water either lost from or entering through the bunds. (Adapted from Rushton, K.R., *Groundwater Hydrology, Conceptual and Computational Models*, Wiley, Chichester, 2003.)

For fields toward the bottom of the terrace, bunds **P** and **Q,** the water table is higher relative to the field water level. For bund **P**, the regional groundwater table passes through the bund with the result that there is a flow from the upstream rice field into the bund, but water flows from the bund to enter the lower rice field. A small seepage face forms on the downstream side of bund **P**. For bund **Q**, due to the relatively higher regional groundwater table, groundwater enters both the upstream and downstream rice fields through the bund.

10.4 Estimating Recharge

There are several methods for estimating recharge, see Scanlon et al. (2002) and also Section 3.2 of Rushton (2003). In some approaches, a direct method is used in which the physical processes are considered; alternatively, indirect methods are available in which changes in conditions in an aquifer are used to estimate the recharge. In the following section, a soil moisture balance method of estimating recharge is presented. This approach involves features, such as soil and crop type, estimates of actual evapotranspiration (AE), bare soil evaporation, and RO. The water table fluctuation method is also considered; this is an indirect method. Finally, the methodology used by the CGWB of India to estimate recharge is reviewed.

10.4.1 Soil Moisture Balance: Conceptual Model

The soil moisture balance method for recharge estimation is based on a physical conceptual model that represents moisture conditions in the soil zone. Excess water in the soil zone can move to the underlying aquifer

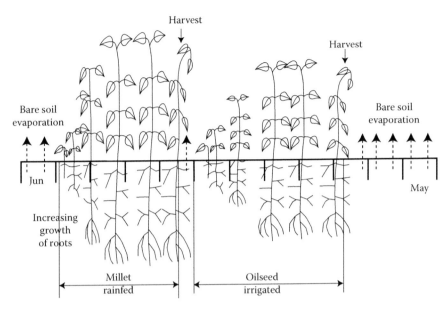

FIGURE 10.9
Diagram illustrating crop and root growth for a wet season crop and a dry season-irrigated crop in a semiarid environment.

as recharge when the soil reaches field capacity and becomes free draining. The soil moisture balance can include a range of climates with different types of soils and crops using parameters presented in FAO (1998).

Detailed descriptions of the soil moisture balance approach are presented in Chapter 3 of Rushton (2003) and Rushton et al. (2006) where algorithms for the calculations are included. Figure 10.9 illustrates crop growth and bare soil conditions in a semiarid environment during a typical year. Important components and processes in the soil moisture balance are described next; three specific situations are illustrated in Figure 10.10 plus detailed daily results in Figure 10.12.

a. *Precipitation and irrigation inflows Pr and Ir*: The inflows to the soil moisture balance are daily rainfall and irrigation (as indicated in Figure 10.10).

b. *Soil*: The ability of the soil to store water is an important property. The maximum and minimum soil moisture contents are field capacity, when the moisture content is high enough for the soil to become free draining, and the wilting point which is the moisture content below which roots are unable to draw water out of the soil. For a sandy soil the difference between the moisture content at field capacity and wilting point is in the range 0.05–0.11 (average 0.08), for clay the range is 0.12–0.20 (FAO 1998).

c. *Crop roots*: The rooting depth of the crop determines the depth from which the roots can draw water. The length of the roots of nonperennial

crops increases with time (Figure 10.9). For wheat the maximum rooting depth is 1.8 m. For permanent grass the rooting depth is 0.5–1.0 m; for apple trees the range is 1.0–2.0 m. For other crops see FAO (1998).

d. *Soil moisture deficit (SMD)*: The water available to the crop roots is expressed as an SMD; this is an equivalent depth of water. Consequently, the SMD is the depth of water required to bring the soil up to field capacity. For the maximum rooting depth of 1.8 m for wheat, in a sandy soil the highest SMD (or total available water [*TAW*]) is $0.08 \times 1.8 = 0.144$ m or 144 mm.

e. *RO*: During rainfall (or due to overirrigation), RO can occur. The magnitude of the RO depends on the rainfall quantity, the wetness of the soil, and the slope of the ground. For each location a matrix is formed of the rainfall quantity and the SMD (which is a measure of the soil surface wetness); the coefficients of the matrix determine the proportion of rainfall that becomes RO. This matrix should be based on field experience where estimates are made of the magnitude of RO for high and low rainfall amounts and differing soil moisture conditions.

f. *Potential and AE or evaporation*: Daily potential reference evapotranspiration can be estimated using a technique such as the Penman–Monteith method (FAO 1998). Alternatively, a methodology requiring less data can be used; the Hargreaves equation is a suitable alternative (Hargreaves and Allen 2003).

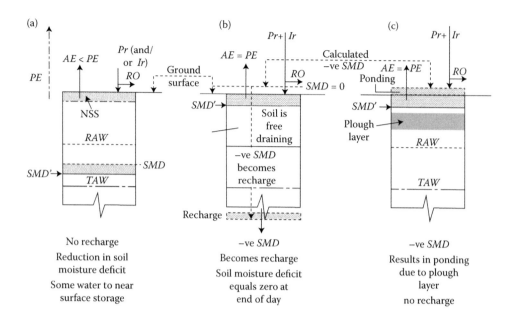

FIGURE 10.10

Examples of water balance calculations: (a) *SMD > RAW* hence reduced evapotranspiration and some water to NSS, (b) and (c) negative *SMD* results in recharge or ponding.

Reference potential evapotranspiration refers to a crop of grass. For other crops, the potential evapotranspiration may be higher or lower than that for grass. The different potential evapotranspiration values are determined using crop coefficients. Approaching harvest the crop coefficient falls below 1.0. Details of crop coefficients and rooting depths of a wide range of crops can be found in FAO (1998).

Bare soil potential evaporation is another important loss from the soil, especially when there is no growing vegetation during a dry season. FAO (1998) includes a methodology for estimating potential evaporation. In a single plot of land, crops can be growing and also bare soil conditions can occur. Consequently, it is necessary to prepare weighted crop coefficients for every day in the year to represent the crop and/or bare soil conditions (see Section 3.3 of Rushton 2003). Bare soil evaporation can only occur from a limited depth; this is represented by the total evaporable water (TEW).

AE is less than the potential value when there is a significant SMD so that the crop is under water stress. This can be represented using the concept of readily available water (*RAW*), which is a fraction of *TAW* (typically 0.5–0.6 of *TAW*). If the SMD is greater than *RAW*, the water stress coefficient is less than 1.0; this determines the rate at which AE can occur. If the SMD is smaller than *RAW*, evapotranspiration occurs at the potential rate.

Estimation of reduced (actual) evapotranspiration or evaporation is illustrated by the diagram of Figure 10.11. The horizontal axis represents the current SMD and the vertical axis the water stress coefficient. Considering the full line which represents a loam soil, once the *SMD* is greater than RAW_L the water stress coefficient is less than 1.0 and evapotranspiration occurs at a reduced rate unless the precipitation is greater than the potential evapotranspiration. When the SMD equals TAW_L evapotranspiration is zero; this corresponds to the wilting point. For sand,

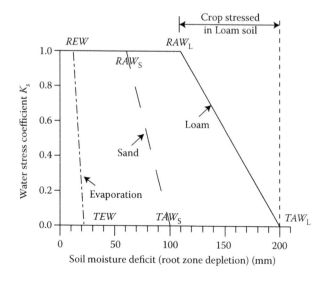

FIGURE 10.11

Reduced evapotranspiration or evaporation when the SMD is greater than the *RAW* or *REW*.

RAW_S and TAW_S are smaller because the difference in moisture content between field capacity and wilting point is smaller. A similar relationship is used for bare soil evaporation; readily evaporable water (*REW*) and *TEW* are significantly smaller.

The occurrence of AE, which is less than the potential value, is illustrated in Figure 10.10a. The components of the soil moisture balance are shown on the diagram; the potential evapotranspiration is represented by a vertical arrow with a broken line to the top left of the diagram. The *SMD* is greater than the *RAW*; consequently, the *AE* (shown by the solid line) is less than the potential evapotranspiration.

g. *Recharge*: When the soil reaches field capacity it becomes free draining and any additional water can move through the soil zone as potential recharge to the underlying aquifer (Figure 10.10b). Since the precipitation is larger than the sum of the potential evapotranspiration and the small SMD at the start of the day *SMD'*, the SMD at the end of the day is negative. Consequently, because the soil is free draining, the recharge equals the quantity of water in excess of that required to bring the soil to field capacity (Figure 10.3b). Alternatively, if there is a low permeability layer in the soil zone, such as due to the plough layer and hard pan in a rice field (Figure 10.3), ponding can occur (Figure 10.3c).

h. *Near-surface storage (NSS)*: A limitation of this conventional SMD approach is that it assumes that any excess water (precipitation minus AE and RO) moves down the soil profile to reduce the SMD toward the bottom of the soil layer. However, in loams, silts, and clays some of the moisture remains close to the soil surface and is transpired by shallow roots during the following days. This is demonstrated by the manner in which vegetation revives for a few days following significant rainfall, despite the fact that at depth the soil may be dry. In the soil moisture balance calculation, this process is represented by *NSS*. In Figure 10.10a, the precipitation is substantially larger than the AE (which is limited because *SMD > RAW*) plus RO; this excess water is divided between a reduction in the SMD and water stored close to the soil surface which is available for evapotranspiration during the following days. This process is of particular importance in semiarid locations where crops can be sown following the early rains even though the deeper soil is still dry.

Examples of the use of a soil moisture balance to estimate recharge for a Miliolite limestone aquifer in western India can be found in Rushton and Rao (1988); also de Silva and Rushton (2007) quantify recharge to a shallow aquifer in Sri Lanka.

10.4.2 Example of the Computational Model for a Soil Moisture Balance

To illustrate the computational model for a soil moisture balance, the results of a daily calculation are presented in Figure 10.12. The calculation represents the conditions for a single year (1993) in a semiarid region where there is one rain-fed crop of millet planted on July 4 and harvested on October 16 and a second irrigated crop of sunflower oilseed planted on November 1 and harvested on March 10 (Figure 10.9). From March 11 to July 3, bare soil conditions apply. The notes at the bottom of Figure 10.12 relate to important findings.

An understanding of the processes described in (a)–(h) above can be gained from a careful examination of the graphs of Figure 10.12. Each of the four graphs is considered in turn.

10.4.2.1 Upper Graph of Figure 10.12

This graph refers to precipitation (open bars) or irrigation (filled green bars), calculated recharge (blue bars above the axis), and estimated RO (blue bars below the axis). The annual rainfall in the study area was 700 mm, with 85% of annual rainfall in less than 2 weeks in July. During this period, recharge is estimated to be more than 200 mm; the high rainfall also resulted in substantial RO. For the dry season crop, it requires about 11 irrigation days of 50 mm each day. On five of the irrigation days, more water was applied than required with the result that the SMD became zero and recharge occurred; the total calculated recharge due to irrigation equaled 67 mm.

10.4.2.2 Second Graph of Figure 10.12

This diagram shows the combined crop coefficient, which represents crop evapotranspiration or bare soil evaporation. From March to May, the crop coefficient is just over 1.0 indicating that the evaporative demand is high. However, unless there is water available, no evaporation will actually occur. Approaching harvest for each crop, the crop coefficient falls below 1.0.

10.4.2.3 Third Graph of Figure 10.12

In this graph potential and actual evaporation or evapotranspiration are plotted. The open bars refer to the potential values; the filled bars refer to actual

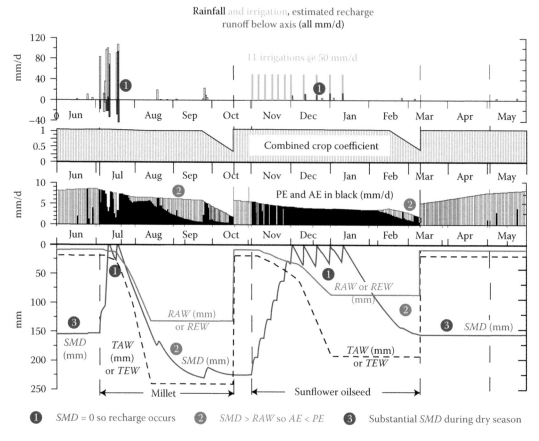

FIGURE 10.12

Daily soil moisture balance for rain-fed and an irrigated dry season crop. (Adapted from Rushton, K.R., *Groundwater Hydrology, Conceptual and Computational Models*, Wiley, Chichester, 2003.)

evaporation or evapotranspiration. In June, on three occasions prior to the sowing of the rain-fed crop, rainfall occurs. Evaporation takes place on four days because the second rainfall event is large enough to meet the evaporative demand and on the following day there is evaporation due to some excess water being transferred to NSS.

In the early days of the millet crop there is high-intensity rainfall, which means that the AE is at the potential rate. However, during the remainder of the growing season rainfall is limited, although there are two occasions when the rainfall is about 20 mm/d and six further days with small rainfall amounts. Consequently, for much of the growing season the AE is well below the potential rate resulting in a poor crop yield.

For the second crop, the first four irrigations at 5-day intervals result in evapotranspiration at the potential rate for 3 days, at a reduced rate for the fourth day, and a calculated zero evapotranspiration on the fifth day. From the fifth irrigation, evapotranspiration occurs at the potential rate. After the sunflower oilseed harvest, evaporation only occurs on 2 days due to small rainfall events.

10.4.2.4 Lowest Graph of Figure 10.12

This graph shows the soil conditions that determine whether the AE is less than the potential; the times when recharge occurs can also be identified. Note that the vertical axis is positive downward to represent a deficit of soil water. The *RAW* (or *REW*) and also the *TAW* (or *TEW*) undergo significant changes with time. This occurs because they depend on the rooting depth or the depth from which evaporation can occur. Consequently, they increase substantially during the initial and development stages of the crop and decrease at harvest. Between the two crops they reduce to the values appropriate for evaporation.

The *SMD* is also plotted on this graph. Whether the *SMD* is smaller or larger than *RAW* (or *REW*) determines whether evapotranspiration (or evaporation) occurs at the potential rate. For the millet crop the *SMD* is larger than *RAW* for all but the early stages of crop growth; consequently, the AE is less than the potential rate. For the second crop which is irrigated (note that the rooting depth of the sunflower oilseed is less than for millet), the *SMD* becomes smaller than *RAW* after the seventh irrigation and evapotranspiration is at the potential rate.

Occasions when recharge occurs can be identified from this graph. Recharge occurs when the *SMD* becomes zero; this arises on four occasions at the start of the rainy season and five occasions due to irrigation of the second crop.

Because this soil moisture balance calculation aims to represent the daily conditions, it is detailed and requires a careful scrutiny to understand the different aspects. Nevertheless, it does lead to an understanding of the complexity of soil moisture conditions and especially the recharge process. Another advantage is that it uses physical parameters, inflows, and outflows, which can be quantified to a reasonable accuracy. Uncertainties can be explored using sensitivity analyses (Eilers et al. 2007, de Silva and Rushton 2007). Alternative methods of estimating recharge are considered below, but they use indirect methods which are not based on the actual physical parameters of the soil moisture balance and the crops.

10.4.3 Water Table Fluctuation Method

In the water-level fluctuation (WLF) method, the rise in water table elevation during the recharge season is multiplied by the specific yield to provide an estimate of the recharge. In a detailed review of the method, Healy and Cook (2002) suggest a number of refinements. The attractiveness of the method is its simplicity and ease of use. It can be viewed as an integrated approach; however, the major difficulty is in identifying a realistic value for the specific yield. Furthermore, it is important to understanding the main causes of changes in the water table elevation. These changes may be due to the inflow of the effective rainfall, but abstraction from the aquifer or groundwater drainage can also have a significant effect.

Insights into the methodology can be gained from a study of recharge by de Silva and Rushton (2007) in which both the water table fluctuation method and a soil moisture balance are used. The study area is in the Kurunagala District of Sri Lanka, a tropical region with a bimodal rainfall pattern but with distinct dry seasons. For the water year of March 2004 to February 2005, the rainfall was 1338 mm and the potential evapotranspiration was 1391 mm. Four wells were monitored with depths of between 4 and 8 m. Taking an average of the water table recovery of the four wells, and using a specific yield of 0.065 (which was determined from analysis of a pumping test in a similar area (de Silva and Rushton 1996)), the recharge estimates for March, October, November, and December 2004 are recorded in the second column of Table 10.2. Column 3 of the table lists monthly recharge values

TABLE 10.2

Comparison of Recharge Estimates from Water-Level Fluctuation Method and Soil Moisture Balance

Month	WLF Approach (mm/month)	SMB for Grass and Aubergine (mm/month)	WLF/SMB (%)
May	26	30	87
October	10	13	77
November	89	111	80
December	30	46	65
Sum	155	200	78

calculated using a daily soil moisture balance with half of the area having a crop of aubergines and half with permanent grass.

The sum of the recharge estimates from the WLF approach for these 4 months is 78% of the corresponding values obtained using a soil moisture balance approach. For each of the months the water table fluctuation estimate is less than that for the soil moisture balance. Three possible reasons for the difference are

a. Some water was taken from wells for domestic purposes during the monsoon season; this reduces the rise in the water table and leads to reduced recharge estimates.

b. Discharge of groundwater is likely to occur through the weathered zone into a valley; in some studies an allowance is made for these discharges.

c. Incorrect estimate of the specific yield; due to the nature of the aquifer system, the specific yield is likely to be different at the individual wells.

Nevertheless, the similarity of the estimates in columns 2 and 3 of Table 10.2 is encouraging.

This study was carried out in an area with a series of rain-fed rice fields in the valley bottom; the observation wells were located on the valley sides. The soil moisture balance technique has been adapted to represent conditions in these rice fields where ponding occurs (de Silva and Rushton 2008).

10.4.4 Pragmatic Approach by the CGWB, India

For more than 50 years, estimates have been made of the groundwater resources of India. The methodology currently used by the CGWB is based on the groundwater resource estimation methodology prepared in 1997; the methodology is outlined by Chatterjee and Purohit (2009). In a review in 2008–2009, it was concluded that the methodology is appropriate and suitable for countrywide groundwater resource estimation.

Recharge during the monsoon season is estimated using the water table fluctuation method; in addition, a rainfall infiltration factor method (RIF) is also employed. Additional contributions or corrections to recharge occur due to the following factors:

Monsoon season: The recharge estimated from the water table fluctuation method is increased by the groundwater draft during the monsoon season.

Nonmonsoon season: Rainfall recharge is estimated using the RIF method (the suitability of this method is questioned because there is usually an SMD during the nonmonsoon season). In addition, recharge due to losses from canals and return flow from irrigation is included as indicated in Table 10.3.

Based on the findings of earlier sections of this chapter, the following comments can be made about the values quoted in Table 10.3:

a. Unlined canal: The assumption that the loss is proportional to the wetted perimeter is incorrect; the results recorded in Table 10.1 indicate that smaller canals lose almost as much water as larger canals. Furthermore, the actual loss is related to the hydraulic conductivity of the underlying soil; consequently, the range of losses is likely to be larger than the values quoted in Table 10.3.

b. Lined canal: As implied by the last four rows of Table 10.1, losses from lined canals are likely to be 60%–80% of unlined canals rather than 20% as recommended.

TABLE 10.3

Additional Components of Recharge Due to Losses from Irrigation According to CGWB, India

a.	Unlined canal	15–30 ha m/d/million m² of wetted perimeter depending on soil type (0.15–0.30 m/d/m² of wetted perimeter)
b.	Lined canal	20% of above
c.	Return flow from surface water irrigation	0.10–0.50 (10%–50%) depending on cropping pattern and depth to water table; shallow water table and paddy crops have higher return flows
d.	Return flow from groundwater irrigation	0.05–0.45 (5%–45%) depending on cropping pattern and depth to water table; shallow water table and paddy crops have higher return flows
e.	Water bodies	1.4 mm/d based on average area of water spread

c. Return flow from surface water irrigation: The range 10%–50% is realistic with 50% being an appropriate value for flooded rice fields (paddy).

d. Return flow from groundwater irrigation: May be as low as 5% if groundwater is only applied when required by the crop.

e. Water bodies: 1.4 mm/d appears to be a low figure which could be appropriate when silt has built up over a long period of time but may be too small for percolation tanks.

Chatterjee and Purohit (2009) also quote the norms used for estimating the draft (abstraction) for various types of wells when used for irrigation or domestic purposes.

Despite some reservations about the CGWB method, it is a realistic attempt at estimating the components that contribute to groundwater recharge; it also provides a plausible approximation of groundwater usage. The resultant water balances published by CGWB demonstrate that overexploitation occurs in many areas and also that there is a possibility of further groundwater exploitation in some localities.

10.5 Artificial Recharge

When water tables fall significantly due to over-exploitation, a solution that is often proposed is to initiate artificial or enhanced recharge. This has been an approach in India for more than 40 years. For an artificial recharge scheme to be effective and economical, it is essential to consider

- Whether there is a reliable source of water
- Whether there is space in an aquifer to store the water
- Whether there are mechanisms to recover the water

In this section, artificial recharge schemes, which were monitored and analyzed thoroughly, will be considered first; subsequently, attention will be turned to schemes currently in progress in India.

10.5.1 Artificial Recharge Pilot Schemes

In the 1980s, pilot schemes for artificial recharge were conducted to explore the potential of artificial recharge in Gujarat, India; this was a joint project between the UNDP (United Nations Development Programme), CGWB, and GWRDC (Gujarat Water Resources Development Corporation). These pilot schemes involved extensive and thorough field work (Bradley and Phadtare 1989), supported by mathematical modeling. Artificial recharge experiments were conducted for both alluvial and limestone aquifers using spreading channels, percolation tanks, and injection wells (Rushton and Phadtare 1989). Three artificial recharge techniques were explored in detail.

a. An injection well in the Mehsana area alluvial aquifer was used to inject water from a river bed into the deeper aquifers. Therefore, the injection well acted as a connection between the shallow and deeper aquifers. The main injection test continued for 250 days; the initial injection rate was 225 m^3/d. As the test continued, the water level in the injection well increased until at the end of the test it was almost 20 m higher than the at-rest levels. Eventually flows into the aquifer almost ceased due to clogging in the vicinity of the injection well. Following thorough cleaning and redevelopment, the well could be used for further recharge. However, the recharge rate of 225 m^3/d is less than the quantity of water already moving vertically from the shallow to the deeper aquifer due to the hydraulic head difference between shallow and deep aquifers. This downward flow is estimated to be about 1500 m^3/d/km^2 based on the estimate of a vertical flow of 1.5 mm/d quoted above as insight a) of Section 10.2.2.1 (see also Rushton and Srivastava 1988, Rushton and Tiwari 1989, Kavalanekar et al. 1992).

b. Spreading channels can be used in unconfined aquifers as a means of increasing the inflow of water above the natural recharge rate. Spreading channels used in the pilot project were typically 100 m long, 2 m deep with a width at the base of the channel of 2 m, and side slopes of 45°; the aquifer is permeable sand. Water was supplied from nearby canals and, although initial recharge rates approaching 1.0 m^3/d/m^2 were achieved, the rates fell to steady values of about 0.25 m^3/d/m^2. To maintain a recharge rate of this order of magnitude for several years, regular cleaning and scraping of the bed and sides of the channel is necessary. Assuming that water is available for artificial recharge for 50 days in a year, the recharge for a single spreading channel is equivalent to about 8000 m^3 in a year. This volume is beneficial for domestic supplies, but it would only be sufficient to irrigate a dry season crop, with a water requirement of 400 mm, covering an area of about 100 m by 200 m.

c. Artificial recharge experiments were conducted in a coastal Miliolite limestone aquifer. Saline intrusion has occurred along much of the coastal limestone belt in western Gujarat; the primary objective of this artificial recharge project was to flush out saline water that had been drawn into the aquifer due to substantial groundwater abstraction close to the coast (Rushton and Rao 1988). The upper part of the Miliolite limestone formation contains major fissures and cavities; therefore, the aquifer can be idealized as having a lower-permeability zone of about 10-m thickness overlain by a higher-permeability zone. Artificial recharge to this highly transmissive aquifer presents no technical difficulties; the water freely enters the aquifer. Three alternative recharge structures were selected: a canal section 10.700 m long in the Miliolite limestone, a recharge basin with a plan area of 80 m^2 and depth 1 m, and an injection shaft of 0.45 m in diameter tapping two major cavities. These studies indicated that instead of flushing out the saline water within the lower-permeability zones, the farmers are the main beneficiaries of the recharge water because they can maintain abstraction from their wells over longer time periods. During extended periods of artificial recharge, water exits the aquifer from springs. Conditions in the coastal region are hardly affected by the artificial recharge.

Another detailed field study and analysis of artificial recharge (managed aquifer recharge), which was conducted by an international team, is described in Gale et al. (2006). Three artificial recharge schemes were studied in detail with extensive field measurements supported by mathematical modeling; socioeconomic surveys were also carried out. The schemes included artificial recharge by means of check dams, recharge ponds, and field bunds; the aquifers were shallow-weathered crystalline hard rocks, weathered–fractured granitic rocks, and Deccan Trap basalts. Field work included monitoring using dug wells and bore wells, flumes for measuring flows, geological mapping, topographical surveys, and data from automatic weather stations.

The focus of these studies was to assess how much water was actually recharged by the structures in comparison to natural recharge. The requirements for successful artificial recharge, a source of water, space in the aquifer to store the water, and mechanisms to recover the water for beneficial use were quantified and put in the overall context of natural recharge and discharge to assess the impact of the artificial recharge in relation to the investment made. They also recommend that every effort should be made to continue collecting and interpreting data for at least 5, and preferably 10 years.

The estimates of recharge for these three sites were found to be about one order of magnitude lower (4–10 mm/yr) than the CGWB estimates. The report by Gale et al. (2006) suggests that managed aquifer recharge is being seen too much as a panacea for water supply problems and groundwater overdraft, without the necessary evaluation of its potential in different climatic, agroecological, hydrogeological, and socioeconomic conditions.

10.5.2 Groundwater Recharge: Master Plan for India

In India, artificial recharge was identified as an important aspect of maintaining groundwater supplies. The National Groundwater Recharge Master Plan for India (Central Ground Water Board 2002) uses two criteria for identifying recharge; the availability of surplus water and the availability of storage space in aquifers. The investments in this program were selected because of the potential available for groundwater recharge, and not by the need for recharge.

A revised Master Plan for Artificial Recharge has been prepared (Central Ground Water Board 2013). This is an update to the Master Plan of 2002; it reports the actual progress made in the implementation targets up to 2010–2011 and also reassesses new areas of recharge to make use of the experience already gained and the input received from the impact assessment of selected schemes. The revised Master Plan states that the monitoring of water levels and water quality is of prime importance in any scheme of artificial recharge to groundwater. Nevertheless there is a failure to recognize that attempts must also be made to estimate how much water is recharged and how much water can be recovered for beneficial use.

Progress of the National Groundwater Recharge Master Plan is described in a publication "Select Case Studies Rain Water Harvesting and Artificial Recharge" (Central Ground Water Board 2011b). In the chapter "Success Stories of Artificial Recharge," numerous examples are given of (usually small scale) artificial recharge projects. Structures suitable for artificial recharge include check dams, gully plugs, nalla bunds, recharge pits, percolation tanks/ponds, subsurface dikes, artificial injection of RO generated from peak storms, recharging of abandoned open wells, and contour trenching or bunding on hilly slopes.

For some of these projects, attempts have been made to evaluate the impact of the schemes, for example, by quantifying the rise in the water table in areas of recharge. Other evidence includes certain wells that no longer become dry between monsoon seasons, additional areas that can be irrigated, and improvements in the quality of the water. Nevertheless, there is very limited information about actual volumes of water recharged and the additional volume of water that has been abstracted. For example, the recharge injection rate for the pilot scheme in the Mehsana alluvial aquifer in the 1980s is quoted (example a) above. However, the fact that the injection rate fell almost to zero at the end of the test is not reported. Although actual volumes of recharged water are difficult to determine, attempts should be made to identify methodologies that have the potential to increase the recharge substantially. In a contribution "The Groundwater Recharge Movement in India" (Sakthivadivel 2007), it is stated that most of the evaluation studies in India are of a qualitative nature with limited objectives.

The challenge of artificial recharge can be summed up by noting that most artificial recharge schemes result in self-clogging of the aquifer. For example, pumping from a well tends to be self-cleansing but an injection well is self-clogging; typically, injection rates into a well are only 10%–20% of discharge rates. An important document on clogging has been prepared by Martin (2013); it recognizes that managed aquifer recharge schemes invariably experience clogging.

10.6 Abstracting Groundwater from Aquifers of Limited Saturated Thickness

Traditional bore wells can be used for deep alluvial aquifers, tapping all the permeable strata. However, when the aquifer is of limited saturated thickness, alternative approaches are required. The traditional large diameter dug well has many advantages for shallow aquifers. Horizontal wells provide an alternative to dug wells. For aquifers overlain by lower permeability strata but with a limited saturated thickness of the productive aquifer, an inverted well, which was developed in Bangladesh, has proved to be successful.

10.6.1 Large-Diameter Dug Wells

Large-diameter dug wells have been used in the Indian subcontinent to abstract water from the weathered zones of aquifers. This chapter explores the reasons for their success, especially when power for operating the pump is only available for a limited time period, and also when irrigation occurs for only a few hours.

The basic features of the operation of a large-diameter dug well during a daily cycle are illustrated in Figure 10.13. Figure 10.13a represents the well full of water before pumping starts. During the initial stages of pumping (Figure 10.13b), most of the water is taken from well storage. In the aquifer, the fall in the water table is

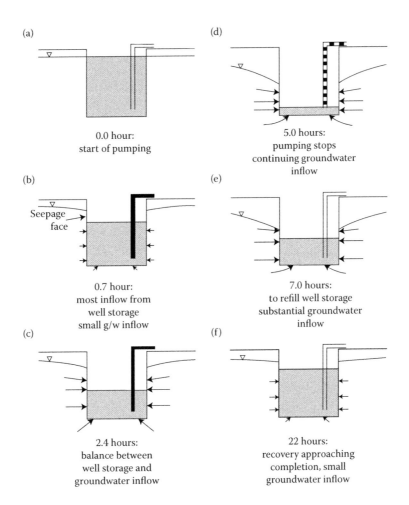

FIGURE 10.13
Daily cycle of abstraction from a large-diameter well. (Adapted from Rushton, K.R., *Groundwater Hydrology, Conceptual and Computational Models*, Wiley, Chichester, 2003.)

small because a seepage face forms on the face of the well; hence, the water table is higher than the well water level. After 2.4 hours (Figure 10.13c), water is taken from both well storage and the aquifer. Pumping ceases after 5 hours (Figure 10.13d).

After pumping stops, the well is refilled from the aquifer; consequently, water continues to be drawn from the aquifer into the well (Figure 10.13e). At 22 hours after pumping started, the well has almost been refilled and the flow from the aquifer is small (Figure 10.13f). This series of diagrams shows that even when the pump has ceased to operate, the well continues to draw water from the aquifer; this behavior is not significant with small-diameter wells.

An example of the well water level and flow components for a representative large-diameter well is included in Figure 10.14; these results are obtained from a radial flow mathematical model (Rushton and Redshaw 1979, Rushton 2003). The analysis represents a single day at the start of the irrigation season. Figure 10.14b indicates that pumping at 300 m³/d

continues for 5 hours. The water level in the well falls by 1.25 m (Figure 10.14a); once pumping stops the well water level recovers slowly.

From the rate at which the water level in the well falls, it is possible to calculate how much water is taken from well storage and therefore how much water is supplied by the aquifer (de Silva and Rushton 1996). Figure 10.14c shows that initially the full discharge of 300 m³/d is taken from well storage; even at the end of the pumping period of 5 hours, 66% of the pumped water is from well storage and 34% is from the aquifer. During recovery, the flow of water from the aquifer to refill well storage decreases only slowly. After 24 hours, 23% of the well storage remains to be replenished. This occurs because the hydraulic conductivity K is quite low and the specific yield S_Y of 0.025 is also small.

Examining the behavior of a dug well over a single day fails to highlight the problems faced by a farmer in deciding how much water should be pumped every day and therefore the area of a crop that can be grown. The critical issue for a farmer is ensuring that there is

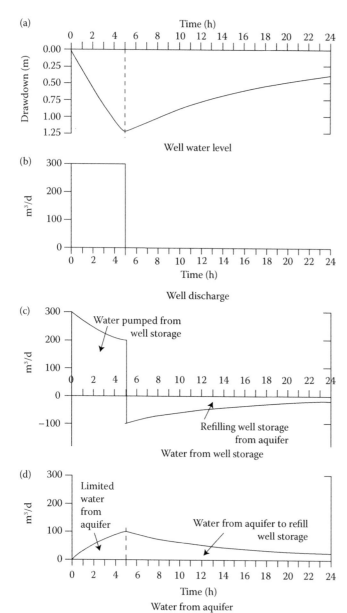

FIGURE 10.14

Example of drawdowns and flows for a well diameter = 7.0 m, initial saturated depth = 8.0 m, $K = 6.0$ m/d, $S_Y = 0.025$, and pumping at 300 m³/d for 5 hours.

water in his well toward the end of the growing season. In a study of the long-term response of abstraction from a large-diameter well, Rushton (2003) found that even when the well water level only falls by 0.8 m at the end of the first day of pumping, by the end of the 84th day little water remains in the well. A rough estimate of the total amount of water that can be withdrawn from a large-diameter well for irrigating a dry season crop can be obtained by assuming that, at most, 50% of the water stored in the aquifer can be abstracted from the well. Frequently, there is insufficient water to grow a dry season crop over the entire land holding of a farmer.

10.6.2 Horizontal Wells

Horizontal wells provide an alternative means of abstracting water from unconfined aquifers of limited saturated thickness. Examples of the successful operation of horizontal wells include a shallow coastal aquifer in Sarawak (Mohamad and Rushton 2006); a cross section of the well in Sarawak is included in Figure 10.15. Also, detailed studies have been made of a 300-m-long horizontal well in a shallow sand aquifer in northwest England, which has operated reliably for more than a decade (Rushton and Brassington 2013).

Within the pipe of a horizontal well, significant hydraulic head loss occurs, especially if the diameter of the pipe is constrained. Therefore, in devising a conceptual model it is necessary to consider the three flow processes: regional groundwater conditions, convergent flows to the well and through the well screen, and also frictional head losses within the well pipe. A schematic diagram of these three components is presented in Figure 10.16a.

Figure 10.16b illustrates schematically how a horizontal well can intercept regional groundwater flow. In this diagram, the regional groundwater flow is from left to right. Flow processes into and within the horizontal well are illustrated in Figure 10.16c. As water is drawn into the well pipe, the quantity flowing through the pipe, and hence the velocity, increases. The head gradient in the pipe is proportional to the square of the velocity; therefore, the hydraulic head gradient steepens significantly toward the discharge end of the pipe. A consequence is that groundwater flows into the well increase markedly toward the discharge end (Rushton and Brassington 2013).

A number of analytical solutions have been devised for horizontal wells but they usually assume either a constant hydraulic head in the well or a constant inflow along the well (Kawecki 2000); neither of these assumptions is valid in practical situations.

Analysis of the operation of horizontal wells in practical field situations is not straightforward. Figure

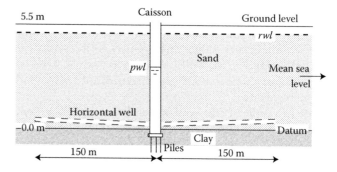

FIGURE 10.15

Details of a horizontal well in a coastal aquifer in Sarawak: *rwl, pwl,* rest, and pwls.

(a)

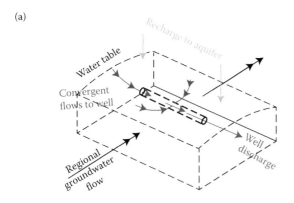

Flow components for horizontal well

(b)

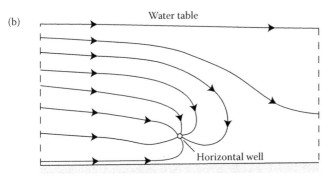

Cross-section with approx. flowlines

(c)

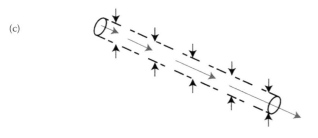

Flow into and through horizontal well

FIGURE 10.16
Components of flow to a horizontal well.

10.16 illustrates how it is necessary to simultaneously consider the regional and local groundwater flow, for which Darcy's law applies, and flow within the well where the hydraulic head loss is proportional to the square of the velocity. The approach adopted for the mathematical modeling of the two case studies referred to above is to solve sequentially for the aquifer and then the pipe. Typically about five iterations are required.

For the horizontal well in the coastal sand aquifer in Sarawak (Figure 10.15), the central caisson of 1.2 m diameter collects water from two arms each extending for 150 m from the caisson. Each horizontal arm consists of two 150-mm slotted pipes. A 6-day period of pumping at 240 m³/d was simulated by the mathematical model which represents all the flow processes indicated in Figure 10.16. Field results for drawdowns in the caisson, the horizontal wells, and several piezometers in the aquifer, were reproduced satisfactorily (Mohamad and Rushton 2006). In a predictive simulation at an abstraction rate of 480 m³/d, the drawdown in the caisson is estimated to be 2.76 m with a head loss along each arm of 0.24 m. The system of a single-suction pump collecting water from horizontal arms into a caisson is more dependable than the alternative of clusters of well points.

Monitoring of the hydraulic heads, at least at either end of a horizontal well, is crucial in developing an understanding of the reliability and efficiency of a horizontal well system (Rushton and Brassington 2013).

10.6.3 Inverted Wells

An initial survey by UNDP of the Baring Tract (about 250 km west–northwest of Dhaka, Bangladesh) concluded that there are no significant aquifers within a depth of 300 m. The UNDP report suggested that in the upper 70 m, the relatively thin zones of fine-grained sand would provide limited groundwater yields for domestic purposes. However, Asad-uz-Zaman of the Barind Multipurpose Development Authority believed that sufficient groundwater could be abstracted for dry season irrigation.

With conventional bore wells, yields are limited when the thickness of the permeable zone is 20 m or less. The left-hand diagram of Figure 10.17 shows a conventional well in an aquifer about 20 m in thickness, drilled to 0.5 m diameter with a solid casing of 0.36 m in diameter extending to a depth of 16 m with a reducer connected to a slotted screen of 0.20 m diameter and 11 m long. The pump is located toward the bottom of the solid casing. With a rest water level (rwl) at 2.8 m and a pwl of 10.2 m (just above the pump) the discharge was 2400 m³/d with a specific capacity of 325 m³/d/m.

After several attempts, a new well arrangement was devised. The solid casing extends nearly to the bottom of the permeable strata with a unique connector at its base, which allows four screens to extend vertically upward as shown in the right-hand diagram of Figure 10.17. This is described as an "inverted well." The pump, which has additional stages, is located toward the bottom of the solid casing and is therefore lower than in a conventional well. This inverted well was constructed close to the original conventional well. Even though the rwl at the time of test pumping was 1.8 m lower, a discharge of 7300 m³/d was achieved with a specific capacity of 960 m³/d/m.

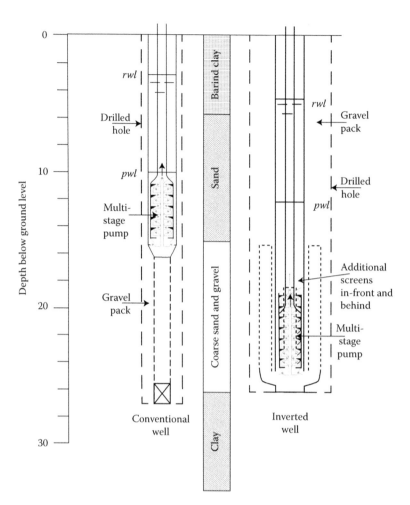

FIGURE 10.17
Comparison of a conventional and an inverted well at the same location.

There are a number of reasons for this dramatic increase in yield:

a. Larger pumped drawdowns are possible because the borehole pump is at a greater depth in the inverted well.

b. The diameter of the drilled hole is double that of the conventional well.

c. With four screens extending vertically upward, the surface area of the screens in the inverted well is 14.6 m² compared with 5.8 m² for the conventional well.

d. For the inverted well, the gravel pack volume is approximately 20 m³ compared to 3.5 m³ for the conventional well.

Further information about the inverted well can be found in Rushton (2003) and Asad-uz-Zaman and Rushton (2006).

Exploitation of the Barind aquifer using bore wells, many of them inverted wells, has continued for more

than 30 years with the number of wells and the area irrigated increasing year by year. Prior to the year 2000, rwls recovered to roughly the same elevation each year. Since then, complete recovery of groundwater levels has not occurred and a new equilibrium appears to have been reached. Continuing careful monitoring is essential to ensure that overexploitation does not occur.

10.7 Conclusions

This chapter illustrates how groundwater exploitation in the Indian subcontinent has changed during the past 40 years. The ability to drill deep wells and the availability of power to pump water from both dug wells and bore wells has resulted in a substantial increase in the volume of groundwater abstracted. In the 1980s and early 1990s, investigations were carried out to understand the wide variety of processes associated with groundwater resources; these investigations were based on extensive field studies which continued for a number of years. The developing understanding of the important groundwater flow processes was confirmed by conceptual and computational modeling. In the past two decades, the number of detailed and thorough field investigations has declined and computer modeling, without the support of detailed fieldwork, has increased.

Changes in groundwater conditions occur only slowly, especially in deep aquifers; it may be many years before the effect of any intervention becomes clear. This is unwelcome news for policymakers who prefer to know quickly the results of interventions. Furthermore, many factors have an impact on groundwater head hydrographs such as changing seasonal rainfall, cropping patterns, groundwater abstraction, and losses from surface water irrigation. Consequently, long-term detailed monitoring is essential. Moreover, obtaining reliable insights from the monitoring results requires careful scrutiny of the field information. Examples in this chapter of vital insights gained from the innovative examination of field results include the flow processes in weathered–fractured aquifers, vertical flows in alluvial aquifers, the substantial water losses through the bunds of rice fields, and the difficulty in maintaining long-term artificial recharge. Well-designed field experiments and careful monitoring allow the formulation of realistic conceptual models. Only when a valid conceptual model has been developed should attempts be made to prepare computational models. The discussion in this chapter has highlighted occasions when inadequate and incorrect computational models have lead to poor guidance and decisions.

Overexploitation is a serious problem, especially in areas with a semiarid climate. It is important to recognize that in many locations there is insufficient rainfall for both a wet season crop and a dry season crop. In hard-rock aquifers, with limited storage, the effect of abstracting too much water rapidly becomes apparent. However, in deep alluvial aquifers, overabstraction can continue for many years without serious consequences. The result is falling water tables and significant increases in the height through which a pump has to raise water due to falling pumping levels. Pragmatic solutions to overexploitation usually require demand-side interventions. A number of successful schemes are described earlier in this chapter; the approaches introduced in these schemes should be implemented widely. The challenge for groundwater specialists is to develop realistic plans to stabilize groundwater abstractions and to help farmers understand that there is a limited amount of groundwater that can be abstracted.

Substantial losses occur from most irrigation schemes; rarely is the amount of water actually used by the crop more than 35% of the water supplied at the head of a major canal irrigation scheme. It is difficult to prevent all losses from canals, especially if they are constructed on relatively permeable ground. Lining of canals has a limited effect; the main advantage of lining is to reduce the hydraulic resistance so that water can be transmitted along a canal with a shallower hydraulic gradient. Although water losses are high, they often add to the groundwater resource; consequently, conjunctive use of surface water and groundwater should be introduced more extensively. However, when developing conjunctive use schemes, the location of the wells must be selected carefully; if they are located in the vicinity of canals, recirculation of water occurs. For rice fields, a more frequent rotational supply of water should be introduced so that the depth of water is never more than 8 cm, thereby reducing the losses through the bunds.

Estimation of recharge to an aquifer system is one of the major challenges in groundwater studies; processes are complex and it is crucial to recognize the significance of soil and crop types. Although the water balance methodology, described in this chapter, may appear to be complicated, the inputs to the model can be derived from documents such as FAO (1998) *Guidelines for Computing Crop Water Requirements*. The computations can be carried out using an Excel spreadsheet. A daily water balance provides an understanding of how conditions change from one location to another and from one season to another. The CGWB approach to recharge estimation provides a useful overall assessment of resources and exploitation. However, detailed field studies supported by water balance modeling and an assessment of other contributions to recharge would improve the reliability of the estimates.

Artificial or managed aquifer recharge is usually difficult to maintain for extended time periods; it is also challenging to assess the success of schemes. The most successful forms of artificial recharge are not planned but result from water losses from irrigation systems and from water distribution networks. The pilot projects in Gujarat in the 1980s and the studies by Gale et al. (2006) demonstrate that artificial recharge rarely achieves the expected targets. Field estimates of the volume of water actually recharged are essential. Since forcing water into an aquifer tends to clog the aquifer, long-term artificial recharge is likely to remain unpredictable.

There are locations where groundwater resources are only partially exploited. In some situations there is no need for additional groundwater. Another reason for limited groundwater utilization is that abstracting water from aquifer systems is difficult, for example, because aquifers are of limited saturated thickness or of low permeability. Vertical small-diameter wells are not efficient at collecting water from an aquifer due to the steep groundwater head gradients as water moves radially toward the well; the presence of a seepage face may also limit their yield. Large-diameter dug wells have advantages not only because there is a larger surface area through which water is drawn into the well but also because they continue to draw in water from the aquifer to refill well storage after the pump is switched off. Horizontal wells also have a considerable potential in shallow unconfined aquifers. Finally, new bore well designs, such as the inverted wells used in Bangladesh, allow the successful exploitation of relatively thin aquifers, which are overlain by a less-permeable material.

References

Ahmad, N. 1974. *Groundwater Resources of Pakistan*. Ripon Printing Press, Lahore.

Asad-uz-Zaman, M. and K.R. Rushton. 2006. Improved yield from aquifers of limited saturated thickness using inverted wells. *Journal of Hydrology* 326:311–24.

Basharat, M. 2012. Spatial and temporal appraisal of groundwater depth and quality in LBDC command—Issues and options. *Pakistan Journal of Engineering and Applied Sciences* 11:14–29.

Bos, M.G. and J. Nugteren. 1990. *On Irrigation Efficiencies*. Publ no 19, International Institute for Land Reclamation and Improvement, Wageningen, the Netherlands.

Bouwer, H. 1969. Theory of seepage from open channels. In: Ven Te Chow (Ed.) *Advances in Hydrosciences*. 5:121–72, Academic Press, New York.

Bradley, E. and P.N. Phadtare. 1989. Paleohydrology affecting recharge to over-exploited semiconfined aquifers in the Mehsana area, Gujarat State, India. *Journal of Hydrology* 108:309–22.

Central Ground Water Board. 2002. *Master Plan for Artificial Recharge to Ground Water in India*. Ministry of Water Resources, Government of India, Faridabad.

Central Ground Water Board. 2011a. *Dynamic Ground Water Resources of India (as on March 31 2009)*. Ministry of Water Resources, Government of India, Faridabad.

Central Ground Water Board. 2011b. *Select Case Studies Rain Water Harvesting and Artificial Recharge*. Ministry of Water Resources, Government of India, Faridabad.

Central Ground Water Board. 2013. *Master Plan for Artificial Recharge to Ground Water in India*. Ministry of Water Resources, Government of India, Faridabad.

Chatterjee, R. and R.R. Purohit. 2009. Estimation of replenishable groundwater resources in India and their status of utilisation. *Current Science of India* 96(12):1581–91.

de Silva, C.S. and K.R. Rushton. 1996. Interpretation of the behaviour of agrowell systems in Sri Lanka using radial flow models. *Hydrological Sciences Journal* 41:825–35.

de Silva, C.S. and K.R. Rushton. 2007. Groundwater recharge estimation using improved soil moisture balance methodology for a tropical climate with distinct dry seasons. *Hydrological Sciences Journal* 52(5):1051–67.

de Silva, C.S. and K.R. Rushton. 2008. Representation of rainfed valley ricefields using a soil–water balance model. *Agricultural Water Management* 93:2271–82.

Eilers, V.M.H., R.C. Carter, and K.R. Rushton. 2007. A single layer soil water balance for estimating deep drainage (potential recharge): An application to cropped land in semi-arid north-east Nigeria. *Geoderma* 140:119–31.

FAO. 1998. *Crop Evapotranspiration: Guidelines for Computing Crop Water Requirements*. FAO Irrigation and Drainage Paper 56, Rome.

Foster, S. 2012. Hard-rock aquifers in tropical regions: Using science to inform development and management policy. *Hydrogeology Journal* 20:659–72.

Gale, I.N., D.M.J. Macdonald, R.C. Calow, I. Neumann, M. Moench, H. Kulkarni, S. Mudrakartha, and K. Palanisami. 2006. *Managed Aquifer Recharge: An Assessment of Its Role and Effectiveness in Watershed Management*. British Geological Survey Commissioned Report CR/06/107N, BGS, Keyworth, UK.

Garduno, H., S. Foster, L. Pradeep-Raj, and F. van Steenberger. 2009. Addressing groundwater depletion through community-based management actions in the weathered granitic basement aquifer of drought-prone Andhra Pradesh-India. World Bank GWMATE Case Profile Collection 19. http://www.worldbank.org/ gwmate.

Hargreaves, G.H. and R.G. Allen. 2003. History and evaluation of Hargreaves evapotranspiration equation. *Journal of Irrigation and Drainage Engineering* 29(1):53–63.

Healy, R.W. and P.R. Cook. 2002. Using groundwater levels to estimate recharge. *Hydrogeology Journal* 10(1):91–109.

Kavalanekar, N.B., S.C. Sharma, and K.R. Rushton. 1992. Over-exploitation of an alluvial aquifer in Gujarat, India. *Hydrological Sciences Journal* 37:329–46.

Kawecki, M.W. 2000. Transient flow to a horizontal water well. *Ground Water* 38:842–50.

Khan, S. and Rushton, K.R. 1996. Reappraisal of flow to tile drains. *Journal of Hydrology* 183:351–95.

Kulkarni, H., P.S. Vijay Shankar, S.B. Deolankar, and M. Shah. 2004. Groundwater demand management at local scale in rural areas of India: A strategy to ensure water well sustainability based on aquifer diffusivity and community participation. *Hydrogeology Journal* 12:184–96.

Latif, M. 2007. Spatial productivity along a canal irrigation system in Pakistan. *Irrigation and Drainage* 56:509–21.

Limaye, S.D. 2010. Groundwater development and management in the Deccan Traps (basalts) of western India. *Hydrogeology Journal* 18:543–58.

Lowe, T.C. and K.R. Rushton. 1990. Water losses from canals and rice fields. *Irrigation and Water Resources Symposium*, Agricultural Engineering Society of Sri Lanka, Peradeniya, Sri Lanka, 107–22.

Martin, R. (Ed.) 2013. Clogging issues associated with managed aquifer recharge methods. IAH Commission on Managing Aquifer Recharge. www.iah.org/recharge/clogging.htm.

Mohamed, A. and K. Rushton. 2006. Horizontal wells in shallow aquifers: Field experiment and numerical model. *Journal of Hydrology* 329:98–109.

Panda, D.K., A. Mishraand, and A. Kumar. 2012. Quantification of trends in groundwater levels of Gujarat in western India. *Hydrological Sciences Journal* 57(7):1325–36.

Qureshi, A.S., H. Turral, and I. Masih. 2004. Strategies for the management of conjunctive use of surface water and groundwater resources in semi-arid areas: A case study from Pakistan. Research Report 86, IWMI, Colombo, Sri Lanka.

Rathod, K.S. and K.R. Rushton. 1991. Interpretation of pumping from two-zone layered aquifer using a numerical model. *Ground Water* 29:499–509.

Rushton, K.R. 2003. *Groundwater Hydrology, Conceptual and Computational Models*. Wiley, Chichester.

Rushton, K.R. 2007. Representation in regional models of saturated river–aquifer interaction for gaining/losing rivers. *Journal of Hydrology* 334:262–81.

Rushton, K.R. and S.C. Redshaw. 1979. *Seepage and Groundwater Flow*. Wiley, Chichester.

Rushton, K.R. and J. Weller. 1985. Response to pumping of a weathered–fractured granite aquifer. *Journal of Hydrology* 80:299–309.

Rushton, K.R. and S.V.R. Rao. 1988. Groundwater flow through a Miliolite limestone aquifer. *Hydrological Sciences Journal* 33:449–64.

Rushton, K.R. and N.S. Srivastava. 1988. Interpretation of injection well tests in an alluvial aquifer. *Journal of Hydrology* 99:49–60.

Rushton, K.R. and P.N. Phadtare. 1989. Artificial recharge pilot projects in Gujarat India. In: A. Sahuquillo, J. Andreu, and T. O'Donnell (Eds.) *Groundwater Management: Quantity and Quality*, IAHS, Wallingford, UK, IAHS Publ. No. 188, pp. 533–45.

Rushton, K.R. and S.C. Tiwari. 1989. Mathematical modelling of a multi-layered alluvial aquifer. *Journal of the Institution of Engineers* (India), 90(CV2):47–54.

Rushton, K.R. and F.C. Brassington. 2013. Hydraulic behaviour and regional impact of a horizontal well in a shallow aquifer: Example from the Sefton Coast, northwest England (UK). *Hydrogeology Journal* 21:1117–28.

Rushton, K.R., V.H.M. Eilers, and R.C. Carter. 2006. Improved soil moisture balance methodology for recharge estimation. *Journal of Hydrology* 318:379–99.

Sakthivadivel, R. 2007. The groundwater recharge movement in India. In: *Comprehensive Assessment of Water Management*. CABI, Agriculture Series, CABI, Wallingford, UK.

Scanlon, B.R., R.W. Healy, and P.R. Cook. 2002. Choosing appropriate techniques for quantifying groundwater recharge. *Hydrogeology Journal* 10(1):18–39.

Shah, T., S. Bhatt, R.K. Shah, and J. Talati. 2008. Groundwater governance through electricity supply management: Assessing an innovative intervention in Gujarat, western India. *Agricultural Water Management* 95:1233–42.

Villholth, K.G. and L.D. Rajasooriyar. 2010. Groundwater resources and management challenges in Sri Lanka—An overview. *Water Resource Management* 24:1489–513.

Wachyan, E. and K.R. Rushton. 1987. Water losses from irrigation canals. *Journal of Hydrology* 92:275–88.

Walker, S.H. and K.R. Rushton. 1984. Verification of lateral percolation losses from irrigated rice fields by a numerical model. *Journal of Hydrology* 71:335–51.

Walker, S.H. and K.R. Rushton. 1986. Water losses through the bunds of irrigated rice fields interpreted through an analogue model. *Agricultural Water Management* 11:59–73.

World Bank. 2010. *Deep Wells and Prudence: Towards Pragmatic Action for Addressing Groundwater Overexploitation in India*. World Bank, Washington, DC.

11
Simulation Optimization Models for Groundwater Management

T. I. Eldho, Boddula Swathi, and Partha Majumdar

CONTENTS

11.1 Introduction

Water is the essential resource for the existence of life on Earth. To meet the various water demands of mankind, there are two major sources, namely, the surface water and the groundwater. As a source of water supply, groundwater is more reliable and it has several inherent advantages over the surface water resources as its availability does not depend directly on the annual rainfall,

assumed to be more pure, and is free from enormous losses due to evaporation. Due to the intensive developments in the surface water resources over the past decades, and significant impacts on the hydrological cycle due to climate change, the groundwater resources are now overstressed leading to various quantity and quality issues. About 97% of the potentially available fresh water, which can be utilized for human use, is beneath the ground surface in the form of groundwater

(Freeze and Cherry, 1979). This shows the importance of groundwater for mankind, which necessitates appropriate management and protection for sustainable utilization. Due to the developments in pumping technologies and power availability over the past few decades, there is widespread drawdown/depletion to major groundwater aquifers all over the world. Further due to unregulated disposal of domestic and industrial wastes, and percolation of pesticides from agricultural fields, severe groundwater contamination has been reported in many parts of the world.

For the appropriate management of the available groundwater resources, it is essential to know the groundwater flow and transport processes in the aquifer systems. Due to the complexity of the groundwater systems, to understand the flow and transport behavior, models are being used, as field investigations in the larger domains are not feasible and are mostly costly. Generally for simulation of groundwater flow and transport processes, models based on numerical methods are used. For groundwater system management, optimization models based on various optimizations tools also can be effectively used. Hence, for solving various groundwater management issues, development of simulation optimization models that realistically model the actual field conditions is necessary. With increase in computational capability of computers and several modeling software being available, a hydrogeologist can effectively simulate different management scenarios for groundwater pumping or remediation and can establish guiding principles and response system for end users. The main focus of this chapter is on the applications of simulation optimization models for various groundwater management problems.

11.2 Challenges in Groundwater Management

As the groundwater moves in the subsurface through the highly anisotropic and heterogeneous porous media, the simulation of flow and transport processes is a highly complex one and hence the appropriate groundwater management poses a number of challenges. In a broad perspective, we have to deal with the following challenges in managing and modeling the groundwater systems.

11.2.1 Complexity of the System

The groundwater flow and transport processes in the aquifer system are very complex and we have to apply a number of simplifications and assumptions to reasonably represent the groundwater systems. Any model will be considered as a simplified version of the actual aquifer system. The complex geological conditions and physical boundaries are often simplified or assumed. Only the most dominant process is considered in the simulation. Few model parameters will be included to satisfy the model calibration, which is based on historical data. All these factors can add up to contribute different patterns of prediction than the realistic scenarios (Hill and Tiedeman, 2007).

11.2.2 Conceptualization of the Groundwater System

The real groundwater system and its behavior are very complicated. Depending on the objective of the groundwater management system, we can conceptualize the groundwater system with respect to dimension, time, anisotropy, heterogeneity, and various other components existing in the system. It is intuitively obvious that no effective management decision can depend on a complete detailed description of a considered system and its behavior. Instead, we need to simplify the description of the considered groundwater system and its behavior to a degree that will be useful for the purpose of planning and making management decisions in specific cases. Generally, these simplifications are implemented through a set of assumptions that expresses our understanding of the groundwater system and its behavior, for the concerned problem. The selection of the appropriate conceptual model for a particular case depends on the objective of the study, available data and resources, and legal and regulatory framework.

11.2.3 Insufficiency of Data

Successful modeling of any groundwater system requires three types of data, namely, static data for constructing hydrogeological framework, dynamic data for assessment of groundwater involving abstraction rates and recharges, and time series data of groundwater levels (Refsgaard et al., 2007; Zhou and Li, 2011). The measurement of all these data is tedious and economically challenging especially in developing countries. Currently there are several local or institutional databases storing these data, but there is an immediate need for a comprehensive, publicly accessible, groundwater database (MOWR, 2001). This can be achieved by starting with regional scale monitoring network. Most of the time, we need to develop groundwater models with the scarce available data and obtain many parameters through calibration and validation processes.

11.2.4 Socioeconomic Factors

In implementing any groundwater management model, we have to deal with several socioeconomic factors. Generally, any new management model attracts inaction from the users due to the politically complex world wherein active modification of already well-established usage patterns is cumbersome and there are monetary concerns. Also, due to the dynamic changes in the nature and climate change and varied usage patterns of resources, the success of any management scenario is uncertain leading to reluctance and a lack of incentives in participating in the long-term management initiatives (FAO, 2003).

11.3 Groundwater Hydrology

Aquifers that are the subsurface units of rock or unconsolidated soil media capable of yielding water are the general sources of groundwater (Bear, 1972). Depending on the geological conditions, aquifers can be either confined or unconfined in nature. The hydrogeological characteristics of a particular aquifer give an idea of the availability of groundwater and its suitability for usage (Fried, 1975). Any model is expressed through a balance of particular extensive quantity, in the case of a groundwater system; it is the principle of conservation of mass of water and/or of solute. The solutions from the models rely on the inclusion of different mechanisms governing the flow and/or solute transport equations. Detailed descriptions of the governing equations describing the groundwater flow and transport processes can be found in many literatures like Fried (1975), Bear (1979), Freeze and Cherry (1979), Anderson and Woessner (1992), etc. From the principle of conservation of mass and Darcy's law, the governing equation for groundwater flow is derived for saturated porous media (Bear, 1972). On solving the flow equation, the head distribution in the entire domain is obtained using which the velocities can also be calculated with Darcy's law. Based on Fick's diffusion laws and mass balance principle, the governing equation for solute-transport equation in a groundwater system in terms of concentration changes is derived. The concentration changes can be due to various transport mechanisms such as advection, dispersion, and retardation to name a few (Bear, 1979). In order to solve the solute transport equation, velocities from the flow regime are required and thus it is essential to have coupled flow and transport groundwater system.

11.3.1 Governing Equations for Groundwater Flow and Transport

The governing partial differential equation describing the flow in a two-dimensional inhomogeneous confined aquifer is given as (Bear, 1979; Wang and Anderson, 1982)

$$\frac{\partial}{\partial x}\left[T_x \frac{\partial h}{\partial x}\right] + \frac{\partial}{\partial y}\left[T_y \frac{\partial h}{\partial y}\right] = S \frac{\partial h}{\partial t} + Q_w \delta(x - x_i)(y - y_j) - q$$

(11.1)

The governing partial differential equation describing the groundwater flow in a two-dimensional inhomogeneous unconfined aquifer is given as (Bear, 1979; Wang and Anderson, 1982)

$$\frac{\partial}{\partial x}\left[K_x h \frac{\partial h}{\partial x}\right] + \frac{\partial}{\partial y}\left[K_y h \frac{\partial h}{\partial y}\right] = S_y \frac{\partial h}{\partial t} + Q_w \delta(x - x_i)(y - y_i) - q$$

(11.2)

The seepage velocity necessary to the solution of the solute transport model is computed using Darcy's law (Bear, 1972) and can be written as: $V_i = -K_{ij}/n \; \partial h/\partial x_i$, $i, j = 1, 2, 3$; where V is the seepage velocity in the x direction [LT^{-1}] and n the porosity.

The governing partial differential equation for transport of a single chemical constituent in saturated porous media in two dimensions, considering the advection, dispersion, and fluid sources/sinks (Freeze and Cherry, 1979; Wang and Anderson, 1982; Zheng and Bennett, 1995) is given as

$$R \frac{\partial C}{\partial t} = \frac{\partial}{\partial x}\left(D_{xx} \frac{\partial C}{\partial x}\right) + \frac{\partial}{\partial y}\left(D_{yy} \frac{\partial C}{\partial y}\right) - \frac{\partial}{\partial x}(V_x C)$$

$$- \frac{\partial}{\partial y}(V_y C) - \frac{c'w}{nb} - R\lambda C - \frac{q_w C}{n}$$

(11.3)

For transient flow and transport analysis, the following initial conditions are used:

$$h(x, y, 0) = h_0(x, y); \quad C(x, y, 0) = f \quad x, y \in \Omega \quad (11.4)$$

The flow and transport equations should be solved with appropriate boundary conditions. Generally, the boundary conditions can be prescribed variable (head or concentration) or gradient of the variable or flux. The boundary conditions can be Dirichlet type written as follows:

$$h(x, y, t) = h_1(x, y, t); \quad C(x, y, t) = g_1 \quad x, y \in \partial\Omega_1 \quad (11.5)$$

Or Neumann's type expressed as, T $\partial h/\partial n = q_1(x, y, t)$ $x, y \in \partial\Omega_2$ for confined aquifer; $Kh \; \partial h/\partial x = q_2(x, y, t)x$, $y \in \partial\Omega_2$ for unconfined aquifer; $(D_{xx}\partial C/\partial x)n_x + (D_{yy}\partial C/\partial y)$ $n_y = g_2(x, y) \in (\Gamma_2)$ for transport problems.

Here $h(x, y, t)$ = Piezometric head [L]; $T(x, y)$ = transmissivity [L^{2}T^{-1}]; $K(x, y)$ = hydraulic conductivity [LT^{-1}];

S = storage coefficient; S_y = specific yield; x, y = horizontal space variables [L]; Q_w = source or sink function ($-Q_w$ = source, Q_w = sink) [LT^{-1}]; t = time [T]; Ω = the flow region; $\partial\Omega$ = the boundary region ($\partial\Omega_1 \cup \partial\Omega_2 = \partial\Omega$); $\partial/\partial n$ = normal derivative; $h_0(x, y)$ = initial head in the flow domain [L]; $h_1(x, y, t)$ = known head value of the boundary head [L]; $q(x, y, t)$ = known inflow rate [L^{2}T^{-1}]; δ = Dirac delta function; x_i, y_i = pumping or recharge well location; where, V_x, V_y = seepage velocity in x and y direction; D_{xx}, D_{yy} = components of dispersion coefficient tensor [L^{2}T^{-1}]; C = dissolved concentration [ML^{-3}]; λ = the reaction rate constant [T^{-1}]; w = the elemental recharge rate with solute concentration c; n = the local porosity; t = time; b = aquifer thickness under the element; R = retardation factor = $1 + \rho_b K_d/n$, where ρ_b is the media bulk density, and K_d the sorption coefficient; q_w = volumetric pumping rate from a source; Γ_1 and Γ_2 are the boundary sections of Ω, f is a given function in Ω, g_1 and g_2 are given functions along boundaries Γ_1 and Γ_2; and n_x and n_y are the components of the unit outer normal vector to the given boundary Γ_2.

11.3.2 Coupled Flow and Transport Mechanism

As groundwater is the most important source of fresh water, we have to protect it from the increasing threat of subsurface contamination. The growth of population and of industrial and agricultural productions coupled with the resulting increased requirements for energy developments have a major impact on the groundwater quality resulting contamination to the aquifer systems. To reduce the groundwater contamination problems and to remediate the already polluted aquifer systems, we need to understand the transport process in the porous media and model the contaminant transport behavior. For the successful groundwater contaminant transport modeling, it is necessary to understand the process very well (Bedient et al., 1999). Broadly, three processes govern the transport of contaminants in the system, namely advection, dispersion, and retardation (Freeze and Cherry, 1979). The study of these processes is useful for predicting the time when an action limit, that is, a concentration limit used in regulations such as drinking water standards, will be reached.

11.3.2.1 Advection

Once a contaminant source is introduced to the aquifer system, the solute is transported due to the bulk motion of the groundwater flow. This process is known as advection (Freeze and Cherry, 1979). The nonreactive solutes are carried at an average rate equal to the average linear velocity, $\bar{v}$, of the water. Therefore, $\bar{v} = v/n$, v being the specific discharge and n is the porosity.

11.3.2.2 Dispersion

Once the contamination is introduced to the aquifer systems, the tendency of the solute to spread out other than following the path as expected with accordance to advection in the system is called hydrodynamic dispersion (mechanical dispersion) (Freeze and Cherry, 1979). The spreading process causes dilution of the solute because of mechanical mixing during fluid advection and because of molecular diffusion of the solute particles. The equation governing the dispersion phenomena is derived using Fick's law (Anderson, 1984).

11.3.2.3 Retardation

In solute transport phenomena, several chemical and physical processes can cause retardation, that is, delay or prolong the transport of solute in the groundwater (Freeze and Cherry, 1979). Four general mechanisms can retard the movement of chemical constituents in groundwater, via dilution, filtration, chemical reaction, and transformation (Bear, 1979).

The constitution of any coupled groundwater flow and transport model involves first solving the governing equation of groundwater flow, which is obtained by combining the flow continuity equation based on conservation of mass. Then velocities are computed with Darcy's law (Bear, 1979), which are used in solving the solute transport equation and which describe the transport of solute (contaminant) in the groundwater system.

11.4 Groundwater Simulation

For appropriate groundwater management, we have to carry out the groundwater simulation for flow and/or transport. Generally, the groundwater simulation can be done by means of physical method, analogs, analytical, or numerical solutions. The physical models to a lab scale or in the field are rarely used nowadays because they are expensive, cumbersome, and difficult in practice. The analog model such as electric resistor analog or viscous analog has limited applications. The analytical method is useful only for a small class of mathematical formulations with simplified governing equations, boundary conditions, and geometry. Hence, the attention was shifted to the prospects of computer modeling in which appropriate solutions for the governing partial differential equations of groundwater flow and transport were found using numerical techniques and soon researchers and engineers were able to successfully simulate and predict the behavior of groundwater systems.

11.4.1 Numerical Modeling

Due to various complexities of groundwater flow and transport problems such as irregularity of domain's boundaries, heterogeneity, nonlinearity, irregular source/sink functions, etc., for the solution of field problems, it is impossible to derive analytical/exact solutions for the mathematical models discussed earlier. Numerical methods are usually employed for solving the mathematical model. In numerical modeling, various numerical methods are employed to transform the mathematical model into a numerical one. The partial differential equations appearing in the mathematical model are represented by their numerical counterparts and solved using appropriate initial and boundary conditions.

In groundwater hydrology, numerical modeling can be used for several purposes like (Wang and Anderson, 1982):

- For the aquifer considered, investigating groundwater system dynamics and understanding the flow patterns.

- As an assessment and planning tool for evaluating recharge, discharge, aquifer storage processes, transport of contaminants, and quantifying sustainable yield.

- As a predictive tool for simulating future scenarios and to recreate the impacts of various activities.

- For planning and designing practical solutions for different development and management scenarios.

- As a groundwater management tool for assessing alternative policies and regulatory guidelines.

- As visualization tools for communicating key messages to the public and decision makers.

For the groundwater simulation, in the last few decades, a variety of numerical methods such as finite difference method (FDM), method of characteristics (MOC), finite volume method (FVM), finite element method (FEM), boundary element method (BEM), analytic element method (AEM), meshless method, etc. have been developed by engineers and scientists. Depending on the groundwater problem to be solved, each of these methods has its own advantages and disadvantages and the choice depends on the complexity of the problem, data available, computational facilities, and investigator's familiarity with the method. Out of the many available numerical methods for groundwater simulation, FDM, FEM, and AEM are the most popular numerical modeling techniques among engineers and scientists. Recently, meshless methods have also been used for groundwater simulation (Mategaonkar and Eldho, 2012). Brief overviews of these techniques are given next.

11.4.1.1 Finite Difference Method

The basis of FDM is to represent the continuous variation of the concerned function by a set of values at points on a grid of intersecting lines. The gradients of the function are then represented by differences in the values at neighboring points and a finite difference version of the equation is formed. In groundwater modeling using FDM, the governing equations are approximated using backward, forward, or central difference schemes (Wang and Anderson, 1982). Initially, the problem domain is discretized to a grid. The nodes can be considered either at the center of grid cells or at the intersection of the grid. Then the governing equation is approximated using a suitable scheme of difference at each node and a system of equations is formed. This system can be solved using direct or iterative solution techniques, after the appropriate application of the boundary conditions, to get the unknown groundwater head or concentration.

Considering homogeneous isotropic confined aquifer in two dimensions, Equation 11.1 can written as

$$\frac{\partial^2 h}{\partial x^2} + \frac{\partial^2 h}{\partial y^2} = \frac{S}{T}\frac{\partial h}{\partial t} - \frac{R(x,y,t)}{T} \qquad (11.6)$$

where S is the storage coefficient, T the transmissivity, and R the recharge or pumping rate. Corresponding FDM form in explicit form (refer to Figure 11.1), Equation 11.6 can be written as (Wang and Anderson, 1982)

$$\frac{h_{I+1,J}^n - 2h_{I,J}^n + h_{I-1,J}^n}{(\Delta x)^2} + \frac{h_{I,J+1}^n - 2h_{I,J}^n + h_{I,J+1}^n}{(\Delta y)^2}$$

$$= \left(\frac{S}{T}\right)\frac{h_{I,J}^{n+1} - h_{I,J}^n}{(\Delta t)} - \frac{R_{I,J}^n}{T} \qquad (11.7)$$

The corresponding FDM form in implicit form can be written as (Wang and Anderson, 1982)

$$\frac{h_{I+1,J}^{n+1} - 2h_{I,J}^{n+1} + h_{I-1,J}^{n+1}}{\Delta x^2} + \frac{h_{I,J+1}^{n+1} - 2h_{I,J}^{n+1} + h_{I,J-1}^{n+1}}{\Delta y^2}$$

$$= \frac{S}{T}\frac{h_{I,J}^{n+1} - h_{I,J}^n}{\Delta t} - \frac{R_{I,J}^{n+1}}{T} \qquad (11.8)$$

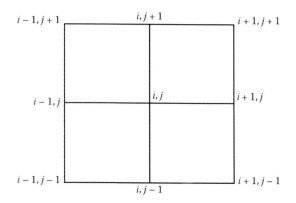

FIGURE 11.1
A typical finite difference grid.

Here $n + 1$ is the current time step and n the previous time step. In Equation 11.7, as there is only one unknown, we can explicitly get the unknown head. In Equation 11.8, as there are more unknowns, the equations are to be formed for all grids and simultaneously solved for all unknowns. In a similar way, the flow equation (11.2) and transport equation (11.3) can be approximated.

In order to apply explicit or implicit schemes, the stability conditions are to be taken into consideration. It is also reported that numerical stability problems arise while simulating solute transport equation using FDM (Bear, 1979). Detailed formulation of FDM for groundwater systems can be found in Bear (1979), Wang and Anderson (1982), Rastogi (2010), etc.

11.4.1.2 Method of Characteristics

In the solution of advection-dispersion equation, the method of characteristics (MOC) technique minimizes numerical diffusion and increases the stability in the solution (Willis and Yeh, 1987). The methodology is based on the characteristic equations of governing partial differential equations. These characteristic equations are solved by discretizing the domain and introducing a set of moving points in each element. The procedure is to track the points through the field/problem domain. The associated transport of contaminant is translated by the distance in accordance with the flow velocity. All points within an element at a given time undergo the same change in concentration due to dispersion. This process then continues for the entire simulation period, where the explicit solution method is to estimate the concentrations at the grid points. Detailed methodology on application of MOC to groundwater problems can be found in Pinder and Cooper (1970), Bear (1979), Willis and Yeh (1987), etc.

11.4.1.3 Finite Element Method

Currently FEM is one of the widely used numerical methods for solving differential and partial differential

equations. It is also widely used for groundwater flow and transport modeling (Pinder and Gray, 1977; Istok, 1989; Desai et al., 2011). The first step in FEM is discretization, where the entire domain is considered as mesh with element shape of our choice. Then, the unknown variable is defined in a piece-wise fashion over the defined individual elements by means of interpolation or shape function. Using the Galerkin approach, the resulting error or residuals due to the trial solution chosen for the governing equation are weighted by a weighting function and set equal to zero over the entire domain. From the elemental equations, matrices are assembled for the global domain, forming a system of algebraic equations. These sets of equations are solved either by iterative or direct techniques after implementation of boundary conditions (Desai et al., 2011). While FEM is used for approximation in space, generally FDM is used for time discretization (Desai et al., 2011).

Using Galerkin's FEM (Desai et al., 2011) and simple two-dimensional triangular element (Figure 11.2) for the approximation of flow Equation 11.1, the first step is to define a trial solution.

$$\hat{h}(x, y, t) = \sum_{L=1}^{NP} h_L(t) N_L(x, y) \qquad (11.9)$$

where h_L is the unknown head, N_L the known basis function at node L, and NP the total number of nodes in the problem domain. A set of simultaneous equations is obtained when residuals weighted by each of the basis functions are forced to be zero and integrated over the entire domain Ω. Thus, Equation 11.1 can be written as

$$\iint_{\Omega} \left[\frac{\partial}{\partial x} \left(T_x \frac{\partial \hat{h}}{\partial x} \right) + \frac{\partial}{\partial y} \left(T_y \frac{\partial \hat{h}}{\partial y} \right) - Q_w + q - S \frac{\partial \hat{h}}{\partial t} \right]$$
$$\times N_L(x, y) dx dy = 0 \qquad (11.10)$$

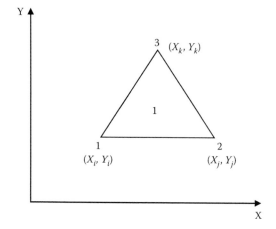

FIGURE 11.2
Linear triangular element used in FEM.

Equation 11.10 can further be written as the summation of individual elements as

$$\sum_e \iint \left(T_x^e \frac{\partial \widehat{h^e}}{\partial x} \left\{ \frac{\partial N_L^e}{\partial x} \right\} + T_y^e \frac{\partial \widehat{h^e}}{\partial y} \left\{ \frac{\partial N_L^e}{\partial y} \right\} \right) dxdy$$

$$+ \sum_e \iint \left(S \frac{\partial \widehat{h^e}}{\partial t} \right) \{N_L^e\} dxdy$$

$$= \int_\Gamma \left(T_x^e \frac{\partial \widehat{h^e}}{\partial x} n_x + T_y^e \frac{\partial \widehat{h^e}}{\partial y} n_y \right) N_L d\Gamma - \sum_e \iint (Q_w) \{N_L^e\} dxdy$$

$$+ \sum_e \iint (q) \{N_L^e\} dxdy$$

$$(11.11)$$

where $\left\{ N_L^e \right\} = \begin{Bmatrix} N_i \\ N_j \\ N_k \end{Bmatrix}$.

For an element, Equation 11.11 can be written in matrix form as

$$\left[G^e \right] \{h_I^e\} + \left[P^e \right] \left\{ \frac{\partial h_I^e}{\partial t} \right\} = \{F^e\} \qquad (11.12)$$

where $I = i, j, k$ are three nodes of triangular elements and G, P, F are the element matrices known as conductance, storage matrices, and recharge vectors, respectively. Summation of elemental matrix Equation 11.12 for all the elements lying within the flow region gives the global matrix as

$$[G]\{h_I\} + [P] \left\{ \frac{\partial h_I}{\partial t} \right\} = \{F\} \qquad (11.13)$$

Applying the implicit finite difference scheme for the $\partial h_I/\partial t$ term in time domain for Equation 11.13 gives

$$[G]\{h_I\}_{t+\Delta t} + [P] \left\{ \frac{h_{t+\Delta t} - h_t}{\Delta t} \right\} = \{F\} \qquad (11.14)$$

The subscripts t and $t + \Delta t$ represent the groundwater head values at earlier and present time steps. By rearranging the terms of Equation 11.14, the general form of the equation can be given as (Istok, 1989; Desai et al., 2011)

$$[[P] + \omega \Delta t[G]]\{h\}_{t+\Delta t} = [[P] - (1-\omega)\Delta t[G]]\{h\}_t$$

$$+ \Delta t (1 - \omega)\{F\}_t + \omega\{F\}_{t+\Delta t} \qquad (11.15)$$

where Δt = time step size, $\{h\}_t$ and $\{h\}_{t+\Delta t}$ are groundwater head vectors at the time t and $t + \Delta t$, respectively, and

ω = relaxation factor which depends on the type of finite difference scheme used. For fully explicit scheme $\omega = 0$, Crank–Nicholson scheme $\omega = 0.5$, fully implicit scheme $\omega = 1$.

Similarly applying Galerkin's FEM for the groundwater solute transport Equation 11.3, we get the following equation (Desai et al., 2011):

$$[[S] + \omega \Delta t[D]]\{C\}_{t+\Delta t} = [[S] - (1-\omega)\Delta t[D]]\{C\}_t$$

$$+ \Delta t (1 - \omega)\{F\}_t + \omega\{F\}_{t+\Delta t} \qquad (11.16)$$

where $[S]$ = element sorption matrix, $[D]$ = element advection–dispersion matrix, ω = relaxation factor, $\{F\}$ = flux matrix, and t and $t + \Delta t$ = beginning and ending time steps.

Detailed formulation of FEM for groundwater flow and transport can be found in Wang and Anderson (1982), Istok (1989), Rastogi (2010), Desai et al. (2011), etc.

11.4.1.4 Analytic Element Method

The AEM was first developed by Strack (1989), wherein the basic idea is that modeling groundwater movement is based on superposition of analytic (standard) solutions to the governing equation, where each element corresponds to a simple hydrogeologic feature (uniform flow, well, stream, lake, inhomogeneity, etc.). Since the solution, in principle, is an analytical one, it has the advantage of the solution being exact and grid independent.

Consider a heterogeneous unconfined aquifer domain having P number of wells and L number of rivers. Let us also assume that the aquifer is being recharged uniformly by precipitation. All the hydrogeological features of the aquifer, that is, wells, river, inhomogeneity boundary (say total number of inhomogeneity boundary is M), uniform recharge, etc. can be represented by analytic elements like well element, head specified line sink, line doublet, and polygonal areal sink, respectively. To simulate the groundwater flow process of the aquifer, L number of unknown strength parameters for L number of head specified line sink, M number of jump in discharge potential across the M number of inhomogeneity boundary and one integration constant have to be found out. Therefore, the total number of unknowns to be solved are $L + M + 1$.

At the control point of the head specified line sink (usually, middle point of line sink) and at the reference point the groundwater head value is known. As the head is known, the discharge potential (Φ) at the middle point of the head line sink and at the reference point (Figure 11.3) can be evaluated using Equations 11.17 and 11.18 (Haitjema, 1995).

$$\Phi(x,y) = kH\phi - \frac{1}{2}kH^2 \quad \text{if } \phi \geq H \qquad (11.17)$$

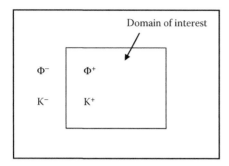

FIGURE 11.3
Inhomogeneity domain used in AEM.

$$\Phi(x,y) = \frac{1}{2}k\phi^2 \quad \text{if } \phi \leq H \qquad (11.18)$$

Expression for discharge potential can also be generated, by adding contribution of all the analytic elements at the reference point and at the control point of the head line sink. By equating the expressions of discharge potential with the Equation 11.17 or 11.18, $L+1$ number of equations are generated as (Haitjema, 1995)

$$\Phi_i(x,y) = \sum_{j=1}^{L} \sigma_{j,LS} \times F_{j,LS}(x,y)$$

$$+ \sum_{j=1}^{M} s_{j,LD} \times F_{j,LD}(x,y) + \Phi_{Areal\_Sink}(x,y)$$

$$+ \sum_{j=1}^{P} \Phi_{j,Well}(x,y) + C \quad where, i = 1, 2, \ldots L+1$$

$$(11.19)$$

where $F_{j,LS}$ is the coefficient function of the jth line sink; $F_{j,LD}$ the coefficient function of the jth line doublet; Φ_{Areal_Sink} the discharge potential contribution from the polygonal areal sink; $\Phi_{j,Well}$ the discharge potential contribution from jth well; C the integration constant; $\sigma_{j,LS}$ the strength of the jth line sink; and $s_{j,LD}$ the jump in the potential across the jth line doublet.

Another M number of equations are generated, by satisfying the continuity of pressure and continuity of the head across the M inhomogeneity boundary (Strack, 1989). Let us consider an inhomogeneous domain (Figure 11.3). Discharge potential and hydraulic conductivity, inside and outside of the rectangular domain, are assumed as Φ^+, k^+ and Φ^-, k^-, respectively. Satisfying continuity of pressure along the boundary of inhomogeneity, a relationship is drawn as (Equation 11.20 [Strack, 1989])

$$\frac{\Phi^+}{k^+} = \frac{\Phi^-}{k^-} \qquad (11.20)$$

Now, jump in the discharge potential (s_n) along the boundary of inhomogeneity can be expressed as (using Equation 11.20)

$$s_n = \Phi^+ - \Phi^- = \Phi^-\frac{k^+}{k^-} - \Phi^- = \frac{k^+ - k^-}{k^-}\Phi^- \quad (11.21)$$

Equation 11.21 further can be written as

$$s_n = \left(\frac{k^+ - k^-}{k^-}\right) \times \left(\sum_{j=1}^{L} \sigma_{j,LS}F_{j,LS}(x,y)\right.$$

$$+ \sum_{j=1}^{M} s_{j,LD}F_{j,LD}(x,y) + \Phi_{Areal\_Sink}(x,y)$$

$$\left. + \sum_{j=1}^{P} \Phi_{j,Well}(x,y) + C\right) \quad Where, n = 1, 2, 3 \ldots N \quad (11.22)$$

Equation 11.22 is used to generate M number of equations across the M line doublets. So there are $L+M+1$ equations for $L+M+1$ unknowns. Solution of these $L+M+1$ equations gives line sink strength ($\sigma_{j,LS}$) at the control points and the integration constant (C) and jumps ($s_{j,LD}$) across the doublet. Discharge potential and head at any point now can be found out using Equations 11.17 through 11.19 as there are no unknown parameters. Detailed derivation of analytic elements (i.e., well, line sink, polygonal areal sink, line doublet, higher order analytic elements, etc.) can be found in Strack and Haitjema (1981a,b), Strack (1989), Haitjema (1995), Fitts (1985), Janković (1997), and Janković and Barnes (1999).

11.4.1.5 Meshless Methods

The idea of getting rid of elements and meshes in the process of numerical treatments has naturally evolved and the concepts of meshless methods have been shaped up (Liu, 2002). The initial idea of meshless methods dates back to the smooth particle hydrodynamics (SPH) method for modeling astrophysical phenomena (Gingold and Monaghan, 1977). The research into meshless methods has become very active, only after the publication of the diffuse element method by Nayroles et al. (1992). Meshless methods use a set of nodes scattered within the problem domain as well as a set of nodes on the boundaries of the domain to represent the problem domain and its boundaries (Figure 11.4). These sets of scattered nodes do not form a mesh and thereby they

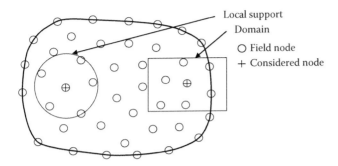

FIGURE 11.4
Local support domain used in meshless to construct shape function.

do not require prior information on the relationship between the nodes for the interpolation or approximation of the unknown functions of the field variables. Many meshless methods have found good applications and shown very good potential to become powerful numerical tools (Liu and Gu, 2005).

While solving Equation 11.1 using a meshless technique, point collocation method (PCM) as described in Mategaonkar and Eldho (2011, 2012), the first step is to define the trial solution:

$$\hat{h}(x,y,t) = \sum_{i=1}^{n} h_i(t)R_i(x,y) \qquad (11.23)$$

where h is the unknown head, $R_i(x,y)$ the shape function at node I, and n the total number of nodes in the support domain (Figure 11.4). The value of the shape function $R_i(x)$ is given as (Liu and Gu, 2005):

$$R_i(x) = \left(r_i^2 + (\alpha_c d_c)^2\right)^q = \left(r_i^2 + C_s^2\right)^q \qquad (11.24)$$

Where,

$$r_i = \sqrt{(x-x_i)^2 + (y-y_i)^2} \quad \text{and} \quad C_s = \alpha_c d_c \qquad (11.25)$$

For multiquadric radial basis function, $q = 0.5$; Equation 11.24 can be written as

$$R_i(x) = \sqrt{(x-x_i)^2 + (y-y_i)^2 + C_s^2} \qquad (11.26)$$

Therefore,

$$\frac{\partial R_i(x,y)}{\partial x} = \frac{(x-x_i)}{R_i(x,y)} \quad \text{and} \quad \frac{\partial^2 R_i(x,y)}{\partial x^2} = \frac{(y-y_i)^2 + C_s^2}{R_i^3(x,y)}$$

$$(11.27)$$

$$\frac{\partial R_i(x,y)}{\partial y} = \frac{(y-y_i)}{R_i(x,y)} \quad \text{and} \quad \frac{\partial^2 R_i(x,y)}{\partial y^2} = \frac{(x-x_i)^2 + C_s^2}{R_i^3(x,y)}$$

$$(11.28)$$

For time discretization, fully implicit finite difference approximation is used (Wang and Anderson, 1982). Substituting all the above terms in Equation 11.6, it can be written as

$$R_i(x,y)h_i^{t+\Delta t} - \frac{T\Delta t}{S}\left(\left(\frac{\partial^2 R_i(x,y)}{\partial x^2}\right) + \left(\frac{\partial^2 R_i(x,y)}{\partial y^2}\right)\right)h_i^{t+\Delta t}$$

$$= R_i(x,y)h_i^t$$

$$(11.29)$$

$R_i(x, y)$, $(\partial^2 R_i(x, y)/\partial x^2)$, and $(\partial^2 R_i(x, y)/\partial y^2)$ values are calculated for each support domain. Shape functions have the Kronecker delta function property (Liu and Gu, 2005), that is,

$$R_i(x) = \begin{cases} 1, i = j, & i = 1,2,\dots n \\ 0, i \neq j, & i,j = 1,2,\dots n \end{cases} \qquad (11.30)$$

Which means the value of shape function outside the support domain of a particular node is zero. All these values of shape functions and derivatives of shape functions are incorporated into the global matrix for the whole problem domain.

Therefore, from Equation 11.29 the system of equations can be written in matrix form as:

$$\left\{[K_1] - \frac{T\Delta t}{S}([K_2] + K_3)\right\}\{h_i\}^{t+\Delta t} = [K_1]\{h_i\}^t \qquad (11.31)$$

where $[K_1]$ is a global matrix for shape function and $[K_2]$ is a global matrix for the second derivative of the shape function with respect to x, $[K_3]$ is a global matrix for the second derivative of a shape function with respect to y. When source or sink terms are to be considered with heterogeneity, then Equation 11.31 can be written as

$$\left\{[K_1] - \left(\frac{\Delta t}{S}\right)\left(T_x[K_2] + T_y[K_3]\right)\right\}\{h_i\}^{t+\Delta t}$$

$$= [K_1]\{h_i\}^t \pm \left(\frac{\Delta t}{S}\right)\left([K_1]Q_w\right) \qquad (11.32)$$

where $Q_w = q_w/a$; q_w is the source or sink term (m³/day); "a" the area of support domain in which the pumping well or recharge well lies (m²); and Q_w the global matrix of the entire source and sink terms. The pumping and recharge wells are considered for source and sink terms. Similarly formulating transport Equation 11.3, the system of equations can be written in matrix form as

$$\left\{[K_1] - \Delta t \left(D_{xx}[K_2] - D_{yy}[K_3]\right)\right\}\{c_i\}^{t+\Delta t}$$

$$= \left([K_1] - \Delta t \left(v_x[K_4] - v_y[K_5]\right)\right)\{c_i\}^t \qquad (11.33)$$

where additionally $[K_4]$ is a global matrix for the first derivative of a shape function with respect to x and $[K_5]$ is a global matrix for the first derivative of a shape function with respect to y.

Detailed description on application of meshless methods to groundwater flow and transport can be found in Li et al. (2003), Kumar and Dodagoudar (2008), Saeedpanah et al. (2011), Li and Mao (2011), Mategaonkar and Eldho (2011, 2012), Kovářík and Muzik (2013), Swathi and Eldho (2014), Pang et al. (2015), etc.

11.5 Modeling Procedure

For groundwater flow and transport modeling, a systematic modeling protocol can be formulated. Figure 11.5 gives the general modeling protocol for the use of coupled flow and transport model. The initial step is to formulate a conceptual model based on simplifying assumptions that best describe the aquifer system, then identify the mechanisms that govern the equations primarily, followed by initial selection of aquifer parameters and then setting of boundary conditions for the particular problem concerned with initial conditions involving various system stresses. Further, the governing equations are solved using the selected numerical technique with initial and boundary conditions and a computer model is developed. The computer model is validated by comparing with available analytical solutions or field measured data. To obtain the appropriate aquifer parameters such as porosity, hydraulic conductivity, storage coefficient, and dispersivity, the calibration process is carried out by adjusting these parameters until the results from the models fall within a certain range of specified accuracy criteria (Willis and Yeh, 1987). Further, the model is applied for the prediction or simulation of the considered scenario.

11.5.1 Important Groundwater Flow and Transport Model Computer Codes

The computer modeling for groundwater flow and transport is well established in the last few decades and a number of software is available to simulate various groundwater problems (http://water.usgs.gov/software/lists/groundwater.). Some of the commonly used flow and transport models are briefly mentioned next (Bedient et al., 1999).

MODFLOW is an FDM groundwater flow simulation model developed by USGS (McDonald and

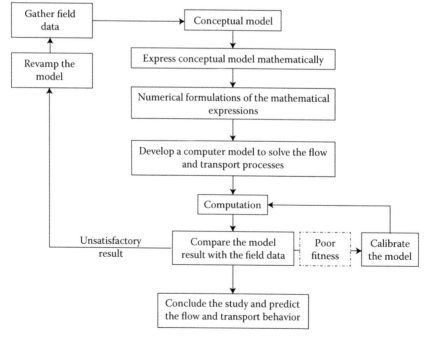

FIGURE 11.5
Numerical modeling procedure for flow and transport simulation.

Harbaugh, 1988). MODFLOW can easily simulate 3D transient/steady-state flow in anisotropic, heterogeneous, layered aquifer system, etc.

MT3D (1990), MT3DMS (1998), and MODPATH are the most widely used solute transport simulators. MT3D/MT3DMS uses finite difference scheme to simulate the transport process and MODPATH uses the particle tracking technique (Shafer, 1987a,b) to simulate advection dominated transport processes. MT3D/MT3DMS and MODPATH can be easily used to simulate anisotropic, heterogeneous, layered aquifer system with complex boundary condition (Zheng and Wang, 1998).

MOC (1998) is a 2D steady and transient solute transport model. The model couples the flow equation with the solute transport equation. The model uses finite difference schemes for solving flow equation and method of characteristics (Bedient et al., 1999) to solve solute the transport equation.

BIOPLUME is a 2D finite difference based model used to simulate the attenuation of organic contaminants in groundwater due to advection, dispersion, sorption, and biodegradation (Rifai et al., 1998). To restore aquifer contaminated with hydrocarbons the biotransformation process is potentially important.

FEMWATER is a software package of the groundwater modeling system (GMS) for different groundwater flow and transport modeling in 2D and 3D (www.epa.gov). FEFLOW is a software package for modeling of different groundwater flow and transport processes in 2D and 3D (www.feflow.info).

3DFEMFAT is 3D FEM model of flow and transport through saturated-unsaturated media. 3DFEMFAT can do simulations of flow or transport, combined sequential flow and transport, or coupled density-dependent flow and transport. (http://www.scisoftware.com/products/3dfemfat_overview.)

AQUA3D is a program developed to solve three-dimensional groundwater flow and transport problems using FEM. AQUA3D solves transient groundwater flow with inhomogeneous and anisotropic flow conditions (http://obinet.engr.uconn.edu/wiki/index.php/Aqua-3D).

AT123D is an analytical groundwater transport model. AT123D computes the spatial-temporal concentration distribution of wastes in the aquifer system and predicts the transient spread of a contaminant plume through a groundwater aquifer (www.seview.com).

GMS is a sophisticated and comprehensive groundwater modeling software package (www.aquaveo.com). It provides tools for every phase of a groundwater simulation including site characterization, model development, calibration, postprocessing, and visualization. GMS supports both FDM and FEM in 2D and 3D including MODFLOW 2000, MODPATH, MT3DMS/RT3D, SEAM3D, ART3D, UTCHEM, FEMWATER, and

SEEP2D. The program's modular design enables the user to select modules in custom combinations, allowing the user to choose only those groundwater modeling capabilities that are required (www.aquavevo.com)

Visual MODFLOW provides professional groundwater flow and contaminant transport modeling using MODFLOW, MODPATH, MT3DMS, and RT3D. The visual MODFLOW 3D-explorer will give a complete and powerful graphical modeling environment (www.swstechnology.com).

11.6 Groundwater Simulation: Hypothetical Case Studies

Here to demonstrate the applications of various simulation models for groundwater flow and transport, two hypothetical case study problems are considered. The first case is for the flow simulation in a confined aquifer and the second case is for the coupled flow and transport simulation in a confined aquifer. The case studies are briefly discussed in the following sections.

11.6.1 Case Study 1

Here a hypothetical homogeneous isotropic confined aquifer in two dimensions of size 1400 m × 1400 m is considered for the flow simulation, as shown in Figure 11.6. The boundary conditions of the problem, various parameters used, and well locations are shown in Figure 11.6. At the center of the aquifer, there is a well pumping at the rate of 10,000 m³/day. FDM (MODFLOW), FEM, and PCM (Mategaonkar, 2012) based flow simulation

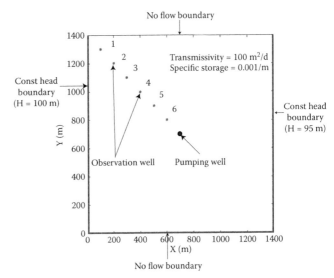

FIGURE 11.6
The hypothetical confined aquifer problem considered (Case study 1).

TABLE 11.1

Head Value (m) at 6 Observation Wells Using FDM, FEM, and PCM Models after 10 Days of Pumping

Observation Point	Analytical	FDM	% Error (Analytical with FDM)	FEM	% Error (Analytical with FEM)	PCM	% Error (Analytical with PCM)
1	97.013	96.751	0.270	96.993	0.02	97.032	0.019
2	93.804	93.259	0.580	93.768	0.036	93.760	0.046
3	90.095	90.111	0.018	90.051	0.044	90.075	0.022
4	85.451	85.312	0.162	85.413	0.038	85.511	0.070
5	78.983	78.184	1.011	78.974	0.009	78.964	0.024
6	67.953	67.680	0.401	67.762	0.191	67.855	0.144

models discussed earlier are applied to analyze the flow pattern for the problem conditions. In the FDM modeling, a grid of 20 m × 20 m is considered. In FEM, 392 numbers of triangular linear elements (100 m sized right-angled triangle) and 225 number nodes were used (Willis and Yeh, 1987). In the PCM modeling (Mategaonkar, 2012), 225 equidistant nodes with a spacing of 100 m is considered. A time step of 0.1 day is used in the simulation.

A comparison of head value at six observation points (Figure 11.6) after 10 days of pumping at a rate of 10,000 m³/day using the different simulation models are shown in Table 11.1. The results are also compared with an analytical solution given by Chan et al. (1976) and FEM model by Willis and Yeh (1987). In comparison with the analytical solution, the meshless PCM models have better accuracy compared to the FDM and FEM models. The contour plot of heads after 10 days of pumping at 10,000 m³/day is shown in Figure 11.7.

In comparison with the grid- or mesh-based methods, the PCM-based meshless model is very easy to simulate with better accuracy.

Further, if a well is pumped for a long period of time with the same discharge value, then after a certain time, the aquifer generally attains steady-state conditions. The AEM can be easily applied under steady-state condition. The problem discussed above is analyzed at steady-state conditions using AEM and MODFLOW models. A comparison of head value at six observation points under steady-state conditions for a pumping rate of 10,000 m³/day using FDM and AEM is shown in Table 11.2. The AEM model is very easy to simulate with better accuracy.

11.6.2 Case Study 2

Here a hypothetical homogeneous isotropic confined aquifer in two dimesions of size 2500 m× 2000 m is considered for the flow and transport simulation, as shown in Figure 11.8. The boundary conditions of the problem, various parameters used, and well locations are shown in Figure 11.8. The aquifer is assumed to have a thickness of 50 m. It is assumed that there are three pumping wells and one injection well. Pumping is taking place at a rate of 4000 m³/day from each of the well. The injection well (at a rate of 2000 m³/day) with a specified TDS concentration of 1000 mg/L provides the source of contamination.

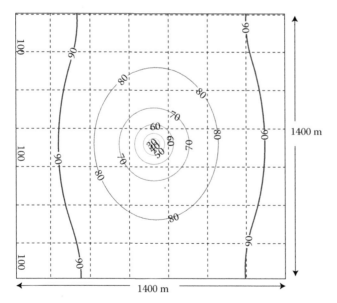

FIGURE 11.7
Contour plot of head after 10 days of continuous pumping (using MODFLOW—Case study 1).

TABLE 11.2

Head Value (m) at Six Observation Wells Using FDM and AEM

Observation Point	FDM	AEM
1	96.71	96.87
2	93.19	93.28
3	90.02	90.53
4	84.54	85.22
5	78.05	78.05
6	67.53	66.30

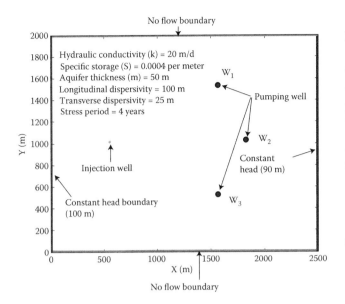

FIGURE 11.8
A hypothetical confined aquifer (Case study 2).

The hypothetical problem has been simulated using MODFLOW and MT3D package available in the GMS software. In the FDM modeling, a grid of 50 m × 50 m is considered. A time step of 0.1 day is used in the simulation. The contamination distribution in the aquifer is found after 1, 2, 3, and 4 years using the MODFLOW and MT3DMS as shown in Figure 11.9. Table 11.3 shows the concentration in wells W_1, W_2, and W_3 after 1, 2, 3, and 4 years of simulation. This problem shows the effectiveness of the simulation model to understand the contamination spreading in an aquifer system.

11.7 Optimization Models

In many of the groundwater problems, we have to make management decisions such as the requirement of a number of pumping wells or rate of pumping to meet specific requirements, appropriate scenarios for aquifer management, or specific remediation schemes such as pump and treat. Also, we have to optimize the aquifer parameters

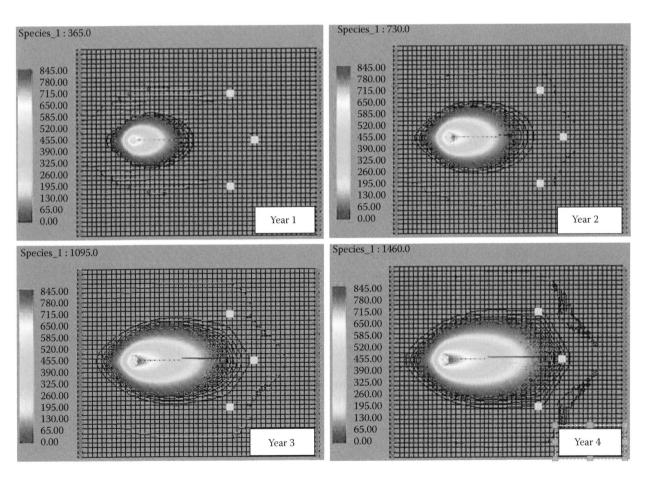

FIGURE 11.9
Contaminant migration after 1, 2, 3, and 4 years in the aquifer domain using MODFLOW-MT3DMS (Case study 2).

TABLE 11.3

Contaminant Concentration at the Pumping Locations at Different Times

Stress Period	W_1 Contaminant Conc. (mg/L)	W_2 Contaminant Conc. (mg/L)	W_3 Contaminant Conc. (mg/L)
1st year	0	0	0
2nd year	0.04	0.09	0.03
3rd year	1.04	2.69	0.88
4th year	5.47	15.61	4.76

using inverse modeling. In all these problems, optimization tools are utilized to facilitate optimal decision making in the planning, design, and operation of the groundwater system considered (Hall and Dracup, 1970). The economics involved with large-scale groundwater projects are indeed very complex as it influences the life and future of many sections of society and different geographical regions in different ways. Further, there is the difficulty of correctly predicting future scenarios, which makes the task of getting optimal decisions based on model forecasts even more challenging. Therefore, the application of optimization techniques is essential and most challenging in groundwater management problems, due to the large number of decision variables involved, stochastic nature of the inputs, and multiple objectives.

Generally, the optimization problem involves finding the maximum or minimum of a given objective function with certain constraints by systematically choosing input values from within an allowed set of range of the variables considered and computing the value of the objective function. The classical optimization techniques such as linear, nonlinear, or dynamic programming are useful in finding the optimum solution or unconstrained maxima or minima of continuous and differentiable functions and they use differential calculus to locate the optimal solutions (Hall and Dracup, 1970; Louck et al., 1981). However, these conventional methods (namely, linear, nonlinear, and dynamic programming) have limited scope in problems of groundwater management where objective functions are nonconvex, nonlinear, and not continuous and/or differentiable, in their domain. The dynamic programming approach faces the problem of dimensionality, whereas the nonlinear programming methods have the limitation of slow rate of convergence, require a large amount of computational storage and time compared with other methods, and they often result in local optimal solutions (Louck et al., 1981).

In order to overcome these limitations, in recent years, meta-heuristic techniques have been used for optimization. Thus, many researchers in the water resources field are now exploring various evolutionary algorithms like GAs (Mckinney and Lin, 1994; Chang and Hsiao, 2002;

Devi Prasad and Park, 2004; Meghna and Minsker, 2006), simulated annealing (Cunha and Sousa, 1999), ant colony optimization (Maier and Simpson, 2003), shuffled complex evolution (Liong and Atiquzzama, 2004), and harmony search (Geem, 2006), and covariance matrix adaptation evolution strategy among others to solve complex water resources problems. Similarly, researchers have also applied different techniques together, like GAs and chaos theory (Cheng et al., 2008), neural networks and genetic programming (Recknagel et al., 2002), and fuzzy logic with GAs (Takagi, 1997) to analyze different categories of water resources engineering problems.

Any optimization problem is defined in terms of an objective function with a set of constraints to be satisfied. A typical objective function can be either a cost minimization or maximization of benefits. The costs can be due to well drilling, well installation, capital and operating cost, contamination remediation process, etc. The typical constraints in groundwater problems on the decision variables can be the number of wells and their location, and rate of pumping/injection rates at individual wells, while the typical constraints on the state variables can be the maintenance of groundwater levels or the permissible contaminant concentrations. Here two of the evolutionary algorithms are briefly discussed.

11.7.1 Genetic Algorithm

Genetic algorithm (GA) is a meta-heuristic global optimization technique based on Darwin's survival of the fittest principle (Holland, 1975). GA generates solution to optimization problems using techniques inspired by evolutionary biology such as selection, crossover, and mutation. Usually the evolution starts with a predefined number of populations within a specified bound of the variable. In each generation, fitness values of all the members of the population are evaluated first. Thereafter, multiple members are selected from the current population based on their dominance (fitness value) and modified by crossover and mutation operators to form a new population set (Mckinny and Lin 1994). The whole process is continued until termination criterion, that is, user defined maximum number of iteration or satisfactory fitness level has been reached for the population. The three operators of genetic algorithm are selection, crossover, and mutation. A detailed description of GA can be found in Goldberg (1989), Goldberg and Deb (1991), Deb (1995, 2012), etc.

11.7.2 Particle Swarm Optimization

Particle swarm optimization (PSO) is a population-based meta-heuristic global optimization technique based on the social behavior of birds and was developed by Eberhart and Kennedy in 1995. In PSO, solutions represented by particles fly through the search space by following the current best individual position and current global best position of particles. Each particle keeps record of its location having best fitness found thus far. This location is called *pbest*. Somehow particles communicate and get information about the location having the best fitness value achieved by any particle of the group. This value is called *gbest*. At each generation, particles update their position by changing their velocity toward *pbest* and *gbest*. A detailed description of the PSO process can be found in Kennedy and Eberhart (2001), Clerc (2010), etc.

11.8 Simulation Optimization Models for Groundwater Management

Simulation Optimization (S/O) models are used in various groundwater management and remediation problems for optimal designs. As discussed earlier, numerical models are used for the simulation of groundwater flow and transport and optimization models are used for selecting the best management scheme. During the past two decades, many groundwater management problems were developed by various researchers using a linked simulation optimization approach (Gill et al., 1984; Chang et al., 1992; Yeh and Chen, 2000; Mattot et al., 2006; Mondal et al., 2010; Mategaonkar and Eldho, 2012). In S/O models, the simulation model attempts to find the state variables in the system (i.e., groundwater head or concentration of contaminant) by solving the partial differential equations that govern the groundwater regime. This information is utilized by the optimization model to obtain optimal solutions to the management problem. In most of these problems, the behavior of the objective function (such as the collective drawdown or the total cost) with respect to the decision variables (such as pumping discharge) is not easily predictable under the appropriate constraints. Researchers have found that meta-heuristic techniques, which have distinct advantages over conventional methods, have a great potential to solve such optimization problems.

A typical flowchart describing the procedure to be followed in an S/O model for groundwater management problems using numerical models and PSO is shown in Figure 11.10.

Following are the important steps used for development of an S/O model for groundwater management.

1. Identification of parameters characterizing the aquifer.
2. Field estimation of relevant hydrogeological parameters at the maximum possible locations, especially at the boundaries.
3. Interpolation/extrapolation of the field parameters for the entire aquifer system under consideration.
4. Identifying the best mathematical governing equation describing the spatio-temporal relationship between the unknowns and parameters.
5. Adopting a numerical technique for solving the mathematical governing equation describing the unknown variable in terms of the state variables such as groundwater head levels or concentrations of pollutants.
6. Calibration of the numerical model for steady and transient conditions.
7. A sensitivity analysis of the model to identify those parameters that need to be estimated more correctly such that the error decreases.
8. Designing the best efficient management policy option for groundwater resources or to remediate the contaminant problem from the aquifer system using an optimization method.
9. Derive the most optimal management scenarios for the problem considered.

11.8.1 Groundwater Management using S/O Models: Case Studies

Here the application of the S/O model for groundwater management is demonstrated using a case study. The above discussed simulation optimization methodology for groundwater management model has been applied to one hypothetical unconfined aquifer problem. The hypothetical unconfined aquifer is of size 6000 m × 12000 m with a thickness of 35 m. The boundary conditions and aquifer properties are as shown in Figure 11.11. It is assumed that there are nine pumping wells. Under continuous pumping policy, the aquifer can be considered under steady-state conditions. To obtain the groundwater head distribution, the aquifer has been simulated using MODFLOW and AEM separately. Thereafter, the simulation model is coupled with GA and PSO to solve two different management problems.

Case a: The objective of the first management problem is to withdraw the maximum amount of water from the aquifer system. It is assumed that the number of pumping wells and location are fixed. Also, hydraulic head value at the well location should be above the permissible limit. Mathematically the management model can be expressed as (Tamer, 2009):

$$\text{Maximize} \quad F = \sum_{i=1}^{N_w} Q_i \qquad (11.34)$$

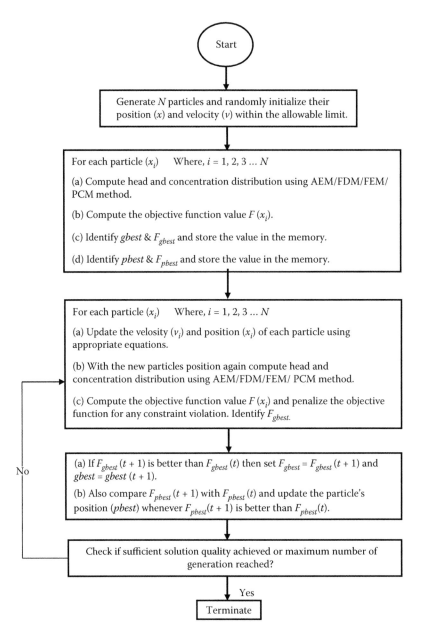

FIGURE 11.10
Flow chart of the PSO-based S/O model for groundwater management.

Subject to, $Q_{i,\min} \leq Q_i \leq Q_{i,\max};\quad i = 1,2,3,\dots N_w$ (11.35)

$$h_i \geq h_{i,\min};\quad i = 1,2,3,4,\dots N_w \qquad (11.36)$$

where F is the objective function; Q_i the pumping rate from the ith well; $Q_{i,\min}$ and $Q_{i,\max}$ are the upper bound and lower bound of the decision variable Q_i; h_i the groundwater head at the ith well; $h_{i,\min}$ the minimum permissible head value at the ith well location; N_w the total number of wells. The constrained optimization problem has been converted to an unconstrained optimization problem using the penalty function method

(Deb, 2012). Whenever a constraint violation happens, a penalty is added to the objective function. The choice of penalty term is somewhat tricky. The larger the value of penalty term for a constraint means more emphasis is given on resolving a constraint violation and vice versa. For the above unconfined aquifer problem, $h_{i,\min}$ is assumed as 5 m at the well location. $Q_{i,\min}$ and $Q_{i,\max}$ are assumed as 0 m^3/d and 15000 m^3/d, respectively.

Maximum withdrawal of water from the aquifer has been obtained by three simulation optimization methods (MODFLOW-GA, MODFLOW-PSO, and AEM-PSO) and is shown in Table 11.4. Contour plot of the head

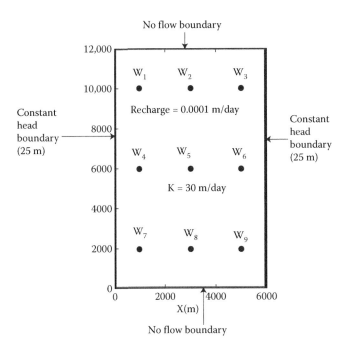

FIGURE 11.11
A hypothetical unconfined aquifer considered (Case a).

using MODFLOW-GA and AEM-PSO are shown in Figure 11.12. Also, a relationship between maximum withdrawals of water with a number of generation for MODFLOW-PSO model is shown in Figure 11.13. From the results, it can be interpreted that PSO yields better results than GA. Also, for a PSO-based model, computational time is less than a GA-based model. An AEM-based model also offers some advantages over MODFLOW. For example, using AEM, it is possible to determine head value at the exact well location very precisely, which in turn increases the efficiency of the simulation optimization model. To obtain an AEM level

of precision using FDM- or FEM-based model, finer discretization of domain near well location is required.

Case b: The objective of the second management problem is to minimize the total pumping cost. The total pumping cost includes well drilling, capital, and operating cost. The decision variables are pumping rates within the specified upper and lower bound. The state variable is the hydraulic head value at the well locations. The hydraulic head value at the well location should be above the specified minimum permissible head value. Also, total amount of pumping from the aquifer should meet/exceed the demand.

Mathematically, the management model can be expressed as follows (Tamer, 2009):

$$\text{minimize } F = \sum_{i=1}^{N_w} \left[c_1 k d_i^{b1} + c_2 Q_i^{b2} (d_i - h_i)^{b3} + c_3 Q_i (d_i - h_i) \right]$$
(11.37)

$$\text{Subject to,} \quad h_i \geq h_{i\min}; \quad i = 1, 2, 3, \ldots N_w \quad (11.38)$$

$$\sum_{i=1}^{N_w} Q_i \geq Q_d \quad (11.39)$$

$$Q_{i,\min} \leq Q_i \leq Q_{i,\max}; \quad i = 1, 2, 3, \ldots N_w \quad (11.40)$$

$$k = \begin{cases} 0 & \text{if } Q_i = 0 \\ & \quad\quad\quad\quad i = 1, 2, 3, \ldots N_w \\ 1 & \text{if } Q_i \neq 0 \end{cases} \quad (11.41)$$

where c_1 is the cost coefficient for well drilling; c_2 the cost coefficient for well and pumping equipments; c_3 the cost coefficient for well operation; $b1$, $b2$, $b3$ are

TABLE 11.4

Maximum Withdrawal of Water by Three Different Simulation Optimization Method

	MODFLOW-GA		MODFLOW-PSO		AEM-PSO	
Well	Pumping Rate (m³/day)	Head at the Well Location (m)	Pumping Rate (m³/day)	Head at the Well Location (m)	Pumping Rate (m³/day)	Head at the Well Location (m)
W_1	11,651.58	7.57	11,543.29	7.66	11,346.82	8.86
W_2	10,130.86	9.60	10,511.37	6.40	10,535.34	5.67
W_3	11,731.02	8.01	11,652.97	7.97	11,768.97	5.0
W_4	9014.51	14.61	11,603.74	5.01	11,374.70	8.61
W_5	10,072.73	10.51	10,455.98	6.63	10,495.16	5.0
W_6	11,861.51	7.26	11,752.69	5.12	11,683.34	7.49
W_7	11,763.25	7.91	11,592.64	5.03	11,485.90	7.73
W_8	8700.61	13.52	10,502.66	6.22	10,436.43	5.03
W_9	10,928.05	11.32	10,732.04	11.13	11,665.69	7.40
Total	95,854.12		100,347.38		100,792.36	

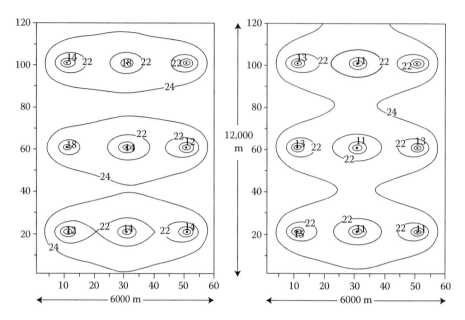

FIGURE 11.12
Contour plot of head distribution after applying S/O model by MODFLOW-GA and AEM-PSO (Case a).

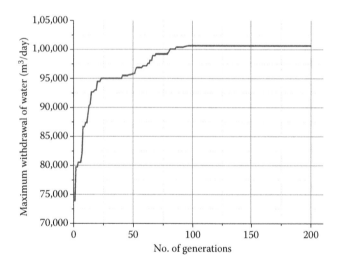

FIGURE 11.13
Relationship between optimal total pumping rates with the number of generations using MODFLOW-PSO (Case a).

optimization models and results are compared as shown in Table 11.6. As can be seen from Table 11.6, AEM-PSO gave the most optimal solution. A relationship between minimum total costs of pumping with the number of generations using MODFLOW-PSO is shown in Figure 11.14.

The case studies presented here shows the effectiveness of the S/O models for appropriate groundwater management. Using the various S/O models discussed, we can generate various scenarios so that the decision makers can make appropriate management decisions in an optimal way.

11.9 Concluding Remarks

In this chapter, groundwater management using simulation optimization models has been discussed in detail from its importance and necessity to application to a few case studies. Groundwater flow and transport governing equations have been introduced and numerical procedures to be followed for solving these equations are explained. The details of evolution of optimization techniques are further explained. Further, the procedure

constants indicating economies of scale; Q_d the water requirement; and d_i the depth at ith well. All the cost coefficients and other constraints are given in Table 11.5. It is assumed that total water demand for the area is 50,000 m³/day. The management problem, described above, has been solved by three different simulation

TABLE 11.5

Cost Coefficients and Constraints

c_1 ($/m)	c_2 ($/m⁴)	c_3 ($/m⁴)	b_1	b_2	b_3	d_i (m)	h_{min} (m)	Q_d (m³/d)
4250	0.13	0.03	0.3	1	1	35	5	50,000

TABLE 11.6

Minimum Pumping Cost to Insure the Water Demand

Well	Pumping Rate (m³/day) (MODFLOW-GA)	Pumping Rate (m³/day) (MODFLOW-PSO)	Pumping Rate (m³/day) (AEM-PSO)
W_1	7471.66	0	0
W_2	0	0	0
W_3	10,772.06	10,924.55	11,575.61
W_4	9144.65	10,348.91	10,579.18
W_5	0	0	7937.62
W_6	8628	9425.95	0
W_7	6142	0	10,500
W_8	9494.67	10,500	0
W_9	0	9007.74	9485.76
Total withdrawal (m³/day)	51,653.04	50,207.15	50,078.17
Total cost ($)	218,279.15	217,703.972	217,065.42

to link simulation and optimization models to get an appropriate S/O model is explained.

The evolution of computing technology offers great potential for simulating the groundwater system scenarios which are helpful for planning and management of resources. By applying an S/O model to a complex problem, the best management policy under a particular set of conditions can be identified. The withdrawal rates without significantly affecting the water quality, restricting and/or regulating the disposal, and possible contamination of aquifer systems can be identified with S/O models. This chapter provides the basis of the groundwater management problems giving details and appropriate methodology to be followed by researchers, planners, and policymakers.

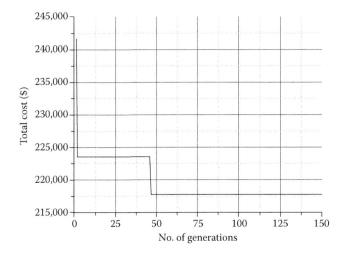

FIGURE 11.14
Relationship between minimum total costs of pumping with the number of generations using MODFLOW-PSO (Case b).

For better efficiency and cost-efficient solutions, different types of management scenarios are to be considered and are to be used in combination. Also, no single problem in groundwater systems is similar to others due to the unique and complex nature of groundwater systems, the management policy or remediation scheme has to be decided on diligently. An effective S/O model will be very useful to solve any of the complex groundwater management problems.

References

Anderson, M.P. 1984. *Movement of Contaminants in Groundwater.* Groundwater Transport Advection and Dispersion, Groundwater Contamination, National Research Council, National Academy Press, Washington, DC.

Anderson, M.P. and W.W. Woessner. 1992. *Applied Groundwater Modeling: Simulation of Flow and Advective Transport.* Academic Press, San Diego, California.

Bear, J. 1972. *Dynamics of Fluids in Porous Media.* Dover Books, New York.

Bear, J. 1979. *Hydraulics of Groundwater.* McGraw-Hill Publishing, New York.

Bedient, P.B., Rifai, H.S. and C.J. Newell. 1999. *Groundwater Contamination-Transport and Remediation*, 2nd ed., Prentice-Hall, Englewood Cliffs, NJ.

Chan, Y.K., Mullineaux, N. and J.R. Reed. 1976. Analytical solutions for draw-downs in rectangular artesian aquifers. *Journal of Hydrology*, 31, 151–160.

Chang, L.C., Shoemaker, C.A. and P.L.-F. Liu. 1992. Optimal time varying pumping rates for groundwater remediation: Application of a constrained optimal algorithm. *Water Resource Research*, 28(12), 3157–3173.

Chang, L.C. and A.T. Hsiao. 2002. Dynamic optimal groundwater remediation including fixed and operation costs. *Groundwater*, 40(5), 481–490.

Cheng, C.T., Wang, W.C., Xu, D.M. and K.W. Chau. 2008. Optimizing hydropower reservoir operation using hybrid genetic algorithm and chaos. *Water Resources Management*, 22(7), 895–909.

Clerc, M. 2010. *Particle Swarm Optimization.* John Wiley & Sons, 93. DOI: 10.1002/9780470612163.

Cunha, M. and J. Sousa. 1999. Water distribution network design optimization: Simulated annealing approach. *Journal of Water Resources Planning and Management*, 125(4), 215–221.

Deb, K. 1995. *Optimization for Engineering Design.* Prentice-Hall of India, New Delhi.

Deb, K. 2012. *Optimization for Engineering Design: Algorithms and Examples.* PHI Learning Pvt. Ltd., New Delhi.

Desai, Y.M., Eldho, T.I. and A.H. Shah. 2011. *Finite Element Method with Applications in Engineering.* Pearson Education, New Delhi.

Devi Prasad, T. and N. Park. 2004. Multi objective genetic algorithms for design of water distribution networks. *Journal of Water Resources Planning and Management*, 130(1), 73–82.

Eberhart, R.C. and J. Kennedy. 1995. A new optimizer using particle swarm theory. In *Proceedings of the Sixth International Symposium on Micro Machine and Human Science*, 1, 39–43. Nagoya, Japan, October 4–6, 1995.

FAO. 2003. Groundwater management: The search for practical approaches. Water Reports No. 25. Rome.

Fitts, C.R. 1985. Modeling aquifer inhomogeneties with analytic elements. Master's thesis, Department of Civil Engineering, University of Minnesota, Minneapolis, MN.

Freeze, R.A. and J.A. Cherry. 1979. *Groundwater*. Prentice-Hall, Englewood Cliffs, NJ.

Fried, J.J. 1975. *Groundwater Pollution*. Elsevier, New York.

Geem, Z.W. 2006. Improved harmony search from ensemble of music players. *Lecture Notes in Artificial Intelligence*, 4251, 86–93.

Gill, P.E., Murray, W., Saunders, A. and M.H. Wright. 1984. Aquifer reclamation design: The use of contaminant transport simulation combined with nonlinear programming. *Water Resource Research*, 20(4), 415–427.

Gingold, R.A. and J.J. Monaghan. 1977. Smooth particle hydrodynamics: Theory and applications to non-spherical stars. *Monthly Notices of the Royal Astronomical Society*, 181, 375–389.

Goldberg, D.E. 1989. *Genetic Algorithms in Search, Optimization and Machine Learning*. Addison-Wesley Pub. Company, Boston, MA.

Goldberg, D.E. and K. Deb. 1991. A comparative analysis of selection schemes used in genetic algorithms. In *Foundations of Genetic Algorithms*, G.J.E. Rawlins (ed.), Morgan Kaufmann, Sanmateo, CA, 69–93.

Haitjema, H.M. 1995. *Analytic Element Modeling of Groundwater Flow*. Academic Press, San Diego, CA.

Hall, W.A. and J.A. Dracup. 1970. *Water Resources Systems Engineering*. McGraw-Hill, New York.

Hill, M.C. and C.R. Tiedeman. 2007. *Effective Groundwater Model Calibration, with Analysis of Sensitivity, Predictions and Uncertainty*. Wiley, New York.

Holland, J.H. 1975. *Adaption in Natural and Artificial Systems*. University of Michigan Press, Ann Arbor, MI.

Istok, J. 1989. *Groundwater Modeling by the Finite Element Method*. American Geophysical Union, Washington.

Janković, I. 1997. High-order analytic elements in modeling groundwater flow, PhD thesis, University of Minnesota, Minneapolis, MN.

Janković, I. and R. Barnes. 1999. High-order line elements in modeling two-dimensional groundwater flow. *Journal of Hydrology*, 226(3), 211–223.

Kennedy, J. and R.C. Eberhart. 2001. *Swarm Intelligence*. Academic Press, CA.

Kovářík, K. and J. Mužík. 2013. A meshless solution for two dimensional density-driven groundwater flow. *Engineering Analysis with Boundary Elements*, 37(2), 187–196.

Kumar, P. and G.R. Dodagoudar. 2008. Two-dimensional modeling of contaminant transport through saturated porous media using the Radial Point Interpolation Method (RPIM). *Journal of Hydrology*, 16, 1497–1505.

Li, J., Chen, Y. and D. Pepper. 2003. Radial basis function method for 1-D and 2-D groundwater. *Computational Mechanics*, 32, 10–15.

Li, Zi and X-z Mao. 2011. Global multiquadric collocation method for groundwater contaminant source identification. *Environmental Modelling and Software*, 26(12), 1611–1621.

Liong, S.-Y. and Md. Atiquzzama. 2004. Optimal design of water distribution network using shuffled complex evolution. *Journal of the Institution of Engineers, Singapore*, 44(1), 93–107.

Liu, G.R. 2002. *Mesh Free Methods: Moving Beyond the FEM*. CRC Press, Boca Raton, FL.

Liu, G.R. and Y.T. Gu. 2005. *An Introduction to Meshfree Methods and Their Programming*. Springer, Dordrecht, Berlin.

Louck, D.P., Stedinger, J.R., and D.A. Haith. 1981. *Water Resources Systems, Planning and Analysis*. Prentice-Hall, London.

Maier, H.R. and A.R. Simpson. 2003. Ant colony optimization for design of water distribution systems. *Journal of Water Resources Planning and Management*, 129(3), 200–209.

Mategaonkar, M. and T.I. Eldho. 2011. Simulation of groundwater flow in unconfined aquifer using meshfree point collocation method. *Engineering Analysis with Boundary Elements*, 35, 700–707.

Mategaonkar, M. and T.I. Eldho. 2012. Groundwater remediation optimization using a point collocation method & particle swarm optimization. *Journal of Environmental Modelling and Software*, 32, 37–48.

Mategaonkar, M. 2012. Simulation-Optimization Model for Groundwater Contamination Remediation Using Meshfree Method and Particle Swarm Optimization, PhD thesis, Department of Civil Engineering, IIT Bombay, India.

Mattot, L.S., Rabideau, A.J. and J.R. Craig. 2006. Pump-and-treat optimization using analytic element method flow models. *Advances in Water Resources*, 29(5), 760–775.

McDonald, M.G. and A.W. Harbaugh. 1988. *A Modular Three Dimensional Finite Difference Groundwater Flow Mode*. MODFLOW, Modeling Techniques, Scientific Software Group, Denver, CO.

McKinney, D.C.M. and M.D. Lin. 1994. Genetic algorithm solution of groundwater management models. *Water Resources Research*, 30(6), 1897–1906.

Meghna, B. and B.S. Minsker. 2006. Groundwater Remediation design using multi scale genetic algorithms. *Journal of Water Resources Planning and Management*, 132(5), 341–350.

MOWR (Ministry of Water Resources). 2001. China agenda for water sector strategy for north China. v. 2: Report No. 22040-CHA, 309 pp.

Mondal, A., Eldho, T.I. and V.V.S. Gurunadha Rao. 2010. Multi-objective groundwater remediation system design using coupled finite element model and non-dominated sorting genetic algorithm II. *Journal of Hydrologic Engineering (ASCE)*, 15(5), 350–359.

Nayroles, N., Touzot, G. and P. Villon. 1992. Generalizing the finite element method: diffuse approximation and diffuse elements. *Computational Mechanics*, 10, 307–318.

Pang, G., Chen, W., and Fu, Z. 2015. Space-fractional advection-dispersion equations by the Kansa method. *Journal of Computational Physics*, 293, 280–296.

Pinder, G.F. and H.H. Cooper. 1970. A numerical technique for calculating the transient position of the salt-water front. *Water Resources Research*, 6(3), 875–882.

Pinder, G.F. and W.G. Gray. 1977. *Finite Element Simulation in Surface and Subsurface Hydrology*. Academic Press, New York.

Rastogi, A.K. 2010. *Numerical Groundwater Hydrology*. Penram Int. Publishing, New Delhi.

Recknagel, F., Bobbin, J., Whigham, P. and H. Wilson. 2002. Comparative application of artificial neural networks and genetic algorithms for multivariate time-series modelling of algal blooms in freshwater lakes. *Journal of Hydroinformatics*, 4(2), 125–133.

Refsgaard, J.C., van der Sluijs, J.P., Hojberg, A.L. and P.A. Vanrolleghem. 2007. Uncertainty in the environmental modeling process—Framework and guidance. *Environmental Modeling and Software*, 22(11), 1543–1556.

Rifai, H.S., Newell, C.J. and J.R. Gonzales et al. 1998. *User's Manual, BIOPLUME III, Natural Attenuation Decision Support System, Version 1.0 U.S. Air Force Center for Environmental Excellence*. Brooks Air Force Base, San Antonio, TX.

Saeedpanah, I., Jabbari, E. and M.A. Shayanfar. 2011. A new approach for analyses of groundwater flow response to tidal fluctuation in a coastal leaky confined aquifer. *Journal of Coastal Research, Special Issue* 64, 1175–1178.

Shafer, J.M. 1987a. Reverse pathline calculation of time related capture zones in non-uniform flow. *Ground Water*, 25(3), 283–289.

Shafer, J.M. 1987b. *GWPATH: Interactive Ground-Water Flow Path Analysis*. Illinois State Water Survey, Champaign, IL.

Strack, O.D. 1989. *Groundwater Mechanics*. Prentice-Hall, Englewood Cliffs, NJ.

Strack, O.D. and H.M. Haitjema. 1981a. Modeling double aquifer flow using a comprehensive potential and distributed singularities: 1. Solution for homogeneous permeability. *Water Resources Research*, 17(5), 1535–1549.

Strack, O.D. and H.M. Haitjema. 1981b. Modeling double aquifer flow using a comprehensive potential and distributed singularities: 2. Solution for inhomogeneous permeabilities. *Water Resources Research*, 17(5), 1551–1560.

Swathi, B. and T.I. Eldho. 2014. Groundwater flow simulation in unconfined aquifers using Meshless Local Petrov-Galerkin (MLPG) method. *Engineering Analysis with Boundary Elements*, 48, 43–52.

Takagi, H. 1997. Introduction to fuzzy systems, neural networks, and genetic algorithms. In: Ruan, D. (ed.), *Intelligent Hybrid Systems*, Kluwer Academic, Norwell, MA, pp. 3–34.

Tamer Ayvaz, M. 2009. Application of harmony search algorithm to the solution of groundwater management models. *Advances in Water Resources*, 32(6), 916–924.

Wang, H. and M.P. Anderson. 1982. *Introduction to Groundwater Modeling Finite Difference and Finite Element Methods*. W. H. Freeman and Company, New York.

Willis, R. and W.G. Yeh. 1987. *Groundwater Systems Planning and Management*. Prentice-Hall, Inc. Englewood Cliffs, NJ.

Yeh, H.D. and J.L. Chen. 2000. Simulations for groundwater remediation using three remedial techniques. *Proceedings of National Science Council. ROC (A)*, 25(5), 309–316.

Zheng, C. and G.D. Bennett. 1995. *Applied Contaminant Transport Modeling—Theory and Practice*. Van Nostrand Reinhold, New York.

Zheng, C. and P.P. Wang. 1998. A Modular Three-Dimensional Transport Model. MT3DMS, Technical Report, US Army Corps of Engineers Waterways Experiment Station, Vicksburg, MS.

Zhou, Y. and Li, W. 2011. A review of regional groundwater flow modeling. *Geoscience Frontiers*, 2(2), 205–214. http://water.usgs.gov/software/lists/groundwater.

12

A New Approach for Groundwater Modeling Based on Connections

B. Sivakumar, X. Han, and F. M. Woldemeskel

CONTENTS

12.1 Introduction

Groundwater makes up more than a quarter of all available freshwater in the world, and it is also one of the purest forms of water on Earth. Despite these, groundwater was one of the least-exploited natural resources until just a few decades ago, as its access was (and still is, to some extent) more difficult and expensive when compared to surface water resources. However, a combination of factors, including explosion in population, dwindling surface water resources per capita, and developments in technology, has changed this situation in recent times. The last few decades have witnessed a significant exploitation of groundwater resources around the world, both in the developed world and in the developing world, including in the United States, Europe, and Asia (e.g., Konikow and Kendy, 2005). Indeed, in many parts of the world, groundwater is the only viable source of water of safe quality, such as the quality required for drinking and household use. With the relative ease in withdrawal of large quantities of water from depths of even hundreds of meters, groundwater has also become the only or primary source of water for irrigation in many countries. While there are already increasing concerns about the recent and current levels of exploitation of groundwater, the situation will likely become even worse in the future, due both to our continued increase in water demands and to the anticipated negative impacts of climate change on our water resources.

As a result, there is a clear and urgent need to improve our understanding and modeling of groundwater flow and transport phenomena.

Flow and solute transport phenomena in natural subsurface formations are influenced by a variety of factors. For instance, they are strongly affected by, among others, the heterogeneity of the groundwater medium and the complexity of the interactions of groundwater with surface and atmospheric water. These, in turn, are dictated by the spatial and temporal scales of the medium and processes of interest. For instance, the hydraulic properties of the medium are often significantly different when different spatial and temporal scales are considered. Similarly, the mechanisms that regulate the soil–water–atmosphere continuum are also different at different scales and often have different levels of complexity. Additional complexity and difficulties arise from natural and anthropogenic biological interferences. Our limited knowledge about the properties of the groundwater medium and its interactions with others severely hampers our ability to develop accurate groundwater models.

Since the pioneering work by Darcy (1856), followed by several other important contributions over the next century (e.g., Boussinesq, 1904; Pennink, 1904, 1905; Hubbert, 1940; Theis, 1940), the past five decades of research have resulted in the proposal of a variety of approaches and development of numerous models for studying groundwater flow and transport phenomena.

The existing approaches may be divided into several groups (or their combinations), including (1) deterministic approaches with aquifers characterized by geologic or hydraulic process properties (e.g., hydraulic head, aquifer porosity, water content, hydraulic conductivity); (2) stochastic approaches mainly based on Gaussian random fields; (3) geostatistical approaches (e.g., kriging, indicator kriging, annealing, transition probabilities, Markov chain-geostatistics); (4) deterministic or stochastic fractal-based approaches (e.g., fractional Brownian motion, fractional Gaussian noise, fractal-multifractal approach); and (5) data-driven approaches (e.g., artificial neural networks, chaos theory, wavelets). Details of these approaches and applications are available in Tóth (1963, 1970), Freeze and Witherspoon (1967), Journel and Huijbregts (1978), Dagan (1979), Delhomme (1979), Kitanidis and Vomvoris (1983), Zimmerman et al. (1998), Neuman (1990), Koltermann and Gorelick (1996), Molz et al. (1997), Coulibaly et al. (2001), Puente et al. (2001), Sivakumar et al. (2005), and Peterson and Western (2014), among others. Despite the progress thus far, there still remain numerous challenges; see, for example, Ojha et al. (2015) and Ramadas et al. (2015) for recent accounts of the challenges in groundwater studies, including in modeling, management, water quality, and climate change impacts.

Regardless of the approach, an important idea in modeling groundwater flow and transport phenomena is to establish connections that generally exist in the dynamics of the underlying system, whether in space or time or space-time. In this context, modern developments in network theory can offer useful ideas—a network is a set of points (or nodes) connected together by a set of lines (or links). In particular, concepts of complex networks (e.g., small-world networks, scale-free networks, network motifs, and community structure) have been found to be very useful in the study of connections in various fields; see, for example, Watts and Strogatz (1998), Barabási and Albert (1999), Milo et al. (2002), and Girvan and Newman (2002) for details of these concepts, and Barabási (2002) and Estrada (2012) for their applications in various fields. Applications of such complex networks-based concepts in hydrology and water resources have recently started, and among the networks studied are rainfall and streamflow monitoring networks as well as river and virtual water networks; see, for example, Rinaldo et al. (2006), Suweis et al. (2011), Scarsoglio et al. (2013), and Sivakumar and Woldemeskel (2014, 2015). Indeed, an argument in favor of network theory as a suitable tool for studying all types of connections in hydrology has also been put forth (e.g., Sivakumar, 2015).

Encouraged by the outcomes of the above studies, we make an attempt here to apply, for the first time, the concepts of complex networks in the field of groundwater

hydrology. Specifically, we examine the spatial connections in a groundwater level monitoring network in a region. For this purpose, daily groundwater levels observed across 125 wells in California are analyzed using two complex networks-based methods: clustering coefficient and degree distribution. The clustering coefficient is a measure of local density and quantifies the tendency of a network to cluster. The degree distribution is a measure of spread and expresses the fraction of nodes in a network with a certain number of links. The results are interpreted to obtain important information about the type of the groundwater level monitoring network.

The rest of this chapter is organized as follows. First, a brief description of the basic concept of a network with two methods employed in this study is provided. Next, details of the study area and groundwater level data used are presented. Then, analysis of groundwater level data and the results are reported. Finally, some closing remarks are made.

12.2 Network and Methods

A *network* is a set of points connected together by a set of lines; see Figure 12.1. The points are called *nodes* or *vertices* and the lines are called *links* or *edges*. Mathematically, a network can be represented as $G = \{P,E\}$, where P is a set of N nodes (P_1, P_2, ..., P_N) and E is a set of n links. The network shown in Figure 12.1 has $N = 7$ (nodes) and $n = 8$ (links), with $P = \{1, 2, 3, 4, 5, 6, 7\}$ and $E = \{\{1,7\}, \{2,4\}, \{2,5\}, \{2,7\}, \{3,7\}, \{4,7\}, \{5,6\}, \{6,7\}\}$. This type of network, consisting of a set of identical nodes connected by identical links, is perhaps the simplest form of network. There are many ways in which networks may be more complex; see Sivakumar (2015) for a few other forms.

There are many different ways to study the properties of a network, such as by its clustering, topology,

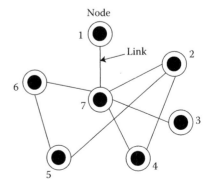

FIGURE 12.1
Concept of a network.

adjacency, and centrality. Accordingly, different measures and methods to represent these properties exist as well, including clustering coefficient, degree distribution, average shortest path length, degree centrality, and community structure. In the present study, clustering coefficient and degree distribution methods are employed to examine the spatial connections in groundwater levels. These two methods are briefly described next.

12.2.1 Clustering Coefficient

The concept of clustering has its origin in sociology, under the name "fraction of transitive triples" (Wasserman and Faust, 1994). The clustering coefficient quantifies the tendency of a network to cluster (Watts and Strogatz, 1998) and, thus, is essentially a measure of local density.

As shown in Figure 12.2 (left), let us consider a network where the node i has k_i links which connect it to k_i other nodes. These k_i nodes are considered the neighbors of node i, and can be identified based on some criterion (say, threshold, T), such as correlation between node i and the other nodes in the network. If the neighbors of i were part of a cluster, then there would be $k_i(k_i - 1)/2$ links between them, as shown in Figure 12.2 (right). The clustering coefficient of node i is then calculated as the ratio between the number E_i of links that actually exist between these k_i nodes (solid lines in Figure 12.2, right) and the total number $k_i(k_i - 1)/2$ (i.e., all lines in Figure 12.2, right),

$$C_i = \frac{2E_i}{k_i(k_i - 1)} \tag{12.1}$$

The clustering coefficient of every other node in the network is calculated by simply repeating the above procedure for the respective nodes. The clustering coefficient of the whole network C is then given by the average of the clustering coefficients C_i's of all the individual nodes.

The clustering coefficient of the individual nodes and of the entire network can be used to obtain useful insights about the network. For instance, it offers important information about the type of network, grouping of nodes, and identification of dominant nodes, among others; for instance, the clustering coefficient $C = 1$ indicates a fully connected network and $C = p$ (where p is the probability of two nodes being connected) indicates a random graph, while C is somewhere between these two (but much higher than the latter) for a small-world network (e.g., Watts and Strogatz, 1998). In the context of spatial groundwater dynamics, information on these is useful for assessment of model complexity, classification of regions exhibiting similar spatial patterns, identification of suitable regions for data interpolation, and many others.

12.2.2 Degree Distribution

As shown in Figure 12.1, different nodes in a network may have a different number of links. The number of links (k) of a node is called the node degree. The spread in the node degrees is characterized by a distribution function, $p(k)$, which expresses the fraction of nodes in a network with degree k. This distribution is called as the degree distribution, and it often serves as a reliable indicator of the type of network. For instance, in a random graph, since the links are placed randomly, most of the nodes have approximately the same degree, and close to the average degree $\langle k \rangle$ of the network. Therefore, the degree distribution of a completely random graph is a Poisson distribution with a peak at $P(\langle k \rangle)$. Similarly, depending on the properties of networks, degree distribution can be Gaussian, exponential, power law, or other.

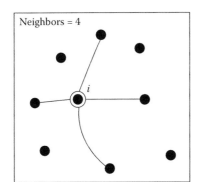
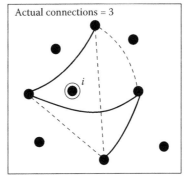

FIGURE 12.2
Calculation of clustering coefficient in a network.

12.3 Study Area and Data

In this study, spatial connections in groundwater levels across California are examined (see Figure 12.3). California is the most populous state (with a population of about 38 million) and the third largest state by area in the United States. California is also the nation's top agricultural state and one of the most important agricultural areas around the world. In particular, the Central Valley in California, with its aquifer system, plays a vital role in agricultural production. Approximately three-fourths of the irrigated land in California and about one-fifth of the irrigated land in the United States are in the Central Valley. From north to south, the Central Valley aquifer system is divided into the Sacramento Valley, the Sacramento-San Joaquin Delta, and the San Joaquin Valley subregions, on the basis of different characteristics of surface-water basins. The Central Valley also supplies about 20% of the nation's groundwater demand from pumping its aquifers, making it the second-most-pumped aquifer system in the United States; see Faunt (2009) for a recent account.

In the past two decades or so, California's population has increased significantly, with the Central Valley also is becoming an important area for the state's expanding urban population. This surge in population has increased the competition for water resources across the state of California and in the Central Valley in particular. The situation is further complicated by the growing competition for the Colorado River water between California, Colorado, Utah, Arizona, and Nevada, not to mention Mexico. As a result, there has been a significant increase in groundwater exploitation across the state. All these necessitate a far better understanding and modeling of groundwater flow and transport phenomena in California toward a more sustainable planning and management of groundwater resources in the system. This offers the motivation for the present study, where spatial connections in groundwater levels are examined.

Groundwater levels (and other data) are measured at numerous locations across California. In this study, groundwater level data observed at 125 wells across the state are studied. Figure 12.3 shows the locations of these 125 wells (in some cases, several wells are located within the same latitude/longitude and, thus, their locations are hard to visualize). The data are at the daily scale, observed during the period January 1, 2013–May 6, 2014. It is important to note that different wells (may) have different periods of data. The period of data and the number of wells in this study are chosen keeping in mind the reliability of the outcomes of the clustering coefficient and degree distribution analysis.

FIGURE 12.3
Study area (California), with locations of the 125 wells considered for analysis.

12.4 Analysis and Results

The clustering coefficient and degree distribution methods are applied to the daily groundwater level data from each of the 125 wells in California. It is important to note that the outcomes of these methods may sometimes be significantly influenced by the correlation threshold (T) used for identification of the neighbors. An optimum threshold level is generally not known *a priori*. While some studies in hydrology have used an arbitrary value of 0.5 (e.g., Scarsoglio et al., 2013), studying the influence of at least a few different threshold levels is often important for more reliable results and interpretations, especially when the outcomes are used for interpolation of data, classification of aquifer, and related purposes. To this end, we consider seven different thresholds in this study: 0.3, 0.4, 0.5, 0.6, 0.7, 0.8, and 0.9.

12.4.1 Clustering Coefficient

Figure 12.4 presents the clustering coefficient (CC) values for the daily groundwater level data from each of the 125 wells in California, with each box corresponding to a specific well (as numbered). Due to space constraints, results for only four thresholds are shown: $T = 0.5$, 0.6, 0.7, and 0.8. For better visualization, different ranges of clustering coefficient values are also shown.

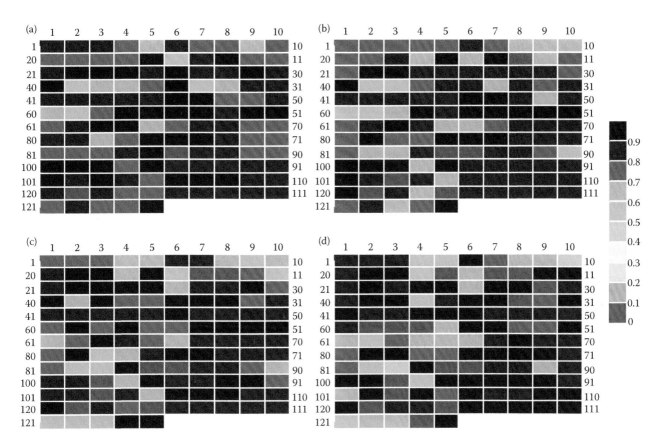

FIGURE 12.4

Clustering coefficients for groundwater level data from California for different thresholds: (a) $T = 0.5$; (b) $T = 0.6$; (c) $T = 0.7$; and (d) $T = 0.8$. A box corresponds to a well.

The clustering coefficient results generally indicate good connections in groundwater levels among the wells, with more than 70% of the wells having clustering coefficient values above 0.60 even up to a threshold level of 0.8; see also Table 12.1, which presents the exact number of wells falling within each clustering coefficient range

TABLE 12.1

Clustering Coefficients for Daily Groundwater Level Data from 125 Wells in California

	Number of Wells within Each Clustering Coefficient (CC) Range for Threshold (*T*)						
CC Range	**0.3**	**0.4**	**0.5**	**0.6**	**0.7**	**0.8**	**0.9**
0.91–1.00	25	27	26	24	30	47	41
0.81–0.90	85	63	51	53	46	22	22
0.71–0.80	6	24	33	18	13	8	3
0.61–0.70	3	5	5	17	14	13	7
0.51–0.60	4	4	6	2	5	9	3
0.41–0.50	0	0	0	3	2	1	3
0.31–0.40	0	0	0	1	4	1	2
0.21–0.30	0	0	0	0	0	1	0
0.11–0.20	0	0	0	0	0	0	0
0.01–0.10	0	0	0	0	0	0	0
0.00	2	2	4	7	11	23	44

for all seven threshold levels considered. The connections are much stronger for the lower side of threshold values, that is, $T = 0.5$ and $T = 0.6$ (and also for thresholds further down—plots not shown), with over 90% of the wells having CC > 0.6. This means most of the wells in the groundwater level monitoring network have high connectivity with the rest of the wells in the network from an overall network perspective. This suggests that interpolation of groundwater level data, at least in specific areas where the clustering coefficient values are generally above 0.6, may offer good results, although identification of such specific areas in itself can be challenging, as not all neighbors can be actual neighbors; see, for example, Sivakumar and Singh (2012) and Sivakumar and Woldemeskel (2014) for details in the context of streamflow and Sivakumar and Woldemeskel (2015) in the context of rainfall.

While Figure 12.4 offers useful information on the extent of connection of each well to the rest of the 124 wells of the network collectively, more useful information may be obtained by comparing the clustering coefficient value of each well with respect to each and every other well in the network on an individual basis. One way to achieve this may be to present the average of clustering coefficients of any two wells for the entire network. This is done in Figure 12.5, again for the four threshold levels shown in Figure 12.4. The results generally show very high connection of each well with respect to each and every other well (light blue, dark blue, and black boxes). This is particularly the case for thresholds up to 0.6, and to some extent even for $T = 0.7$; see Figure 12.5a–c. Only in a very few cases, and that too only at a very high threshold level of 0.8, the connection is considerably weak (yellow, orange, and red boxes); see Figure 12.5d.

The clustering coefficient values for each of the 125 wells (Figure 12.4) and their comparison with each and every other well (Figure 12.5) indeed provide useful information about individual connections in the network. However, what is of an even broader interest is the identification of the nature of the entire network of 125 wells as a whole, for development of an appropriate model to represent the structure and dynamics of the entire network, including in the identification of the complexity of the model. To this end, the clustering coefficient of the entire network, calculated as the average of the clustering coefficients for all the 125 wells, is useful.

The clustering coefficient values of the entire network for the seven different thresholds are 0.827, 0.819,

0.807, 0.778, 0.780, 0.743, and 0.669. Figure 12.6 presents the variation of the clustering coefficient value with respect to threshold level. Overall, the clustering coefficient value decreases with an increase in the threshold value, as normally expected, but the decrease is more pronounced at higher thresholds (i.e., from $T = 0.8$). The high clustering coefficient values seem to suggest that the network is not a purely random graph, as the clustering coefficient values for random networks are typically very low, essentially due to random distribution of links; see, for example, Watts and Strogatz (1998). As the clustering coefficient for the groundwater level monitoring network is much higher than that for the random network but lower than the ones expected for fully connected networks (for which CC should be equal to 1.0), it may be reasonable to interpret that the network is a small-world network. However, such an interpretation needs further evidence. In what follows, we present the results from the degree distribution method to obtain some additional information about the type of network, especially from the point of view of spread of the network. We will specifically address

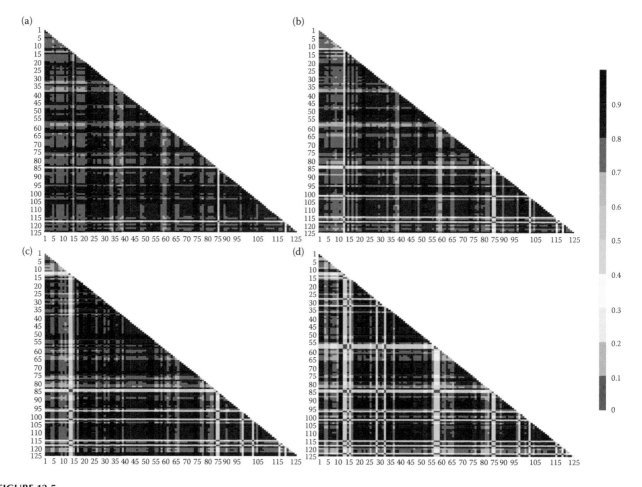

FIGURE 12.5
Clustering coefficients for groundwater level data from California for different thresholds: (a) $T = 0.5$; (b) $T = 0.6$; (c) $T = 0.7$; and (d) $T = 0.8$. The value in any box (well) indicates the average of clustering coefficients of the two corresponding boxes.

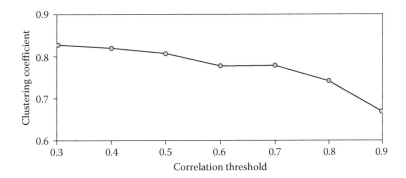

FIGURE 12.6
Relationship between clustering coefficient and correlation threshold for the groundwater level monitoring network of 125 wells in California.

the small-world nature of the groundwater level monitoring network further in a future study, by exploring also the concept of shortest path length (e.g., Watts and Strogatz, 1998).

12.4.2 Degree Distribution

Figure 12.7 presents the results from the degree distribution analysis of groundwater levels from the 125 wells in California, for the seven threshold levels considered in this study (*threshold values are in increasing order from right to left in each figure*). The degree distribution results are shown both in the normal scale (Figure 12.7a) and in the log–log scale (Figure 12.7b), and are the values of complementary cumulative distribution,

defined as the fraction of nodes with degree at least k and denoted as $p(K \geq k)$.

While the degree distribution for the groundwater level data network changes with respect to the correlation thresholds, there also seem to be some patterns in the way it changes (i.e., shape of the curve). For instance, when the threshold is 0.3, there are about 80% of the wells with at least 40 neighbors, but the number of wells with at least 40 neighbors is zero when the threshold is 0.9 (see Figure 12.7a). These observations have important implications for groundwater data monitoring, including in the identification of nearest neighbors for data interpolation purposes.

The degree distribution curves in Figure 12.7 seem to resemble exponential distribution over some parts (left,

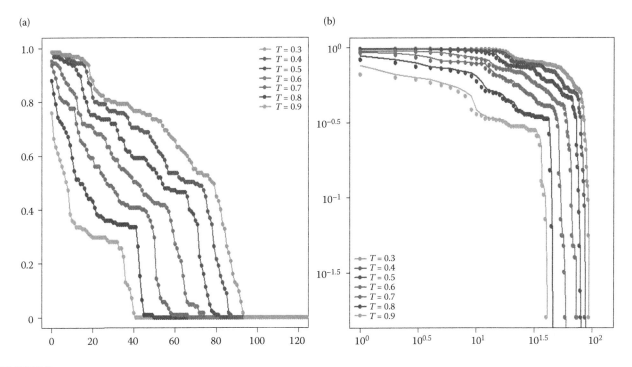

FIGURE 12.7
Degree distribution results for the groundwater level data monitoring network of 125 stations in California: (a) normal scale; and (b) log–log scale; threshold values are in increasing order from right to left.

especially for higher thresholds) and power-law distribution over others (right, especially for lower thresholds), although there are also changes with respect to thresholds. It may, therefore, be reasonable to interpret that the groundwater level monitoring network is a combination of exponential distribution and power-law distribution, with significant resemblance to exponentially truncated power-law distribution. This is an important observation in the context of groundwater modeling studies because it is a common practice to assume Gaussian distribution for groundwater flow and transport phenomena.

12.5 Closing Remarks

A fundamental requirement in studying the dynamics of groundwater flow and transport processes is the identification of the nature and extent of connections in space and time. This study introduced some of the modern developments in the field of complex systems science, especially concepts of complex networks, for studying groundwater. The study employed the concepts of clustering coefficient and degree distribution to examine the spatial connections in groundwater levels in a region. Analysis of daily groundwater level data observed in a monitoring network of 125 wells in California, considering also different threshold levels (of spatial connection), offers some interesting observations about the strength of connection of each well with the rest of the network collectively and with each and every other well individually, as well as about the nature of the network as a whole. The results suggest that the groundwater level monitoring network is not a purely random network, but is more likely a small-world network, with a combination of exponential and power-law distribution, rather than Gaussian distribution. These results have important implications for modeling and prediction of groundwater levels, including for interpolation/extrapolation of data.

Recent developments in network theory offer new avenues for studying complex systems, such as hydrologic systems. Applications of the ideas of network theory in the field of hydrology and water resources are in the very early stages, with only a very few networks studied until now; rainfall and streamflow monitoring networks, river networks, and virtual water networks. There is indeed a great potential for employing network theory for exploring various types of connections in hydrology, as recently highlighted by Sivakumar (2015). It is in this context we have made the very first attempt to explore the utility of the concepts of network theory for groundwater studies. It is indeed our hope that this study will lead to many more network theory-based studies to further advance our understanding of connections in groundwater flow and transport phenomena.

Acknowledgments

This study is supported by the Australian Research Council (ARC) Future Fellowship grant (FT110100328). Bellie Sivakumar acknowledges the financial support from ARC through this Future Fellowship grant.

References

Barabási, A.-L. 2002. *Linked: The New Science of Networks*, Perseus, Cambridge, MA.

Barabási, A.-L. and R. Albert. 1999. Emergence of scaling in random networks, *Science*, 286, 509–512.

Boussinesq, J. 1904. Recherches th´eoriques sur ´ecoulement des nappes d'eau infiltr´ees dans le sol", *J. de math´ematiques pures et appliqu´ees*, 5(X(1)(5–78), 363–394.

Coulibaly, P., Anctil, F., Aravena, R., and B. Bobée. 2001. Artificial neural network modeling of water table depth fluctuations, *Water Resources Research.*, 37(4), 885–896.

Dagan, G. 1979. Models of groundwater flow in statistically homogeneous porous formations, *Water Resources Research*, 15(1), 47–63.

Darcy, H. P. G. 1856. *Les Fontaines Publiques de la Ville de Dijon*, V. Dalmont, Paris.

Delhomme, J. P. 1979. Spatial variability and uncertainty in groundwater flow parameters: A geostatistical approach, *Water Resources Research*, 15(2), 269–280.

Estrada, E. 2012. *The Structure of Complex Networks: Theory and Applications*, Oxford University Press, Oxford, UK.

Faunt, C. C. (ed.) 2009. *Groundwater Availability of the Central Valley Aquifer*, U.S. Geol. Surv. Prof. Pap., California, 1766, 1–225.

Freeze, R. A. and P. A. Witherspoon. 1967. Theoretical analysis of regional groundwater flow, II: Effect of water table configuration and subsurface permeability variations, *Water Resources Research*, 3(2), 623–634.

Girvan, M. and M. E. J. Newman. 2002. Community structure in social and biological networks, *Proceedings of the National Academy of Sciences USA*, 99, 7821–7826.

Hubbert, M. K. 1940. The theory of groundwater motion, *Journal of Geology*, 48, 785–944.

Journel, A. G. and C. J. Huijbregts. 1978. *Mining Geostatistics*, Academic, San Diego, CA.

Kitanidis, P. K. and E. G. Vomvoris. 1983. A geostatistical approach to the inverse problem in groundwater modeling (steady state) and one-dimensional simulations, *Water Resources Research*, 19(3), 677–690.

Koltermann, C. E. and S. M. Gorelick. 1996. Heterogeneity in sedimentary deposits: A review of structure-imitating, process-imitating, and descriptive approaches, *Water Resources Research*, 32, 2617–2658.

Konikow, L. F. and E. Kendy. 2005. Groundwater depletion: A global problem, *Hydrogeology Journal*, 13(1), 317–320.

Milo, R., Shen-Orr, S., Itzkovitz, S., Kashtan, N., Chklovskii, D., and U. Alon. 2002. Network motifs: simple building blocks of complex networks, *Science*, 298, 824–827.

Molz, F. M., Liu, H. H., and J. Szulga. 1997. Fractional Brownian motion and fractional Gaussian noise in subsurface hydrology: A review, presentation of fundamental properties, and extensions, *Water Resources Research*, 33(10), 2273–2286.

Neuman, S. P. 1990. Universal scaling of hydraulic conductivities and dispersivities in geologic media, *Water Resources Research*, 26, 1749–1758.

Ojha, R., Ramadas, M., and R. S. Govindaraju. 2015. Current and future challenges in groundwater. I: Modeling and management of resources, *Journal of Hydrologic Engineering*, 20(1), A4014007.

Pennink, J. M. K. 1904. Investigations for ground-water supplies. *American Society of Civil Engineers, Transactions*, V 54(D), 169–181.

Pennink, J. M. K. 1905. Over de beweging van groundwater, *De Ingenieur*, 20(30), 482–492.

Peterson, T. J. and A. W. Western. 2014. Nonlinear time-series modeling of unconfined groundwater head, *Water Resources Research*, 50, 8330–8355.

Puente, C. E., Robayo, O., Diaz, M. C., and B. Sivakumar. 2001. A fractal-multifractal approach to groundwater contamination. 1. Modeling conservative tracers at the Borden site, *Stochastic Environmental Research and Risk Assessment*, 15(5), 357–371.

Ramadas, M., Ojha, R., and R. S. Govindaraju. 2015. Current and future challenges in groundwater. II: Water quality modeling, *Journal of Hydrologic Engineering*, 20(1), A4014008.

Rinaldo, A., Banavar, J. R. and A. Maritan. 2006. Trees, networks, and hydrology, *Water Resources Research*, 42, W06D07, doi: 10.1029/2005WR004108.

Scarsoglio, S., Laio, F., and L. Ridolfi. 2013. Climate dynamics: A network-based approach for the analysis of global precipitation, *PLoS ONE*, 8(8), e71129, doi: 10.1371/journal.pone.0071129.

Sivakumar, B. 2015. Networks: A generic theory for hydrology? *Stochastic Environmental Research and Risk Assessment*, 29, 761–771.

Sivakumar, B., Harter, T. and H. Zhang. 2005. Solute transport in a heterogeneous aquifer: A search for nonlinear deterministic dynamics, *Nonlinear Processes in Geophysics*, 12, 211–218.

Sivakumar, B. and V. P. Singh. 2012. Hydrologic system complexity and nonlinear dynamic concepts for a catchment classification framework, *Hydrology and Earth System Sciences*, 16, 4119–4131.

Sivakumar, B. and F. M. Woldemeskel. 2014. Complex networks for stream flow dynamics, *Hydrology and Earth System Sciences*, 18, 4564–4578.

Sivakumar, B. and F. M. Woldemeskel. 2015. A network-based analysis of spatial rainfall connections, *Environmental Modelling and Software*, 69, 55–62.

Suweis, S., Konar, M., Dalin, C., Hanasaki, N., Rinaldo, A., and I. Rodriguez-Iturbe. 2011. Structure and controls of the global virtual water trade network, *Geophysical Research Letters*, 38, L10403, doi: 10.1029/2011GL046837.

Theis, C. V. 1940. The source of water derived from wells: Essential factors controlling the response of an aquifer to development, *Civil Engineering*, 10(5), 277–280.

Tóth, J. 1963. A theoretical analysis of groundwater flow in small drainage basins, *Journal of Geophysical Research*, 68, 4785–4812.

Tóth, J. 1970. A conceptual model of the groundwater regime and the hydrogeological environment. *Journal of Hydrology*, 19, 164–176.

Wasserman, S. and K. Faust. 1994. *Social Network Analysis*, Cambridge University Press, Cambridge, UK.

Watts, D. J. and S. H. Strogatz. 1998. Collective dynamics of 'small-world' networks, *Nature*, 393, 440–442.

Zimmerman, D. A., de Marsily, G., C. A. Gotway et al. 1998. A comparison of seven geostatistically based inverse approaches to estimate transmissivities for modeling advective transport by groundwater flow, *Water Resources Research*, 34(6), 1373–1413.

13

Mathematical Modeling to Evolve Predevelopment Management Schemes: A Case Study in Boro River Valley, Okavango Delta, Botswana, Southern Africa

M. Thangarajan

CONTENTS

13.1 Introduction

Overdevelopment of groundwater resources to meet the increase demand for drinking, industries, irrigation, and other purposes is ever-growing in developed and developing countries. The overdevelopment of groundwater resources leads to the decline of water level and thereby causes socioeconomic and environmental degradation. It is, thus, imperative to manage the groundwater resources in an optimal manner. Management schemes can be evolved, only if the groundwater potential is assessed in a more realistic manner. Mathematical modeling in conjunction with detailed field investigations has been proved to be a potential tool for this purpose. Evolving predevelopment management schemes still

work out to be a better choice. One such study was carried out in Boro River valley, Okavango Delta, Botswana. The River Boro emerges from the Okavango swamps in the northwestern part of Botswana.

The groundwater resource in the valley has been quantified through exploratory drilling, test pumping, long-term water level monitoring, and hydrochemical analysis of groundwater samples by the Department of Water Affairs, Government of Botswana. A preliminary mathematical model with six layers was conceptualized by making use of available data. The main aquifer is the fourth layer and bottom most (sixth layer) layer is the saline one. The model of the basin was constructed and calibrated for steady-state conditions. A number of prognostic runs were made and an optimal one was identified which will ensure a minimum upward leakage from the bottom saline unit to the pumping aquifer. The interflows across the aquifers were quantified for different scenario runs. This study indicates that the aquifer can sustain a pumping rate of 8000 m³/day, if River Boro receives regular flow as it receives now.

13.2 Study Area

The Boro River valley is located in Okavango Delta (Figure 13.1), Botswana (Southern Africa). The Okavango River originates in the Angolan highlands reaches in northwestern Botswana and terminates as a huge inland

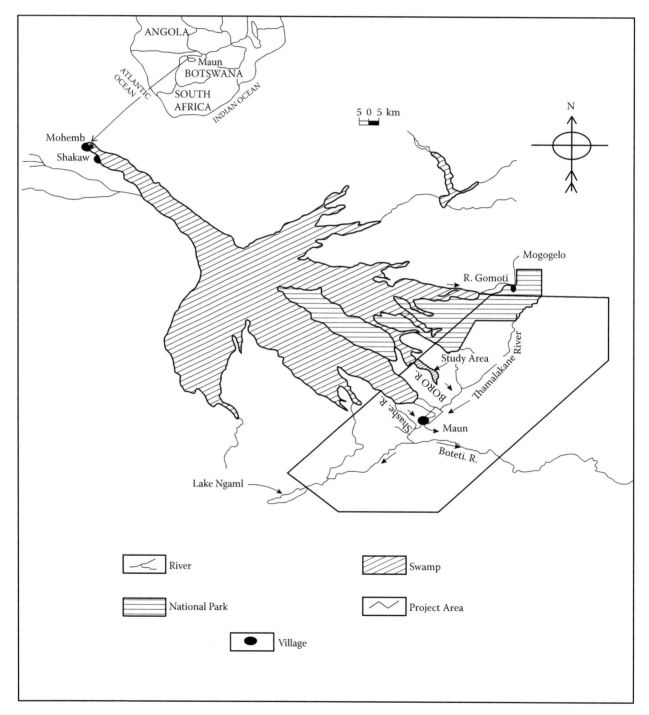

FIGURE 13.1
Location map of Okavango Delta. (Adapted from Thangarajan, M. et al. *Journal of Geological Society of India*, 55, 623–648, June 2000.)

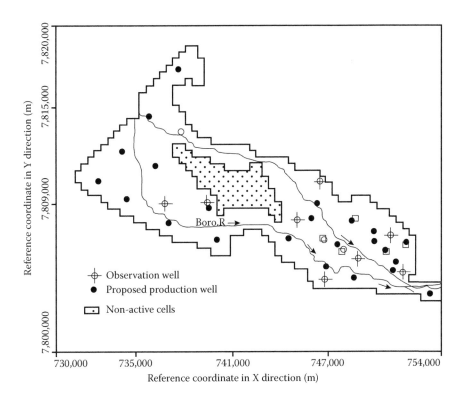

FIGURE 13.2
Location map of Boro River Valley in Okavango Delta.

delta, namely the Okavango Delta. The River Boro runs between delta perennial swamps and the Thamalakane River (Figure 13.2) and drains the delta to Lake Ngami through River Thamalakane. Maun, a small town that is located on the bank of River Thamalakane, is emerging as a famous wildlife tourist center in Botswana and the demand for water is ever-increasing. The demand during the year 1997 was about 3540 m³/day and it is likely to increase to about 4200 m³/day during the year 2000. The present demand is met from the well fields located in the valleys of Shashe and Thamalakane and the future demand has to be met from new resources. Therefore, the government of Botswana formulated a project in 1996 to explore and assess the new groundwater structures through integrated geophysical and geohydrological studies in the neighborhood of Maun. The Boro River valley (Figure 13.2) covering an area about 170 km² is one such system wherein geohydrological, geophysical, and chemical quality studies were combined to find suitable structures for exploiting groundwater. The data collected under this project was used to conceptualize the groundwater flow regime and a preliminary mathematical model was developed to study the aquifer response and thereby to evolve predevelopmental management schemes. The model was constructed and calibrated for steady-state condition. The model calibration has clearly indicated that middle and lower aquifers are semiconfined in nature. The calibrated model was then used to

study the aquifer response under four possible scenarios for evolving optimal well field locations in the lower semiconfined aquifer.

13.3 Hydrogeological Setting

The aquifer system belongs to Kalahari beds. There are three main aquifer systems in the river valley separated by two aquitards (Figure 13.3). The top zone (surface soil) is the low permeable zone with average thickness of 10 m. The top two aquifers are freshwater bearing units and the bottom one is saline. The vertical subsurface geological section along the valley is shown in Figure 13.3. An outline of the hydrogeological conditions in each of these units is given next.

13.3.1 Shallow Semiconfined Aquifer

The shallow semiconfined aquifer consists of coarse to medium grained sand. The thickness of sand varies from a few meters at the valley margins to as much as 30 m in the valley center and followed by 10-m thick interlayered sands and clays, clayey sands, and clay of the upper confining layer. The overlying bed is the sandy clay with low permeability. The aquifer receives

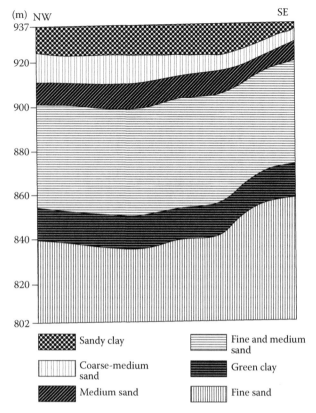

FIGURE 13.3
Vertical subsurface section along Boro River valley. (Adapted from Shashe Well Field Conceptual Model, Interim Report (Appendix G). 1996. Eastend Investments Pvt. Ltd., Gaborone, Botswana.)

recharge through river infiltration. The valley gets surface water during August–October of every year through the River Boro. The groundwater gradient in this unit is from the northwest to southeast.

Recharge to the semiconfined aquifer has been predominantly from river water infiltration as downward leakage. Major rainfall events also recharge the top aquifer to a limited extent and it is derived from model calibration as 1.8 mm per annum. This is well within the range of values recommended by the recharge committee of Botswana. The mean annual rainfall for 1986–1996 from the Maun airport weather station is about 380 mm. Without any abstraction, the main discharge from this aquifer is the evapotranspiration, which works out to be 2.4 mm/year and subsurface outflow toward the River Thamalakane.

13.3.2 Lower (Semiconfined) Aquifer

The aquifer, which occurs approximately between 40 and 50 m below ground surface, is termed a lower semiconfined aquifer, which is comprised of fine to medium grained sand. This unit is separated from the top unconfined aquifer by 10 m sandy silts, silty clays, and minor clays, which form the confining layer to the middle

aquifer. Fifteen-meter poor permeable green clays below follow the bottom of the middle aquifer, which the underlying sands contain brackish to saline water. Groundwater in this unit is under semiconfined conditions.

The transmissivity values estimated through pumping tests range from 10 to 50 m²/day. The storativity values estimated are 8.3×10^{-4}–4.3×10^{-3}. The vertical hydraulic conductivity of the middle aquifer ranges from 6.2×10^{-4} to 2.0×10^{-2} m/day. The average yield of boreholes completed in this aquifer is 130 m³/day with a range from 60 to 360 m³/day.

13.3.3 Bottom Aquifer (Saline)

The aquifer, which occurs approximately between 75 and 80 m below the ground level, is comprised of fine to medium sands and contains brackish to saline water. There is not much information available regarding hydraulic characteristics of this unit. The hydraulic conductivity value is assumed to be 2 m/d and its storativity value is 5×10^{-4}. The bottom aquifer is hydraulically connected with the overlying middle semiconfined aquifer.

13.4 Model Design

13.4.1 Methodology

Groundwater models are the simplified representations of the subsurface aquifer systems. The calibrated and validated models may be used to predict aquifer response to pumping stresses. Groundwater flow in three dimensions in a porous media of constant density can be expressed by the following partial differential equation (Rushton and Redshaw, 1979):

$$\frac{\delta}{\delta x}\left(K_{xx}\frac{\delta h}{\delta x}\right) + \frac{\delta}{\delta y}\left(K_{yy}\frac{\delta h}{\delta y}\right) + \frac{\delta}{\delta z}\left(K_{zz}\frac{\delta h}{\delta z}\right) = S_s\frac{\delta h}{\delta t} \pm W$$

(13.1)

where

K_{xx}, K_{yy}, K_{zz} = the hydraulic conductivity along x, y, and z coordinates, which are assumed to be parallel to the major axes of hydraulic conductivity (LT⁻¹)
h = potentiometric head (L)
W = volumetric flux per unit volume and represents sources and/or sinks of water (T⁻¹)
S_s = the specific storage of the porous material (L⁻¹)
t = time (T)

Equation 13.1 describes groundwater flow under nonequilibrium conditions in a heterogeneous and anisotropic medium, provided the principal axes of hydraulic conductivity are aligned with the x–y Cartesian coordinate axes. The groundwater flow equation together with specification of flow and/or initial head conditions at the boundaries constitutes a mathematical representation of the aquifer system. Numerical methods are used in general to solve the groundwater flow equation.

The computer software program Modflow developed by the United States Geological Survey (USGS, 1988) was used for this study. The Block centered finite difference approach is used in this program to solve the groundwater flow equation. A pre- and postmodel processor, namely, Visual-Modflow developed by Nilson Guigner and Thomas Franz of Waterloo Hydrologic Software Inc., Waterloo, Ontario, Canada (1996) was used for graphical data input, and for analysis and presentation of the output data.

13.4.2 Conceptual Model

The Boro River valley aquifer system was conceptualized as a six-layer system with three aquifers separated by two confining layers (confining/semiconfining) as follows:

Top soil with low permeable zone

Shallow semiconfined aquifer

Upper confining unit

Lower semiconfined aquifer

Lower semiconfining unit

Lower brackish/saline aquifer

The northwestern boundary of this narrow river valley is taken as the inflow boundary. The southeastern boundary is the line joining the confluence point of Rivers Boro and Kunyere Fault. The lateral boundary represents the freshwater zone in the top semiconfined and middle semiconfined aquifers. This delineation is an approximate one and is based on available data. The lateral boundaries were delineated based on the lithology of recently drilled lateral boundary definition boreholes, the airborne EM survey, and TEM sounding data. The area considered for this model study is 170 km². The study area was divided into square grids and the map is shown in Figure 13.4.

13.4.2.1 Initial Condition and Boundary Conditions of the Model

The following boundary conditions were used in the present model.

The northwestern boundary of the valley, close to the vicinity of a perennial swamp in the top aquifer and middle aquifer were taken as the subsurface inflow boundary. The quantum of inflow flux was calculated

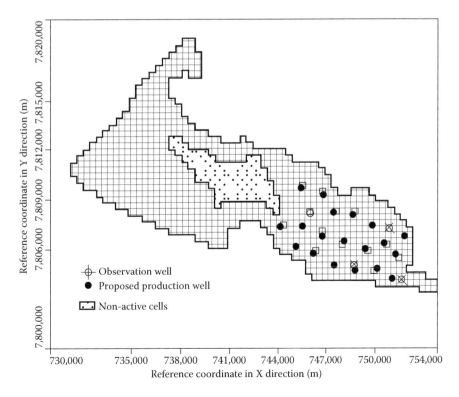

FIGURE 13.4
Grid map of the modeled area.

using transmissivity values and the hydraulic gradient. It was estimated that approximately 80 m³/d is received as inflow in the top aquifer and about 240 m³/day is received in the middle aquifer. Layers 3 and 5 (silty sands and clays) were taken as aquitards. Since the bottom aquifer (saline unit below green clay) is laterally extended, the northwest and southeast boundaries were assumed as inflow and outflow boundaries, respectively. The eastern and western lateral boundaries were treated as no flow boundaries as the flow is predominantly from northwest to southeast.

The subsurface outflow toward the southeastern direction was simulated as fixed heads at the junction point of River Boro and Kunyere Fault. The water levels monitored in the presently drilled bore wells are very useful in fixing the boundary heads in all three layers.

The aquifer parameter's hydraulic conductivity (K (m/day)), specific yield (Sy), and specific storage (Ss (L-1)) were assigned zonewise for each layer. Specific yield values for the top unconfined and middle semiconfined aquifers were established through model calibration. A specific yield of 0.015 was uniformly assumed for the middle semiconfined aquifer. The vertical permeability for each layer is assigned as one-tenth of horizontal conductivity. The upper semiconfining unit and lower confining unit were assumed to have storativity values of 0.0021 and 0.0017, respectively. The hydraulic conductivity value of 0.0055 m/day was set for both the upper confining and lower confining units.

The water levels monitored during March 1997 were assumed to be the initial condition of the steady-state model.

13.4.3 Model Calibration

13.4.3.1 Steady-State Calibration

The aquifer condition of March 1997 was assumed to be the initial condition for the steady-state model

calibration. The model could not be initialized to an early date due to the nonavailability of water level data before January 1997. Minimizing the difference between the computed and the field water level for each observation point started the steady-state model calibration. A number of trial runs were made by varying the input/output stresses and the hydraulic conductivity values of the top and middle aquifers in order to keep the root mean square (RMS) error below 0.2 m and mean error below 0.1 m. The computed versus observed head for selected observation points is shown in Figure 13.5.

This figure indicates that there is a fairly good agreement between the calculated and observed water levels. The calibrated zonal hydraulic conductivity (K) values for the top and middle aquifers and upper aquitard are shown in Table 13.1.

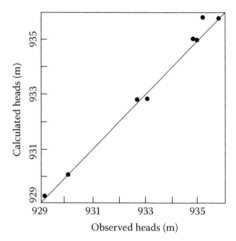

Mean error = 0.209
Mean absolute error = 0.116
Root mean square error = 0.267

FIGURE 13.5
Scatter plot for steady-state model.

TABLE 13.1

Model Parameters Derived from Steady-State Model Calibration

Layer	Unit Description	Average Thickness (m)	Field Derived Horizontal Hydraulic Conductivity (K, m/d)	Model Calibrated Horizontal Hydraulic Conductivity (Kₓ, m/d)	Storativity (Specific Yield/Aquifer Thickness)
1	Surficial low permeability zone	13.5	NA	0.0055	0.0024
2	Shallow semiconfined aquifer	6	NA	2	0.0024
3	Middle semiconfining unit	7.5	NA	0.0055	0.00, 215
4	Lower semiconfined aquifer	30	1 to 2 m/d in northwestern sector 4 m/d in southeastern sector	5 northwestern sector 4 southeastern sector 0.005 at Kunyere fault 5 southeast of Kunyere fault	0.0011 (Field derived) 0.0024 northwestern sector 0.00066 southeastern sector
5	Lower semiconfining unit	9.5	NA	0.0055	0.0017
6	Bottom brackish/saline unit	40 (assumed)	NA	2	0.001

13.5 Prediction Scenario Runs

Four preliminary model prediction scenarios were run utilizing simulated production bore wells completed in the lower semiconfined aquifer. The area was divided into two sectors, northwestern and southeastern for modeling purpose. Scenario 1 involved well fields in both the northwestern and southeastern parts of the area. Location maps of the proposed boreholes are shown in Figure 13.3. In Scenario 2, 3, and 4, the pumping boreholes were all placed in the southeastern part of this exploration closer to the Kunyere Fault. Scenarios 1 and 2 were run without any recharge from river infiltration and as such are worst-case scenarios. Scenarios 3 and 4 were run with 3 months per year of recharge from river flooding with the remaining 9 months being dry. The flood event was modeled as emanating solely from the River Boro. The width of the flood zone was set at 500 m and the depth of floodwaters at 0.5 m. The unsaturated vertical permeability of the riverbed was taken as 0.15 m/d. This value is based on the falling head permeability test. The river infiltration was calculated based on the following equation:

$$Q = C \times (h1 - h2)$$

$$C = k \times l \times w / m$$

where
  K = vertical permeability of river bed
  L = length of river bed
  W = width of river bed
  H1 = head in river bed
  H2 = head in aquifer

13.5.1 Prediction Scenario 1

In this scenario, 20 simulated production boreholes were placed in the 170 km² area (3 km spacing). The boreholes were split between the larger northwest sector and smaller southeastern sector. Each production borehole was pumping 200 m³/day from the lower semiconfined aquifer. The downward feed from the top aquifer and top aquitard and upward feed from the bottom aquifer is given in Table 13.2. Figures 13.6 and 13.7 are predicted water level contour maps for the lower semiconfined aquifer at the end of year 1 and 10 of simulation. Table 13.2 provides a summary of the computed inflows and outflows. The table indicates a decreasing trend in upward leakage from the underlying brackish system.

13.5.2 Prediction Scenario 2

This scenario likewise involves pumping from 20 simulated production boreholes in the lower semiconfined aquifer, but in the smaller 45 km² southeastern sector of the model area. The boreholes were spaced about 1.5 km apart with each borehole pumping at a rate of 200 m³/day for a total withdrawal of 4000 m³/day. Predicted water level contour map after one year of pumping is shown in Figure 13.8. The modeling indicates that the upper semiconfined aquifer begins to become unconfined after year 8 of the simulation. Table 13.3 provides a summary of the computed inflow/outflow quantities from the lower semiconfined aquifer system. Most of the input to the pumped aquifer is from downward leakage and subsurface inflow. Upward leakage from the bottom brackish aquifer system is minimal.

13.5.3 Prediction Scenario 3

This scenario was run with the same borehole configuration and pumping abstraction as Scenario 2. In addition, a very conservative estimate of 3 months of river infiltration was simulated each year. The average width of the river flow was taken as 800 m on the basis of aerial photograph interpretation. The stream bed thickness was considered to be 3 m with a water column of 0.5 m on the top of the

TABLE 13.2

Input/Output Quantities for Model Prediction Scenario 1 (no River Recharge)

Year	Pumpage (m³/d)	Subsurface Outflow (m³/d)	Upward Leakage from the Pumping Aquifer (m³/d)	Total Output from the Pumping Aquifer (m³/d)	Downward Leakage to the Pumping Aquifer (m³/d)	Upward Leakage to the Pumping Aquifer (m³/d)	Input from Storage (m³/d)	Inflow from the North-western Boundary to the Pumping Aquifer (m³/d)	Total Input to the Pumping Aquifer (m³/d)
1	4000	236	112	4348	2379	604	218	1147	4348
2	4000	300	160	4460	2448	522	180	1310	4460
4	4000	303	216	4519	2311	511	162	1535	4519
8	4000	260	293	4553	2055	507	139	1852	4553
10	4000	208	322	4530	1927	500	129	1974	4530

Total number of bore holes = 20 with 3 km spacing and average pumping rate 200 m³/d.

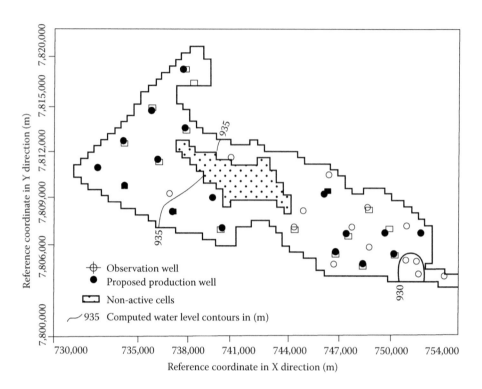

FIGURE 13.6
Predicted water level contours in lower semiconfined aquifer for January 1998 (prediction scenario 1).

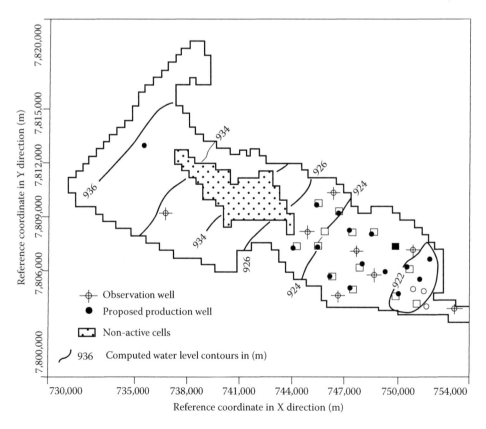

FIGURE 13.7
Predicted water level contours in lower semiconfined aquifer for January 2007 (prediction scenario 1).

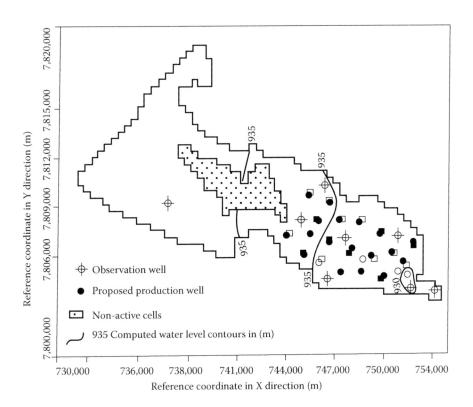

FIGURE 13.8
Predicted water level contours in lower semiconfined aquifer for January 1998 (prediction scenario 2).

TABLE 13.3

Input/Output Quantities for Model Prediction Scenario 2 (with no River Recharge)

Year	Pumpage (m³/d)	Subsurface Outflow (m³/d)	Upward Leakage from the Pumping Aquifer (m³/d)	Total Output from the Pumping Aquifer (m³/d)	Downward Leakage to the Pumping Aquifer (m³/d)	Upward Leakage to the Pumping Aquifer (m³/d)	Input from Storage (m³/d)	Inflow from the North-western Boundary to the Pumping Aquifer (m³/d)	Total Input to the Pumping Aquifer (m³/d)
1	4000	200	97	4297	2954	752	343	248	4297
2	4000	235	119	4354	3135	660	296	263	4354
4	4000	188	−138	4050	3106	361	270	313	4050
8	4000	84	−284	3968	2804	312	210	473	3800
10	4000	62	−284	3778	2665	303	200	610	3778

(−) represents change in direction of leakage to the pumping aquifer.
Total number of bore holes = 20 with 1.5 km spacing and average pumping rate 200 m³/d in southeastern sector only.

stream bed. Predicted water level contours at the end of Year 1 simulation is shown in Figure 13.9.

Table 13.4 provides a summary of computed inflows/outflows from the pumping aquifer. The input of river infiltration has resulted in increased downward leakage and less upward leakage from the underlying brackish aquifer system.

13.5.4 Prediction Scenario 4

Scenario 4 involved doubling the pumping rate of each simulated production borehole (400 m³/day). All other parameters were kept as in Scenario 3. The predicted water level contours for the lower semiconfined aquifer system are shown in Figures 13.10 and 13.11 after year 1 and 10 years of pumping. Table 13.5 summarizes computed inflows and outflows from the pumping aquifer. The increased abstraction during the simulation resulted in increased river infiltration and downward leakage. Input to the pumped aquifer from upward leakage and lateral flow did not change markedly from Scenario 3. Overall, the withdrawal of 8000 m³/day appears possible from this 45-km² area with a conservative estimate input for river recharge.

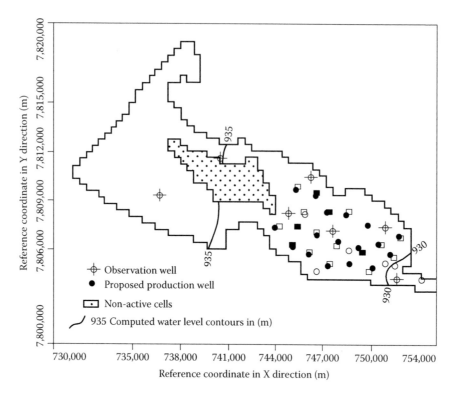

FIGURE 13.9

Predicted water level contours in lower semiconfined aquifer for January 1998 (prediction scenario 3).

TABLE 13.4

Input/Output Quantities for Model Prediction Scenario 3 (with River Recharge)

Year	Pump-age (m³/d)	Subsur-face Outflow (m³/d)	Upward Leakage from the Pump-ing Aquifer (m³/d)	Total Output from the Pump-ing Aquifer (m³/d)	River Infil-tra-tion (m³/d)	Down-ward Leakage to the Pump-ing Aquifer (m³/d)	Upward Leakage to the Pump-ing Aquifer (m³/d)	Input from Stor-age (m³/d)	Inflow from the Northwest-ern Bound-ary to the Pumping Aquifer (m³/d)	Total Input to the Pumping Aquifer (m³/d)
During first year river flow (3 months)	4000	509	3646	8155	3665	3818	294	174	204	8155
End of first year	4000	466	−49	4417	0	3732	273	245	167	4417
During second year river flow (3 months)	4000	629	3937	8566	3930	4280	129	90	137	8566
End of second year	4000	557	−48	4509	0	3930	206	204	119	4509
During fourth year river flow (3 months)	4000	678	4620	9298	4568	4548	54	37	91	9298
End of fourth year	4000	601	−46	4555	0	4157	151	165	82	4555
During eighth year river flow (3 months)	4000	714	5680	10,394	5468	4799	20	17	90	10,394
End of eighth year	4000	614	−33	4581	0	4267	99	128	87	4581
During tenth year river flow (3 months)	4000	662	6055	10,717	5764	4840	0	10	103	10,717
End of tenth year	4000	578	−8	4570	0	4311	55	104	100	4570

(−) represents change in direction of leakage.

Total number of bore holes = 20 with 1.5 km spacing and average pumping rate 200 m³/d (southeastern sector only).

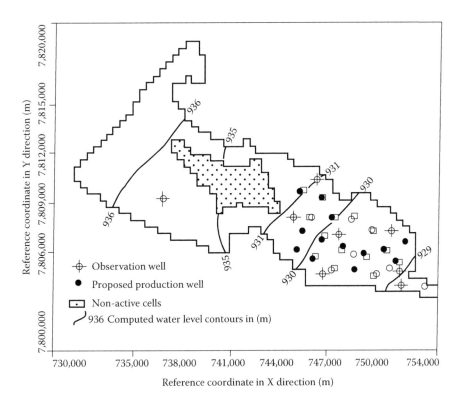

FIGURE 13.10
Predicted water level contours in lower semiconfined aquifer for January 1998 (prediction scenario 4).

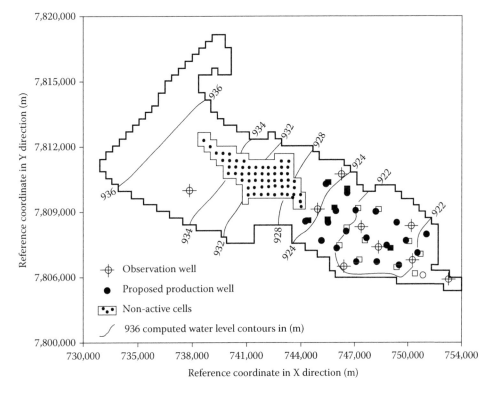

FIGURE 13.11
Predicted water level contours in lower semiconfined aquifer for January 2007 (prediction scenario 4).

TABLE 13.5

Input/Output Quantities for Model Prediction Scenario 4 (with River Recharge)

Year	Pump-age (m³/d)	Subsur-face Outflow (m³/d)	Upward Leakage from the Pump-ing Aquifer (m³/d)	Total Output from the Pump-ing Aquifer (m³/d)	River Infil-tration (m³/d)	Down-ward Leakage to the Pump-ing Aquifer (m³/d)	Upward Leakage to the Pump-ing Aquifer (m³/d)	Input from Stor-age (m³/d)	Inflow from the Northwest-ern Bound-ary to the Pumping Aquifer (m³/d)	Total Input to the Pump-ing Aquifer (m³/d)
During first year river flow (3 months)	8000	483	3782	12,265	4258	6575	613	613	206	12,265
End of first year	8000	422	−273	8149	0	6634	686	657	172	8149
During second year river flow (3 months)	8000	573	4598	13,171	5127	7344	361	188	151	13,171
End of second year	8000	498	−588	7910	0	6953	466	354	137	7910
During fourth year river flow (3 months)	8000	589	6194	14,783	6650	7737	190	66	140	14,783
End of fourth year	8000	506	−540	7966	0	7183	349	297	137	7966
During eighth year river flow (3 months)	8000	569	8557	17,126	8814	8031	50	14	217	17,126
End of eighth year	8000	481	−462	8019	0	7415	185	201	218	8019
During tenth year river flow (3 months)	8000	504	9383	17,887	9548	8040	26	5	268	17,887
End of tenth year	8000	412	−429	7983	0	7444	117	155	267	7983

(−) represents change in direction of leakage.

Total number of bore holes = 20 with 1.5 km spacing and average pumping rate 400 m³/d (southeastern sector only).

13.6 Uncertainties in the Model

Due to limited data and/or data gaps, some assumptions and estimates were made during the conceptualization of the Boro River valley system. Any error associated with these assumptions will be reflected in the model calibration. The following uncertainties are inherent in the model and should be verified through additional field investigations:

- The demarcation of lateral boundary was based on the transient electromagnetic data and lithologs of a few boreholes. This database could be improved by drilling a few more boundary definition boreholes.

- The vertical permeability of upper and lower aquitards (2×10^{-4} m/day) arrived through pumping test and model calibration needs further field verification.

- The transient calibration for the middle aquifer could not be carried out for want of time-dependent water level data.

- The assumption of inflow and outflow across the boundaries in the lower aquifer has to be verified through field experiments.

13.7 Conclusions

The preliminary modeling supports the future consideration of this area for development. It appears that the area can support a considerable amount of freshwater withdrawal. In all of the model cases, upconing of higher TDS water does not appear to be a concern. The preliminary modeling study indicates that substantial development of potential even in the smaller 45 km² southeastern modeled sector is possible. The simulation run with modest estimates of river-derived recharge to the lower semiconfined aquifer system indicates the development quantities on the order of 5000–10,000 m³/day are possible. The present modeling is very preliminary and subject to data density constraints. Prior to any development plans for the area, a more intense investigation/characterization program is required followed by much more intensive predictive modeling.

Acknowledgments

The author wishes to thank the Department of Water Affairs, Government of Botswana for permitting publication of this paper. The following hydrologists of Eastend Invest Ltd., Gaborone, Botswana; M. Masie, T. Rana, Vincent Uhl, and T.B. Bakaya are thanked for their help in providing the hydrogeological data for the present model study as well to G.G. Gabaakae of Department of Water Affairs Government of Botswana, Gaborone, Botswana for useful discussion during model calibration. My special thanks to Dr. H.K. Gupta, director, NGRI, Hyderabad at the time of this study and T.B. Bakaya for arranging my visit to Botswana as modeling expert and carry out this study.

References

Guiguer, N. and T. Franz. 1996. *Visual Modflow*, Waterloo Hydrogeologic Software, Waterloo, Ontario, Canada.

Mcdonald, M.G. and A.W. Harbaugh. 1988. A modular three-dimensional finite difference groundwater flow model, USGS Technical Report on MODELING Techniques, Book 6.

Rushton, K.R. and S.C. Redshaw. 1979. *Seepage and Groundwater Flow*, John Wiley & Sons Ltd., London, pp. 1–330.

Shashe Well Field Conceptual Model, Interim Report (Appendix G). 1996. Eastend Investments Pvt. Ltd., Gaborone, Botswana.

Thangarajan, M., Masie, M., Rana, T., Vincent, U., Bakaya, T.B., and G.G. Gabaakae. 2000. Simulation of multilayer aquifer system to evolve optimal management schemes: A case study in Shashe river valley, Okavango Delta, Botswana, *Journal of Geological Society of India*, 55, 623–648.

14

Tackling Heterogeneity in Groundwater Numerical Modeling: A Comparison of Linear and Inverse Geostatistical Approaches— Example of a Volcanic Aquifer in the East African Rift

Moumtaz Razack

CONTENTS

14.1 Introduction

Groundwater represents the major, even sometimes unique source of freshwater in many parts of the world. A recent UN report (UN World Water Development Report, 2012) stated that groundwater resources accounts for nearly half of all drinking water worldwide. Moreover, due to climate changes effects and the reduced availability of surface water, people are increasingly turning to groundwater for agricultural, industrial, and drinking water needs. However, due to these increasing pressures and demands linked to population growth and economic development, groundwater is not being used sustainably, according to needs and demands. In many countries, to respond to growing demands, groundwater is overexploited, leading to critically low levels in aquifers. Proper and sustainable management of aquifers is a key environmental challenge of this century. Sustainable management of groundwater involves decision making, to achieve goals, under certain constraints. These decisions require prior comprehensive knowledge of the systems to be able to anticipate their responses to planned actions. In the real world, knowledge and understanding of hydrogeological systems are not obvious to the extent that these systems are often very complex and heterogeneous. Water resources managers are faced with disparate and fragmented information about the underlying geology, water balance, climate change impact, etc. of such complex systems, which results in risky decision making processes.

The tool to understanding the system and its behavior and to predict its response is modeling. Efficient numerical modeling may constitute a valuable mean to tackle and solve this issue. An important and unavoidable issue in modeling consists of taking into account in numerical models meant to represent real systems, the natural heterogeneity of these underground systems. Even aquifer systems that are characterized by relatively simple geology can still contain a high degree of natural heterogeneity.

In recent decades, much research has been devoted to this theme. One can find comprehensive scientific reviews on these methods in the literature (de Marsily et al., 1998, 2005; Dagan, 2002). These approaches can be grouped into three broad categories: geostatistical (or stochastic) models; Boolean models (Cacas et al., 1990; Haldorsen and Chang, 1986; Long and Billaux, 1987);

and genetic models (Teles, 2004). The last two categories (Boolean and genetic models) are more oriented toward reservoir engineering and have been developed in the context of the oil industry. Their applications in hydrogeology are rather confidential. These models are not yet within reach of practitioner hydrogeologists. Advantages and issues related to these methods are widely analyzed by de Marsily et al. (2005). On the other hand, the category of geostatistical models has considerably expanded in hydrogeology. However, a rather paradoxical point was raised recently by Renard (2007) concerning the ignorance and rather reduced use of geostatistical modeling in operational hydrogeology. Yet practical tools currently exist to effectively implement some of the geostatistical methods in hydrogeology. In this chapter, we focus on two geostatistical methods, for developing forward modeling (kriging) and inverse modeling (pilot points). The modeling is limited to the resolution of the flow equation. The transport problem is not covered here.

The example selected to illustrate this theme is the Upper Awash volcanic aquifer system, located in Ethiopia in the East African Rift region. This aquifer system has a very complex structure and is highly heterogeneous, due to the fracturing and weathering of the basalts and intercalation of scorious layers. However, it is of vital importance for the country. Understanding the functioning of this volcanic aquifer system has become a vital necessity for the sustainability of water resources.

14.2 The Forward and Inverse Approaches in Hydrogeology

The groundwater 2D flow equation, in transient state and for heterogeneous media, combining mass conservation and Darcy's law, can be expressed as

$$\frac{\partial}{\partial x}\left(K_x \frac{\partial H}{\partial x}\right) + \frac{\partial}{\partial y}\left(K_y \frac{\partial H}{\partial y}\right) = S_s \frac{\partial H}{\partial t} + W \quad (14.1)$$

where K_x, K_y is the hydraulic conductivity (LT^{-1}); S_s is the specific storage (L^{-1}); H is the hydraulic head (L), and W is the sink/source term (inflow/outflow to/from the system per unit volume of system per unit of time T^{-1}). When the medium is isotropic, then $K_x = K_y$. For steady state flow, $\delta H/\delta t = 0$, and Equation 14.1 is written as

$$\frac{\partial}{\partial x}\left(K_x \frac{\partial H}{\partial x}\right) + \frac{\partial}{\partial y}\left(K_y \frac{\partial H}{\partial y}\right) = W \quad (14.2)$$

Solving the groundwater flow equation in steady state requires knowledge of the hydraulic parameters (K_x, K_y) spatial distribution, of the boundary conditions and other components as recharge and sink/source terms. When such knowledge is available, the hydraulic head H is therefore the main variable to solve for the model and the approach is termed a forward model.

However, as very often in practice, knowledge of hydraulic parameters is very limited and in any case insufficient to grasp the spatial heterogeneity of the underground environment, the forward approach cannot be implemented. We then proceed by model calibration. One seeks to match the calculated H values by the model to the observed H values by adjusting hydraulic parameters (hydraulic conductivity K or transmissivity T in steady flow) and other aspects of the model, which are considered then unknown. This process is called model calibration and can be performed through nonautomated trial and error approach or automated inverse approach. In essence, both calibration processes are the same. However, when dealing with highly heterogeneous media, the nonautomated trial and error approach turns very quickly quite tedious, time-consuming, and approximate.

Inverse modeling has the ability to automatically determine hydraulic parameter values that give the best match between observed and simulated hydraulic heads. A number of reviews have been published in the literature, pointing out the multiple benefits of inverse models in hydrogeology, but also discussing the issues linked with the use (or misuse) of these models (Carrera, 1987; Carrera et al., 2005; de Marsily et al., 2000; McLaughlin and Townley, 1996; Poeter and Hill, 1997; Yeh, 1986; Zhou et al., 2014). By using this approach, one must be aware of a major issue pertaining to the nonuniqueness of the solution (i.e., the parameter field) derived from inverse modeling. How this problem is solved with the pilot points method used in this work is outlined next.

14.3 Geostatistics-Kriging

Geostatistics comes from the mining industry (Matheron, 1963, 1965). Its early applications in hydrogeology date back to the 1970s (Delhomme, 1978, 1979). Kriging is one of the geostatistical methods to carry out the estimation of a spatial variable on sound bases. The primary interest of geostatistics is the fact that this discipline, before performing an estimation of the concerned parameter, first characterizes the spatial variability of this parameter. This property is fundamental to characterize heterogeneous aquifers. Matheron (1963) had called "regionalized variable" any variable distributed in space, in a structured way. Analysis of the spatial structure is achieved by calculating the variogram, which is defined by the following equation for a given regionalized variable Z:

$$\gamma(h) = \frac{1}{2} E[(Z(x+h) - Z(x))^2] \qquad (14.3)$$

where $h = x_i - x_j$; E is the mathematical expectation. The variogram is the basic tool for geostatistical analysis. The assumptions underlying the definition of the variogram are detailed in basic textbooks on geostatistics (Isaak and Srivastava, 1989; Kitanidis, 2000). On this basis, geostatistics provides the best estimation of the variable at unsampled points. Contrary to other estimation techniques, geostatistics takes into account the observable spatial correlation between sample points to predict the variable at unsampled points. The geostatistical estimator is a BLUE (Best Linear Unbiased Estimator) as it is characterized by the two following properties: (1) the average of the errors of estimate is zero, and (2) the variance of the errors of estimate is a minimum (see Equations 14.5 and 14.6).

Kriging (Universal Kriging) is a linear geostatistical estimation method, which enables the estimation of a regionalized variable (Z) at any point in space, based on its measured values at other locations. A few noteworthy features regarding this interpolation method are highlighted hereunder. A detailed presentation can be found in basic geostatistics references (Isaaks and Srivastava, 1989; Journel and Huijbregts, 1978; Kitanidis, 2000).

Let us consider a regionalized variable Z which variogram $\gamma(h)$ is defined by the expression 14.3. The variogram $\gamma(h)$, as stated above, shows the spatial variability of the variable. The kriging estimator Z_0^* at a point x_0 is written:

$$Z^*(x_0) = Z_0^* = \sum_{i=1}^{N} \lambda_i Z(x_i) \qquad (14.4)$$

Z_0^* is a linear estimator, N is the number of values involved in the estimation, and λ_i are weights. The kriging estimator satisfies two major conditions, called the unbiasedness and optimality conditions. These conditions stipulate that the errors of estimate (differences between the true values and the estimated values) should have a mean equal to zero and a minimum variance. They are written, respectively:

$$E(Z^* - Z_0) = 0 \qquad (14.5)$$

$$Var(Z^* - Z_0) \text{ minimum} \qquad (14.6)$$

Developing Equation 14.4 to satisfy these two conditions leads to a system of kriging equations written in terms of the variogram. Solving this system yields the N weights λ_i to be used in Equation 14.4 and the kriged estimate Z_0^* can be calculated. The variance of the error of estimate (kriging variance) is given as:

$$\sigma^2(x_0) = Var(Z^* - Z_0) = \sum_{i=1}^{N} \lambda_i \gamma(x_i - x_0) + \mu_0 \qquad (14.7)$$

where μ_0 is the Lagrange multiplier (Isaak and Srivastava, 1989). One should note that kriging takes into account: (i) the spatial positions of the point to be estimated and the known points; and (ii) the spatial variability of the variable through the variogram. The variance of the errors of estimate is an indicator of the confidence to be granted to the kriging estimates. If the variance is lower, then the confidence level is higher.

In this work, we applied the variogram analysis to characterize the spatial variability of the hydrogeologic parameters over the study area. Then kriging technique was used to estimate the parameter values over the study area. The estimated kriged values were then used as an input to the digital model to carry out a forward modeling of the site.

14.4 Pilot Point Principles

The pilot point method is an inverse modeling approach, back-calculating suitable parameter values through the comparison of measured and simulated observations. In most cases, the adjusted parameter is the horizontal hydraulic conductivity (or the transmissivity). But any parameter playing a role in the domain hydrodynamics (river conductance, vertical hydraulic conductivity, recharge, fluxes, etc.) can be adjusted in inverse modeling. The observations are usually the groundwater hydraulic heads (piezometric data). The principles and mathematical background related to the pilot points method in the context of groundwater model calibration can be found in Doherty (2003) and Doherty et al. (2010).

The following description is intended as a brief reminder of the major concepts, free of any mathematical difficulty, so that the method can be understood by a hydrogeologist without extensive mathematical culture and especially that the method is correctly applied to practical modeling cases of heterogeneous aquifers. The pilot points method was proposed first by de Marsily (1978), who introduced this method in order to minimize uncertainty with enhanced heterogeneity recognition. The pilot point method is a numerical models parameterization method that actually constitutes an interesting alternative to the traditional zone wise parameterization method. Pilot points can be seen as a 2D scatter point set. Instead of creating a zone and having the inverse model estimate one value for the entire zone, the value of the parameter is interpolated from the pilot points. The pilot point values are considered as unknowns and are adjusted to make the model fit available head observations. This calibration is done using an inverse procedure. The inverse model optimizes the parameter values at the pilot points. The optimized parameter values are then interpolated from the pilot points to the grid cells using kriging.

However, this method raises some numerical, practical, and plausibility issues. The inverse problem is qualified as well-posed when a unique solution exists and as ill-posed when there are multiple solutions, the latter always being the case when the number of parameters exceeds the number of observations. This means that when the number of unknowns, that is, parameters, is not kept smaller than the number of head observations comprising the calibration dataset, the resulting inverse problem is ill-posed and can lead to numerical instability of the inversion algorithm (McLaughlin and Townley, 1996). At the same time, one is faced with the problem of locating the pilot points.

The use of strategies meant to transform an ill-posed problem into a well-posed problem is generally known as regularization. Regularization is a way of alleviating the ill-posedness of the inverse problem through incorporation of prior information into the objective function (Christensen and Doherty, 2008; Doherty, 2003). The mathematical background of this procedure is detailed in Doherty (2003), Doherty and Hunt (2010), Doherty et al. (2010), and Tonkin and Doherty (2005). With regularization, the number of parameters can exceed the number of observations. The pilot point approach and regularization can be used in conjunction with the well known parameter estimation software PEST (Doherty, 2004). As a result, complex hydraulic conductivity distributions can be defined, resulting in extremely low residual error.

A key concern of the pilot point approach is how to determine the number of pilot points and their locations. A number of strategies have been proposed in the literature. Gomez-Hernandez et al. (1997) recommended two or three pilot points per correlation length. LaVenue and Pickens (1992) placed the pilot points in the highest sensitivity regions. Wen et al. (2002) proposed to randomly locate the pilot points such that the spacing between the pilot points is one correlation length.

In this work, a simple strategy was applied to distribute the pilot points uniformly throughout the flow domain. To take account of the geostatistical properties of the hydraulic conductivity field and as its spatial structure is known through the variogram, several pilot points distributions were tested (<1, 2, and 3 pilot points per correlation length). Modeling was performed using the MODFLOW and PEST software, included in the package Groundwater Vistas (ESI, 2011).

14.5 Study Area and Available Data

The study area is located in the East African Rift region, which is found in a particular geodynamic context, related to the separation of African and Arabian plates since about 30 million years ago. The East African Rift is an active continental rift zone. It includes the Main Ethiopian Rift, which continues south as the Kenyan Rift Valley. Therefore, the volcanic formations resulting from plate tectonics outcrop over a major part of the territory. The volcanic aquifer system, which is the focus of this chapter, is located in Ethiopia (Upper Awash aquifer).

The Upper Awash volcanic system is situated in central Ethiopia at the western margin of the Main Ethiopian Rift (MER). Addis Ababa, the capital city, is located in the northern part of the basin (Figure 14.1). The Upper Awash volcanic aquifer system is one of the most productive systems in Ethiopia providing water for agriculture, industry, and both for the rural and urban communities including the capital city Addis Ababa. The major productive units are located within the volcanic rock sequences, particularly basalts and associated volcano clastic and alluvial deposits. The catchment area is 10,841 km^2. The Upper Awash Basin has three main physiographic zones, namely the plateau starting from the In toto mountain range in the north, the escarpment in the middle, and the rift, extending up to Lake Koka in the south.

The formation of the Ethiopian Rift during the Miocene period separated the eastern and western highlands. The Upper Awash Basin is exclusively located within the north-central plateau and the adjacent escarpment and rift. The central plateau is drained westward by the Blue Nile River drainage system and north-eastwards by the Awash River drainage system.

The aquifer properties in the basin are controlled by the litho-stratigraphy of the volcanic rocks and the structures that affect them. Three aquifer units have been differentiated in the area (Figure 14.2): (i) The Upper Basaltic Aquifer: This unit is composed of quaternary basalt flows, scoria and spatter cones, and tertiary-Neogene basalts. The upper basalt aquifer is overlain by ignimbrites and tuffs at places. (ii) The Lower Basaltic Aquifer. This deep unit is composed of lower tertiary basalts, dominantly scoraceous, and (iii) localized and regional low permeability sequences. Quaternary rhyolites, trachytes, and tertiary rhyolites have low permeability, except along weathered and fractured zones.

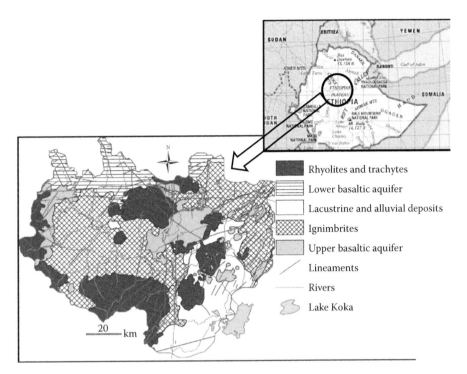

FIGURE 14.1
Location of the Upper Awash basin, East Africa, Ethiopia.

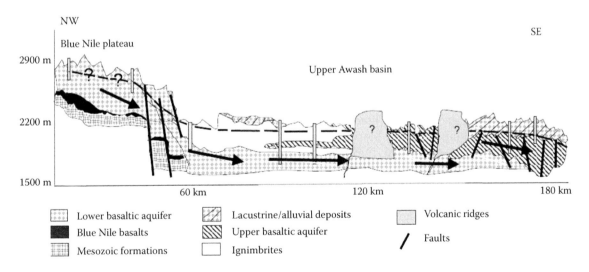

FIGURE 14.2
Conceptual NW-SE cross-section of the Upper Awash aquifer system.

14.6 Available Hydraulic Conductivity Data

Hydraulic properties (transmissivity, hydraulic conductivity) of the Upper Awash aquifers have been estimated through pumping tests which lasted between 24 and 72 hours. The transmissivity of the aquifers has been first calculated from the time-drawdown and recovery data of the constant rate tests. Hydraulic conductivity was then estimated from the wells with known screen length. A set of 81 values was thus available, distributed almost evenly over the study site (Figure 14.3). Few data points are outside the basin limits. But as the geological formations are similar, these data were kept for further analyses. Statistics of the K data set are given in Table 14.1.

The available hydraulic conductivity data varies between 1.6E–02 m/d and 3.09E+02 m/d, spanning

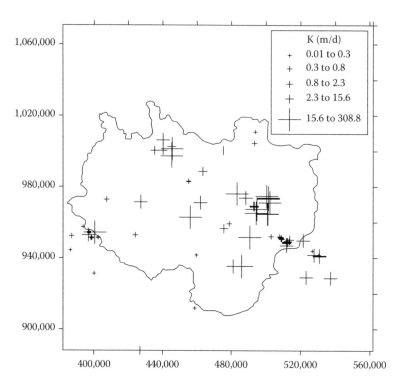

FIGURE 14.3
Distribution of observed hydraulic conductivity (K, m/d) data over the study area.

several orders of magnitude. The frequency distribution of the K data, plotted on a probability diagram, is shown in Figure 14.4.

The plot shows that K data can be fitted by a lognormal distribution. The log-normality of aquifers' hydraulic properties is well recognized in the literature, as well for porous aquifers (Aboufarissi and Marino, 1984; Ahmed and de Marsily, 1987; Delhomme, 1978; Razack and Huntley, 1991) as for fractured aquifers (Bracq and Delay, 1997; Huntley et al., 1992; Jalludin and Razack, 2004; Razack and Lasm, 2006). It is recommended in the geostatistical literature to perform kriging using normal data rather than data that have skewed distributions. A normal distribution improves the geostatistical estimation (Ahmed et al., 1988). A lognormal transformation is often used to this end. Accordingly, the variographic

analysis (i.e., calculation of the experimental variogram; adjustment of an analytical function to the experimental variogram) is performed using the logarithmic transform of the data. The variogram of log (K) is reported in Figure 14.5.

The experimental variogram shows no nugget effect. It increases from the origin and at large distances and it fluctuates around the sill, which indicates that log (K) is stationary. These characteristics indicate that the data display a well-defined spatial structure. The theoretical model fitted to the experimental log (K) variogram is an exponential model, whose general equation is

$$\gamma(h) = C_0 + C_1 \left[1 - \exp\left(-\frac{3h}{a} \right) \right] \qquad (14.8)$$

TABLE 14.1

Statistics of Various Hydraulic Conductivity K Data Sets

K data sets	Average	Minimum	Maximum	SD	CV	N
Observed K (m/d)	19.4	1.6 E–02	3.09E+02	52.9	2.7	81
Estimated K (m/d) by backtransform of krigedlogK	3.12	2.60E–02	2.82E+02	8.9	2.9	10,888
Simulated K field with <1 PP per CL	22.4	1.40E–02	9.75E+02	50.6	2.2	10,888
Simulated K field with 2 PP per CL	116.8	1.10E–03	6.02E+03	375.4	3.2	10,888
Simulated K field with 3 PP per CL	272.9	1.00E–03	1.00E+04	949.3	3.5	10,888

Note: SD: standard deviation; CV: coefficient of variation; PP: pilot point; CL: correlation length; N: sample size.

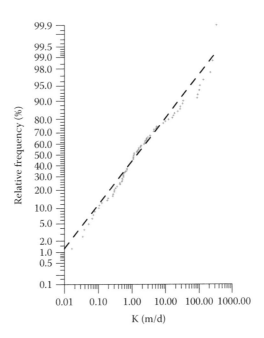

FIGURE 14.4
Frequency distribution of observed hydraulic conductivity values (K, m/d) on a probability diagram.

where C_0 is the nugget effect, $C = C_0 + C_1$ is the sill, a is the practical range (distance at which 95% of the sill has been reached), and h is the distance between sampling points. The nugget effect represents any small-scale data variability or possible sampling (measurement and/or location) errors. The sill indicates the total variance. The range is the distance between sampling points at which the sill is reached. Beyond the range, the variance measured between the data points is independent from the respective data points and there is no longer a correlation between the points. The best fit is given for the following parameter values: $C_0 = 0$; $C = 1.15$; $a = 21,600$ m. These parameters (nugget effect, sill, range) characterize the spatial

structure (or regionalization) of the logarithm of the hydraulic conductivity. The exponential variogram is thus written as:

$$\gamma(h) = 1.15\left[1 - \exp\left(-\frac{3h}{21600}\right)\right] \quad (14.9)$$

This variogram model shown in Figure 14.5 and the available log (K) sample data (81 values) were used to perform a kriged estimation of log (K) over the whole domain on a grid with a mesh size of 1 km × 1 km. The kriged log (K) values are used afterward to make an estimation of the hydraulic conductivity over the study area through a back transform of log (K) to K. Some

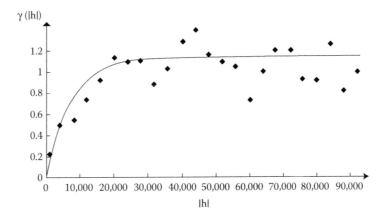

FIGURE 14.5
Experimental variogram of hydraulic conductivity and fitted exponential model.

noteworthy points should be emphasized at this stage. The logarithmic transform K to log (K) is a nonlinear transform. In this case, when the kriged unbiased estimates log (K) are back transformed, then their unbiasedness property is lost. The hydraulic conductivity values estimated using this procedure are no longer unbiased. The estimated hydraulic conductivity values are mapped in Figure 14.6a. This geostatistical procedure makes it possible to obtain an assessment of the hydraulic conductivity over the whole study domain.

The frequency distribution of the estimated K values, obtained using the above procedure, is given in Figure 14.7a. This graph shows that the estimated K values follow also a lognormal distribution. Their statistics are reported in Table 14.1.

Comparing the statistics of observed K values and those of the estimated K values shows that the average and the standard deviation of estimated values are smaller than those of the observed values. The coefficient of variation, however, is equivalent in the two cases. Note also that the range of estimated values of K (Min K = 2.60E–02; Max K = 2.82E+02 m/d) is roughly

equivalent to that of observed K (Min K = 1.60E-02 m/d; Max K = 3.09E+02 m/d).

14.7 Modeling the Upper Awash Volcanic System with the Use of Linear and Inverse Geostatistical Models

14.7.1 Conceptual Groundwater Model of the Upper Awash

The volcanic aquifer system of Upper Awash includes several volcanic sequences. Information on the exact location of these sequences with respect to each other is not sufficient to give a clear view of the multilayer structure of this system. Moreover, no isolation between the sequences was made in the wells before performing pumping tests. Therefore, in the present state of knowledge, it is unrealistic to consider a multilayer modeling of this system. Given the intense fracturing affecting the

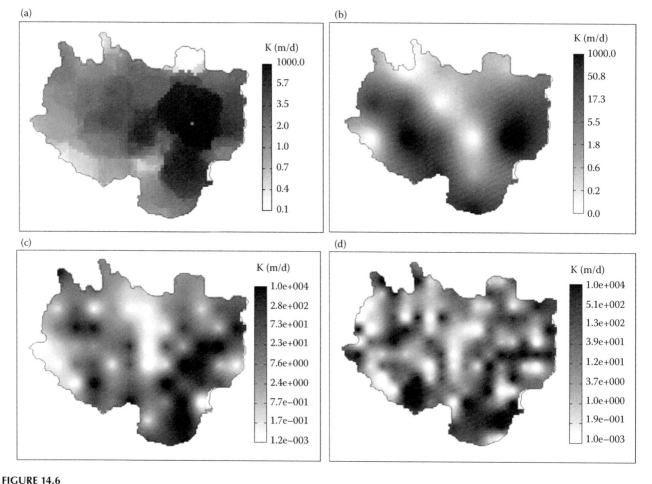

FIGURE 14.6
Hydraulic conductivity (K, m/d) maps obtained using various procedures. (a) Estimated K by backtran form of kriged logK. (b) Simulated K field with <1 PP per CL. (c) Simulated K field with 2 PP per CL. (d) Simulated K field with 3 PP per CL.

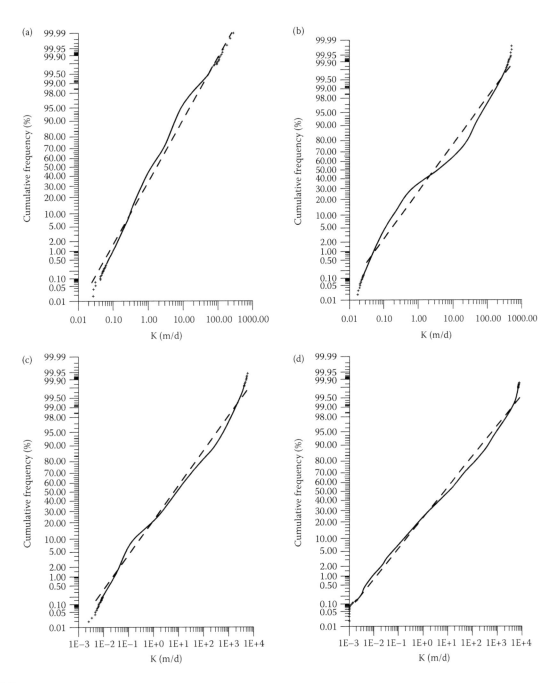

FIGURE 14.7
Frequency distributions of K data sets obtained using various procedures. (a) Estimated K by backtran form of kriged logK. (b) Simulated K field with <1 PP per CL. (c) Simulated K field with 2 PP per CL. (d) Simulated K field with 3 PP per CL.

volcanic sequences, there is a strong hydraulic interconnection between these different sequences. It follows that the hydraulic head is the same for all sequences. Moreover, a significant relationship is found between the hydraulic head and topography, regardless of the depth of the wells. Thus, one can plausibly assume that all volcanic sequences form a global unconfined single layer aquifer system. The numerical models were developed on this basis. The boundary conditions are illustrated in Figure 14.8.

The northern limit between the basins of Awash and Blue Nile has been modeled as an inflow boundary (Neumann boundary). Flows imposed between the two basins result from previous works (Yitbarek, 2009) and have not been modified or recalibrated. The east, west, and southwest limits are no-flow boundaries. These limits mainly include massive rhyolitic intrusions. The limit with Lake Koka was considered as a constant head boundary (Dirichlet boundary). A general head boundary was considered at either side of the lake. The

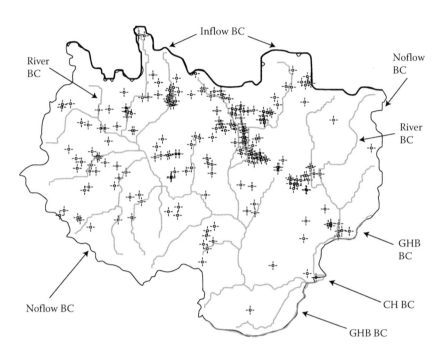

FIGURE 14.8
Boundary conditions of the Upper Awash numerical model and display of the hydraulic heads data used for model calibration.

exchanges between the rivers and the aquifer have been simulated using Cauchy type boundary (river boundary). An average value of 2.10^5 m²/d was affected to the conductance of all head dependant boundaries (river, GHB). The conductance was not calibrated, as the work is focused on hydraulic conductivity.

Groundwater recharge is a fundamental component in the water balance of any aquifer and a key component in any model of groundwater flow. Its accurate quantification is crucial to proper management and protection of groundwater resources. The groundwater recharge in the study area has been estimated by Yitbarek (2009) using groundwater table fluctuation (Healy and Cook, 2002) and base flow (Riser et al., 2005) methods. This outcome was introduced as such in the models developed here.

The numerical models were calibrated on the basis of 250 hydraulic head values almost evenly spread over the entire area of study (Figure 14.8). The accuracy of the calibration of the groundwater flow models was assessed using the following quantitative measures: Plot of simulated heads versus observed heads and computation of R^2 (determination coefficient), residual mean, absolute residual mean, RMS error, minimum residual, and maximum residual.

14.7.2 Forward Model of Upper Awash

The first model that was developed is a forward model in the sense that no model adjustment was sought. The conductivity field obtained by back transform of the kriged log (K) values was input in the numerical model and K values were not modified to improve the model. The calibration diagram is shown in Figure 14.9 and the residuals statistics in Table 14.2.

The calibration diagram indicates that the observed values of H are well simulated by the forward model. A determination coefficient of $R^2 = 0.972$ is obtained. The RMS error is equal to 38.3 m. The residuals statistics show that model calibration remains satisfactory. The residuals are distributed following a normal law, indicating the correctness of the calibration.

14.7.3 Geostatistical Inverse Models of Upper Awash Using the Pilot Points Approach

The three distributions of pilot points over the study domain, used in inverse models, that is, <1, 2, and 3 pilot points per correlation length, are plotted in Figure 14.10.

The parameter pace imposed in this inverse approach is 10^{-3} m/d $< K < 10^{+4}$ m/d. This parameter pace is larger than the range of K values deduced from pumping tests (Table 14.1). The limits of this parameter pace have been shifted for several reasons; (i) not to restrict the degree of heterogeneity of the volcanic system; and (ii) based on values published in the literature. Calibration diagrams of the three inverse models are plotted in Figure 14.9 and calibration statistics are given in Table 14.2. The conductivity fields resulting from these inverse approaches are plotted in Figures 14.6b, 14.6c, and 14.6d.

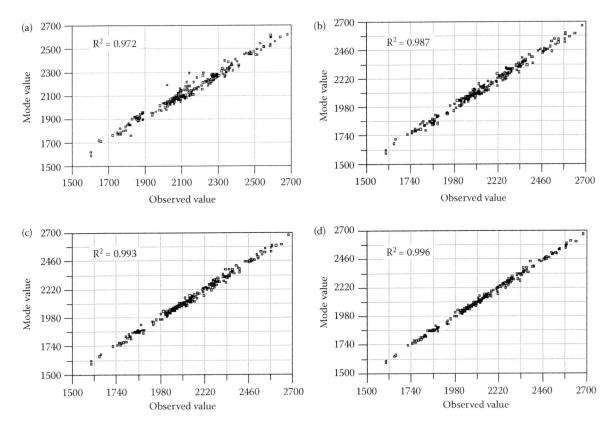

FIGURE 14.9
Simulated heads versus observed heads of the different numerical models of Upper Awash. (a) Forward model. (b) Inverse model with <1 PP per CL. (c) Inverse model with 2 PP per CL. (d) Inverse model with 3 PP per CL.

TABLE 14.2

Residuals Statistics of the Forward and Inverse Models of Upper Awash

	Residual Mean (m)	Residual SD (m)	Absolute Residual Mean (m)	Minimum Residual (m)	Maximum Residual (m)	RMS Error (m)
Forward model	−6	37.8	29	−172.7	71.7	38.3
Inverse model with <1 PP per CL	0.8	24.7	19.4	−76.8	64.4	24.7
Inverse model with 2 PP per CL	1.47	17.4	13.3	−53.9	56.6	17.4
Inverse model with 3 PP per CL	1.2	13.3	9.27	49.3	54.7	13.4

Note: PP: pilot point; CL: correlation length; SD: standard deviation.

14.7.4 Discussion

The four models of the Upper Awash aquifer system developed in this work produce a result with acceptable accuracy. Table 14.2 shows the calibration results of the four models. The following points are worth noting:

- The coefficient of determination of the relationship H simulated versus H observed of the forward model is R² = 0.972, those of the three inverse models are higher, respectively, R² = 0.987, R² = 0.993, and R² = 0.996. The coefficient of determination is quite significant in all four cases.

- The four models produce quite satisfactory calibration accuracy. However, the accuracy of inverse models remains greater than that of the forward model.

- The ARM (absolute residual mean) is 29 m for the forward model, and it reduces drastically to 9.2 m for the inverse model with 3 pilot points correlation length.

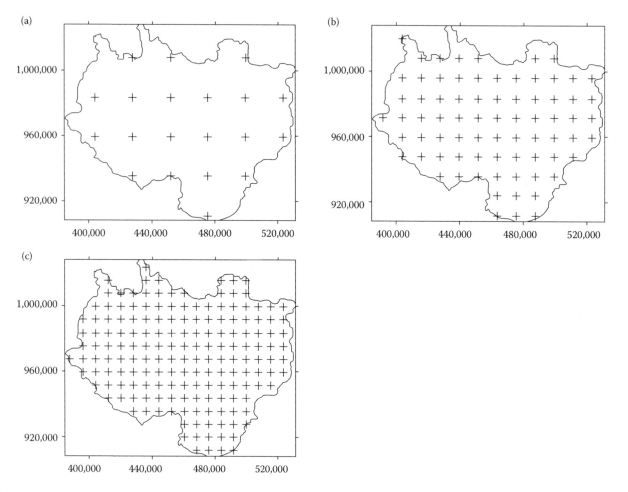

FIGURE 14.10
Distributions of pilot points (PP). (a) <1 PP per correlation length. (b) 2 PP per correlation length. (c) 3 PP per correlation length.

- With the forward model, the residuals range is quite large (–172.7 m; 71.7 m). This interval is reduced with the successive inverse models. With 3 PP/CL, the residuals range is quite small (49.3 m, 54.7 m).

- The RMS error greatly reduces between the forward model (RMSE = 38.3 m) and inverse models (RMSE equal, respectively, to 24.7 m, 17.4 m and 13.4 m).

At this stage, without considering the plausibility of the models, the results highlight very clearly the following:

- The forward approach provides a model whose accuracy is acceptable.
- The implementation of the pilot points substantially improves the accuracy of the models.
- The accuracy increases with the number of pilot points.

The plausibility of the models is discussed below by considering statistical properties of the K field. The visual patterns of the K fields (Figure 14.6) allow making the following comments:

- Between Figures 14.6a (forward model) and 14.6d (3 PP/CL), the structure of the K field is very different.
- Scale of the K heterogeneities is roughly equivalent for the forward model and the model built with <1 PP/CL, while this scale becomes much finer when increasing the number of PP. With 3 PP/CL, Figure 14.6d shows a highly heterogeneous system.

Figure 14.7 shows the distributions of K data sets in a probability plot. As noted above, the values of K estimated by back transform of kriged log (K) values follow a lognormal distribution (Figure 4.7a). Note that the K field simulated with <1 PP/CL slightly deviates from a lognormal distribution. On the other hand, K fields simulated with 2 and 3 PP/CL preserve their lognormal properties. This finding suggests that when implementing the pilot points method, the fact of using a small number of pilot points can lead to degrading the statistical distribution of the parameter.

The statistics of the various K fields (Table 14.1) show that:

- The dispersion of values around the arithmetic mean remains similar (CV of same magnitude in all four cases).
- When the number of PP increases, the arithmetic mean of K becomes higher, because probably the inverse approach uses a wider parameter space.
- The model with 3 PP/CL uses the entire parameter space that was assigned (10–3; 10 + 4 m/d). It seems likely that with a wider parameter space, the inverse model would have expanded the range of simulated values of K. In this case, the model would have been overparameterized and the plausibility of the model could be questioned.

14.8 Concluding Remarks

Considerable advances have been made in recent decades, from a scientific point of view, in order to develop numerical models more representative of actual aquifer systems and taking into account their degree of heterogeneity, which in some cases can be very high. It can be stated that the development of the geostatistics theory in the 1960s and its applications some years later in hydrogeology were a decisive breakthrough in the field of hydrogeological modeling. However, operational hydrogeology did not follow this remarkable theoretical advance. Yet the needs are real and urgent in order to efficiently model complex aquifer systems containing significant water resources, which require sustainable and rational management for their preservation as pressures and demands are increasingly higher.

Currently the tools are available to implement by practitioner some of the methods that were developed to model complex and heterogeneous aquifer systems. In this work, two readily available methods, namely, kriging and pilot points, were compared to model a highly complex aquifer system, the Upper Awash system in Ethiopia, East Africa. Kriging was used to elaborate a forward numerical model and pilot points for inverse models. The number of pilot points to be used in the inverse approach was also tested. To this end, <1, 2, and 3 pilot points per correlation length were used.

The results show that all four models (forward model, three inverse models with <1, 2, and 3 PP/CL) lead to an acceptable calibration and generate varying degrees of heterogeneity in the system. However, the quality of calibration by taking into account the degree of heterogeneity is significantly better with the inverse approach and with increased pilot points. The results of the three inverse models highlight the following points:

- The use of too few pilot points (<1 PP/CL) can alter the statistical properties of K field generated by this approach.
- There is a real risk to overparameterize the model when the number of pilot points is too high. The result reached here shows that with 3 PP/CL, the inverse approach uses the entire parameter space and thereby probably reached the limit of overcalibration of the model.
- The use of 2 PP/CL seems to be a good compromise.

Note that the inverse modeling was conducted without using the pumping test data as a subset of pilot points with fixed K values. In the continuation of this work, several forthcoming operations are planned:

1. To test the use of the available K values as a subset of PP with fixed values.
2. To perform sensitivity/uncertainty analysis of the models.
3. To move to transient models and transport models to assess the validity of the conductivity fields generated by these different approaches.
4. To improve the models by first assessing the conductance of the 3rd type boundary conditions (River, GHB) in the field and then using PEST for optimization. This will require further field woks.

In practice, quite often, no sufficient data of the parameters are available to analyze and define their structural properties. The work here has pointed out the importance of knowing the correlation length to implement the pilot points method. It may be interesting or even necessary for the community in hydrogeology to establish databases containing orders of magnitude of variables like the correlation length of hydrogeological parameters, according to the geological properties of the aquifers, to implement these relatively novel methods, such as the pilot points, on a sound basis.

References

Aboufirassi, A. and M.A. Marino. 1984. Cokriging of aquifer transmissivity from field measurements of transmissivity and specific capacity. *Mathematical Geology*, 16(1): 19–35.

Ahmed, S. and G. de Marsily. 1987. Comparison of geostatistical methods for estimating transmissivity using data on transmissivity and specific capacity. *Water Resource Research*, 23(9): 1717–1737.

Ahmed, S., de Marsily, G., and A. Talbot. 1988. Combined use of hydraulic and electrical properties of an aquifer in a geostatistical estimation of transmissivity. *Ground Water* 26(1): 78–86.

Bracq, P. and F. Delay. 1997. Transmissivity and morphological features in chalk aquifer: a geostatistical approach of their relationship. *Journal of Hydrology*, 191: 139–160.

Cacas, M.C., Ledoux, E., de Marsily, G. et al. 1990. Modelling fracture flow with a discrete fracture network: calibration and validation of the flow model. *Water Resource Research*, 26(1): 479–489.

Carrera, J. 1987. State of the art of the inverse problem applied to the flow and solute transport equations. In: Custodio, E., Gurgui, A., Ferreira, J. L. (Eds.), *Analytical and Numerical Groundwater Flow and Quality Modelling*. Series C, Mathematical and physical sciences. 224. D. Reidel, Norwell, MA, pp. 549–583.

Carrera, J., Alcolea, A., Medina, A., Hidalgo, J., and L.J. Slooten. 2005. Inverse problem in hydrogeology. *Hydrogeology Journal*, 13(1): 206–222.

Christensen, S. and J. Doherty. 2008. Predictive error dependencies when using pilotpoints and singular value decomposition in groundwater model calibration. *Advanced Water Resources*, 31(4): 674–700.

Dagan, G. 2002. An overview of stochastic modeling of groundwater flow and transport: from theory to applications. *EOS, Transactions of the American Geophysical Union*, 83(53): 31–12.

Delhomme, J.P. 1978. Kriging in hydrosciences. *Advanced Water Resources*, 1(5): 251–266.

Delhomme, J.P. 1979. Spatial variability and uncertainty in groundwater flow parameters: A geostatistical approach. *Water Resource Research*, 15(2): 269–280.

de Marsily, G. 1978. De l'identification des syst_meshydrologiques [On the calibration of hydrologic systems]. Doctoral thesis, University Paris VI.

de Marsily, G., Delhomme, J.P., Coudrain-Ribstein, A., and A.M. Lavenue. 2000. Four decades of inverse problems in hydrogeology. In: Zhang, D., Winter, C. L. (Eds.), Theory, modeling, and field investigations in hydrogeology: a special volume in honor of S.P. Neuman's 60th birthday. Geological Society of America Special Paper, Boulder, Colorado, 348, pp. 1–17.

de Marsily, G., Delay, F., Gonzalves, J., Renard, P., Teles, V., and S. Violette. 2005. Dealing with spatial heterogeneity. *Hydrogeology Journal*, 13: 161–183.

de Marsily, G., Schafmeister, M.T., Delay, F., and V. Teles. 1998. On some current methods to represent the heterogeneity of natural media in hydrogeology. *Hydrogeology Journal*, 6: 115–130.

Doherty, J. 2003. Ground water model calibration using pilot points and regularization. *Ground Water*, 2(41): 170–177.

Doherty, J. 2004. *PEST: Model-Independent Parameter Estimation. User Manual*, Watermark Numerical Computing, Brisbane, QLD, Australia.

Doherty, J.E. and R.J. Hunt. 2010. Approaches to highly parameterized inversion: a guide to using PEST for groundwater-model calibration. US Department of the Interior, US Geological Survey.

Doherty, J.E., Fienen, M.F., and R.J. Hunt. 2010. Approaches to highly parameterized inversion: pilot-point theory, guidelines, and research directions. US Geological Survey Scientific Investigations Report 2010–5168. p.36.

Environmental Simulations, Inc. 2011. Groundwater Vistas. Reinholds.

Gomez-Hernandez, J.J., Sahuquillo, A., and J.E. Capilla. 1997. Stochastic simulation of transmissivity fields conditional to both transmissivity and piezometric data. 1. *Theory. Journal of Hydrology*, 204(1–4): 162–74.

Haldorsen, H.H. and D.M. Chang. 1986. Notes on stochastic shales from outcrop to simulation models. In: Lake L.W., Carol H.B. Jr. (Eds.) *Reservoir Characterization*. Academic, New York, pp 152–167.

Healy, R.W. and P.G. Cook. 2002. Using ground-water levels to estimate recharge. *Hydrogeology Journal*, 10:91–109.

Huntley, D., Nommensen, R., and D. Steffey. 1992. The use of specific capacity to assess transmissivity in fractured rock aquifers. *Ground Water*, 30(3): 396–402.

Isaaks, E.H. and M.R. Srivastava. 1989. *An introduction to Applied Geostatistics*. Oxford University Press, New York.

Jalludin, M. and M. Razack. 2004. Assessment of hydraulic properties of sedimentary and volcanic aquifer systems under arid conditions in the Republic of Djibouti (Horn of Africa). *Hydrogeology Journal*, 12, 159–170.

Journel, A.G. and C.J. Huijbregts. 1978. *Mining Geostatistics*. Academic Press, New York.

Kitanidis, P. K. 2000. *Introduction to Geostatistics: Application in Hydrogeology*. Cambridge University Press, Cambridge, UK.

Lavenue, A.M. and J.F. Pickens. 1992. Application of a coupled adjoint sensitivity and kriging approach to calibrate a groundwater flow model. *Water Resources Research*, 28(6): 1543–69.

Long, J.C. and D. Billaux. 1987. From field data to fracture network modelling: an example incorporating spatial structure. *Water Resources Research*, 23(7): 1201–1216.

Matheron, G. 1963. Principles of Geostatistics. *Economic Geology*, 58: 1246–1266.

Matheron, G. 1965. *Les variables régionalisées et leur estimation [Regionalized variables and their estimation]*. Masson, Paris, 185 pp.

McLaughlin, D. and L.R. Townley. 1996. A reassessment of the groundwater inverse problem. *Water Resources Research*, 32(5): 1131–1161.

Poeter, E.P. and M.C. Hill. 1997. Inverse models: A necessary next step in ground-water modeling. *Ground Water*, 35(2): 250–260.

Razack, M. and D. Huntley. 1991. Assessing transmissivity from specific capacity in a large and heterogeneous alluvial aquifer. *Ground Water*, 29(6): 856–861.

Razack, M. and T. Lasm. 2006. Geostistical estimation of the transmissivity in a highly fractured metamorphic and crystalline aquifer (Man-Danane Region, western Ivory Coast). *Journal of Hydrology*, 325(1–4): 164–178.

Renard, Ph. 2007. Stochastic Hydrogeology: what professional really need? *Ground Water*, 45(5): 531–541.

Riser, D.W., Gburek, W.J., and G.J. Folmer. 2005. Comparison of methods for estimating groundwater recharge and base flow at a small watershed underlain by fractured bed rock in the eastern United States. U.S Geological Survey Scientific Invest. Report 2005–5038, 31.

Teles, V., Delay, F., and G. de Marsily. 2004. Comparison between different methods for characterizing the heterogeneity of alluvial media: groundwater flow and transport simulations. *Journal of Hydrology*, 294(1–3): 103–121.

Tonkin, M.J. and J. Doherty. 2005. A hybrid regularized inversion methodology for highly parameterized environmental models. *Water Resources Research*, 41(10): 1–16.

UN World Water Development Report. 2012. 4th edition. http://www.unesco.org/new/en/natural-sciences/environment/water/.

Wen, X.H., Deutsch, C.V., and A.S. Cullick. 2002. Construction of geostatistical aquifer models integrating dynamic flow and tracer data using inverse technique. *Journal of Hydrology*, 255(3): 151–168.

Yeh, W.W. 1986. Review of parameter identification procedures in groundwater hydrology: The inverse problem. *Water Resources Research*, 22(2): 95–108.

Yitbarek, A. 2009. Hydrogeological and hydrochemical framework of complex volcanic system in the Upper Awash River basin, Central Ethiopia: With special emphasis on inter-basins groundwater transfer between Blue Nile and Awash rivers. PhD thesis, University of Poitiers, France.

Zhou, H., Jaime Gómez-Hernández, J., and L. Liangping. 2014. Inverse methods in hydrogeology: Evolution and recent trends. *Advanced Water Research*, 63: 22–37.

Section IV

Transport Modeling

15

Transform Techniques for Solute Transport in Ground Water

Mritunjay Kumar Singh, Vijay P. Singh, and Shafique Ahamad

CONTENTS

15.1 Introduction

In daily life, the importance of water is realized only when one can face the scarcity of water. The challenge for any country is to have sustainable water security. In most countries, water resources are mainly surface and/or groundwater. Groundwater is part of the hydrological cycle that flows through the saturated zone at rates that are influenced by gravity, pressure, and the characteristics of geologic formations. Groundwater is one of the main sources of water supply for meeting agricultural, domestic, environmental, and industrial demands in many areas. Indeed, it is the source of water supply for many various cities and towns, especially in the United States and several western countries. Groundwater is also an important environmental asset that provides base flow to streams and supports wetlands and other groundwater-dependent ecosystems. In India, it is an important resource for potable water for both urban and rural areas and is the source of drinking water for more than 90% of the rural population that does not have access to a public water supply system. About 42% of the water used for irrigation comes from groundwater. Groundwater is an integral part of the environment. Its availability depends on rainfall and recharge conditions.

The quality of groundwater has been declining rapidly since the beginning of the twentieth century, primarily due to urbanization and various human activities. There has been a lack of adequate attention to water conservation, efficiency in water use, water reuse and recycle, groundwater recharge, and ecosystem sustainability. An uncontrolled use of the bore well technology has led to the extraction of groundwater at such a high rate that often recharge is far less than withdrawal. The causes of low water availability in many regions of India are linked directly to the diminishing forest cover and soil degradation.

Pollution of groundwater resources has become a major problem around the world today. The pollution of air, water, and land has adversely affected the public. The solid, liquid, and the gaseous wastes generated, not regularly and properly treated, result in environmental as well as groundwater pollution. For example, when air is polluted pollutants during rainfall fall on the ground, seep into, and contaminate groundwater. Leaching of pollutants from pesticides and fertilizers into the aquifers has polluted groundwater, and water extraction without proper recharge of groundwater resources has further exacerbated groundwater pollution. Further, agricultural waste, industrial waste, and the municipal solid waste have also polluted surface water and groundwater. In particular, groundwater pollution is caused when products, like gasoline, oil, road salts, chemicals, etc., get into the groundwater reservoir. These products render groundwater unsafe for human consumption. Pesticides and fertilizers find their way into groundwater over a period. Road salt, toxic substances from mining sites, and used motor oil also may seep into groundwater reservoirs (Fried, 1975; Sharma and Reddy, 2004; Rausch et al., 2005; Thangarajan, 2006). It is also possible for untreated wastes from septic tanks and toxic chemicals from underground storage tanks and leaky landfills to pollute groundwater reservoirs (Batu, 2006; Singh et al., 2010c). The terms *contaminant* and *pollutant* are two different scientific terms, but many environmentalists, in recent times, are using them as synonyms. The term "contaminant" is usually defined as a substance, human or naturally produced, that is found in a place where the quantity should not be or in amounts greater than it should be (Javandel et al., 1984). However, the term "pollutant" refers to any substance introduced into the environment that adversely affects the usefulness of a resource (Bear and Verruji, 1987).

Addition of any undesirable substances to groundwater either by human activities or by natural occurrence is termed groundwater pollution. Pollution problems vary from simple inconvenience, such as taste, odor, color, hardness, or foaming, to serious health hazards due to pathogenic organisms, flammable or explosive substances, or toxic chemicals and their by-products. The adverse effects of groundwater pollution are the result of human activity at the ground surface, unintentionally by agriculture, domestic and industrial effluents, or unexpectedly by subsurface or surface disposal of sewage and industrial wastes. Solute transport in groundwater generally necessitates special attention due to (1) very slow velocity of water and (2) interaction of solute particle with the solid matrix.

From a geo-environmental point of view, the source of groundwater pollution is divided into the following three groups:

- Sources originating on the ground surface
- Sources originating above the water table
- Sources originating below the water table

On the ground surface, various water-soluble products stored or spread cause groundwater pollution, such as infiltration of polluted surface water, land disposal of solid and liquid wastes, accidental spills, fertilizers and pesticides, disposal of sewage and water treatment plant sludge, salt storage and spreading on roads, animal feedlots, etc. There are varieties of substances deposited or stored on the ground above the water table that cause groundwater pollution, such as waste disposal in excavations, landfills, and septic tanks, leakage from underground storage tanks, pipelines, etc. There are numerous situations where materials are stored or disposed of below the water table, such as waste disposal in wet excavations, deep well injection and mines, agricultural drainage wells and tiles, abandoned or improperly constructed wells, etc., leading to serious groundwater pollution.

For mapping groundwater pollution and developing appropriate technologies for remediation, mathematical models are applied. These models are capable of simulating a wide range of hydrogeologic conditions. In recent years, models used to predict the fate and transport of pollutants in groundwater reservoirs. Groundwater models broadly are divided into two categories: (1) groundwater flow models and (2) solute transport models. Groundwater flow models solve for the distribution of head or groundwater flow, whereas solute transport models solve for the concentration of solute as affected by advection (movement of the solute with the average groundwater flow), dispersion (spreading and mixing of the solute), and chemical reactions, which slow down or transform solutes. A model is any device that represents an approximation of a field situation. Physical models, such as laboratory sand tanks, simulate groundwater flow directly. A mathematical model simulates groundwater flow indirectly by means of governing equations to represent the physical processes that occur in the system, together with equations that describe heads or flows along the boundaries of the model.

Mathematical models may be solved analytically, numerically, or both. When the assumptions used to derive an analytical solution are adjudged too simple and/or inappropriate for the problem under consideration, a numerical model is employed. For example, analytical solutions usually assume a homogeneous porous medium. A numerical model may be easier to apply than an analytical model if the analytical problem involves a complex superposition of solutions, for example, with image well theory. Because most of the physical problems in nature are highly irregular, numerical models rather than analytical models usually address them. The analytical solution, however, can be used to benchmark numerical codes and solutions. Input parameters may also be validated by the analytical solution. Unfortunately, for most practical problems, because of the heterogeneity of the solution domain, the irregular shape of domain boundaries, and the nonanalytical form of various functions, analytical solutions of mathematical models are intractable. Indeed, this may be an open research problem in solute transport modeling that needs to be explored further. Conveniently, researchers transform a mathematical model into a numerical one and then solve it by means of computational techniques.

We need groundwater models to make predictions about a groundwater system's response to a given stress, to understand the physical system, to design field studies, and to make policy decisions. A transport model consists of two parts: (1) a computer program to solve for groundwater flux and (2) a computer program to solve for solute concentration.

15.2 Solute Transport Processes

The solute transport process controls the extent of pollutant migration in the subsurface region. It mainly involves three processes, such as advection, diffusion, and dispersion. These processes are available only to deal with the transport of nonreactive pollutants in the subsurface. Nonreactive pollutants are dissolved pollutants, not influenced by chemical reactions or microbiological processes. In the case of reactive pollutants, these transport processes are considered along with various mass transfer and microbial degradation processes. Advection is the movement of the dissolved pollutant with groundwater, and it refers to the average linear flow velocity of the bulk of pollutant. Due to advection, the pollutant moves with the flow at a velocity equal to the seepage velocity of groundwater in the porous medium. The groundwater seepage velocity is defined as follows:

$$v = \frac{Kh}{\eta}$$

where K is the hydraulic conductivity, h is the hydraulic gradient, and η is the porosity of the porous material.

Diffusion is a microscale process, which causes movement of a solute in water from an area of higher concentration to an area of lower concentration. The difference in concentration is the concentration gradient. Diffusion ceases when there is no concentration gradient. It can occur even when the fluid is not flowing or is flowing in the direction opposite the pollutant movement. At the macroscale level, the solute transport is investigated by considering the average groundwater velocity. However, at the microscale level, the actual velocity of water may vary from point to point and can be either lower or higher than the average velocity. The difference in microscale water velocities arises due to pore size, path length, and friction in pores. Due to these differences in velocities, mixing occurs along the flow path. This mixing is nothing but mechanical dispersion, hydrodynamic dispersion, or simply dispersion. The mixing that occurs along the direction of the flow path is the longitudinal dispersion and along the direction, normal to the direction of the flow path, it is transverse dispersion (Fried and Combarnous, 1971; Bear, 1972; Sharma and Reddy, 2004; Batu, 2006; Singh et al., 2010c).

15.2.1 Review of Solute Transport Literature

Groundwater pollution remediation usually requires a quantitative knowledge about the distribution and fate of pollutants. The advection-dispersion equation represents a standard model to predict the solute movement in groundwater, based on the conservation of mass and Fick's law of diffusion (Fried and Combarnous, 1971; Bear, 1972; Chrysikopoulos et al., 1990).

To study the effect of initial and boundary conditions on the distribution of solute in time and space (or distance), Gehrson and Nir (1969) solved a one-dimensional advection-dispersion equation. The effects of hydrodynamic dispersion, diffusion, radioactive decay, and simple chemical interactions of the solute have been included. Many of the studies have been carried out for infinite, semi-infinite, and finite aquifer lengths. Day (1977) conducted a series of experiments to investigate longitudinal dispersion of reactive contaminants through natural channels and mountain streams in particular. Considering the Cauchy-type boundary condition, analytical solutions derived for chemical transport with simultaneous adsorption, zero order production, and first order decay. van Genuchten (1981) solved a one-dimensional advection-dispersion equation with adsorption, zero order production, and first order decay, considering constant dispersion parameters. van Genuchten and Alves (1982) obtained analytical solutions with the use of Laplace transform for first as well as third type boundary conditions.

Using Fourier transform and the method of superposition, Valocchi and Roberts (1983) derived analytical solutions of a one-dimensional advection-dispersion equation in semi-infinite and infinite domains for continuous input of a periodically fluctuating concentration and finite-duration input of a pulse type concentration.

Parker and van Genuchten (1984) presented the transformation between volume-averaged pore fluid concentration and flux-averaged concentration for a one-dimensional nonreactive solute transport equation. Considering uniform and time-dependent dispersion coefficients, Barry and Sposito (1989) investigated analytical solutions of a one-dimensional convection-dispersion model using Green's function. Using the Laplace transform, Chrysikopoulos et al. (1990) obtained an analytical solution for a one-dimensional solute transport through porous media in a semi-infinite domain with flux type, inlet boundary condition. Fry et al. (1993) derived an analytical solution, using an eigenfunction integral equation method, for the advection-dispersion equation with rate-limited desorption and first-order decay. Considering the first and third type inlet boundary conditions and arbitrary initial conditions, Toride et al. (1993) presented a comprehensive set of analytical solutions for one-dimensional non-equilibrium solute transport through semi-infinite soil systems with the use of Laplace transform.

Leij and van Genuchten (1995) studied approximate analytical solutions for solute transport in two-layer porous media. Using the principle of superposition, Ge and Lu (1996) presented a semi-analytical solution of one-dimensional advection-dispersive solute transport equation with an arbitrary concentration boundary condition. The solution was obtained for the third type boundary condition with the use of time-dependent source decay constant and Heaviside function. Using a generalized integral transform technique (GITT), Liu et al. (2000) derived the solution of one-dimensional solute transport equation with Cauchy-type boundary condition in a finite domain. Serrano (2001) obtained a series solution for one-dimensional nonlinear advection-dispersive equation using the method of decomposition. The nonlinearity introduced in the equation was due to nonlinear degradation or decay or by a chemical constituent, which follows a nonlinear sorption isotherm. Considering sine and cosine variations of waste discharge concentration at the upstream boundary and nonzero initial condition throughout the aquifer, Shukla (2002) obtained an analytical solution of one-dimensional advection-dispersion equation for unsteady transport dispersion of nonconservative pollutant/biochemical oxygen demand with first order decay.

Smedt (2006) presented analytical solutions for solute transport in rivers, including the effect of transient storage and first order decay, using the modified Bessel function of first kind for different boundary conditions. Using the Laplace transform, Yeh and Yeh (2007) obtained an analytical solution of one-dimensional groundwater transport equation. Srinivasan and Clement (2008a, b) mathematically derived the analytical solution of one-dimensional multi-species reactive transport equation

coupled with sorption and sequential first order reaction. Jaiswal et al. (2009) solved analytically a linear advection-diffusion equation in one-dimensional semi-infinite medium with temporally and spatially dependent variables, using the Laplace transform. Guerrero et al. (2009) solved a linear advection-diffusion transport equation with constant coefficient using a generalized integral transform technique for both first and third type boundary conditions. Chen and Liu (2011) derived the generalized analytical solution for one-dimensional advection-dispersion equation with a constant dispersion coefficient in a finite domain, subject to an arbitrary time-dependent inlet boundary condition. Sometimes, due to external causes, the amount of contaminant at the intermediate portion may vary with time and, therefore, Singh et al. (2011) solved a one-dimensional solute transport equation for a pulse type boundary condition, considering time-dependent dispersion coefficient and seepage velocity. Assuming initial concentration as an exponentially decreasing function of space, Singh et al. (2012) solved analytically a one-dimensional advection-dispersion equation in a homogeneous semi-infinite aquifer with the use of Laplace transform for both Dirichlet and Cauchy-type time-dependent boundary conditions. Guerrero et al. (2013) presented a one-dimensional advection-dispersion transport equation through multilayered media using the classical integral transform technique. Singh et al. (2014a) developed an analytical solution for one-dimensional uniform and time varying solute dispersion along transient groundwater flow in a semi-infinite aquifer.

Using the effect of molecular diffusion, adsorption, and first order decay, Basha and El-Habel (1993) solved a one-dimensional advection-dispersion equation with the method of superposition. Considering uniform dispersion coefficient and periodic boundary conditions, Logan and Zlotnik (1995) explored analytical solutions of convective-diffusive equation with decay in a semi-infinite domain and derived a formula for representing closed form analytical solutions. The same work extended further in the form of hypergeometric functions, assuming scale-dependent dispersion coefficient and periodic type time-dependent boundary conditions (Logan, 1996). Assuming dispersivity in the form of separable power law dependence on both time and scale, Su et al. (2005) derived an explicit closed form solution in a semi-infinite domain with an instantaneous point-source (Dirac delta function) for constant concentration and constant flux type boundary conditions. Sander and Braddock (2005) developed analytical solutions for transient, unsaturated transport of water and contaminants through horizontal porous media. Singh et al. (2009b) presented a one-dimensional advection-dispersion equation with time-dependent dispersion coefficients, that is, sinusoidally and exponentially varying.

Singh et al. (2009a) further developed a solute transport model for one-dimensional homogeneous porous formations with time dependent point-source concentration. Singh et al. (2010a,b,c) also presented an analytical solution for solute transport along and against time dependent source concentration in a homogeneous finite aquifer. Wang et al. (2011) proposed a stepwise superposition approximation approach to solve for nonzero initial concentration for first type and third type boundary conditions using the existing zero initial concentration solutions. Singh et al. (2013) derived an analytical approach to one-dimensional solute dispersion along and against transient groundwater flow in aquifers. The solutions mentioned thus far have been one-dimensional when the length of the aquifer is very large as compared to the width and depth of the aquifer. However, for comparatively small groundwater reservoirs, the width and depth of the aquifer are explicitly considered and hence two- or three-dimensional problems are accordingly addressed.

Considering a wide range of Reynolds number in a two-dimensional isotropic porous medium, Harleman and Rumer (1963) analytically and experimentally investigated the longitudinal and lateral dispersion coefficient. Hoeksema and Kitanidis (1984) developed a geo-statistical approach to estimate the transmissivity from head and transmissivity measurements in a two-dimensional steady flow without pumping or acceleration terms for the Dirichlet boundary condition. Hantush and Marino (1990) analyzed the hydraulic head variability in space and time in a two-dimensional transient groundwater flow along a finite heterogeneous aquifer within the framework of a Galerkin finite element procedure and an eigenvalue technique. Using the Laplace transform and Fourier analysis technique, Tsai and Chen (1995) developed a finite analytical-numerical solution for the migration of groundwater contamination in a two-dimensional semi-infinite aquifer. Huang and van Genuchten (1995) developed an approximate analytical solution for simulating solute transport in unsaturated soil by infiltration of water and a dissolved tracer during ponding. Assuming time-dependent dispersion along uniform flow, the two-dimensional advection-diffusion equation, Aral and Liao (1996) solved analytically for an instantaneous point injection and a continuous point source. The dispersion coefficient was considered a uniform, linear, and asymptotically and exponentially temporally dependent term.

Shan and Javandel (1997) developed analytical solutions for two-dimensional solute transport in a vertical section of a homogeneous aquifer with uniform groundwater flow for two different source conditions: (1) a constant concentration of known value at the source, and (2) a constant flux rate of known value from the source. Hantush and Marino (1998) developed

analytical solutions for the first-order rate model in an infinite porous medium using the Fourier and Laplace transforms and superposition principle. The solutions considered a rectangular source with (1) an instantaneous release of a contaminant mass and (2) an exponentially decaying source applied at a fixed rate. Chen et al. (2003) presented a two-dimensional Laplace-transformed power series solution for solute transport in a radially convergent flow field. Massabo et al. (2006) presented an analytical solution of a two-dimensional convection-dispersion equation by considering the anisotropic dispersion coefficient and the effect of chemical decay or adsorption like reaction inside the liquid phase. Park and Baik (2008) solved a two-dimensional advection-diffusion model in closed form with the use of the superposition method for a ground level finite area source. The dependence of ground level concentration on important parameters was also examined and the finite area source solutions compared with the solution for a laterally infinite area source and a point source. Using the Laplace transform, Zhan et al. (2009) studied two-dimensional solute transport in an aquifer-aquitard system. Aniszewski (2009) presented a mathematical model and practical verification of groundwater and contaminant transport in a natural aquifer. Considering the dispersion coefficient and seepage velocity varying with time, Singh et al. (2010a, b, c) presented a two-dimensional solute transport in a homogeneous finite aquifer using the Hankel transform. Using a cylindrical coordinate system, Chen et al. (2011) obtained an analytical solution of a two-dimensional advection-dispersion equation, subject to first- and third-type inlet boundary conditions. Kumar and Yadav (2013) extensively explored pollutant dispersion in groundwater: its degradation and rehabilitation. The finite Hankel transform technique of second kind and the generalized integral transform techniques were employed. Singh and Kumari (2014) presented different sets of initial and boundary conditions in confined and unconfined aquifers available in the literature. Singh et al. (2015) recently presented pollutant's horizontal dispersion along and against sinusoidally varying velocity from a pulse type point source.

Mathematical modeling of pollutant concentration patterns in unsteady groundwater flow has made a significant contribution to the development and assessment of groundwater systems. The intensive use of natural resources and the large production of wastes by modern society are posing a great threat to groundwater quality and have resulted in many incidents of groundwater pollution (Kumar et al., 2005). This chapter may help develop measures for minimizing the pollutant concentration in aquifers and predict the time lapse to reach the minimum concentration.

15.2.2 Use of Solute Transport Models in Practice

There are many ways to classify conceptual or mathematical groundwater flow models. One-, two-, or three-dimensional (spatial) models considered are either transient or steady state, confined or unconfined. The maximum amount of groundwater flow in an aquifer commonly used is from local flow systems. Groundwater levels and flow in local flow systems are affected by seasonal variations in recharge because recharge areas are relatively shallow transient groundwater flow systems. Regional flow systems are found to be less transient than local systems and are intermediate flow systems.

A conceptual model of an aquifer system is a simplified, qualitative description of the physical system. A conceptual model may include a description of the aquifer and confining units that make up the aquifer system, boundary conditions, flow regimes, sources and sinks of water, and general directions of flow. Then the calibrated model is used to quantify various aspects of the conceptual model and to draw conclusions about regional groundwater flow in the aquifer. The calibrated model represents the movement of water through the parts of the aquifer system not greatly affected by seasonal variations in groundwater recharge from precipitation, which may be approximately 10% of the total flow in the aquifer. Numerical groundwater flow models are approximate representations of aquifers and have some limitations. These limitations are usually associated with (1) the purpose of the groundwater flow model, (2) the understanding of the aquifer, (3) the quantity and quality of data used to constrain parameters in the flow model, and (4) the assumptions made during model development.

15.3 Mathematical Models

The advection-dispersion equation along with different initial and boundary conditions are known as advection-dispersion models or solute transport models. Analysis of a contaminant transport problem requires the use of a mathematical model that is commensurate with the application. In the literature, there are several mathematical techniques available to solve the advection-dispersion model. These techniques involve generally two main approaches, such as analytical and numerical (Singh and Das, 2015).

15.3.1 Analytical Model

Analytical methods are useful for providing initial and approximate solutions of alternative pollution scenarios, conducting sensitivity analyses to investigate the effects of various parameters or processes on contaminant transport, and extrapolating results over large time and spatial scales where numerical solutions become impractical. The literature contains many analytical solutions for advection-dispersion type transport problems in one, two, and three dimensions. There are several methods available to find analytical solutions of advection-dispersion models, such as Laplace transform technique, Fourier transform technique, generalized integral transform technique, Green's function method, power series method, Hankel transform technique, and method of superposition principle. The methods used here are according to the physical model, dimensions, types of initial and boundary conditions, and temporal and spatial dependency of the coefficients in the equation of the physical problem (Mahato, 2013).

15.3.2 Numerical Model

Except for applications to well hydraulics, analytical solutions may not be widely used in practical applications. Numerical solutions are much more versatile and with widespread and easy availability of computers are now easier to use than the more complex analytical solutions. The numerical methods, such as finite difference, finite elements, integrated finite difference, boundary integral equation method, boundary-fitted coordinate method, and analytical elements, are widely used in groundwater modeling. The boundary integral equation method (Liggette and Liu, 1983; Liggette, 1987) and analytic elements (Strack, 1987, 1988) are relatively new techniques and have not yet been widely used. The integrated finite difference (IFD) technique is closely related with the finite element method. Finite difference and finite element methods are commonly used to solve groundwater-modeling problems closely related with physical systems (van Genuchten, 1982; Anderson and Woessner, 2002; Rastogi, 2007).

15.3.3 Finite Difference Method (FDM)

The advection-dispersion equation, subject to different initial and boundary conditions, is usually solved by approximating the first and second order derivatives contained in the equation by forward, backward, and central difference approximations. The numerical solution is obtained commonly by finite-difference techniques, such as Crank-Nicolson implicit method and an explicit finite difference method.

In the Crank-Nicolson implicit method, the time derivative of advection-dispersion equation is approximated by forward difference approximation; the first-order space derivative is approximated by the average of central difference approximation of the nth level and the $(n + 1)$th level and the second order space derivative is approximated as the average of the central difference approximation of the nth level and the $(n + 1)$th level. After computing the set of approximations, the method gives a system of equations solved by a tri-diagonal method or the Gauss elimination method, which is assumed unconditionally stable. However, the determination of the system of equations at each time step, especially in a two-dimensional case, makes the problem complex.

Explicit finite-difference method is a second-order accurate method. Let us consider a one-dimensional domain on which the problem is formulated. This is rectangular where x ranges from x_{min} to x_{max} and time domain t ranges from 0 to T. Divide the interval $[x_{min}, x_{max}]$ into M equal subintervals with length Δx, indexed by $i = 0, 1, 2, …, M$ and interval $[0, t]$ into equal subintervals with length Δt, indexed by $j = 0, 1, 2, ….$ Let $c_{i,j}$ denote the approximation of grid point at $x_i = x_0 + i\Delta x$ and $t_j = t_0 + j\Delta t$. The partial order derivatives become:

$$\frac{\partial c}{\partial t} \approx \frac{c_{i,j+1} - c_{i,j}}{\Delta t} \quad \text{(Forward difference formula)} \quad (15.1)$$

$$\frac{\partial c}{\partial x} \approx \frac{c_{i+1,j} - c_{i-1,j}}{2\Delta x} \quad \text{(Central difference formula)} \quad (15.2)$$

$$\frac{\partial^2 c}{\partial x^2} \approx \frac{c_{i+1,j} - 2c_{i,j} + c_{i-1,j}}{\Delta x^2} \quad \text{(Central difference formula)}$$
$$(15.3)$$

Using these approximations in a one-dimensional solute transport model, all the values of $c_{i,j+1}$ can be determined for the entire grid at each time level. This method is popularly known as explicit finite-difference method. It is second-order accurate in the x-direction and first-order accurate in the t-direction, and is easy to implement. The explicit finite-difference solution is unstable unless the ratio $\Delta t/\Delta x^2$ is sufficiently small. This means that small errors may come either due to arithmetic inaccuracies or due to the approximate nature of the derivative expressions, which tends to accumulate and grow as one proceeds rather than dampen out.

15.4 Solute Transport in Groundwater

The general equation governing the hydrodynamic dispersion of a homogeneous fluid is the fundamental advection-dispersion equation, which is based on the conservation of mass and Fick's first law of diffusion.

15.4.1 Fick's First Law

Fick's first law relates the diffusive flux to the concentration gradient. It states that the rate of transfer of diffusive substance through the unit area of a section is proportional to the concentration gradient normal to that section.

$$J_D = -D^* \frac{\partial c}{\partial x} \quad (15.4)$$

where J_D is the diffusive mass flux, and D^* is the diffusion coefficient. The negative sign indicates that contaminant moves from the zone of higher concentration to the zone of lower concentration.

15.4.2 Advection-Dispersion Equation

Let $c(x, y, z)$ be the concentration of the fluid at any point (x, y, z) and u_x, u_y, u_z be the velocity components parallel to the coordinates axes. Consider a small element of volume $dxdydz$ in the form of a rectangular parallelepiped, as shown in Figure 15.1 whose sides are parallel to the axes of coordinates.

The dispersive mass flux $\vec{J}_{diff}(J_x, J_y, J_z)$ from Fick's first law of diffusion

$$J_x = -D_x \frac{\partial c}{\partial x}, J_y = -D_y \frac{\partial c}{\partial y}, J_z = -D_z \frac{\partial c}{\partial z} \quad (15.5)$$

and, the convective mass flux $\vec{J}_{conv}$ through the element can be written as:

$$\vec{J}_{conv} = \vec{u}c \quad (15.6)$$

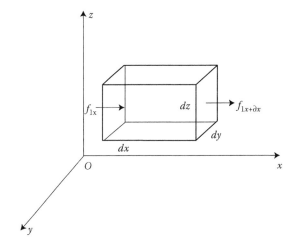

FIGURE 15.1
A small rectangular parallelepiped (Adapted from Kumari, P. 2013. *Study of analytical and numerical approach of solute transport modeling.* PhD thesis, ISM Dhanbad.)

where D_x, D_y, and D_z are the dispersion coefficients and $\vec{u}(u_x, u_y, u_z)$ is the flow velocity.

Therefore, the total flux, including advection and diffusion transport fluxes, is

$$\vec{J}(J_1, J_2, J_3) = \vec{J}_{conv} + \vec{J}_{diff} \tag{15.7}$$

The total flux entering the element along the x-direction is

$$J_1 = \left(u_x c - D_x \frac{\partial c}{\partial x} \right) \partial y\, \partial z = f_1(x, y, z) \quad \text{(say)} \tag{15.8}$$

and the total flux leaving the element from the x-direction is

$$f_1(x + \partial x, y, z) = f_1(x, y, z) + \partial x \frac{\partial}{\partial x} f_1(x, y, z) + \cdots \tag{15.9}$$

The excess of solute flow along the x-axis is written as follows:

Difference of mass per unit time = mass entering the element—mass leaving the element

$$= -\partial x \frac{\partial}{\partial x} f_1(x, y, z)$$

$$= -\partial x \frac{\partial}{\partial x} \left(u_x c - D_x \frac{\partial c}{\partial x} \right) \partial y\, \partial z \quad \text{[Using Equation 15.8]} \tag{15.10}$$

Similarly, the excess of solute flow along the y-axis is

$$= -\frac{\partial}{\partial y} \left(u_y c - D_y \frac{\partial c}{\partial y} \right) \partial x\, \partial y\, \partial z \tag{15.11}$$

and the excess of solute flow along the z-axis is

$$= -\frac{\partial}{\partial z} \left(u_z c - D_z \frac{\partial c}{\partial z} \right) \partial x\, \partial y\, \partial z \tag{15.12}$$

The total excess of solute flow in and out from all the coordinate axes of the element, that is, the difference of mass of solute per unit time, is:

$$= - \left[\begin{array}{l} \dfrac{\partial}{\partial x} \left(u_x c - D_x \dfrac{\partial c}{\partial x} \right) + \dfrac{\partial}{\partial y} \left(u_y c - D_y \dfrac{\partial c}{\partial y} \right) \\[2ex] + \dfrac{\partial}{\partial z} \left(u_z c - D_z \dfrac{\partial c}{\partial z} \right) \end{array} \right] \partial x\, \partial y\, \partial z \tag{15.13}$$

Since the dissolved substance is assumed nonreactive, the difference between the flux into the element and the flux out of the element equals the amount of dissolved substance accumulated in the element. Therefore, the rate of mass change per unit time becomes

$$\frac{\partial c}{\partial t} \partial x\, \partial y\, \partial z \tag{15.14}$$

From the principal of conservation of mass, the volume of the aquifer per unit time is equal to the rate of solute concentration in the aquifer.

Therefore, rate of change of solute concentration = rate of solute flow in – rate of solute flow out

$$\frac{\partial c}{\partial t} \partial x\, \partial y\, \partial z = - \left[\begin{array}{l} \dfrac{\partial}{\partial x} \left(u_x c - D_x \dfrac{\partial c}{\partial x} \right) + \dfrac{\partial}{\partial y} \left(u_y c - D_y \dfrac{\partial c}{\partial y} \right) \\[2ex] + \dfrac{\partial}{\partial z} \left(u_z c - D_z \dfrac{\partial c}{\partial z} \right) \end{array} \right] \partial x\, \partial y\, \partial z$$

$$\Rightarrow \frac{\partial c}{\partial t} = \left[\begin{array}{l} \dfrac{\partial}{\partial x} \left(D_x \dfrac{\partial c}{\partial x} - u_x c \right) + \dfrac{\partial}{\partial y} \left(D_y \dfrac{\partial c}{\partial y} - u_y c \right) \\[2ex] + \dfrac{\partial}{\partial z} \left(D_z \dfrac{\partial c}{\partial z} - u_z c \right) \end{array} \right]$$

$$\tag{15.15}$$

Equation 15.15 is known as the hydrodynamic dispersion equation or advection-dispersion equation in a three-dimensional coordinate system.

Here D_x, D_y, and D_z are the dispersion coefficients, considered as functions of time or position or constant depending on the type of aquifer.

For a one-dimensional case, the advection-dispersion equation with constant coefficients becomes

$$\frac{\partial c}{\partial t} = D_x \frac{\partial^2 c}{\partial x^2} - u_x \frac{\partial c}{\partial x} \tag{15.16}$$

15.4.3 Initial and Boundary Conditions

The advection-dispersion equation represents the expression of mass balance and dimensions. It does not contain any information related to any specific case of flow, not even the shape of the domain within which the flow occurs. Therefore, each equation has an infinite number of possible solutions, each of which corresponds to a particular case of flow through a porous medium. To obtain one particular solution corresponding to a certain specific problem of interest, it is necessary to provide supplementary information that is not contained in the equation. This supplementary information defines the initial and boundary conditions of the problem.

The initial condition describes the distribution of values of the considered state variables at some initial time, usually taken as $t = 0$, at all points within the considered domain. The initial condition, representing the concentration of contaminant transport in a general form for a three-dimensional system, becomes:

$$c(x, y, z, t) = f_2(x, y, z); \quad t = 0$$

where $c(x, y, z, t)$ is the required state variable, and $f_2(x, y, z)$ is a known function or constant or zero, depending on the physical model of the problem.

Boundary conditions represent the way the considered domain interacts with its adjacent environment and there are three types of mathematical conditions written as follows:

1. First type boundary condition or Dirichlet type boundary condition or specified head boundary: A specific head boundary, simulated by setting the head at the relevant boundary nodes, which is equal to known head values. If the boundary is a river, the head along the boundary may vary spatially, whereas for lakes and reservoirs, the boundary is described by a constant head condition. It prescribes the concentration along a portion of the boundary. Mathematically, it becomes

$$c = f_3(x, y, z, t)$$

where $f_3(x, y, z, t)$ may be constant or a given function of space or time or both for that particular portion of the boundary.

2. Second type boundary condition or Neumann type boundary condition or specific flow boundary: In the case of specified flow boundary, the derivative of head (flux) across the boundary is given. It is used to describe fluxes for surface water bodies, spring flow, underflow, and seepage to or from bedrock underlying the modeled system. It prescribes the normal gradient of concentration over a certain portion of the boundary. Mathematically it becomes

$$\left(D_{ij} \frac{\partial c}{\partial x_j} \right) n_i = f_4(x, y, z, t)$$

where f_4 is a known function and n_i is the directional cosine. For impervious boundaries, f_4 becomes zero.

3. Mixed type boundary condition or Cauchy-type boundary condition or head dependent flow boundary: The head dependent flow boundaries relate with boundary heads and boundary flows. The flux across this type of boundary is dependent on the difference between a user-supplied specified head on one side of the boundary and the model calculated head on the other side. The Robin type boundary condition prescribes a linear combination of concentration and its gradient along the boundary. Mathematically, it becomes

$$\left(D_{ij} \frac{\partial c}{\partial x_j} - v_i c \right) n_i = f_5(x, y, z, t)$$

where f_5 is a known function. The first term on the left-hand side represents flux by dispersion and the second term represents the effect of advection. These descriptions of boundary conditions are explored by Aral (1990).

15.4.4 Dispersion Theory

Sl. No.	Author	Relation between D and u
1.	Taylor (1953)	D proportional to u^2
2.	Bear and Todd (1960), Kumar (1983)	D is proportional to u
3.	Freeze and Cherry (1979)	D is proportional to a power n of velocity u which ranges between 1 and 2
4.	Scheidegger (1957)	1. $D = \alpha u^2$, where α is a constant of the porous medium 2. $D = \beta u$, where β is another constant of the porous medium
5.	Ghosh and Sharma (2006)	D is directly proportional to the velocity u with a power ranging from 1 to 1.2

15.4.5 Velocity Assumptions

We consider three different time-dependent forms of velocity expressions. The first two and last one are followed by Aral and Liao (1996). The velocity expression is considered, based on the properties of algebraic sigmoid, which includes the error function. It starts to progress from a small beginning and accelerates in the rainy season and reaches a limit over a period. These expressions are as follows:

1. Exponentially decreasing form of velocity

$$u = u_0 f(t), \ f(t) = 1 - \exp\left(\frac{-mt}{K} \right)$$

$$\Rightarrow T^* = \frac{1}{m} \left(mt + K \left(\exp\left(-\frac{mt}{K} \right) - 1 \right) \right) \quad (15.17)$$

2. Asymptotic form of velocity

$$u = u_0 f(t), \; f(t) = \frac{mt}{(mt+K)} \Rightarrow T^* = \frac{1}{m}\left(mt - K\frac{mt}{(mt+K)}\right)$$

$$\tag{15.18}$$

3. Algebraic sigmoid form of velocity

$$u = u_0 f(t), \; f(t) = \frac{mt}{\sqrt{(mt)^2 + K^2}}$$

$$\Rightarrow T^* = \frac{1}{m}\left(\sqrt{(mt)^2 + K^2} - K\right) \tag{15.19}$$

4. Sinusoidal form of velocity

$$u = u_0 f(t), \; f(t) = 1 - \sin(Kmt)$$

$$\Rightarrow T^* = \frac{1}{m}\left(mt + \frac{1}{K}(\cos(Kmt) - 1)\right) \tag{15.20}$$

where K is the arbitrary constant. Considering $K = 0$ in Equations 15.17 through 15.20, results in $f(t) = 1$ and accordingly the expression of T^* may change. It represents the problem with uniform velocity and dispersion coefficient.

15.5 Two-Dimensional Advection-Dispersion Problem and Solution

The advection-dispersion equation, popularly known as ADE, is derived using the principle of conservation of mass and Fick's law of diffusion. It is a partial differential equation and it satisfies Darcy's law if the medium is porous. The advection-dispersion equation is widely used in describing the pollutant distribution in aquifers, oil reservoirs, rivers, lakes, and air. This equation may also be used in describing similar phenomena in biophysics and biomedical sciences.

The advection-dispersion equation in two-dimensional form for an isotropic horizontal medium is given as follows:

$$\frac{\partial C}{\partial t} = \frac{\partial}{\partial x}\left[D_x(x,t)\frac{\partial C}{\partial x} - u(x,t)C\right]$$

$$+ \frac{\partial}{\partial y}\left[D_y(y,t)\frac{\partial C}{\partial y} - v(y,t)C\right] \tag{15.21}$$

where C is the solute concentration of the pollutant at position (x, y) of the plane at time t.

Now the solute dispersion components in both the longitudinal and lateral directions are considered here which are proportional to the square of the respective velocity components. The flow domain is considered temporally dependent. Therefore, we consider (Jaiswal et al. 2011; Kumar et al., 2012; Singh et al. 2012):

$$u(x,t) = u_0 f(mt); \quad v(y,t) = v_0 f(mt) \tag{15.22}$$

and

$$D_x(x,t) = D_{x0} f^2(mt); \quad D_y(y,t) = D_{y0} f^2(mt) \tag{15.23}$$

where u_0, v_0 in Equation 15.22 may be referred to as uniform longitudinal and lateral velocity components, respectively, each of dimension [LT^{-1}]; and coefficients D_{x0}, D_{y0} in Equation 15.23 may be referred to as the initial longitudinal and lateral dispersion coefficients, respectively, each of dimension [L^{2}T^{-1}]. Also, m is a flow resistance coefficient whose dimension is inverse of the dimension of time t, that is, of dimension [T^{-1}]. Hence, $f(mt)$ is a nondimensional expression, $f(mt)$ is chosen as a sinusoidal form of temporally dependent dispersion, $f(mt) = 1 - \sin(mt)$, or an exponential form of temporally dependent dispersion, $f(mt) = \exp(mt)$, in such a way, that $f(mt) = 1$ for $m = 0$ or $t = 0$.

15.5.1 Analytical Solution

Using Equations 15.22 and 15.23, Equation 15.21 is expressed as follows:

$$\frac{1}{f(mt)}\frac{\partial C}{\partial t} = \frac{\partial}{\partial x}\left[D_{x0} f(mt)\frac{\partial C}{\partial x} - u_0 C\right]$$

$$+ \frac{\partial}{\partial y}\left[D_{y0} f(mt)\frac{\partial C}{\partial y} - v_0 C\right] \tag{15.24}$$

Now the system is assumed to be polluted initially, that is, there is some initial contaminant existing in the aquifer at $t = 0$. Also, the time-dependent source contaminant concentration is injected at the origin of aquifer and let the contaminant concentration gradient at the infinite distance in both the directions away from the source be zero at all times. The system is assumed to be of semi-infinite extent along both the directions.

The set of initial and boundary conditions stated above is expressed as follows:

$$C(x,y,t) = C_i \exp\{-\gamma(x+y)\}; \quad x \ge 0, y \ge 0, t = 0 \tag{15.25}$$

$$-D_x(x,t)\frac{\partial C}{\partial x}+u(x,t)C=\frac{uC_0}{2}[1+\exp(-qt)]; \quad x=0, t>0$$

(15.26)

$$-D_y(y,t)\frac{\partial C}{\partial y}+v(y,t)C=\frac{vC_0}{2}[1+\exp(-qt)]; \quad y=0, t>0$$

(15.27)

$$\frac{\partial C}{\partial x}=0, \frac{\partial C}{\partial y}=0; \quad x\to\infty, y\to\infty, t\geq 0 \quad (15.28)$$

where C_i is the initial solute concentration [ML^{-3}] that is included with an exponentially decreasing function of space describing the distribution of concentration at all points in both directions of the flow domain, that is, at $t=0$, and γ is a constant coefficient parameter whose dimension is the inverse of space variable, that is, [L^{-1}]. Here, C_0 is the solute concentration [ML^{-3}], and q is the decay rate coefficient [T^{-1}].

Using Equations 15.22 and 15.23, in Equations 15.25 through 15.28, we obtain the following:

$$C(x,y,t)=C_i\exp\{-\gamma(x+y)\}; \quad x\geq 0, y\geq 0, t=0 \quad (15.29)$$

$$-D_{x0}f(mt)\frac{\partial C}{\partial x}+u_0C=\frac{u_0C_0}{2}[1+\exp(-qt)]; \quad x=0, t>0$$

(15.30)

$$-D_{y0}f(mt)\frac{\partial C}{\partial y}+v_0C=\frac{v_0C_0}{2}[1+\exp(-qt)]; \quad y=0, t>0$$

(15.31)

$$\frac{\partial C}{\partial x}=0, \frac{\partial C}{\partial y}=0; \quad x\to\infty, y\to\infty, t\geq 0 \quad (15.32)$$

Introducing new independent variables, X and Y, using the transformations (Jaiswal et al., 2011; Singh et al., 2012):

$$X=\int\frac{dx}{f(mt)}=\frac{x}{f(mt)} \quad (15.33)$$

and

$$Y=\int\frac{dy}{f(mt)}=\frac{y}{f(mt)} \quad (15.34)$$

Equation 15.24 becomes

$$\frac{\partial C}{\partial t}=D_{x0}\frac{\partial^2 C}{\partial X^2}+D_{y0}\frac{\partial^2 C}{\partial Y^2}-u_0\frac{\partial C}{\partial X}-v_0\frac{\partial C}{\partial Y} \quad (15.35)$$

Also, initial and boundary conditions given by Equations 15.29 through 15.32 become

$$C(X,Y,t)=C_i\exp\{-\gamma(X+Y)\}; \quad X\geq 0, Y\geq 0, t=0 \quad (15.36)$$

$$-D_{x0}\frac{\partial C}{\partial X}+u_0C=\frac{u_0C_0}{2}[1+\exp(-qt)]; \quad X=0, t>0 \quad (15.37)$$

$$-D_{y0}\frac{\partial C}{\partial Y}+v_0C=\frac{v_0C_0}{2}[1+\exp(-qt)]; \quad Y=0, t>0 \quad (15.38)$$

$$\frac{\partial C}{\partial X}=0, \frac{\partial C}{\partial Y}=0 \quad X\to\infty, Y\to\infty, t\geq 0 \quad (15.39)$$

Again, introducing another transformation

$$\psi=X+Y \quad (15.40)$$

that reduces Equation 15.35 to a one-dimensional equation

$$\frac{\partial C}{\partial t}=D_0\frac{\partial^2 C}{\partial\psi^2}-U_0\frac{\partial C}{\partial\psi} \quad (15.41)$$

where $D_0=(D_{x0}+D_{y0})$ and $U_0=(u_0+v_0)$.

In addition, the initial and boundary conditions 15.36 through 15.39 become

$$C(\psi,t)=C_i\exp(-\gamma\psi); \quad 0\leq\psi<\infty, t=0 \quad (15.42)$$

$$-D_0\frac{\partial C}{\partial\psi}+U_0C=\frac{U_0C_0}{2}[1+\exp(-qt)]; \quad \psi=0, t>0 \quad (15.43)$$

$$\frac{\partial C}{\partial\psi}=0; \quad \psi\to\infty, t\geq 0 \quad (15.44)$$

Further, the following transformation is considered:

$$C(\psi,t)=k(\psi,t)\exp\left(\frac{U_0}{2D_0}\psi-\frac{U_0^2}{4D_0}t\right) \quad (15.45)$$

Using Equation 15.45 along with the Laplace transform and its inverse, the solution of transformed initial and boundary value problem is obtained as follows:

$$C(\psi,t)=F(\psi,t)-G(\psi,t)-H(\psi,t)+I(\psi,t) \quad (15.46)$$

where

$$C(X,t) = F(X,t) - G(X,t) - H(X,t) + I(X,t) \qquad (15.54)$$

where

$$F(\psi,t) = \frac{C_0}{2}\left[2U_0\sqrt{\frac{t}{\pi D_0}}\exp\left\{\frac{U_0\psi}{2D_0} - \frac{\psi^2}{4D_0 t} - \frac{U_0^2 t}{4D_0}\right\}\right.$$

$$+ erfc\left\{\frac{\psi - U_0 t}{2\sqrt{D_0 t}}\right\} - \left(1 + \frac{U_0\psi}{D_0} + \frac{U_0^2 t}{D_0}\right)$$

$$\left. \times \exp\left(\frac{U_0\psi}{D_0}\right)erfc\left\{\frac{\psi + U_0 t}{2\sqrt{D_0 t}}\right\}\right] \qquad (15.47)$$

$$F(X,t) = \frac{C_0}{2}\left[2u_0\sqrt{\frac{t}{\pi D_0}}\exp\left\{\frac{u_0 X}{2D_0} - \frac{X^2}{4D_0 t} - \frac{u_0^2 t}{4D_0}\right\}\right.$$

$$+ erfc\left\{\frac{X - u_0 t}{2\sqrt{D_0 t}}\right\} - \left(1 + \frac{u_0 X}{D_0} + \frac{u_0^2 t}{D_0}\right)$$

$$\left. \times \exp\left(\frac{u_0 X}{D_0}\right)erfc\left\{\frac{X + u_0 t}{2\sqrt{D_0 t}}\right\}\right] \qquad (15.55)$$

$$G(\psi,t) = \frac{qC_0}{2}\left[\frac{1}{2U_0}\sqrt{\frac{t}{\pi D_0}}\left(2D_0 + U_0\psi + U_0^2 t\right)\right.$$

$$\times \exp\left\{\frac{U_0\psi}{2D_0} - \frac{\psi^2}{4D_0 t} - \frac{U_0^2 t}{4D_0}\right\}$$

$$+ \frac{1}{2U_0^2}\left(U_0^2 t - U_0\psi - D_0\right)erfc\left\{\frac{\psi - U_0 t}{2\sqrt{D_0 t}}\right\}$$

$$- \frac{1}{2U_0^2}\left\{U_0^2 t - D_0 + \frac{U_0^2}{2}\left(\frac{\psi + U_0 t}{\sqrt{D_0}}\right)^2\right\}$$

$$\left. \times \exp\left(\frac{U_0\psi}{D_0}\right)erfc\left\{\frac{\psi + U_0 t}{2\sqrt{D_0 t}}\right\}\right] \qquad (15.48)$$

$$G(X,t) = \frac{qC_0}{2}\left[\frac{1}{2u_0}\sqrt{\frac{t}{\pi D_0}}\left(2D_0 + u_0 X + u_0^2 t\right)\right.$$

$$\times \exp\left\{\frac{u_0 X}{2D_0} - \frac{X^2}{4D_0 t} - \frac{u_0^2 t}{4D_0}\right\} + \frac{1}{2u_0^2}\left(u_0^2 t - u_0 X - D_0\right)$$

$$\times erfc\left\{\frac{X - u_0 t}{2\sqrt{D_0 t}}\right\} - \frac{1}{2u_0^2}\left\{u_0^2 t - D_0 + \frac{u_0^2}{2}\left(\frac{X + u_0 t}{\sqrt{D_0}}\right)^2\right\}$$

$$\left. \times \exp\left(\frac{u_0 X}{D_0}\right)erfc\left\{\frac{X + u_0 t}{2\sqrt{D_0 t}}\right\}\right] \qquad (15.56)$$

$$H(\psi,t) = \frac{1}{2}I(\psi,t)\left[erfc\left\{\frac{\psi - (2\gamma D_0 + U_0)t}{2\sqrt{D_0 t}}\right\} - \left(\frac{\gamma D_0 + U_0}{\gamma D_0}\right)\right.$$

$$\left. \times \exp\left\{\left(2\gamma + \frac{U_0}{D_0}\right)\psi\right\} \times erfc\left\{\frac{\psi + (2\gamma D_0 + U_0)t}{2\sqrt{D_0 t}}\right\}\right]$$

$$+ \frac{U_0 C_i}{2\gamma D_0}\exp\left(\frac{U_0\psi}{D_0}\right)erfc\left\{\frac{\psi + U_0 t}{2\sqrt{D_0 t}}\right\} \qquad (15.49)$$

$$H(X,t) = \frac{1}{2}I(X,t)\left[erfc\left\{\frac{X - (2\gamma D_0 + u_0)t}{2\sqrt{D_0 t}}\right\} - \left(\frac{\gamma D_0 + u_0}{\gamma D_0}\right)\right.$$

$$\left. \times \exp\left\{\left(2\gamma + \frac{u_0}{D_0}\right)X\right\} \times erfc\left\{\frac{X + (2\gamma D_0 + u_0)t}{2\sqrt{D_0 t}}\right\}\right]$$

$$+ \frac{u_0 C_i}{2\gamma D_0}\exp\left(\frac{u_0 X}{D_0}\right)erfc\left\{\frac{X + u_0 t}{2\sqrt{D_0 t}}\right\} \qquad (15.57)$$

$$I(\psi,t) = C_i\exp\{(\gamma^2 D_0 + \gamma U_0)t - \gamma\psi\} \qquad (15.50)$$

$$\psi = \frac{x+y}{f(mt)} \qquad (15.51)$$

$$I(X,t) = C_i\exp\{(\gamma^2 D_0 + \gamma u_0)t - \gamma X\} \qquad (15.58)$$

$$D_0 = (D_{x0} + D_{y0}) \qquad (15.52)$$

$$X = \frac{x}{f(mt)} \qquad (15.59)$$

$$U_0 = (u_0 + v_0) \qquad (15.53)$$

When we put $y = 0$ in the above problem, it is converted into a one-dimensional problem, except it takes on an independent time variable that was earlier employed by Crank (1975). The solution obtained in this case is as follows:

15.6 Results and Discussions

The analytical solution given in Equation 15.46 for the Cauchy-type input condition, is computed for the input values $C_i = 0.01$, $C_0 = 1.0$, $u_0 = 0.009$ (km/year), $v_0 = 0.0009$ (km/year), $D_{x0} = 0.9$ (km²/year), $D_{y0} = 0.09$ (km²/year), $m = 0.1$ (/year), $\gamma = 0.1$ (/km), and $q = 0.001$ (/year) in transient groundwater flow in a finite domain $0 \le x \le 1$ (km) and $0 \le y \le 1$ (km) along longitudinal and transverse directions, respectively, and time domain $0.4 \le t \le 1.6$ (years) with a sinusoidal form of temporally dependent dispersion and an exponential form of temporally dependent dispersion. The solution is shown graphically in Figure 15.2, which depicts a two-dimensional contaminant concentration distribution. The transverse mixing may be considered in the case of shallow aquifers and surface water bodies; otherwise, it can be ignored because in comparison of the respective longitudinal components of velocity in water bodies of substantial depth, the transverse components of velocity and dispersion coefficients are approximately found one-tenth of the respective longitudinal components in the field conditions. The concentration value increases with time but decreases with the distance travelled for both forms of temporally dependent dispersion, that is, sinusoidal and exponential, represented by solid and dotted lines, respectively. One can easily observe that the concentration values near the source are the same for both the forms but as we move away from the origin, the concentration values for the sinusoidal form of dispersion are smaller than that for the exponential form of dispersion for all times. The solution obtained with the Cauchy-type boundary condition is more realistic

because it includes a linear combination of concentration and its gradient is prescribed on the boundary.

15.7 Sensitivity Analysis

The analytical solution given by Equation 15.46 is computed using a MATLAB® program and the concentration values, thus obtained, are shown in Tables 15.1 through 15.4 with input values of sensitive parameters, like dispersion, seepage velocity, and decay rate coefficients in the range of $D_{x0} = 0.7$–1.0 (km²/year), $D_{y0} = 0.07$–0.1 (km²/year), $u_0 = 0.007$–0.01 (km/year), $v_0 = 0.0007$–0.001 (km/year), and $q = 0.0008$–0.002 (/year). The concentration values at different positions obtained in rows (i) and (ii) are for the exponential form of velocity and sinusoidal form of velocity, respectively. It is observed that the concentration values for the sinusoidal form of velocity are less than that for the exponential form of velocity at each position with respect to time and space. Also it can be observed that the concentration values increase initially when we increase the values of sensitive parameters D_{x0}, D_{y0}, u_0, v_0, and q; however, the concentration values decrease with respect to distance.

When we decrease or increase the values of D_{x0}, D_{y0}, u_0, v_0, and q beyond the range mentioned above, the concentration values show anomalies, that is, the

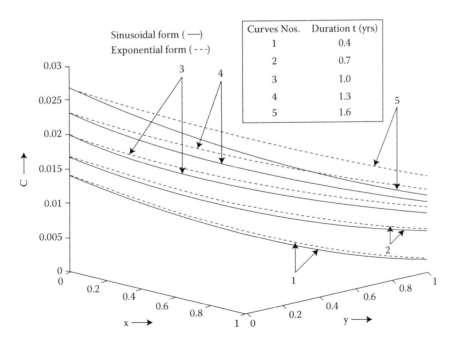

FIGURE 15.2
Two-dimensional concentration distribution for exponential form of velocity (dotted line), and sinusoidal form of velocity (solid line) for temporally dependent input source of concentration with the Cauchy-type boundary condition along transient flow.

TABLE 15.1

Contaminant Concentration with $D_{x0} = 0.7$, $D_{y0} = 0.07$, $u_0 = 0.007$, $v_0 = 0.0007$, and $q = 0.0008$ for (1) Exponential Form of Velocity and (2) Sinusoidal Form of Velocity

x	0	0.1	0.2	0.3	0.4	0.5	0.6	0.7	0.8	0.9	1.0
y	0	0.1	0.2	0.3	0.4	0.5	0.6	0.7	0.8	0.9	1.0
t = 0.4 year											
(i)	0.0113	0.0098	0.0083	0.0068	0.0054	0.0041	0.0029	0.0020	0.0013	0.0007	0.0003
(ii)	0.0113	0.0097	0.0081	0.0065	0.0049	0.0036	0.0024	0.0015	0.0009	0.0004	0.0002
t = 0.7 year											
(i)	0.0134	0.0119	0.0105	0.0092	0.0081	0.0070	0.0061	0.0054	0.0048	0.0044	0.0041
(ii)	0.0134	0.0117	0.0101	0.0087	0.0074	0.0063	0.0054	0.0048	0.0043	0.0040	0.0039
t = 1.0 year											
(i)	0.0157	0.0142	0.0129	0.0117	0.0106	0.0097	0.0089	0.0083	0.0077	0.0073	0.0070
(ii)	0.0157	0.0139	0.0123	0.0109	0.0098	0.0088	0.0081	0.0075	0.0070	0.0068	0.0066
t = 1.3 year											
(i)	0.0181	0.0167	0.0153	0.0142	0.0132	0.0123	0.0115	0.0108	0.0102	0.0098	0.0093
(ii)	0.0181	0.0162	0.0146	0.0132	0.0121	0.0111	0.0103	0.0097	0.0092	0.0087	0.0084
t = 1.6 years											
(i)	0.0207	0.0193	0.0179	0.0168	0.0157	0.0148	0.0139	0.0131	0.0125	0.0118	0.0113
(ii)	0.0207	0.0187	0.0170	0.0155	0.0143	0.0132	0.0122	0.0114	0.0107	0.0101	0.0095

distribution of concentration in space and time for this particular range of values of sensitive parameters is altered when we go beyond the range mentioned above. Hence, it obviously distorts the nature of the solution, which is bounded with some suitable values.

With this consideration, we fixed a range of values for sensitive parameters mentioned above and found that our solution for the present problem is bounded and acceptable for this particular range of values of sensitive parameters.

TABLE 15.2

Contaminant Concentration with $D_{x0} = 0.8$, $D_{y0} = 0.08$, $u_0 = 0.008$, $v_0 = 0.0008$, and $q = 0.0009$ for (1) Exponential Form of Velocity and (2) Sinusoidal Form of Velocity

x	0	0.1	0.2	0.3	0.4	0.5	0.6	0.7	0.8	0.9	1.0
y	0	0.1	0.2	0.3	0.4	0.5	0.6	0.7	0.8	0.9	1.0
t = 0.4 year											
(i)	0.0127	0.0110	0.0093	0.0077	0.0062	0.0049	0.0037	0.0028	0.0020	0.0015	0.0012
(ii)	0.0127	0.0108	0.0090	0.0073	0.0058	0.0044	0.0032	0.0023	0.0016	0.0012	0.0010
t = 0.7 year											
(i)	0.0151	0.0135	0.0119	0.0106	0.0093	0.0083	0.0073	0.0066	0.0060	0.0055	0.0052
(ii)	0.0151	0.0132	0.0115	0.0100	0.0087	0.0075	0.0066	0.0059	0.0054	0.0051	0.0050
t = 1.0 year											
(i)	0.0177	0.0161	0.0147	0.0134	0.0123	0.0113	0.0105	0.0098	0.0092	0.0087	0.0083
(ii)	0.0177	0.0158	0.0141	0.0127	0.0114	0.0104	0.0095	0.0089	0.0084	0.0080	0.0077
t = 1.3 year											
(i)	0.0206	0.0190	0.0176	0.0164	0.0152	0.0142	0.0134	0.0126	0.0119	0.0113	0.0107
(ii)	0.0206	0.0186	0.0168	0.0153	0.0140	0.0129	0.0120	0.0112	0.0105	0.0099	0.0094
t = 1.6 years											
(i)	0.0236	0.0221	0.0207	0.0194	0.0182	0.0171	0.0161	0.0151	0.0143	0.0135	0.0127
(ii)	0.0236	0.0215	0.0197	0.0180	0.0165	0.0151	0.0140	0.0129	0.0120	0.0111	0.0104

TABLE 15.3

Contaminant Concentration with $D_{x0} = 0.9$, $D_{y0} = 0.09$, $u_0 = 0.009$, $v_0 = 0.0009$, and $q = 0.001$ for (1) Exponential Form of Velocity and (2) Sinusoidal Form of Velocity

x	0	0.1	0.2	0.3	0.4	0.5	0.6	0.7	0.8	0.9	1.0
y	0	0.1	0.2	0.3	0.4	0.5	0.6	0.7	0.8	0.9	1.0
t = 0.4 year											
(i)	0.0141	0.0122	0.0104	0.0087	0.0072	0.0058	0.0045	0.0036	0.0028	0.0023	0.0020
(ii)	0.0141	0.0121	0.0102	0.0083	0.0067	0.0052	0.0040	0.0031	0.0024	0.0020	0.0018
t = 0.7 year											
(i)	0.0169	0.0151	0.0135	0.0120	0.0107	0.0096	0.0086	0.0078	0.0072	0.0067	0.0063
(ii)	0.0169	0.0148	0.0130	0.0114	0.0100	0.0088	0.0079	0.0071	0.0066	0.0062	0.0060
t = 1.0 year											
(i)	0.0199	0.0182	0.0167	0.0153	0.0141	0.0131	0.0121	0.0113	0.0106	0.0100	0.0095
(ii)	0.0199	0.0178	0.0160	0.0145	0.0131	0.0120	0.0111	0.0103	0.0096	0.0091	0.0086
t = 1.3 year											
(i)	0.0232	0.0216	0.0201	0.0187	0.0174	0.0163	0.0153	0.0144	0.0135	0.0127	0.0120
(ii)	0.0232	0.0211	0.0192	0.0175	0.0161	0.0148	0.0136	0.0126	0.0117	0.0110	0.0103
t = 1.6 years											
(i)	0.0268	0.0252	0.0236	0.0222	0.0208	0.0195	0.0182	0.0171	0.0160	0.0150	0.0141
(ii)	0.0268	0.0245	0.0225	0.0205	0.0187	0.0171	0.0156	0.0143	0.0131	0.0121	0.0112

TABLE 15.4

Contaminant Concentration with $D_{x0} = 1.0$, $D_{y0} = 0.1$, $u_0 = 0.01$, $v_0 = 0.001$, and $q = 0.002$ for (1) Exponential Form of Velocity and (2) Sinusoidal Form of Velocity

x	0	0.1	0.2	0.3	0.4	0.5	0.6	0.7	0.8	0.9	1.0
y	0	0.1	0.2	0.3	0.4	0.5	0.6	0.7	0.8	0.9	1.0
t = 0.4 year											
(i)	0.0157	0.0136	0.0117	0.0099	0.0082	0.0067	0.0054	0.0044	0.0036	0.0031	0.0027
(ii)	0.0157	0.0135	0.0114	0.0094	0.0077	0.0061	0.0049	0.0039	0.0032	0.0028	0.0026
t = 0.7 year											
(i)	0.0188	0.0169	0.0152	0.0136	0.0122	0.0110	0.0100	0.0091	0.0084	0.0078	0.0074
(ii)	0.0188	0.0166	0.0147	0.0130	0.0115	0.0102	0.0092	0.0083	0.0077	0.0072	0.0069
t = 1.0 year											
(i)	0.0222	0.0204	0.0188	0.0173	0.0160	0.0148	0.0138	0.0129	0.0121	0.0114	0.0107
(ii)	0.0222	0.0200	0.0181	0.0164	0.0149	0.0137	0.0126	0.0116	0.0108	0.0102	0.0096
t = 1.3 year											
(i)	0.0260	0.0243	0.0226	0.0211	0.0197	0.0184	0.0172	0.0161	0.0151	0.0142	0.0133
(ii)	0.0260	0.0238	0.0217	0.0198	0.0182	0.0166	0.0153	0.0140	0.0129	0.0120	0.0111
t = 1.6 years											
(i)	0.0301	0.0284	0.0267	0.0251	0.0235	0.0219	0.0204	0.0190	0.0177	0.0165	0.0154
(ii)	0.0301	0.0277	0.0254	0.0232	0.0210	0.0190	0.0172	0.0156	0.0142	0.0130	0.0119

15.8 Conclusions

For the analytical solution of two-dimensional solute transport along transient groundwater flow with time-dependent dispersion in a semi-infinite aquifer, the Laplace transform is used. The solutions describe the nature of contaminant concentration with respect to space and time for the Cauchy-type input condition. The results obtained for two expressions of temporally dependent dispersion, such as sinusoidally and exponentially increasing forms, are more realistic because the time-dependent input concentration is considered at the source. The time-dependent source of input concentration is a little more significant than the uniform source of input concentration. In the solution obtained for the Cauchy-type input condition, if we put $y = 0$, then the solution is converted to the solution of one-dimensional problem discussed by Singh et al. (2012) and it validates the solution of the present work in this case. In this chapter, the significance of transform techniques in solute transport modeling/groundwater was briefly presented. The occurrence of ADE and its methodology was explored to solve ADEs. The dispersion theory was employed here for solving ADE. Different types of boundary conditions applicable to represent the pollutant sources were discussed in this chapter to model the physical system. The sensitivity analysis was explored for input parameters taken into consideration for the physical system. The need for solute transport modeling in groundwater through transform techniques was presented here that can help manage groundwater resources, and remedial measures may then be accordingly developed.

Acknowledgment

The authors are thankful to the editors for their constructive comments and valuable suggestions, which have helped improve the quality of the chapter.

Appendix

The Laplace transform of a function $F(t)$ is as follows:

$$f(p) = L[F(t)] = \int_0^\infty e^{pt} F(t) dt$$

Its inverse is as follows:

$$F(t) = L^{-1}[f(p)]$$

Here, a series of inverse Laplace transforms of some functions is given in the form of a table. To find the inverse Laplace transform of different functions, the following abbreviations are used:

$$A = \frac{1}{\sqrt{\pi t}} \exp\left(-\frac{x^2}{4t}\right), \quad B = erfc\left(\frac{x}{2\sqrt{t}}\right),$$

$$C = \exp(a^2 t - ax) erfc\left(\frac{x}{2\sqrt{t}} - a\sqrt{t}\right) \quad \text{and,}$$

$$D = \exp(a^2 t + ax) erfc\left(\frac{x}{2\sqrt{t}} + a\sqrt{t}\right)$$

SI. No.	$f(p)$	$F(t)$
1.	$\dfrac{e^{-x\sqrt{p}}}{p - a^2}$	$\dfrac{1}{2}(C + D)$
2.	$\dfrac{e^{-x\sqrt{p}}}{\sqrt{p}(p - a^2)}$	$\dfrac{1}{2a}(C - D)$
3.	$\dfrac{\sqrt{p}\,e^{-x\sqrt{p}}}{(p - a^2)^2}$	$tA + \dfrac{1}{4a}(1 - ax + 2a^2 t)C - \dfrac{1}{4a}(1 + ax + 2a^2 t)D$
4.	$\dfrac{e^{-x\sqrt{p}}}{(p - a^2)^2}$	$\dfrac{1}{4a}(2at - x)C + \dfrac{1}{4a}(2at + x)D$
5.	$\dfrac{e^{-x\sqrt{p}}}{\sqrt{p}(p - a^2)^2}$	$\dfrac{t}{a^2}A - \dfrac{1}{4a^3}(1 + ax - 2a^2 t)C + \dfrac{1}{4a^3}(1 - ax - 2a^2 t)D$
6.	$\dfrac{e^{-x\sqrt{p}}}{(p - a^2)(\sqrt{p} + a)}$	$tA + \dfrac{1}{4a}C - \dfrac{1}{4a}(1 + 2ax + 4a^2 t)D$

(Continued)

SI. No.	$f(p)$	$F(t)$
7.	$\dfrac{e^{-x\sqrt{p}}}{(p-a^2)^2(\sqrt{p}+a)}$	$\dfrac{t}{4a^2}(1+ax+2a^2t)A+\dfrac{1}{16a^3}(4a^2t-2ax-1)C-\dfrac{1}{16a^3}[4a^2t-1+2a^2(x+2at)^2]D$
8.	$\dfrac{e^{-x\sqrt{p}}}{(p-a^2)(\sqrt{p}+a)^2}$	$\dfrac{1}{8a^2}C-\dfrac{t}{2a}(1+ax+2a^2t)A+\dfrac{1}{8a^2}[-1+2ax+8a^2t+2a^2(x+2at)^2]D$
9.	$\dfrac{e^{-x\sqrt{p}}}{(p-a^2)^2(\sqrt{p}+a)^2}$	$\dfrac{1}{16a^4}\left[1+a(4a^2t-1)(x+2at)+\dfrac{4}{3}a^3(x+2at)^3\right]D-\dfrac{1}{16a^4}(1+ax-2a^2t)C$ $-\dfrac{t}{12a^3}[-3+4a^2t+a^2(x+2at)^2]A$
10.	$\dfrac{e^{-x\sqrt{p}}}{(p-a^2)(\sqrt{p}+a)^3}$	$\dfrac{t}{12a^2}[-3+3ax+14a^2t+2a^2(x+2at)^2]A+\dfrac{1}{16a^3}[C+(2ax-1)D]$ $-\dfrac{1}{24a}[3(x+2at)(x+6at)+2a(x+2at)^3]D$
11.	$\dfrac{e^{-x\sqrt{p}}}{(p-a^2)(\sqrt{p}+b)}$	$\dfrac{1}{2(a+b)}C-\dfrac{1}{2(a-b)}D+\dfrac{b}{a^2-b^2}\exp(b^2t+bx)erfc\left(\dfrac{x}{2\sqrt{t}}+b\sqrt{t}\right)$
12.	$\dfrac{e^{-x\sqrt{p}}}{(p-a^2)(\sqrt{p}+b)^2}$	$\dfrac{1}{2(a+b)^2}C+\dfrac{1}{2(a-b)^2}D+\dfrac{2bt}{a^2-b^2}A$ $-\left[\dfrac{a^2+b^2}{(a^2-b^2)^2}+\dfrac{2b^2t+bx}{(a^2-b^2)}\right]\exp(b^2t+bx)erfc\left(\dfrac{x}{2\sqrt{t}}+b\sqrt{t}\right)$
13.	$\dfrac{e^{-x\sqrt{p}}}{(p-a^2)^2(\sqrt{p}+b)}$	$\dfrac{1-(2at-x)(a+b)}{4a(a+b)^2}C+\dfrac{(2at+x)(a-b)-1}{4a(a-b)^2}D$ $-\dfrac{t}{a^2-b^2}A+\dfrac{b}{a^2-b^2}\exp(b^2t+bx)erfc\left(\dfrac{x}{2\sqrt{t}}+b\sqrt{t}\right)$

References

Anderson, M. P. and W. W. Woessner. 2002. *Applied Groundwater Modeling: Simulation of Flow and Advective Transport*. Academic Press, London, England.

Aniszewski, A. 2009. Mathematical modeling and practical verification of groundwater and contaminant transport in a chosen natural aquifer. *Acta Geophysica* 57(2):435–453.

Aral, M. M. 1990. *Groundwater Modeling in Multilayer Aquifers: Unsteady Flow*. Lewis Publishers, Chelsea, Michigan.

Aral, M. M. and B. Liao 1996. Analytical solutions for two-dimensional transport equation with time-dependent dispersion coefficients. *Journal of Hydraulic Engineering* 1(1):20–32.

Barry, D. A. and G. Sposito. 1989. Analytical solution of a convection—Dispersion model with time-dependent transport coefficients. *Water Resources Research* 25(12):2407–2416.

Basha, H. A. and F. S. El-Habel. 1993. Analytical solution of the one-dimensional time dependent transport equation. *Water Resources Research* 29(9):3209–3214.

Batu, V. 2006. *Applied Flow and Solute Transport Modeling in Aquifers*. CRC Press, Taylor & Francis Group, Boca Raton, Florida.

Bear, J. 1972. *Dynamics of Fluids in Porous Media*. Elsevier, New York.

Bear, J. and D. K. Todd. 1960. The transition zone between fresh and saltwater in coastal aquifers. University of California, Water Resources Centre Contribution No. 29.

Bear, J. and A. Verruji. 1987. *Modeling Groundwater Flow and Pollution*. D. Reidel Publishing Company, Kluwer Academic Publishing Group, Holland.

Chen, J. S., Chen, J. T., Liu, C. W., Liang, C. P., and C. M. Lin. 2011. Analytical solutions to two-dimensional advection-dispersion equation in cylindrical coordinates in finite domain subject to first- and third-type inlet boundary conditions. *Journal of Hydrology* 405(3–4):522–531.

Chen, J. S. and C. W. Liu. 2011. Generalized analytical solution for advection-dispersion equation in finite spatial domain with arbitrary time-dependent inlet boundary condition. *Hydrology and Earth System Sciences* 15:2471–2479.

Chen, J. S., Liu, C. W., and C. M. Liao. 2003. Two-dimensional Laplace-transformed power series solution for solute transport in a radially convergent flow field, *Advanced Water Resources* 26:1113–1124.

Chrysikopoulos, C. V., Roberts, P. V., and P. K. Kitanidis. 1990. One-dimensional solute transport in porous media with partial well-to-well recirculation: Application to field experiment. *Water Resources Research* 26(6):1189–1195.

Crank, J. 1975. *The Mathematics of Diffusion*. Oxford University Press, Oxford, U.K.

Day, T. J. 1977. Longitudinal dispersion of fluid particles in mountain streams: 1. Theory and field evidence. *Journal of Hydrology* 16(1):7–25.

Freeze, R. A. and J. A. Cherry. 1979. *Groundwater*. Prentice-Hall, Upper Saddle River, NJ.

Fried, J. J. 1975. *Groundwater Pollution*, Elsevier Scientific Pub. Comp., Amsterdam.

Fried, J. J. and M. A. Combarnous. 1971. Dispersion in porous media. *Advances in Hydroscience* 7:169–281.

Fry, V. A., Istok, J. D., and R. B. Guenther. 1993. An analytical solution to the solute transport equation with rate-limited desorption and decay. *Water Resources Research* 29(9):3201–3208.

Ge, S. and Lu, N. 1996. A semi-analytical solution of one-dimensional advective-dispersive solute transport under an arbitrary concentration boundary condition. *Groundwater* 34(3):501–503.

Gehrson, N. D. and A. Nir. 1969. Effects of boundary conditions of model on tracer distribution in flow through porous mediums. *Water Resources Research* 5(4):830–839.

Ghosh, N. C. and K. D. Sharma. 2006. *Groundwater Modelling and Management*. Capital Publishing Company, New Delhi, 442–478.

Guerrero, J. S. P., Pimentel, L. C. G., Skaggs, T. H., and M. T. van Genuchten. 2009. Analytical solution of the advection–diffusion transport equation using a change-of-variable and integral transform technique. *International Journal of Heat and Mass Transfer* 52:3297–3304.

Guerrero, J. S. P., Pontedeiro, E. M., van Genuchten, M. T., and T. H. Skaggs. 2013. Analytical solutions of the one-dimensional advection-dispersion solute transport equation subject to time-dependent boundary conditions. *Chemical Engineering Journal* 221:487–491.

Hantush, M. M. and M. A. Marino, 1990. Temporal and spatial variability of hydraulic heads in finite heterogeneous aquifers: Numerical modeling. *Model CARE 90: Calibration and Reliability in Groundwater Modeling*, IAHS, 195:23–31.

Hantush, M. M. and M. A. Marino. 1998. Interlayer diffusive transfer and transport of contaminants in stratified formation. II: analytical solutions. *Journal of Hydrologic Engineering* 3(4):241–247.

Harleman, D. R. F. and R. R. Rumer. 1963. Longitudinal and lateral dispersion in an isotropic porous media. *Journal of Fluid Mechanics* 16:385–394.

Hoeksema, R. J. and P. K. Kitanidis. 1984. An application of the Geostatistical approach to the inverse problem in two-dimensional groundwater modeling. *Water Resources Research* 20(7):1003–1020.

Huang, K. and M. Th. van Genuchten. 1995. An analytical solution for predicting solute transport during ponded infiltration. U.S. Salinity Laboratory, USDA, ARS, Riverside, 159(4):217–223.

Jaiswal, D. K., Kumar A., Kumar, N., and M. K. Singh. 2011. Solute transport along temporally and spatially dependent flows through horizontal semi-infinite media: Dispersion proportional to square of velocity. *Journal of Hydrologic Engineering* 16(3):228–238.

Jaiswal, D. K., Kumar, A., Kumar, N., and R. R. Yadav. 2009. Analytical solutions for temporally and spatially dependent solute dispersion of pulse type input concentration in one-dimensional semi-infinite media. *Journal of Hydro-Environment Research* 2:254–263.

Javandel, I., Doughty, C., and C. F. Tsang. 1984. *Groundwater Transport: Handbook of Mathematical Models*. American Geophy. Union, Washington, D.C.

Kumar, A., Jaiswal, D. K., and N. Kumar. 2012. One-dimensional solute dispersion along unsteady flow through a heterogeneous medium, dispersion being proportional to the square of velocity. *Hydrological Sciences Journal* 57(6):1223–1230.

Kumar, N. 1983. Dispersion of pollutants in semi-infinite porous media with unsteady velocity distribution. *Nordic Hydrology* 14:167–178.

Kumar, N. and S. K. Yadav. 2013. Pollutant dispersion in groundwater: Its degradation and rehabilitation. *Proceedings of the National Conference Recent Advances in Mathematics and Applied Mathematics* 1:1–25.

Kumar, R., Singh, R. D., and K. D. Sharma. 2005. Water resources of India. *Current Science*, 89(5): 794–811.

Kumari, P. 2013. Study of Analytical and Numerical Approach of Solute Transport Modeling. PhD thesis, ISM Dhanbad.

Leij, F. J. and M. Th. van Genuchten. 1995. Approximate analytical solutions for solute transport in two-layer porous media. *Transactions of Porous Media* 18:65–85.

Liggette, J. A. 1987. Advances in the boundary integral equation method in subsurface flow. *Water Resources Bulletin* 23(4):637–651.

Liggette, J. A. and P. L. F. Liu. 1983. *The Boundary Integral Equation Method for Porous Media Flow*. Allen and Unwin, London.

Liu, C., Szecsody, J. E., Zachara, J. M., and P. B. William. 2000. Use of the generalized integral transform method for solving equations of solute transport in porous media. *Advances in Water Research* 23:483–492.

Logan, J. D. 1996. Solute transport in porous media with scale dependent dispersion and periodic boundary conditions. *Journal of Hydrology* 184(3–4):261–276.

Logan, J. D. and V. Zlotnik. 1995. The convection-diffusion equation with periodic boundary conditions. *Applied Mathematics Letters* 8(3):55–61.

Mahato, N. K. 2013. Study of Solute Transport Modeling along Unsteady Groundwater flow in Aquifer. PhD thesis, ISM Dhanbad.

Massabo, M., Cianci, R., and O. Paladino. 2006. Some analytical solutions for two-dimensional convection-dispersion equation in cylindrical geometry. *Environmental Modelling and Software* 21(5):681–688.

Park, Y. S. and J. J. Baik. 2008. Analytical solution of the advection-diffusion equation for a ground-level finite area source. *Atmospheric Environment* 42:9063–9069.

Parker, J. C. and M. Th. Van Genuchten. 1984. Flux-averaged and volume-averaged concentrations in continuum approaches to solute transport. *Water Resources Research* 20(7):866–872.

Rastogi, A. K. 2007. *Numerical Groundwater Hydrology*. Penram International Publishing Pvt. Ltd., India.

Rausch, R., Schafer, W., Therrien, R., and C. Wagner. 2005. *Solute Transport Modeling: An Introduction to Models and Solution Strategies*. Gebr. BorntragerVerlagsbuchhandlung Science Publishers, Berlin, Germany.

Sander, G. C. and R. D. Braddock. 2005. Analytical solutions to the transient, unsaturated transport of water and contaminants through horizontal porous media. *Advances in Water Resources* 28:1102–1111.

Scheidegger, A. E. 1957. *The Physics of Flow through Porous Media*. University of Toronto Press, Toronto.

Serrano, S. E. 2001. Solute transport under non-linear sorption and decay. *Water Research* 35(6):1525–1533.

Shan, C. and I. Javandel. 1997. Analytical solutions for solute transport in a vertical aquifer section. *Journal of Contaminant Hydrology* 27:63–82.

Sharma, H. D. and K. R. Reddy. 2004. *Geo-Environmental Engineering*. John Wiley & Sons, New York.

Shukla, V. P. 2002. Analytical solutions for unsteady transport dispersion of non-conservative pollutant with time-dependent periodic waste discharge concentration. *Journal of Hydraulic Engineering* 128(9):866–869.

Singh, M. K., Ahamad, S., and V. P. Singh. 2012. Analytical solution for one-dimensional solute dispersion with time-dependent source concentration along uniform groundwater flow in a homogeneous porous formation. *Journal of Engineering Mechanics* 8138(8):1045–1056.

Singh, M. K., Ahamad, S., and V. P. Singh. 2014. One-dimensional uniform and time varying solute dispersion along transient groundwater flow in a semi-infinite aquifer. *Acta Geophysica* 62(4):872–892.

Singh, M. K., and P. Das. 2015. Scale dependent solute dispersion with linear isotherm in heterogeneous medium. *Journal of Hydrology* 520:289–299.

Singh, M. K. and P. Kumari. 2014. Contaminant concentration prediction along unsteady groundwater flow. *Modelling and Simulation of Diffusive Processes, Series: Simulation Foundations, Methods and Applications,* Springer XII:339.

Singh, M. K., Mahato, N. K., and N. Kumar. 2015. Pollutant's horizontal dispersion along and against sinusoidally varying velocity from a pulse type point source. *Acta Geophysica* 63(1):214–231.

Singh, M. K., Mahato, N. K., and V.P. Singh. 2011. Longitudinal dispersion with constant source concentration along unsteady groundwater flow in finite aquifer: Analytical solution with pulse type boundary condition. *Natural Science* 3(3):186–192.

Singh, M. K., Mahato, N. K., and V. P. Singh. 2013. An analytical approach to one-dimensional solute dispersion along and against transient groundwater flow in aquifers. *Journal of Groundwater Research* 2(1):65–78.

Singh, M. K., Singh, P., and V. P. Singh. 2009a. Solute transport model for one-dimensional homogeneous porous formations with time dependent point-source concentration. *Advances in Theoretical and Applied Mechanics* 2(3):143–157.

Singh, M. K., Singh, P., and V. P. Singh. 2010a. Analytical solution for solute transport along and against time dependent source concentration in homogeneous finite aquifer. *Advances in Theoretical and Applied Mechanics* 3(3):99–119.

Singh, M. K., Singh, P., and V. P. Singh. 2010b. Analytical solution for two-dimensional solute transport in finite aquifer with time-dependent source concentration. *Journal of Engineering Mechanics* 136(10):1309–1315.

Singh, M. K., Singh, V. P., Singh, P., and D. Shukla. 2009b. Analytical solution for conservative solute transport in one-dimensional homogeneous porous formation with time-dependent velocity. *Journal of Engineering Mechanics* 135(9):1015–1021.

Singh, P., Singh, M. K., and V. P. Singh. 2010c. *Contaminant Transport in Unsteady Groundwater Flow: Analytical Solutions.* LAP LAMBERT Academic Publishing, Germany.

Smedt, F. D. 2006. Analytical solutions for transport of decaying solutes in rivers with transient storage. *Journal of Hydrology* 330:672–680.

Srinivasan, V. and T. P. Clement. 2008a. Analytical solutions for sequentially coupled one-dimensional reactive transport problems-part I: Mathematical derivations. *Advances in Water Resources* 31:203–218.

Srinivasan, V. and T. P. Clement. 2008b. Analytical solutions for sequentially coupled one-dimensional reactive transport problems-part II: Apecial cases, implementation and testing. *Advances in Water Resources* 31:219–232.

Strack, O. D. L. 1987. *The Analytic Element Method for Regional Groundwater Modeling in: Solving Ground Water Problems with Models.* National Water Well Assoc., Columbus, OH, 929–941.

Strack, O. D. L. 1988. *Groundwater Mechanics.* Prentice-Hall, Enlgewood Cliffs, NJ, 732.

Su, N., Sander, G. C., Fawang, L., Anh, V., and D. A. Barry. 2005. Similarity solutions for solute transport in fractal porous media using a time and scale-dependent dispersivity. *Applied Mathematical Modeling* 29:852–870.

Taylor, G. 1953. Dispersion of soluble matter in the solvent flowing slowly through a tube. *Proceedings of the Royal Society of London, A* 219:186–203.

Thangarajan, M. 2006. *Groundwater: Resource Evaluation, Augmentation, Contamination, Restoration, Modeling and Management.* Capital Publishing Company, New Delhi, India.

Toride, N., Leij, F. J., and M. Th. van Genuchten. 1993. A comprehensive set of analytical solutions for non-equilibrium solute transport with first-order decay and zero-order production. *Water Resources Research* 29(7):2167–2182.

Tsai, W. F. and C. J. Chen. 1995. Unsteady finite-analytic method for solute transport in ground-water flow. *Journal of Engineering Mechanics* 121(2):230–243.

Valocchi, A. J. and P. V. Roberts. 1983. Attenuation of groundwater contaminant pulses. *Journal of Hydraulic Engineering* 109(12):1665–1682.

van Genuchten, M. Th. 1981. Analytical solution for chemical transport with simultaneous adsorption, zero order production and first order decay. *Jouranl of Hydrology* 49:213–233.

van Genuchten, M. Th. 1982. A comparison of numerical solutions of the one-dimensional unsaturated-saturated flow and mass transport equations. *Advances in Water Resources* 5:47–55.

van Genucheten, M. Th. and W. J. Alves. 1982. Analytical solution of one dimensional convective-dispersion solute transport equation. Technical Bulletin No. 1661, 1–51. U.S. Department of Agriculture, Washington DC.

Wang, H., Han, R., Zhao, Y., Lu, W., and Y. Zhang. 2011. Stepwise superposition approximation approach for analytical solutions with non-zero initial concentration using existing solutions of zero initial concentration in contaminate transport. *Journal of Environmental Science* 23(6):923–930.

Yeh, H. D. and G. T. Yeh, 2007. Analysis of point-source and boundary-source solutions of one-dimensional groundwater transport equation. *Journal of Environmental Engineering* 133(11):1032–1041.

Zhan, H., Wen, Z., Huang, G., and D. Sun. 2009. Analytical solution of two-dimensional solute transport in an aquifer-aquitard system. *Journal of Contaminant Hydrology* 107:162–174.

16

Precise Hydrogeological Facies Modeling for Vertical 2-D and 3-D Groundwater Flow, Mass Transport, and Heat Transport Simulation

Naoaki Shibasaki

CONTENTS

16.1 Introduction

Groundwater simulation using computer technology has been employed since the 1960s. After the first application of digital computer simulation by Tyson and Weber (1964), a similar groundwater simulation model was applied to the aquifer system in the western part of Tokyo by Shibasaki et al. (1969). The model was called a horizontal two-dimensional model, but the leakage term from the upper unconfined aquifer through the upper confining layer was taken into account in the model considering the hydrogeological settings of the Musashino Terrace in western Tokyo. In the 1970s, serious damages caused by land subsidence became significant social problems in Japan (Shibasaki and

Research Group for Water Balance, 1995). To simulate occurrence of land subsidence in the Quaternary aquifer system, the quasi-three-dimensional aquifer model (Kamata et al., 1973) was developed. In the model, leakage from the upper confined aquifer and the squeeze from the confining layers were simulated. Based on the simulated heads in the confining clay layer, the clay compaction was calculated using Terzaghi's one-dimensional consolidation theory.

The importance of hydrogeological facies classification was recognized in mid-1970s when the land subsidence caused by the deep extraction of water-soluble natural gas became serious issues in the Funabashi gas field, Japan. Because the quasi-three-dimensional model could not simulate piezometric heads in multilayered geologic structures, the vertical two-dimensional multiple aquifers model (Kamata et al., 1976) was developed. The model was applied to the hydrogeological cross-section up to a depth of 2000 m in the Funabashi gas field to simulated head distribution and occurrence of soil compaction in different geologic layers. In the late 1970s, the quasi-three-dimensional multi-aquifers model (Fujinawa, 1977; Fujisaki et al., 1979) was developed to simulate groundwater flow in a multilayered groundwater basin. The model was used to find the optimal groundwater pumpage in the groundwater basin without causing additional land subsidence.

The author used the quasi-three-dimensional multi-aquifers model (Fujisaki et al., 1979) in the 1980s to early part of the 1990s to simulate and predict groundwater flow and land subsidence in several groundwater basins in Japan. The model code was also translated for a notebook-type computer by the author and applied to the Metro Manila groundwater basin in the Philippines during the project carried out by the Japan International Cooperation Agency (JICA, 1992). In JICA's groundwater management and land subsidence control project in Bangkok, Thailand, the author employed the MODFLOW code (McDonald and Harbaugh, 1988) to simulate a three-dimensional groundwater flow system (Kokusai Kogyo and JICA, 1995). The groundwater levels and land subsidence in the Bangkok metropolitan area were carefully monitored (Ramnarong et al., 1995) and a methodology of practical parameter estimation for 3-D groundwater modeling based on hydrogeological classification was developed (Shibasaki, 1999).

In this chapter, the author reviews the methods of parameter estimation for vertical 2-D and 3-D groundwater modeling. Then a more precise method of parameter estimation based on hydrogeological facies modeling is introduced. The detailed method will help reliable simulations of groundwater flow, mass transport, and heat transport studies.

16.2 Practical Parameter Estimation in Japan Aquifer

Prior to creating groundwater models, it is necessary to understand the subsurface geologic conditions in a groundwater basin. Especially in Japan, most groundwater basins are composed of Quaternary sediments, so the hydrostratigraphic units should be identified based on various geologic and hydrogeologic data. As already pointed out by the Research Group for Water Balance (1973,1976), integrated geological investigations and interpretations utilizing the knowledge and methods

of Quaternary geology, hydrogeology, applied geology, geophysics, civil engineering, etc. are necessary to demarcate hydrostratigraphic units.

In this section, a case study of practical parameter estimation for the vertical 2-D groundwater simulation in the Kanto groundwater basin, Japan is reviewed.

16.2.1 Outline of Kanto Groundwater Basin

The Kanto Plain is the most populated area in Japan, where Tokyo Metropolis and six prefectures are situated. The Kanto groundwater basin is the largest groundwater basin in Japan. The first benchmark in Tokyo was installed in 1891 and the maximum land subsidence reached 4.5 m in the 1970s (Shibasaki and Research Group for Water Balance, 1995). From the 1980s, the depression of the piezometric head had moved to the north and the land subsidence more than 2 cm/year recorded in the wide areas of the northern Kanto Plain.

The monitoring network of groundwater level and land subsidence was expanded in the whole Kanto groundwater basin, having 292 sites and 450 observation wells as of March 1986 (Shibasaki and Research Group for Water Balance, 1995). However, by the time of the mid-1980s, groundwater simulation studies were separately carried out by prefecture and ministry. From the mid-1980s, the Kanto Regional Agricultural Administration Office of the Ministry of Agriculture, Forestry and Fisheries, Japan started detailed land subsidence studies in the central and the northern parts of the Kanto Plain (Figure 16.1). The author participated in the project and the geohydrologic parameters were estimated (Shibasaki, 1989; Matsumoto et al., 1994). Here the brief method of practical parameter estimation for vertical 2-D model is reviewed.

16.2.2 Establishment of Basic Subsurface Stratigraphy

In the 1980s, several groundwater and land subsidence observations wells with depths up to 600 m were drilled by continuous core boring (Shibasaki, 1989). Although the previous subsurface geological profiles were made based on the viewpoints of continuity of gravel layers that were obtained from cutting samples and resistivity logs of existing wells, the Geological Research Group of Central Kanto Plain (1994) carried out integrated core sample analyses including volcanic ash, composition of heavy minerals and rock fragments, gravel composition, paleomagnetism, total sulfur content, molluscan fossils, foraminiferal fossils, diatom fossils, pollen fossils, and fission track dating. Based on the detailed core sample analyses with identification of sedimentation cycles at five locations in the central Kanto Plain, the basic subsurface stratigraphy up to 600 m in depth was

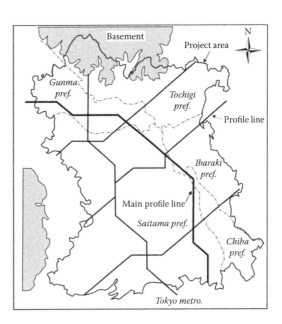

FIGURE 16.1
Location of Kanto groundwater basin, Japan.

established. Then the basic stratigraphy was expanded to 14 other existing core drillings with depths from 185 to 700 m located in the central to northern parts of the Kanto Plain. As a result, nine stratigraphic units were identified. Each stratigraphic unit basically consists of one sedimentary cycle starting from gravelly layers.

16.2.3 Preparation of Subsurface Geological Profile

To clarify the subsurface geology of the Kanto groundwater basin, a total of five subsurface geological profiles were created (Matsumoto et al., 1994). For each geological profile, columnar sections of the analyzed core borings and existing wells were projected. The stratigraphic boundaries identified at the core borings were carefully traced among the columnar sections of existing wells, paying attention to the sedimentary cycles found in the core borings.

In general, the accuracy of lithological descriptions of existing deep wells are poor because the descriptions are made based on the observation of cutting samples and the results of electric loggings. Furthermore, the level of lithological description varies by the drilling operators. However, it was assumed that the recognition of aquifer facies (coarse sediments such as gravel layers) should be identical because that is the most important factor to make the well successful. Therefore, it was judged that the brief facies classification, such as gravel, sand, and clay of the columnar section of existing wells, was reliable.

As shown in Figure 16.2, the created subsurface geological profile shows clear extent and structure of each

stratigraphic unit (Matsumoto et al., 1994). In addition, the subsurface geology was clearly correlated with the surface geology outcropping at the western marginal area of the Kanto groundwater basin. The several faults along the Motoarakawa tectonic belt (Shimizu and Horiguchi, 1981) having NW-SE orientation were also shown in the profile. These faults affect the continuities and the structures of those stratigraphic units in the profile.

16.2.4 Preparation of Hydrogeological Profile

Based on the established basic subsurface stratigraphy and the created subsurface geological profiles said above, hydrogeological profiles were prepared (Matsumoto et al., 1994). In the hydrogeological profiles, facies classification was made considering the aquifer productivities. The well depths, locations of screen pipes, transmissivity values, specific storage values, and the height of the piezometric head are also shown in the hydrogeological profiles.

As mentioned earlier, nine stratigraphic units were identified in the Kanto groundwater basin. These stratigraphic units were classified into four hydrogeologic layers. The hydrogeological classification was made by the values of specific storage reflecting the dynamic properties of soil by geologic age. The hydrogeologic layers were named C layer (the Lower to Middle Pleistocene), B layer (the Middle Pleistocene), A layer (the Upper Pleistocene), and H layer (the Late Pleistocene to Holocene) in ascending order. In the profile as shown in Figure 16.3, the discontinuities of aquifers by the faults are well expressed and the elevations of the layer

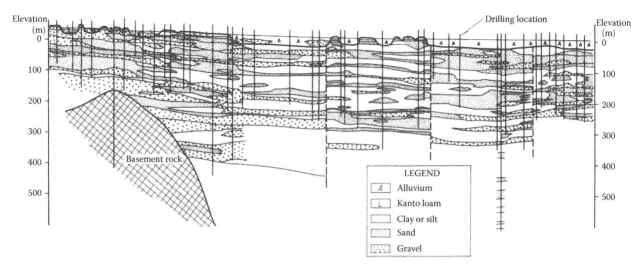

FIGURE 16.2
Subsurface geological profile along NW-SE line.

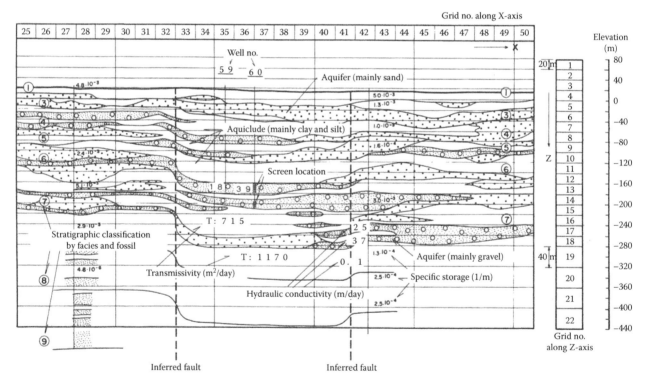

FIGURE 16.3
Hydrogeological profile along SW-NE line.

boundaries are lowered showing graben-shape along the Motoarakawa tectonic belt and the Arakawa River, which is the main river flowing through the central Kanto Plain.

16.2.5 Creation of Vertical 2-D Model

The vertical 2-D multiple aquifers models were created in the Kanto groundwater basin to simulate groundwater

flow and land subsidence (Shibasaki, 1989; Matsumoto et al., 1994). The vertical 2-D model is suitable to simulate hydraulic heads in a multilayered aquifer system and to compute layer compactions in different layers particularly in a groundwater basin having complexity of aquifer system and facies change.

The largest model was created along one of the hydrogeological profiles located from the northwestern to the southeastern parts in the basin (Figure 16.4).

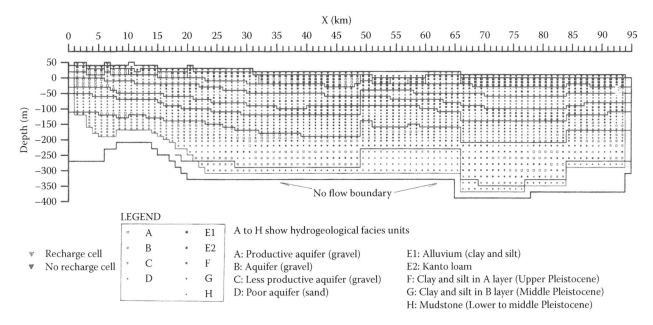

FIGURE 16.4
Vertical 2-D model along NW-SE line.

The length of the model is 92 km. The orientation of the model is approximately the same with the main groundwater flow direction. Considering the thickness of the geologic layers and the geologic structures expressed in the hydrogeological profile, the model cell has 1 km in size along the x-direction and 10 m in thickness along the z-direction up to the 26th model layers (= − 200 m in elevation) where changes in facies are significant. Below −200 m in elevation, the thickness of the model layers was set as 20 m.

The vertical model grid system was superimposed on the hydrogeological profile, and then the hydrogeological facies unit was identified at each cell. The hydrogeological facies were briefly classified into (1) sand and gravel facies and (2) clay and silt facies, and then the facies were subclassified into nine units based on the results of aquifer parameter estimation mentioned later.

Since the created vertical 2-D model was crossing the central part of the Kanto groundwater basin, three kinds of boundary conditions were assigned to the model considering the actual hydrogeological conditions. It was assumed that the left (NW) and right (SE) perimeters of the model have inflow/outflow of groundwater so that constant head boundaries were assigned. Near the constant head boundaries so-called "cushion zones" where greater values of hydraulic conductivity and specific storage than the actual ones were assigned to avoid extreme rise/decline of hydraulic heads. At the ground surface, the constant head boundaries were assigned to express the occurrence of groundwater recharge. It was assumed that the groundwater table in the unconfined aquifers

resembles the river bed elevations; therefore, the prescribed heads in the terrace areas and alluvial lowland were set as 5–10 m lower and a few meters lower than the ground elevation, respectively. No flow boundaries were assigned at the model bottom because the geologic layer was assumed to have very low hydraulic conductivity.

16.2.6 Parameter Estimation Using Well Production Test and Soil Test

Reliable parameter estimation is one of the important points for groundwater simulation studies. Especially for the regional groundwater simulation, it is common that the hydrogeologic structures and aquifer parameters vary within the model domain. Therefore, it is necessary to collect reliable data at as many as possible different locations for estimating aquifer parameters.

16.2.6.1 Hydraulic Conductivity

Generally hydraulic conductivity values are obtained from the results of pumping tests and laboratory soil tests. In the study, the data of as many as possible pumping tests and laboratory soil tests were collected; however, it was not sufficient to grasp a regional distribution of hydraulic conductivity. Therefore, the empirical method developed by Logan (1964) was employed to estimate transmissivity from specific capacity using well production test data. Then the hydraulic conductivity values were calculated using the screen lengths of wells.

According to Logan (1964), the empirical relationship between transmissivity and specific capacity in a confined aquifer can be expressed by:

$$T = 1.22Sc \tag{16.1}$$

where T is the transmissivity (m²/day) and Sc is the specific capacity (m³/day/m). The hydraulic conductivity values can be computed by the definition of the transmissivity, that is:

$$K = T/b \tag{16.2}$$

where K is the hydraulic conductivity (m/day) and b is the thickness (m) of the confined aquifer. Therefore, if the well inventories having groundwater pumpage, drawdown, and a total length of screen pipes are available, the transmissivity value can be computed at each well using Equation 16.1. Furthermore, if it can be assumed that the total screen length is the same as the thickness of the aquifer(s), hydraulic conductivity values can simply be computed using Equation 16.2. For clay and silt facies, hydraulic conductivity values obtained from existing soil consolidation tests were used.

However, it is common that the hydraulic conductivity values obtained by the said methods generally take a wide range of variation. A statistical method was hence employed to identify typical values of hydraulic conductivity by facies and stratigraphy. Since the statistical distributions of geohydrologic parameters including hydraulic conductivity show logarithmic normal distributions (Shibasaki et al., 1967), the logarithmic average and logarithmic standard deviation by facies and stratigraphy were computed after collecting thousands of exiting well records.

As a result, it was revealed that the hydraulic conductivities of aquifer facies (gravel and sand facies) show unique values by facies, but the values were regardless of stratigraphic unit. In addition, the variation of hydraulic conductivity was small in each facies (Figure 16.5). On the other hand, the distribution of hydraulic conductivities in clay and silt facies and mudstone facies of C layer take different ranges by stratigraphic unit (Figure 16.6). It was found that the hydraulic conductivities decreased with depth, showing that the geohydrologic properties of clay to silt facies reflect their geohistory. Therefore, it is necessary for estimating hydraulic conductivities of clay and silt to identify the differences in consolidation history and degree of diagenesis by stratigraphic unit.

16.2.6.2 Specific Storage

It is common that the values of storage coefficient obtained by pumping tests take a wide range of variation even if the test is precisely and carefully performed. For that reason, the specific storage values of clay to silt facies were estimated from the volume compressibility values obtained by soil consolidation test (Figure 16.7).

According to the consolidation theory with neglecting the compressibility of water, the specific storage can be written as

$$Ss = m_v \cdot \rho_w g = m_v \cdot \gamma_w \tag{16.3}$$

where Ss is the specific storage, m_v is the volume compressibility, ρ_w is the density of water, g is the acceleration of gravity, and γ_w is the unit weight of water. When the units are assigned as Ss (1/m), m_v (cm²/kg), and γ_w (g/cm³), the Ss can be computed by

$$Ss = 1/10 \cdot m_v \cdot \gamma_w \tag{16.4}$$

$$S = Ss \cdot b \tag{16.5}$$

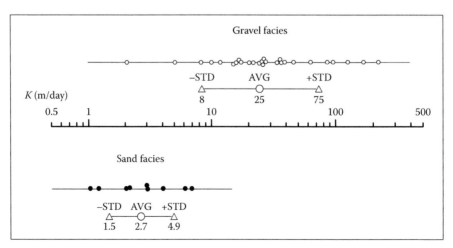

FIGURE 16.5
Hydraulic conductivity of gravel and sand facies.

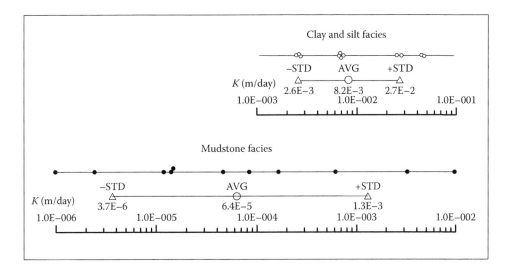

FIGURE 16.6
Hydraulic conductivity of clay to silt facies and mudstone facies.

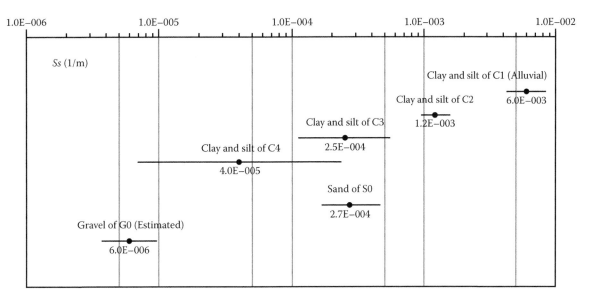

FIGURE 16.7
Hydraulic conductivity of clay to silt facies and mudstone facies.

where S is storage coefficient and b the thickness of the confined aquifer.

For the aquifer facies (gravel and sand facies), typical values of specific storage presented by Domenico and Mifflin (1965) were assigned considering the facies descriptions.

16.3 Practical Parameter Estimation in Thailand Aquifer

The method of practical parameter estimation was applied to the Bangkok groundwater basin where severe land subsidence has occurred since the 1970s. The JICA conducted an integrated study project on management of groundwater and land subsidence in the Bangkok metropolitan area and its vicinity from 1992–1995 (Kokusai Kogyo and JICA, 1995). In this section, a case study of practical parameter estimation for the 3-D groundwater simulation in the Bangkok groundwater basin is reviewed.

16.3.1 Outline of Bangkok Groundwater Basin

The Bangkok groundwater basin is situated beneath the flood plain of the Chao Phraya River, in the Lower Central Plain of Thailand, and it extends southward beneath the Gulf of Thailand. The basin is also known as the Lower Chao Phraya basin. The basin is bounded on the east and west by mountains and on the north by small hills that divide it from the Upper Central Plain.

About the subsurface hydrogeology of the Bangkok groundwater basin, the Department of Mineral Resources (DMR) of Thailand identified and named 8 aquifers within 550 m in depth based on logs of groundwater wells. These aquifers consist mainly of sand and gravel separated by virtually impervious strata of clay (Ramnarong, 1976). After drilling about 250 monitoring wells at about 100 monitoring stations, the underlying unconsolidated deposits such as sand, gravel, and clay were subdivided into 8 principle aquifer units as follows (Ramnarong and Buapeng, 1992):

1. Bangkok aquifer (BK) (50 m zone)
2. Phra Pradaeng aquifer (PD) (100 m zone)
3. Nakhon Luang aquifer (NL) (150 m zone)
4. Nonthaburi aquifer (NB) (200 m zone)
5. Sam Khok aquifer (SK) (300 m zone)
6. Phaya Thai aquifer (PT) (350 m zone)
7. Thon Buri aquifer (TB) (450 m zone)
8. Pak Nam aquifer (PN) (550 m zone)

16.3.2 Facies Modeling for Each Aquifer Unit

Transmissivity is the most important parameter for the groundwater models including the MODFLOW (McDonald and Harbaugh, 1988). Generally, transmissivity values are obtained from pumping tests. In the Bangkok Metropolitan area, pumping tests records of existing wells were collected and detailed pumping tests were performed at the newly constructed 18 monitoring wells by JICA (Kokusai Kogyo and JICA, 1995). However, the number of pumping tests and obtained aquifer parameters were still inadequate to evaluate

regional aquifer characteristics. Therefore, the transmissivity values by aquifer unit were estimated from the hydrogeological information and specific capacity data of the production wells listed on the groundwater database. Fortunately, Kokusai Kogyo and JICA (1995) made the groundwater database having more than 13,000 existing well records. About a half of the well records had lithologic logs and well production test data. Although the accuracy of existing well data may be poorer than those of the monitoring wells' data, the regional aquifer characteristics could be figured by aquifer unit (Shibasaki, 1999).

Due to the limitation of calculation speed and memory capacity of personal computers in the 1990s, the 3-D groundwater simulation model having eight model layers was created for the Bangkok groundwater basin (Kokusai Kogyo and JICA, 1995; Shibasaki, 1999). The concept of the facies modeling adapted to the Bangkok groundwater basin is shown in Figure 16.8. It was revealed from the hydrogeological profiles that several clay layers occur in each aquifer unit and these clay layers have different continuities. Therefore, it was judged that the isopach map of clay facies for each aquifer unit should be prepared and those clay layers should be treated as a single clay layer.

After the preparation of subsurface geological profiles, the subsurface hydrostratigraphic units were identified at each columnar section of the exiting well. Then the XYZ data of the top and bottom elevations of each aquifer unit as well as the total thickness of clay facies in each aquifer unit, which is mentioned later, were prepared for the Kriging interpolation. It can be assumed that most horizontal groundwater flow occurs in the aquifer facies, which is composed of gravel facies and sand facies. On the other hand, contribution of the clay portion to the total transmissivity of the aquifer unit is considerably small because of its small hydraulic conductivity. Therefore, if the thickness of aquifer facies

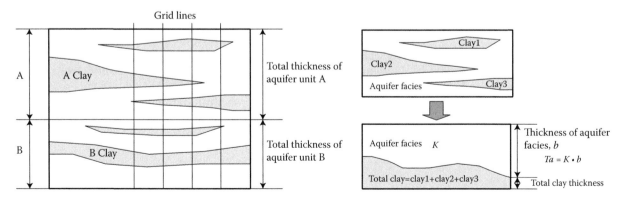

FIGURE 16.8
Schematic profiles showing concept of facies modeling by aquifer unit. (*Note*: The left figure shows an illustration of actual distributions of aquifer facies and clay facies. The right figure shows the differentiation of aquifer facies and clay facies by the facies modeling method developed by Shibasaki (1999).)

and its hydraulic conductivity are known, the transmissivity of aquifer facies can be computed.

16.3.3 Preparation of Clay Content Map

After the identification of subsurface hydrostratigraphic units at each columnar section, the total thickness of clay facies in the aquifer unit was also computed. In the Bangkok case, the thickness of clay was computed by the multipliers shown in Table 16.1. The multiplier values were assumed based on the general soil classification. In case a geologic layer was described as sandy clay in the columnar section, a value of 0.7 was multiplied to the thickness. Similarly, a value of 0.3 was multiplied to the thickness when the layer was described as clayey sand. The total thickness of clay facies in the aquifer unit was then computed by adding up those thicknesses.

After gridding the top and bottom elevations of each aquifer unit by the Kriging method, the isopach map of each aquifer unit was created. Then the clay content of each aquifer unit was computed by Equation 16.6 and the clay content map was prepared:

$$(\text{Clay content}) = \frac{\left(\begin{array}{c} \text{total thickness of clay in} \\ \text{the aquifer unit} \end{array}\right)}{\left(\text{thickness of the aquifer unit}\right)} \times 100\,(\%)$$

(16.6)

16.3.4 Estimation of T and K from Well Production Test

In the Bangkok groundwater basin, detailed pumping tests were carried out at 18 monitoring wells drilled in the three monitoring stations constructed by Kokusai Kogyo and JICA (1995). Transmissivity and storage coefficient values of each aquifer unit were obtained by a series of pumping tests consisting of step-drawdown tests, continuous pumping tests, and recovery tests. However, it was very difficult to estimate parameter distribution in the entire Bangkok groundwater basin from the limited data. Further, it was still difficult to assign reliable parameters to each model cell from

existing pumping test results because the number of existing pumping tests was inadequate for covering the groundwater basin. However, the 3-D simulation model required input parameters at every 2×2 km cell for each aquifer unit.

Generally pumping tests have been conducted at monitoring wells and some production wells in most groundwater basins; the number of pumping tests and obtained aquifer parameters are still inadequate to evaluate regional aquifer characteristics in most cases. Further, even if precise pumping tests with several observation wells are conducted, the obtained aquifer parameters still take wide ranges of variation and the reliability of the parameters are limited (Research Group for Water Balance, 1976). In practice, it is not easy to carry out many accurate pumping tests due to economical constraint. Therefore, the following methods of estimating transmissivity (T) and hydraulic conductivity (K) developed by Shibasaki (1999) were employed in the study.

Classify production wells by aquifer unit using screen depths: The production wells having screen length and production test data were used to compute specific capacity. The tapped aquifer of each well was identified by the top and bottom elevation maps of the aquifer units. If the well has screens located at two or more aquifer units, the specific capacity value was not used for the statistical analysis.

Compute specific capacity value from production test: Specific capacity values were simply computed using the production test data. Then the frequency distributions of specific capacity by aquifer unit with a logarithmic average and a range of the logarithmic standard deviations were statistically investigated. Figure 16.9 summarizes the frequency distribution of specific capacity values with the results of the statistical analysis by aquifer unit.

Estimation of apparent transmissivity from specific capacity using well data: Transmissivity values of the production wells were estimated by following three methods examined by Shibasaki (1996, 1999):

- Unsteady-state estimation method by Shibasaki (1996)

- Steady-state estimation method by Shibasaki (1996)

- Estimation method by Logan (1964)

1. *Unsteady-state estimation method by Shibasaki (1996)*: If the well has the data of discharge,

TABLE 16.1

Multiplier to Compute Clay Thickness

Actual Facies	Multiplier	Equation to Compute Clay Thickness
Clay	1.0	Thickness of clay = 1.0 × (thickness of clay)
Sandy clay	0.7	Thickness of clay = 0.7 × (thickness of sandy clay)
Clayey sand	0.3	Thickness of clay = 0.3 × (thickness of clayey sand)

drawdown, pumping time, and well radius, the specific capacity is given by the approximate nonequilibrium equation without well loss in confined aquifers presented by Cooper and Jacob (1946), that is,

$$Sc = \frac{Q}{s} = \frac{4\pi T}{2.30\log\left(2.25Tt/r^2S\right)} \qquad (16.7)$$

where Sc is specific capacity (m²/d), Q is well discharge (m³/d), s is drawdown (m), T is transmissivity (m²/d), t is pumping duration (day), r is well radius (m), and S is storage coefficient (dimensionless).

It is known that large changes in S cause comparatively small changes in specific capacity (Domenico and Schwartz, 1990). In addition,

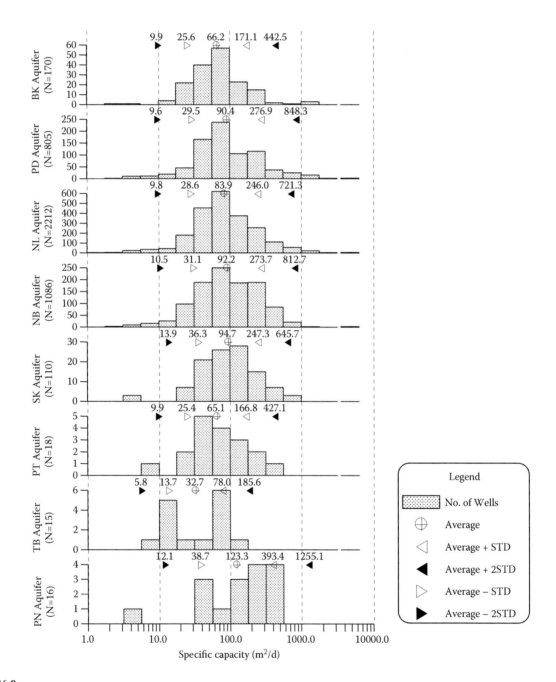

FIGURE 16.9
Frequency distribution and results of statistical analysis of specific capacity. (Adapted from Shibasaki, N., *Journal of Geosciences, Osaka City University*, 42, (Art. 2), 21–43, 1999.)

although the values of storage coefficient obtained from pumping tests generally take a wide range of variation, Walton (1970) suggested the range of S values in confined aquifers varies between 1.0E–5 and 1.0E–3 for all soils and the range of S values varies between 5.0E–5 and 1.0E–2 in productive aquifers. Therefore, for confined aquifers, the initial value of storage coefficient can be assumed to be 1.0E–3, and then T can be obtained by solving Equation 16.7 numerically substituting t and r (Shibasaki, 1996).

2. *Steady-state estimation method by Shibasaki (1996)*: The steady-state estimation was used for the wells not having pumping time. The equilibrium equation for confined aquifers developed by Thiem (1906) is written as

$$T = \frac{Q \ln(r_2/r_1)}{2\pi(s_1 - s_2)} \qquad (16.8)$$

where s_1 (m) and s_2 (m) are drawdowns of piezometric level at distances r_2 (m) and r_1 (m), respectively. Changing from natural to common log base and reducing constants, Equation 16.8 can be written as

$$T = \frac{0.366Q \log(r_2/r_1)}{(s_1 - s_2)} \qquad (16.9)$$

When r_1 moves to the well and r_2 moves outward along the cone of depression to a point at which the pumping influence is not measurable, the $(s_1 - s_2)$ becomes the drawdown at the well, s.

$$T = \frac{0.366Q \log(r_2/r_1)}{s} \qquad (16.10)$$

There are some empirical equations using hydraulic conductivity or transmissivity to obtain radius of influence. Klimentov and Pykhachev (1961) gave an empirical equation for computing the radius of influence as below:

$$r_2 = 2s\sqrt{T} \qquad (16.11)$$

where s (m) is drawdown in a well having well radius r (m). Substituting Equation 16.11 in Equation 16.10, T can be determined by

$$T = \frac{0.366Q \log(2s\sqrt{T}/r)}{s} = \frac{0.183Q \log(4Ts^2/r^2)}{s} \qquad (16.12)$$

where r is the well radius. Thus, Sc can be expressed as

$$Sc = \frac{Q}{s} = \frac{5.46T}{\log(4Ts^2/r^2)} \qquad (16.13)$$

By solving Equation 16.13 with applying r and s with some numerical methods, T values can be obtained (Shibasaki, 1996).

3. *Estimation method by Logan (1964)*: If the well has only the value of specific capacity, the following equation presented by Logan (1964) was used to estimate transmissivity:

$$T = 1.22 Sc \qquad (16.14)$$

Logan (1964) used Equation 16.8 to estimate transmissivity. He found that if a value for the quantity $\log(r_2/r_1)$ in a given area is obtained, Equation 16.10 is solvable from the results of routine production tests. He empirically employed a "typical" value of 3.32 for the $\log(r_2/r_1)$, and then substituted the value for the log ratio in Equation 16.10.

16.3.5 Estimate Hydraulic Conductivity from Transmissivity and Screen Length

Normally it can be said that production wells are partially wells and the screen length is shorter than the aquifer thickness as shown in Figure 16.10. The estimated transmissivity by the above methods can be called "transmissivity by pumping test" Tp because the estimated transmissivity describes the ability of the perforated portion of the aquifer to transmit water. After obtaining Tp of the well by the said methods, the hydraulic conductivity K value was estimated based on the definition of transmissivity given by

$$K = Tp/l \qquad (16.15)$$

where l is the screen length and Tp is the transmissivity by pumping test. If it is assumed that the aquifer facies has a uniform value of hydraulic conductivity from the top to the bottom of aquifer facies at a

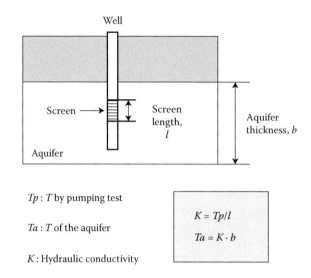

FIGURE 16.10
Schematic profile showing transmissivity by pumping test and transmissivity of aquifer.

location, transmissivity of the aquifer facies can be computed by

$$Ta = K \cdot b \qquad (16.16)$$

where Ta is the transmissivity of the aquifer facies and b is the thickness of the aquifer facies. In the Bangkok groundwater basin, the distribution of estimated hydraulic conductivity was statistically analyzed by aquifer unit, and then the estimated hydraulic conductivity map for each aquifer unit was prepared.

16.3.6 Preparation of Apparent Hydraulic Conductivity for MODFLOW Input

The MODFLOW (McDonald and Harbaugh, 1988) requires transmissivity or horizontal hydraulic conductivity depending on the selection of "user specified" or "calculated" options in the layer property menu. When the "user specified" option is selected, the transmissivity of the aquifer facies Ta can be directly input. However, it is common to select the "calculated" option in the MODFLOW data input. In this case the MODFLOW computes transmissivity value from the model layer

thickness and the horizontal hydraulic conductivity input to the model. Therefore, it is necessary to compute apparent hydraulic conductivity of the aquifer unit for the MODFLOW input.

Figure 16.11 shows the concept of apparent hydraulic conductivity of an aquifer unit for MODFLOW input. The apparent hydraulic conductivity Ka can be computed by

$$Ka = Ta/L \qquad (16.17)$$

where L is the thickness of the model layer, which the MODFLOW computes from the top and the bottom elevations of the model layer.

The following is the summary of procedures to estimate horizontal hydraulic conductivity employed for the modeling of the Bangkok groundwater basin.

1. Estimate K from Tp and screen length.
2. Compute clay content of each aquifer unit.
3. Compute Ta from K, clay content, and thickness of the aquifer unit.
4. Compute Ka for MODFLOW by dividing Ta by the thickness of the aquifer unit.

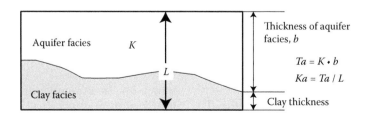

FIGURE 16.11
Schematic profile showing the concept of apparent hydraulic conductivity of an aquifer unit for MODFLOW input. (*Note*: Ta is the transmissivity of aquifer facies, K is the hydraulic conductivity of aquifer facies, and Ka is the apparent hydraulic conductivity of an aquifer unit.)

16.4 Hydrogeological Facies Modeling for Parameter Estimation

From the 2000s, the necessity of precise facies modeling has been increased to simulate 3-D mass transport for the various groundwater fields. The progress of computer technology and the development of numerical solutions have helped create fine groundwater simulation models.

Particularly for mass transport and heat transport simulations, it is necessary to divide the model domain into fine cells to reflect actual hydrogeologic conditions. The author created several groundwater models based on the idea of precise hydrogeological facies modeling.

16.4.1 Concept of Precise Hydrogeological Facies Modeling

As mentioned earlier, identification of hydrostratigraphic units is essential for groundwater modeling. Furthermore, facies classification is also necessary to create fine groundwater simulation models. In general, the hydrogeologic facies are identified based on the lithological logs of geologic borings and/or well drilling records including geophysical loggings. The model layers are thinner for the fine model to obtain reasonable resolution of simulated heads and/or simulated concentration.

Figure 16.12 shows the schematic profile for the concept of precise hydrogeological facies modeling. When the model layers have uniform top and bottom elevations, the horizontal hydraulic conductivity should be computed based on the facies distribution in the layer. For example, the horizontal hydraulic conductivity of Layer 1 is uniformly 10 m/day; however, the hydraulic

conductivity of Layer 2 at No. 1 location is smaller than 10 m/day because some portion is occupied by clay. In this way, it is required to consider the thickness of the facies in the model layer and its hydraulic conductivity.

16.4.2 Method of Precise Hydrogeological Facies Modeling

Figure 16.13 shows an example of computing horizontal hydraulic conductivity in a model layer. Considering the facies distributions in the model layer and those hydraulic conductivity values, the apparent transmissivity of the layer can be computed. Then the horizontal hydraulic conductivity of the model layer can be obtained by the apparent transmissivity divided by the thickness of the model layer.

There are six distribution patterns of a particular facies at the target model layer as shown in Figure 16.14. Considering the relationships among the top and bottom elevations of model layer and top and bottom elevations of the facies, the distribution pattern can be identified and the thickness of the facies in the model layer can be computed. Based on this method, the facies ratio map can be prepared for each model layer.

16.4.3 Application of Precise Hydrogeological Facies Modeling

The precise hydrogeological facies modeling was applied to southern Sendai Plain, Japan (Ouchi et al., 2011). A total of 195 columnar sections from existing geologic boring and existing well drilling records were collected and arranged for hydrogeological interpretations. The hydrogeologic facies were classified into 11 categories to assign hydrogeologic parameters.

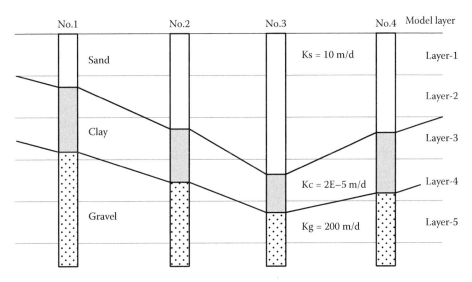

FIGURE 16.12
Schematic profile showing the concept of precise hydrogeological facies modeling.

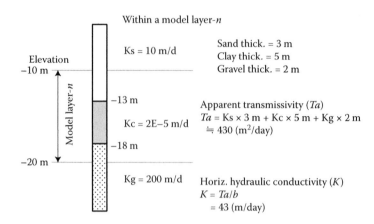

FIGURE 16.13
Example of computing horizontal hydraulic conductivity in a model layer.

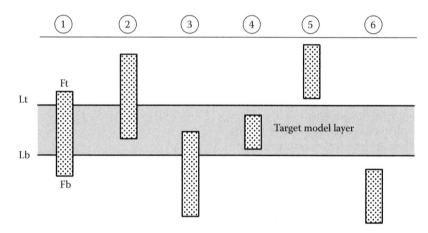

FIGURE 16.14
Distribution patterns of particular facies at the target model layer. (*Note*: Lt: layer top, Lb: layer bottom, Ft: facies top, Fb: facies bottom.)

Based on the typical hydraulic conductivity values presented by Bair and Lahm (2006), the initial values of the horizontal hydraulic conductivity were assigned. It was assumed that the effective porosity and specific yield showed the same value, the typical value of effective porosity for each facies was obtained from Bair and Lahm (2006).

After computing facies ratios for each model layer being 2 m in thickness, XYZ data files consisting of UTM Easting values, UTM Northing values, and horizontal hydraulic conductivity values at existing boring points for the target model layers were prepared. The hydraulic conductivity values were obtained by the previously mentioned method in which the apparent transmissivity was divided by the thickness of the model layer. To avoid negative values of horizontal hydraulic conductivity by the Kriging interpolation, the conductivity values were once converted into logarithmic values and then gridded. After creating the logarithmic grid values, those values were again converted into the

arithmetic values. The horizontal hydraulic conductivity value at the central point of each cell was obtained by the slice command of Surfer (Golden Software, Inc.).

Figure 16.15 shows the 3-D distribution of horizontal hydraulic conductivity in southern Sendai Plain, Japan. It was evaluated that the estimated horizontal hydraulic conductivity distribution showed a good agreement with the manually prepared facies profile. Similarly, the 3-D distribution of effective porosity was also prepared as shown in Figure 16.16. These parameters were input to the 3-D groundwater simulation model using MODFLOW. During the historical model calibration, the simulated heads showed a good agreement with the actual heads after slight modification of parameter values. The calibrated model was used to simulate seawater intrusion using SEAWAT (Langevin et al., 2008).

Figure 16.17 shows the temperature profile of groundwater in Kitakata City, Fukushima, Japan. The gradient of groundwater temperature is not uniform; relatively higher gradient is observed at depths where clay facies

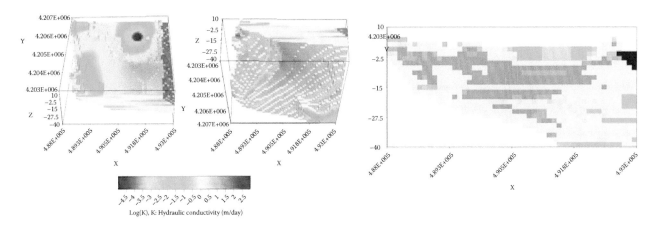

FIGURE 16.15
3-D distribution of horizontal hydraulic conductivity in southern Sendai Plain. (*Note*: The right figure shows the W-E profile at UTM-N = 4,206,000 m.)

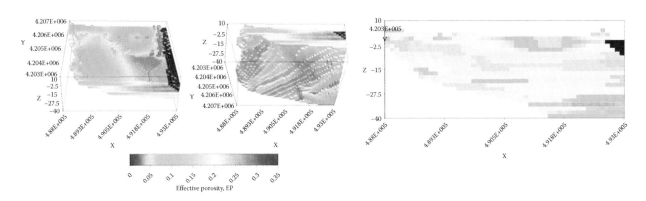

FIGURE 16.16
3-D distribution of effective porosity in southern Sendai Plain. (*Note*: The right figure shows the W-E profile at UTM-N = 4,206,000 m.)

exists. On the other hand, lower gradient is observed at thick gravelly layers and the screen portion. This indicates that the groundwater temperature is controlled by facies distributions and subsequent groundwater flow conditions. Therefore, it is also necessary for heat transport simulation to carry out precise hydrogeological facies modeling introduced in this chapter.

16.5 Conclusions and Recommendations

Identification of hydrogeologic unit is essential for groundwater modeling. Particularly for the Quaternary groundwater basins, integrated investigations on Quaternary geology, hydrogeology, applied geology, geophysics, civil engineering, and others are necessary to identify hydrogeologic units and to estimate geohydrologic parameters. For the vertical 2-D and 3-D groundwater modelings having many model layers, not

only the identification of hydrogeologic unit but also the precise hydrogeological facies modeling is necessary to carry out reliable groundwater simulation.

In the case study of the Kanto groundwater basin in Japan, preparation of accurate geological profiles and hydrogeological profiles based on the facies classification helped create reliable vertical 2-D multiple aquifer models. Necessary geohydrologic parameters were estimated from the existing well production tests and soil tests. The hydraulic conductivities of aquifer facies show unique values by facies; however, those of clayey facies showed different ranges of variation by hydrostratigraphic unit. Specific storage values of clayey facies were assigned based on the volume compressibility obtained from soil consolidation tests.

For the groundwater modeling of the Bangkok groundwater basin in Thailand, facies modeling was carried out in the eight principle aquifer units. Based on the subsurface geological profiles, subsurface hydrologic units were identified at columnar sections of the existing wells. Then the aquifer facies

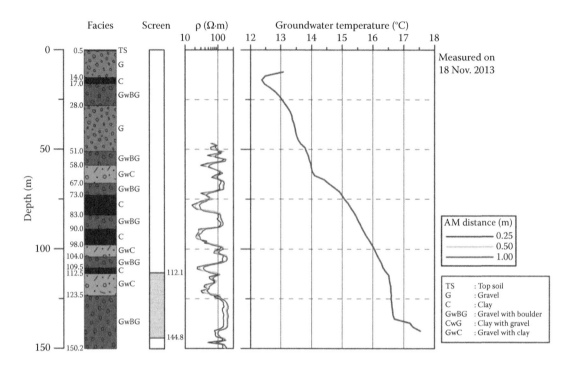

FIGURE 16.17

Example of temperature profile of groundwater in Kitakata City, Fukushima, Japan. (Adapted from Kaneko, S. and N. Shibasaki. 2014. Analysis on groundwater flow and groundwater temperature in urban area of Kitakata City, Japan. *Research Project for Regeneration of Harmonies between Human Activity and Nature in Bandai-Asahi National Park, Symbiotic Systems Science, Fukushima University*, 14, 96–105. (In Japanese.))

and clay facies were identified to prepare clay content maps. Transmissivity values were estimated from the specific capacity values of existing production wells. Depending on the available information of an existing well's production test data, a suitable method for estimating transmissivity from specific capacity was selected. Then the apparent hydraulic conductivity by aquifer unit for MODFLOW input was computed cell by cell.

From the 2000s, the author employed precise hydrogeological facies modeling for groundwater flow, mass transport, and heat transport simulations. Considering the facies distributions in the model layer and those hydraulic conductivity values, the apparent transmissivity of the layer can be computed. Then the horizontal hydraulic conductivity of the model layer was computed considering the thickness of the model layer. The precise hydrogeological facies modeling was applied to the southern Sendai Plain, Japan to simulate seawater intrusion in the coastal aquifer. The precise hydrogeological facies modeling is also necessary for heat transport simulation because the actual profile of groundwater temperature is controlled by facies distributions and subsequent groundwater flow conditions. It is recommended that the hydrogeologic unit should be identified first, and then the precise hydrogeological facies modeling should be carried out in the identified hydrogeologic unit.

References

Bair, E.S. and T.D. Lahm. 2006. *Practical Problems in Groundwater Hydrology*. Pearson Prentice Hall, New Jersey..

Cooper, H.H., Jr. and C.E. Jacob. 1946. A generalized graphical method for evaluating formation constants and summarizing well-field history. *Transactions American Geophysical Union*, 27, 526–534.

Domenico, P.A. and M.D. Mifflin. 1965. Water from low-permeability sediments and land subsidence. *Water Resources Research*, 1, 563–576.

Domenico, P.A. and F.W. Schwartz. 1990. *Physical and Chemical Hydrogeology*. John Wiley & Sons, New York.

Fujinawa, K. 1977. Finite element analysis of groundwater flow in multiaquifer systems, 2. A quasi three-dimensional flow model. *Journal of Hydrology*, 33, 349–362.

Fujisaki, K., H. Oka, and A. Kamata. 1979. A Galerkin finite element analysis of unsteady groundwater flow by a quasi three-dimensional multiaquifer model and three-dimensional model. *Earth Science*, 33, 73–84. (In Japanese with English abstract.)

Geological Research Group of Central Kanto Plain. 1994. Subsurface Geology in Central Kanto Plain. Monograph, 42, The Association for the Geological Collaboration in Japan. (In Japanese with English abstract.)

JICA. 1992. Study for the groundwater development in Metro Manila. *Main Report, Japan International Cooperation Agency*. URL: http://libopac.jica.go.jp/images/report/P0000075186.html.

Kamata, A., K. Harada, and H. Nirei. 1976. Analysis of land subsidence by vertical two dimensional multi-aquifer model. *Publication No. 121 of the IAHS, Proceedings of the Second International Symposium on Land Subsidence*, Anaheim, California, December 13–17, 1976, 201–210.

Kamata, A., M. Murakami, and K. Harada. 1973. Application of a quasi three-dimensional aquifer model to analysing land subsidence. *Earth Science*, 27, 131–140. (In Japanese with English abstract.)

Kaneko, S. and N. Shibasaki. 2014. Analysis on groundwater flow and groundwater temperature in urban area of Kitakata City, Japan. *Research Project for Regeneration of Harmonies between Human Activity and Nature in Bandai-Asahi National Park, Symbiotic Systems Science, Fukushima University*, 14, 96–105. (In Japanese.)

Klimentov, P.P. and G.B. Pykhachev. 1961. Groundwater dynamics (in Russian). Translated by Hokao, Z. and M. Nagai. 1967. Lattice, Tokyo, 523p. (In Japanese.)

Kokusai Kogyo and JICA. 1995. The study on management of groundwater and land subsidence in the Bangkok metropolitan area and its vicinity. *Final Report, Japan International Cooperation Agency*. URL: http://libopac.jica.go.jp/images/report/P0000084486.html.

Langevin, C.D., D.T. Thorne, Jr., A.M. Dausman, M.C. Sukop, and W. Guo. 2008. SEAWAT Version4: A Computer Program for Simulation of Multi-Species Solute and Heat Transport: U.S. Geological Survey Techniques and Methods Book 6, Chapter A22, 39p.

Logan, J. 1964. Estimating transmissibility from routine production tests of water wells. *Ground Water*, 2, 35–37.

Matsumoto, T., N. Shibasaki, M. Fujihara, and T. Tomita. 1994. Hydrogeological Modeling for Kanto groundwater basin–Importance of Establishing Aquifer Units based on the Subsurface Stratigraphy–Subsurface Geology in Central Kanto Plain, Monograph, 42, The Association for the Geological Collaboration in Japan, 165–176. (In Japanese with English abstract.)

McDonald, M.G. and A.W. Harbaugh. 1988. A modular three-dimensional finite-difference ground-water flow model. Techniques of Water-Resources Investigations 06-A1, USGS, 576p.

Ouchi, T., N. Shibasaki, K. Mori, and T. Takahashi. 2011. Mechanism of groundwater flow and simulation analysis of saline water intrusion in southern Sendai Plain, Japan. *Fourth International Groundwater Conference (IGWC-2011) on the Climate Change on Groundwater Resources with Special Reference to Hard Rock Terrain*, September 27–30, 2011, Yadava College, Madurai, India, 416–423.

Ramnarong, V. 1976. Pumping tests for Nakhon Luang and Bangkok aquifers. *Open File Report No. 90*, DMR, Bangkok.

Ramnarong, V. and S. Buapeng. 1992. Groundwater resources of Bangkok and its vicinity: Impact and management. *Proceedings of National Conference on Geologic Resources of Thailand: Potential for Future Development*, Bangkok, Thailand, November 17–24, 172–184.

Ramnarong, V., S. Buapeng, N. Shibasaki, and A. Kamata. 1995. Groundwater levels and land subsidence in the Bangkok metropolitan area. In Barends, F.B.J., F.J.J. Brouwer, and F.H. Schroder, Eds., *Land Subsidence*, Balkema, Rotterdam, 197–205.

Research Group for Water Balance. 1973. *Groundwater Resources Research*. Kyoritsu Shuppan, Tokyo. (In Japanese.)

Research Group for Water Balance. 1976. *Groundwater Basin Management*. Tokai University Press, Tokyo. (In Japanese.)

Shibasaki, N. 1989. Estimating geohydrologic parameters for groundwater management and land subsidence control in Kanto plain, Japan. In Gupta, C.P. ed., *Proceedings of International Workshop on Appropriate Methodologies for Development and Management of Groundwater Resources in Developing Countries*, Oxford & IBH Publishing, New Delhi, 2, 539–544.

Shibasaki, N. 1996. Relationship between transmissivity and specific capacity for evaluating aquifer characteristics. *Journal Geological Society of Japan*, 102, 419–430.

Shibasaki, N. 1999. Study on methodology of practical parameter estimation for groundwater modeling based on hydrogeological classification. *Journal of Geosciences, Osaka City University*, 42(Art. 2), 21–43.

Shibasaki, T., S. Azemoto, and T. Kotsuma. 1967. A various investigation on coefficient of aquifer. *Soils and Concrete*, 48, 5–11. (In Japanese.)

Shibasaki, T., A. Kamata, and S. Shindo. 1969. The hydrologic balance in the land subsidence phenomena. Land Subsidence 1, IAHS-UNESCO, 201–215.

Shibasaki, T. and Research Group for Water Balance. 1995. *Environmental Management of Groundwater Basins*. Tokai University Press, Tokyo.

Shimizu, Y. and M. Horiguchi. 1981. Motoarakawa tectonic belt in northern part of Omiya upland, Kanto plain, central Japan. *Seismicity and Tectonics of the Kanto District, Central Japan, the Memoirs of the Geological Society of Japan*, 20, 95–102. (In Japanese with English abstract.)

Thiem, G. 1906. *Hydrologische Methoden*. Gedhardt, Leipzig.

Tyson, H.N. and E.M. Weber. 1964. Groundwater management for nation's future—computer simulation of groundwater basins. *Journal Hydraulics Division, Proceedings of the American Society of Civil Engineers*, 90, HY4, 57–77.

Walton, W.C. 1970. *Groundwater Resource Evaluation*. McGraw-Hill, New York.

17

Simulation of Flow and Transport in Fractured Rocks: An Approach for Practitioners

Christopher J. Neville and Vivek S. Bedekar

CONTENTS

17.1 Introduction

The abundant literature on the simulation of flow and transport in fractured rocks is full of important insights; however, what is conspicuously absent is guidance for applied groundwater modelers. In this chapter, a "textbook" treatment of an approach for the simplified simulation of solute transport in fractured rocks is presented. The approach is directed specifically at practitioners charged with developing analyses that incorporate the transport processes of advection along the fractures and diffusion into the porous matrix blocks. The preparation of this chapter has been motivated by occasional requests the authors have received for guidance in applying the widely used solute transport simulator MT3DMS (Zheng and Wang, 1999; Zheng, 2010) for the analysis of solute transport in fracture-coupled porous media. Although the material presented is not new, as far as the authors are aware it has never been presented in a form that is intended to be applied.

Several comprehensive numerical simulation codes are available that are capable of representing coupled transport between a network of discrete fractures and porous blocks. These simulators include FracMan (Miller et al., 1994), TRAFRAP (Huyakorn et al., 1987), STAFF3D (Huyakorn et al., 1983a,b,c), FRACTRAN (Sudicky and McLaren, 1992), and FRAC3DVS (Therrien and Sudicky, 1996). In an ideal world, the characterization of fractured rock sites would yield sufficient data to constrain these numerical simulators. However, the available data for typical sites do not include the data that are most fundamental to the analysis of solute transport in systems of discrete fractures: the locations, orientations, and extents of the fractures, the variations of apertures of the individual fractures, and the connections between fractures. It is generally not feasible to characterize typical sites at the level of detail that would support estimation of these data.

The simplified approach presented here is intended to support screening-level calculations of solute transport in fractured rocks considering the important attenuation mechanism of diffusion into the porous matrix blocks. The approach incorporates established techniques to estimate the bulk-average transmissivity and the average fracture apertures. The approach then proceeds to the simulation of solute transport. Matrix diffusion is simulated with the first-order mass transfer coefficient approach. Physically-based expressions are presented to estimate the mass transfer coefficient, extending beyond the treatment of the mass transfer coefficient as a "fitting parameter."

17.2 Conceptual Model

Photographs of a physical system that motivates the analysis are shown in Figure 17.1. The complex rock mass is idealized as a set of near-parallel, horizontally extensive fractures that conduct groundwater. The fractures are separated by slabs of porous rock that are much less permeable than the fractures.

Summer view

Winter view

FIGURE 17.1
Dolostone quarry, southern Ontario (authors' photographs).

17.3 Inference of the Properties of the Fracture System

The first step in characterizing a fractured rock aquifer that is idealized as a set of parallel fractures separated by porous slabs is the estimation of the average aperture of the fractures. A constant-rate pumping test is conducted to accomplish this objective. A constant-rate pumping test is conducted to estimate the bulk-average transmissivity of the formation. The bulk transmissivity represents the cumulative transmissivity of the fractures. With an estimate of the fracture spacing, the average transmissivity and aperture of the individual fractures is back-calculated from the bulk transmissivity.

17.3.1 Estimation of the Bulk Transmissivity of the Aquifer

The conceptual model for the pumping test is shown schematically in Figure 17.2. The upper portion of the borehole is cased against the confining strata. The open-hole section of the borehole is assumed to penetrate the full thickness of the aquifer, with negligible groundwater flow above and below the open interval. The open-hole section of the well has a radius r_w and the casing through the confining strata has a radius r_c. The borehole is pumped at a constant rate and changes in groundwater levels are monitored only in the pumping well.

For demonstration purposes, the drawdowns are calculated with an analytical solution. The solution used here is an adaptation for pumping at a constant rate

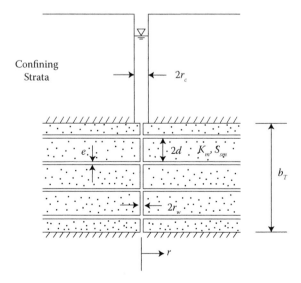

FIGURE 17.2
Conceptual model for the open-hole pumping test.

TABLE 17.1

Parameters for the Demonstration Analysis

Parameter	Value
Aquifer thickness, b_T	40.0 m
Casing radius, r_c	0.05 m
Borehole radius, r_w	0.05 m
Matrix hydraulic conductivity, K_m	1.157×10^{-6} m/s
Matrix specific storage, S_{sm}	10^{-5}/m
Spacing between fractures, S	1.0 m
Number of fractures, $n = b_T/S$	40
Fracture aperture, e	10^{-4} m
Pumping rate, Q	10^{-3} m³/s

assumed that there are neither skin losses nor nonlinear losses in the formation adjacent to the well or in the well itself.

The first plot of the simulated drawdowns in the idealized pumping well is presented in Figure 17.3. This is a "conventional" log–log plot, prepared in anticipation of a Theis (1935) analysis. The early portion of the response, which extends to about 100 seconds, approximates a straight line. At first glance, this linear response might be attributed to the effects of wellbore storage. However, the slope on the log–log plot exceeds the unit slope that is characteristic of a pure wellbore storage response. In fact, the early time response is a complex combination of the effects of wellbore storage, drawdown in the formation, and leakage from the porous matrix slabs.

To highlight the complexity of the early-time response, in Figure 17.4 the drawdowns calculated for impermeable slabs are superimposed on the simulated drawdowns. In contrast to the results for permeable slabs, the early-time drawdowns lie on a straight line of unit slope. Without knowing the fracture and matrix properties in advance, it is never obvious what portion of the response should be matched to estimate the bulk transmissivity of the aquifer.

In contrast to the log–log plot, the estimation of the bulk transmissivity of the aquifer is straightforward with the Cooper and Jacob (1946) semi-log analysis. The

of the Barker and Black (1983) solution for a slug test. In addition to considering transient flow in the fractures and the porous matrix slabs, the solution considers wellbore storage. Kazemi (1969) introduced the conceptual model that underlies the solution, and analytical solutions have been derived by de Swaan (1976), Boulton and Streltsova (1977), and Moench (1984). Details of the analytical solution are presented in Appendix A. The final results are obtained with numerical inversion of the Laplace-transform solution using the de Hoog et al. (1982) algorithm.

The parameter values specified for the calculation of the synthetic drawdowns are listed on Table 17.1. It is

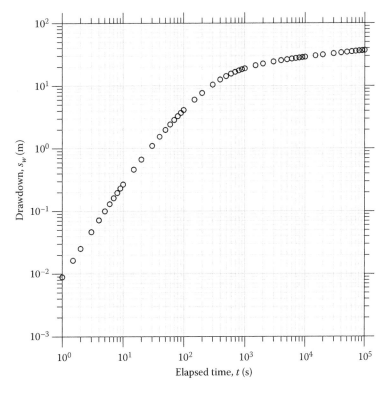

FIGURE 17.3
Log–log plot of simulated drawdowns in the pumping well.

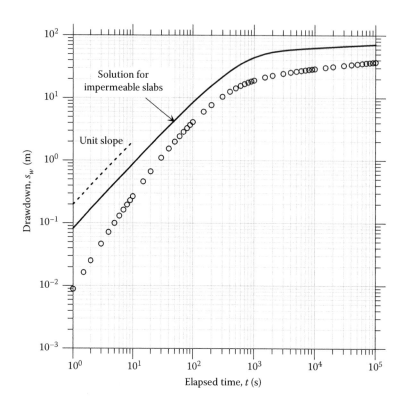

FIGURE 17.4
Log–log plot of simulated drawdowns in the pumping well, with solution for impermeable slabs.

key to the analysis is recognizing the portion of the response that is consistent with the conceptual model underlying the Cooper–Jacob (and Theis) model. The appropriate portion is referred to as the period of Infinite Acting Radial Flow (*IARF*) (Horne, 1995, 2009). Referring to Figure 17.5, this period corresponds to the interval between 10^4 and 10^5 seconds, during which the drawdowns approximate a straight line *and* the derivative of the drawdown reaches a plateau.

The slope on the semi-log plot of the drawdowns between 10^4 and 10^5 seconds is

$$\Delta s = 36.97 \text{ m} - 29.13 \text{ m} = 7.84 \text{ m/log cycle } t$$

The total transmissivity of the rock mass estimated with the Cooper–Jacob analysis is therefore

$$T_T = 2.303 \frac{Q}{4\pi} \frac{1}{\Delta s} = 2.303 \frac{(10^{-3} \text{ m}^3/\text{s})}{4\pi} \frac{1}{(7.84 \text{ m})}$$

$$= 2.34 \times 10^{-5} \text{ m}^2/\text{s}$$

To assess whether this estimate of the transmissivity is reliable, the bulk transmissivity of the aquifer given by the cumulative transmissivity of the fractures is calculated:

$$T_{bulk} = n \times \left[\frac{\rho g}{\mu} \frac{e^3}{12} \right]$$

$$= (40) \times \left[\frac{(1000 \text{ kg/m}^3)(9.807 \text{ m/s}^2)}{(1.40 \times 10^{-3} (\text{kg/m} \cdot \text{s}))} \frac{(10^{-4} \text{ m})^3}{12} \right]$$

$$= 2.34 \times 10^{-5} \text{ m}^2/\text{s}$$

Here n is the number of fractures across the thickness of the aquifer, μ is the dynamic viscosity of water [ML^{-1} T^{-1}], ρ is the density of water [ML^{-3}], and g is the acceleration due to gravity [LT^{-2}].

The two transmissivity estimates are essentially identical. Without knowing the "answer at the back of the book," the Cooper–Jacob analysis has yielded a reliable estimate of the bulk transmissivity of the fractures. This is an important result. When applied correctly, that is, over the appropriate late time response, the Cooper–Jacob analysis yields a representative estimate of the bulk-average transmissivity without having to diagnose and simulate the features that give rise to the complex early-time response. To emphasize this point, in Figure 17.5, the log–log plot is revisited with the results of the Cooper–Jacob analysis. As shown in Figure 17.6, this solution can match the synthetic drawdowns; however, special care must be taken to restrict the match to the late-time portion of the data. This is certainly not an

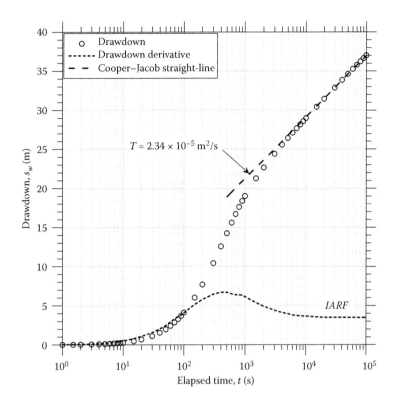

FIGURE 17.5
Semi-log plot of simulated drawdowns.

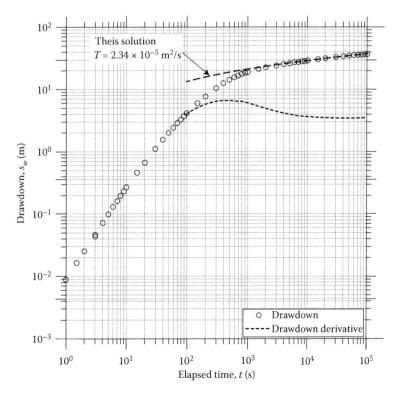

FIGURE 17.6
Log–log plot with the transmissivity estimated from the Cooper–Jacob analysis.

instance where a computer-assisted regression analysis of the complete drawdown record would yield a reliable estimate of the transmissivity.

17.3.2 Estimation of the Average Aperture of the Fractures

The results of the preceding calculations suggest that when applied appropriately, the Cooper–Jacob analysis yields a reliable estimate of the bulk-average transmissivity. However, the present analysis is conducted *not* to estimate the bulk-average transmissivity but instead to estimate the average fracture aperture. Each fracture is idealized as smooth-walled parallel plates separated by an aperture e. The transmissivity of an individual fracture for this model is given by the cubic flow law (Snow, 1965):

$$T_f = \frac{\rho g}{\mu} \frac{e^3}{12} \tag{17.1}$$

The bulk-average transmissivity estimated from the constant-rate pumping test represents the cumulative transmissivity of the fractures:

$$T_T = n \times T_f \tag{17.2}$$

Therefore, given the bulk transmissivity and the number of fractures across the aquifer thickness, the average aperture is given by:

$$e = \left[\frac{1}{n} T_T \frac{12\mu}{\rho g} \right]^{1/3} \tag{17.3}$$

Rather than work in terms of the number of fractures, it is more practical to estimate the average spacing between the fractures, S [L]. The spacing can be estimated from a variety of sources, including rock outcrops, geophysical logs, flowmeter profiles, and borehole cameras. For a total aquifer thickness b_T [L] and a fracture spacing S, the average aperture is calculated from:

$$e = \left[\frac{S}{b_T} T_T \frac{12\mu}{\rho g} \right]^{1/3} \tag{17.4}$$

The back-calculation of the average fracture aperture for the demonstration analysis is illustrated in Figure 17.7. The aquifer thickness and bulk transmissivity are 40 m and 2.34×10^{-5} m²/s, respectively, and the average fracture spacing is 1.0 m. This yields an average fracture aperture of 10^{-4} m, identical to the value specified for the calculations of the drawdowns due to pumping.

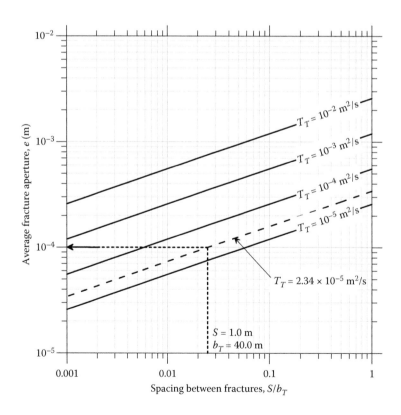

FIGURE 17.7
Estimation of average fracture aperture from bulk transmissivity.

17.4 Simulation of Solute Transport

17.4.1 Overview of Simulation Approaches for Fractured–Porous Media

Fractured–porous media have two distinct pore spaces: the void space between the fracture walls and the pore space in the intact matrix. The void space of the fractures may represent only a relatively small fraction of the total volume of the medium, with typical fractions ranging from 0.01% to 0.1% of the total volume. In contrast, the porosity of the intact matrix blocks may be relatively high. For example, the matrix porosities of intact samples of dolostone from a site in southern Ontario range from 1% to 10% (Najiwa, 2003; Burns, 2005).

The time scales for active storage and release of water from the porous matrix are generally brief. Therefore, unless the specific focus of an application is the transient hydraulic response, distinct fracture and matrix porosities generally do not have to be considered in the context of groundwater flow simulation. The effective hydraulic conductivity of an open fracture is generally much larger than the conductivity of the intact matrix so flow will occur primarily in the fracture network. As shown in the previous example calculations, the matrix does affect the early-time transient response during a pumping test conducted in fractured rock; however, the long-term response is controlled by the properties of the fractures. As only the fracture network is relevant with respect to long-term flow conditions, if the network is sufficiently dense, flow in a fractured–porous medium can be represented using an equivalent porous medium (EPM) approach. In this approach, the fracture network is replaced by porous medium with the same effective hydraulic conductivity, defined as the bulk hydraulic conductivity that yields the same discharge under the same bulk hydraulic gradient. This approach has been adopted to simulate groundwater flow for many applications.

In contrast to groundwater flow, the porous matrix may play a significant role with respect to solute transport. The pore space in the matrix represents a significant reservoir into and out of which solute can diffuse from the fractures. Conventional numerical solute transport simulators do not represent the diffusive mass exchange between the fractures and the matrix; therefore, EPM simulations conducted with these simulators miss a fundamental aspect of solute transport in fractured–porous media. Several numerical simulators of solute transport are capable of representing the coupled transport between a network of discrete fractures and porous blocks, including the previously mentioned FracMan, TRAFRAP, STAFF3D, FRACTRAN, and FRAC3DVS.

The simulation codes that are available for fractured–porous media provide important insights into the processes that control the migration of solutes in groundwater. However, two fundamental problems arise when trying to apply these codes in practice. First, the geometry and properties of the fracture network are generally not known. Second, the computational demands are very high and simulation of large-scale sites is frequently not feasible. An alternative approach is described here: the *dual-domain* conceptualization approximated with a first-order mass transfer (FOMT) approach. The FOMT approach that is implemented in the widely available solute transport simulator MT3DMS retains the simplicity of the EPM approach, but provides an approximate approach for representing the diffusive exchange of solutes between the fractures and the porous matrix.

17.4.2 The FOMT Approach

With the dual-domain conceptualization, the subsurface is idealized as two overlapping continua, a *mobile* domain in which there is active groundwater flow and advective–dispersive transport is dominant, and an *immobile* domain in which groundwater flow is not significant and diffusive transport is dominant. In the context of fractured–porous media, the mobile domain corresponds to the fractures and the immobile domain corresponds to the porous matrix.

The FOMT implementation of the dual-domain conceptualization is perhaps simplest to understand by examining the one-dimensional forms of the governing equations for the concentrations in the mobile and immobile domains. For simplicity, sorption and transformation reactions are neglected.

The statement of mass conservation for solute in the mobile domain is

$$\theta_m \frac{\partial C_m}{\partial t} = -q \frac{\partial C_m}{\partial x} + \theta_m D_{Lm} \frac{\partial^2 C_m}{\partial x^2} - \varsigma(C_m - C_{im}) \quad (17.5)$$

The statement of mass conservation for solute in the immobile domain is

$$\theta_{im} \frac{\partial C_{im}}{\partial t} = \varsigma(C_m - C_{im}) \quad (17.6)$$

Here

C_m: concentration in the mobile domain [ML^{-3}]
C_{im}: concentration in the immobile domain [ML^{-3}]
q: Darcy flux [LT^{-1}]
D_{Lm}: longitudinal dispersion coefficient [L^{2}T^{-1}]
θ_m: porosity of the mobile zone (fractures) [–]
θ_{im}: porosity of the immobile zone (matrix blocks) [–]
ς: first-order mass transfer coefficient [T^{-1}]

The linking term ζ $(C_m - C_{im})$ appears in both Equations 17.5 and 17.6, but with opposite signs. This reflects the fact that the mass transfer from the mobile region to the immobile region represents a sink from the perspective of the mobile region, while at the same time representing a source from the perspective of the immobile region.

The mass transfer coefficient in Equations 17.5 and 17.6, ζ, is frequently treated phenomenologically, that is, as a "curve-fitting" parameter with little underlying physical significance. At first glance it appears that the value of ζ cannot be determined *a priori*, and therefore the whole FOMT approach has limited predictive capability. However, as shown in the next section, the mass transfer coefficient can be estimated from physical quantities, at least for idealized systems.

17.4.3 Application of the FOMT Approach for Simplified Geometrical Representations of Fractured–Porous Media

Exact relations for the diffusive exchange between fractures and matrix blocks consisting of uniform slabs or spheres have been developed. Van Genuchten and Dalton (1986) and Sudicky (1990) have shown that with some simplifying assumptions, the FOMT coefficient, ζ, can be estimated with simple closed-form expressions such that the FOMT approach approximates the exact relations for the slab and sphere models. Van Genuchten and Dalton's expressions for the FOMT coefficient are given next.

For Parallel Slabs:

$$\zeta = \frac{3\theta_{im}D^*_{im}}{d^2} \qquad (17.7)$$

For Spheres:

$$\zeta = \frac{15\theta_{im}D^*_{im}}{r_0^2} \qquad (17.8)$$

Here

θ_{im}: the porosity of the matrix slabs and spheres [L^3L^{-3}]
D^*_{im}: effective diffusion coefficient for the immobile zone [L^2T^{-1}]
d: slab half-thickness [L]
r_0: sphere radius [L]

The demonstration analysis introduced in Section 17.3 is extended to examine quantitatively the reliability of the FOMT approach for simulating solute transport in a fractured–porous medium idealized as a set of parallel fractures separated by porous slabs. The conceptual model for the transport analysis is shown schematically

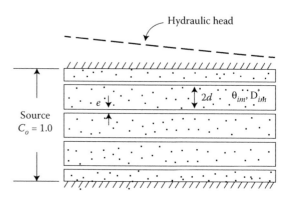

FIGURE 17.8
Conceptual model for solute transport in a fractured–porous matrix system.

in Figure 17.8. Concentration profiles along a fracture are calculated with three simulation approaches. In the first analysis, transport in the fractured–porous system is simulated with an exact solution. In the second analysis, the results of the exact solution are compared with the results of an FOMT model as implemented in an exact Laplace-transform solution. In the third analysis, the results of the exact solution are compared with the results of the FOMT as implemented in the widely used numerical simulator MT3DMS.

The analytical approach for transport along a set of discrete fractures separated by porous slabs is presented in Barker (1982) and Sudicky and Frind (1982). The calculations presented here are obtained by numerical inversion of the Laplace-transform solution (de Hoog et al., 1982). Details of the Laplace transform solution are presented in Appendix B.

The following parameter values are specified in the demonstration analysis:

- Fracture aperture, $e = 10^{-4}$ m
- Slab half-thickness = Half of the spacing between fractures, $d = S/2 = 0.5$ m
- Porosity of the matrix blocks, $\theta_{im} = 0.01$
- Effective diffusion coefficient in the matrix, $D_{im}^* = 1.380 \times 10^{-5}$ m^2/day
- Velocity along the fractures, $v = 0.1$ m/day
- Longitudinal dispersion coefficient, $D_{Lm} = 0.01$ m^2/day

The fracture aperture was estimated in Section 17.2, from an estimate of the bulk transmissivity derived from the single-well pumping test and an estimate of the fracture spacing.

The immobile zone porosity, θ_{im}, corresponds to the porosity of the matrix blocks. Compilations of typical values of the matrix porosities of rocks are available from several sources. Particularly good sources of literature values for a range of rock types are Davis (1969) and Wolff (1982). There are abundant laboratory-determined values of the porosities of intact igneous rocks, obtained from studies conducted to support the evaluation of potential geologic repositories (see, e.g., Bradbury et al., 1982; Skagius and Neretnieks, 1986).

The effective diffusion coefficient in the matrix, D_{im}^{*}, is typically estimated by multiplying the free-solution diffusion coefficient, D_0, by the matrix tortuosity, τ (Bear, 1972):

$$D_{im}^{*} = \tau \times D_0 \qquad (17.9)$$

Tabulations of the free-solution diffusion coefficient are widely available (see, e.g., U.S. EPA, 1996; Lide, 2005). The tortuosity is included to account for the tortuous diffusion pathways due to the presence of solid minerals comprising the matrix. Our experience suggests that tortuosity values back-calculated from diffusion experiments lie within the following range:

$$\theta_{im}^{2} < \tau < \theta_{im} \qquad (17.10)$$

Alternatively, the effective diffusion coefficient, D_{im}^{*}, can be specified directly based on literature values or the results of diffusion tests. An excellent compilation of effective diffusion coefficients is presented in Rowe et al. (2004).

The longitudinal dispersion coefficient accounts for variations in the velocities along the fractures that are smaller than the resolution of the analysis. Here the dispersion coefficient is specified as the product of the velocity along the fractures and a longitudinal dispersivity, α_L, of 0.1 m. A relatively small value of the longitudinal dispersivity is specified for the demonstration calculations, consistent with the idealized conceptual model (Gelhar, 1987).

17.4.3.1 Exact Analytical Solution

The concentration profiles along the fractures calculated with the exact Laplace-transform solution for transport along a set of discrete fractures with porous matrix slabs are shown in Figure 17.9.

17.4.3.2 Exact Analytical Solution and Analytical Implementation of the FOMT Model

For the second analysis, the discrete fracture system is simulated as an equivalent bi-continuum, using the FOMT approach implemented in the exact Laplace-transform solution MPNE1D (Neville et al., 2000). The MPNE1D solution requires specification of the following parameters:

- Total porosity, θ
- Fraction of the total porosity that is mobile, ϕ
- The first-order mass transfer coefficient, ζ

The mobile porosity for a set of fractures separated by porous slabs is

$$\theta_m = \frac{e}{S+e}$$

$$= \frac{(10^{-4}\,\mathrm{m})}{(1.0\,\mathrm{m}) + (10^{-4}\,\mathrm{m})} = 9.999 \times 10^{-5}$$

The total porosity, equal to the sum of the mobile and immobile region porosities, is

$$\theta = \theta_m + \theta_{im}$$

$$= (9.999 \times 10^{-5}) + (0.01) = 1.010 \times 10^{-2}$$

The ratio of the mobile porosity to the total porosity is

$$\phi = \frac{\theta_m}{\theta}$$

$$= \frac{(9.999 \times 10^{-5})}{(1.010 \times 10^{-2})} = 9.90 \times 10^{-5}$$

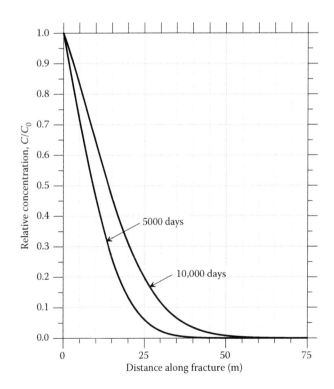

FIGURE 17.9
Results from the discrete fracture analytical solution.

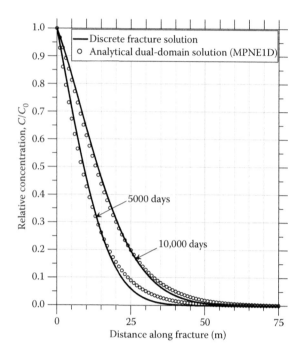

FIGURE 17.10
Comparison of discrete fracture and analytical FOMT solutions.

The FOMT coefficient for transport in a set of parallel slabs is calculated as

$$\zeta = \frac{3\theta_{im}D_{im}^*}{d^2}$$

$$\zeta = \frac{3(0.01)(1.38\times10^{-5}\,\mathrm{m^2/d})}{(0.5\,\mathrm{m})^2} = 1.656\times10^{-6}/\mathrm{day}$$

The results of the exact solution and the MPNE1D solution are shown in Figure 17.10. The results confirm that the FOMT implementation of the dual-domain formulation provides a relatively good match to the results from an exact solution for transport along a set of parallel fractures. The mismatch observed at the leading front of the concentration profile between the two results is examined in the next section.

17.4.3.3 Exact Analytical Solution and Numerical Implementation of the FOMT Model

For the third analysis, the discrete fracture system is simulated with the widely used three-dimensional solute transport simulator MT3DMS (Zheng and Wang, 1999; Zheng, 2010). Since 1999, MT3DMS has supported the dual-domain conceptual model with the FOMT approach. For MT3DMS, the mobile and immobile porosities along with the FOMT coefficient are specified:

- $\theta_m = 9.999 \times 10^{-5}$

- $\theta_{im} = 0.01$
- $\zeta = 1.656 \times 10^{-6}/\mathrm{day}$

The MT3DMS model is one-dimensional, with 101 grid blocks of dimension $1 \times 1 \times 1$ m. The hydraulic conductivity and constant-head conditions at either end of the model are specified such that the Darcy flux is 9.999×10^{-6} m/day. Dividing the Darcy flux by the mobile porosity yields an average linear groundwater velocity of 0.1 m/day.

The results of the MT3DMS simulations are shown in Figure 17.11. Although there is some additional tailing in the MT3DMS solution, the general trend of the results confirms that the FOMT approach as implemented in MT3DMS is capable of approximating transport in a simple system of discrete parallel fractures.

The source of the mismatch between the exact analytical solution for parallel fractures and the MPNE1D and MT3DMS results is not immediately obvious. The mismatch may be due to the approximations inherent in the FOMT approach described here, or to the approximations inherent in the MT3DMS discrete numerical solution. An additional analysis has been conducted to examine this issue. The results of the analytical and numerical implementations of the FOMT approach are superimposed in Figure 17.12. The results are essentially identical, confirming that the inability to match the discrete-fracture analytical results exactly is due to approximations inherent in the FOMT approach. It is possible that an improved match might be obtained by

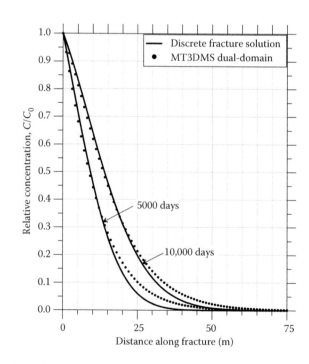

FIGURE 17.11
Results from MT3DMS dual-domain analysis.

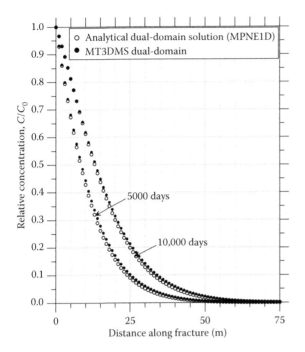

FIGURE 17.12
Comparison of FOMT solutions.

adjusting the parameters specified for the FOMT calculations. However, the objective here has been to assess the performance of an approach based on the independent specification of parameters rather than through calibration.

An analytical solution implementing the FOMT dual-domain approach is sufficient for the demonstration analysis. The flexibility of a numerical solution such as the MT3DMS approach is required when flow conditions are three-dimensional, material properties are variable, and there are complex distributions of sources and sinks. In these instances, variations in fracture spacing and the properties of the intact rock can be represented by specifying FOMT coefficients that vary in space.

17.5 Conclusions

Practicing hydrogeologists are charged frequently with providing quantitative analyses of solute transport in fractured rock settings. In practice, it is generally not feasible to characterize typical sites at the level of detail that would support the estimation of the data that are most fundamental to the analysis of solute transport in systems of discrete fractures: the locations, orientations, and extents of the fractures, the variations of apertures of the individual fractures, and the connections between

fractures. In this chapter, a simplified approach has been presented for simulating solute transport in fractured rocks. The approach is directed specifically at practitioners charged with developing analyses that incorporate the transport processes of advection along the fractures and diffusion into the porous matrix blocks. The approach may be particularly useful for screening-level assessments and to guide the design of site characterization programs and more sophisticated numerical modeling efforts.

Hydraulic testing data for fractured-rock sites are often conducted over relatively long open-hole intervals. A major portion of the development has been devoted to an approach for interpreting hydraulic testing that yields representative estimates of the bulk-average transmissivity. Coupled with an estimate of the average spacing between fractures, the bulk-average transmissivity provides a basis for inferring average fracture apertures.

An exact solution for transport in a set of ideal discrete fractures separated by porous matrix blocks has been used to assess the reliability of results obtained using a simplified FOMT approach to represent the interaction between the fractures and the porous matrix. The key results of the analyses are as follows:

- The results of the demonstration analyses confirm that the FOMT formulation approximates relatively well the results from an exact analytical solution for transport along a set of parallel fractures separated by porous slabs.

- The general trends of the results confirm that the FOMT representation of the dual-domain formulation as implemented in MT3DMS offers the capability of accounting for the important attenuation process of matrix diffusion.

Ma and Zheng (2011) indicate that the immobile porosity and FOMT coefficient are fitting parameters that generally cannot be estimated independently. In this chapter, we have demonstrated that the FOMT approach is not necessarily merely phenomenological when applied to represent transport in fractured–porous media—at least for idealized systems. van Genuchten and Dalton (1986) and Sudicky (1990) have shown that with some simplifying assumptions, simple closed-form expressions can be derived for ζ such that the FOMT coefficient approach effectively mimics the slab and sphere models.

All of the solutions used to generate the results presented in this chapter are freely available from the authors on request. Detailed notes on a complete development of closed-form expressions for the FOMT coefficient are also available on request.

Appendix A: Details of the Analytical Solution for a Pumping Test in a System of Parallel Fractures Separated by Porous Slabs

The notation follows Barker and Black (1983).

1. Governing equation for the drawdown in the fractures:

$$T_f\left(\frac{\partial^2 h_f}{\partial r^2} + \frac{1}{r}\frac{\partial h_f}{\partial r}\right) = S_f\frac{\partial h_f}{\partial t} - q_L \quad r_w \le r < \infty \quad \text{(A.1)}$$

subject to the following boundary and initial conditions:

$$h_f(r_w, t) = H(t) \tag{A.2}$$

$$h_f(\infty, t) = 0 \tag{A.3}$$

$$h_f(r, 0) = 0 \tag{A.4}$$

2. Governing equation for the drawdown in the porous slabs:

$$K_m\frac{\partial^2 h_m}{\partial z^2} = S_{sm}\frac{\partial h_m}{\partial t} \quad 0 \le z \le d \tag{A.5}$$

subject to the following boundary and initial conditions:

$$\frac{\partial h_m}{\partial z}(r, 0, t) = 0 \tag{A.6}$$

$$h_m(r, d, t) = h_f(r, t) \tag{A.7}$$

$$h_m(r, z, 0) = 0 \tag{A.8}$$

3. Leakage term between the fractures and the porous slabs:

$$q_L = -2nK_m\frac{\partial h_m}{\partial z}(r, d, t) \tag{A.9}$$

4. Linking term between the drawdown in the fracture network, h_f, and the drawdown in the pumping well, H:

$$2\pi r_w T_f\frac{\partial h_f}{\partial r}(r_w, t) - \pi r_c^2\frac{dH}{dt} = Q \tag{A.10}$$

subject to the following initial conditions:

$$H(0) = 0 \tag{A.11}$$

5. Laplace-transform solution for drawdown in the pumping well:

$$\bar{H}(p) = \frac{Q}{p}\frac{1}{\pi r_c^2[p + 2(r_w/r_c^2)T_f q(K_1(qr_w)/K_0(qr_w))]}\frac{K_0(qr)}{K_0(qr_w)} \tag{A.12}$$

Appendix B: Details of the Analytical Solution for Solute Transport along a System of Parallel Fractures Separated by Porous Slabs

The notation follows Sudicky and Frind (1982).

1. Governing equation for solute transport along the fractures:

$$-vb\frac{\partial c}{\partial z} + Db\frac{\partial^2 c}{\partial z^2} - Rb\lambda c = Rb\frac{\partial c}{\partial t} + q \quad 0 \le z < \infty \tag{B.1}$$

subject to the following boundary and initial conditions:

$$c(0, t) = c_0 \tag{B.2}$$

$$c(\infty, t) = 0 \tag{B.3}$$

$$c(z, 0) = 0 \tag{B.4}$$

2. Governing equation for solute transport in the porous slabs

$$D'\frac{\partial^2 c'}{\partial x^2} - R'\lambda c' = R'\frac{\partial c'}{\partial t} \quad b \le x \le B \tag{B.5}$$

subject to the following boundary and initial conditions:

$$c'(b, z, t) = c(z, t) \tag{B.6}$$

$$\frac{\partial c'}{\partial x}(B, z, t) = 0 \tag{B.7}$$

$$c'(x, z, 0) = 0 \tag{B.8}$$

3. Linkage term between the fractures and the porous slabs:

$$q = -\theta' D \frac{\partial c'}{\partial x}(b, z, t) \qquad \text{(B.9)}$$

4. Laplace-transform solution for concentration in the fractures:

$$\bar{c}(z, p) = \frac{c_0}{p} EXP\{m^- z\} \qquad \text{(B.10)}$$

$$m^- = \upsilon \left[1 - \sqrt{1 + K^2 \left(P + \frac{P^{1/2}}{A} TANH\{\sigma P^{1/2}\} \right)} \right] \qquad \text{(B.11)}$$

$$P = p + \lambda \qquad \text{(B.12)}$$

$$\upsilon = \frac{v}{2D} \qquad \text{(B.13)}$$

$$\sigma = \left(\frac{R'}{D'} \right)^{1/2} (B - b) \qquad \text{(B.14)}$$

$$A = \frac{bR}{\theta'(R'D')^{1/2}} \qquad \text{(B.15)}$$

$$K = \frac{2(RD')^{1/2}}{v} \qquad \text{(B.16)}$$

References

Barker, J.A. 1982. Laplace transform solutions for solute transport in fissured aquifers, *Advances in Water Resources*, 5(2), 98–104.

Barker, J.A. and J.H. Black. 1983. Slug tests in fissured aquifers, *Water Resources Research*, 19(6), 1558–1564.

Bear, J. 1972. *Dynamics of Fluids in Porous Media*, American Elsevier, New York.

Boulton, N.S. and T.D. Streltsova. 1977. Unsteady flow to a pumped well in a fissured waterbearing formation, *Journal of Hydrology*, 35, 257–269.

Bradbury, M.H., Lever, D., and D. Kinsey. 1982. Aqueous diffusion in crystalline rock, in *Proceedings of the Fifth International Symposium on the Scientific Basis of Nuclear Waste Management*, W. Lutze (Ed.), Elsevier, Berlin, pp. 569–578.

Burns, L.S. 2005. Fracture Network Characteristics and Velocities of Groundwater, Heat and Contaminants in a Dolostone Aquifer in Cambridge, Ontario, MSc thesis,

University of Waterloo, Department of Earth Sciences, Waterloo, Ontario.

Cooper, H.H., Jr. and C.E. Jacob. 1946. A generalized graphical method for evaluating formation constants and summarizing well-field history, *Transactions of the American Geophysical Union*, 27(4), 526–534.

Davis, S.N. 1969. Porosity and permeability of natural materials, in *Flow Through Porous Media*, R.J.M. De Wiest (Ed.), Academic Press, New York, pp. 53–89.

de Hoog, F.R., J.H. Knight, and A.N. Stokes. 1982. An improved method for numerical inversion of Laplace transforms, *SIAM Journal of Scientific and Statistical Computing*, 3(3), 357–366.

de Swaan, A. 1976. Analytic solutions for determining naturally fractured reservoir properties by well testing, *Society of Petroleum Engineers Journal*, 16(3), 117–122.

Gelhar, L.W. 1987. *Applications of Stochastic Models to Solute Transport in Fractured Rocks*, SKB Technical Report 87–07, Swedish Nuclear Fuel and Waste Management Co., Stockhold, Sweden.

Horne, R.N. 1995. *Modern Well Test Analysis*, 2nd ed., Petroway, Inc., Palo Alto, California.

Horne, R.N. 2009. Basic interpretation—Homogeneous reservoirs, in *Transient Well Testing*, M.M. Kamal (Ed.), Society of Petroleum Engineers Monograph, Volume 23, Society of Petroleum Engineers, Richardson, TX, pp. 69–90.

Huyakorn, P.S., B.H. Lester, and C.R. Faust. 1983a. Finite element techniques for modeling groundwater flow in fractured aquifers, *Water Resources Research*, 19(4), 1019–1035.

Huyakorn, P.S., B.H. Lester, and J.W. Mercer. 1983b. An efficient finite element technique for modeling transport in fractured porous media, 1. Single species transport, *Water Resources Research*, 19(3), 841–854.

Huyakorn, P.S., B.H. Lester, and J.W. Mercer. 1983c. An efficient finite element technique for modeling transport in fractured porous media, 2. Nuclide decay chain transport, *Water Resources Research*, 19(5), 1286–1296.

Huyakorn, P.S., H.O. White, Jr., and T.D. Wadsworth. 1987. TRAFRAP-WT, A Two-Dimensional Finite Element Code for Simulating Fluid Flow and Transport of Radionuclides in Fractured Porous Media with Water Table Boundary Conditions, IGWMC-FOS33, Colorado School of Mines, Golden, CO.

Kazemi, H. 1969. Pressure transient analysis of naturally fractured reservoirs with uniform fracture distribution, *Society of Petroleum Engineers Journal*, 9(4), 451–462.

Lide, D.R. 2005. *CRC Handbook of Chemistry and Physics, Internet Version 2005*, http://www.hbcnetbase.com, CRC Press, Boca Raton, FL.

Ma, R. and C. Zheng. 2011. Not all mass transfer coefficients are created equal, *Ground Water*, 49(6), 772–774.

Miller, I., G. Lee, T. Kleine, and W. Dershowitz. 1994. *MAFIC Fracture/Matrix Flow and Transport Code: User Documentation*, Golder Associates, Inc., Seattle, WA.

Moench, A.F. 1984. Doubleporosity models for a fissured groundwater reservoir with fracture skin, *Water Resources Research*, 20(7), 831–846.

Najiwa, H.M.S. 2003. Geological Characterization of the Middle Silurian Dolomite Aquifer in Cambridge, Ontario, MSc.

thesis, University of Waterloo, Department of Earth Sciences, Waterloo, Ontario.

Neville, C.J., M. Ibaraki, and E.A. Sudicky. 2000. Solute transport with multiprocess nonequilibrium: A semi-analytical solution approach, *Journal of Contaminant Hydrology*, 44, 141–159.

Rowe, R.K., R.M. Quigley, R.W.I. Brachman, and J.R. Booker. 2004. *Barrier Systems for Waste Disposal*, Spon Press, New York.

Skagius, K. and I. Neretnieks. 1986. Porosities and diffusivities of some nonsorbing species in crystalline rocks, *Water Resources Research*, 22(3), 389–398.

Snow, D.T. 1965. A Parallel Plate Model of Fractured Permeable Media, PhD thesis, University of California, Berkeley.

Sudicky, E.A. 1990. The Laplace transform Galerkin technique for efficient time-continuous solute of solute transport in double-porosity media, *Geoderma*, 46, 209–232.

Sudicky, E.A. and E.O. Frind. 1982. Contaminant transport in fractured porous media: Analytical solutions for a system of parallel fractures, *Water Resources Research*, 18(6), 1634–1642.

Sudicky, E.A. and R.G. McLaren. 1992. The Laplace transform Galerkin technique for largescale simulation of mass transport in discretely fractured porous formations, *Water Resources Research*, 28(2), 499–514.

Theis, C.V. 1935. The relation between the lowering of the piezometric surface and the rate and duration of discharge of a well using ground-water storage, *Transactions of the American Geophysical Union, 16th Annual Meeting*, Part 2, pp. 519–524.

Therrien, R. and Sudicky, E.A. 1996. Three-dimensional analysis of variably-saturated flow and solute transport in discretely-fractured porous media, *Journal of Contaminant Hydrology*, 23, 1–44.

U.S. EPA. 1996. Soil screening guidance: Technical background document, second edition, *EPA/540/R95/128*, United States Environmental Protection Agency, Office of Solid Waste and Emergency Response, Washington, DC.

van Genuchten, M.T. and F.N. Dalton. 1986. Models for simulating salt movement in aggregated field soils, *Geoderma*, 48, 165–183.

Wolff, R.G. 1982. Physical Properties of Rocks—Porosity, Permeability, Distribution Coefficients and Dispersivity, United States Geological Survey, OpenFile Report 82–166.

Zheng, C. 2010. *MT3DMS v5.3 Supplemental User's Guide*, Technical Report to the U.S. Army Engineer Research and Development Center, Department of Geological Sciences, University of Alabama.

Zheng, C. and P.P. Wang, 1999. MT3DMS—A Modular Three-Dimensional Multi-Species Transport Model for Simulation of Advection, Dispersion and Chemical Reactions of Contaminants in Groundwater Systems: Documentation and User's Guide, U.S. Army Engineer Research and Development Center, Contract Report SERDP-99-1.

18

Modeling of Radionuclide Transport in Groundwater

L. Elango, M. Thirumurugan, Faby Sunny, and Manish Chopra

CONTENTS

18.1 Introduction

Radionuclides undergo radioactive decay through emission of alpha and beta particles or gamma rays. There are about 2000 known radionuclides present in the earth. These naturally occurring radionuclides are ubiquitous trace elements found in rocks and soils. In general, radionuclides can be categorized in two ways: (1) by type of radioactive decay (alpha, beta, or gamma emission) and (2) by naturally occurring or manufactured. The radioactive elements are also present in water. The presence of radioactive elements in the groundwater depends on the geology and geochemistry of the rock formation. Groundwater is one of the likely pathways for radionuclide transport during accidental releases from nuclear facilities, such as power plants, fuel processing plants, and mining or milling operations. This can be estimated by direct measurement of activity or using mathematical models or a combination of these techniques.

Groundwater model is a computational method that represents an approximation of a groundwater system (Anderson and Woessner, 1992), which involves simulation of flow in an aquifer and its response to various input/output systems. Groundwater models are used as a tool to investigate a wide variety of hydrogeologic conditions. It can provide the estimates of water balance and travel time along flow paths. The groundwater models are also used to calculate the fate and transport of contaminants for dose evaluation. These models can also simulate the migration of solutes through the subsurface media and the boundaries of the aquifer system. Thus, the groundwater models play a major role in the study of problems associated with migration of radionuclides.

18.2 Applications of Radionuclide Models

Radionuclides may enter into groundwater systems due to leaching from nuclear storage facilities, near surface burials of wastes, mining and milling operations, and also due to unfortunate accidents in nuclear plants.

18.2.1 Geologic Storage of High-Level Waste

Geological disposal of nuclear waste is being planned in several parts of the world in order to isolate radionuclides from the environment. Hence, performance assessments rely on numerical models that use data collected over a comparatively short period of time (John and Helen, 2008). The deep geological conditions of the repository, the characteristics of nuclear waste, and geotechnical aspects control the confinement of wastes within the repository. The long-term performance and safety assessment of the system is carried out by simulating cumulative effects of changes in the properties of the repository system and the effects of the repository on the environment. Such a study was carried out by Claudia et al. (2013) in the fractured medium, where the radionuclide migration was simulated and the effective dose to an adult by ingestion of groundwater was estimated by numerical methods.

18.2.2 Near-Surface Disposal

Shallow land burial is a relatively simple disposal method for low level nuclear wastes. The basic principle of near-surface disposal is to keep the radiation dose due to the disposal practice to a level as low as possible. The assessment of radionuclide migration from near-surface disposal bears many similarities and some differences from isolation of high-level waste. The major distinctions are (1) wastes are not heat producing and (2) activity level is very low. The analysis procedure for developing a source term through corrosion or breaching of waste containers, an appropriate leach rate for the waste form, and estimating groundwater flow and radionuclide transport was carried out by Aikens et al. (1979). The performance assessment of the consequences of a breach of isolation and prediction of the long-term stability of the site was carried out. Nair and Krishnamoorthy (1998) have studied the probabilistic safety assessment of a near surface radioactive waste disposal and simulated the annual effective dose. Bugai et al. (2002) studied the radionuclide migration from a near surface waste disposal site in the Chernobyl zone to the geo-environment for safety assessment of waste dumps and predicted the future scenario.

18.2.3 Uranium Mining and Milling

Most of the groundwater contamination problems are associated with mining, milling, and waste disposal operations involved in the use of radioactive materials (NRC, 1979; Shepherd and Cherry, 1980; John and Helen, 2008). The least probability event of release from mill tailings (solid residues and the associated liquids) that contains both radioactive and nonradioactive materials may lead to contamination of groundwater. Tailings are the waste material that remains after uranium extraction from the ore. The solids consist primarily of the finely ground bulk of the original ore and a variety of chemicals precipitated from the tailings liquids (Thomas, 1990). The behavior of the radioactive contaminants in the tailings varies from simple to complex. Such tailings ponds may become a probable source for groundwater contamination. Hence, the long-term radiological impact assessment of a uranium tailings pond is being made a part of an impact assessment exercise in the uranium-mining industry. Elango et al. (2012) and Nair et al. (2010) have studied the radionuclide migration into groundwater from a tailings pond including the incorporation of decay chain transport.

18.2.4 Nuclear Power Plants

The nuclear power plant could have a very low probable accidental release, which ranges over small leaks from contaminated water streams in nuclear plants to major releases caused by a core meltdown accident (NRC, 1975, 1978; Niemczyk et al., 1981; John and Helen, 2008). In such cases, the nuclear wastes may get mixed with groundwater from the disposal facilities. The migration of radionuclide release into groundwater is much less as compared to that released through the air. Several researchers have estimated the transport of radionuclides released from an accident at the Fukushima Daiichi Nuclear Power Plant (Chino et al., 2011; Morino et al., 2013). Christoudias et al. (2014) estimated the probability of contamination at a proposed nuclear power plant in the eastern Mediterranean and Middle East region that is prone to natural disasters.

18.3 Parameters of Transport and Flow Equations

Various parameters of the aquifer medium control the groundwater flow and transport of contaminants. They are porosity, hydraulic conductivity, advection, dispersion, and sorption and retardation factor. The equation that governs the contaminant is based on the parameters, which are discussed in this section.

18.3.1 Porosity and Effective Porosity

Porosity and effective porosity are necessary to solve the flow and solute transport equations. The volume of void spaces in the rock is connected and they contribute to fluid flow or permeability in a reservoir or aquifer. These connected pores constitute effective porosity, which excludes the isolated pores and pore volumes occupied by water adsorbed on clay minerals or other grains. Thus, the effective porosity is less than total porosity.

18.3.2 Hydraulic Conductivity

The hydraulic conductivity (K) is one of the hydraulic properties of soil and it can be defined as a measure of the soil's ability to transmit water. This parameter typically determines potential of an aquifer for contamination. Soils with high hydraulic conductivities and large pore spaces are likely candidates for transporting contaminants. The hydraulic conductivity is defined by Darcy's law, which is explained later in this chapter.

18.3.3 Advection

The term advection is defined as the process by which moving groundwater carries dissolved solutes, where direction and rate of mass transport coincide with groundwater flow. The advective velocity is also called the seepage or pore water velocity, which is responsible for the transport of solutes through the groundwater medium. The one-dimensional flux of a solute through a porous medium is expressed by the following equation (Schulze and Makuch, 2004):

$$J = V_x C n_e \tag{18.1}$$

where
J = Mass flux per unit area per unit time
V_x = Advective velocity in the direction of flow
C = Concentration in mass per unit volume of solution
n_e = Effective porosity

18.3.4 Hydrodynamic Dispersion or Dispersion

Dispersion is a process of fluid mixing which acts to dilute the solute and lower its concentration. Dispersion leads to spreading of solutes in the direction of flow (longitudinal dispersion as shown in Figure 18.1) and perpendicular to it (transverse dispersion as shown in Figure 18.2).

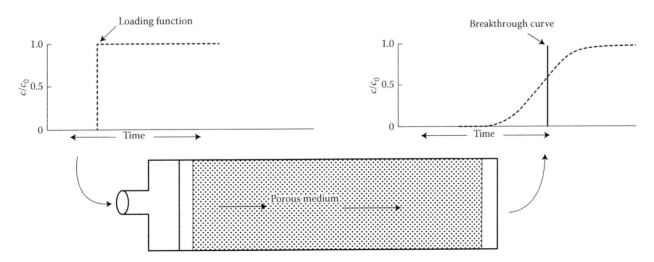

FIGURE 18.1
Longitudinal dispersion: Flow through column experiment: Continuous input of tracer at the inflow end and relative concentration versus time at the outflow end.

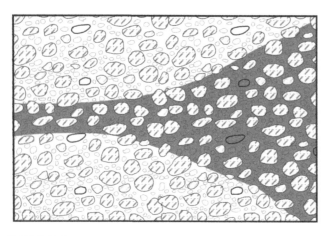

FIGURE 18.2
Transverse dispersion: Mixing in the direction perpendicular to the groundwater flow.

Dispersion occurs in a porous medium because of two processes:

- Mechanical dispersion (dispersion)
- Molecular diffusion (diffusion)

$$D_L = D' + D_m \qquad (18.2)$$

where
D_L = Longitudinal coefficient of hydrodynamic dispersion, (L^2T^{-1})
D' = Molecular diffusion coefficient, (L^2T^{-1})
D_m = Mechanical dispersion coefficient (L^2T^{-1})

In general, the longitudinal dispersivity is about 10 times larger than transverse dispersivity due to the local variation in the velocity field being more dominant in the direction of flow (Schulze and Makuch, 2004). Transverse dispersivity is because of branching out of flow paths from the center of the solute transport due to tortuosity of the medium.

18.3.4.1 Mechanical Dispersion

Mechanical dispersion is caused by the friction on pore walls, variations in pore size, and difference in flow paths. Some of the flow paths are faster because they follow a more direct path or because they are going through the center of void spaces in which water moves faster, hence less friction is involved (Schulze and Makuch, 2004). Other flow paths may be slower because they are closer to the pore walls, thus being exposed to more friction and hence leading to slowing down the movement of water molecules. The different flow paths of the water cause the mechanical dispersion.

Mechanical dispersion in flow direction (x) is given by (Schulze and Makuch, 2004)

$$D_m = \alpha_x v_x \qquad (18.3)$$

where
D_m = Mechanical dispersion coefficient (L^2T^{-1})
α_x = Longitudinal dispersivity (m)
v_x = Advective velocity along x direction (m s^{-1})

18.3.4.2 Molecular Diffusion

Molecular diffusion is caused by solutes kinetic energy of random motion. The coefficient of molecular diffusion is smaller for liquids because of the impact of solids with groundwater. The value of the coefficient of molecular diffusion depends on the type of solute in the groundwater medium, but for major anions and cations it usually ranges between 10^{-10} and 10^{-11} m^2 s^{-1} (Schulze and Makuch, 2004). The mathematical expression of molecular diffusion is given in following equation (Fick's law):

$$J = -D\frac{dC}{dx} \qquad (18.4)$$

where
J = Solute mass flux per unit area and per unit time $(ML^{-2}T^{-1})$
D = Molecular diffusion coefficient $(L^{-2}T^{-1})$
C = Solute concentration (ML^{-3})
dC/dx = Concentration gradient (ML^{-4})

18.3.5 Sorption and Retardation Factor

Van der waals forces, hydrogen bonding, hydrophobic forces, coulomb forces, ligand exchange, and covalent bonding between solute and geological medium are major mechanisms that cause sorption (Schulze and Makuch, 2004). In general, two types of sorption are distinguished: reversible and irreversible. Reversible sorption causes the solute concentration front to trail behind the advective front, while irreversible sorption effectively removes the solute concentration from the groundwater. Thus, the effect of reversible sorption on solute transport in groundwater is described by the retardation factor, which is defined using the average linear velocity of the medium and advective velocity of the solute.

18.4 Groundwater Flow Equations and Radioactive Transport

The migration of radionuclides in groundwater can be described by two equations, namely, groundwater flow and mass transport of the radionuclides. The movement of groundwater in the area must be studied before the transport equation can be solved.

18.4.1 Groundwater Flow

Groundwater moves in the pores of soil or rock under gravity, capillary forces, and pressure difference. Figure 18.1 conceptually illustrates a flow experiment through a sand-packed column and the relationship between the flows to the properties of the medium. Water enters the column under pressure through one end. The pressure can be measured and observed by means of a thin vertical pipe open in the sand at inlet and outlet. Darcy found that the discharge, Q (V^*) area of cross-section (A) of column or velocity (V) is proportional to the difference in the height of the water, h (hydraulic head), between the ends and inversely proportional to the flow length L (Darcy, 1856), the proportionality constant K is known as hydraulic conductivity.

$$V_x = -K \frac{\Delta h}{\Delta L} \qquad (18.5)$$

Equation 18.5 is known as Darcy's law (Darcy, 1856) and is the basis for groundwater flow equations in the saturated zone. An approximation of the flux in the major flow direction can be obtained using Darcy's law (Darcy, 1856):

$$V_x = -K \frac{\Delta h}{\Delta x} \cong -K \frac{dh}{dx} \qquad (18.6)$$

where dh/dx is the hydraulic gradient in the direction of flow. This assumes a homogeneous isotropic medium in which the gradient is constant over the augmentation. Water is moving only in the pore spaces, hence the actual groundwater velocity would be greater than the Darcy velocity, V_x. The groundwater velocity (U) can be estimated from V_x by using the effective porosity, n_e (Darcy, 1856)

$$U = \frac{V_x}{n_e} \qquad (18.7)$$

18.4.2 Mass Transport

Mass transport in the groundwater is strongly influenced by the geological structure of the porous medium. Radioactive wastes may move in the unsaturated region before entering into the saturated zone. The flow direction in the unsaturated region is downward until the flow reaches the zone of saturation, where the flow is

mostly along a lateral direction. The mass transport equation (ANS, 1980) is obtained by assuming local equilibrium of the dissolved species between water and rock:

$$R_d n \frac{\partial c}{\partial t} - \nabla . \; nD \; (\nabla c) + \nabla .(Vc) + \left[R_d \frac{\partial \theta}{\partial t} + \lambda n R_d \right] c = 0$$

$$(18.8)$$

where
   R_d = Retardation factor
   c = Concentration of dissolved constituent (g cm^{-3})
   D = Dispersion tensor (cm^2 s^{-1})
   V = Darcy velocity vector (cm s^{-1})
   λ = Radioactive decay constant (s^{-1}), which is given as
   n = Porosity

$$\lambda = \frac{\ln 2}{t_{1/2}}$$

where, $t_{1/2}$ is the half-life of a radioactive nuclide.

θ Moisture content

If dispersion tensor is homogeneous and isotropic, the equation can be written for a saturated medium as (ANS, 1980)

$$R_d \frac{\partial c}{\partial t} - D(\nabla . \nabla c) + \frac{V}{n} .(Vc) + \lambda R_d c = 0 \lim_{x \to \infty} \qquad (18.9)$$

If fluid flux is assumed as uniform along the x-axis, the equation can be written as (ANS, 1980)

$$\frac{\partial c}{\partial t} - \frac{D_x}{R_d} \frac{\partial^2 c}{\partial x^2} - \frac{D_y}{R_d} \frac{\partial^2 c}{\partial y^2} - \frac{D_z}{R_d} \frac{\partial^2 c}{\partial z^2} + \frac{U}{R_d} \frac{\partial c}{\partial x} + \lambda c = 0 \quad (18.10)$$

where U is the pore water velocity, D_x, D_y, and D_z are the dispersion coefficients in the x, y, and z directions, respectively. The above equations are valid for isotropic media only, but they may be applied to anisotropic formations while dispersivities are obtained in the field.

18.4.3 Decay Chain

Naturally occurring radioactive materials and many fission products undergo radioactive decay through a series of transformations (loss of particles or electromagnetic energy from an unstable nucleus) rather than in a single step (USEPA). These radionuclides emit energy or particles until the last step while each transformation becomes another radionuclide. Human-made elements, which are all heavier than uranium and unstable, undergo decay in this way. This decay chain or decay series ends in a stable nuclide. In a constant one-dimensional velocity field, the general equation can be written as (Burkholder and Rosinger, 1980)

$$R_{d1}\frac{\partial c_1}{\partial t} + U\frac{\partial c_1}{\partial x} = D\frac{\partial^2 c_1}{\partial x^2} - R_{d1}\lambda_1 c_1 \qquad (18.11)$$

$$R_{di}\frac{\partial c_i}{\partial t} + U\frac{\partial c_i}{\partial x} = D\frac{\partial^2 c_i}{\partial x^2} - R_{di}\lambda_i c_i + R_{di-1}\lambda_{i-1}c_{i-1}(i \geq 2)$$

$$(18.12)$$

where

R_{di} = Retardation factor for species i
U = Pore water velocity
c_i = Concentration of species i
D = Dispersion coefficient in the direction of flow
λ_i = Radioactive decay constant for species i

The mobility of contaminants is analyzed using the velocity at which contaminants move through the subsurface media. However, contaminants may be sorbed on or bonded to the surfaces of solids, particularly organic molecules which slow the contaminant plume relative to the movement of the groundwater itself, and spread it out (Willing, 2007). Sorption characteristics are expressed as distribution coefficient (K_d), a ratio of the amount of a solute in sorbed form on solids to the amount in dissolved form in water at equilibrium.

If the progeny are assumed to have equal retardation, then the concentration of the ith progeny in terms of parent concentration (Burkholder and Rosinger, 1980) is written as:

$$c_i = \frac{\lambda_i c_1}{\lambda_1}\prod_{m=1}^{i-1}\lambda_m\sum_{j=1}^{i}\frac{e^{-\lambda_j t}}{\prod_{\substack{k=1 \\ k \neq j}}^{i}(\lambda_k - \lambda_j)} \qquad (18.13)$$

If retardation values are different with long-chain decays for the parent and the progeny, the numerical model is a better option.

18.4.4 Uranium Decay Series

Radioactive decay occurs when an unstable isotope transforms by emitting alpha (α) and beta (β) particles and gamma (γ) rays. There are three natural decay series with ^{238}U, ^{232}Th, and ^{235}U as the parent radionuclides. In general, the radionuclides in these series are approximately in a state of secular equilibrium in nature, in which the activities of all radionuclides within each series are nearly same.

Two conditions are necessary for secular equilibrium. First, the half-life of the parent radionuclides must be much longer than that of any other radionuclide in the series. This condition is generally met for the ^{238}U, ^{232}Th, and ^{235}U decay series in naturally occurring ores. Second, a sufficiently long period of time must have

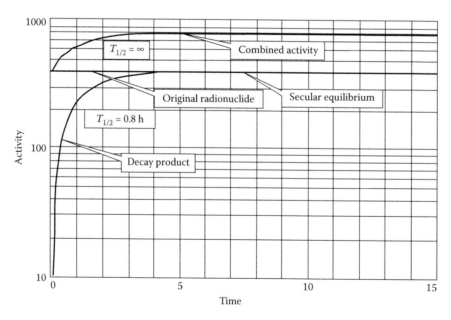

FIGURE 18.3
Relation between the radioactivities of a parent and daughter nuclide (secular equilibrium).

elapsed, for example, 10 half-lives of the decay product to allow for in-growth of the decay products. Under secular equilibrium as shown in Figure 18.3, the activity of the parent radionuclides decays infinitely slow as compared to the daughter radionuclide (half-lives of ∞ and 0.8 h, respectively).

The radionuclides of ^{238}U, ^{232}Th, and ^{235}U decay series are shown in Figures 18.4, 18.5, and 18.6 along with the major mode of radioactive decay for each.

18.5 Groundwater Model Calibration and Validation

A model is a set of parameters and validations of comparing the solution of these parameters with field measured data. Collecting data and monitoring concentrations of contaminants in the subsurface is mandatory but it is costly because the media in which

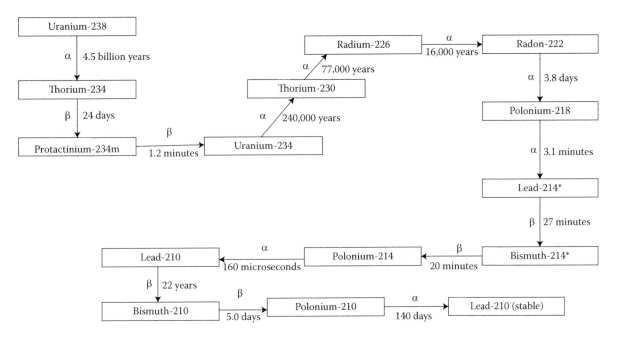

FIGURE 18.4
Natural decay series of ^{238}U.

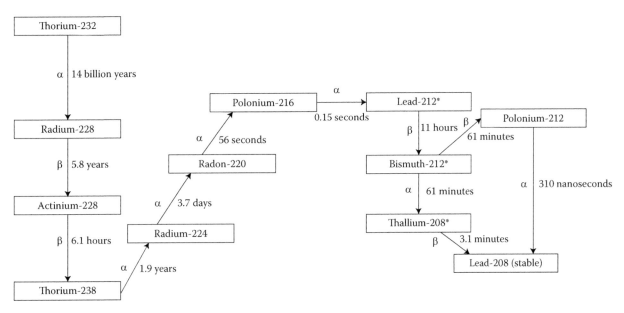

FIGURE 18.5
Natural decay series of ^{232}Th.

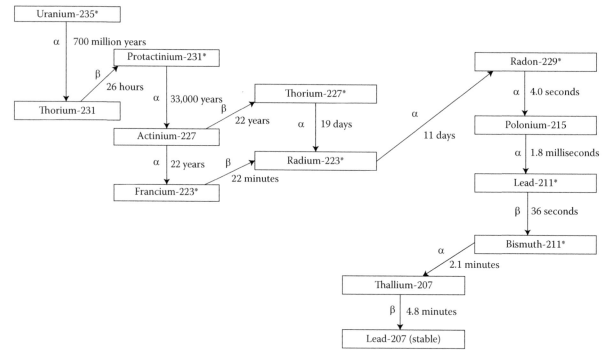

FIGURE 18.6
Natural decay series of ^{235}U.

the radionuclide migration is taking place can only be measured directly from wells. Groundwater contamination models are calibrated for a problem by defining field and laboratory measured parameters, running the model, and comparing the results with observed data. If the results are poor, the parameters need to be modified until a desired level of match is obtained (Mercer and Faust, 1980). Modifying boundary conditions and parameters for final simulation must still be in agreement with the knowledge of the geology and hydrogeology of the model area.

Validation of the groundwater models is defined as the process of demonstrating that a given site-specific model is capable of making sufficiently accurate predictions (Henriksen et al., 2003). The transport of contaminant and its progeny from the source need to be considered for a long time period. This period is considered as a representative time during which the geology and climate may not change to any great extent (Roberts, 1990). Over such a timescale, modeling results should be interpreted only as indicators of potential impacts and trends, rather than absolute values (Camus et al., 1998). The effective dose rate gives the impact of the transport of radionuclides in the study area. The simulated radionuclide concentration has to be translated into effective dose rates for people who consume the groundwater in the region. The guideline value for dose rate through drinking water

pathway is defined by the World Health Organisation (WHO-2004) as 0.10 mSv/yr.

18.5.1 Sensitivity and Uncertainty Analysis

A sensitivity analysis is the process of varying inputs over a sensible range and observing the relative change in model response. Uncertainty analysis is used for quantification of the uncertainty in the model output due to uncertainty in the input parameter values. Typically hydraulic conductivity (K) and distribution coefficient (K_d) are the most important parameters in groundwater flow and contaminant transport modeling. Any variation in hydraulic conductivity is expected to change the groundwater flow-field significantly (Elango et al., 2012). This change in the flow velocity will affect the contaminant concentration and, thus, the estimation of total effective dose on humans, due to consumption of groundwater.

18.6 Radionuclide Transport Modeling in Groundwater: A Case Study

The radionuclide migration and its radiological impacts due to a proposed uranium tailings pond was assessed

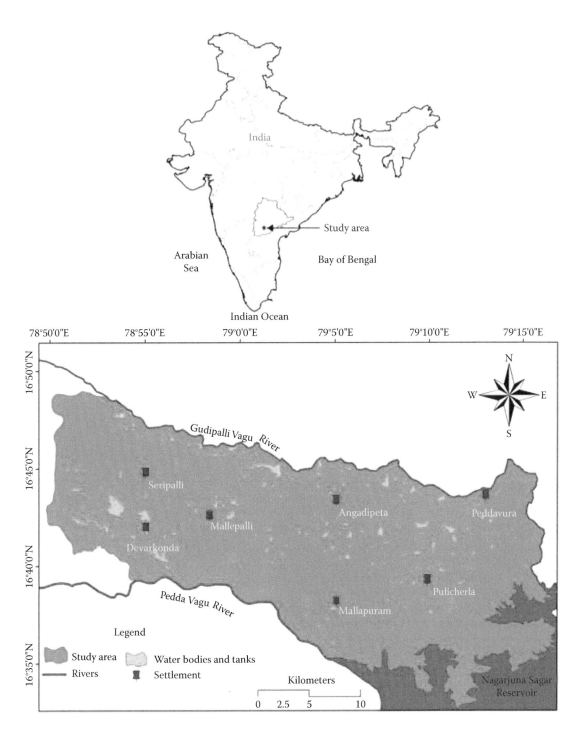

FIGURE 18.7
Location of the study area.

using numerical modeling by Elango et al. (2012). The tailings pond is located in the northern part of Cuddapah basin in the Nalgonda district, Telangana state, about 80 km E-SE of Hyderabad, India (Figure 18.7). The average annual rainfall in this area is about 1000 mm. This region has a well-defined watershed and covers an area of about 750 km² (Elango et al., 2012). The southeastern side of the study area is bounded by the Nagarjuna Sagar reservoir. The southern and northern sides were bounded by the Pedda Vagu River and Gudipalli Vagu River, respectively, which flow seasonally during the southwest monsoon from July–September and the northeast monsoon from October–December. Elevation of this area ranges from 160 to 350 m above mean sea level.

18.6.1 Geology and Hydrogeology

Geologically, this region is comprised of granite/granitic gneisss, pink biotite granite, grey hornblende biotite gneiss, migmatite granite, and metabasalt belonging to the late Archean (Brindha et al., 2012). These granites/gneisses are commonly medium to coarse grained, forming the basement and are traversed by a number of dolerite dykes and quartz veins (GSI, 1995). The youngest member in the Cuddapah supergroup (i.e., Srisailam formation) overlies the basement rock with a discrete unconformity (Brindha et al., 2012). The Srisailam formations are mainly arenaceous and include pebbly-gritty quartzite shale with dolomitic limestone, intercalated sequence of shale-quartzite, and massive quartzite (Brindha et al., 2012).

Hydrogeological conditions of the area were assessed by the observation of large diameter unlined wells. The temporal variation in groundwater level was assessed in several wells during the study carried out from March 2003 to December 2009 (Elango et al., 2012). Most of the wells in this area penetrate into the fractured layer. Groundwater occurs under unconfined condition in the topsoil and weathered/fractured rocks. Rainfall is the principle source of recharge apart from irrigation return flow and seepage from surface water bodies.

18.6.2 Model Conceptualization and Boundary Condition

Initially, the study area was conceptualized into a three-layer system. The soil cover is underlain by a weathered layer that extends downward and its thickness various spatially. This weathered zone was subdivided into a highly weathered layer and a moderately weathered/fractured layer (Elango et al., 2012). The two rivers namely Gudipalli Vagu (River) and Pedda Vagu (River) form the boundaries of northern and southern end, and they were considered as river head boundaries (Cauchi/Robbins condition). The southeastern part of the study area bounded by Nagarjuna sagar reservoir was considered as temporally variable head boundary. The western part of the study area was considered as variable head boundary and the head in this boundary was assigned based on the measured groundwater level in the wells closer to this boundary (Elango et al., 2012).

18.6.3 Governing Equations for Groundwater Flow and Radionuclide Transport Modeling

In general, during groundwater flow modeling the computer code numerically solves a system (matrix) of algebraic equations and this matrix represents the approximation of the partial differential equation of groundwater flow system (Kresic, 2007). The governing groundwater flow

equation (Equation 18.14) through porous media under unconfined condition given by Rushton (2003) is used in this study.

$$\frac{\partial}{\partial x}\left(K_x h \frac{\partial h}{\partial x}\right) + \frac{\partial}{\partial y}\left(K_y h \frac{\partial h}{\partial y}\right) + \frac{\partial}{\partial z}\left(K_z h \frac{\partial h}{\partial z}\right) = S_y \frac{\partial h}{\partial t} + q$$

(18.14)

where K_x, K_y, and K_z are the hydraulic conductivity (L/T) along x, y, and z coordinates; h is the hydraulic head (L); S_y is the specific yield (dimensionless); q is the source/sink term (L/T); and t is the time (T).

Equation 18.14 was numerically approximated to simulate the spatial and temporal variation of groundwater head and also to estimate the seepage velocity in the aquifer formation. Seepage velocity of the formation forms the basis to study the movement of radionuclides in groundwater. The spatial and temporal variation of the uranium nuclides over the area was studied using the decay-chain transport equation (Equation 18.15) given by Nair et al. (2010).

$$R_i \frac{\partial N_i}{\partial x} = \frac{\partial}{\partial x}\left(D_x \frac{\partial N_i}{\partial x}\right) + \frac{\partial}{\partial y}\left(D_y \frac{\partial N_i}{\partial y}\right) + \frac{\partial}{\partial z}\left(D_z \frac{\partial N_i}{\partial zx}\right)$$

$$- u \frac{\partial N_i}{\partial x} - v \frac{\partial N_i}{\partial y} - w \frac{\partial N_i}{\partial z} - R_i \lambda_i N_i + R_{i-1} \lambda_{i-1} N_{i-1}$$

when $i > 1$ to M (18.15)

where M is the number of total nuclides in the decay-chain; N_i is the concentration of the parent in groundwater (atoms L^{-3} or mol); D_x, D_y, and D_z are the hydrodynamic dispersion coefficients in x, y, and z directions ($L^2 T^{-1}$); x, y, and z are the longitudinal, lateral, and vertical distances from the source area (L); u, v, and w are the components of groundwater seepage velocity along x, y, and z directions (LT^{-1}); R_i is the retardation factor of the radionuclide for the linear isotherm, and λ_i is the radioactive decay constant or chemical reaction rate constant of the parent radionuclide (T^{-1}). Further, R_i can be computed using the following equation:

$$R_i = 1 + \frac{K_{dip b}}{n}$$

(18.16)

where K_{di} is the distribution coefficient of nuclide "i" ($L^3 M^{-1}$); ρb is the bulk density of the aquifer ($M L^{-3}$); and is the porosity.

The numerical solution for Equations 18.14 and 18.15 was obtained by discretizing the region into 3D finite element cells using the FEFLOW 6.0 (Elango et al., 2012).

18.6.4 Model Input Parameters

The hydraulic conductivity and specific yield/specific storage values were assigned from pumping tests carried out in the area. The groundwater level measured in March 2003 was considered as the initial hydraulic head and it ranged from 349.5 m (amsl) near the northwestern boundary to 164.5 m (amsl) at the southeastern part of the area. The rainfall recharge was estimated from rain gauge stations using the Theissen-Polygon method based on the Groundwater Resource Estimation Committee (GEC, 1997) and assigned to the model (Elango et al., 2012).

18.6.5 Flow Model Calibration and Simulation

Calibration was carried out by varying the model input parameters within the allowable range. The calibration was performed initially under steady-state condition by considering the groundwater head measured in March 2003 (Figure 18.8a). Then the calibration was carried out under transient conditions for the period from March 2003 to December 2009. Figure 18.8b shows the comparison of observed and simulated head over this period. During this calibration, the hydraulic conductivity and

specific yield values were varied within an allowable range during the number of trials until a reasonable match between observed and simulated groundwater head was achieved (Figure 18.9).

18.6.6 Modeling Radionuclide Transport from the Tailings Pond

The potential effect of the mine-waste storage on the groundwater was estimated by using the source term. The radionuclide source term was treated as a decaying-concentration boundary condition, where the concentration of the ith species at the source area was computed using the following equation given by Nair et al. (2010):

$$\varphi_i(t) = N_i(t)K_{li} \exp[-(\lambda i + K_{li})t] \qquad (18.17)$$

where $\varphi_i(t)$ is the release rate of the ith nuclide (atoms T^{-1}); $N_i(t)$ is the total of the ith nuclide (atoms) at time (t); λ_i is the radioactive decay constant or chemical reaction rate constant of the parent radionuclide (T^{-1}); K_{li} is the leach rate or fractional release rate of the ith nuclide (T^{-1}); and t is the time (T). The exponential term in above equation represents the source

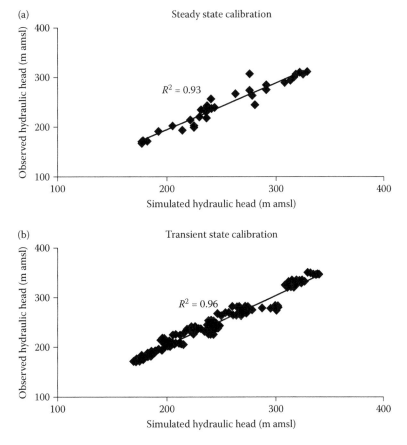

FIGURE 18.8
Comparison between simulated and observed head at (a) steady and (b) transient state condition.

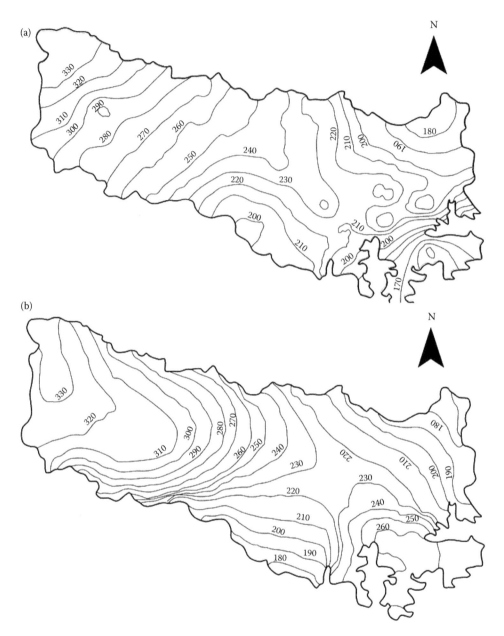

FIGURE 18.9
Contours of (a) observed and (b) simulated regional groundwater levels in September 2009.

depletion due to radioactive decay and leaching from the tailings pond.

The leach rate from the tailing ponds was calculated as a function of the infiltration velocity (Nair et al., 2010).

$$K_{li} = \frac{vS}{nVR_{is}} \tag{18.18}$$

where v is the infiltration rate of water from the tailings pond ($L\ T^{-1}$), S is the surface area of the tailings pond (L^2), V is the volume of the tailings pond (L^3), n is the porosity of the tailings (dimensionless), and R_{is} is

the retardation factor (dimensionless) of nuclide i in the tailings pond. The ingrowth of the progeny in the tailings pond was evaluated using the Bateman equations (Elango et al., 2012). The following first-order differential equation was used to calculate the number of atoms of the parent radionuclide at the source area (Nair et al., 2010):

$$\frac{dN_1}{dt} = -(\lambda_1 + K_{l1})N_1; \quad \text{with initial condition } N_1(t=0)$$

$$= N_1^0 \tag{18.19}$$

where $N_1(t)$ is the number of atoms of the parent radionuclide at time t (atoms) and N_1^0 is the number of atoms of the parent radionuclide present in the tailings pond initially (other terms as previously defined).

The activity of the ith daughter nuclide was calculated using the following equation (Nair et al., 2010):

$$\frac{dN_i}{dt} = \lambda_{i-1}N_{i-1} - (\lambda_i + K_{li})N_i \qquad (18.20)$$

with initial condition $N_i(t = 0) = N_i^0$

where $N_i(t)$ is the number of atoms of the ith daughter radionuclide at time t (atoms) and N_i^0 is its number of atoms present in the tailings pond initially.

The concentrations of the four long-lived radionuclides ^{238}U, ^{234}U, ^{230}Th, and ^{226}Ra in the tailings pond were calculated as a function of time based on their initial inventories. The source properties of half-life, distribution coefficient, and ingestion-dose coefficient of the radionuclides in the tailings pond inventory are given in Table 18.1. The effect of leaching from the tailings pond was simulated by considering an array of injection wells in the source area (Elango et al., 2012). Each node in the source was considered to have an initial concentration of 0.129 Bq/l (238U/234U) discharging into the field with an exponential decay-rate equivalent to the decay constant plus the leach rate of the contaminant radionuclide (Elango et al., 2012). The initial activities of ^{230}Th and ^{226}Ra are ten times higher than the $^{238}U/^{234}U$ as 90% of the uranium has been extracted from the mined ore material and 10% goes to the tailings pond as waste (Elango et al., 2012). The quantity of recharge of the tailings water from the pond based on the design infiltration velocity of 1.0×10^{-9} m/s was estimated to be 69.19 m³/day (Elango et al., 2012).

18.6.7 Simulation of Effect of Uranium Tailings Pond

Simulation of transport of uranium and its long-lived progeny was carried out using the properties given in Table 18.1. The variation in concentration of radionuclides versus distance from the edge of the tailings pond, along

the direction of groundwater flow for different distances after a time period of 1000, 3000, 5000, 7000, and 10,000 years was plotted and given in Figure 18.10. This figure shows the concentrations of ^{238}U, ^{234}U, and ^{230}Th decrease with distance and become less than 0.025 Bq/l at a distance of 340 m from the edge of the tailings pond along the groundwater flow direction (Elango et al., 2012). The concentrations of ^{238}U and ^{234}U are identical. In general, the ^{226}Ra, with its lower distribution coefficient (K_d) value, higher initial activity (almost 10 times that of ^{238}U and ^{234}U) at the tailings pond, shorter half-life, and greater in situ production, migrates further and the concentration increases compared to the other three radionuclides (Elango et al., 2012).

In order to study the possible impact of the transport of radionuclides, the simulated concentrations were translated into effective dose rates for people who may consume the groundwater on the downstream side of the tailings pond. The effective dose rate is equal to the product of concentration, drinking water consumption rate (taken to be 2.2 l/day), and the ingestion dose coefficient (Nair et al., 2010). The total annual effective dose to the public through possible use of groundwater for drinking along its flow path is shown in Figure 18.11. The observed maximum effective dose rate at an elapsed time of 10,000 years is about 2.5 times lower than 0.1 mSv/y (WHO, 2004) (Elango et al., 2012).

18.6.8 Conclusion

Groundwater modeling has become an intense indispensible tool to understand the possible migration of radionuclides from nuclear storage facilities, milling operations, tailings ponds, and unfortunate accidents in nuclear plants. A case study on the application of radionuclide transport modeling to understand implications of a proposed uranium tailings pond was presented. This study indicates that the maximum dose that would be received by the members of the public who consume the affected groundwater at 340 m from the tailings pond is predicted to be about 0.04 mSv/year after 10,000 years of transport. This value is about 2.5 times lower than the WHO drinking water guideline of 0.1 mSv/year (Elango et al., 2012).

TABLE 18.1

Properties of Radionuclides Used in the Model

Radionuclides	Half-Life (years)	Distribution Coefficient for Soil (mL/g)	Ingestion-Dose Coefficient (Sv/Bq)
^{238}U	4.5×10^9	500	4.5×10^{-8}
^{234}U	2.44×10^5	500	4.9×10^{-8}
^{230}Th	7.70×10^4	2000	2.1×10^{-7}
^{226}Ra	1.60×10^3	500	2.8×10^{-7}

Acknowledgment

The authors would like to thank the Board of Research in Nuclear Sciences, Department of Atomic Energy (GOI) for their financial support to carry out the case study under Grant No. 2007/36/35.

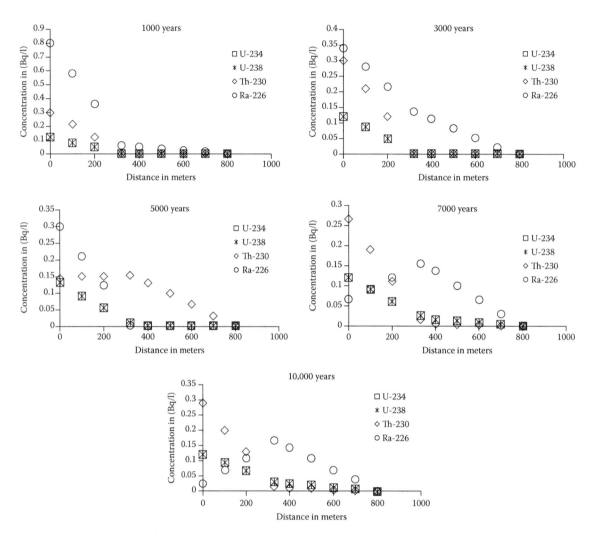

FIGURE 18.10
Simulated concentration of radionuclides versus distance from the edge of the tailings pond along the direction of groundwater flow.

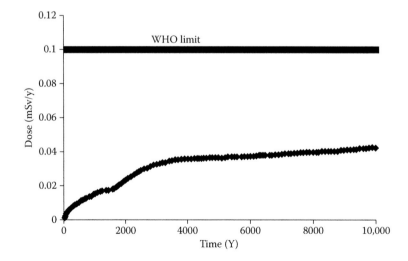

FIGURE 18.11
Total effective annual dose rates from all the long-lived radionuclides at a distance of 340 m from the edge of the tailings pond, along the direction of groundwater flow, compared to the WHO limit. (Adapted from Elango, L. et al. *Hydrogeol. J.*, 20, 797–812, 2012.)

References

Aikens, A.E., Berlin, R.E., Clancy, J., and O.I. Oztunali. 1979. *Generic Methodology for Assessment of Radiation Doses from Groundwater Migration of Radionuclides in LWR Wastes in Shallow Land Burial Trenches.* Atomic Industrial Forum, Washington, DC.

Anderson, M.P. and W.W. Woessner. 1992. *Applied Groundwater Modelling, Simulation of Flow and Advective Transport.* Academic Press, San Diego, CA.

ANS. 1980. American Nuclear Society: Evaluation of radionuclide transport in groundwater for nuclear power sites. Technical report: 2.17.

Brindha, K., Rajesh, R., Murugan, R., and L. Elango. 2012. Nitrate pollution in groundwater in some rural areas of nalgonda district, Andhra Pradesh, India. *Journal of Environmental Science and Engineering,* 54(1), 64–70.

Bugai, D., Dewiere, L., Kashparov, V., Skalskyy, A., Levchuk, S., and V. Barthes. 2002. Chernobyl case study increases confidence level in radionuclide transport assessments in the geosphere. *Technical Report*: 30, 1–14.

Burkholder, H.C. and E.L.J. Rosinger. 1980. A model for the transport of radionuclides and their decay products through geologic media. *Nuclear Technology,* 49(1): 150–158.

Camus, H., Little, R., Acton, D., Aguero, A., and D. Chambers et al. 1998. Long-term contaminant migration and impacts from uranium mill tailings. *Journal of Environmental Radioactivity,* 42(2–3), 289–304.

Chino, M., Nakayama, H., Nagai, H., Terada, H., Katata, G., and H. Yamazawa. 2011. Preliminary estimation of release amounts of 131I and 137Cs accidentally discharged from the Fukushima Daiichi Nuclear Power Plant into the atmosphere. *Journal of Nuclear Science and Technology,* 48, 1129–1134.

Christoudias, T., Protestos, Y., and J. Lelieveld. 2014. Atmospheric dispersion of radioactivity from nuclear plant accidents: Global assessment and case study for the Eastern Mediterranean and Middle East. *Energies,* 7(12), 8338–8354.

Claudia, S., Antonio, C.M.A., and R.O. Jose de Jesus. 2013. Radionuclide transport in fractured rock: Numerical assessment for high level waste repository. *Science and Technology of Nuclear Installations,* 2013, 1–17.

Darcy, H. 1856. *Les Fontaines Publiques de la Ville de Dijon.* Victor Dalmont, Paris.

Elango, L., Brindha, K., Kalpana, L., Sunny, F., Nair, R.N., and R. Murugan. 2012. Groundwater flow and radionuclide decay-chain transport modelling around proposed uranium tailings pond in India. *Hydrogeology Journal,* 20, 797–812.

GEC. 1997. *Groundwater Resource Estimation Committee: Report of the Groundwater Resources Estimation Methodology.* Government of India, New Delhi.

GSI. 1995. *Geology and Minerals Map of Nalgonda District, Andhra Pradesh, India.* Geological Survey of India, New Delhi.

Henriksen, H.J., Troldborg, L., Nyegaard, P., Sonnenborg, T.O., Refsgaard, J.S., and B. Madsen. 2003. Methodology for construction, calibration and validation of a national hydrological model for Denmark. *Journal of Hydrology,* 280, 52–71.

John, E.T. and A.G. Helen. 2008. *Radiological Risk Assessment and Environmental Analysis.* Oxford University Press, Oxford.

Kresic, N. 2007. *Hydrogeology and Groundwater Modelling.* (2nd ed.) CRC Press, Taylor & Francis Group, Boca Raton, FL.

Mercer, J.W. and C.R. Faust. 1980. Groundwater modelling: An overview. *Groundwater,* 18(2), 108–115.

Morino, Y., Ohara, T., Watanabe, M., Hayashi, S., and M. Nishizawa. 2013. Episode analysis of deposition of radiocesium from the Fukushima Daiichi Nuclear Power Plant accident. *Environmental Science and Technology,* 47, 2314–2322.

Nair, R.N. and T.M. Krishnamoorthy. 1998. Probabilistic safety assessment for near surface radioactive waste disposal facilities. *Environmental Modeling and Software,* 14(1999), 447–460.

Nair, R.N., Sunny, F., and S.T. Manikandan. 2010. Modelling of decay chain transport in groundwater from uranium tailings ponds. *Applied Mathematical Modeling,* 34, 2300–2311.

Niemczyk, S.J., Adams, K., Murfin, W.B., Ritchle, L.T., Eppel, E.W., and J.D. Johnson. 1981. *The Consequence from Liquid Pathways After a Reactor Meltdown Accident.* NUREG/CR-1598. U.S. Nuclear Regulatory Commission, Washington, DC.

NRC. 1975. *Nuclear Regulatory Commission: Reactor Safety Study, An Assessment of Accident Risks in U.S. Commercial Nuclear Power Plants.* WASH-1400. U.S. Nuclear Regulatory Commission, Washington, DC.

NRC. 1978. *Nuclear Regulatory Commission: Liquid Pathways Generic Study.* NUREG-0440. U.S. Nuclear Regulatory Commission, Washington, DC.

NRC. 1979. *Nuclear Regulatory Commission: Draft Generic Environmental Impact statement on Uranium Milling.* NUREG-0511. U.S. Nuclear Regulatory Commission, Washington, DC.

Roberts, L.E.J. 1990. Radioactive waste management: Annual review of nuclear particle. *Science* 40, 79–112.

Rushton, K.R. 2003. *Groundwater Hydrology, Conceptual and Computational Models.* Wiley, Chichester, UK.

Schulze, D. and Makuch. 2004. *Advection, Dispersion, Sorption, Degradation, Attenuation,* Eolss Publisher (www.eolss.net). University of Texas at El Paso.

Shepherd, T.A. and J.A. Cherry. 1980. Contaminant migration in seepage from Uranium mill tailings impoundments—an overview. In Uranium Tailings Management, Proceedings of the Third Symposium. Civil Engineering Department, Colorado State University, Fort Collins.

Thomas, K.T. 1990. Management of wastes from uranium mines and mills. *IAEA Bulletin,* 23(2), 33–35.

USEPA. Available from "http://www.epa.gov/radiation/understand/chain.html”

WHO. 2004. *Guidelines for Drinking-Water Quality.* (3rd ed.), Recommendations: Vol.1. World Health Organization, Geneva.

Willing, P. 2007. A nanotechnical guide to groundwater modelling. *NRDC,* 1–36.

19

Benchmarking Reactive Transport Codes for Subsurface Environmental Problems

Dipankar Dwivedi, Bhavna Arora, Sergi Molins, and Carl I. Steefel

CONTENTS

19.1 Introduction

Reactive transport models (RTMs) can be applied to investigate and understand a broad range of environmental problems such as CO_2 sequestration, acid mine drainage, aquifer contamination, geochemical weathering, as well as carbon and nutrient cycling at regional and global scales. In this respect, RTMs are essential tools to investigate important environmental issues by improving our mechanistic understanding, facilitating data interpretation, and predicting system evolution.

The past few years have witnessed extraordinary advances in the development of new and the improvement of existing capabilities of RTMs. Advancements in measurement technology (e.g., spectroscopic methods, genomic characterization, and isotope fractionation) have expanded the role of RTMs to simulate pore scale processes (Steefel et al., 2013; Molins et al., 2014), include microbial dynamics (Scheibe et al., 2009; Steefel et al., 2014b), and discern the contribution of different sources and/or processes based on isotope ratios. The rapid advancement of computer architectures and computational methods have enabled RTMs to simulate reactive processes in heterogeneous and complex multidimensional systems (Li et al., 2010), incorporate high-resolution information (Trebotich et al., 2014), and improve code performance (Hammond et al., 2014). Certain reactive transport models have also included capabilities for coupling hydrological and reactive transport processes at watershed and catchment scales (Gwo and Yeh, 2004; Srivastava et al., 2007; Beisman et al., 2015; Dwivedi et al., 2015a) and/or bridging across scales (Scheibe et al., 2014; Arora et al., 2015b).

Despite the availability of a large number of codes that essentially solve the same governing equations, substantial differences exist among them. Users can differentiate codes based on (1) the types of capabilities they offer for simulating flow and transport, biogeochemical processes, and coupling mechanisms (e.g., multiphase multicomponent flow, kinetic isotope fractionation, biomass growth, porosity–permeability coupling, atomistic or transition theory-based rate law formulations), (2) differences in numerical schemes and formulations (e.g., global implicit or operator splitting), and (3) the ease and flexibility of their use (e.g., user support groups, graphical user interface). With the recent advances in computer hardware and performance, the ability to parallelize the codes, and the availability of open source software have become important issues for some users. Thus, a variety of RTMs and numerical codes are available to users to describe the interaction of several complex and competing biogeochemical processes at a range of spatial and time scales. For complex physical and chemical

scenarios, such as those involving multiple interacting components or multiple phases, the only way to verify codes and build confidence is through benchmarking activities (Keyes et al., 2013; Steefel et al., 2015a).

Benchmarking can be defined in different ways based on the reasons for their use (Figure 19.1). First, benchmarking can be used to compare conceptual and numerical capabilities of models and identify needs for further improvement. Second, benchmarks can be developed as standardized references to test existing or new models. Third, benchmarking activities can build confidence in models and provide measures of model performance. Overall, benchmarking can be defined as an important scientific approach to verify, validate, develop, and compare RTMs. There are several ways in which benchmark problems can be developed. For example, a benchmark can be built on a hypothetical scenario or actual field data. Similarly, a benchmark can have a single problem or a set of simple to complex problems. Considerable attention has been given to different types of benchmarking studies in the recent literature. In this chapter, we review recent benchmarking efforts that compare RTMs on relevant subsurface environmental problems.

The chapter is organized as follows. Section 19.2 summarizes the governing equations for hydrological flow, solute transport, and chemical reactions. Section 19.3 provides a description of the numerical approaches used in RTMs. Section 19.4 describes relevant benchmarking and code comparison activities within the past 5 years (2010–2015). Section 19.5 concludes the chapter with a discussion of the need for new benchmarking activities. The focus of this chapter is on benchmarking activities in reactive transport modeling in the geosciences and not on the RTMs themselves. Consequently, it is not the objective of the current chapter to provide a comprehensive description of all the participating RTMs in the benchmark studies described here. We refer readers interested in these models to user manuals or relevant review studies that provide a detailed description of the different RTMs (Zhang et al., 2012; Steefel et al., 2015b).

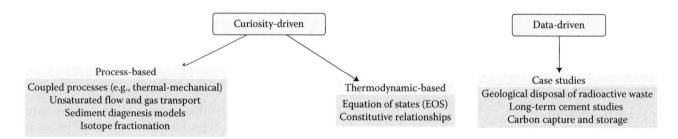

FIGURE 19.1
Benchmarking activities can be defined based on how they are used. (Modified from Kolditz, O. et al. *Environ. Earth Sci.*, 67(2), 613–632, 2012, doi:10.1007/s12665-012-1656-5.)

19.2 Governing Equations

Although this book addresses flow and reactive transport processes primarily in groundwater, we need to consider these processes in the unsaturated zone also. Physical features of the unsaturated zone, like macropores and fractures, contribute to 90% of contaminant transport to the deeper zones. Similarly, geochemical processes in the unsaturated zone (also known as the vadoze zone) can affect the mobility of contaminants through different mechanisms of precipitation, sorption–desorption, ion-exchange, redox reactions, complexation, and colloidal interactions. In the following section, we will describe how these processes can be solved mathematically at the continuum scale by using the continuity equation together with momentum, energy, and reactive or nonreactive species balances (Steefel and Lasaga, 1994; Molins et al., 2012; Hammond et al., 2014; Steefel et al., 2015b). Equations of flow and transport at the continuum scale utilize various constitutive relationships to parameterize the heterogeneous subsurface system under variably saturated conditions, for example, permeability, dispersivity, and specific mineral surface area (Lichtner and Kang, 2007). Variably saturated flow is represented either by solving a set of multiphase flow and conservation equations—where local equilibrium is assumed between phases—or the Richards equation—which assumes single phase, variably saturated, isothermal systems (Neuman, 1973; Panday et al., 1993). Furthermore, for describing reactive transport in the variably saturated subsurface environment, we also need to combine the transport and biogeochemical reaction network. The most important governing equations for describing and simulating reactive flow and transport at the continuum subsurface environmental are described next.

19.2.1 Fluid Flow

19.2.1.1 Single-Phase Flow

The general formulation of single-phase flow that is also applicable to variable density flow is typically described by the continuity equation as given by (Bear, 1975):

$$\frac{\partial [\phi \rho_f]}{\partial t} = \nabla \cdot [\rho_f \mathbf{q}] + \rho_f Q_a, \tag{19.1}$$

where ∇ denotes the divergence operator; ρ_f [kg m^{-3} water] represents the fluid density, $\mathbf{q}$[m^3 H$_2$O m^{-2} medium s^{-1}] is the volumetric or Darcy flux of water, Q_a[m^3 H$_2$O m^{-3} medium s^{-1}] is the volumetric source or sink term, t[s] is the time, and ϕ[m^3 void m^{-3} medium] is

a dimensionless quantity, which represents the porosity of the porous media.

Darcy's law provides the constitutive relationship between the volumetric or Darcy flux and fluid pressure gradient. Darcy's law is used to describe the single-phase flow in permeable subsurface porous media (Bear, 1975), as:

$$\mathbf{q} = \phi \mathbf{v} = -\frac{\mathbf{k}_{sat}}{\mu} [\nabla P - \rho_f g e_z], \tag{19.2}$$

where $\mathbf{v}$[m s^{-1}] denotes the pore velocity, z[m] is the depth, ∇P[kg m^{-1} s^{-2}, or Pa m^{-1}] is the gradient of the fluid pressure, g[m s^{-2}] is the acceleration due to gravity, $\mathbf{k}_{sat}$[m^2] is the permeability tensor for fully saturated conditions, e_z is a unit vector in the vertical direction, and μ[kg s^{-1} m^{-1}] is the dynamic viscosity, respectively. Alternatively, Darcy's law as described in Equation 19.2 can also be written as the product of the hydraulic head, h[m] and the hydraulic conductivity tensor, $\mathbf{K}$[m s^{-1}], as:

$$q = -\mathbf{K}\nabla h. \tag{19.3}$$

19.2.1.2 Variable Saturated Flow

Variably saturated flow can be described either by a set of equations for multiphase flow (e.g., liquid and gas) or by the Richards equation that assumes only one active phase (e.g., liquid), while the gas phase is assumed to be passive.

19.2.1.2.1 Richards Equation

The Richards equation (Richards, 1931) has extensively been used in the literature to describe the variably saturated flow under the assumption of a passive air phase that is at the atmospheric pressure (e.g., Neuman, 1973; Arora et al., 2011; Riley et al., 2014; Dwivedi et al., 2015b). The Richards equation can be formulated either based on the saturation of the aqueous phase or based on the fluid pressure. The saturation-based formulation is given as:

$$S_a S_s \frac{\partial h}{\partial t} + \phi \frac{\partial S_a}{\partial t} = \nabla \cdot [k_{ra} \mathbf{K} \nabla h] + Q_a, \tag{19.4}$$

where S_a[m^3 H$_2$O m^{-3} void] and k_{ra}[–] are dimensionless quantities, which denote saturation of the aqueous phase and the relative permeability, respectively, and S_s[m^{-1}] is the specific storage coefficient. The equivalent pressure-based formulation is given as:

$$\frac{\partial [S_a \phi \rho_f]}{\delta t} = \nabla \cdot \left[-\rho_f \frac{k_{ra} k_{sat}}{\mu} (\nabla P - \rho_f g e_z) \right] + \rho_f Q_a. \tag{19.5}$$

19.2.2 Multiphase Flow

The mass conservation equation of multiphase transport, in general, requires the solution of each component j in phase α:

$$\frac{\partial \left[\phi \sum_\alpha \rho_\alpha S_\alpha Y_{j\alpha} \right]}{\delta t} = \nabla \cdot \left[-\sum_\alpha \rho_\alpha Y_{j\alpha} \frac{k_{r\alpha} k_{sat}}{\mu} (\nabla P - \rho_\alpha g e_z) \right]$$
$$+ \rho_\alpha Q_j, \qquad (19.6)$$

where $Y_{j\alpha}$ denotes the mass fraction of component j in phase α, and all other variables (e.g., $k_{r\alpha}$, etc.) are the same as defined earlier. Thus, the occurrence of components like CO_2 or H_2O in various phases, gas, liquid, and solid can be described using the multiphase flow. Equation 19.6 is applicable in a wide spectrum of problems ranging from hydrothermal to nuclear waste disposal to problems addressing geological CO_2 sequestration in deep saline aquifers.

19.2.2.1 Relative Permeability and Saturation Formulations

Although several models can be used to describe the relationship between the capillary pressure and aqueous phase saturation, this relationship is often described using van Genuchten or Brooks-Corey formulations. Additionally, the relative permeability functions are typically based on the Mualem and Burdine formulations (Brooks and Corey, 1964; Mualem, 1976; van Genuchten, 1980). The effective saturation, $S_{ea}[-]$, of the aqueous phase can be related to the capillary pressure, $P_c[Pa]$, using the van Genuchten relationship (van Genuchten, 1980):

$$S_{ea} = \left[1 + (\alpha \, | \, P_c \, |)^n \right]^{-m}. \qquad (19.7)$$

The effective saturation is defined as:

$$S_{ea} = \frac{S_a - S_{ra}}{1 - S_{ra}} \qquad (19.8)$$

The relative permeability based on the Mualem relative permeability and van Genuchten saturation relationships can be expressed (Mualem, 1976; van Genuchten, 1980), as:

$$S_a = S_{ra} + \frac{1 - S_{ra}}{(1 + \alpha \psi_a^n)^m}, \qquad (19.9)$$

$$k_{ra} = S_{ea}^l \left[(1 - S_{ea}^{1/m})^m \right]^2, \qquad (19.10)$$

where $S_{ra}[-]$ is the residual saturation of the aqueous phase, ψ_a denotes the pressure head [m], $\alpha[m^{-1}]$, n, m, and l are soil hydraulic function parameters, also known as Mualem-van Genuchten parameters, with $m = 1 - 1/n$.

The other commonly used relationship between the capillary pressure and aqueous phase saturation is given by the Brooks-Corey curve (Brooks and Corey, 1964):

$$S_{ea} = \left[\frac{| P_c |}{P_e} \right]^\lambda, \qquad (19.11)$$

where P_e defines the entry pressure [Pa] and λ is a fitting parameter.

For the gas phase, the relative permeability is calculated from the following formulation:

$$k_{rg} = 1 - k_{ra}, \qquad (19.12)$$

Equation 19.12 provides a simple formulation to calculate the relative permeability of the gas phase; however, a more detailed approach suggested by Luckner et al. (1989) is often used:

$$k_{rg} = (1 - S_{eg})^{1/3} \left[1 - S_{eg}^{1/m} \right]^{2m}, \qquad (19.13)$$

where effective gas saturation S_{eg} is related to the liquid saturation S_a, and the residual gas saturation S_{rg}, as:

$$S_{eg} = \frac{S_a}{1 - S_{rg}}. \qquad (19.14)$$

So far in this chapter, we have talked about fluid flow in porous media under variably saturated conditions. In the following sections, we will talk about processes that change concentrations of solutes (or chemical components, used interchangeably in this section) in space and time. Natural processes that change concentrations in the porous media can be broadly categorized as transport or transformation. The passive chemical substances (also known as nonreactive tracers) transport by advection or diffusion, or both. Advective transport is driven by fluid flow, while diffusive transport is due to the random motion of a substance within the fluid. In addition, advective and diffusive transport are also associated with the mechanical dispersion that indicates spreading and mixing of substances due to molecular diffusion and variations in fluid flow velocity (Bear, 1975). In general, the dispersion term is used to account for the mechanical dispersion and the diffusion (Gelhar et al., 1992). Molecular diffusion and mechanical dispersion are examples of processes that occur at different spatial scales. Molecular diffusion and dispersion occur at microscopic and macroscopic scales, respectively.

In contrast to a passive chemical substance, a reactive chemical substance can also undergo a transformation, which can change the substance of interest into another substance, during the transport process. The substance experiences transformation either by physical processes, for example, radioactive decay or biogeochemical processes, for example, denitrification.

19.2.3 Molecular Diffusion

One of the modes of transport of solute species in porous media is through molecular diffusion. Fick's First Law states that the molecular diffusive flux of species in solution is proportional to the concentration gradient (Bear, 1975):

$$J_j = -D_j \nabla C_j, \tag{19.15}$$

where D_j denotes the diffusion coefficient, which is specific to chemical species considered as indicated by the subscript j. Fick's Second Law, derived by integrating the diffusive fluxes over a control volume, states that the divergence of the diffusive flux is the accumulation rate of the solute species, as (Bear, 1975):

$$\frac{\partial C_j}{\partial t} = -\nabla \cdot [J_j] = \nabla \cdot [D_j \nabla C_j]. \tag{19.16}$$

Fick's Law is a remarkable relationship that is used to describe transport processes in porous media driven by concentration gradients. However, molecular diffusive transport can also occur due to gradients caused by chemical or electrostatic potential (i.e., electrochemical migration); the Nernst–Planck equation considers gradients due to chemical or electrostatic potential apart from concentration gradients (Bear, 1975; Steefel and Maher, 2009). Additionally, it is also important to include a tortuosity correction, as well, in the case of diffusion in porous media (Steefel and Maher, 2009). The Nernst–Planck equation and a discussion of how to incorporate tortuosity correction is available in (Steefel et al., 2015b).

19.2.4 Mechanical Dispersion

Mechanical dispersion is defined in terms of the fluid velocity and dispersivity, α, with longitudinal and transverse components. Dispersivity describes the spreading in the longitudinal and transverse direction of a plume due to the spatially variable velocity. Dispersivity is the property of the porous media and is a scale dependent parameter (Dagan, 1990; Gelhar et al., 1992).

$$\begin{aligned} D_L &= \alpha_L V_i, \\ D_T &= \alpha_T V_i, \end{aligned} \tag{19.17}$$

where V_i is the average velocity in the principal direction of flow, and the subscripts L (longitudinal) and T (transverse) denote the dispersion coefficient parallel and perpendicular to the principal direction of flow, respectively (Bear, 1975).

19.2.5 Reactive Transport Equations

The reactive transport modeling includes combining the transport and biogeochemically driven transformation processes in the continuity equation; the reactive transport equation can be described as follows:

$$\frac{\partial(\phi S_L C_i)}{\partial t} = \nabla \cdot (\phi S_L D_i^* \nabla C_i) - \nabla \cdot (\mathbf{q} C_i) - \sum_{r=1}^{Nr} \nu_{ir} R_r - \sum_{m=1}^{Nm} \nu_{im} R_m$$

$$- \sum_{l=1}^{Ng} \nu_{il} R_l \tag{19.18}$$

The accumulation term (mol m^{-3} medium s^{-1}) appears on the left-hand side of the Equation (19.18), which is calculated by multiplying the porosity and liquid saturation, S_L, and the concentration. The first and second right-hand side terms account for diffusive and advective transport; third, fourth, and fifth right-hand side terms describe various reactions that are partitioned between aqueous phase reactions, R_r, mineral reactions, R_m, and gas reactions, R_l, respectively. These reaction terms are typically assumed to be kinetically controlled, although an equilibrium treatment is possible as well.

Reactions are geochemical processes that lead to the transformation of reactant species to product species. Reactions can be classified into homogeneous if they occur within the same phase (usually, the aqueous phase) or heterogeneous if they involve the transformation of species from or into a different phase, for example, the solid phase in the case of mineral dissolution–precipitation reactions. These reactions can also be classified into equilibrium or kinetic depending on whether local equilibrium can be assumed. Which treatment is justified depends on the relative time scales of the processes involved, for example, if the rate of dissolution of a mineral is much faster than the rate of transport of reactants and products the reaction may be assumed to be in equilibrium. However, a kinetic treatment is always more general (Steefel and Lasaga, 1994).

The transformation from reactant to product species occurs according to a stoichiometric relationship. The stoichiometry of a generic reaction involving N_c reactants with chemical formulae A_j and a product with chemical formula A_i can be written as

$$\sum_{j=1}^{N_c} \nu_{ij} A_j = A_i, \tag{19.19}$$

where v_{ij} are the stoichiometric coefficients of reactant j in reaction i.

When a reaction is in equilibrium, the relationship between the concentrations of products and reactants can be expressed with the law of mass action:

$$\log K_i = \sum_{j=1}^{N_c} v_{ij} \log(\gamma_j c_i) - \log(\gamma_i c_i), \quad (19.20)$$

where K_i is the equilibrium constant and γ_i and γ_j are the activity coefficients of product and reactant species.

Geochemical equilibrium between aqueous species makes it possible to calculate the concentration of a set of secondary species as a function of a set of primary species. Further, one can define a set of total component concentrations, one for each primary species, the masses of which are not affected by geochemical aqueous equilibrium reactions. The total mass of a component (ψ_j) is defined as the sum of the mass of a primary species and the mass of a number of secondary species multiplied by the corresponding stoichiometric coefficient:

$$\psi_j = c_j + \sum_{i=1}^{N_r} v_{ij} c_i, \quad (19.21)$$

where N_c is the number of aqueous equilibrium reactions.

Definition of a set of total component or primary species concentrations that includes both individual primary species and a set of secondary species assumed to be at equilibrium is an approach that is widely used in geochemical models (Lichtner, 1985; Yeh and Tripathi, 1989; Steefel and Lasaga, 1994; Steefel and MacQuarrie, 1996) to reduce the size of the system to solve, and more specifically the number of transport equations. Because equilibrium aqueous complexation reactions always exist, Equation 19.18 can be modified as follows to account for equilibrium reactions by defining a total concentration as:

$$\frac{\partial(\phi S_L \psi_i)}{\partial t} = \nabla \cdot (\phi S_L D_i^* \nabla \psi_i) - \nabla \cdot (\mathbf{q} \psi_i) - \sum_{r=1}^{N_r} v_{ir} R_r$$

$$- \sum_{m=1}^{N_m} v_{im} R_m - \sum_{l=1}^{N_g} v_{il} R_l \quad (19.22)$$

When local equilibrium cannot be assumed, reaction rates need to be calculated explicitly. The reaction rates are in general a nonlinear function of concentrations of the geochemical species. Additionally, for heterogeneous reactions, the rates can be a function of material properties (symbolically: π), for example,

reactive surface area in the case of mineral dissolution–precipitation. The particular expression depends on the reaction type. For example, mathematical formulations include the transition state theory rate law for mineral dissolution–precipitation, a first order dependence on concentration for radioactive decay or a Monod-type formulation for microbially mediated reduction–oxidation reactions. Readers are referred to additional sources for more detailed formulations. However, for the sake of completeness, the equilibrium mass action law and kinetic rate expression for reaction (Equation 19.23) are provided here only as a generic function (Table 19.1):

$$R_k = f(c_i, \pi) \quad (19.23)$$

For multiphase and multicomponent reactive transport problems, Equations 19.18 and 19.22 can be further extended to consider transport in different phases (in addition to liquid phase) by incorporating the corresponding accumulation and transport terms for each phase (Mayer et al., 2001; Molins and Mayer, 2007). Furthermore, depending on the level of complexity desired, the Fickian treatment of dispersion or the Nernst–Planck equation can also be incorporated in Equations 19.18 and 19.19 (Mayer et al., 2001; Rasouli et al., 2015).

19.2.6 Thermal Processes

The reactive transport processes are typically temperature dependent. An additional governing equation used for nonisothermal problems are based on energy conservation of the energy balance. The continuity equation for energy conservation in a porous medium is given by:

$$\frac{\delta}{\delta t}\left[\phi \sum_{\alpha} S_\alpha \rho_\alpha U_\alpha + (1-\phi)\rho_r c_r T\right] + \nabla \cdot [\mathbf{q}_\alpha \rho_\alpha H_\alpha - \kappa \nabla T] = Q_e,$$

$$(19.24)$$

where T is the temperature and the subscript α denotes a fluid phase with Darcy velocity q_α, density ρ_α, saturation S_α, internal energy U_α, and enthalpy H_α. The coefficient κ is the thermal conductivity of the medium and c_r and ρ_r represent the specific heat and density of the porous medium. The quantity Q_e accounts for a source/sink term. The relationship between internal energy and enthalpy is given here:

$$U_\alpha = H_\alpha - \frac{P_\alpha}{\rho_\alpha}. \quad (19.25)$$

Thermal conductivity is defined in terms of fully saturated, κ_{sat}, and dry thermal conductivities κ_{dry}, and

TABLE 19.1

Examples of Chemical Equilibrium and Kinetic Reaction Equations[a]

	Reaction	Equilibrium Mass Action Law
Equilibrium Reactions		
Ion exchange	$vACl_u(aq) + uBX_v(s) \leftrightarrow uBCl_v(aq) + vAX_u(s)$	$K_{eq} = \dfrac{[BCl_v]^u [AX_u]^v}{[ACl_u]^v [BX_v]^u}$
Surface complexation	$XOH + M^{Z+} = XOM^{Z+1} + H^+$	$K_{eq} = \dfrac{[XOH^{Z+1}][H^+]}{[XOH][M^{2+}]}$
		$K_{app} = K_{int} e^{\frac{zF\psi_0}{RT}}$ (b)
Mineral dissolution		$K_m = \displaystyle\prod_{j=1}^{Nc} (\gamma_j c_j)^{v_{mj}}$ (c)
Kinetic Reactions		
Monod rate law	$R_m = \mu_{max} B \left[\dfrac{[E_D]}{K_{E_D} + [E_D]} \right] \left[\dfrac{[E_A]}{K_{E_A} + [E_A]} \right]$ (d)	
Mineral dissolution	$R_m^{\cdot} = sgn[\Omega] A_m k_m \left(\prod a^n \right) \left\| \exp \left[\dfrac{\eta \Delta G}{RT} \right] - 1 \right\|^m = sgn[\Omega] A_m k_m \left(\prod a^n \right) \left\| \left[\dfrac{Q_m}{K_{eq}} \right]^\eta - 1 \right\|^m$	
	$sqn[\Omega] = sgn[\log(Q_m / K_m)]$ (e)	

[a] The brackets [] refer to activities, K_{eq} is the equilibrium constant, and R_m is the reaction rate (mol kg$_{water}^{-1}$ s^{-1}).

[b] K_{int} is the intrinsic equilibrium constant that does not depend on surface charge, z is the charge of the surface species, π_0 is the mean surface potential (V), F is Faradays' constant (96,485 C mol^{-1}), R is the gas constant (8.314 J mol^{-1} K^{-1}), and T is the absolute temperature (K).

[c] v_{ij} are the stoichiometric coefficients of primary species j in mineral m.

[d] μ_{max} is the maximum intrinsic rate (mol mol^{-1} biomass s^{-1}) and B is the biomass concentration. Symbols E_A and E_D represent the electron acceptor and donor, respectively (mol kg$_{water}^{-1}$), and K_s is the half saturation constant (mol kg$_{water}^{-1}$).

[e] A_m is the reactive surface area of the reacting mineral (m^2 m^{-3} porous medium), k_m is the rate constant (mol m^{-2} s^{-1}), and the term Πa^n is product of all of the far from equilibrium effects on the reaction, n can be interpreted as the reaction order, and Q_m and K_{eq} are the ion activity product and equilibrium constant, respectively.

the liquid saturation S_l, as suggested by Somerton et al. (2013):

$$\kappa = \kappa_{dry} + \sqrt{S_l}(\kappa_{sat} - \kappa_{dry}) \qquad (19.26)$$

19.2.7 Remarks

In this chapter, we have described flow and reactive transport processes in variably saturated subsurface porous media for multiphase, multicomponents, and nonisothermal processes. We have also described kinetically controlled and equilibrium-based biogeochemical reaction processes that are typically considered in modern reactive transport modeling. The kinetically controlled biogeochemical reaction processes typically include mineral dissolution and precipitation reactions, homogeneous (aqueous phase) reactions, and microbially mediated homogeneous and heterogeneous reactions. The equilibrium-based biogeochemical reaction processes include sorption or surface complexation. An in-depth review of kinetically controlled and equilibrium-based biogeochemical reactions can be found in

the literature (e.g., Steefel and Maher, 2009; Li et al., 2010; Steefel et al., 2015b). Furthermore, the surface complexation models have been described as multisite surface complexation, constant capacitance model, diffuse layer model, triple layer model, gas-aqueous phase exchange, and multisite ion exchange reactions. A detailed description of various surface complexation models is available elsewhere (Davis and Kent, 1990; Dzombak, 1990; Goldberg et al., 2007).

Geochemical transformation of substances such as mineral precipitation and dissolution can lead to changes in porosity and permeability. Several studies have shown that modest reductions in porosity from mineral precipitation can cause large reductions in permeability (Vaughan, 1987). Changes in permeability and porosity are likely to modify the unsaturated flow properties of the rock. Additionally, it is essential to address flow and transport in fractured porous media as well as for reliable predictions of several issues, such as groundwater contamination, nuclear waste cleanup activities, carbon capture and storage, etc. The prevalence of fractures and heterogeneity in porous media has led to a full suite of the multicontinuum class of models. Some examples include (1) equivalent continuum model (ECM) (Faybishenko et al., 2000); (2) dual permeability model (DPM), dual or multiporosity model (Barenblatt et al., 1960; Warren and Root, 2013), multiple interacting continua approach (MINC) (Pruess, 2013), and (3) discrete fracture and matrix model (Snow, 1965). Broadly speaking, the dual-continuum approach considers two interacting regions, one associated with the less permeable intraaggregate pore region, or the soil matrix domain, and the other associated with the interaggregate pore region, or the fracture domain. A review of the different multicontinuum approaches including the governing equations and exchange functions are available elsewhere (Faybishenko et al., 2000; Berkowitz, 2002; Simunek et al., 2003).

19.3 Numerical Schemes

As described in the previous section, reactive transport modeling in its most general case entails the solution of a coupled, nonlinear set of differential and algebraic equations describing a number of physical and geochemical processes over time in a spatial domain. Finite element (FE) and finite volume (FV) methods have been used to discretize the differential equations describing flow, transport, and energy balance (Equations 19.1, 19.4, or 19.6, 19.18, and 19.24) in reactive transport problems. Both methods allow for the use of structured and unstructured meshes. Unstructured meshes make it possible to handle complex domain geometries, but obtaining quality meshes is often a rather involved process.

19.3.1 Coupling and Solution Approaches

One can envision solving this large set of mixed (flow, energy balance, transport, and chemical) equations simultaneously, accounting for the feedback mechanisms between processes directly. This approach is commonly termed *global implicit* or *one step* method in that the primary unknowns (and thus their derived quantities) are evaluated at the new time step ($n + 1$). Often, each of the processes or group of processes is solved separately, rather than simultaneously, in a sequential manner over the same time step. This alternative approach is referred to as *operator splitting*, wherein the operators corresponding to each process are applied to the primary unknowns sequentially. An operator splitting error is associated with this approach especially when there is a strong coupling between the *split* processes, although this error may be controlled by choosing sufficiently small time steps. An iterative procedure may also be used to minimize this error.

In RTMs, it is common to solve flow and energy balance (Equations 19.1, 19.4 or 19.6, and 19.24) separately from the reactive transport equations (Equation 19.18). The feedback processes from reactive transport processes, for example, evolution of fluid densities and medium porosity, are accounted for sequentially at the end of the reactive transport step. The updated flow field resulting from this evolution is passed back to the reactive transport problem. In contrast, the coupling between transport and geochemical processes is considered both using global implicit and operator splitting approaches. The global implicit and operator implicit approaches for reactive transport are discussed in more detail next.

While saturated flow, energy balance, and advective–diffusive transport can be described by linear differential equations, the equations describing the nonlinear multiphase and variably saturated flow equations (Equations 19.6 and 19.4), respectively, as well as the geochemical processes (Equations 19.20 and 19.23) are nonlinear. The Newton–Raphson method is widely employed to solve nonlinear equations. In this method, the set of equations ($F(x) = 0$) is linearized near an approximation of the solution (x^k) using the Jacobian ($J \equiv F'(x)$) to obtain a correction to the approximation (δx^{k+1}):

$$J(x^k)\delta x^{k+1} = -F(x^k) \qquad (19.27)$$

This process is iterated until the approximation of the solution converges to a set of values within a given tolerance.

19.3.2 Geochemical Problem Approach

Since the mass action law (Equation 19.20), which describes geochemical equilibrium, makes it possible to calculate the concentration of geochemical species as a function of a subset of species, referred to as primary or component species, the degrees of freedom of the geochemical problem equals the number of these primary species. The geochemical problem consists of a series of conservation equations one for each primary species, which need to be solved together with the mass action law and the kinetic reactions equations (Equations 19.20 and 19.23) (Steefel and MacQuarrie, 1996). Each conservation equation defines the total component concentration, which includes the primary species and the secondary species in equilibrium with it. Given that the mass action law and kinetic rate law are algebraic, nonlinear equations, the resulting system needs to be solved with the Newton–Raphson method. The size of this nonlinear system of equations can be reduced if the mass action law equations are substituted directly into conservation equations, that is, the secondary species are formally eliminated from the equations in the so-called direct substitution approach (DSA).

19.3.3 Operator Splitting Approach for Reactive Transport

In the operator splitting approach, the transport equations are solved separately from the geochemical reactions. This makes it possible to take advantage of numerical properties of the transport equations. Advective–diffusive transport can then be described by a set of linear equations which are also decoupled, that is, each total component concentration can be solved for independently from the rest. Further, high order explicit methods then can be used for advection, thus minimizing numerical dispersion.

An additional advantage of the operator splitting approach is that the reaction step becomes an embarrassingly parallel problem, trivial to code for distributed memory systems, where geochemistry is solved grid cell by grid cell. Further, the size of the Jacobian matrix needed in each cell equals the number of primary species. The main drawback of the operator splitting approach is the need to limit the time step size such that the Courant number is less than one. The Courant condition guarantees that mass is not transported more than a single grid cell in a time step (Steefel and MacQuarrie, 1996).

19.3.4 Global Implicit Approach for Reactive Transport

In the global implicit approach, the reactive transport problem consists of a set of nonlinear equations that solves transport and reaction over the entire simulation domain simultaneously. This allows for arbitrarily large time steps, although time discretization errors are not necessarily completely eliminated in this case. Following the DSA, the set of equations can be formulated so that the concentrations of the primary species (rather than the total component concentrations) are the unknowns of the system (Lichtner, 1992). The size of the resulting Jacobian matrix is the size of the domain times the number of component species, in which now the chemical problem in each cell does not depend only on the concentrations in that cell, but also on the concentrations in the adjacent cells, through the coupling introduced by the transport processes.

19.4 Benchmarking Studies from 2010 to 2015

As described in the previous sections, RTMs can have different geochemical capabilities, numerical schemes, and added architecture. This section provides an overview of recent reactive transport benchmarks in the geosciences field. Benchmarking efforts in the past have

targeted relevant reactive transport problems such as the performance of geological radioactive waste repositories (Pruess et al., 2004; Tsang et al., 2008), uranium migration (De Windt et al., 2003), and biogeochemical processes affecting root water uptake (Nowack et al., 2006). Recent literature has seen an increase in model verification and intercomparison studies highlighting the interest in and need for these activities. Literature on these recent benchmarking activities was amassed by searching existing refereed literature published within the past 5 years (2010–2015) using Web of Science and Google Scholar with search terms such as "model intercomparison," "reactive transport benchmarks," "benchmarking reactive transport," "model verification," and specific environmental problem names. This literature showed both community-based and individual benchmarking efforts. As suggested in the introduction section, each benchmarking initiative has its own process and style. Thus, in this section, we describe the benchmarking strategy adopted, the results achieved and the nature of environmental problems targeted for both the individual benchmarking studies and community-based initiatives undertaken during this time frame.

19.4.1 Individual Benchmarking Studies

In this subsection, we describe the individual benchmarking studies that have been conducted within the past 5 years. In reviewing these examples, we emphasize the motivation and the need behind these benchmarking efforts. These categories are by no means an exhaustive list but serve as examples of the basis for individual benchmarking efforts.

19.4.2 Benchmarking for Developing Standardized References

The aim of these benchmark problems is to design a test problem that is reproducible and can serve as a validation tool for other codes. For example, Carrayrou et al. (2009) presented the reactive transport benchmark of MoMaS, a challenging test for numerical methods used in modern RTMs. MoMaS refers to modeling, mathematics, and numerical simulations related to nuclear waste management problems. The motivation behind this benchmarking exercise was to simulate the development of new numerical methods and present a standard benchmark for other codes. The benchmark was presented in three levels—easy, medium, and hard in one- and two-spatial dimensions. This exercise resulted in several publications with different codes simulating the different levels of the MoMaS benchmark (De Dieuleveult and Erhel, 2009; Lagneau and van der Lee, 2009; Mayer and MacQuarrie, 2009; Carrayrou et al., 2010). This benchmark exercise highlighted that although several RTMs were able to reproduce the MoMaS benchmark test cases of different levels with both sequentially iterative and global approaches, the 2D problem emphasized the need for parallel computations to decrease computation time for simulations. This benchmarking exercise still fosters collaboration through conferences and meetings.

19.4.3 Benchmarking for Code and Conceptual Model Comparisons

There have been several benchmarking studies within the past 5 years that are designed to test different aspects of the code capabilities. A few of these studies are described here.

Kempka et al. (2013) employed three codes (TOUGH2-MP/ECO2N, DuMu^X, and OpenGeoSys) to improve their understanding of carbon capture and storage at the Ketzin pilot site located in Eastern Germany. This intercomparison study revealed that matching downhole pressure data and CO_2 arrival times requires appropriate anisotropic permeability information (e.g., the introduction of distinct near-well and far-field permeability tensors in the codes).

In another study on carbon capture and storage, Tambach et al. (2013) compared two codes— TOUGHREACT and a Shell reservoir simulator MoReS coupled with PHREEQC—to simulate CO_2 storage in a deep saline aquifer. The purpose of this benchmarking exercise was to obtain confidence and quality control on different types of reservoir simulators.

There have been other benchmarking studies that focus on coupling mechanisms. A historical description of several benchmarking examples dealing with both single and coupled thermal-hydro-mechanical/chemical processes (THM/C) processes have been presented in Kolditz et al. (2015). This book is specifically designed for code developers and describes verification tests that deal with common applications of THM/C coupled models like geothermal energy utilization, nuclear waste disposal, and carbon dioxide storage in the deep geological formation.

19.4.4 Benchmarking for Model Performance and Reliability

In the absence of closed-form solutions for realistically complex subsurface problems, some benchmarks are designed to test model implementation and performance. For example, He et al. (2015) established a coupling interface for two open source codes OpenGeoSys and IPhreeqc and proposed a flexible MPI-based parallelization scheme, which allows an optimized allocation of computer resources for the node-wise calculation of chemical reactions as well as for solving flow and solute transport processes (Figure 19.2a). The motivation

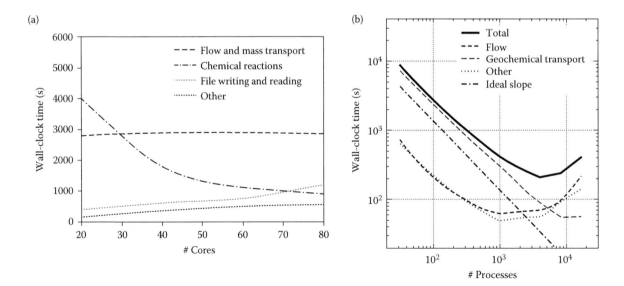

FIGURE 19.2
Breakdown of the wall clock time into time spent in flow, transport, geochemical reactions, and other (i.e., initialization, writing input output files, and interface coupling) for (a). The 3D test case using OGS-IPhreeqc coupling when 20 cores are used for domain decomposition related processes (Modified from He, W. et al. 2015. *Geoscientific Model Development*, 8: 2369–2402.) and (b). the Hanford 300 Area test problem using PFOTRAN (Hammond, G. E., P. C. Lichtner, and R. T. Mills: Evaluating the performance of parallel subsurface simulators: An illustrative example with PFLOTRAN. *Water Resources Research*, 2014. 50(1). 208–228. doi:10.1002/2012WR013483. Figure 10, Copyright Wiley-VCH Verlag GmbH & Co. KGaA. Reproduced with permission.)

behind this benchmarking exercise was to test and verify the new coupling interface and parallelization scheme for precision and performance. Nardi et al. (2014) also provide a benchmark problem to test the reliability of a newly developed interface between the general purpose COMSOL multiphysics framework and the geochemical code PHREEQC.

In another benchmarking study, Hammond et al. (2014) provide a reference for the scalability and performance of parallel subsurface simulators. The benchmark is set up on three realistic problem scenarios dealing with (a) copper leaching from a mineral ore deposit, (b) variably saturated flow and transport with a doublet well configuration, and (c) uranium surface complexation at the Hanford 300 Area (Figure 19.2b). The study indicated strong scaling of PFLOTRAN when the number of degrees of freedom is maintained at or above 10K per process.

19.4.5 Community-Based Benchmarking Efforts

In this subsection, we describe benchmarking activities initiated within the past 5 years by the geosciences community as a group. This includes the SSBench initiative described in detail next. There are other community-based benchmarking initiatives that have continued over the past several years and are still ongoing, including the DECOVALEX project (Tsang et al., 2008), long-term cement studies (Savage et al., 2011; Soler et al., 2015), and Sim-Seq (Pruess et al., 2004; Mukhopadhyay et al., 2014), but these efforts have been adequately summarized elsewhere. In addition, there have been recently initiated community-based efforts. For example, geothermal code comparison efforts have been recently initiated by the DOE Geothermal Technologies Office (Scheibe et al., 2013). These efforts have fostered the customization of an open-source framework called GTO-Velo for sharing data, input and output files, and other benchmarking relevant information (White et al., 2015).

Another ongoing initiative is the CO_2Bench. This is a benchmarking initiative considering the problem of CO_2 injection and sequestration in subsurface geological formations (Kolditz et al., 2012; Kühn et al., 2012). This benchmarking project is an important research collaboration intended to advance the understanding and mathematical modeling of coupled thermal-hydro-mechanical/chemical processes (THM/C). This benchmarking initiative has thus far provided a systematic multistep approach for developing benchmarks for CO_2 injection problems such as establishing process coupling, checking thermodynamic status, and capturing site characteristics. However, these efforts are only now underway and will not be discussed further here.

19.4.6 Benchmarking for Environmentally Relevant Topics: The SSBench Initiative

The Subsurface Environmental Simulation Benchmarking Workshop (or SSBench workshop) was initiated in 2011 as a workshop at Lawrence Berkeley National Laboratory for advancing the understanding and mathematical modeling of subsurface reactive transport processes (Steefel et al., 2015a). As the name suggests, the objective of SSBench was to present environmentally relevant benchmark problems that test conceptual model capabilities, numerical implementation, process coupling, and the accuracy and efficiency of the RTMs. SSBench has included the participation of several international research teams and code developers, and has been held annually at different locations since its inception. The benchmarking strategy of SSBench is to have at least three different codes simulate each benchmark with the intent to provide these comparisons to the geochemical community and to improve on existing modeling frameworks. The primary focus in the manuscripts is on the development of a benchmark that provides rigorous tests of one or more essential features of a numerical reactive transport simulator. The objective is not to verify this or that specific code—the reactive transport codes play a supporting role in this regard—but rather to use the codes to verify that a common solution of the problem can be achieved. The SSBench community has thus far produced several interesting benchmarking problems, which are presented next within the specific environmental problem they target.

19.4.6.1 Acid Rock Drainage

Reactive transport modeling of metal cycling associated with sediment, tailings, and waste rock is often needed for the management of abandoned mines and remediation of mining-impacted areas (Caruso, 2004; Arora et al., 2014). The SSBench presented two benchmark problems dealing with (1) the generation and attenuation of acid rock drainage (ARD) (Mayer et al., 2015) and (2) sediment–metal interactions in mining-impacted lake sediments (Arora et al., 2015a).

The first benchmark was designed as a three-component problem that compares RTMs on simulating relevant ARD processes such as pyrite oxidation in mine tailings, pH-buffering by primary and secondary mineral phases, and metal release above and below the water table. The study indicated that all participating codes were able to capture system evolution and long-term (10 years) mass loading predictions. A second benchmark study focuses on the cycling of metal contaminants in lake sediments downstream of mining operations. This study compares four RTMs in their

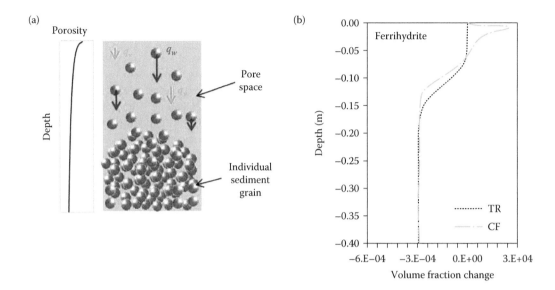

FIGURE 19.3
(a) Differences in compaction schemes for two reactive transport models where ToughReact (TR) simply advects mineral and aqueous components at a given solid (q_w) and fluid (q_v) burial rates but CrunchFlow (CF) readjusts porosity internally to honor conservation of mass at each grid cell following Berner's approach (Berner, 1980) (Adapted from Arora, B. et al. 2015a. *Computers & Geosciences*, 19(3): 613–633.) and (b) results of simulating compaction in a 1D column using these two RTMs. (With kind permission from Springer Science + Business Media: *Computers & Geoscicience*, A reactive transport benchmark on heavy metal cycling in lake sediments, 19(3), 2015a, 613–633, doi: 10.1007/s10596-014-9445-8, Arora, B. et al., Figure 10.)

handling of microbial reductive dissolution of iron hydroxides, the release of sorbed metals into lake water, reaction of these metals with biogenic sulfide to form sulfide minerals, and sedimentation driving the burial of ferrihydrite and other minerals. This study highlights that results of benchmark implementation by different codes is dominated by the differences in conceptual models of sediment compaction (Figure 19.3).

19.4.6.2 Metal/Radionuclide Fate and Transport

The use of reactive transport modeling is often required to obtain a predictive understanding of the coupled processes affecting the transformation and bioreduction of toxic metals and radionuclides in the subsurface (Jardine, 2008). The SSBench presented two benchmarks dealing with chromium bioreduction and three with uranium fate and transport processes.

The first benchmark on chromium focuses on the simulation of microbially mediated redox reactions with the inclusion of biomass growth and decay as well as explicit consideration of catabolic and anabolic pathways (Molins et al., 2015). This study indicated excellent agreement between results obtained with all participating codes. A second benchmark study compares RTMs in their simulation of processes affecting chromium isotope fractionation such as the dissolution of chromium bearing minerals and abiotic aqueous chromium reduction from iron (Wanner et al., 2015). Results of certain

components of this benchmark were also compared with analytical Rayleigh-type fractionation model. Overall, model intercomparison of this benchmark resulted in a good agreement (Wanner et al., 2014a).

The first benchmark on uranium focuses on an acetate amendment field experiment and compares RTMs on simulating a complex biogeochemical reaction network that considers microbially mediated reduction and immobilization of uranium (Yabusaki et al., 2015). This study indicated that even though a large number of reactions, different rate law formulations, multisite surface complexation model, as well as interdependent transport and reaction processes were considered, the participating codes were able to reproduce observed uranium behavior. This study also highlighted that discrepancies at the micromolar level for certain components (e.g., U(VI), Fe(II), sulfide) were pH-related and might be due to differences in numerical solution schemes of the different codes. A second benchmark on uranium focuses on comparing RTMs on simulating multirate surface complexation and dual domain mass transfer in one- and two-spatial dimensions (Greskowiak et al., 2015). This study also emphasized that pH and carbonate activity should be accurately calculated as even slight deviations in their values can have a significant impact on simulation results. A third benchmark focuses on mixing controlled reactions and the importance of Fe-hydroxides in reoxidizing biogenically reduced uranium (Şengör et al., 2015). This study showed that comparing mixing controlled reactions provides an excellent test for analyzing numerical dispersion and comparing numerical solution schemes across codes (Figure 19.4a).

19.4.6.3 Geological and Engineered Waste Repositories

Reactive transport models are important tools to provide accurate assessments of risk and engineering performance for issues associated with CO_2 storage, geothermal energy generation, and nuclear waste disposal in geological and engineered repositories (Bildstein et al., 2014). The SSBench presented three benchmarks dealing with this topic.

The first benchmark focuses on the geochemical evolution of a cement–clay interface as might be expected in an engineered barrier system and compares RTMs in simulating steep pH and Eh gradients and highly complex mineralogy (Marty et al., 2015). Despite differences in the activity correction models and other nuances of the codes, code intercomparison showed only minor discrepancies in simulating the benchmark problem. The second benchmark also focuses on cementitious materials, but with a focus on discrete fractures and their effect on system evolution (Perko et al., 2015). This benchmark study highlighted that differences in simulation results of codes could be due to different solution approaches (finite volume versus finite element), different geochemical solvers, and/or different numerical

schemes (sequential iterative versus noniterative) (Figure 19.4b). A third benchmark relevant to nuclear waste repositories considers the evolution of porosity, permeability, as well as tortuosity and compares RTMs on six hypothetical scenarios (including clogging) with increasing geochemical or physical complexity (Xie et al., 2015). The permeability–porosity relationship was described using the Carman–Kozeny formulation and tortuosity-porosity (and pore diffusion-porosity) relationship was described using Archie's law. Although the codes showed good agreement in general, this benchmark problem highlights the differences in simulation results when pore clogging or phase disappearance is considered.

19.4.6.4 Diffusion-Dominated Systems

Diffusion-dominated systems are important aspects of contaminant storage in waste containment barriers and contaminant retardation in groundwater (Shackelford and Daniel, 1991; Steefel et al., 2014a). The SSBench presented two benchmarks that deal with the problem of multicomponent diffusion.

The first benchmark focuses on multicomponent diffusion and electrochemical migration based on the Nernst–Planck equation rather than Fick's Law (Figure 19.5a) (Rasouli et al., 2015). This study showed that comparing multicomponent diffusion and species-dependent transverse dispersion formulations across

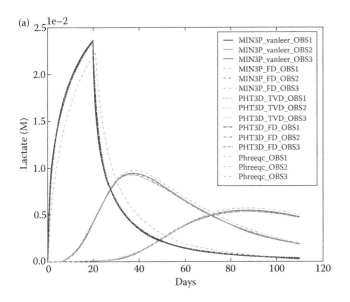
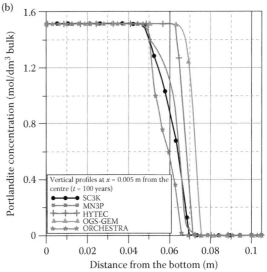

FIGURE 19.4
Differences in numerical schemes cause minor discrepancies in simulation results for (a) tracer concentrations at observation (OBS) points 1, 2, and 3 considering dispersion using 1 m longitudinal dispersivity. TVD stands for total variation diminishing scheme and FD for finite difference method. (With kind permission from Springer Science + Business Media: *Computers & Geoscience*, A reactive transport benchmark on modeling biogenic uraninite re-oxidation by Fe(III)-(hydr)oxides, 19(3), 2015, 569–583, Şengör, S. S. et al., Figure 5.) and (b) portlandite concentrations at 100 years especially near the concrete crack interface. (With kind permission from Springer Science + Business Media: *Computers & Geoscience*, Decalcification of cracked cement structures, 19(3), 2015, 673–693, doi:10.1007/s10596-014-9467-2, Perko, J. et al., Figure 6.)

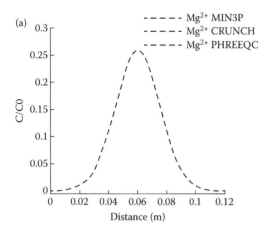

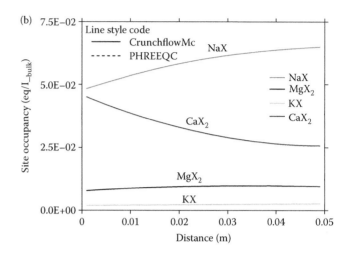

FIGURE 19.5
Simulating multispecies diffusion using Nernst–Planck equation. (a) Results showing 1D simulation of transverse multicomponent diffusion for the case of transport of mixed electrolytes in pure water. (With kind permission from Springer Science + Business Media: *Computers & Geoscience*, Benchmarks for multicomponent diffusion and electrochemical migration, 19(3), 2015, 523 -533, Rasouli, P. et al., Figure 6.) and (b) implementation of an EDL and a Stern layer. Simulated results from both codes show the preferred uptake of Ca^{2+} onto montmorillonite surfaces in exchange for Na^+, Mg^{2+}, and K^+. (With kind permission from Springer Science + Business Media: *Computers & Geoscience*, Benchmark reactive transport simulations of a column experiment in compacted bentonite with multispecies diffusion and explicit treatment of electrostatic effects, 19(3), 2015, 535–550, Alt-Epping, P. et al., Figure 11.)

codes provides an excellent test for verifying conceptual model treatments. A second benchmark also uses the Nernst–Planck equation to simulate multispecies diffusion through compacted bentonite and further considers an explicit treatment of the electrical double layer including anion exclusion effects (Figure 19.5b) (Alt-Epping et al., 2015). This study suggests that the codes show slight discrepancies in simulation results under transient conditions, but display a good agreement for results of the simulated species concentrations under steady-state conditions.

A brief summary of these can be found in Steefel et al. (2015a). A summary of the capabilities of all the participating codes used in SSBench has been provided in Steefel et al. (2015b).

19.5 Summary and Outlook

Applications of reactive transport models are growing in part due to significant advances in computational methods and measurement techniques. There is a growing need to benchmark reactive transport codes with the aim of verifying conceptual models, validating numerical schemes and coupling mechanisms, and evaluating the robustness and performance of RTMs. This chapter summarizes these benchmarking efforts initiated by the geosciences community as a group or as individual studies within the past 5 years.

In this context, the SSBench is an ongoing international benchmarking initiative that has produced several benchmarks on environmentally relevant problems such as acid rock drainage, chromium bioreduction, uranium fate and transport, geological and engineered waste repositories, and diffusion-dominated systems. Some of these benchmarks were designed to compare codes on specific capabilities like kinetic isotope fractionation (Wanner et al., 2015), microbially mediated reactions with evolving biomass (Molins et al., 2015), reaction-induced porosity and permeability changes (Xie et al., 2015), multirate and multisite models (Greskowiak et al., 2015; Yabusaki et al., 2015), and others were designed to test conceptual models like multicomponent diffusion (Rasouli et al., 2015) and transport in electrical double layers (Alt-Epping et al., 2015). This benchmarking initiative highlighted the robustness of the RTMs when considering long-term performance and general geochemical system evolution and related small discrepancies in results mainly to differences in numerical simulation approaches. The major discrepancies were related to differences in conceptual modeling framework. Efforts on individual benchmarking studies also showed significant advances in developing benchmarks as standardized references for other codes, code comparison activities, and model reliability and performance issues.

Although the geosciences community has made significant advances in these benchmarking activities, certain challenges related to benchmarking process coupling and model structures remain. Little progress has been made in the use of benchmarking activities within the context of designing and evaluating new experiments. Opportunities abound to advance the benchmarking activities to thermodynamic databases, isotope fractionation studies, as well as reactive transport modeling at both pore and watershed scales.

Acknowledgments

This material is based on work supported as part of the Genomes to Watershed Scientific Focus Area at Lawrence Berkeley National Laboratory, under Award Number DE-AC02-05CH11231 and Interoperable Design of Extreme-Scale Application Software (IDEAS), under Award Number DE-AC02-05CH11231 funded by the U.S. Department of Energy, Office of Science, Office of Biological and Environmental Research.

References

Alt-Epping, P., C. Tournassat, P. Rasouli, C. I. Steefel, K. U. Mayer, A. Jenni, U. Mäder, S. S. Sengor, and R. Fernández. 2015. Benchmark reactive transport simulations of a column experiment in compacted bentonite with multispecies diffusion and explicit treatment of electrostatic effects. *Computers & Geosciences,* 19(3): 535–550.

Arora, B., K. U. Mayer, C. I. Steefel, N. F. Spycher, S. S. Şengör, D. Jacques, and P. Alt-Epping. 2014. Reactive transport benchmarks on heavy metal cycling. *Goldschmidt Abstracts,* 47(4): 69.

Arora, B., B. P. Mohanty, and J. T. McGuire. 2011. Inverse estimation of parameters for multidomain flow models in soil columns with different macropore densities. *Water Resources Research,* 47(4): 1–17, doi:10.1029/2010WR009451.

Arora, B., B. P. Mohanty, and J. T. McGuire. 2015b. An integrated Markov chain Monte Carlo algorithm for upscaling hydrological and geochemical parameters from column to field scale. *Science of the Total Environment,* 512–513, 428–443, doi:10.1016/j.scitotenv.2015.01.048.

Arora, B., S. S. Şengör, N. F. Spycher, and C. I. Steefel. 2015a. A reactive transport benchmark on heavy metal cycling in lake sediments. *Computers & Geosciences,* 19(3): 613–633, doi: 10.1007/s10596-014-9445-8.

Barenblatt, G., I. Zheltov, and I. Kochina. 1960. Basic concepts in the theory of seepage of homogeneous liquids in fissured rocks [strata]. *Journal of Applied Mathematics and Mechanics,* 24(5): 1286–1303, doi:10.1016/0021-8928(60)90,107-6.

Bear, J. 1975. Dynamics of fluids in porous media. *Soil Science* 120(2): 162–163, doi:10.1097/00,010,694-197,508,000-00,022.

Beisman, J. J., R. M. Maxwell, A. K. Navarre-Sitchler, C. I. Steefel, and S. Molins. 2015. ParCrunchFlow: An efficient, parallel reactive transport simulation tool for physically and chemically heterogeneous saturated subsurface environments. *Computers & Geosciences,* 19(2): 403–422. doi:10.1007/s10596-015-9475-x.

Berkowitz, B. 2002. Characterizing flow and transport in fractured geological media: A review. *Advances in Water Resources,* 25(8–12): 861–884, doi:10.1016/S0309-1708(02)00,042-8.

Berner, R. A. 1980. *Early Diagenesis: A Theoretical Approach.* Princeton University Press, Princeton, NJ.

Bildstein, O., F. Claret, J. Perko, D. Jacques, and C. I. Steefel. 2014. Benchmark for reactive transport codes with application to concrete alteration. *Goldschmidt Abstracts,* 199, Goldschmidt Conference, Sacramento, CA.

Brooks, R. and A. Corey. 1964. Hydraulic properties of porous media. *Hydrology Paper,* Colorado State University, Fort Collins. 3, 37.

Carrayrou, J., J. Hoffmann, P. Knabner, S. Kräutle, C. Dieuleveult, J. Erhel, J. Lee, V. Lagneau, K. U. Mayer, and K. T. B. MacQuarrie. 2010. Comparison of numerical methods for simulating strongly nonlinear and heterogeneous reactive transport problems—The MoMaS benchmark case. *Computers & Geosciences,* 14(3), 483–502, doi:10.1007/s10596-010-9178-2.

Carrayrou, J., M. Kern, and P. Knabner. 2009. Reactive transport benchmark of MoMaS. *Computers & Geosciences,* 14(3), 385–392, doi: 10.1007/s10596-009-9157-7.

Caruso, B. S. 2004. Modeling metals transport and sediment/water interactions in a mining impacted mountain stream. *Journal of the American Water Resources Association,* 40(6), 1603–1615, doi:10.1111/j.1752-1688.2004.tb01609.x.

Dagan, G. 1990. Transport in heterogeneous porous formations: Spatial moments, ergodicity, and effective dispersion. *Water Resources Research* 26(6), 1281–1290, doi:10.1029/WR026i006p01281.

Davis, J. A. and D. B. Kent. 1990. Surface complexation modeling in aqueous geochemistry. *Reviews in Mineralogy,* 23: 176–260.

De Dieuleveult, C. and J. Erhel. 2009. A global approach to reactive transport: Application to the MoMas benchmark. *Computers & Geosciences,* 14(3): 451–464, doi:10.1007/s10596-009-9163-9.

De Windt, L., A. Burnol, P. Montarnal, and J. van der Lee. 2003. Intercomparison of reactive transport models applied to UO₂ oxidative dissolution and uranium migration. *Journal of Contaminant Hydrology,* 61(1-4): 303–312, doi:10.1016/S0169-7722(02)00,127-4.

Dwivedi, D., C. I. Steefel, B. Arora, and G. Bisht. 2015a. *Impact of Hyporheic Zone on Biogeochemical Cycling of Carbon,* presented at 2015 Goldschmidt, Prague, Czech Republic, August 16–21.

Dwivedi, D., W. J. Riley, and J. Y. Tang. 2015b. Carbon saturation affects soil C dynamics? *Proceedings, TOUGH Symposium 2015,* Lawrence Berkeley National Laboratory, Berkeley, California, September 28–30, 2015, 476–482.

Dzombak, D. A. 1990. *Surface Complexation Modeling: Hydrous Ferric Oxide.* John Wiley & Sons, New York.

Faybishenko, B., P. A. Witherspoon, and S. M. Benson (Eds.) 2000. *Dynamics of Fluids in Fractured Rock, Geophysical Monograph Series.* American Geophysical Union, Washington, DC.

Gelhar, L. W., C. Welty, and K. R. Rehfeldt. 1992. A critical review of data on field-scale dispersion in aquifers. *Water Resources Research*, 28: 1955–1974, doi:10.1029/92WR00607.

Goldberg, S., L. J. Criscenti, D. R. Turner, J. A. Davis, and K. J. Cantrell. 2007. Adsorption–desorption processes in subsurface reactive transport modeling. *Vadose Zone Journal* 6(3): 407, doi:10.2136/vzj2006.0085.

Greskowiak, J., J. Gwo, D. Jacques, J. Yin, and K. U. Mayer. 2015. A benchmark for multi-rate surface complexation and 1D dual-domain multi-component reactive transport of U(VI). *Computers & Geosciences*, 19(3): 585–597.

Gwo, J. P. and G. T. Yeh. 2004. High-performance simulation of surface-subsurface coupled flow and reactive transport at watershed scale, in *Proceedings of the International Conference on Computational Methods*, pp. 1–4, Singapore.

Hammond, G. E., P. C. Lichtner, and R. T. Mills. 2014. Evaluating the performance of parallel subsurface simulators: An illustrative example with PFLOTRAN. *Water Resources Research*, 50(1): 208–228, doi:10.1002/2012WR013483.

He, W., C. Beyer, J. H. Fleckenstein, E. Jang, O. Kolditz, D. Naumov, and T. Kalbacher. 2015. A parallelization scheme to simulate reactive transport in the subsurface environment with OGS# IPhreeqc. *Geoscientific Model Development*, 8: 2369–2402.

Jardine, P. M. 2008. Chapter 1 influence of coupled processes on contaminant fate and transport in subsurface environments. *Advances in Agronomy*, 99: 1–99.

Kempka, T., H. Class, U.-J. Görke, B. Norden, O. Kolditz, M. Kühn, L. Walter, W. Wang, and B. Zehner. 2013. A dynamic flow simulation code intercomparison based on the revised static model of the Ketzin pilot site. *Energy Procedia* 40: 418–427.

Keyes, D. E. et al. 2013. Multiphysics simulations: Challenges and opportunities. *International Journal of High Performance Computing Applications*, 27(1): 4–83, doi:10.1177/1094342012468181.

Kolditz, O. et al. 2012. A systematic benchmarking approach for geologic CO$_2$ injection and storage. *Environmental Earth Sciences*, 67(2): 613–632, doi:10.1007/s12665-012-1656-5.

Kolditz, O., H. Shao, W. Wang, and S. Bauer (Eds.) 2015. *Thermo-Hydro-Mechanical-Chemical Processes in Fractured Porous Media: Modelling and Benchmarking*. Terrestrial Environmental Sciences, Springer International Publishing, Cham.

Kühn, M., U.-J. Görke, J. T. Birkholzer, and O. Kolditz. 2012. The CLEAN project in the context of CO$_2$ storage and enhanced gas recovery. *Environmental Earth Sciences*, 67(2): 307–310.

Lagneau, V. and J. van der Lee. 2009. HYTEC results of the MoMas reactive transport benchmark. *Computers & Geosciences*, 14(3), 435–449, doi:10.1007/s10596-009-9159-5.

Li, L., C. I. Steefel, M. B. Kowalsky, A. Englert, and S. S. Hubbard. 2010. Effects of physical and geochemical heterogeneities on mineral transformation and biomass accumulation during biostimulation experiments at Rifle, Colorado. *Journal of Contaminant Hydrology*, 112(1-4): 45–63, doi:10.1016/j.jconhyd.2009.10.006.

Lichtner, P. C. 1985. Continuum model for simultaneous chemical reactions and mass transport in hydrothermal systems, *Geochimica et Cosmochimica Acta*, 49(3): 779–800.

Lichtner, P. C. 1992. Time-space continuum description of fluid/rock interaction in permeable media. *Water Resources Research*, 28(12): 3135–3155.

Lichtner, P. C. and Q. Kang. 2007. Upscaling pore-scale reactive transport equations using a multiscale continuum formulation. *Water Resources Research*, 43(12): 1–19, doi:10.1029/2006WR005664.

Luckner, L., M. T. Van Genuchten, and D. R. Nielsen. 1989. A consistent set of parametric models for the two-phase flow of immiscible fluids in the subsurface. *Water Resources Research*, 25(10): 2187–2193, doi:10.1029/WR025i010p02187.

Marty, N. C. M. et al. 2015. Benchmarks for multicomponent reactive transport across a cement/clay interface. *Computers & Geoscience*, 19(3): 635–653.

Mayer, K. U. and K. T. B. MacQuarrie. 2009. Solution of the MoMaS reactive transport benchmark with MIN3P—Model formulation and simulation results. *Computers & Geoscience*, 14(3): 405–419, doi:10.1007/s10596-009-9158-6.

Mayer, K. U., D. W. Blowes, and E. O. Frind. 2001. Reactive transport modeling of an *in situ* reactive barrier for the treatment of hexavalent chromium and trichloroethylene in groundwater. *Water Resources Research*, 37(12): 3091–3103, doi:10.1029/2001WR000234.

Mayer, K. U., P. Alt-Epping, D. Jacques, B. Arora, and C. I. Steefel. 2015. Benchmark problems for reactive transport modeling of the generation and attenuation of acid rock drainage. *Computers & Geoscience*, 19(3): 599–611.

Molins, S. and K. U. Mayer. 2007. Coupling between geochemical reactions and multicomponent gas and solute transport in unsaturated media: A reactive transport modeling study. *Water Resources Research*, 43(5): n/a–n/a, doi:10.1029/2006WR005206.

Molins, S., D. Trebotich, C. I. Steefel, and C. Shen. 2012. An investigation of the effect of pore scale flow on average geochemical reaction rates using direct numerical simulation. *Water Resources Research*, 48: 1–11, doi:10.1029/2011WR011404.

Molins, S., J. Greskowiak, C. Wanner, and K. U. Mayer. 2015. A benchmark for microbially mediated chromium reduction under denitrifying conditions in a biostimulation column experiment. *Computers & Geoscience*, 19(3): 479–496.

Molins, S., D. Trebotich, L. Yang, J. B. Ajo-Franklin, T. J. Ligocki, C. Shen, and C. I. Steefel. 2014. Pore-scale controls on calcite dissolution rates from flow-through laboratory and numerical experiments. *Environmental Science & Technology*, 48(13): 7453–60, doi:10.1021/es5013438.

Mualem, Y. 1976. A new model for predicting the hydraulic conductivity of unsaturated porous media. *Water Resources Research*, 12(3): 513–522, doi:10.1029/WR012i003p00513.

Mukhopadhyay, S., C. Doughty, D. Bacon, J., Li, L. Wei, H. Yamamoto, S. Gasda, S. A. Hosseini, J.-P. Nicot, and J. T. Birkholzer. 2014. The Sim-SEQ Project: Comparison of selected flow models for the S-3 site. *Transport in Porous Media*, 108(1): 207–231, doi:10.1007/s11242-014-0361-0.

Nardi, A., A. Idiart, P. Trinchero, L. M. de Vries, and J. Molinero. 2014. Interface COMSOL-PHREEQC (iCP), an efficient numerical framework for the solution of coupled multiphysics and geochemistry. *Computers & Geoscience*, 69: 10–21.

Neuman, S. P. 1973. Saturated–unsaturated seepage. *Journal of the Hydraulics Division*, American Society of Civil Engineers, New York, 99: HY12.

Nowack, B. et al. 2006. Verification and intercomparison of reactive transport codes to describe root-uptake. *Plant Soil*, 285(1-2): 305–321, doi:10.1007/s11104-006-9017-3.

Panday, S., P. S. Huyakorn, R. Therrien, and R. L. Nichols. 1993. Improved three-dimensional finite-element techniques for field simulation of variably saturated flow and transport. *Journal of Contaminant Hydrology*, 12(1-2): 3–33, doi:10.1016/0169-7722(93)90,013-I.

Perko, J., K. U. Mayer, G. Kosakowski, L. De Windt, J. Govaerts, D. Jacques, D. Su, and J. C. L. Meeussen. 2015. Decalcification of cracked cement structures. *Computers & Geoscience*, 19(3): 673–693, doi:10.1007/s10596-014-9467-2.

Pruess, K. 2013. A practical method for modeling fluid and heat flow in fractured porous media. *Society of Petroleum Engineering Journals*, 25(01): 14–26, doi:10.2118/10,509-PA.

Pruess, K., J. García, T. Kovscek, C. Oldenburg, J. Rutqvist, C. Steefel, and T. Xu. 2004. Code intercomparison builds confidence in numerical simulation models for geologic disposal of CO_2. *Energy* 29(9-10): 1431–1444, doi:10.1016/j.energy.2004.03.077.

Rasouli, P., C. I. Steefel, K. U. Mayer, and M. Rolle. 2015. Benchmarks for multicomponent diffusion and electrochemical migration. *Computers & Geoscience*, 19(3): 523–533.

Richards, L. A. 1931. Capillary conduction of liquids through porous mediums. *Journal of Applied Physics*, 1: 318–333, doi:10.1063/1.1,745,010.

Riley, W. J., F. M. Maggi, M. Kleber, M. S. Torn, J. Y. Tang, D. Dwivedi, and N. Guerry. 2014. Long residence times of rapidly decomposable soil organic matter: Application of a multi-phase, multi-component, and vertically resolved model (BAMS1) to soil carbon dynamics. *GeoscientiRic Model Development*, 7: 1335. doi: 10.5194/gmd-7-1335-2014.

Savage, D. et al. 2011. A comparative study of the modelling of cement hydration and cement–rock laboratory experiments. *Applied Geochemistry*, 26(7): 1138–1152.

Scheibe, T. D., R. Mahadevan, Y. Fang, S. Garg, P. E. Long, and D. R. Lovley. 2009. Coupling a genome-scale metabolic model with a reactive transport model to describe *in situ* uranium bioremediation. *Microbial Biotechnology*, 2(2): 274–86, doi:10.1111/j.1751-7915.2009.00,087.x.

Scheibe, T. D., E. M. Murphy, X. Chen, A. K. Rice, K. C. Carroll, B. J. Palmer, A. M. Tartakovsky, I. Battiato, and B. D. Wood. 2014. An analysis platform for multiscale hydrogeologic modeling with emphasis on hybrid multiscale methods. *Groundwater*, 1–19, doi:10.1111/gwat.12,179.

Scheibe, T., M. White, and S. White. 2013. *Outcomes of the 2013 GTO Workshop on Geothermal Code Comparison*. Pacific Northwest National Laboratory, Battelle.

Şengör, S. S., K. U. Mayer, J. Greskowiak, C. Wanner, D. Su, and H. Prommer. 2015. A reactive transport benchmark on modeling biogenic uraninite re-oxidation by Fe(III)-(hydr)oxides. *Computers & Geoscience*, 19(3): 569–583.

Shackelford, C. D. and D. E. Daniel. 1991. Diffusion in saturated soil. I: Background. *Journal of Geotechnical Engineering*, 117(3): 467–484.

Simunek, J., J. Simunek, N. J. Jarvis, N. J. Jarvis, M. T. van Genuchten, M. T. van Genuchten, A. Gardenas, and A. Gardenas. 2003. Review and comparison of models for describing non-equilibrium and preferential flow and transport in the vadose zone. *Journal of Hydrology*, 272: 14–35, doi:10.1016/S0022-1694(02)00,252-4.

Snow, D. T. 1965. A parallel plate model of fractured permeable media. University of California, Berkeley.

Soler, J. M., J. Landa, V. Havlova, Y. Tachi, T. Ebina, P. Sardini, M. Siitari-Kauppi, J. Eikenberg, and A. J. Martin. 2015. Comparative modeling of an *in situ* diffusion experiment in granite at the Grimsel Test Site. *Journal of Contaminant Hydrology*, 179: 89–101.

Somerton, W. H., A. H. El-Shaarani, and S. M. Mobarak. 2013. High temperature behavior of rocks associated with geothermal type reservoirs, in *SPE California Regional Meeting*, Society of Petroleum Engineers.

Srivastava, P., K. W. Migliaccio, and J. Simunek. 2007. Landscape models for simulating water quality at point, field, and watershed scales, *Transactions of ASABE* 50(5): 1683–1693.

Steefel, C. I. and A. C. Lasaga. 1994. A coupled model for transport of multiple chemical species and kinetic precipitation/dissolution reactions with application to reactive flow in single phase hydrothermal systems. *American Journal of Science*, 294, 529–592, doi:10.2475/ajs.294.5.529.

Steefel, C. I., P. Alt-Epping, P. Rasouli, K. U. Mayer, A. Jenni, M. Rolle, and C. Tournassat. 2014a. Reactive transport benchmarks for multicomponent diffusion and electrical double layer transport. *Goldschmidt Abstracts*, 2372.

Steefel, C. I. and K. T. B. MacQuarrie. 1996. Approaches to modeling of reactive transport in porous media. *Reviews in Mineralogy and Geochemistry*, 34(1): 85–129.

Steefel, C. I. and K. Maher. 2009. Fluid-rock interaction: A reactive transport approach. *Reviews in Mineralogy and Geochemistry* 70(1): 485–532, doi:10.2138/rmg.2009.70.11.

Steefel, C. I., S. Molins, and D. Trebotich. 2013. Pore scale processes associated with subsurface CO_2 injection and sequestration. *Reviews in Mineralogy & Geochemistry* 77(1): 259–303.

Steefel, C. I. et al. 2014b. The GEWaSC framework: Multiscale modeling of coupled biogeochemical, microbiological, and, *Goldschmidt Abstracts*, p. 2373, Sacramento, CA.

Steefel, C. I., S. B. Yabusaki, and K. U. Mayer. 2015a. Reactive transport benchmarks for subsurface environmental simulation, *Computers & Geoscience*, 19(3): 439–443.

Steefel, C. I. et al. 2015b. Reactive transport codes for subsurface environmental simulation. *Computers & Geoscience*, 19: 445–478, doi:10.1007/s10596-014-9443-x.

Tambach, T. J., C. H. Pentland, G. Zhang, H. Huang, and J. R. Snippe. 2013. A comparative study of reactive transport modelling using toughReact and MoReS for modelling CO_2 sequestration, *Second EAGE Sustainable Earth Sciences (SES) Conference and Exhibition*.

Trebotich, D., M. F. Adams, S. Molins, C. I. Steefel, and C. Shen. 2014. High-resolution simulation of pore-scale reactive transport processes associated with carbon sequestration. *Computer Science Engineering*, 16(6): 22–31, doi:10.1109/MCSE.2014.77.

Tsang, C.-F., O. Stephansson, L. Jing, and F. Kautsky. 2008. DECOVALEX Project: From 1992 to 2007. *Environmental Geology*, 57(6): 1221–1237, doi:10.1007/s00254-008-1625-1.

Van Genuchten, M. T. 1980. A closed-form equation for predicting the hydraulic conductivity of unsaturated soils. *Soil Science Society of America Journal*, 44: 892, doi:10.2136/sssaj1980.03615995004400050002x.

Vaughan, P. J. 1987. Analysis of permeability reduction during flow of heated, aqueous fluid through westerly granite. in *Coupled Processes Associated with Nuclear Waste Repositories*, C.-F. Tsang, Ed., Academic Press, New York.

Wanner, C., J. L. Druhan, R. T. Amos, P. Alt-Epping, and C. I. Steefel. 2015. Benchmarking the simulation of Cr isotope fractionation. *Computers & Geoscience*, 19(3): 497–521.

Wanner, C., S. Molins, J. L. Druhan, J. Greskowiak, R. T. Amos, K. U. Mayer, P. Alt-Epping, and C. I. Steefel. 2014a. Benchmarking the simulation of microbial Cr(VI) reduction and Cr isotope fractionation, *Goldschmidt Abstracts*, p. 2655, Sacramento, CA.

Warren, J. E. and P. J. Root. 2013. The behavior of naturally fractured reservoirs. *Society of Petroleum Engineering Journals*, 3(03): 245–255, doi:10.2118/426-PA.

White, S. K., S. Purohit, and L. Boyd. 2015. Using GTO-Velo to facilitate communication and sharing of simulation results in support of the geothermal technologies office code comparison study. *Proceedings, Fourtieth Workshop on Geothermal Reservoir Engineering*, pp. 1–10, Stanford University, Stanford, CA.

Xie, M., K. U. Mayer, F. Claret, P. Alt-Epping, D. Jacques, C. Steefel, C. Chiaberge, and J. Simunek. 2015. Implementation and evaluation of permeability–porosity and tortuosity-porosity relationships linked to mineral dissolution–precipitation, *Computers & Geoscience*, 19(3): 655–671.

Yabusaki, S. B., S. S. Şengör, and Y. Fang. 2015. A uranium bioremediation reactive transport benchmark. *Computers & Geoscience*, 19(3): 551–567.

Yeh, G. T. and V. S. Tripathi. 1989. A critical evaluation of recent developments in hydrogeochemical transport models of reactive multichemical components. *Water Resources Research*, 25(1): 93–108.

Zhang, F. et al. 2012. *Groundwater Reactive Transport Models*, Bentham Science Publishers Ltd., Oak Park.

Section V

Pollution and Remediation

20

Toward Quantification of Ion Exchange in a Sandstone Aquifer

John Tellam, Harriet Carlyle, and Hamdi El-Ghonemy

CONTENTS

20.1 Introduction

Ion exchange is a very important process in determining the migration of a number of solutes, including pollutants, through aquifer systems. It affects the relative proportions of ions, and by doing so can induce other reactions to occur resulting in the loss or gain of solute to the groundwater and, often, a change in pH: for example, displacement of fresh, calcite-saturated groundwater by an Na-rich solution (e.g., landfill leachate or sea water) will usually result in the increase of Ca in solution and hence, possibly, the precipitation of calcite and a fall in pH. Such changes in chemistry will often have implications for migration of pollutants, for example, transition metals. Ion exchange is also directly involved with attenuating a number of pollutant species (e.g., transition metals and NH_4^+).

Despite the importance of the process, few attempts have been made at quantification of ion exchange in the hydrogeological context: the studies associated with the Borden site in Canada (e.g., Dance and Reardon (1983) and Reardon et al. (1983)) and those undertaken

by Appelo and co-workers (e.g., Appelo, 1994; van Breukelen et al., 1998) are exceptions, and a great deal more work has been undertaken on soils (e.g., see Sposito, 1981; McBride, 1994).

This chapter outlines three studies aimed at quantifying the ion exchange behavior of the English Triassic sandstones, a fluviatile/Aeolian red-bed sequence whose main exchange phases are clay minerals, though Mn and Fe oxyhydroxides are also present (cf. Dzombak and Morel, 1990).

The first study described undertook laboratory determination of ion exchange parameters for sandstone samples from northwest England (Carlyle, 1991). The data were interpreted using a standard soil science approach, and then used in reactive transport computer modeling studies of the intrusion of estuary water into the aquifer over a 40-year period (Carlyle et al., 2004). Although surprisingly successful in reproducing the breakthroughs of the various cations at the pumping wells, the study showed that although the standard ion exchange model proved satisfactory, it provided a far from perfect description of the exchange processes as observed in the laboratory. As a result, a second study examined the exchange behavior of sandstone samples in detail (El-Ghonemy, 1997), and this resulted in a new empirical description of exchange for the sandstones. A third study attempted to determine how the various exchange phases contribute to the overall exchange behavior of the rock, and hence whether knowledge of the mineralogy of the sandstones can enable exchange properties to be predicted (Parker et al., 2000). In all cases, only major cation (Ca, Mg, Na, K) ion exchange has been considered.

20.2 The English Triassic Sandstone

In the areas studied, the Triassic sandstone typically contains (e.g., Gillespie, 1987): quartz clasts—50%–70% of whole rock volume; feldspar clasts (mainly K-feldspar)—3%–6%; calcite and dolomite clasts and cement—1%–3%; and clay minerals (illite, chlorite, kaolite, smectite, mixed layers)—<5%. Iron oxyhydroxides coat almost all the grains, with Fe comprising 1%–2% of the rock, and Mn oxyhydroxides are also common (around 100–400 ppm of the rock). Pyrite and various accessory minerals are also occasionally present (Travis and Greenwood, 1911), though pyrite was not observed in any of the samples used in the studies.

The hydraulic conductivity as measured in the laboratory typically averages about 1 m/d, but ranges from <10^{-3} to >10 m/d. Porosity averages around 25%, and specific yield around 10%–15%, though good data are few. The sandstone is fractured, but often the regional hydraulic conductivity is not much greater than the intergranular conductivity.

20.3 The Application of the Gaines–Thomas Exchange Model in Predicting Groundwater Chemistry in a Zone of Estuary Intrusion

This study (Carlyle, 1991; Carlyle et al., 2004) attempted to apply an ion exchange model widely used in soil science in a region of northwest England where estuary water has been intruding the sandstone aquifer since around 1930. The aquifer underlies the town of Widnes. Abstraction from industrial boreholes in the town and industrial and public supply boreholes to the north have resulted in a northward-directed head gradient and the intrusion of brackish water (Cl < 10,000 mg/L) from the Mersey Estuary (Tellam and Lloyd, 1986; Howard, 1988).

20.3.1 Determination of Exchange Parameters

Sixteen cylindrical plugs, representing the main lithologies present, were obtained from a 239-m borehole core in Widnes. The plugs were analyzed intact using the method of Reardon et al. (1983), vacuum saturation with deionized water, removal of porewater using centrifugation, analysis to determine porewater concentrations in equilibrium with the exchangers, saturation with 1 M Li–Cl, centrifugation, and analysis to determine cation exchange capacity (CEC) and the composition of the exchange sites. From the porewater concentrations and the compositions of the exchange sites, selectivity coefficients were obtained assuming the Gaines–Thomas (GT) convention (Gaines and Thomas, 1953):

$$K_{A2+/B+} = \frac{(B^+)^2 A^{2+}X}{(A^{2+})B^+X^2}$$

where ( ) represents activities calculated using the concentrations and the extended Debye-Hückel equation and $A^{2+}X$ is the equivalent fraction of ion A^{2+} on the exchanger (= sorbed A^{2+}/CEC, both in equivalents). Two "runs" of experiments were completed (Table 20.1) and between the runs the cores were saturated with a saline solution, and hence the exchange conditions are different in Run 1 and Run 2 as indicated in Table 20.2. Hence, the experiments cover a very wide range of conditions. No pH control was attempted in the experiments, but was always in the range 6–6.5. Field sample pHs for 1979–1980 were in the range 6.0–8.0, with most values lying between 6.7 and 7.4.

A flushing method was also used to determine ion exchange parameters (cf. Kool et al., 1989). This involved passing a set of solutions through the core plugs and recording the breakthrough curves. From these data the mass transfers between solution and exchangers was calculated using a direct search method as the resulting equation is a fifth-order polynomial. CEC estimates obtained using the 1 M Li–Cl method described above were then used with the mass transfer data to calculate selectivity coefficients assuming a particular ion exchange model (in this case GT). This approach has the advantages that less drying and porewater extraction is needed, and the validity of the chosen ion exchange model is directly determined.

20.3.2 Results

The results from the standard method are shown in Table 20.1. They have similar ranges to those given by Appelo and Postma (1993). The flushing method produced similar parameters on average. Close examination of the data sets indicated that the ion exchange parameters vary with exchange site composition. In the

TABLE 20.1

CEC (meq/100 g) and K Values for Standard Method Experiments (Sample Numbering Scheme: First 3 Alphanumeric Characters Indicate Depth; A, B Signify Different Samples from the Same Depth)

Sample	CEC (meq 1⁻¹)	Run	$K_{Ca/Mg}$	$K_{Ca/Na}$	$K_{Ca/K}$	$K_{K/Na}$	CaX_2	NaX
10a	2.30	1	(5.63)	0.22	0.10	1.51	0.73	0.02
10b	1.44	1	0.87	0.08	0.22	0.60	0.49	0.01
10cA	1.83	1	0.95	0.43	(23.68)	0.13	0.58	0.03
10cB	1.79	1	0.77	0.74	1.17	0.79	0.59	0.02
20a	1.51	1	1.63	0.03	0.02	1.25	0.50	0.03
20bA	1.15	1	0.87	0.16	0.04	2.05	0.57	0.03
20bB	1.71	1	0.43	0.94	0.03	5.79	0.59	0.02
35b	1.21	1	1.56	(7.34)	0.16	6.82	0.46	0.01
40aA	0.85	1	1.03	0.53	0.06	3.10	0.59	0.02
40a13	0.97	1	0.96	0.28	0.11	2.18	0.60	0.01
40b	1.00	1	0.84	0.32	0.11	1.73	0.57	0.03
55aA	0.98	1	0.83	0.59	0.04	4.04	0.62	0.03
55aB	0.84	1	1.10	0.68	0.11	2.51	0.61	0.02
59a	0.95	1	1.01	0.60	0.11	2.34	0.59	0.02
78a	0.92	1	(2.84)	0.03	0.04	0.90	0.53	0.04
78b	0.09	1	0.47	0.00	0.00	1.56	0.29	0.23
10cA	2.20	2	0.66	0.45	0.09	2.23	0.37	0.40
10cB	1.62	2	0.88	0.76	0.05	4.11	0.33	0.47
20	1.17	2	0.89	0.67	0.19	1.87	0.34	0.42
20bB	2.00	2	0.68	0.59	0.14	2.07	0.37	0.42
35b	0.69	2	0.77	0.31	1.00	1.79	0.27	0.45
40aA	0.71	2	0.39	1.17	0.05	4.70	0.23	0.42
40aB	0.83	2	0.38	0.64	0.14	2.11	0.24	0.44
40b	1.44	2	2.10	(30.41)	0.39	8.85	0.79	0.12
55aA	0.60	2	1.27	(75.7)	0.26	(17.24)	0.53	0.11
55aB	0.92	2	1.02	(12.76)	0.17	8.61	0.57	0.17
59aB	0.69	2	(3.67)	(476)	0.51	(30.6)	0.80	0.13
78	0.81		1.29	0.88	0.17	2.31	0.35	0.44
78b	0.81	2	1.37	1.24	0.23	2.33	0.33	0.45
Mean[a]	1.20		0.96	0.51	0.20	2.90	–	–
St. Dev.[a]	0.49		0.40	0.35	0.28	2.27	–	–
Number	19		26	26	28	27	–	–
Geom. mean[a]	1.13		0.89	0.39	0.12	2.15	–	–
Total Range[a]	0.6–2.3		0.38–2.1	0.0–1.2	0.0–1.2	0.13–8.9	–	–
Range[b]			0.4–4.0	2.8–11.1	0.06–0.7	4.0–6.7	–	–

[a] Ignoring outliers indicated by italics. Zeroes ignored for geometric mean calculations.

[b] Range indicated by Appelo and Postma (1993).

TABLE 20.2

Porewater Concentration Ranges in Experiment Runs 1 and 2

Run	Ca (mg/L)	Mg (mg/L)	Na (mg/L)	K (mg/L)
Run 1	6–24	3–11	5–65	4–12
Run 2	22–533	6–213	313–2058	13–70

case of the flushing method it was sometimes necessary to make changes (mean 13%, minimum 3%, and maximum 37%) to the cation exchange capacity from one flush to another in order to obtain an interpretation.

This problem suggests that the GT model, although often provides a satisfactory description, is not perfect, and this observation led to the second study described below (see Section 20.4).

20.3.3 Modeling of the Field System

Pumped water Cl and hardness data spanning a period of around 40 years were available for a number of boreholes in Widnes, together with full major ion analyses for 1980/1981. Using the exchange parameters obtained from the laboratory experiments, a one-dimensional reactive transport mixing cell model was used to predict breakthrough at the field scale. The breakthrough patterns thus obtained, assuming GT/constant CEC exchange with calcite equilibrium, significantly overestimated the concentrations of the divalent ions and the 1980 sulfate concentration, and underestimated the 1980 alkalinity concentration. Inclusion of a representation of sulfate reduction in the estuarine alluvium failed to reproduce the 1980 HCO_3 and pH values. However, [34]S and [18]O evidence from a previous study (Barker et al., 1998) indicates that partial CO_2 degassing occurs from the alluvium following sulfate reduction. By including these mechanisms, and using GT exchange with averaged laboratory parameters, good matches for SO_4, HCO_3, pH, and cation concentrations were obtained. It was concluded that the GT/constant CEC model with averaged laboratory parameter values can be used successfully when predicting estuary water intrusion in this aquifer at a regional scale and over long time scales, despite the numerous assumptions necessary. This result agrees with the conclusion of Appelo (1994) who also studied a large-scale flow system.

20.4 Investigation of the Exchange Properties of Sandstone Samples in the Laboratory

The results of the investigation described in the last section suggest that the GT model, although appearing to perform satisfactorily when applied in a regional investigation, clearly did not describe the exchange processes fully. Hence, a second study was undertaken to examine the exchange behavior of the sandstones in more detail. Four intact core plug samples were analyzed to determine how their exchange properties varied as a function of exchange site composition. A search was then made to obtain an empirical model that would provide a better description for the ion exchange properties of the samples.

20.4.1 Methods

The samples were from core from the same aquifer as the previous study. The basic approach followed that of Jensen and Babcock (1973). The plugs were flushed with a solution of known cation ratio until equilibrium was attained: solutions containing only two cations were used. The plug was then flushed to equilibrium with a concentrated solution in order to displace the previously sorbed cations. From the data collected, the selectivity coefficients were calculated for a wide range of exchange site compositions.

20.4.2 Results and Interpretation

Selectivity coefficients were found often to vary considerably as a function of exchange site composition, a result that helps explain the variations seen in Table 20.1. Figure 20.1 shows example results for GT selectivity coefficients.

One way to interpret the data is to assume that the variation in selectivity coefficients is due to variation in sorbed ion activity coefficients. If this is done, then the variation in sorbed ion activity coefficient is of the form shown in Figure 20.2. Similar activity coefficient variation was found for most of the data sets, and in principle this approach might be used with a constant selectivity coefficient. However, it was also found that a simple empirical power function fitted the data very well:

$$K_{A2+/B+} = \frac{(B^+)^2[A^{2+}X]^n}{(A^{2+})[B^+X^2]^n}$$

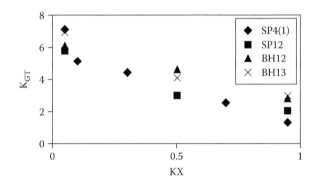

FIGURE 20.1
The variation of the GT selectivity coefficient (K_{GT}) as a function of the equivalent fraction of K on the exchange sites (KX) of four Triassic sandstone samples undergoing K/Na exchange.

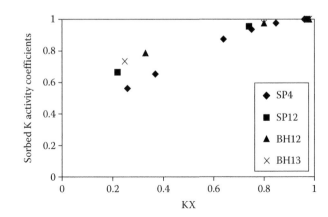

FIGURE 20.2
The variation of sorbed phase activity coefficients as a function of the equivalent fraction of K^+ on the exchange sites (KX) of four Triassic sandstone samples undergoing K/Na exchange. Values were calculated using the method of Argersinger et al. (1950).

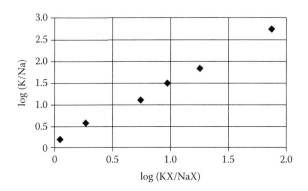

FIGURE 20.3

The power function relationship for K/Na exchange on Triassic sandstone sample SP4.

Figure 20.3 shows some example data: a power function relationship is clearly indicated. Power functions have been used before to describe ion exchange systems, notably by Rothmund and Kornfeld (1918, 1919), Walton (1949), Langmuir (1981, 1997), and Bond (1995), but have rarely been used in the hydrogeological context. The thermodynamic justification is limited (e.g., see Langmuir, 1997), but power functions can derived by assuming that the sorbed cations are effectively represented by an ideal solid solution (Garrels and Christ, 1965). They also appear to be appropriate in exchange systems comprising one type of exchanger, a topic investigated in the study described in the next section.

20.5 The Exchange Behavior of Individual Exchange Phases in the Sandstone

To investigate the reasons for the exchange behavior and to attempt to see if exchange properties could be predicted for the given knowledge of the mineralogy of the rock, a study was undertaken to examine the exchange properties of some of the main exchange phases in the sandstone. The study also undertook a preliminary modeling investigation of issues associated with the up-scaling of results from the laboratory scale to the field scale.

20.5.1 Laboratory Methods

The exchange phases in the sandstone are clays and possibly to some extent oxides/hydroxides. In this study (Parker, 2004), the clay phases only were considered. Because it is not feasible to extract enough pure clay from the sandstones, samples of kaolinite, chlorite, montmorillonite, and illite were purchased from the Clay Minerals Repository, Missouri. The exchange properties of the clays were determined again using the Jensen and Babcock (1973) method, but this time in batch reactors. Experiments involved examining single clays, and binary and ternary mixtures of the clays. XRD indicated that the clay samples contained no recognizable amounts of other minerals. Again, pH was not adjusted, but was measured at all stages.

20.5.2 Laboratory Investigation Results

In all cases the GT, Gapon, and Vanselow selectivity coefficients varied as a function of the exchange site composition (Figure 20.4). However, the exchange on all the pure and all the mixed clay samples was well described by a power function. Figure 20.5 shows example results. The power function parameters vary smoothly as a function of mixture proportions, as shown in Figure 20.6. It would therefore appear that there is a possibility that the exchange properties of rocks containing only clay exchange phases may be estimated using data on the proportions of the clay components.

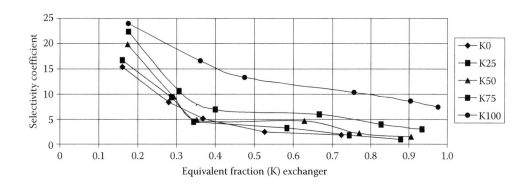

FIGURE 20.4

The variation of GT selectivity coefficients as a function of equivalent fraction of potassium for various mixtures of kaolinite and illite (K25 = 25% kaolinite).

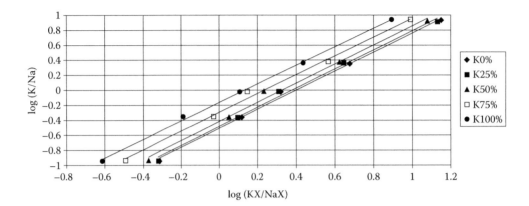

FIGURE 20.5
Power function relationships for K/Na exchange for various mixtures of kaolinite and illite (K25 = 25% kaolinite).

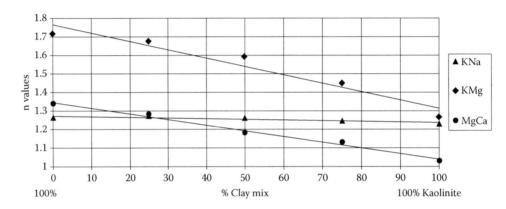

FIGURE 20.6
The variation of the power function exponent as a function of mix % for kaolinite/illite mixtures for three binary exchange systems—K/Na, K/Mg, and Mg/Ca.

20.6 Preliminary Modeling Investigations

Scoping modeling investigations have been undertaken in order to investigate the effect of multiple GT exchangers in contact with each other and also to investigate the effect of sampling groundwater—such as might occur when sampling well water—that have been in contact with rock of different exchange properties.

20.6.1 Equilibrium between Multiple Exchangers

A simple spreadsheet calculation can be used to investigate the case where all sites are in equilibrium with the same binary solution. The steps of the calculation are as follows:

1. Assign for each of N different types of site values for equilibrium constants and CEC.

2. Calculate the composition of the exchange sites for each exchanger when in equilibrium with a set of solutions of given compositions.

3. For each solution, sum the sorbed masses, weighted according to the CEC of each site type, to obtain sorbed equivalent fractions for the exchanger as a whole.

4. Determine the relationship between these sorbed equivalent fractions and the known solution compositions.

It was found that power function relationships provide a good approximation of the overall behavior of the assemblages. There is some variation in the parameters n and K with solution concentration, but this is small compared with the variation in Gaines–Thomas (and also Vanselow) selectivity coefficients of the various component phases, except when solutions with extreme concentration ratios are included.

20.6.2 Preliminary Up-Scaling Investigation

To investigate the issues associated with the up-scaling of laboratory results to the field scale, modeling work using PHREEQM (Appelo and Postma, 1993) was undertaken. The basic approach adopted was to model the breakthrough from two stream tubes with different ion exchange properties, to combine the effluent stream, and then to interpret the mixed effluent. The combined effluent water is taken to represent the water obtained from, for example, a sampling well. This study has shown that for a series of perfect G–T exchangers:

1. The chemical changes observed in the "sampling well" can be reproduced adequately using a GT description with averaged exchange capacities and selectivity coefficients but only if the exchange properties of each stream tube are very similar.

2. For stream tubes where the exchange properties differ significantly from each other, the GT description no longer is satisfactory, whatever the exchange properties assumed.

3. A power function description will describe the breakthrough curves very well for cases where the geochemical properties of the stream tubes are very different.

4. The power function description breaks down where the difference in the exchange properties of the stream tubes is extremely large (usually unrealistically large)—that is, the power function description appears to be only approximate for heterogeneous systems.

At the field scale, it proved possible to model the breakthrough of estuary water over a 40-year period using the laboratory data and a Gaines–Thomas convention description (Carlyle et al., 2004). However, it is suspected that the range of exchange site occupancy in the field system was limited and that the variation in selectivity coefficient value would therefore also have been limited. Modeling suggests that well water chemistry may be described using a power function approach if the water has arrived at the well through different pathways with different significantly Gaines–Thomas–Vanselow exchanger properties. It seems likely, therefore, that a power function/Rothmund–Kornfeld approach would have provided just as satisfactory a description of the field data as the Gaines–Thomas model did. We conclude that the Rothmund–Kornfeld approach is a very convenient means of quantifying major cation exchange in the sandstones studied and valid over a larger exchanger composition range than other, more usual approaches. The Rothmund–Kornfeld power relationship appears to arise when either the water is in equilibrium with an exchanger assemblage that is made up of a series of Gaines–Thomas or Vanselow exchangers of different properties, or when the water is a mixture of waters that have been in contact with exchangers of different properties. In general, it appears that where systems have only a limited range of exchange properties or exchange compositions, the GT approach works well, but for more heterogeneous systems Rothmund–Kornfeld power relationship works better. We suggest that these findings will be valid for any system where clay minerals are the dominant exchange phase.

20.7 Conclusions

At the laboratory sample scale, none of the usual models of exchange behavior provide a really good description of ion exchange in the Triassic sandstones of England. However, a power (Rothmund–Kornfeld) function description is significantly better than the other usual relationships. The main exchange phases are almost certainly clay minerals and it has been shown that ion exchange on "pure" clay minerals can also be well described by a power function. Preliminary modeling experiments suggest that a power function description may arise in cases where several Gaines–Thomas or Vanselow exchangers with rather different properties are present. Mixed exchangers would be likely to be the case in sandstones; in the case of clays, exchange properties are likely to vary with location on a crystal surface (e.g., edge versus face sites).

Acknowledgments

We thank the Natural Environment Research Council and MJCA Ltd. for financial support.

References

Appelo, C.A.J. 1994. Cation and proton exchange, pH variations, and carbonate reactions in a freshening aquifer. *Water Resources Research*, 30, 2793–2805.

Appelo, C.A.J. and D. Postma. 1993. *Geochemistry, Groundwater and Pollution*. Balkema, Rotterdam.

Argersinger, W.J., Davidson, A.W., and O.D. Bonner. 1950. Thermodynamics of ion exchange phenomena. *Transactions of the Kansas Academy of Science*, 53, 404–410.

Barker, A., Newton, R., Bottrell, S.H., and J.H. Tellam. 1998. Processes affecting groundwater chemistry in a zone of saline intrusion into an urban aquifer. *Applied Geochemistry*, 6, 735–750.

Bond, W.I. 1995. On the Rothmund–Kornfeld description of cation exchange. *Soil Science Society of America Journal* 59, 436–443.

Carlyle, H.F. 1991. The hydrochemical recognition of ion exchange during seawater intrusion at Widnes, Merseyside, UK. Unpublished PhD thesis, Earth Sciences, University of Birmingham.

Carlyle, H.F., Tellam, J.H., and K. Parker. 2004. The use of laboratory-determined ion exchange parameters in the prediction of field-scale major cation migration over a 40 year period. *Journal of Contaminant Hydrology*, 68, 55–81.

Dance, J.T. and E.J. Reardon. 1983. Migration of contaminants in groundwater at a landfill: A case study, 5. Cation migration in the dispersion test. *Journal of Hydrology*, 63, 109–130.

Dzombak, D.A. and F.M.M. Morel. 1990. *Surface Complexation Modelling: Hydrous Ferric Oxide*. John Wiley & Sons, New York, 393.

El-Ghonemy, H.M.R. 1997. Laboratory experiments for quantifying and describing cation exchange in UK Triassic Sandstones. Unpublished PhD Thesis, University of Birmingham, UK.

Gillespie, K.W. 1987. The sedimentology and diagenetic history of the sandstones in the Merseyside Permo-Triassic aquifer. Unpublished MSc Thesis, University of Birmingham.

Gaines, G.L. and H.C. Thomas. 1953. Adsorption studies on clay minerals, II. A formulation of the thermodynamics of exchange adsorption. *Journal of Chemical Physics*, 21, 714–718.

Garrels, R.M. and C.L. Christ. 1965. *Solutions, Minerals and Equilibria*. Freeman, Cooper & Company, California.

Howard, K.W.F. 1988. Beneficial aspects of sea-water intrusion. *Groundwater*, 25, 398–406.

Jensen, H.E. and K.L. Babcock. 1973. Cation exchange equilibria on a Yolo loam. *Hilgardia*, 41, 475–487.

Kool, J.B., Parker, J.C.V., and L.W. Zelazny. 1989. On the estimation of cation exchange parameters from column displacement experiments. *Soil Science Society of American Journal*, 53, 1347–1355.

Langmuir, D. 1981. The power exchange function: A general model for metal adsorption onto geological materials. In: Tewari, D.H. (ed.) *Adsorption from Aqueous Solutions*. Plenum Press, New York, pp. 1–17.

Langmuir, D. 1997. *Aqueous Environmental Geochemistry*. Prentice Hall, New Jersey.

McBride, M.B. 1994. *Environmental Chemistry of Soils*. Oxford University Press, Oxford, UK.

Parker, K.E. 2004. Calculating ion exchange parameters from pure clay assemblages. Unpublished PhD Thesis, University of Birmingham, UK.

Parker, K.E., Tellam, J.H., and M.I. Cliff. 2000. Predicting ion exchange parameters for Triassic sandstone aquifers. In: Sililo, O.T.N. et al. (Eds.), *Groundwater: Past Achievements and Future Challenges*. Balkema, Rotterdam, pp. 581–586.

Reardon, E.J., Dance J.T., and J. L. Lolcama. 1983. Field determination of cation exchange properties for calcareous sand. *Groundwater*, 21, 421–428.

Rothmund, V. and G. Kornfeld. 1918. Der Basenaustauschim Permutit. *Zeitschrift für anorganische und allgemeine Chemie* 108, 215–225.

Rothmund, V. and G. Kornfeld. 1919. Der Basenaustauschim Permutit. *Zeitschrift für anorganische und allgemeine Chemie* 103, 129–163.

Sposito, G. 1981. *The Thermodynamics of Soil Solutions*. Oxford Clarendon Press, Oxford.

Tellam J.H. and J.W. Lloyd. 1986. Problems in the recognition of seawater intrusion by chemical means: An example of apparent chemical equivalence. *Quarterly Journal of Engineering Geology*, 19, 389–398.

Travis, C.B. and H.W. Greenwood. 1911. The mineralogical and chemical constitution of the Triassic rocks of Wirral. *Proceedings of the Liverpool Geological Society*, 11, 116–139.

Van Breukelen, B.M., Appelo, C.A.J., and T.N. Olsthoom. 1998. Hydrogeochemical transport modelling of 24 years of Rhine water infiltration in the dunes of the Amsterdam Water Supply. *Journal of Hydrology*, 209, 281–296.

Walton, H.F. 1949. Ion exchange equilibria. In: Nachod, F.C. (Ed.), *Ion Exchange Theory and Practice*. Academic Press, New York.

21

Evolution of Arsenic Contamination Process and Mobilization in Central Gangetic Plain Aquifer System and Its Remedial Measures

Manoj Kumar, AL. Ramanathan, Alok Kumar, and Shailesh Kumar Yadav

CONTENTS

21.1 Introduction

Readily accessible drinking water is a chief factor that has an adverse impact on health and life expectancy of the population in the present scenario (Kumar et al., 2007; WHO, 2004). Arsenic (As) contamination in the groundwater of the central Gangetic plain was first reported in the year 2002; since then, a number of studies on the groundwater have been conducted by different researchers in this region. Severe health problems caused by naturally occurring As contamination in groundwater triggered extensive studies on arsenic distribution and mobilization in different domains worldwide. Identification of chemical and geological processes responsible for As contamination in the groundwater of the region might be useful to install new tubewells and may allow tapping As-free aquifers as easily available drinking water sources. Approximately 30–35 million people are estimated to be vulnerable of As contamination >50 ppb in Bangladesh and 6 million in West Bengal (Chakraborti et al., 2002; Mandal and Suzuki, 2002; Srivastava and Sharma, 2013). In Bihar state alone, located in the eastern part of the Gangetic plain, 0.9 million inhabitants from 15 districts are reported to be living in an As risk zone (Saha et al., 2009) thus posing a serious risk to the population at large.

21.2 Arsenic: Sources and Occurrence

Arsenic is a notorious element, also known as the "king of poison." It is found in profusion in the earth crust, a trace concentration is also present in soil, rock, water, and air. Arsenic has become a menace to the lives of several hundred million people in different parts of the world and has become the world's greatest environmental calamity (Ravenscroft et al., 2009; Smedley and Kinniburgh, 2002). Arsenic has been used in the ancient world around 3000 bc for human welfare like to harden bronze. Hippocrates suggested the use of As compounds as an ulcer treatment.

Average As concentration in earth crust has been reported to be 1.8 ppm (Mason, 1966) but it could occur up to five times in shale and alluvium. Arsenic is present as a major constituent in more than 200 minerals, including elemental As, arsenides, sulfides, oxides, arsenates, and arsenites. Table 21.1 shows the important minerals with their percentage of As concentration. The most abundant As ore mineral is arsenopyrite, FeAsS. It is generally believed that arsenopyrite, with the other dominant As-sulfide minerals realgar and orpiment, are only formed under high-temperature conditions in the earth crust. It has also been observed that high As concentrations are also found in many oxide minerals and hydrous metal oxides, either as part of the mineral structure or as sorbed species, which includes sorption on the edges of clays and on the surface of the calcite (Goldberg and Glaubig, 1988; Smedley and Kinniburgh, 2002) a common mineral in many sediments. Sand and sandstones mainly having constituent minerals quartz and feldspar tend to have the lowest As concentration while mud and clay usually have higher concentrations (Ravenscroft et al., 2009).

Besides natural, there are anthropogenic sources of As to the environment like application of As-based pesticides, combustion of coal, mining activities and use of preservatives of wood like chromated copper arsenate (Smedley and Kinninburg, 2002). Arsenic enters groundwater through the weathering processes of As mineral-bearing rocks followed by runoff, deposition, and leaching (NIH and CGWB (Government of India [GoI]), 2010). Average As concentration in soils has been reported in the order of 5–10 mg kg^{-1}. Boyle and Jonasson (1973) and Shacklette et al. (1974) quoted an average value of 7.2 and 7.4 mg kg^{-1}, respectively, for American soils. Average As concentration of the stream sediment was found in the range of 5–8 mg/kg^{-1} in England and Wales (AGRG, 1978). Similar concentration

has also been reported in sediments of the Ganges with a range of 1.2–2.6 mg kg^{-1}, from Brahmaputra River with a range of 1.4–5.9 mg kg^{-1} and from the Meghna River with a range of 1.3–5.6 mg kg^{-1} (Datta and Subramanian, 1997). There is little evidence suggesting that atmospheric As poses a real health threat to drinking water sources. But through atmosphere the threat is from direct inhalation of domestic coal-fired smoke and particularly from consumption of foods dried over domestic coal fires, rather than from drinking water affected by atmospheric inputs (Finkelman et al., 1999; Smedley and Kinniburgh, 2002).

The concentration of As in freshwater strongly depends on the source of As, local geochemical environment, and the amount available. It can vary by more than 10^3 orders of magnitude. The main factors for mobilization and accumulation of As are rock-water interactions and availability of favorable physical and geochemical conditions in the aquifers (Smedley and Kinniburgh, 2002). Geochemistry of As is controlled by many factors that include redox potential, adsorption/desorption, precipitation/dissolution, As speciation, pH, and biological transformation.

21.3　Aqueous Speciation and Toxicity of Arsenic

Arsenic behaves differently among the heavy metalloids and oxyanion-forming elements (e.g., As, Sb, Mo, Cr, Re, Se, and V) in its sensitivity and mobilization at the typical values of pH (6.5–8.5) found in natural aquifers. Arsenic can be found in several oxidation states (–3, +3, 0, and +5) and forms organic and inorganic species in the environment (Smedley and Kinniburgh, 2002). Inorganic forms are dominant in natural water while organic forms are found in surface water where biological activity takes place or where waters are significantly affected with industrial waters. Inorganic forms are oxyanions of trivalent arsenite As(III) and or pentavalent arsenate As(V) and major organic species are dimethylarsinic acid (DMA) and monomethylarsonic acid (MMA). Inorganic species are more toxic than organic (NIH and CGWB [GoI], 2010).

TABLE 21.1

Some Arsenic Bearing Minerals

Mineral Arsenic	Chemical Formula	Content (%)
Lollingite	(FeAs$_2$)	73
Arsenopyrite	(FeAsS)	46
Orpiment	As$_2$S$_3$	61
Realgar	As$_4$S$_4$ or AsS	70
Native arsenic	As	90–100

Source: Modified from National Institute of Hydrology (NIH), Roorkee, Central GroundWater Board (CGWB) and Ministry of Water Resources, Government of India (GoI). 2010. Mitigation and remedy of groundwater arsenic menace in India: A vision document by New Delhi, pp. 1–7.

21.4　Geological Setting and Aquifer Types of Central Gangetic Plain

The Ganga River basin, also known for the world's largest alluvial sedimentation and deposition, is an end result of the India-Asia plate collision that had started

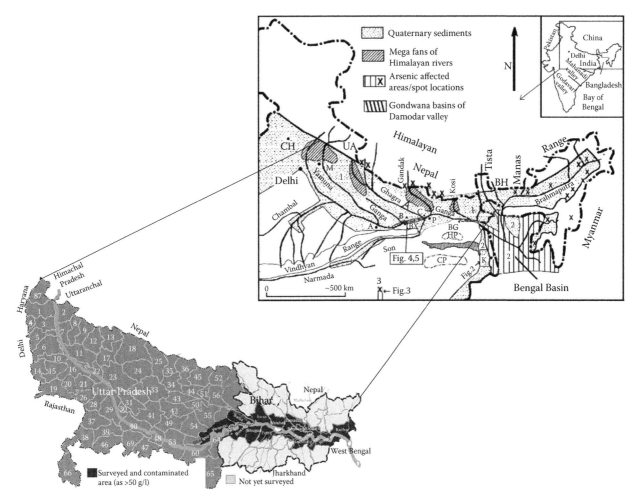

FIGURE 21.1
Central Gangetic plain. (Modified from National Institute of Hydrology (NIH), Roorkee, Central Groundwater Board (CGWB) and Ministry of Water Resources, Government of India (GoI). 2010. Mitigation and remedy of groundwater arsenic menace in India: A vision document by New Delhi, pp. 1–7 and Acharyya, S. K. and B. A. Shah, *J. Environ. Sci. Health A*, 42, 1795–1805, 2007.)

in the Palaeogene geological period. In central Gangetic plain (Figure 21.1), the sediment deposition is divided into Pleistocene and Holocene deposition, which comprises flood plains and piedmont plains of eastern Uttar Pradesh and Bihar (Revenscroft, 2001). A flat topography exists with a north to south slope in the Holocene alluvial central Gangetic plain. Acharyya and Shah (2007) have discussed geomorphologic and quaternary morphostratigraphy of central Gangetic plain on 1:50,000 scale, along with field observations to identify fluvial landforms and soil characteristics. The terrain is divided into two types of deposition: the older alluvium (Pleistocene) is characterized by the presence of yellow brown clay with profuse calcareous and ferruginous concentrations. The newer alluvium (Holocene) is characterized by unoxidized, organic-rich sand, silt, and clay and restricted to low-lying fluvial and fluviolacustrine settings. The major part of the Ganga plain consists of interfluves upland terrace surface (Singh,

1996). The plain has been incised by dendritic drainage and channels containing a good amount of organic mud of the Holocene age (Ravenscroft et al., 2005). Shah (2008) revealed a mineral assemblage (quartz, muscovite, chlorite, kaolinite, feldspar, amphibole, and goethite) with the help of XRD studies on soil samples of As-safe older alluvial and As-contaminated newer alluvium from the central Gangetic plain. In the central Gangetic plain tubewells are tapped mostly in shallow aquifers, which hold 30% of total replenishable groundwater. Shallow aquifers are the main source of drinking water to fulfill the daily requirements of the local population and remain as the major input for societal development.

Aquifers of Holocene sandy sediments are unconfined or semiconfined in the central Gangetic plains, which are mainly used for water extraction. The area has high fertility and a large area (84%) is used for agriculture purposes. A drilling was done by using the drill-cut method up to 300 m and found shallow aquifers at a depth of 120–140 m and deeper aquifers found at 225–240 m below ground level (Saha and Shukla, 2014).

21.5 Environmental Conditions and Groundwater Quality of the Central Gangetic Plain

The quality of water plays an important role in the management of water resources. A number of authors have worked over the years to add to the inventory of water quality in the central Gangetic plain and prominent among them have been highlighted in this text. Saha (1999) had suggested that in Bihar entire drinking water and a major portion of water to irrigate the agricultural fields is extracted from quaternary aquifers. The water-table of Ballia district in Uttar Pradesh (U.P.) varies between 27 and 48 m below ground level (Chandrasekharam et al., 2007) while the water level in Bihar in premonsoon and postmonsoon remained at 3.22–5.82 and 4.7–7.82 m bgl, respectively (Saha et al., 2010). The average value of groundwater quality parameters of different districts reported in different literatures is shown in Table 21.2. The groundwater of central Gangetic plain is neutral to alkaline in nature. Higher average pH values have been observed in Bhagalpur (8.31 ± 0.13) (Kumar et al., 2010a) and Samastipur (8.20 ± 06.4) (Saha and Shukla, 2014). Bicarbonate shows a very wide range in average concentration from premonsoon (79 ± 42) and postmonsoon (105 ± 41.19) ppm in Bhagalpur (Kumar et al., 2010b) to 468 ± 96 ppm in Ballia (Kumar et al., 2015). Higher concentration of HCO_3^- in the postmonsoon period may be due to intense weathering of carbonaceous sandstones followed by subsequent precipitation of HCO_3^- with other cations (Kumar et al., 2010b). The average total dissolved solids (TDS) varies between 423 ± 103 to 793 ± 220 ppm in the central Gangetic plain. Higher concentration of Fe and lower concentration of Mn is a good indicator of the reducing environmental conditions in the central Gangetic plain.

There is unevenness in As concentration in the entire central Gangetic plain due to different sedimentary settings. Average As concentration of 331 ± 156 ppb has been reported in Ballia and Ghazipur (U.P.) (Srivastava and Sharma 2013). Inorganic As(III) was observed in higher percentages in most of the studies in the central Gangetic plain except Ballia. The reason may be the variation in regional geology and unevenness of the covered thick sediments with varying content of organic matter and organic carbon (Chandrasekharam et al., 2007).

21.6 Factors Controlling Arsenic Mobilization in Central Gangetic Plain

There is a misunderstanding that the composition of groundwater does not change naturally. However, a common cause of change in water quality is interaction between aquifer material and the water flowing through them. Factors that control the dissolved minerals in groundwater include (1) the types of minerals that make up the aquifer, (2) the length of time that the water is in contact with the minerals, and (3) the chemical state of the groundwater (NIH and CGWB [GoI], 2010). Different rocks have different minerals and groundwater in contact with those materials will have different compositions. The longer the contact time with minerals, the greater the extent of its reaction with those minerals and the higher will be the content of dissolved minerals. According to Saha et al. (2010) dissolution of calcite and dolomite and infiltration of rain water are responsible for shaping the groundwater chemistry in the central Gangetic plain. Kumar et al. (2010b) observed arsenic enrichment to the river proximity indicating fluvial input as a major contributor of As in the central Gangetic plain. The As contamination may result from a combination of natural and anthropogenic processes like weathering, biological and mining activities, combustion of fossil fuels, use of arsenical pesticides, herbicides, and crop desiccants, and use of As as an additive to livestock, particularly for poultry feed (Huq et al., 2001). Carbonate and silicate weathering plays an important role in groundwater quality and As concentration along with surface-groundwater interactions, ion exchange, and anthropogenic activities. Temporal and seasonal variations also affect the As concentration. Concentration of As is slightly higher in premonsoon than postmonsoon times (Kumar et al., 2010b). Kumar et al. (2010a) observed that types of weathering i.e., carbonate and silicate along with anthropogenic activities control the solute chemistry in premonsoon while anthropogenic activities control solute chemistry in postmonsoon in the central Gangetic plain. The correlation between different parameters in the central Gangetic plains suggests reductive dissolution of Fe (III) oxyhydroxides driven by microbial degradation of younger alluvial sedimentary organic matter under strongly reducing conditions is the probable cause for arsenic mobilization as shown in Figure 21.2.

In this process, consumption of O_2 and NO_3^- occur, which causes redox-alteration in the aquifer (Nickson et al., 2000). Shah (2014) stated that younger tubewells have comparatively low As concentration than older ones. Tubewells of shallow aquifers have higher concentration than deeper aquifers. Although it was not

TABLE 21.2

Average Solute Chemistry of the Groundwater in Central Gangetic Plain

S.No.	Study Area	No.s	pH	HCO$_3$– (ppm)	TDS (ppm)	Fe (ppm)	Mn^{++} (ppm)	As(III) (ppb)	As(V) (ppb)	As(t) (ppb)	References
1.	Ballia (U.P.), India	12	7.5 ± 0.3	–	–	3.53 ± 1.40	0.36 ± 0.33	16 ± 20	33 ± 26	49 ± 34	Chandrasekharam et al. (2007)
2.	Ballia (U.P.), India	65	8.0	323 ± 13	–	3.19 ± 2.39	–	29 ± 37	33 ± 52	49 ± 68	Chauhan et al. (2009)
3.	Ballia and Ghazipur (U.P.), India	36	–	–	–	–	–	–	–	331 ± 156	Srivastava and Sharma (2013)
4.	Ballia (U.P.) and Buxar (Bihar)	121	–	–	–	3.22 ± 2.43	–	–	–	100 ± 150	Shah (2008)
5.	Balia (U.P.) and Patna (Bihar), India	89	–	–	–	–	–	–	–	137.5 ± 232.6	Chandra et al. (2011)
6.	Ghazipur (U.P.), India	PreM-30	7.8	79 ± 42	–	0.61 ± 0	–	34.3 ± 37	22.1 ± 29	48.4 ± 63	Kumar et al. (2010a)
		PostM-30	7.7 ± 0.3	380 ± 212	–	0.10 ± 0.09	–	44.5 ± 43.9	17.6 ± 13.4	38.3 ± 49.9	
7.	Varanasi (U.P.), India	Shallow 29	7.5 ± 0.4	403 ± 138	628 ± 203	0.8 ± 0.48	–	–	–	5.8 ± 13.6	Raju (2012)
		Deep 22	7.6 ± 0.3	346 ± 55	423 ± 103	1.4 ± 0.39	–	–	–	16 ± 24.9	
8.	Bhojpur (Bihar), India	77	7.1 ± 0.4	288.4 ± 99	425 ± 124	1.1 ± 1.6	–	–	–	84.52 ± 120.4	Saha et al. (2009)
9.	Bhojpur (Bihar), India	23	7 ± 0	280 ± 115	–	–	–	–	–	123 ± 113	Saha et al. (2011)
10.	Bhagalpur (U.P.), India	PreM-36	8.3 ± 0.1	105 ± 42.2	455 ± 133	3.19 ± 2.23	0.66 ± 0.56	34.43 ± 21.04	16.81 ± 10.28	51.23 ± 27.64	Kumar et al. (2010b)
		PostM-36	8.0 ± 0.1	129 ± 57.7	441 ± 128	2.82 ± 1.98	0.63 ± 0.55	32.73 ± 18.04	10.24 ± 11.49	48.97 ± 25.76	
11.	Samastipur (Bihar), India	57	8.2 ± 0.4	276.3 ± 73	469 ± 229	1.1 ± 1.2	–	–	–	62.2 ± 66.4	Saha and Shukla (2014)
12.	Samastipur (Bihar), India	23	–	–	–	–	0.45 ± 0.34	–	–	32.14 ± 32.85	Kumar et al. (2016)
13.	Ballia (U.P.), India		7.3 ± 0.2	468 ± 96	736 ± 191	1.4 ± 1.4	0.51 ± 0.27	–	–	35.1 ± 23.8	Kumar et al. (2015)
	Buxar (Bihar), India	84	7.0 ± 0.3	465 ± 86	793 ± 220	1.6 ± 1.9	0.56 ± 0.22	–	–	20.3 ± 13.4	
	Bhojpur (Bihar), India		7.3 ± 0.2	393 ± 85	667 ± 167	2.3 ± 1.9	0.56 ± 0.28	–	–	24.3 ± 27.2	

Note: U.P. = Uttar Pradesh; No.s = number of samples; PreM = pre-monsoon; PostM = post-monsoon; Shallow = shallow aquifers; Deep = deep aquifers.

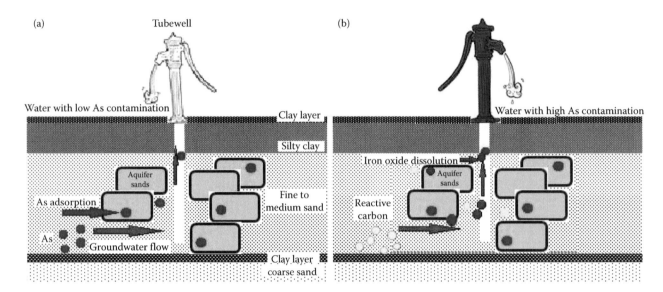

FIGURE 21.2
Conceptual representation of processes regulating the As content of groundwater. A distinction is drawn between low-As aquifer that is contaminated by As released elsewhere (a) and locally (b). Groundwater flow could play a role by supplying As directly or indirectly by supplying reactive organic carbon that triggers reductive dissolution of iron oxides. (Modified from van Geen, A. 2011. International Drilling to Recover Aquifer Sands (IDRAs) and Arsenic Contaminated Groundwater in Asia Scientific Drilling, No. 12, September.)

applicable in all cases, As concentration in newer and older tubewells depends on the lithocharacters of quaternary sediments and also on the release mechanisms (Shah, 2014). Other theories that have been put forth inferring As pollution in groundwater at different regions are oxidation of As-bearing pyrite minerals and competitive exchange with phosphate ($H_2PO_4^-$), but in the central Gangetic plain reductive dissolution of Fe(III) oxyhydroxides fairly explains the mobilization of As mechanism thus far.

21.7 Socio-Economic Effects of Arsenic Contamination of Central Gangetic Plain

The absence of information about the quality of groundwater with respect to As contamination has grave implications owing to dependence of considerable population on groundwater in the central Gangetic plains. Arsenic poisoning has prevalence with respect to age, gender, education, and the economic status of the household. Of these factors, age and economic status of the individual play a crucial role in controlling As exposure as has been observed in the studies conducted in Bangladesh (Hadi and Parveen, 2004) and West Bengal (Mukherjee et al., 2006). Arsenic has been extensively covered in the published literature over the years, the consumption of As contaminated water leads to various pathological manifestations on the body of the user, for example, hyperpigmentation of the skin, black spots in the initial

stages to keratosis to skin cancer, to liver and kidney failure in the final stages (BRAC, 2000). These clinical manifestations are responsible for social exclusion (e.g., no marital relations are formed in the affected families) in the region due to the stigma attached to the diseased individual (Nasreen, 2003; Sarker, 1999).

The economic status of a family influences dissemination of the information about the As poisoning. The As exposure in the region has attained serious proportion and may be due to incessant increase in the number of wells at a cheap cost, which is due to the fact that the owner of the land is also the owner of the groundwater with no state guidelines on pumping. In light of an increased number of wells but no knowledge of the quality of groundwater available, the problem is aggravating day by day. A study conducted by Van Geen and Singh (2013) suggests that the economic status of the household had a role in the purchase parity of the As testing kits in the central Gangetic plain. They also suggested that given the cost involved in the As testing of groundwater, it becomes imminent that the state comes forward with some policy intervention on As and water quality testing in the region.

21.8 Remedial Measures for Arsenic Contamination

The source of As in groundwater is of geogenic origin (Kumar et al., 2016; Shah, 2008) in the central Gangetic plain, so its immediate restoration is not feasible until you have sufficient knowledge of all physiochemical processes controlling its release from the sediments. Thus, the alternate sources or adoption of suitable technological options should be promoted to ensure supply of potable water in As affected areas. The As remedial options can broadly be grouped as:

- Use of surface water sources
- Exploring alternate As-free aquifer
- Use of As removal treatment plants or filters
- Rainwater harvesting
- As remediation

21.8.1 Use of Surface Water Sources

Surface water can be used for drinking purposes after suitable purification by conventional methods of treatment, namely, flocculation, coagulation, rapid sand filtration, and disinfection, as an alternate option have been used by some government organizations. Availability of surface water with technological and economic feasibility may prove successful in the central Gangetic plain.

21.8.2 Exploring Alternate Arsenic Safe Aquifers

The distribution of As in the groundwater is very heterogeneous in the central Gangetic plain as has been found in different studies that not all shallow aquifers are As contaminated but contamination depends on their age and usage (Kumar et al., 2010a,b). Many studies have reported that shallow aquifers have higher concentrations of As than the deeper aquifers. Shallow aquifers are from alluvium and colluvium deposits while deeper alluviums are from carbonate deposits (Williams et al., 1996). The As-contaminated zones are mostly reported within a depth of 100 m below ground level (bgl) in the central Gangetic plain.

Therefore, shallow aquifers should be continuously monitored and marked for their As contamination status, the knowledge of which should be disseminated through public means such that safe aquifers may be used by the population at large. The deep aquifers underneath shallow aquifers are reported normally As-free from West Bengal and Bangladesh plains (Dhar et al., 2008). A thick clay layer with suitable composition separated deep aquifers from shallow aquifers. From the isotopic studies carried out in West Bengal, it is observed that there is no hydraulic connection between deep and shallow aquifers as they belong to different ages (Saha et al., 2011).

21.8.3 Arsenic Removal Treatment Plant and Filters

A number of As removal filters and devices, developed by various organizations, are being used for removing arsenic contamination from groundwater. Working principles of these filters and devices depend on the scientific proposal of the organizations. These techniques are widely being used in West Bengal and there are many organizations like central government, state government, institutions, and private organizations that have come forward for implementation of these devices and provide safe As-free drinking water in As affected areas. There is a need for such a policy intervention in the central Gangetic plain as well.

21.8.4 Rainwater Harvesting

Rainwater harvesting is one the best worldwide methods for those places where other conventional methods are not easily available. It is a simple, affordable, socially acceptable, and technically feasible method that may be popularized in As affected areas but will depend on the success of monsoons in the region.

21.8.5 Arsenic Remediation

Removal of As from groundwater used for drinking purposes has been accomplished using many technologies. In the last few years, many small-scale technologies have been developed and tested in various countries including India. All the available technologies are mainly based on five principles:

1. Co-precipitation
2. Adsorption
3. Oxidation and filtration
4. Ion exchange
5. Membrane technology

1. Co-precipitation

   In this process, As(III) oxidized into As(V) by suitable oxidizing agents like bleaching powder, alum of ferric sulfate, etc. Co-precipitation of As with ferric chloride is an effective and economical technique for removing As from water (Cheng et al., 1994; Gulledge and O'Connor, 1973). Iron hydroxide formed from the salt of ferric chloride, having a high absorption capacity of As(V). Absorption of As(III) is more difficult than As(V), so it rapidly oxidized to As(V) by oxidizing agents such as hydrogen peroxide, potassium permanganate, and sodium hypochlorite. The following steps are involved in the co-precipitation process of As removal:

   a. Addition of bleaching powder/oxidizing agents
   b. Addition of alum/ferric chloride
   c. Mixing of chemical and slow stirring

d. Sedimentation

e. Filtration

2. Adsorption

For purification of water and wastewater, adsorption solids such as activated carbon, activated metal hydride, and synthetic resin are being used widely by industrial applications. A good adsorbent having some surface properties such as high-surface area and polarity is used.

a. Activated Carbon

It is an organic sorbent, commonly used to remove organic and metal contamination from water and wastewater. Activated carbon is a crude form of graphite having an amorphous structure, highly porous, with visible cracks, crevices, and slits of molecular dimensions (Hamerlinck et al., 1994). Adsorption capacity of activated carbon can be improved by impregnation of some of the metals such as copper and ferrous ions.

b. Activated Alumina

Activated Alumina (AA) is prepared from aluminum hydroxide by the process of thermal dehydration. UNEP classified AA adsorption as the best technology among all available technologies for As removal from water. Sorption of As(V) is pH dependent, it adsorbed between pH 6 to pH 8 while As(III) is highly pH dependent and adsorbed at pH 7.6 with high affinity toward AA (Singh and Pant, 2004). Adsorption capacity of AA is improved if iron oxide is impregnated with AA (Singh and Pant, 2004, 2006). Adsorption of As(III) is exothermic when adsorbed on AA and endothermic when adsorbed on impregnated AA (Kuriakose et al., 2004).

c. Iron-based sorbents

Iron hydroxide is the best sorbent in As adsorption because it has high-surface area and strong chemical affinity toward As (Zeng et al., 2007). Most iron oxides are fine powders that are difficult to separate from solution. Therefore, the EPA has proposed iron oxide-coated sand filtration as an emerging technology for As removal at small water facilities (Thirunavukkarasu et al., 2003; USEPA, 1999). Beside iron, other metals are also used as an adsorbent such as titanium oxide, clay mineral, zeolite, manganese dioxide, lanthanum hydroxide, etc.

3. Oxidation and Filtration

Oxidation and filtration processes are mainly designed to remove naturally occurring iron and manganese from water. Arsenite can be oxidized into arsenate by the action of oxygen, ozone, chlorine, permanganate, and hydrogen peroxide. Air oxidation of arsenite is very slow, can take weeks for oxidation (Pierce and Moore, 1982), but chemicals like chlorine and permanganate oxidized rapidly. The removal efficiency of As is strongly dependent on the initial ratio of iron to the As.

4. Ion Exchange

Ion exchange is a process in which ions of one insoluble substance are replaced by other solubles having a similar charge. Arsenic removal is accomplished by passing water continuously through a resin column under pressure. Arsenic(V) can be removed by using a strong base anion exchange resin in either chloride or hydroxide form.

5. Membrane Technology

Membrane technology is the best technology for removing a wide range of contaminants from water. A major drawback of this technique is that it involves high cost. It is a pressure-driven process. Membrane filtration is best over adsorption because the removal efficiency is less affected by the chemical composition and the pH of the water. They are categorized into four categories:

a. Microfiltration (MF)

b. Ultrafiltration (UF)

c. Nanofiltration (NF)

d. Reverse osmosis (RO)

21.8.6 Innovative Technologies

Some innovative technologies are also used in As removal from drinking water such as permeable reactive barrier, phytoremediation, biological treatment, and electrochemical treatment. But all these methods have their own advantages as well as limitations.

21.9 Summary

Arsenic is emerging as a serious health hazard in the central Gangetic plain causing multiple disorders of skin, lungs, liver, kidney, mental disorder, and cancers of multiple organs in severe conditions. Major parts of central Gangetic plain come under newer Holocene alluvium sediment deposition containing mud clay rich in organic matter. Shallow aquifers are the main source of water for irrigation and drinking purposes for the people in the central Gangetic plain. Dissolution of calcite and dolomite and subsequent infiltration of rainwater is controlling the groundwater quality in the central Gangetic plain. Reductive dissolution of Fe(III) oxyhydroxide driven by microbial degradation of younger alluvium containing organic matter is the widely accepted mechanism for mobilization of As in groundwater. Younger and deeper tube wells contain less As or are free from As and vice versa. There are many socio-economic impacts on the residing population even social exclusion (e.g., no marital relations are formed in the affected families). The As contamination is of geogenic in origin in central Gangetic plain, hence immediate restoration is not possible. Some effective steps that may be taken up include use of surface water sources, tapping alternate As-free aquifers, use of As removal treatment plants or filters, rainwater harvesting, and As remediation techniques. In all these steps, rainwater harvesting is the best, simple, affordable, socially acceptable, technically feasible and worldwide acceptable method.

21.10 Recommendations

It is clear that there is unevenness in As concentration throughout the central Gangetic plain, hence remedial steps should be adopted on the basis of efficiency and affordability. There is a need for policy intervention with regard to providing safe drinking water to the people by keeping a constant check on the status of available drinking water quality in the region. The management of aquifer resources is the need of the hour and transboundary aquifer management should be taken up in the region. The groundwater as a resource is precious so its incessant wastage should be checked by fixing the number of tube wells in an area based on population. Information is an important aspect in the exploitation of the groundwater as a resource. Campaigning should be done at the Panchayat level by experts to save the very precious groundwater and for adopting green technologies like rain water harvesting by the residing population. There should be bodies at the Panchayat level to monitor water quality at regular intervals. Due to shortage of time and limitations of the text it was not possible to include all the studies in this chapter. So it is recommended that there should be a record of all the studies done in entire Gangetic plain. It would be helpful to compose proper guidelines, which should be disseminated to the people of the region through various government bodies.

Acknowledgments

The first author is highly thankful to CSIR for providing stipend for PhD work.

References

Acharyya, S. K. and B. A. Shah. 2007. Groundwater arsenic contamination affecting different geologic domains in India—A review: Influence of geological setting, fluvial geomorphology and quaternary stratigraphy, *Journal of Environmental Science and Health, Part A*, 42, 1795–1805.

AGRG. 1978. *The Wolfson Geochemical Atlas of England and Wales*, Clarenson Press, Oxford.

Boyle, R. W. and I. R. Jonasson. 1973. The geochemistry of As and its use as an indicator element in geochemical prospecting, *Journal of Geochemical Exploration*, 2, 251–296.

BRAC. 2000. BRAC Annual Report. BRAC, Dhaka.

Chakraborti, D., Rahman, M. M. Paul, K. Chowdhury, U. K. Sengupta, M. K., Lodh, D. et al. 2002. Arsenic calamity in the Indian sub-continent—What lessons have been learned? *Talanta*, 58, 3–22.

Chandra, S., Ahmed, S., Nagaiah, E., Singh, S. K., and P. C. Chandra. 2011. Geophysical exploration for lithological control of arsenic contamination in groundwater in Middle Ganga Plains, India, *Physics and Chemistry of the Earth*, 36, 1353–1362.

Chandrasekharam, D., Joshi, A., and V. Chandrasekhar. 2007. *Arsenic Content in Groundwater and Soils of Ballia, Uttar Pradesh*, Taylor & Francis Group, London, UK, pp. 1021–1024.

Chauhan, V. S., Nickson, R. T., Chauhan, D., Iyengar, L., and N. Sankararamakrishnan. 2009. Ground water geochemistry of Ballia district, Uttar Pradesh, India and mechanism of arsenic release, *Chemosphere*, 75, 83–91.

Cheng, R. C., Liang, S., Wang, S., and M. D. Beuhler. 1994. Enhanced coagulation for arsenic removal, *Journal of American Water Works Association*, 86(9), 79–90.

Datta, D. K. and V. Subramanian. 1997. Texture and mineralogy of sediments from the Ganges–Brahmaputra–Meghna river system in the Bengal basin, Bangladesh and their environmental implications, *Environmental Geology*, 30, 181–188.

Dhar, R. K., Zheng, Y., Stute, M., Van Geen, A., Cheng, Z., Shanewaz, M., and K. M. Ahmed. 2008. Temporal variability of groundwater chemistry in shallow and deep aquifers of Araihazar, Bangladesh, *Journal of Contaminant Hydrology*, 99(1), 97–111.

Finkelman, R. B., Belkin, H. E., and B. Zheng. 1999. Health impacts of domestic coal use in China, *Proceedings of the National Academy of Sciences, India Section A: Physical Science, USA*, 96, 3427–3431.

Goldberg, S. and R. A. Glaubig. 1988. Anion sorption on a calcareous, montmorillonitic soil—Arsenic, *Soil Science Society of American Journal*, 52, 1297–1300.

Gulledge, J. H. and J. T. O'Connor. 1973. Removal of arsenic (V) from water by adsorption on aluminum and ferric hydroxides (PDF), *Journal of American Water Works Association*, 65(8), 548–552.

Hadi, A. and R. Parveen. 2004. Arsenicosis in Bangladesh: Prevalence and socio-economic correlates, *Public Health*, 118(8), 559–564.

Hamerlinck, Y., Mertens, D. H., and E. F. Vansant. 1994. *Activated Carbon Principles in Separation Technology*, Elsevier, New York.

Huq, S. M. I., Ara, Q. A. J., Islam, K., Zaher, A., and R. Naidu. 2001. The possible contamination from arsenic through food chain. In: Bhattacharya, P., Jacks, G., and Khan, A. A. (Eds.), *Groundwater Arsenic Contamination in the Bengal Delta Plain of Bangladesh. Proceedings of the KTH-Dhaka University Seminar*, KTH Special Publication, TRITA-AMI Report 3084, pp. 9–96.

Kumar, M., Kumar, P., Ramanathan, A., L., Bhattacharya, P., Thunvik, R., Singh, U. K., Tsujimura, M., and O. Sracek, 2010a. Arsenic enrichment in groundwater in the middle Gangetic plain of Ghazipur district in Uttar Pradesh, *Journal of Geochemical Exploration*, 105, 83–94.

Kumar, P., Kumar, M., Ramanathan, A. L., and M. Tsujimura. 2010b. Tracing the factors responsible for arsenic enrichment in groundwater of the middle Gangetic plain, India: A source identification perspective, *Environmental Geochemistry and Health*, 32, 129–146.

Kumar, M., Kumari, K., Ramanathan, A. L., and R. Saxena. 2007. A comparative evaluation of groundwater suitability for irrigation and drinking purposes in two agriculture dominated districts of Punjab, India, *Environmental Geology*, 53, 553–574.

Kumar, M., Kumar, M., Kumar, A., Singh, V. B., Kumar, S., Ramanathan, AL., and Bhattacharya, P. 2015. Arsenic distribution and mobilization: A case study of three districts of Uttar Pradesh and Bihar (India). In: Ramanathan, AL., Johnston, S., Mukherjee, A., and Nath, B. (eds.), *Safe and Sustainable Use of Arsenic-Contaminated Aquifers in the Gangetic Plain: A Multidisciplinary Approach*, Co-published by Springer, International Publishing, Cham, Switzerland with Capital Publishing Company, New Delhi, India. pp. 121–135.

Kumar, M., Rahman, M.M., Ramanathan, AL., and Naidua, R. 2016. Arsenic and other elements in drinking water and dietary components from the middle Gangetic plain of Bihar, India: Health risk index, *Science of the Total Environment*. 539, 125–134.

Kuriakose, S., Singh, T. S., and K. K. Pant. 2004. Adsorption of As(III) from aqueous solution onto iron oxide impregnated activated alumina, *Water Quality Research Journal Canada*, 39(3), 258–266.

Mandal, B. K. and K. T. Suzuki. 2002. Arsenic round the world: A review, *Talanta*, 58, 201–235.

Mason, B. 1966. *Principles of Geochemistry*, 2nd ed., McGraw-Hill, New York.

Mukherjee, A., Sengupta, M. K., Hossain, M. A. et al. 2006. Arsenic contamination in groundwater: A global perspective with emphasis on the Asian scenario, *Journal of Health Population and Nutrition*, 24(2), 142–163.

Nasreen, M. 2003. Social impacts of arsenicosis. In: Ahmed, M. F. (ed.) *Arsenic Contamination: Bangladesh Perspective*. ITN-Bangladesh, Dhaka, Bangladesh, pp. 340–353.

National Institute of Hydrology (NIH), Roorkee, Central Groundwater Board (CGWB) and Ministry of Water Resources, Government of India (GoI). 2010. Mitigation and remedy of groundwater arsenic menace in India: A vision document by New Delhi. pp. 1–7.

Nickson, R. T., McArthur, J. M., Ravenscroft, P. Burgess, W. G., and K. M. Ahmed. 2000. Mechanism of arsenic release to groundwater, Bangladesh and West Bengal, *Applied Geochemistry*, 15, 403–413.

Pierce, M. L. and C. B. Moore. 1982. Adsorption of arsenite and arsenate on amorphous iron hydroxide, *Water Research*, 16(7), 1247–1253.

Raju, N. J. 2012. Evaluation of hydrogeochemical processes in the Pleistocene aquifers of Middle Ganga plain, Uttar Pradesh, India, *Environmental Earth Science*, 65, 1291–1308.

Ravenscroft, P. 2001. Distribution of groundwater arsenic in Bangladesh related to geology. In: Jack, G., Bhattacharya, P., Khan, A. A. (eds.), *Groundwater Arsenic Contamination in the Bengal Delta Plain of Bangladesh*, KTH Special Publication. TRITA-AMI report 3084, pp. 41–56.

Ravenscroft, P., Brammer, H., and K. Richards. 2009. *Arsenic Pollution: A Global Synthesis*, Wiley-Blackwel, Chichester, U.K.

Ravenscroft, P., Burgess, W. G., Ahmed, K. M., Burren, M., and J. Perrin. 2005. Arsenic in groundwater of the Bengal basin, Bangladesh: Distribution, filed relations, and hydrological setting, *Hydrogeology Journal*, 13, 727–751.

Saha, D. 1999. Hydrogeological framework and groundwater resources of Bihar state. Unpubl. Report, Central Groundwater Board, Patna, India.

Saha, D., Sarangam, S. S., Dwivedi, S. N., and K. G. Bhartariya. 2010. Evaluation of hydrogeochemical processes in arsenic-contaminated alluvial aquifers in parts of mid-Ganga Basin, Bihar, eastern India, *Environmental Earth Science*, 61, 799–811.

Saha, D. and R. R. Shukla. 2014. Genesis of arsenic-rich groundwater and the search for alternative safe aquifers in the Gangetic plain, India, *Water Environment Research*, 85, 2254–2264.

Saha, D., Sinha, U. K., and U. K. Dwivedi. 2011. Characterization of recharge processes in shallow and deeper aquifers using isotopic signatures and geochemical behavior of groundwater in an arsenic enriched part of the Ganga plain, *Applied Geochemistry*, 26, 432–443.

Saha, D., Sreehari, S. M., Dwivedi, S. N., and K. G. Bhartariya. 2009. Evaluation of hydrogeochemical processes in arsenic contaminated alluvial aquifers in parts of mid-Ganga basin, Bihar, Eastern India, *Environmental Earth Science*, 61, 799–811.

Sarker, P. C. 1999. Beliefs and arsenicosis and their impact on social disintegration in Bangladesh: Challenges to social work interventions. In: *Proceedings of Joint Conference of AASW, IFSW, AASW and AAWWF*, Brisbane, Australia, pp. 217–221.

Shacklette, H. T., Boerngen, J. G., and J. R. Keith. 1974. *Selenium, Fluorine, and Arsenic in Superficial Materials of the Conterminous United States*, US Government Printing Office, Washington, DC, U.S. Geological Survey, Circ. 692.

Shah, B. A. 2008. Role of quaternary stratigraphy on arsenic-contaminated groundwater from parts of Middle Ganga Plain, UP–Bihar, India, *Environmental Geology*, 35, 1553–1561.

Shah, B. A. 2014. Arsenic in groundwater, quaternary sediments, and suspended river sediments from the middle Gangetic plain, India: Distribution, field relations, and geomorphological setting, *Arabian Journal of Geosciences*, 7, 3525–3536.

Singh, I. B. 1996. Late quaternary sedimentation of Ganga plain foreland basin, *Geological Survey of India*, Special Publication, 21, 161–172.

Singh, T. S. and K. K. Pant. 2004. Equilibrium, kinetics and thermodynamic studies for adsorption of As(III) on activated alumina, *Separation and Purification Technology*, 36(2), 139–147.

Singh, T. S. and K. K. Pant. 2006. Kinetics and mass transfer studies on the adsorption of arsenic onto activated alumina and iron oxide impregnated activated alumina. *Water Quality Research Journal of Canada*, 41(2), 147–156.

Smedley, P. L. and D. G. Kinniburgh. 2002. A review of the source, behavior and distribution of arsenic in natural waters, *Applied Geochemistry*, 17, 517–568.

Srivastava, S. and Y. K. Sharma. 2013. Arsenic occurrence and accumulation in soil and water of eastern districts of Uttar Pradesh, India, *Environmental Monitoring and Assessment*, 18, 4995–5002.

Thirunavukkarasu, O. S., Viraraghavan, T., and T. Subramanian. 2003. Arsenic removal from drinking water using iron oxide-coated sand, *Water, Air, & Soil Pollution*, 142(1–4), 95–111.

USEPA. 1999. Technologies and Costs for Removal of Arsenic from Drinking Water, Draft Report, EPA-815-R-00-012, Washington, DC.

Van Geen, A. 2011. International Drilling to Recover Aquifer Sands (IDRAs) and Arsenic Contaminated Groundwater in Asia Scientific Drilling, No. 12, September.

Van Geen, A. and Singh, C.K. 2013. Piloting a novel delivery mechanism of a critical public health service in India: Arsenic testing of tubewell water in the field for a fee, *International Growth Centre*, Policy Note 13/0238.

WHO. 2004. Guidelines for Drinking Water Quality, 3rd ed. Vol. 1 Recommendations.

Williams, M., Fordyce, F., Paijitprapapon, A., and P. Charoenchaisri. 1996. Arsenic contamination in surface drainage and groundwater in part of the southeast Asian tin belt, Nakhon Si Thammarat Province, southern Thailand, *Environmental Geology*, 27(1), 16–33.

Zeng, H., Fisher, B., and D. E. Giammar. 2007. Individual and competitive adsorption of arsenate and phosphate to a high-surface-area iron oxide-based sorbent, *Environmental Science and Technology*, 42(1), 147–152.

22

Fluoride in Groundwater: Mobilization, Trends, and Remediation

G. Jacks

CONTENTS

22.1 Introduction

The first reports on adverse effects of high-fluoride groundwater used as drinking water date back to the late 1930s. Since then, this has been recognized as a nationwide problem in India. The mechanisms of mobilization are largely known; the problem exists in 20 out of 35 Indian states and is most pronounced in the semi-arid parts of the country. There is a close correlation between the concentration of fluoride in groundwater and soil conditions, which is more pronounced in areas with alkaline–sodic soils. This means that high-fluoride groundwater is also closely related to irrigation practices. An urgent task at hand is to decrease the alkalinization and salinization of soils in a way that will also benefit agricultural production in terms of soil conditions and nutrient supply, notably trace element availability. Water harvesting and improving irrigation practices will help in lowering fluoride concentrations in groundwater. Even if such remedial actions are inadequate, they will at least lower the demands placed on ex situ treatment of water in terms of water filters, etc. The past four decades have largely revealed the mechanisms behind the mobilization which in turn has helped in the development and implementation of combined flow and reaction modeling. However, there is still a need to investigate which mechanism limits the concentration of fluoride. In this regard, it has been suggested that fluorite constitutes a secondary solid phase determining

the higher concentrations. However, in many cases, the high-fluoride groundwater turns out to be undersaturated with respect to fluorite. Moreover, ion-exchange is a possible mechanism that could explain the commonly seen accumulation of fluoride in calcrete.

Excess fluoride in groundwater is a worldwide problem, notably in semi-arid countries (Figure 22.1). It is seen in a belt from North Africa through the Middle East via India to China as well as in Central America and Australia (UNICEF 2014) (Figure 22.1). Another high-fluoride area is along the Rift Valley in East Africa. Fluoride has a very narrow therapeutic spectrum—it protects against tooth decay at levels around 1 mg/L in drinking water but causes dental fluorosis at slightly higher levels, notably in warmer climates like India where the need for fluid is elevated and mostly satisfied by the locally available groundwater. Viswanathan et al. (2009) came to the conclusion that 0.5–0.65 mg/L is the optimal concentration in drinking water taking into account food intake and analyses of the accumulation of fluoride in skeletal bones.

The first reports regarding fluorosis date back to 1937 (Shortt et al. 1937a,b) in which skeletal fluorosis was described from Nellore District of Earlier Madras Province and in present day Andhra Pradesh. However, cases with "mottled teeth" were noted in annual reports a few years prior to this (Mahajan 1934; Visvanathan 1935). Recent reports (Tang et al. 2008; Choi et al. 2015) indicate that chronic exposure of children to excess

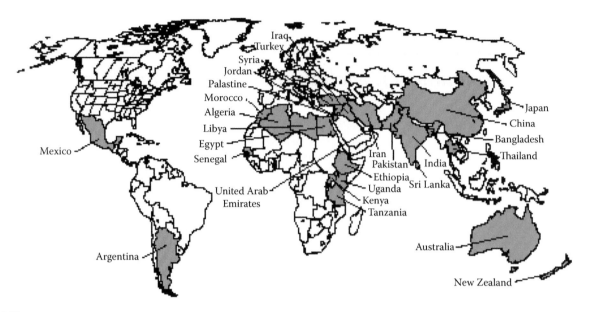

FIGURE 22.1
Countries with endemic fluorosis due to fluoride in groundwater. (From UNICEF, Water Front Issue 13, 1999.)

fluoride affects intellectual capacity. Excess fluoride may also give gastrointestinal manifestations (Susheela et al. 1993). It has been shown that drinking water is by far the main source of exposure to excess fluoride (Amalraj and Pius 2013) even though some varieties of tea could have elevated fluoride content (Susheela 1999; Gupta and Sandesh 2012). It is worth noting that the uptake of fluoride is less when combined with a higher calcium intake, for example, from food and water (Teotia and Teotia 1994; Kravchenko et al. 2014).

Fluorosis is endemic in 20 out of 35 Indian states (FR & RDF 2014) with very high levels recorded in Haryana and Punjab where 48 and 42 mg/L have been detected, respectively (FR & RDF 2014). Even a higher concentration, 86 mg/L, is given by Brindha and Elango (2011) in their extensive review on the same (Figure 22.2). The prevalence of dental and skeletal fluorosis is given in a comprehensive review by Garg and Singh (2007) and Arlappa et al. (2013). Table 22.1 provides the information regarding fluoride concentrations in groundwater in India, site, aquifer lithology, and ranges (Madhavan and Subramanian 2007).

An estimated 60 million people are considered to be exposed to excessive fluoride in their drinking water in India as given in Table 22.2 (Muralidharan et al. 2011). Besides certain hot spots like Andhra Pradesh, Haryana and part of Rajasthan, excess fluoride is also found in most states in India. In an extensive investigation in three districts in Tamil Nadu, the permissible level was exceeded in 89% of the samples (Karthikeyan et al. 2010).

To underline the severity of the situation, it can be mentioned that there are a handful of sites where the

prevalence of skeletal fluorosis is 25%–47% and a dozen sites where the dental fluorosis has a prevalence of up to 90% (Arlappa et al. 2013). An estimated 800 publications in international journals deal with fluoride in

FIGURE 22.2
Ranges of fluoride in groundwater in India. (From Brindha, K. and L. Elango. 2011. *Fluoride: Properties, Applications and Environmental Management.* S. D. Monroy, Ed., Nova Publishers, pp. 111–136.)

TABLE 22.1

Fluoride Concentration in Groundwater in India, Site, Aquifer Lithology, and Ranges

Place	Lithological Characteristics	F⁻ mg/L
Cuddapah (AP)	Granite	0.3–8.0
Guntur (AP)	Charnockites	0.6–2.5
Medak (AP)	Archean crystalline	0.4–6.9
Nalgonda (AP)	Granite gneisses	0.4–20.0
Pennar river basin (AP)	Granite shist and gneisses	<1.0–1.0
Visakhapatnam (AP)	Khondalite, lepynite, and charnockite	0.2–8.4
Karbi Anglong (Assam)	Cretaceous sandstone	0.4–20.6
Bihar	Shist and gneisses	0.1–2.5
Delhi	Amphibolites and quartzites	0.2–32.5
Gujarat	Basalt quaternary sediments	0.1–40.0
Karnataka	Granite and hornblende shist, gneisses	0.4–7.4
Kerala	Charnockites and quaternary	0.1
Shivpuri (MP)	Gneisses	0.2–6.4
Orissa	Granite shist and gneisses	0.1–16.4
Punjab	Recent alluvium	0.1–16.5
Ajmer (RJ)	Calcshist, biotite shist, and gneisess	0.1–12.0
Barmer (RJ)	Sandstone and alluvial deposits	Traces-13.2
Churu (RJ)	Sandstone and alluvial deposits	Traces-30.0
Dungarpur (RJ)	Shist and gneisses, fluorapatite	0.1–10.0
Jaipur (RJ)	Mica shist and quartzite	4.5–28.1
Jaisalmer (RJ)	Sandstone and alluvial deposits	0.2–4.6
Nagaur (RJ)	Limestone, biotite shist	Traces-44.0
Pali (RJ)	Limestone, biotite shist	5.6
Puskhar valley	Calcshist and gneisses, biotite shist	0.19–13.5
Sirohi (RJ)	Shist and gneisses, fluorspar	<1.0–6.0
SE Rajasthan	Calöcshist and gneisses, biotite shist	0.2–16.2
Udaipur (RJ)	Amphibolites, gneisses, and shist	0.1–11.7
Tamil Nadu	Charnockite, gneisses, and shist	0.51–5.0
Uttar Pradesh	Recent alluvium	0.1–17.5
West Bengal	Recent alluvium	0.1–2.2
Indian rivers	Various lithology	~0.5

Source: From Springer Science+Business Media: *Groundwater; Resource Evaluation, Augmentation, Contamination, Restoration, Modeling and Management,* Environmental impact assessment and remediation of contaminated groundwater systems including evolution of fluoride and arsenic contamination process in groundwater, 2007, Madhavan, N. and V. Subramanian.

Note: AP = Andhra Pradesh, MP = Madhya Pradesh, and RJ = Rajasthan.

groundwater in India, including hydrogeology, hydrochemistry, health aspects, and removal technologies.

TABLE 22.2

Fluoride Exposure by Groundwater in India

Region	Population Influenced by Fluoride 1–1.5 mg/L (in Millions)	Population Influenced by Fluoride > 1.5 mg/L (in Millions)
Uttar Pradesh	110	11.6
Andhra Pradesh	63	1.27
Rajasthan Haryana	23	14.8
Total in India	739	54.4

Source: After Muralidharan, D., Rangarajan, R., and B. K. Shankar, *Current Science,* 100(5), 638–640, 2011.

22.2 Hydrogeochemistry of Fluoride

Fluoride is the 24th most common element in Earth's crust, and the most common mineral containing fluorine as a major component is fluorite (CaF_2). However, the bulk of fluorine is present in hydroxy-minerals in which fluoride partly replaces hydroxyl-ions. Such a mineral is hydroxyapatite, called fluorapatite when fluoride is a major component. Many other minerals with hydroxyl-groups like amphiboles, pyroxenes, and micas also contain appreciable amounts of fluorine. Hallett et al. (2015) have budgeted the fluorine in rocks and regoliths in two sites in Andhra Pradesh and one site in Sri Lanka where both areas have elevated fluoride in groundwater. Their work shows that fluoride in groundwater has multiple mineralogical sources both in rocks and regoliths.

In India, a sizeable amount of fluoride is circulated by rainfall; approximately 0.2–0.5 mg/L is found in rainwater (Chandrawanshi and Patel 1999; Satsangi et al. 2002; Tiwari et al. 2012). With evaporation and transpiration this contributes to the generally elevated fluoride concentrations in groundwater. Furthermore, use of phosphate fertilizers manufactured from fluoride containing phosphate rock may add fluoride to the soil–water cycle (Srinivasa Rao 1997; Kundu et al. 2009). According to Srinivasa Rao (1997), 0.34 mg/L of fluoride could be traced back to application of phosphate fertilizer. Pauwels et al. (2015) could also trace some of the fluoride back to phosphate fertilizers.

Groundwater with excess fluoride is commonly of the Na–HCO$_3$ type. Some researchers consider that fluorite (CaF$_2$) is a phase that controls the solubility in groundwater (e.g., Handa 1975; Jacks et al. 1995, 2005).

$$CaF_2 \leftrightarrow Ca^{2+} + 2F^-$$

However, many investigations in high-fluoride areas indicate that fluorite is undersaturated. Saxena and Ahmed (2003) collected groundwater from 58 sites spread across almost all of India and their data showed undersaturation with respect to fluorite. Similarly, Tirumalesh et al. (2007), Pettenati et al. (2014), and Sajil Kumar et al. (2015) found undersaturation with respect to fluorite. Under semi-arid conditions, calcite (CaCO$_3$) is precipitated in the soil zone lowering Ca^{2+} in solution and allowing fluoride concentrations to increase. This occurs when evaporation and evapotranspiration decrease the soil moisture. In a slope in Tamil Nadu, about 700 m long, the calcrete (soil concretions of CaCO$_3$) changes composition downslope from pure calcite to Mg–calcite, dolomite being found at the foot of the slope (Figure 22.3). Although it is rather uncommon for dolomite to form as a new phase, it can occur when the water movements are slow and there is enough time to form the rather complicated crystallography of dolomite (Jacks et al. 2005; Sajil Kumar et al. 2015). In such calcretes up to 8000 mg/kg of fluorine was found (Jacks et al. 2005). Thus, it remains an open question whether the elevated fluoride content in calcrete is due to a specific phase, fluorite, or that it occurs, instead, after being adsorbed onto clay minerals like zeolite in the calcrete (Reddy et al. 2010). Calcrete does contain clay minerals in the fabric itself, in pores and outside the nodules (Wright and Peeters 1989). The development of the groundwater along a slope from Ca–HCO$_3$–water to an Na–HCO$_3$ or even Na–HCO$_3$–Cl water could occur at different scales, along less than a kilometer, as mentioned above (Jacks et al. 2005) or over stretches up to 17 km (Reddy et al. 2010). The soil processes over an even longer stretch and the development from nonsaline conditions via alkaline conditions to saline are described for the Upper Gangetic Plains by Singh et al. (2006).

Whether fluorite is saturated or not is a crucial question. If it is undersaturated there must be some other sink for the fluoride such as adsorption. The high contents of fluoride in calcrete, for example, could be explained in this way. Reddy et al. (2010) and Mondal et al. (2014) consider that zeolites contained in the calcrete could act as adsorbents. In this regard, it has been hypotetized that sepiolite could act as a sink for fluoride (Jacks et al. 1995). However, it is worth noting that sepiolite, palygorskite, and zeolite could all act as adsorbants.

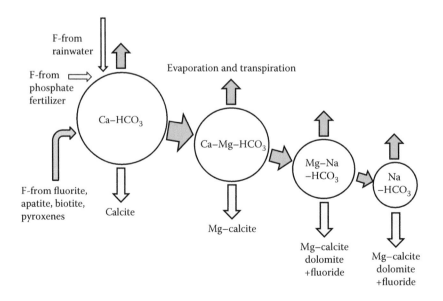

FIGURE 22.3
Conceptual model of the formation of high-fluoride groundwater in India visualizing processes from recharge to discharge areas.

Palygorskite, for example, tends to form in connection with soil concretions under strong evaporation (Hojati et al. 2013). Banerjee (2014) considers that weathering of primary minerals, such as fluorapatite, is a long-term process. Desorption from zeolites and similar clay minerals could be a faster adaptation to the current water conditions. The regolith in sites in Andhra Pradesh and Sri Lanka yielded often a faster release than that from bedrock in leaching tests (Hallett et al. 2015). The excess fluoride groundwater is commonly found in hard rock areas in acid to intermediate gneisses (Tirumalesh et al. 2007; Pettenati et al. 2014; Singaraja et al. 2014). However, it is also common in sedimentary areas, like in a valley in West Bengal (Mondal et al. 2014) and in the Ganga Plain (Misra and Mishra 2007). Misra (2013) has documented the presence of generally elevated fluoride in groundwater in an area with sandstone and limestone and in which there are numerous stone quarries with groundwater exposed for evaporation. The groundwater bears the sign of being concentrated by evaporation and evapotranspiration and enriched in sodium along with mobilization of fluoride.

As already mentioned, the rocks associated with high-fluoride groundwater are usually granitic while basic rocks in hard rock areas are seldom found to contain high-fluoride groundwater (Mondal et al. 2014). However, it is noteworthy that even in a humid tropical area, like in Assam, there could be found very high concentrations of fluoride up to 15 mg/L when acidic granites provide both a soft groundwater and a readily soluble source of fluoride in the form of fluorite (Sahadevan and Chandrasekharam 2008).

While some investigations have found fluorite to be close to saturation and even oversaturated, there are no reports about the formation of secondary fluorite. Several publications have even found fluorite to be undersaturated albeit fluoride concentrations are high (e.g., Reddy et al. 2010). This publication has investigated a long slope from recharge areas to discharge areas and has found that although fluoride is released to the water during recharge, it is then precipitated or adsorbed onto calcrete downslope. Due to calcrete formation, the fluoride increases slightly downslope while a portion of the fluoride is adsorbed on the calcrete. Furthermore, a high content of fluorine is found downslope in the calcrete, often above 2–3000 mg/kg. However, an even higher content, up to 9000 mg/kg was found by Jacks et al. (2005) and Jacks and Sharma (1995) at a site in Tamil Nadu. Dolomite has been found to be a portion of the calcrete both by Reddy et al. (2010) and Jacks et al. (2005). Dolomitic calcrete had a higher content of fluorine indicating a downslope position. Reddy et al. (2010) and Sajil Kumar (2015) found a sizeable oversaturation in the groundwater with respect to dolomite, while fluorite was generally undersaturated even with a trend of larger undersaturation downslope, due to loss of calcium precipitated as calcrete. This is an indication that fluorite is not a limiting phase for the dissolution of fluoride. However, this elevated content in calcrete may not be a defined phase but rather adsorbed to clay minerals like zeolite, palygoskite, and sepiolite incorporated during the calcrete formation. Many other anions are coprecipitated along with calcite like phosphate and arsenate. In Ethiopia, for example, natural zeolites are tested for the elimination of fluoride from drinking water (Gómez-Hortigüela et al. 2013).

22.3 Modeling

The modeling of the hydrogeochemistry could be done either by taking hydrochemistry into account or by combining this with a hydrogeological flow model. Common hydrochemical programs are MINTEQ and PHREEQC. Banerjee (2014) has modeled the dissolution of fluorapatite in the process of forming high-fluoride groundwater. One of the results is that a very long time is needed for the mobilization of fluoride by this mechanism. This may indicate that other mechanisms like desorption from clays like zeolite are important ways of forming high-fluoride groundwater (Mondal et al. 2014). Recently combined models dealing with hydrochemistry as well as flow dynamics (Packhurst and Appelo 1999) are used to study the behavior of fluoride in groundwater (Pettenati et al. 2013). This study focused the effect that rice cultivation had on fluoride concentration as a result of irrigation with groundwater. The result showed that out of different land uses, irrigation had a pronounced elevating effect on fluoride concentration in groundwater. This is an important finding that stresses the risk that an increasing abundance of high-fluoride groundwater can occur.

22.4 Future Development

High-fluoride groundwater is associated with Na–HCO$_3$ type water, common in areas with saline and sodic soils. The extent of saline soil is 7 M ha (Pal et al. 2003). Out of this, 2.32 M ha in Punjab, Haryana, and Uttar Pradesh are sodic soils with a high pH, 9.5–10.5 and ESP (exchangeable sodium percentage) in the order of 80%–95%. In these areas, high-fluoride groundwater is common. According to El-Swaify et al. (1983), the extent of alkaline soil was only 0.6 M ha in 1979. The current extent of sodic/alkaline soil in the whole of India is given as 3.79 M ha (CSSRI 2014). As mentioned earlier, irrigation tends to speed up the process of increasing fluoride in groundwater as once the groundwater has become sodic, it tends to decrease the hydraulic conductivity of soil increasing evaporation, runoff, and soil loss (Mandal et al. 2008). Even conjunctive use of an alkaline groundwater and a nonalkaline surface water for irrigation renders the soil more alkaline (Choudhary et al. 2006). Furthermore, it is worth noting that the formation of alkaline soil is a problem for cultivation resulting in a decreasing availability of phosphorus, zinc, and a risk for molybdenosis in grazing animals (Gupta and Lipsett 1981). Approximately 50% of the agricultural soils in India are zinc deficient (Cakmak 2009) and one of the reasons is this is a high soil pH, rendering zinc less available for plant uptake. Finally, regarding groundwater, there is often a correlation between soil pH and fluoride concentrations (Jacks et al. 2005), which is still another reason to combat soil alkalinity.

Groundwater recharge on a national scale is quite well known in India. In a nationwide investigation, Rangarajan and Athavale (2000) have used injected tritium for the recharge assessments and have given a figure of 148 mm recharge for a national mean, while Raju (1998) using groundwater fluctuations, has come to a recharge of 130 mm. It is noteworthy that these two studies deviate by little more than 10% from each other which is quite an achievement. The recharge, in the semi-arid parts of India where the high-fluoride groundwater is most common, is in the order of 30 mm (Jacks and Sharma 1983). The depletion of groundwater occurring in the peninsula, and not least in NW India, is a threat to both agriculture and economy in terms of energy subsidies needed for pumping and costs for the individual farmer (Shah 2014). In this respect, a very large number of investigations have been carried out when it comes to promoting recharge (e.g., Singh 2014). The current overuse of groundwater for irrigation, especially in peninsular India, is a threat not only to the groundwater as such because it may also change the crop quality. Perrin et al. (2011) have constructed scenarios with different rates of pumping for a site in Andhra Pradesh, indicating that the total salinity will increase 3- to 5-fold. However, other parameters may also change, like soil alkalinity, which may promote dissolution of fluoride and desorption of fluoride from clay minerals (Mondal 2014). It has been noted that a rapid increase in soil pH, half a pH-unit over a 2–3-year period, was recorded from the Indira Gandhi canal (Jaglan and Qureshi 1996). Thus, it is likely that fluoride concentrations will tend to increase, needing countermeasures. The tendency of drilling deeper wells may also contribute to increased fluoride content in groundwater as weathering over longer time contributes to release of more fluoride (Pauwels et al. 2015).

A concern in this connection, albeit within a longer time perspective, is climate change. There are several projections regarding the coming decades (Kumar et al. 2006; Mall et al. 2006; Turner and Annamalai 2012). The climate is expected to be warmer but most of India is also expected to receive more rainfall. Runoff will remain more or less as per now except for in the northwest (Mall et al. 2006; Turner and Annamalai 2012). Rivers like Narmada and Sabarmati are expected to have decreased runoff (Mall et al. 2006). These areas already gave groundwater with excess fluoride and endemic fluorosis and this will probably be an increasing problem in the northwestern part of India.

22.5 Remedial Actions

22.5.1 Water Treatment

The measures used to decrease the exposure of people to high fluoride concentrations are usually filters. However, filters do not perform very well for the removal of fluoride due to the fact that they function best at a low pH, while fluoride-rich waters are generally alkaline with a pH over 8. Furthermore, filters in India are usually expected to be handled by women, who are often too burdened with other tasks to be able to service them properly. This was seen in Bangladesh, where less than 10% of arsenic removal filters were in use 2 years after their introduction (Jakariya et al. 2007). Although fluoride removal plants have been developed and built, their maintenance is often poor. Moreover, plants using the so-called Nalgonda technique, based on addition of alum and lime leaves aluminum complexes in the water that could be toxic (Meenakshi 2006). However, using stable aluminum compounds could offer a good solution (Karthikeyan and Elango 2009). Furthermore, ferric compounds like laterite have been found useful (Sarkar et al. 2007). In Sri Lanka, crushed red bricks are used. Although this is a cheap solution, the overall capacity is low. Other media tested for fluoride removal is the shell from tamarind fruits (Suvsankar et al. 2012), cotton nut shell carbon (Mariappan et al. 2015), and nanohydroxyapatite (Pandi and Viswanathan 2015). Although membrane technologies could offer a possible solution, the problem of remineralizing the water still remains. For an extensive review regarding fluoride removal, see Bhatnagar et al. (2011). Based on their research, although they consider activated alumina as the most established treatment, they also mention modified chitosan as another promising alternative. However, successful batch tests at the laboratory may fail when applied to field conditions. Moreover, the regeneration of filter materials needs to be studied further (Bhatnagar et al. 2011).

22.5.2 Artificial Recharge

Reddy and Raj (1997) observed low fluoride groundwater in the vicinity of recharge ponds. Muralidharan et al. (2011) subsequently recommend water harvesting in the upstream areas of catchments to dilute the groundwater. It is worth noting that extensive experience from such an approach has been gained at NGRI (National Institute of Geophysical Research) in Hyderabad. Furthermore, recent research has studied the efficiency of recharge from small percolation tanks in hard rock terrain in India (Bhagavan and Raghu 2005; Glendenning and Vervoort 2010; Boisson et al. 2014; Massuel et al. 2014; Pettenati et al. 2014), which is typical for areas with elevated fluoride levels in groundwater. It turns out that around 60% of the water collected water percolate to the groundwater. The recharged water is found in close vicinity to the tank. However, tanks such as these need to be properly managed, for example, desilting (Muralidharan et al. 1995; Madhnure et al. 2015). A modeling approach using a watershed model taking into account crop growth has been used to evaluate the recharge through percolation ponds (Selvarajan et al. 1995). A field study on the catchment scale is reported by Glendenning and Vervoort (2010). For the percolation ponds they reported between 12 and 52 mm/day while the areal recharge was 2–7 mm for the studied period. Pettenati et al. (2014) have used modeling with reactive transport models to visualize different scenarios. Even if rainwater harvesting does not result in acceptable concentrations of fluoride in the groundwater, it is important as it will decrease the demand on the capacity of filters and other ex-situ technologies.

22.5.3 Remediation of Soil Alkalinity

There is often a direct relationship between soil pH and fluoride content in groundwater (Jacks et al. 2005). Saline soils occupy about 9.1 million hectares (Mha) in India and sodic-alkaline soils about 3.6 Mha (Dagar et al. 2001). Compared to data from approximately a decade earlier, which gave the extent of saline soils at 7 Mha and sodic soils at 2.5 mHa (Dahiya and Anlauf 1990) there seems to have been a rapid increase in the extent of these affected soils. Moreover, the extent of sodic soils was only 0.6 Mha in 1979 (El-Swaify et al. 1983). Along with promoting recharge, it is also necessary to remediate the alkalinity/sodicity of soils. Amelioration of sodic soils is often done by application of gypsum (Chhabra 1995; Minhas 1996). The gypsum application is often amended by the addition of organic matter such as cow dung or straw (Prapagar et al. 2012). Another solution is bioremediation, which has also been extensively tested, for example, by tree plantation (Stille et al. 2011). Tree plantations have been used to decrease the alkalinity of soils by increasing the carbon dioxide pressure in the root zone and promoting dissolution of soil calcrete (Prakash and Hasan 2011). Phytoremediation takes place with a number of tree species, among them the easily cultivated *Prosopis juliflora* (Qadir et al. 2007). This will also affect groundwater quality as there is usually a correlation between soil pH and fluoride concentration in groundwater (Jacks et al. 2005). Alveteg and Jönsson (1991) have demonstrated that application of gypsum lowers the alkalinity affects the fluoride concentrations in groundwater to some extent. It was found that the fluoride content was lowered from about 5 mg/L to about 2–3 mg/L in wells situated in the gypsum treated area. An extensive overview of possible methods for remediating alkaline/sodic soils is given by Minhas and Sharma (2003). They also report that a sizable portion of alkaline soil in the Indo Gangetic Plain has been remediated (Minhas and Sharma 2003). In this respect a total of 1.4 Mha of alkali land has been reclaimed in India up to the year 2005 (Tripathi 2009); nevertheless, the area of alkaline soil is given as 3.8 Mha by CSSRI (2014). As a result of this changing policy, new cropping patterns may have to be introduced (Ambast et al. 2006). One such cropping pattern is the much debated system for rice intensification (SRI). However, tests in India show saving of about 35%–40% of the water demand and a higher productivity (Singh et al. 2012). To date, large portions of rice cultivation in Tamil Nadu and Bihar, drought-prone states, are now under SRI regime (Uphoff 2012). Moreover, although remediation of degraded soils is a global problem, examples from several countries including India, reveal that it is an economically feasible task (Qadir et al. 2014).

22.6 Conclusions

The common occurrence of high-fluoride groundwater in India is a serious health problem. This is because groundwater is by far the largest source of fluoride in terms of human intake. In India, the permissible level of fluoride in drinking water is lower than elsewhere due to the climate and a generally low calcium intake. The past 40 years of research has, by and large, explained the mechanisms of mobilization regarding high fluoride concentrations in groundwater. If there is a pioneering publication in this respect, it would be Handa (1975). However, there is still some uncertainty regarding what is the main sink for fluoride in groundwater systems, whether under some conditions fluorite is a concentration limiting specie or whether adsorption/desorption is the major mechanism. Many of the high fluoride groundwater is undersaturated with respect to fluorite while it seems that adsorption/desorption depending on the current chemical groundwater conditions may be very important. In recent years, modeling has been applied to the problem using models like PHREEQC. This has been used to shed light on the effect of artificial recharge. As such, there is a need for a more detailed investigation into the mechanisms that determine the fluoride levels in groundwater, as it seems that fluorite (CaF_2), as mentioned, is undersaturated in high-fluoride groundwater in many cases, notably what phases in calcrete, with its elevated content of fluorine, govern the release and uptake of fluoride.

A large number of articles have been published on fluoride in groundwater, a problem that has been detected in the majority of Indian states. As fluoride concentrations in groundwater are related to soil conditions, it is likely that the areal extent of alkaline–sodic soils mirrors the extent of the problem pertaining to excess fluoride in groundwater. Although the figures regarding the area with alkaline soils vary considerably between different sources, there seems to be a trend that the areal extent is increasing albeit remedial actions have had a limiting effect.

When considering remedial actions, there are indications that artificial recharge or water harvesting may be helpful, even though it does not seem to solve the problem. While lower fluoride concentrations are seen in connection to check dams and similar structures, the concentrations are usually still in excess of the permissible limit. Even if this is the case, lowering the fluoride concentrations *in situ* is recommendable as it places less stress on the extent of ex situ treatment like filters. In other words, feeding filters with lower concentrations of fluoride will lessen the need for replacing or regenerating the filters.

It appears that alkalinity in soils has been increasing over the last few decades, which is likely to cause increase in groundwater fluoride concentrations. Thus, all possible efforts to combat the development of alkalinity in soils should be taken especially as there are so many other problems associated with it. A good example of this is the fact that soil alkalinity results in lower availability of many nutrients like phosphate and trace elements, especially zinc. A lack of such elements may lower crop yields and decrease the nutritional value of the food produced. Soil alkalinity may also cause deterioration in the physical conditions of the soil, like permeability. Thus, irrigation efficiency needs to be improved and new cropping patterns may be introduced. An example of a new cropping pattern is the so-called system for rice intensification (SRI), which is gaining ground on the Indian peninsula. This new cropping pattern may save a lot of water and decrease the effects of soil alkalinity.

Acknowledgments

Sincere thanks to Dr. B. K. Handa, who was the superintending chemist at Central Ground Water Board of India during my first participation in a project in India. He visited us in Coimbatore every other month and was a fantastic source of knowledge and inspiration. During his stay, the workdays used to be long but full of ideas and proposals that remained with us. Also, thanks to all the authors of the numerous articles written about fluoride in groundwater in India from which this author hopefully learned a lot.

References

Alveteg, T. and M. Jönsson. 1991. Amendment of High Fluoride Groundwaters. M.Sc. thesis. Royal Inst of Technology (KTH), Stockholm, Sweden.

Amalraj, A. and A. Pius. 2013. Health risk from fluoride exposure of a population in selected areas of Tamil Nadu South India. *Food Science and Human Wellness*, 2(2), 73–86.

Ambast, S. K., Tyagi, N. K., and S. K. Raul. 2006. Management of declining groundwater in the Trans Indo-Gangetic Plain (India): Some options. *Agricultural Water Management*, 82(3), 279–296.

Arlappa, N., Aatif Queshi, I., and R. Srinivas. 2013. Fluorosis in India: An overview. *International Journal of Research and Development*, 1(1), 97–100.

Banerjee, A. 2014. Groundwater fluoride contamination: A reappraisal. *Geoscience Frontiers* 6(2), 277–284.

Bhagavan, S. V. B. K. and V. Raghu. 2005. Utility of check dams in dilution of fluoride concentration in ground water and the resultant analysis of blood serum and urine of villagers, Anantapur District, Andhra Pradesh, India. *Environmental Geochemistry and Health*, 27(1), 97–108.

Bhatnagar, A., Kumar, E., and M. Sillanpää. 2011. Fluoride removal from water by adsorption—A review. *Chemical Engineering Journal*, 171(3), 811–840. Rajasthan, India. Nordic Soc. of Clay Res. Report No 10, pp. 5–6.

Boisson, A., Baisset, M., Alazard, M. et al. 2014. Comparison of surface and groundwater balance approaches in the evaluation of managed aquifer recharge structures: Case of a percolation tank in a crystalline aquifer in India. *Journal of Hydrology*, 518 (Part B), 1620–1633.

Brindha, K. and L. Elango. 2011. Fluoride in groundwater: Causes, implications and mitigation measures. In *Fluoride: Properties, Applications and Environmental Management*. S. D. Monroy, Ed. Nova Publishers, New York, pp. 111–136.

Cakmak, I. 2009. Enrichment of fertilizers with zinc: An excellent investment for humanity and crop production in India. *Journal of Trace Elements in Medicine and Biology*, 23(4), 281–289.

Chandrawanshi, C. K. and K. S. Patel. 1999. Fluoride deposition in Central India. *Environmental Monitoring and Assessment,* 55(2), 252–265.

Chhabra, R. 1995. Nutrient requirement for sodic soils. *Fertilizer News* 40, 13–21.

Choi, A. L., Zhang, Y., Sun, G. et al. 2015. Association of lifetime exposure to fluoride and cognitive functions in Chinese children: A pilot study. *Neurotoxicology and Teratology* 47, 96–101.

Choudhary, O. P., Ghuman, B. S., Josan, A. S., and M. S. Bajwa. 2006. Effect of alternating irrigation with sodic and non-sodic waters on soil properties and sunflower yield. *Agricultural Water Management,* 85(1–2), 151–156.

CSSRI (Central Soil Salinity Research Institute). 2014. http://cssri.nic.in/. Accessed May 10, 2015.

Dagar, J. C., Singh, G., and N. T. Singh. 2001. Evaluation of forest and fruit trees used for rehabilitation of semi-arid alkali-sodic soils in India. *Arid Land Research and Management,* 15(2), 115–133.

Dahiya, I. S. and R. Anlauf. 1990. Sodic soils in India, the reclamation and management. *Zeitschrift für Kulturtechnik und Landentwicklung.* 31, 26–34.

El-Swaify, S. A., Arunin, S. S., and I. P. Abrol. 1983. Soil salinization: Development of the salt affected soils. In *Natural Systems for Development: What Planners Need to Know.* C.A. Carpenter, Ed. McMillan Publ. Co., London, UK.

FR & RDF (Fluorosis Research & Rural Development Foundation). 2014. www.fluorideandfluorosis.com/ Organisation. Accessed May 21, 2015.

Garg, V. K. and B. Singh. 2007. Fluoride in drinking water and fluorosis. www.eco-web.com/edi/070207.html. Accessed April 20, 2015.

Glendenning, C. J. and E. W. Vervoort. 2010. Hydrological impacts of rainwater harvesting (RWH) in a case study catchment: The Aravan River, Rajasthan, India. Part 1: Field-scale impacts. *Agricultural Water Management,* 98(2), 331–342.

Gómez-Hortigüela, L., Pérez-Pariente, J., Garcia, R., Chebude, Y., and I. Diaz. 2013. Natural zeolite from Ethiopia for elimination of fluoride from drinking water. *Separation and Purification Technology* 120, 224–229.

Gupta, P. and N. Sandesh. 2012. Estimation of fluoride in tea infusions, prepared from different forms of teas, commercially available in Mathura city. *Journal of International Society of Preventive and Community Dentistry,* 2(2), 64–68.

Gupta, U. and J. Lipsett. 1981. Molybdenum in soils, plants and animals. *Advances in Agronomy,* 4, 73–115.

Hallett, B. M., Dharmagunawardhane, H. A., Aytal, S., Valsami-Jones, E., Ahmed, S., and W. C. Burgess. 2015. Mineralogical sources of groundwater fluoride in Archean bedrock/roegloth aquifers: Mass balances from southern India and north-central Sri Lanka. *Journal of Hydrology: Regional Studies,* 4, 111–130.

Handa, B. K. 1975. Geochemistry and genesis of fluoride-containing ground waters in India. *Ground Water* 13, 275–281.

Hojati, S., Khademi, H., Faz Cano, A., Ayobi, S., and A. Landi. 2013. Factors affecting the occurrence of palygorskite in Central Iranian soils developed on Tertiary sediments. *Pedosphere* 23(3), 359–371.

Jacks, G., Bhattacharya, P., Chaudhary, C., and K. P. Singh. 2005. Controls on the genesis of some high-fluoride groundwaters in India. *Applied Geochemistry,* 20(2), 221–228.

Jacks, G., Rajagopalan, K., Alveteg, T., and M. Jönsson. 1995. Genesis of high-F groundwaters, southern India. *Applied Geochemistry,* Suppl. Issue No. 2, 241–244.

Jacks, G. and V. P. Sharma. 1983. Nitrogen circulation and nitrate in groundwater in Southern India. *Environmental Geology,* 5(2), 61–64.

Jacks, G. and V. P. Sharma. 1995. Geochemistry of calcic horizons in relation to hillslope processes, southern India. *Geoderma* 67(3–4), 203–214.

Jaglan, M. S. and M. H. Qureshi. 1996. Irrigation development and its environmental consequences in arid regions in India. *Environmental Management,* 20(3), 323–336.

Jakariya, M., von Brömssen, M., Jacks, G. et al. 2007. Searching for sustainable arsenic mitigation strategy in Bangladesh: Experiences from two upazilas. *International Journal of Environment and Pollution,* 31, 415–430.

Karthikeyan, K., Nanthakumar, K.,Velmurugan, P., Tamilarasi, S., and P. Lakshmanaperumalsamy. 2010. Prevalence of certain inorganic constituents in groundwater samples of Erode district, Tamil Nadu, India, with special emphasis on fluoride, fluorosis and its remedial measures. *Environmental Monitoring and Assessment,* 160(1–4), 141–155.

Karthikeyan, M. and K. P. Elango. 2009. Removal of fluoride from water using aluminium containing compounds. *Journal of Environmental Science,* 21, 1513–1518.

Kravchenko, J., Rango, T., Akushevich, I. et al. 2014. The effect of non-fluoride factors on risk of dental fluorosis: Evidence from rural populations of the Main Ethiopian Rift. *Science of the Total Environment,* 466–489, 595–606.

Kumar, K. R., Sahai, A. K., Krishna Kumar, K. et al. 2006. High resolution climate change scenarios for India for the 21st century. *Current Science* 90, 334–345.

Kundu, M. C., Mandal, G., and C. Hazra. 2009. Nitrate and fluoride contamination in groundwater of an intensively managed agroecosystem: A functional relationship. *Science of the Total Environment,* 407(8), 2771–2782.

Madhavan, N. and V. Subramanian. 2007. Environmental impact assessment and remediation of contaminated groundwater systems including evolution of fluoride and arsenic contamination process in groundwater. In *Groundwater; Resource Evaluation, Augmentation, Contamination, Restoration, Modeling and Management.* M. Thangarajan, Ed. Springer Verlag and Capital Publishing Company, Heidelberg, Germany.

Madhnure, P., Rao, P. S., and A. D. Rao. 2015. Establishing strategies for sustainable groundwater management plan for typical granitic aquifers—A pilot study near Hyderabad, India. *Aquatic Procedia* 4, 1307–1314.

Mahajan. 1934. Annual report. VIO Hyderabad State, 3, Indian council of Agricultural Science, New Delhi.

Mall, R. K., Gupta, A., Singh, R., Dingh, R. S., and L. S. Rathore. 2006. Water resources and climate change: An Indian perspective. *Current Science* 90, 1610–1625.

Mandal, U. K., Bhardway, A. K., Warrington, D. N., Goldstein, D., Bar Tal, A., and G. J. Levy. 2008. Changes in soil hydraulic conductivity, runoff, and soil loss due to irrigation with different types of saline sodic water. *Geoderma* 144(3–4), 509–516.

Mariappan, R., Vairamuthu, R., and A. Ganapathy. 2015. Use of chemically activated cotton nut shell carbon for the removal of fluoride contaminated drinking water: Kinetics evaluation. *Chinese Journal of Chemical Engineering*, 23, 710–721.

Massuel, S., Perrin, J., Mascre, C., Mohamed, W. et al. 2014. Managed aquifer recharge in South India: What to expect from small percolation tanks in hard rock? *Journal of Hydrology*, 512, 157–167.

Meenakshi, R. C. M. 2006. Fluoride in drinking water and its removal. *Journal of Hazardous Material*, B137, 456–463.

Minhas, P.S. 1996. Saline water management for irrigation in India. *Agricultural Water Management*, 30, 1–24.

Minhas, P.S. and O. P. Sharma, 2003. Management of soil salinity and alkalinity problems in India. *Journal of Crop Production*, 7, 181–230.

Misra, A. K. 2013. Influence of stone quarries on groundwater quality and health in Fatehpur Sikri, India. *International Journal of Sustainable Built Environment*, 2, 73–88.

Misra, A. K. and A. Mishra. 2007. Study of quaternary aquifers in Ganga Plain, India: Focus on groundwater salinity, fluoride and fluorosis. *Journal of Hazardous Material*, 144(1–2), 438–448.

Mondal, D., Gupta, S. D., Reddy, D. V., and P. Nagabhushanam 2014. Geochemical controls on fluoride concentrations in groundwater from alluvial aquifers of the Birbhum district, West Bengal, India. *Journal of Geochemical Exploration*, 145, 190–206.

Muralidharan, D., Rangarajan, R., Hodlur, G.K., and U. Sathyanarayana 1995. Optimal desilting for improving the efficiency of tanks in semi-arid regions. *Journal of the Geological Society of India*, 65, 83–88.

Muralidharan, D., Rangarajan, R., and B. K. Shankar. 2011. Vicious cycle of fluoride in semi-arid India—A health concern. *Current Science*, 100(5), 638–640.

Packhurst, D. L. and J. Appelo. 1999. *Users Guide to PHREEQC (version 2)—A Computer Program for Speciation, Batch-Reaction, One Dimensional Transport Inverse Geochemical Calculation*. U.S. Geological Survey Water Resources Investigation Report, pp. 99–4259.

Pal, D. K., Srivastava, P., Durge, S. J., and T. Bhattacharyya, 2003. Role of microtopography in the formation of sodic soils in the semi-arid part of the Indo-Gangetic Plains, India. *Catena* 51(1), 3–31.

Pandi, K. and N. Viswanathan. 2015. *In situ* precipitation of nano-hydroxyapatite in gelatin polymatrix towards specific fluoride sorption. *International Journal of Biological Macromolecules* 74, 351–359.

Pauwels, H., Négrel, P., Dewandel, B. et al. 2015. Hydrochemical borehole logs characterizing fluoride contamination in a crystalline aquifer (Maheshwaram, India). *Journal of Hydrology*, 525, 302–312.

Perrin, J., Mascré, C., Pauwels, H., and S. Ahmed. 2011. Solute recycling: An emerging threat to groundwater quality in southern India. *Journal of Hydrology*, 398(1–2), 144–154.

Pettenati, M., Perrin, J., Pauwels, H., and S. Ahmed. 2013. Simulating fluoride evolution in groundwater using a reactive multicomponent transient transport model: Application to a crystalline aquifer of Southern India. *Applied Geochemistry,* 29, 102–116.

Pettenati, M., Picot-Colbeaux, G., Thiery, S. et al. 2014. Water quality during managed aquifer recharge (MAR) in Indian crystalline basement aquifers: Reactive transport modelling in the critical zone. *Procedia Earth Planetary Science*, 10, 82–87.

Prakash, J. and S. Hasan. 2011. Reclamation of sodic soil in North India through *Acacia nilotica* and *Dalbergia sissoo*. *Journal of Natural Science, Biology and Medicine*, 2(3 Suppl.), 132.

Prapagar, K., Indraratne, S. P., and P. Premanandharajah. 2012. Effect of soil amendment on reclamation of sodic-saline soil. *Tropical Agricultural Research*, 23(3), 168–176.

Qadir, M., Oster, J. D., Schubert, S., Noble, A. D., and K. L. Sahrawat. 2007. Phytoremediation of sodic and saline soils. *Advances in Agronomy*, 96, 197–247.

Qadir, M., Quillérou, E., Nangia, V. et al. 2014. Economics of salt-induced land degradation and restoration. *Natural Research Forum* 38(4), 282–296.

Raju, K. C. B. 1998. Importance of recharging depleted aquifers: State of the art of artificial recharge in India. *Journal of the Geological Society of India*, 51, 429–454.

Rangarajan, R. and R. N. Athavale. 2000. Annual replenishable ground water potential of India—An estimate based on injected tritium studies. *Journal of Hydrology*, 234(1), 38–53.

Reddy, D. V., Nagabhushanam, P., Sukhija, B. S., Reddy, A. G. S., and P. L. Smedley. 2010. Fluoride dynamics in the granitic aquifer of the Wailapally watershed. *Chemical Geology*, 269(3–4), 278–289.

Reddy, T. N. and P. Raj. 1997. Hydrogeological conditions and optimum well discharges in granitic terrain in parts of Nalgonda district, Andhra Pradesh, India. *Journal of the Geological Society of India*, 49, 61–74.

Sahadevan, S. and D. Chandrasekharam. 2008. High fluoride groundwater of Karbi-Anglong district, Assam, Northeastern India: Source characterization. In *Groundwater for Sustainable Development: Problems, Perspectives and Challenges*. P. Bhattacharya, A.L. Ramanathan, A.B. Mukherjee, J. Bundschuh, D. Chandrasekharam, and A.K. Keshari, Eds. Taylor & Francis Group, London, pp. 301–310.

Sajil Kumar, P. J., Jegathambal, P., Nair, S., and E. J. James. 2015. Temperature and pH dependant geochemical modeling of fluoride mobilization in the groundwater of a crystalline aquifer in southern India. *Journal of Geochemical Exploration*, 156, 1–9.

Sarkar, M., Banerjee, A., Pramanick, P. P., and A. R. Sarkar. 2007. Design and operation of fixed bed laterite column for the removal of fluoride from water. *Chemical Engineering Journal*, 131(1–3), 329–335.

Satsangi, G. R., Lakhani, A., Khare, P., Singh, S.P., Kumari, S. S., and S. S. Srivastava. 2002. Measurement of ion concentrations in settled coarse particles and aerosols at a semiarid rural site in India. *Environment International*, 28(1–2), 1–7.

Saxena, V. K. and S. Ahmed. 2003. Inferring the chemical parameters for the dissolution of fluoride in groundwater. *Environmental Geology*, 43(6), 731–736.

Selvarajan, M., Bhattacharya, A. K., and F. W. T. Penning de Vries. 1995. Combined use of watershed, aquifer and crop simulation models to evaluate groundwater recharge through percolation ponds. *Agricultural Systems* 47(1), 1–24.

Shah, T. 2014. Towards a managed aquifer recharge strategy for Gujarat, India: An economist's dialogue with hydrogeologists. *Journal of Hydrology,* 518(Part A), 94–107.

Shortt, H. E., McRobert, G. T., Barrand, T. W., and A. S. M. Nayyar. 1937b. Endemic fluorosis in the Madras Presidency. *Indian Journal of Medical Research,* 25, 353–358.

Shortt, H. E., Pandit, C. G., and Raghavachari. 1937a. Endemic fluorosis in the Nellore District of South India. *Indian Gazette* 72, 396–398.

Singaraja, C., Chidambaram, S., Anandhan, P. et al. 2014. Geochemical evaluation of fluoride contamination groundwater in the Thoothukudi District of Tamil Nadu, India. *Applied Water Science,* 4(3), 241–250.

Singh, A. 2014. Groundwater resources management through the application of modelling: A review. *Science of the Total Environment,* 499(1), 414–423.

Singh, S., Parkash, B., and B. Bhosle. 2006. Pedogenic processes on the Ganga Deoha-Ghaghara Interfluve, Upper Gangetic Plains, India. *Quaternary International,* 159(1), 57–73.

Singh, Y. V., Singh, K. K., and S. K. Sharma. 2012. Influence of crop nutrition on grain yield, seed quality and water productivity under two rice cultivation systems. *Rice Science* 20(2), 129–138.

Srinivasa Rao, N. 1997. The occurrence and behaviour of fluoride in the groundwater of the Lower Vamsadhara River basin, India. *Hydrological Sciences Journal,* 42(6), 877–892.

Stille, L., Sineets, E., Wicke, B., Singh, R., and G. Singh. 2011. The economic performance of four (agro-) forestry systems on alkaline soils in the state of Haryana in India. *Energy for Sustainable Development,* 15(4), 388–397.

Susheela, A. K. 1999. Fluorosis management programme in India. *Current Science,* 77(10), 1250–1256.

Susheela, A. K., Kumar, A., Bhatnagar, M., and R. Bahadur. 1993. Prevalence of endemic fluorosis with gastrointestinal manifestations in people living in some north-Indian villages. *Fluoride* 26(2), 97–104.

Suvsankar, V., Rajkumar, S., Murugesh, S., and A. Darchen. 2012. Tamarind (*Tamarindus indica*) fruit shell carbon: A calcium-rich promising adsorbent for fluoride removal from groundwater. *Journal of Hazardous Material,* 225–226, 164–172.

Tang, O. Q., Du, J., Ma, H. H., Jiang, S. J. J., and X. J. Zhou. 2008. Fluoride and children's intelligence: A meta-analysis. *Biological Trace Element Research,* 126(1–3), 115–120.

Teotia, S. P. S. and M. Teotia. 1994. Dental fluorosis—A disorder of high fluoride and low dietary calcium interactions. *Fluoride* 27, 59–66.

Tirumalesh, K., Shivanna, K., and A. A. Jalihal. 2007. Isotope hydrochemical approach to understand fluoride release into groundwaters of Ilkal area, Bagalkot District, Karnataka, India. *Hydrogeology Journal,* 15(3), 589–598.

Tiwari, S., Chate, D. S., Bisht, D. S., Srivastava, M. K., and B. Padmanabhamurty, 2012. Rainwater chemistry in the North Western Himalayan Region, India. *Atmospheric Research,* 104–105, 128–138.

Tripathi, R. S. 2009. *Alkali Land Reclamation.* Mittal Publishers, New Delhi.

Turner, A. G. and H. Annamalai. 2012. Climate change and the South Asian summer monsoon. *Nature Climate Change* 2, 587–595.

UNICEF. 1999. Water Front, Issue 13.

UNICEF. 2014. UNICEF's Position on Water Fluoridation. www.nofluoride.com/Unicef_fluor.cfm. Accessed May 10, 2015.

Uphoff, N. 2012. Comment to "The system of rice intensification: Time for and empirical turn." *NJAS—Wageningen Journal of Life Science* 59(1–2), 53–60.

Viswanathan, G., Jaswanth, A., Gopalakrishnan, S., Sivailango, S., and G. Aditya. 2009. Determining optimal fluoride concentration in drinking water for fluoride endemic regions in South India. *Science of the Total Environment,* 407(19), 5298–5307.

Visvanathan, G. R. 1935. Annual report Madras. Indian council of Agricultural Research, New Delhi (quoted from Indian Institute of Science, 33A, 1, 1951).

Wright, V. P. and C. Peeters. 1989. Origins of some early Carboniferous calcrete fabrics revealed by cathodoluminiscence: Implications for interpreting the sites of calcrete formation. *Sedimentary Geology,* 65(3–4), 345–353.

23

Compound Wells for Skimming Freshwater from Fresh Saline Aquifers

Deepak Kashyap, K. Saravanan, M. E. E. Shalabey, and Anupma Sharma

CONTENTS

23.1 Introduction

Groundwater development provides assured water supply for agricultural, municipal, and industrial activities. The agricultural groundwater development not only augments the canal water supply, but also facilitates timely irrigation at critical times. However, several aquifers worldwide contain fresh usable groundwater only in a not-so-thick layer toward the top. This freshwater layer is underlain by a relatively thick layer of unusable saline water. Such *Fresh–Saline* aquifers (termed henceforth as F–S aquifers) occurring invariably in coastal regions are quite common in inland aquifers also. In coastal regions, groundwater salinity is mostly of marine origin such as salinity originating from marine transgressions, seawater intrusion, incidental flooding by seawater, and groundwater enriched in salts by seawater sprays. In inland areas and parts of coastal areas,

groundwater salinity is of terrestrial origin that can be attributed to natural or anthropogenic factors. Natural factors include groundwater enrichment in salts by evaporation at or near land surface or by dissolution of naturally occurring soluble minerals underground while anthropogenic factors include groundwater enrichment in salts by irrigation and subsurface waste disposal.

In India, F–S aquifers are frequently encountered in the fertile alluvial Indo-Gangetic Plains of North India, and deltaic formations on the east coast of India and Saurashtra coast of Gujarat. Globally the problem exists in several parts of the world, namely, basins of West and Central Asia, lowlands of South America and Europe, parts of North America, Northwestern Pacific margin, and eastern Australia. Most of the affected parts fall in the category of fertile agricultural areas, along the coast and in deltas.

The design challenge posed by the F–S aquifers is to arrive at such a well configuration that permits *skimming* of freshwater without drawing the saltwater into the pumped discharge. Such wells termed *skimming wells* have traditionally been partially penetrating wells tapping only the upper portion of the freshwater layer. The partially penetrating skimming wells, in spite of being simple to install and design, are not suitable if the freshwater thickness is less than 30 m (Asghar et al., 2002). Two other well systems (scavenger well, recirculation well) that seem to work well even when the freshwater layer is thin have been in vogue lately. These compound wells essentially reduce the rise of underlying saltwater toward the pumping well by way of additional innovative pumping/recharge.

Recognizing that hydraulically, the compound wells are essentially an extension of the partially penetrating well, the present chapter commences with a section on the partially penetrating well and goes on to build up the theory of the two compound wells.

23.2 Traditional Development of Fresh–Saline Aquifers: Partially Penetrating Wells

Traditionally, groundwater is developed in F–S aquifers through partially penetrating wells tapping only the freshwater layer and leaving out adequate cushion between the screen-bottom and the static interface between the freshwater and saltwater (Figure 23.1). The cushion is necessary because the interface tends to rise (upcone) as a consequence of pumping. With prolonged pumping, the upconed interface may reach the screen and the well may start yielding groundwater of enhanced salinity. When the pumping stops, the upconed heavier saltwater starts falling and over time may reach its initial (static) position.

The phenomenon of upconing is mainly attributed to advective transport of saltwater. However, apart from the upconing, there is some upward movement of saltwater due to dispersion also, which leads to formation of a dispersed interface instead of a sharp interface between freshwater and saltwater. The dispersed interface, in which the fluid concentration varies from that of freshwater to that of saltwater, enlarges during the upconing process and this significantly affects the salinity of pumped water. Schmorak and Mercado (1969) observed that wells become contaminated with saltwater long before undiluted saltwater reaches them, a phenomenon ascribed to the miscible nature of freshwater and saltwater.

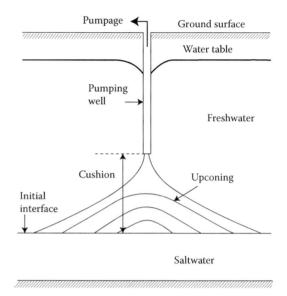

FIGURE 23.1
Saltwater upconing below a pumping well.

23.2.1 Simulation

The design of wells in freshwater aquifers is mostly focused on the discharge and drawdown only. However, the design of partially penetrating wells in F–S aquifers requires due attention to the upward movement of the interface during pumping and the downward movement in the recovery stage. Various strategies for quantification of this saltwater movement are discussed in the following paragraphs.

23.2.1.1 Sharp Interface Approach

Saltwater upconing below a partially penetrating well may be analyzed using the sharp interface method, which assumes freshwater and saltwater to be immiscible fluids. This method is applicable when the thickness of the dispersed interface is relatively small compared to the thickness of the aquifer. A dispersed interface is considered thin if it is less than one-third the thickness of the freshwater zone (Reilly and Goodman, 1985). Critical rise of a sharp interface is the rise to a location above which only an unstable cone can exist (Muskat, 1937).

Bear and Dagan (1964) presented the following analytical solution of the upconing in an anisotropic aquifer of infinite thickness.

$$\zeta(r,t) = \frac{Q}{2\pi(\Delta\gamma/\gamma_f)K_rD}\left[\frac{1}{(1+R^2)^{1/2}} - \frac{1}{[(1+T)^2+R^2]^{1/2}}\right]$$

(23.1)

Here

$$R = \frac{r}{D}\left(\frac{K_z}{K_r}\right)^{1/2} \quad \text{and} \quad T = \left(\frac{\Delta\gamma}{\gamma_f}\right)\frac{tK_z}{2\phi D} \qquad (23.2)$$

where $\zeta(r,t)$ = rise of interface above its position at a radial distance r from the center of well at time t; Q = time invariant discharge; γ_s, γ_f = saltwater and freshwater specific weights; $\Delta\gamma = \gamma_s - \gamma_f$; K_r, K_z = hydraulic conductivities in r and z directions, respectively; D = vertical distance between the initial position of interface and bottom of well; and ϕ = aquifer porosity. The major assumptions in this solution are that water is abstracted from a point sink, the aquifer is of infinite thickness, and upconing at the well center is small, that is, $\zeta(0,t) \leq 0.25D$.

Dagan and Bear (1968) presented another analytical solution for an aquifer of finite thickness. The solution, based on the assumptions of a point sink and small upconing [$\zeta(0,t) \leq 0.33D$], is as follows:

$$\zeta(r,t) = \frac{\gamma_f Q}{2\pi \Delta\gamma (K_r K_z)^{1/2}} \int_0^\infty \frac{\cosh[\lambda(A-D)]}{\sinh(\lambda A)}$$

$$\times \left\{ 1 - \exp\left(\frac{-\lambda K_z \Delta\gamma t}{\phi[\gamma_f \coth(\lambda A) + \gamma_s \coth(\lambda B)]}\right) \right\} J_0(\lambda r) d\lambda \qquad (23.3)$$

where A, B = initial freshwater and saltwater layer thicknesses; J_0 = Bessel's function of first kind and order zero; and λ = Fourier function.

23.2.1.2 Dispersed Interface Approach

With the advent of modern electronic computers, it is perfectly possible to simulate numerically the *total* upward saltwater movement below a pumping partially penetrating well accounting for both the advective and dispersive transport—and without making the assumptions inherent in the analytical solutions. The end-product from such a simulation may comprise the spatial distribution of salt concentration in a well's vicinity and the salt concentration in the pumped water at advancing times (Shalabey et al., 2006). The contour of 0.5 concentration may be deemed to be the average interface.

Shalabey (1991) developed a numerical model of vertical saltwater movement below a partially penetrating well. The model incorporating a finite difference-based solution of the coupled differential equations governing two-dimensional axis-symmetric flow of variable density fluid was subsequently applied to the pumping and recovery tests carried out by Schmorak and Mercado (1969) on wells in the Ashqelon region in the coastal plain

of Israel. The numerical solution reproduced the upconing and settlement of interface (0.5 isochlor) reasonably well uniformly at all values of ζ/D, even when the analytical solutions fail to reproduce the observed upconing (Figure 23.2). It also reproduced well the observed salt concentration in the pumped water (Figure 23.3) at advancing time. A parametric study on the model revealed that the upconing reduces as the screen length is increased. Further, it is found to decrease as the thickness of the saltwater decreases.

23.3 Design Aspects

The design of a skimming partially penetrating well has to satisfy all the general requirements of well design. The additional design variables are the cushion, duration of a pumping spell, and finally the rest period between two pumping spells. Incorporation of these design variables in design would apparently require simulation of vertical saltwater movement either by sharp interface or dispersed interface solution. The corresponding design criteria may be stated as follows.

23.3.1 Sharp Interface Approach

The design criteria with this approach could be as follows:

1. The cushion should be large enough to keep the upconed interface adequately below the screen-bottom at the end of a pumping spell of the design duration.
2. Rest period between two pumping spells must be long enough to permit the interface to fall back to its static position before the commencement of the next spell.

These criteria can be implemented by invoking the analytical solutions (Equations 23.1 and 23.3) for the sharp interface upconing described earlier. The residual upconing subsequent to the closure of pumping can be computed through superposition. This approach may lead to overestimation of upconing in case the screen length is not small enough. However, this may still provide "safe" but conservative design of the pumping discharge. The other issue as discussed earlier is that the upconing beyond the threshold level (0.33 times the cushion; Equation 23.3) may be underestimated leading to "unsafe" design of discharge. However, it may be recalled (refer Figure 23.2) that beyond 0.33D the upconed interface may become unstable and rise rather quickly. As such, it may be desirable to restrict the permissible

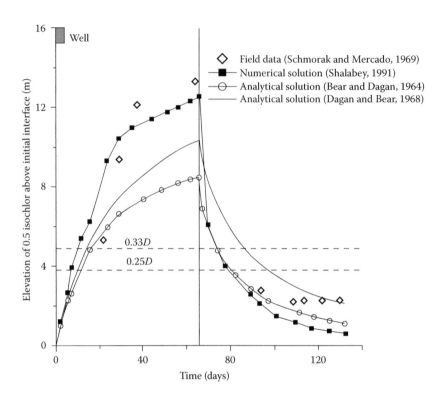

FIGURE 23.2
Upconing and settlement of dispersed interface versus time. (From Shalabey, M.E.E., Kashyap, D., and A. Sharma, *Journal of Hydrologic Engineering*, ASCE, 11(4), 306–318, 2006.)

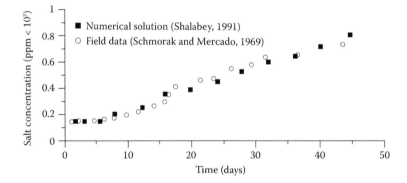

FIGURE 23.3
Illustration of increase in observed and simulated saltwater concentration in pumped water with time for given discharge in Ashqelon aquifer, Israel. (From Shalabey, M.E.E., Kashyap, D., and A. Sharma, *Journal of Hydrologic Engineering*, ASCE, 11(4), 306–318, 2006.)

upconing to 0.33*D*—for which the analytical solution (Equation 23.3) holds (the threshold limit for analytical solution given by Equation 23.1 is 0.25*D*). Thus, it may be concluded that the analytical solutions, in spite of rather restrictive assumptions, can be used for the designs.

23.3.2 Dispersed Interface Approach

As pointed out earlier, as a consequence of dispersion, the wells may start yielding saltwater long before the upconed interface encroaches on the screen. Thus, a more rigorous design requirement could be to ensure

that salt concentration in the pumped water remains below an acceptable level. Thus, invoking a dispersion interface model, the design criteria could be as follows:

1. The cushion should be large enough to keep the salt concentration in the pumped water below the permissible value at the end of a pumping spell of the design duration.
2. Rest period between two pumping spells must be long enough to permit the 0.5 isochlor to fall back to its static position before the commencement of the next spell.

This approach, though more credible theoretically, may be more difficult to implement because it would require setting up a numerical model which always would have its own uncertainties like numerical dispersion, poorly known dispersion parameter, etc. As such, the sharp interface approach involving an easy-to-implement analytical solution may be preferable. The chances of saltwater entry into the pumping well may be small enough when the upconing is restricted to 0.33*D*. Nevertheless, the dispersed interface models are useful for the advancement of scientific knowledge and also for providing insight. For example, the desirability of restricting the upconing to 0.33*D* is derived from the dispersed interface modeling (Shalabey et al., 2006).

23.3.3 Suitability

The partially penetrating skimming wells, although simple to design and install, are effective in limiting saltwater upconing if the freshwater layer is not thin, that is, the freshwater thickness is more than 30 m. In case of thin freshwater lenses, two other well systems, namely the scavenger well and recirculation well can be usefully employed by incorporating additional innovative pumping/recharge mechanisms in the partially penetrating skimming wells. Details are as provided in the following sections.

23.4 Scavenger Well System

The scavenger well system consists of two wells, namely, production well and scavenger well, located side by side—usually located in a single bore hole. The production well taps the freshwater zone while the scavenger well taps the saline water zone. These wells pump fresh and saline waters from the same site simultaneously without mixing, through two separate discharge systems as shown in Figure 23.4.

The concept of the scavenger was developed independently by different workers in different parts of the world in the 1960s. C.E. Jacob in 1965 took out a patent on the "Doublet Well" (Wickersham, 1977), designed to recover the upper fluid, while recirculating the lower fluid and keeping the interface as a flow line, a concept essentially the same as the scavenger well. Long (1965) studied feasibility of scavenger well application in Louisiana, USA. Zack and Candelario in 1984 reported the effectiveness of scavenger wells in coastal areas of Puerto Rico, where many wells were abandoned because they were inadvertently screened in the saltwater part of the aquifer. By installing scavenger wells in these abandoned wells, freshwater could be extracted from the thin freshwater lenses occurring at the surface of the water table. The scavenger wells were tested further in the lower Indus basin and have shown their usefulness in skimming of freshwater (Stoner and Bakiewicz, 1992). More than 400 scavenger wells have been installed in the lower Indus basin to tackle water logging and soil salinity problems. More recently, Alam and Olsthoorn (2014) have numerically shown that scavenging is the only long-term option to solve the longstanding problem of sustainable groundwater extraction and overcome the salinization problem in F–S aquifers of the Indus Basin, Pakistan. The efficacy of scavenger wells has also been investigated for stopping saltwater intrusion in Louisiana (Tsai, 2011).

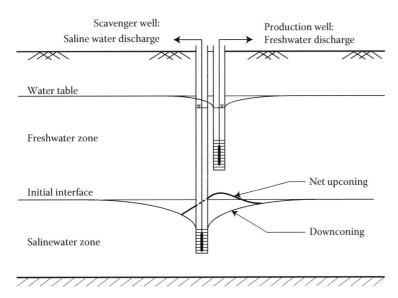

FIGURE 23.4
Saltwater upconing below a scavenger well system.

Basically, in a scavenger well system, the rise of saline water due to upconing caused by pumping from production well (Q_1) is countered by the downconing of the interface caused by pumping from a scavenger well (Q_2). The pumping rates from the two wells are adjusted in such a way that the underlying saline water does not intrude into the production well. While an under-pumping from the scavenger well may cause salinization of the production well, an over-pumping, apart from increasing the costs of pumping and the problem of saline water disposal, also creates downward gradient in the freshwater zone leading to wastage of freshwater.

23.4.1 Simulation of Scavenger Well System

The scavenging well system may be mathematically viewed as a well with two screens—first through which the production discharge is implemented, and the other through which the scavenging discharge occurs. As such, the system may be simulated invoking either the sharp interface approach, or the dispersed interface approach discussed earlier. Field and Critchley (1993) presented a solution for the upconing invoking Bear and Dagan's (1964) analytical solution for a partially penetrating well, and the principle of superposition.

$$\zeta_{net}(r,t) = \frac{Q_1}{2\pi(\Delta\gamma_1/\gamma_f)K_r D_1}\left[\frac{1}{(1+R_1'^2)^{1/2}} - \frac{1}{[(1+T_1')^2 + R_1'^2]^{1/2}}\right]$$

$$- \frac{Q_2}{2\pi(\Delta\gamma_2/\gamma_s)K_r D_2}\left[\frac{1}{(1+R_2'^2)^{1/2}} - \frac{1}{[(1+T_2')^2 + R_2'^2]^{1/2}}\right]$$

$$(23.4)$$

Here, $\zeta_{net}(r, t)$ = net interface position at a radial distance r from the center of the pumping well at a time t since the beginning of pumpage. The subscripts 1 and 2 stand for production and scavenger well screens, respectively, and R' and T' are dimensionless distance and time parameters defined as follows.

$$R_1' = \frac{r}{D_1}\left(\frac{K_z}{K_r}\right)^{1/2} \quad R_2' = \frac{r}{D_2}\left(\frac{K_z}{K_r}\right)^{1/2}$$

$$T_1' = \left(\frac{\Delta\gamma_1}{\gamma_f}\right)\frac{tK_z}{2\phi D_1} \quad T_2' = \left(\frac{\Delta\gamma_2}{\gamma_s}\right)\frac{tK_z}{2\phi D_2}$$

$$(23.5)$$

where $Q_{1,2}$ = time invariant discharge rates, K_r, K_z = radial and vertical hydraulic conductivities, D_1 = distance between the interface and the bottom of the production well screen, D_2 = distance between the interface and the top of the scavenger well screen, γ_f, γ_s = specific weights

of fresh and saline waters, $\Delta\gamma_1 = (\gamma_s - \gamma_f)$, $\Delta\gamma_2 = (\gamma_f - \gamma_s)$, and ϕ = aquifer porosity.

The analytical solution (Equation 23.4) is based on several assumptions discussed earlier. In case the assumptions are severely violated, the more rigorous approach accounting for dispersive transport of saltwater may be adopted for design of a scavenger well system (e.g., Saravanan et al., 2014). Such models based on the numerical solution of partial differential equations describing unsteady state two-dimensional axis-symmetric groundwater flow and transport in cylindrical coordinates can be used for studying the response of compound wells and arrive at the time variation of production well salinity and interface position.

23.4.2 An Insight into Scavenging Well Mechanics

Saravanan (2011) and Saravanan et al. (2014) conducted a detailed simulation study of the scavenging well system using the dispersed interface approach—revealing the mechanics of the system. The typical velocity fields are shown in Figure 23.5a and b. Figure 23.5a shows the velocity field with no scavenging discharge, that is, a partially penetrating well. For this kind of well operation, the vertical and lateral movement of the saline water is toward the production well that leads to upconing of saltwater. However, on introducing scavenging discharge, the vertical and lateral movement of the saline water toward the production well attenuates. In fact, as Q_2/Q_1 increases and scavenger discharge becomes equal to production well discharge (i.e., $Q_2/Q_1 = 1.0$) the vertically upward movement of the saline water is completely arrested across the initial interface and there is a minor downward movement of the freshwater toward the scavenger well (Figure 23.5b). These figures also reveal the necessity of optimal design of a scavenger well system to minimize both the freshwater wastage and excessive saltwater pumpage, which in turn leads to the problem of saline water disposal.

Figure 23.6 illustrates how the upconing attenuates as the scavenging Q_2/Q_1 is enhanced. For $Q_2/Q_1 \geq 0.6$, the upconing becomes insignificant. The time variation of the interface position below the well during pumping and recovery phases is shown in Figure 23.7. Figure 23.8 shows the variation in pumped water salinity at advancing times. The impact of introducing scavenging discharge in the form of significant reduction in the production well salinity is very much visible in this figure. It also reveals that as the scavenging discharge increases, the production well salinity decreases. The pumped water becomes practically salt-free as the scavenging discharge gets equal to the production well discharge.

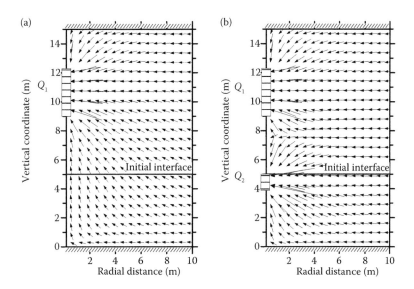

FIGURE 23.5
Velocity fields at the end of 24 hours for scavenger well system for $k_z/k_r = 0.3$ (a) $Q_2/Q_1 = 0.0$ and (b) $Q_2/Q_1 = 1.0$.

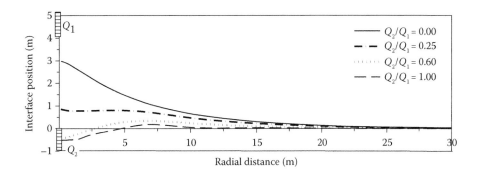

FIGURE 23.6
Impact of scavenger well on upconing: Interface position at the end of 24 hours.

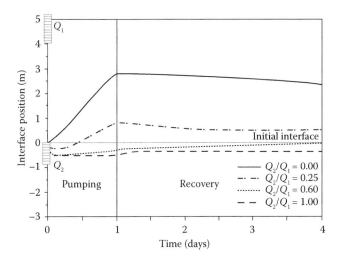

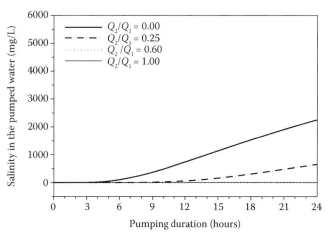

FIGURE 23.7
Scavenger well system: Time series of interface position during pumping and recovery ($r = r_w$).

FIGURE 23.8
Scavenger well system: Time series of production well salinity.

23.4.3 Sensitivity Analysis

The simulation study on scavenging well mechanics also determines the sensitivity of the system performance with respect to Q_2/Q_1, k_z/k_r ratios, and thickness of freshwater zone. Figure 23.9 shows three isochlors (0.1, 0.2, and 0.3) at the end of 24 hours. Taking salinity in the saline layer (C_s) to be 30,000 mg/L, these isochlors represent variation of absolute salinity from 3000 to 9000 mg/L. It may be seen that at $Q_2/Q_1 = 0$, these isochlors terminate into the production well screen leading to some salinity in the production well. However, as Q_2/Q_1 increases the isochlors are downturned away from the production well.

The production and scavenger well salinities for various levels of the scavenging discharge at different k_z/k_r values at the end of 24 hours are presented in Table 23.1. The table shows that the production well salinity decreases as the scavenging discharge increases or vertical anisotropy decreases. This decrease is apparently on account of the reduced drawdown and hence reduced upconing as the scavenging discharge increases or vertical anisotropy decreases. Similarly, scavenger well salinity also slightly decreases as the

scavenging discharge increases or vertical anisotropy decreases. This response is due to increase in the downward flow of freshwater toward the scavenger well and reduction in upconing as the scavenging discharge increases or vertical anisotropy decreases (refer to Figures 23.5 and 23.6).

23.4.4 Design Aspects

Saravanan et al. (2014) employed a numerical model to establish the optimum parameters for scavenger well design. It was established that optimal scavenger requirement (expressed as a percentage of the production discharge) varies from 30% to 140% with the permissible production well salinity as 2000 mg/L. The scavenging requirement reduces substantially as the permissible salinity level is increased to 3000 mg/L. At this level, the optimal scavenging requirement varies from 0.1% to 90% of the production discharge. In general, the optimal scavenging requirement is quite sensitive to production discharge, vertical anisotropy, and radial intrinsic permeability. It increases as the production discharge increases or vertical anisotropy increases or radial intrinsic permeability decreases.

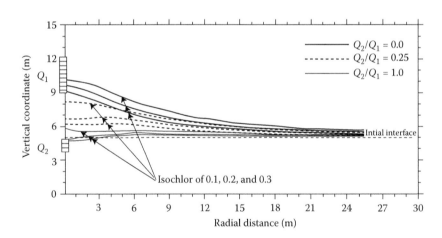

FIGURE 23.9
Scavenger well system: Dispersed interface isochlors (0.1, 0.2, and 0.3).

TABLE 23.1

Production and Scavenger Well Salinities for Varying Discharge Ratio and Vertical Anisotropy

	Production Well Salinity (C_1) in mg/L				Scavenger Well Salinity (C_2) in mg/L				Q_2/Q_1 [a]
	Q_2/Q_1				Q_2/Q_1				Required for
k_z/k_r	0	0.25	0.6	1.0	0	0.25	0.6	1.00	$C_1 = C^*$
1.0	5666.53	2628.75	154.89	3.22	–	28,436.1	23,962.0	20,656.2	0.33
0.6	4242.05	1770.45	67.52	2.69	–	27,716.2	23,411.7	20,560.3	0.22
0.3	2286.67	664.42	7.22	1.99	–	26,343.1	22,402.2	20,406.1	0.04

[a] Required discharge ratio with permissible production well salinity (C^*) = 2000 mg/L.

For low/moderate production discharge, the production well screen is optimally located quite close to the upper boundary of the saturated domain. However, for high production discharge, the production well screen needs to be lowered to accommodate the drawdown. In the numerical experiments performed by Saravanan et al. (2014), the production well screen was found to be optimally located at a depth varying from 12% to 26% of the initial freshwater thickness for both the permissible salinity levels (2000 mg/L and 3000 mg/L). The scavenger well screen was optimally located near the initial interface position irrespective of the magnitude of the production discharge.

23.5 Recirculation Well System

A recirculation well system comprises two closely spaced wells, a production well and a recirculation (recharge) well as shown in Figure 23.10. The screens of both the wells are located in the freshwater zone with an objective of reducing the effective upconing. The freshwater is pumped through the production well, and a portion of it is injected back into the aquifer through the recirculation well.

In 1963, Smith and Pirson applied this technique to reduce the mixing of saline water with oil. MacDonald and Kitanidis (1993) examined flow in an unconfined aquifer near a recirculation well, with emphasis on understanding the behavior of the free surface. Recirculation well system has also been used to remove the volatile organic compounds from groundwater aquifers (Lesage et al., 2003). A physical model of recirculation well system was developed by Sufi et al. (1998) and was used to calibrate a density dependent 3D finite

element numerical model (Sakr, 1995). Using MATLAB®-based numerical models, Alam and Olsthoorn (2014) have shown that recirculation wells can substantially delay salinization due to upconing in F–S aquifers in the Indus Basin of Pakistan.

In a recirculation well system, the rise of saline water due to pumping of freshwater through a production well is countered by the downconing of the interface caused by recharging through a recirculation well. The discharge of a production well and a freshwater recharge through a recirculation well are adjusted so that the underlying saline water may not intrude into the production well. The saline water disposal problem present in the scavenger well system is overcome in the recirculation well design. Still, a flow pattern from the deeper recharge screen to the shallower production well depth may get established within the aquifer causing some intermixing of saline water with the circulating freshwater.

23.5.1 Simulation of Recirculation Well System

The recirculation well system may be mathematically viewed as a single well with two screens—first through which the production discharge is implemented, and the other through which the recharge into the aquifer takes place. The time variation of production well salinity and the position of the interface in response to the recharge from the recirculation well can be simulated using a numerical model based on the numerical solution of partial differential equations describing unsteady state two-dimensional axis-symmetric groundwater flow and transport in cylindrical coordinates (Saravanan, 2011). Such a model would facilitate realistic simulation of the recirculation well system accounting for both advective and dispersive components of saltwater movement in an F–S aquifer. Solutions based on a sharp interface approach have not been attempted.

23.5.2 Flow Mechanics of a Recirculation Well System

Saravanan et al. (2007) and Saravanan (2011) conducted a detailed simulation study of the recirculation well system using the dispersed interface approach—revealing the mechanics of the system. In a recirculation well system, the recirculation well screen can be placed (1) close to the initial interface—Configuration I or (2) close to the production well screen—Configuration II. Taking Q_1 as the production discharge and Q_2 as the recirculation well recharge, the typical velocity fields for different values of discharge ratio Q_2/Q_1 are shown in Figure 23.11a–c for both Configurations I and II. Figure 23.11a shows the velocity field for $Q_2/Q_1 = 0$ (corresponds to zero recharge) wherein the velocity is almost vertical

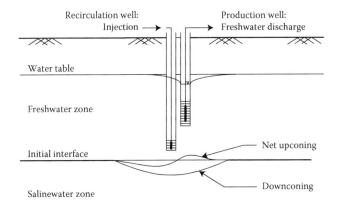

FIGURE 23.10
Saltwater upconing below a recirculation well system.

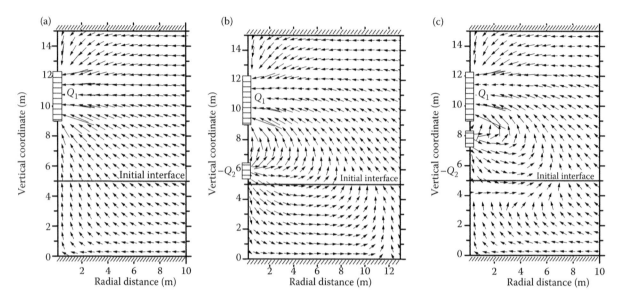

FIGURE 23.11
Velocity fields at the end of 24 hours for recirculation well system for $k_z/k_r = 0.3$ (a) $Q_2/Q_1 = 0.0$, (b) Configuration I: $Q_2/Q_1 = 0.5$, and (c) Configuration II: $Q_2/Q_1 = 0.5$.

just above and below the production well screen and almost horizontal across it. The vertical velocity below the production well is transporting the saline water toward the production well by advection. However, on introducing recharge, the vertical and lateral movement of the saline water toward the production well is cut down and instead a recirculating flow regime is created between the production and recirculation well. This prevents the advective transport of saline water toward the production well. The recirculating flow regime increases as the ratio Q_2/Q_1 increases. The regime of recirculating flow extends vertically below the initial interface in case the recirculation well is placed just above it (Figure 23.11b). In case the recirculation well is placed close to the production well (Figure 23.11c), a major portion of the recharge may be recaptured by the production well.

Figure 23.12a and b shows the model computed interface (i.e., 0.5 isochlor) at the end of 24 hours of recharging for the two positions of the recirculation well. The time variation of the interface position just below the recirculation well system during pumping and recovery phases is shown in Figure 23.13a and b for the two positions of the recirculation well. It may be seen that as the recirculation recharge (Q_2/Q_1) increases, the upconing reduces in both cases. The reduction is maximum in the vicinity of the well (when the interface position becomes even negative, i.e., *downconing*) and diminishes away from the well. It may further be seen that the reduction of the upconing on account of the recirculation is more pronounced in Configuration I (Figure 23.12a). This implies that as the recirculation screen is lowered and

placed just above the initial interface position, the recirculation get more efficient.

The variation of production well salinity (C_1) at advancing times is shown in Figure 23.14a and b for the two positions of the recirculation well. The figures reveal that as the ratio Q_2/Q_1 increases the production well salinity reduces. The reduction is quite significant as the ratio Q_2/Q_1 increases beyond 0.4 for both positions of the recirculation well. A comparison of corresponding C1 time series presented in Figure 23.14a and b reveals that generally attenuation of C_1 is more pronounced for Configuration II of the recirculation well. However, if the production well salinity increases beyond 2500 mg/L, then the salinity of recharging water (which is a portion of production discharge) also increases, resulting in ineffective control of upconing through Configuration II. In such a case, Configuration II yields slightly higher C_1. It is thus inferred that while Configuration I is more efficient in controlling the upconing, Configuration II generally controls the production well salinity (C_1) more efficiently.

23.5.3 Sensitivity Analysis

The simulation study discussed above determines the sensitivity of the system performance with respect to Q_2/Q_1, k_z/k_r ratios, and thickness of freshwater zone for both configurations I and II. Figure 23.15a and b shows three isochlors (0.1, 0.2, and 0.3) at the end of 24 hours. Taking $C_s = 30,000$ mg/L, these isochlors represent variation of absolute salinity from 3000 to 9000 mg/L. It may

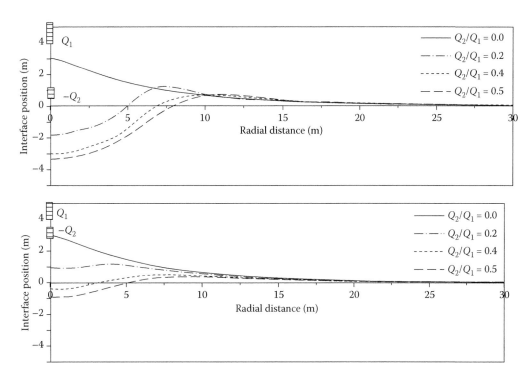

FIGURE 23.12
Impact of recirculation well system on upconing for $k_z/k_r = 0.3$: Interface position at the end of 24 hours (a) Configuration I and (b) Configuration II.

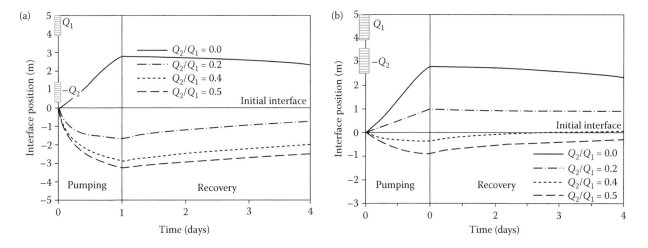

FIGURE 23.13
Recirculation well system: Time series of interface position during pumping and recovery $(r = r_w)$ for $k_z/k_r = 0.3$ (a) Configuration I and (b) Configuration II.

be seen that at $Q_2/Q_1 = 0$, these isochlors terminate into the production well screen leading to some salinity in the production well. However, as Q_2/Q_1 increases, the isochlors are downturned.

The salinity levels in the pumped water for various Q_2/Q_1 and k_z/k_r ratios are presented in Table 23.2. Further, assuming the permissible salinity in pumped water to be 2000 mg/L, the necessary recirculation recharge rates are also interpolated and presented in the table. This salinity level is suitable for irrigating salt-sensitive

crops (Ayers and Westcot, 1985). It may be seen that the placement of the recirculation screen in Configuration II leads to a lower requirement of recirculation recharge.

The scavenging discharge/recirculating recharge effectively reduces the vertically upward velocity component near the interface and hence reduces the resultant upconing. Irrespective of well configuration and discharge/recharge ratio (Q_2/Q_1), increase in k_z/k_r increases the upconing and consequently leads to more salinity in the production well. The upconing and production well

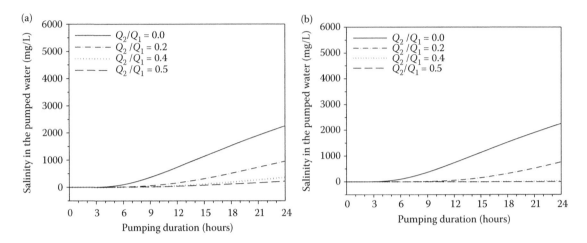

FIGURE 23.14
Recirculation well system: Time series of production well salinity for $k_z/k_r = 0.3$ (a) Configuration I and (b) Configuration II.

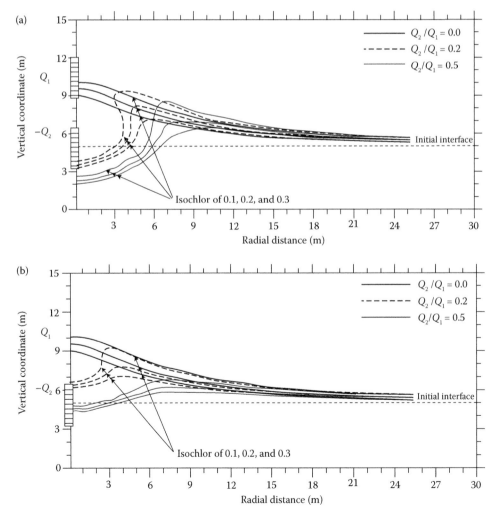

FIGURE 23.15
Recirculation well system: Dispersed interface isochlors (0.1, 0.2, and 0.3) for $k_z/k_r = 0.3$ (a) Configuration I and (b) Configuration II.

TABLE 23.2

Production Well Salinity for Varying Recharge Ratio and Vertical Anisotropy

Recirculation Well Position	k_z/k_r	Production Well Salinity (C_1) in mg/L				Q_2/Q_1[a] Required for $C_1 = C^*$
		Q_2/Q_1				
		0	0.2	0.4	0.5	
Close to the initial interface (Configuration I)	1.0	5666.53	4107.85	2698.5	2122.2	0.52
	0.6	4242.05	2620.27	1435.32	1026.78	0.30
	0.3	2286.67	968.07	369.01	227.4	0.04
Close to the production well (Configuration II)	1.0	5666.53	4377.96	2138.33	1113.27	0.41
	0.6	4242.05	2677.1	738.91	256.68	0.27
	0.3	2286.67	774.42	45.27	5.31	0.04

[a] Required recharge ratio with permissible production well salinity (C^*) = 2000 mg/L.

salinity are more sensitive to the ratio Q_2/Q_1. As Q_2/Q_1 increases, the upconing reduces and even completely vanishes at large Q_2/Q_1. However, in a scavenger well system, an excessively large Q_2/Q_1 can cause downconing, which is due to flow of freshwater toward the scavenger well. Similarly, in a recirculation well system excessively large Q_2/Q_1 may cause downconing, which is due to significant mixing of freshwater with saline water. The excessively large Q_2/Q_1 causes wastage of freshwater and increases the overall pumping cost.

23.5.4 Design Aspects

Placing the recirculation well screen close to the initial position of the interface is effective in controlling the upconing. However, this screen position leads to widening of the dispersed interface causing enhanced production well salinity. On the other hand, placement of the recirculation well screen close to the production well, though not so effective in controlling the upconing, restricts the dispersed interface more effectively and hence reduces the production well salinity more significantly.

Numerical experiments have shown that the vertically upward movement of the saline water toward the production well screen is attenuated with the production well salinity reduced by 60% by a recirculation recharge equaling 40% of the production discharge while placing the recirculation well screen close to the production well. However, when the recirculation well screen is placed close to the interface, a recirculation recharge equaling 50% of the production well discharge is found to show similar attenuation/reduction.

23.6 Conclusion

Freshwater is underlain by saline water in many aquifers worldwide. In case of thin freshwater lenses, the compound wells can prove to be more effective than the traditional partially penetrating skimming well in controlling the saltwater upconing and pumping water of permissible salinity. Depending on the existing field conditions at a given site, the compound wells may be developed as a scavenger well or a recirculation well system. However, excessive scavenging discharge/recirculation recharge could easily be counterproductive. This calls for a credible simulation of the flow/transport mechanisms that attenuate the upconing, and evolving optimal designs.

References

Alam, N. and T.N. Olsthoorn. 2014. Punjab scavenger wells for sustainable additional groundwater irrigation. *Agricultural Water Management*, 138, 55–67.

Asghar, M.N., Prathapar, S.A., and M.S. Shafique. 2002. Extracting relatively-fresh groundwater from aquifers underlain by salty groundwater. *Agricultural Water Management*, 52, 119–137.

Ayers, R.S. and D.W. Westcot. 1985. Water Quality for Agriculture, Food and Agriculture Organization of the United Nations. Irrigation and Drainage Paper No. 29, Rev. 1, Rome.

Bear, J. and G. Dagan. 1964. Some exact solutions of interface problems by means of the hodograph method. *Journal of Geophysical Research*, 69, 1563–1572.

Dagan, G. and J. Bear. 1968. Solving the problem of local interface upconing in a coastal aquifer by the method of small perturbations. *Journal of Hydraulic Research*, 6(1), 15–44.

Field, M. and M. Critchley. 1993. *Multi Level Pumping Wells as a Means for Remediating a Contaminated Coastal Aquifer*. U.S. Environmental Protection Agency, Washington, D.C., EPA/600/R-93/209 (NTIS PB95138301).

Lesage, S., Brown, S., Millar, K., and H. Steer. 2003. Simulation of a ground water recirculation well with a dual-column laboratory setup. *Ground Water Monitoring and Remediation*, 23(2), 102–110.

Long, R.A. 1965. Feasibility of a scavenger well system as a solution to the problem of vertical salt water encroachment, Louisiana Geological Survey, Water Resources Pamphlet Number 15.

Macdonald, T.R. and P.K. Kitanidis, 1993. Modeling the free-surface of an unconfined aquifer near a recirculation well. *Ground Water*, 31(5), 774–780.

Muskat, M. 1937. *The Flow of Homogeneous Fluids through Porous Media*. McGraw-Hill, Inc., New York.

Reilly, T. E. and A. S. Goodman. 1985. Quantitative analysis of saltwater–freshwater relationships in ground-water systems—A historical perspective. *Journal of Hydrology*, 80, 125–160.

Sakr, S. 1995. *Variable Density Groundwater Contaminant Transport Modeling (VDGWTRN)*. Department of Civil Engineering, Colorado State University, Colorado.

Saravanan, K. 2011. Modeling and design of scavenging skimming well system in fresh–saline aquifers, PhD Thesis (Unpublished), IIT Roorkee, Roorkee, India.

Saravanan, K., Kashyap, D., and A. Sharma. 2007. Numerical simulation of flow and mass transport in the vicinity of recirculation wells. *3rd Int. Groundwater Conference (IGS-2007) on Water, Environment & Agriculture—Present Problems & Future Challenges*, Feb. 7–10, 2007, Tamil Nadu Agricultural University, Coimbatore.

Saravanan, K., Kashyap, D., and A. Sharma, 2014. Model assisted design of scavenger well system. *Journal of Hydrology*, 510, 313–324.

Schmorak, S. and A. Mercado. 1969. Upconing of fresh water seawater interface below pumping wells-field study. *Water Resource Research*, 5(6), 1290–1311.

Shalabey, M.E.E. 1991. A study on saltwater transport towards a pumping well. Ph.D. thesis (Unpublished), University of Roorkee, Roorkee, India.

Shalabey, M.E.E., Kashyap, D., and A. Sharma. 2006. Numerical model of saltwater transport toward a pumping well. *Journal of Hydrologic Engineering, ASCE*, 11(4), 306–318.

Smith, C.E. and S.J. Pirson. 1963. Water coning control in oil wells by fluid injection. *Society Petroleum Engineers Journal*, 158, 314–326.

Stoner, R.F. and W. Bakiewicz. 1992. Scavenger wells. 1 Historic Development. *Proceedings of the 12th Saltwater Intrusion Meeting*, Barcelona, Spain, 545–556.

Sufi, A.B., Latif, M., and G.V. Skogerboe. 1998. Simulating skimming well techniques for sustainable exploitation of groundwater. *Irrigation and Drainage Systems*, 12, 203–226.

Tsai, F.T.-C. 2011. Stop saltwater intrusion toward water wells using scavenger wells, *Proc. World Environmental and Water Resources Congress*, Palm Springs, California, May 22–26, 2011, 904–913.

Wickersham, G. 1977. Review of C. E. Jacob's doublet well. *Ground Water*, 15(5), 344–347.

Zack, A. and R.M. Candelario, 1984. A hydraulic technique for designing scavenger-production well couples to withdraw freshwater from aquifers containing saline water. Final Technical Report, Project A-075-PR.

24

An Overview of Electrochemical Processes for Purification of Water Contaminated by Agricultural Activities

S. Vasudevan and Mu. Naushad

CONTENTS

24.1 Introduction

Water, the generous gift of nature, is sure to become scarce unless the ever-growing population is enlightened enough in handling the increasing stress and to avoid the crisis due to the expanding demand on this precious commodity. Management of water and its resources by conservation and its judicious use help to preserve the available water. Even then, whether it is from surface or underground sources, it has become impossible to obtain good quality water for human consumption. Thus, the dwindling quantity and lessening quality of water require effective steps to be taken urgently for the sustenance of the living beings of today and tomorrow.

"What does this glass of water contain?" worries one before he starts quenching his thirst. The extent of water contamination is so much and so varied that organic, inorganic, and biological impurities are present in the water due to natural as well as induced reasons. The responsibility fell on scientists and engineers to provide appropriate technologies not only to prevent pollution at the source itself but also to treat the user end.

Agriculture is the single largest user of fresh water on a global basis despite more rapid growth in other sectors' usage. Agriculture activity is the one of the main route causes for contamination of surface water and groundwater through leaching and surface runoff; for example, sediments from eroded soil, salts from irrigation, nutrients from organic and mineral fertilizers, pathogens from livestock, and chemicals from pesticides and farm machinery. In India, 70% of the total population depends on agriculture. India ranks second in world farming; irrigation has acquired increasing importance in agriculture worldwide.

Coping with these developments requires various tactics to overcome the water shortage and satisfy the needs of all. The main activity in this direction is to decrease the pollution level of discharged effluents and treatment of contaminated water to acceptable quality. Conventionally, water purification methods involve physicochemical separation of contaminants or biological treatment. The physicochemical methods aim at shifting the pollutants (landfill), concentrate the pollutants (adsorption), transfer the pollutant to another medium (air stripping), or cause secondary pollution (chemical precipitation leading to sludge). Biological technologies require a narrow range of operating conditions.

In this context, the role of electrochemistry to environmental applications is expanding due to the characteristics of electrochemical processes, namely, versatility, energy efficiency, environmental compatibility, cost effectiveness, etc. (Martínez-Huitle and Brillas 2008, 2009; Butler et al. 2011; Kabdasli et al. 2012; Kuokkanen et al. 2013; Vasudevan et al. 2013; Sirés et al. 2014; Vasudevan and Oturan 2014). The growth of electrocatalysts and electrochemical engineering in the past two decades has made a total revolution in the electrode and cell design technologies that are particularly important for water and effluent treatment. In this review, brief accounts of a few successful electrochemical technologies are presented.

24.2 Sources of Contamination through Agricultural Activities

In most countries, water has been polluted by fertilizers and pesticides used during agricultural activities. This pollution is due to excess use of fertilizers and pesticides in agricultural activities to enhance productivity due to rapid population increase. Development of technology threatens the groundwater and surface water on a large scale. This shows there is an obvious risk for humans in the future (Barcelona and Naymik 1984; Agrawal 1999; Novotny 1999; Jimoh and Ayodaeji 2003; Pimentel et al. 2004; Divya and Belagali 2012).

The use of toxic pesticides and herbicides in agricultural lands to manage pest problems has become a common practice around the world. Pesticide/herbicide leaching occurs when pesticides mix with water and move through the soil, ultimately contaminating groundwater. The leaching of pesticide/herbicide is not only originated from applied agricultural lands but also from pesticide mixing areas, pesticide application machinery washing sites, or disposal areas. Leaching happens, most likely, by water soluble pesticides. Their presence in water causes toxicity toward humans and animals. The short-term impacts are headaches and nausea and chronic impacts are cancer, reproductive harm, and endocrine disruption. In addition, acute dangers such as nerve, skin, and eye irritation and damage, headaches, dizziness, nausea, fatigue, and systemic poisoning can sometimes be dramatic, and even occasionally fatal. Children are particularly vulnerable to these health effects due to their less-developed metabolism and the on-going maturation of their organ systems (Arcury et al. 2006; Anderson et al. 2011; Sugeng et al. 2013; Kamaraj et al. 2014a,b).

Nitrate (NO_3^-) leaching is a common issue in most irrigated agricultural regions of the world, especially where the crops/grass with high requirements of water and nitrogen (N) tend to increase the potential risk of NO_3^- pollution to the groundwater. Nitrate is very mobile in the environment and cannot be held up well in any type of material. The strong oxidizing power of nitrates causes detrimental effects in the environment. In humans, increasing NO_3^- concentrations in drinking water causes two adverse health effects: induction of "blue-baby syndrome" (methemoglobinemia), especially in infants, and the potential formation of carcinogenic nitrosamines. Second, formation of nitrosoamines from nitrite can give rise to cancers of the digestive tract since nitrosamines are the most efficacious carcinogens in mammals (Vasudevan et al. 2010a,b; Abirami et al. 2011; Mook et al. 2012).

The phosphate applied to agricultural land (via synthetic fertilizers, composts, manures, biosolids, etc.) can provide valuable plant nutrients. However, if not managed correctly, excess phosphate can have negative environmental consequences. Eutrophication is one of the main problems encountered due to high levels of phosphate in water (Vasudevan et al. 2008, 2009a,b; Vasudevan and Lakshmi 2012).

To increase the nutrient content in agricultural lands, manures and biosolids are used. But the manures and biosolids contain not only nutrients (namely, carbon, nitrogen, and phosphorus), but they may also contain organic contaminants, including pharmaceuticals and personal care products (PPCPs). Heavy metal (e.g., lead, cadmium, arsenic, and mercury) contamination is also possible due to the usage of fertilizers, organic wastes such as manures, and industrial byproduct wastes in agricultural lands.

24.3 Electrochemical Water Treatment Processes

Electrochemical technologies, for the removal of contaminants due to agricultural activities, can be broadly divided two categories, namely, physical (e.g., electrodialysis, electrocoagulation, and electroflotation) and chemical involving direct reaction at the electrode or an indirect reaction with a reagent generated electrolytically (Grimm et al. 1998; Martínez-Huitle and Brillas 2008; Martínez-Huitle and Brillas 2009; Butler et al. 2011; Kabdasli et al. 2012; Kuokkanen et al. 2013; Vasudevan et al. 2013; Chaplin 2014; Sirés et al. 2014; Vasudevan and Oturan 2014).

24.3.1 Electrodialysis

Electrodialysis (ED) is a membrane-based process involving the separation and concentration of electrolytes (dissolved salt concentration) by the electromigration of the species through ion exchange membranes by means of direct current (DC) (Banasiak and Schafer 2009). The process uses a driving force to transfer ionic species from the source water through cathode (positively charged ions) and anode (negatively charged ions) to a concentrate wastewater stream, creating a more dilute stream. Anion and cation exchange membranes are arranged alternatively between two electrodes, which enable ions to pass through one type of membrane while the other membrane blocks the passage thus creating a set of "concentrating" cells and "dilution/purification" cells. Cellulose acetate membranes are conventionally used in ED stacks and polysulfone, polyacrylonitrie, and polyfluorosulphonic acid type membranes are some of the new membranes presently in use. The water to be purified is circulated through the dilution compartment and the dissolved salts are concentrated in the stream through the concentration compartment.

In the present context, this process is very useful to remove anions and cations from water contaminated by agricultural activities (Banasiak and Schafer 2009; Gherasim et al. 2014). Further, the electrodialysis process has important applications: (1) production and supply of potable water from brackish water (with typical salt concentration of 1000–3000 mg/L, (2) recovery and recycling of metal ions such as in plating industry effluents for water recycling, (3) salt removal from industrial effluents for water recycling, (4) removing electrolytes from nonelectrolytes, and (5) salt splitting, for example, in the recovery of NaOH and H_2SO_4 (this requires one pair of electrodes for each pair of membrane). Brackish water ED systems can operate over wide pH ranges and up to 43°C (depending on the membrane type) at a typical energy consumption of 1.6–2.6 kWh/m³ of purified water.

24.3.2 Electrocoagulation and Electroflotation

Electrocoagulation is the process of destabilizing suspended, emulsified, or dissolved contaminants in an aqueous medium by introducing an electric current

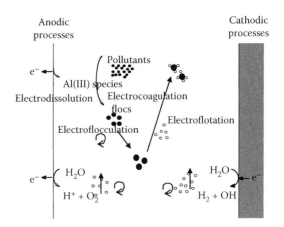

FIGURE 24.1
The working principle of the electrocoagulation process.

into the medium (Figure 24.1). In other words, electrocoagulation is the electrochemical production of destabilization agents that brings about neutralization of electric charge for removing pollutant. Once charged, the particles bond together like small magnets to form a mass. In its simplest form, an electrocoagulation unit is made up of an electrolytic cell with one anode and one cathode. The conductive metal plates/rods are commonly known as sacrificial electrodes, made up of the same or different materials used for the anode and cathode. The materials employed in electrocoagulation processes are usually aluminum or iron. Recently, magnesium and zinc are also applied as sacrificial anodes (Vasudevan and Oturan 2013; Vasudevan et al. 2010a,b, 2013) for electrocoagulation. Electroflotation is a simple process that floats pollutants to the surface of a water body by tiny bubbles of hydrogen and oxygen gases generated from water electrolysis. Dimensionally stable anodes (Ti/IrO_2, Ti/RuO_2, Ti/TiO_2-RuO_2, and Ti/Pt) are also used. Recently, highly effective $Ti/IrO_x-Sb_2O_5-SnO_2$ is used and its life is predicted to be about 20 years.

This process involves applying an electric current to sacrificial electrodes inside an electrolytic cell where the current generates a coagulating agent and gas bubbles. For example, in the case of aluminum, magnesium, zinc, or iron electrodes, aluminum in the form of Al^{3+}, magnesium in the form of Mg^{2+}, zinc in the form of Zn^{2+}, and iron in the form of Fe^{2+} will be produced by anodic dissolution during electrocoagulation process. Then, Al^{3+}, Mg^{2+}, Zn^{2+}, and Fe^{2+} react immediately with hydroxide ions in solution to produce $Al(OH)_3$, $Mg(OH)_2$, $Zn(OH)_2$, and $Fe(OH)_2$, respectively. The metallic hydroxide was produced up to a sufficient concentration to initiate polymerization or condensation reactions. The polymerization reactions for aluminum and ferrous hydroxides are as follows:

$$Al(OH)_3 + Al(OH)_3 \rightarrow (OH)_2Al - O - Al(OH)_2 + H_2 \quad (24.1)$$

$$Fe(OH)_2 + Fe(OH)_2 \rightarrow (OH)Fe - O - Fe(OH) + H_2O \quad (24.2)$$

The polymeric complexes $[Al_2(O)(OH)_4]$ and $[Fe_2(O)(OH)_2]$ allow removal of pollutants from water mainly by either adsorption, surface complexation, or coprecipitation. The following reaction explains how the pollutant is removed by surface complexation:

$$M - H_{(aq)} + (HO)OFe_{(S)} \rightarrow M - OFe_{(S)} + H_2O \quad (24.3)$$

where M represents the metal pollutant, which acts as a ligand. In general, polymeric aluminum or iron complexes can have both positive and negative charges capable of attracting the opposite charges of pollutants and remove them from water. For example, boron will be removed as follows:

$$Al^{3+} + BO_2^- + 2OH^- + nH_2O \leftrightarrow Al(OH)_2BO_2 . nH_2O \quad (24.4)$$

This process can also be used for the removal of nitrate, phosphate, cadmium, lead, etc. Further, the electrocoagulation process has been widely used in textile wastewater, refractory oily wastewater, urban wastewater, restaurant wastewater, wasters loaded with phenol, and laundry effluent. Likewise, it has been used to remove color, clay, toxic metals, inorganic metals, and drinking water treatment (Vasudevan and Oturan 2014).

24.3.3 Direct Electrochemical Processes

In an electrochemical cell, the anodic reaction is oxidation (dissolution in the case of soluble anodes) and the cathodic reaction is reduction. Under appropriate conditions, these two reactions can be effectively exploited to remove pollutants from water or effluents (Comninellis 1994).

24.3.3.1 Anodic Processes

Pesticide and herbicide destruction is an important example of direct electrochemical oxidation used in agricultural lands. The process is more suitable for water with high concentrations of organic impurities. The oxidation of Cr^{3+} to Cr^{6+} is commercially adopted for the recycling of the pharmaceutical, electronic, and aerospace industries. The oxidation is carried out at a lead-dioxide anode and continuous flow cells with Nafion membranes to separate the anolyte and catholyte in use.

24.3.3.2 Cathodic Processes

The permissible metal ion concentration in water is in the range of 0.02–5 mg/L, whereas water from various sources contains unacceptable levels of metallic impurities. The reduction of these metal ions to their elemental state by electrochemical deposition is a convenient and cheap method for purification. The metals are deposited in the strippable form so that they can be recovered as a saleable product. For solutions of fairly low concentrations, electrolyzers with two-dimensional and three-dimensional electrode structures and a variety of mass transfer enhancement methods have been developed and successfully operated. Thus, processes for removal and recovery of copper, lead, mercury, zinc, silver, chromium, etc. are in practice.

24.3.4 Indirect Electrochemical Processes

Direct electrochemical oxidation or reduction may not be possible in certain cases due to limitations, namely, the required potential for driving the desired reaction. Under these circumstances it is advisable to identify a suitable reversible or irreversible redox system that can be efficiently generated and allowed to react with the pollutant (Panizza and Cerisola 2009). Examples of such redox pairs are Fe^{2+}/Fe^{3+}, Mn^{2+}/Mn^{3+}, $Ag^+/^{2+}$, Cl^-/ClO^-, etc. These reactions can be made to occur *in situ* in the electrolytic cell itself by maintaining a low concentration of the reactant redox intermediate or externally, by generating high concentration of the reactant in an electrochemical cell and reacting in an external reactor with the effluent to be treated.

The generation of short-lived intermediates such as hydroxyl radical, peroxide ion, and hydro-peroxide radical is another area of electrochemical treatment for the destruction of toxic chemicals. The oxidation of phenols and discoloration of dye effluents at tin-oxide electrode involves the generation of OH^-. This is produced by reduction of hydrogen peroxide with Fe^{2+}. With this system, oxidation of substituted benzene derivatives, formaldehyde, and cyanide has been effectively carried out.

Apart from these, electrochemical pH adjustment is an efficient and cheap method of neutralization of acid and alkaline waters and effluents, which simultaneously help in precipitating certain metal ions such as iron, calcium, chromium, magnesium, etc., without the need for the addition of any external chemicals.

24.3.5 Electrochemical Disinfection of Water

In most places, water disinfection was used for water discharged from agricultural lands. Usually disinfection is carried out with chlorine, bleaching powder, or hypochlorite. The basic chemical, chlorine, itself is an electrochemical product from which the others are produced. Even if chlorination is done hypochlorite is the effective reactant in most instances. Electrolytic chlorination or hypochlorination is a simple, cost-effective, and onsite process that eliminates the transport and storage of gaseous chlorine. User-friendly portable chlorine and hypochlorite generators are available for water and effluents treatment (Panizza and Cerisola 2009).

Recent findings indicate that generation of toxic dioxin and AOX due to the usage of chlorine or hypochlorite and alternatively chlorine dioxide, ozone, and hydrogen peroxide are recommended. Chlorine dioxide generators employing a membrane cell approach has been developed in which a sodium chlorate solution as anolyte is electrolyzed in a cell with Nafion membrane. The anolyte becomes acidic and the chlorate is decomposed to give chlorine dioxide. Electrochemical ozone generator employing a proton exchange membrane (PEM) electolyzer generates ozone at an efficiency of over 15% and at a concentration in the range of 60–75 mg/L, which is much higher than the ozone generator based on the air oxidation of electric discharge. This high concentration of ozone is capable of destructing any pollutants and bacteria and combined with ultraviolet radiation even destroying viruses and pathogens. Hydrogen peroxide of 2%–3% concentration can be conveniently produced onsite using a trickle-bed electrolyzer. Oxygen is reduced under alkaline conditions at a highly catalytic carbon bed electrode over which the electrolyte trickles down generating hydrogen peroxide.

24.3.6 Photoelectrochemical Methods

In recent years, photoelectrochemistry has led to a new and interesting possibility for treatment of pollutants from wastewater. In this case, suspensions of semiconductor particles (mostly TiO_2) can be used to harness the light with production of electrons and holes in the solid, which can destroy pollutants by means of reduction and oxidation, respectively. In this way, water containing organic, inorganic, or microbiological pollutants can be effectively treated (Fresno et al. 2014).

24.3.7 Electrochemical Advanced Oxidation Processes (EAOPs)

In the case of agricultural wastewater treatment, chemical oxidation is often necessary to remove organic matter (biodegradable or not) that consumes oxygen dissolved in water. The oxidations by ozone or hydrogen peroxide (H_2O_2) are methods used as complement or in competition with the activated carbon filtration or nanofiltration. But in some cases, the conventional oxidation is still inadequate and remains inefficient. Moreover, the conventional oxidation treatments as well as biological treatments prove ineffective against certain types of organic micropollutants. To eliminate these kind of pollutants from water, more powerful processes, namely advanced oxidation processes (AOPs), have been developed. The AOPs are based on the *in situ* generation of strong oxidizing agent hydroxyl radicals (•OH) and their high oxidation power. The use of •OH in the treatment/remediation of contaminated water is justified by a number of advantages: (1) they are not toxic (very short lifetime); (2) they are simple to produce and use; (3) they are very efficient to remove organic pollutants: the main feature of these powerful species is their ability to transform refractory inorganic compounds to biodegradable products; (4) they are not corrosive to the equipment; and (5) they do not induce secondary pollution: the final products of oxidation are CO_2 and H_2O, and inorganic ions (mineralization).

Electrochemistry constitutes one of the clean and effective ways to produce *in situ* hydroxyl radical (•OH), a highly strong oxidizing agent of organic matter in waters. Due to its very high standard oxidation power (E°(•OH/H_2O) = 2.80 V/SHE), this radical species is able to react nonselectively with organic or organometallic pollutants yielding dehydrogenated or hydroxylated derivatives, which can be in turn completely mineralized, that is, converted into CO_2, water, and inorganic ions.

Recently, the electrochemical advanced oxidation processes (EAOPs) have received great attention by their environmental safety and compatibility (operating at mild conditions), versatility, high efficiency, and amenability of automation (Brillas et al. 2009).

24.4 Electrochemical Technologies Developed at CSIR-Central Electrochemical Research Institute (CECRI), Karaikudi, Tamil Nadu, India

The CSIR-Central Electrochemical Research Institute with its vast experience in the field of electrochemical technologies has developed a number of processes mentioned above. For example, electrochemical removal of nitrate, electrochemical removal of inorganic ions like phosphate, boron, arsenic, lead, chromium, etc., electrochemical removal of organic contaminants like pesticides, herbicides, etc.

24.4.1 Electrochemical Dearsenator

A 2A and 20A capacity electrochemical dearsenator employing aluminum alloy/magnesium alloy/mild steel anodes and GI/mild steel cathodes and which can treat 2.0/40 L of drinking water per hour was developed. The unit will remove both As(III) and As(V) from 3000 to 20 ppb. The electrochemical unit can be scaled up to any capacity if needed. The dearsenator was used with solar energy systems. In this electrochemical dearsenator, mild steel and galvanized iron is used as anode and cathode, respectively. The reactions are as follows:

At the cathode:

$$2H_2O + 2e^- \rightarrow H_2(g) + 2OH^- \qquad (24.5)$$

At the anode

$$Fe \rightarrow Fe^{2+} + 2e^- \qquad (24.6)$$

$$Fe^{2+} + 2H_2O \rightarrow Fe(OH)_2 + H_2 \qquad (24.7)$$

$$Fe(OH)_2 \rightarrow Fe(OH)_3 \qquad (24.8)$$

$$As + Fe(OH)_3 \rightarrow Fe - As \text{ (complex)} + H_2O \qquad (24.9)$$

24.4.2 Electrochemical Defluoridator

This process works on the principle of electrochemical dissolution of aluminum alloy anode followed by coagulation and electroflotation. Based on the above principal, both domestic (20 L/h) and community (200 L/h) model electrochemical defluoridators are developed. These units will reduce fluoride content from 5 ppm to less than 1.5 ppm of fluoride from drinking water at a flow rate of 20 L/h. Aluminum hydroxide is generated in an electrochemical cell fitted with soluble aluminum alloy electrodes. When a DC current is applied, the anode dissolves and provides the required aluminum hydroxide as per the reactions (10)–(15) and that adsorbs and removes fluoride from the water. The process provides effective defluoridation media that has selective high efficiency for fluoride removal. The total system is simple in operation and maintenance. The reaction mechanisms are as follows:

At the cathode:

$$2H_2O + 2e^- \rightarrow H_2(g) + 2OH^- \qquad (24.10)$$

At the anode:

$$Al \rightarrow Al^{3+} + 3e^- \qquad (24.11)$$

In the electrolyte:

$$Al^{3+}(aq) + 3H_2O \rightarrow Al(OH)_3 + 3H^+(aq) \qquad (24.12)$$

$$nAl(OH)_3 \rightarrow Al_n(OH)_{3n} \qquad (24.13)$$

$$Al^{3+} + 6F^- \rightarrow AlF_6^{3-} \qquad (24.14)$$

$$AlF_6^{3-} + 3Na^+ \rightarrow Na_3AlF_6 \qquad (24.15)$$

24.4.3 Electrochemical Removal of Nitrate

A prototype portable electrochemical nitrate removal unit was developed and this unit will remove the nitrate from 1000 mg/L to less than 40 mg/L with the acceptable level of nitrite and ammonia after treatment. The units will treat 40 L of drinking water per hour. This technology will remove nitrate to environmentally friendly nitrogen. The removal mechanism is as follows:

At cathode

$$NO_3^- + 6H_2O + 8e^- \rightarrow NH_3 + 9OH^- \qquad (24.16)$$

At anode

$$2Cl^- \rightarrow Cl_2 + 2e^- \qquad (24.17)$$

$$Cl_2 + H_2O \rightarrow HClO + HCl \qquad (24.18)$$

Net reaction

$$2NH_3 + 3HClO \rightarrow N_2 + 3HCl + 3H_2O \qquad (24.19)$$

24.4.4 Removal of Other Heavy Metals

Electrochemical removal of other inorganic ions like chromium, mercury, copper, boron, etc. is successfully completed at the laboratory scale level. The flow cell for domestic and community model is under development.

24.5 Conclusions

Electrochemical technologies are a vital and enabling discipline in many areas of environmental treatment, including (a) clean synthesis, (b) monitoring of process efficiency and pollutants, (c) removal of contaminants, (d) recycling of pollutants, (e) water sterilization, and (f) clean energy conversion. The field of environmental electrochemistry has witnessed many progresses and has demonstrated many successes, as evidenced by the increasing literature. In this review, a variety of selected electrochemical processes for environmental protection of water affected by agricultural activities has been presented through electrodialytic processes, electrocoagulation and electroflotation, anodic and cathodic processes, photoelectrochemical methods, and electrochemical advanced oxidation processes. Finally, the technologies developed in CSIR-CECRI, Karaikudi, India with respect to removal of contaminants due to agricultural activities are also discussed.

Acknowledgment

S. Vasudevan expresses his sincere thanks to Director, CSIR-CECRI, Karaikudi for his encouragement and permission to publish this review chapter.

References

Abirami, D., S. Vasudevan, and E. Florence. 2011. Nitrate reduction in water: Influence of the addition of a second metal on the performances of the Pd/CeO$_2$ catalyst. *Journal of Hazardous Material*, 185: 1412–1417.

Agrawal, G.D. 1999. Diffuse agricultural water pollution in India. *Water Science Technology*, 39: 33–47.

Anderson, LM, B.A. Diwan, N.T. Fear, and E. Roman. 2011. Critical windows of exposure for children's health: Cancer in human epidemiological studies and neoplasms in experimental animal models. *Environmental Health Perspectives*, 108: 573–594.

Arcury, T.A., S.A. Quandt, D.B. Barr, J.A. Hoppin, L. McCauley, and J.G. Grzywacz. 2006. Farmworker exposure to pesticides: Methodologic issues for the collection of comparable data. *Environmental Health Perspectives*, 114: 923–928.

Banasiak, L.J. and A.I. Schafer. 2009. Removal of boron, fluoride and nitrate by electrodialysis in the presence of organic matter. *Journal of Membrane Science*, 334: 101–109.

Barcelona, M.J. and T.G. Naymik. 1984. Dynamics of a fertilizer contaminant plume in ground water. *Environmental Science & Technology*, 18: 257–261.

Brillas, E., I. Sirés, and M.A. Oturan. 2009. Electro-Fenton process and related electrochemical technologies based on Fenton's reaction chemistry. *Chemical Reviews*, 109: 6570–6631.

Butler, E., Y.T. Hung, R. Yu-Li Yeh, and M. S. Al Ahmad. 2011. Electrocoagulation in wastewater treatment. *Water*, 3: 495–525.

Chaplin, B.P. 2014. Critical review of electrochemical advanced oxidation processes for water treatment applications. *Environmental Science: Processes Impacts*. 16: 1182–1203.

Comninellis, C. 1994. Electrocatalysis in the electrochemical conversion/combustion of organic pollutants for waste water treatment. *Electrochimica Acta*, 39: 1857–1862.

Divya, J. and S.L. Belagali. 2012. Impact of chemical fertilizers on water quality in selected agricultural areas of Mysore district, Karnataka, India. *International Journal of Environmental Science*, 2: 1449–1458.

Fresno, F., R. Portela, S. Suarez, and J. M. Coronado. 2014. Photocatalytic materials: Recent achievements and near future trends. *Journal of Materials Chemistry A*, 2: 2863–2884.

Gherasim, C-V., J. Krivcik, and P. Mikulasek. 2014. Investigation of batch electrodialysis process for removal of lead ions from aqueous solutions. *Chemical Engineering Journal*, 256: 324–334.

Grimm, J., D. Bessarabov, and R. Sanderson. 1998. Review of electro-assisted methods for water purification. *Desalination*, 115: 285–294.

Jimoh, O.D. and M.A. Ayodaeji. 2003. Effect of agrochemicals on surface and ground waters in the Tunga-Kawo (Nigeria) irrigation scheme. *Journal of Hydrological Science*, 48: 1013–1023.

Kabdasli, I., I. Arslan-Alaton, T. Ölmez-Han, and O. Tünay. 2012. Electrocoagulation applications for industrial wastewaters: A critical review. *Environmental Technology Reviews*. 1: 2–45.

Kamaraj, R., D.J. Davidson, G. Sozhan, and S. Vasudevan. 2014a. Adsorption of 2,4-dichlorophenoxyacetic acid (2,4-D) from water by *in situ* generated metal hydroxides using sacrificial anodes. *Journal of Taiwan Institute of Chemical Engineering*, 45: 2943–2949.

Kamaraj, R., D.J. Davidson, G. Sozhan, and S. Vasudevan. 2014b. An *in situ* electrosynthesis of metal hydroxides and their application for adsorption of 4-chloro-2-methylphenoxyacetic acid (MCPA) from aqueous solution. *Journal of Environmental Chemical Engineering*, 2: 2068–2077.

Kuokkanen, V., T. Kuokkanen, J. Rämö, and U. Lassi. 2013. Recent applications of electrocoagulation in treatment of water and wastewater—A review. *Green and Sustainable Chemistry*. 3: 89–121.

Martínez-Huitle, C.A. and E. Brillas. 2008. Electrochemical alternatives for drinking water disinfection. *Angewandte Chemie International Edition*, 47: 1998–2005.

Martínez-Huitle, C.A. and E. Brillas. 2009. Decontamination of wastewaters containing synthetic organic dyes by electrochemical methods: A general review. *Applied Catalysis B: Environmental*, 87: 105–145.

Mook, W.T., M.H. Chakrabarti, M.K. Aroua, G.M.A. Khan, B.S. Ali, M.S. Islam, and M.A. Abu Hassan. 2012. Removal of total ammonia nitrogen (TAN), nitrate and total organic carbon (TOC) from aquaculture wastewater using electrochemical technology: A review. *Desalination*, 285: 1–13.

Novotny, V. 1999. Diffuse pollution from agriculture—A worldwide outlook. *Water Science and Technology*, 39: 1–137.

Panizza, M. and G. Cerisola. 2009. Direct and mediated anodic oxidation of organic pollutants. *Chemical Reviews*, 109: 6541–6569.

Pimentel, D., B. Berger, D. Filiberto, M. Newton, B. Wolfe, E. Karabinakis, S. Clark, E. Poon, E. Abbett, and N. Sudha. 2004. Water resources: Agricultural and environmental issues. *Bio Science*. 54: 909–918.

Sirés, I., E. Brillas, M.A. Oturan, M.A. Rodrigo, and M. Panizza. 2014. Electrochemical advanced oxidation processes: Today and tomorrow. A Review. *Environmental Science and Pollution Research*, 21: 8336–8367.

Sugeng, A.J., P.I. Beamer, E.A. Lutz, and C.B. Rosales. 2013. Hazard-ranking of agricultural pesticides for chronic health effects in Yuma County, Arizona. *Science of the Total Environment* 463–464: 35–41.

Vasudevan, S., Florence Epron, J. Lakshmi, S. Ravichandran, S. Mohan, and G. Sozhan. 2010a. Removal of NO_3^- from drinking water by electrocoagulation—An alternate approach. *Clean—Soil, Air, Water*, 38: 225–229.

Vasudevan, S. and J. Lakshmi. 2012. The adsorption of phosphate by graphene from aqueous solution. *RSC Advances*, 2: 5234–5242.

Vasudevan, S., J. Lakshmi, J. Jayaraj, and G. Sozhan. 2009a. Remediation of phosphate-contaminated water by electrocoagulation with aluminium, aluminium alloy and mild steel anodes. *Journal of Hazardous Material*, 164: 1480–1486.

Vasudevan, S., J. Lakshmi, and R. Packiyam. 2010b. Electrocoagulation studies on removal of cadmium using magnesium electrode. *Journal of Applied Electrochemistry*, 40: 2023–2032.

Vasudevan, S., J. Lakshmi, J, and G. Sozhan. 2009b. Optimization of the process parameters for the removal of phosphate from drinking water by electrocoagulation. *Desalination and Water Treatment*, 12: 407–414.

Vasudevan, S., J. Lakshmi, and G. Sozhan. 2013. Electrochemically assisted coagulation for the removal of boron from water using zinc anode. *Desalination*, 310: 122–129.

Vasudevan, S. and M.A. Oturan. 2013. Electrochemistry and water pollution. E. Lichtfouse, J. Schwarzbauer, and D. Robert, Eds. *Environmental Chemistry for a Sustainable World*, Vol. 3. Springer Publishers, London, UK. pp. 27–68.

Vasudevan, S. and M.A. Oturan. 2014. Electrochemistry: As cause and cure in water Pollution—An Overview. *Environmental Chemistry Letters*,12: 97–108.

Vasudevan, S, G. Sozhan, S. Ravichandran, J. Jayaraj, J. Lakshmi, and S. Margrat Sheela. 2008. Studies on the removal of phosphate from drinking water by electrocoagulation process. *Industrial & Engineering Chemistry Research*, 47: 2018–2013.

25

Fashionable Techniques for Assessment of Groundwater Potential, Pollution, Prevention, and Remedial Measures

R. Annadurai and Sachikanta Nanda

CONTENTS

25.1 Introduction

Groundwater is the most important resource for human-kind. Keeping view of this, the hydrological decade has been celebrated internationally from the year 1965 to 1974, which aimed at hydrological education and research, development of analytical techniques, and collection of hydrological information. During the last four decades, a lot of new techniques have been developed to understand the assessment, utilization, and management of water resources. This chapter aims at understanding the possible development in locating groundwater as well as its quality studies and also targeting the sources of groundwater contamination and their remedial measures.

Groundwater constitutes 0.6% of the total fresh water, which accounts for 2.8% of all water. The basic idea of studying groundwater is important as unlike other sources of fresh water this is naturally suitable and economically exploitable. Therefore, groundwater has suitable composition in most cases and is free from turbidity, objectionable colors, and mostly free from pathogenic organisms whose presence may require further treatment. It is also relatively safe from hazards of chemical, radiogenic, and biological pollution unlike surface water. Moreover, it is slowly affected by drought and other climatic changes and is available locally in many cases, which demands less cost for exploration. The major sources of groundwater are meteoric water, connate water, and juvenile water. It may occur in solid rocky confined or unconfined aquifers (Singh, 2009).

25.2 Scientific Methods

To locate groundwater accurately and to determine the depth, quantity, and quality of the water, several techniques must be used, and possible locations must be tested and studied to identify the hydrologic and geologic features; those are important for study, planning, and management of the resource. Conditions for large quantities of water of shallow groundwater are more favorable under valleys than under hills. In some regions, the presence of water-loving plants such as cottonwoods or willows indicates groundwater at shallow depth. Areas where water is at the surface such as springs, seeps, swamps, or lakes reflect the presence of groundwater, although not in large quantities necessarily or of usable quantities.

Rocks are the most valuable clues of all. They are the primary consideration of a hydrologist to prepare a geological map and cross section of the lithographic distribution and positions, both surface and subsurface, to locate groundwater favorably. Suitable occurrences of sedimentary rocks for several kilometers ensure the chances of more permeability. Similarly, the fractures or weathered zones in different rocks contain openings for large amounts of water to pass through. Types and orientation of joints and occurrences of lineaments and lineament cross sections may be clues for obtaining suitable information regarding the groundwater. The hydrologists locate the wells in the target area and collect necessary information like depth to water, amount of water pumped, and types of rocks penetrated by the well. Wells are tested to determine the amount of water moving through the aquifer, the volume of water that can enter a well, and the effects of pumping on the water levels in that area. A careful chemical analysis also gives information about the quality of water in the aquifer.

Dowsing is a method of using a forked stick, rod, pendulum, or similar device to locate groundwater, minerals, or other hidden or lost substances, and has been a subject of discussion for years. Although tools and methods vary widely, many dowsers probably use the forked stick which may come from varieties of trees including the willow, peach, and witch-hazel. Other dowsers use key, wire coat hangers, pliers, wire rods, pendulums, or various kinds of elaborate boxes or electrical instruments. In the classic form of using forked sticks, one fork is held in each hand with the palms upward. The bottom or butt end of the "Y" is pointed skyward at an angle of about 45°. The dowser then walks back and forth on the area to be tested. When he or she passes over a source of water, the butt end of the stick is supposed to rotate or be attracted downward. Water dowsers practice mainly in rural and suburban areas where localities are uncertain about the groundwater location and supply (Water dowsing, U.S. Department of the interior/geological survey, 2003). Because drilling and development of a well cost a lot, residents gamble on a dry hole and turn to a water dowser for advice. Some water exists under the surface almost everywhere. This explains why the water dowsers are appearing to be successful. To locate groundwater accurately, however, as to depth, quantity, and quality, a number of techniques must be used. Hydrologic, geologic, and geophysical knowledge are a must to determine the depth and extent of different groundwater bearing strata and the quantity and quality of water found in each (Zohdy et al., 1974).

25.2.1 Groundwater Exploration Techniques

The studies to be carried out for a groundwater development project may be subdivided into:

- A preliminary survey at a regional scale
- A detailed survey to select the best well location

The main objective of a preliminary survey is to minimize the cost of groundwater development. Planning of a preliminary survey includes the total cost of the study and implementation of the groundwater development project. The importance of the exploration should not be neglected; particularly in areas of discontinuous aquifers where little information is available. The major expectations from a preliminary survey will be the following:

- Identification of aquifers and estimation of their characteristics
  - The geological setting
  - Hydraulic continuity
  - Groundwater quantity and quality assessment and expected water demand
  - Depth of water from the ground
  - Physical characteristics of the formation to be penetrated to reach water
- Identify appropriate development methods

- Performance of dug or drilled wells
- Well depth, diameter, and where wells are open to aquifer
- Construction methods
- Technical specifications
- Use of local or imported techniques
- Cost
- Selection of pumps or water lifting devices
- Assessment of risk failure

Therefore, a preliminary survey may identify the groundwater conditions and, taking these and other conditions into account, a reasonable basis for a choice of groundwater development may be obtained. Identifying the groundwater conditions includes the determination of groundwater locations and physical properties of the aquifer, estimating the quantity and quality, and determining depth to the water table from the surface. The groundwater conditions will determine what kinds of wells may be constructed. Depending on these possibilities and on the local resources and economics, costs are compared and a choice may be made.

In different parts of the world, geological and hydrological maps of 1/200,000 to 1/1,000,000 exist or are in the process of preparation and can provide essential information on aquifers, their extension, their boundaries and their lithology, and on the depth to water level. This information is usually sufficient to determine whether the aquifers are continuous within the area considered for the livestock water supply project. In addition to the classical geological and hydrogeological maps, satellite images may provide complementary information on geological formations and structures. Aerial photographs may also be available and geophysical surveys have been made for many regions. Information from an inventory of water wells is the basis for a more accurate identification of the aquifer, which will then be tapped in the framework of the future groundwater development project. From the data collected on the existing dug and drilled wells, it will be possible to establish the hydraulic continuity of the aquifer and to map the depth to water, the distribution of the good wells (the discharge of which exceeds a minimum value admissible for the considered purpose), and water quality. In the case of discontinuous aquifers, the data collected should also include dry wells (dug or drilled) in order to establish a correlation between as many parameters as possible. Well discharge is usually correlated with

- The nature of the water-bearing formation
- The formation fracture characteristics, if any (from air photo interpretation)

- The distance to important tectonic structures (from the satellite imagery and air photo interpretation)
- The depth of penetration of the wells into the aquifer

In many countries, this formation has already been collected by the service responsible for the water resource inventory in which case the additional field investigations required will be limited to updating the inventory or checking questionable information.

In many developed countries there is not only a heavy reliance on groundwater as a primary drinking supply but also as a supply of water for both agriculture and industrial use. The reliance on groundwater is such that it is necessary to ensure that there are significant quantities of water and that the water is of a high quality. The use of geophysics for both groundwater resource mapping and for water quality evaluations has increased dramatically over the last 10 years in large part to the rapid advances in microprocessors and associated numerical modeling solutions. However, despite its sometimes spectacular success, for the majority of groundwater studies, the use of geophysics is still often not considered. In part, it is poor publicity of the potential use of geophysics and poor dissemination of some of the more complex technical issues. It is also due to practical implementation difficulties and cost limitations. Unfortunately, it is sometimes because geologists and engineers may have experienced an inappropriate use of geophysics in the past or more unfortunately had some geophysics oversold to them under false pretenses of its possibilities thus leading to not just the poor use of geophysics but to the delivery of misleading or wrong results. This chapter attempts to guide the reader to same.

Groundwater applications of near-surface geophysics include mapping the depth and thickness of aquifers, mapping aquitards or confining units, and locating preferential fluid migration paths such as groundwater and that from saltwater intrusion.

25.3 Modern Techniques

As far as the remote sensing applications in groundwater hydrology is concerned, aerial photographs and visible and near-infrared satellite images have been used for groundwater exploration experimentally since the 1960s with only limited success (Engman and Gurney, 1991). The absence of spectral resolution did not allow effective use in intrasite groundwater prospecting. However, with the advent of high resolution multispectral satellite sensors, the use of satellite imagery (including micro-wave imagery) for groundwater prospecting dramati-cally increased in the late 1980s (Engman and Gurney, 1991; Welch and Remillard, 1991; Meijerink, 2000; Jackson, 2002). The use of a remote sensing technique has been proved a very cost-effective approach in pros-pecting and preliminary surveys, because of high cost of drilling. Generally, the analysis of aerial photographs or satellite imagery is recommended prior to ground surveys and fieldwork because it may eliminate areas of potentially low water-bearing strata and may also indi-cate promising areas for intensive field investigations (Revzon et al., 1988). It should be noted, however, that the adoption of remote sensing does not eliminate the *in situ* data collection, which is still essential to verify the accuracy of remote sensing data and their interpre-tation. Of course, remote sensing helps minimize the amount of field data collection Jha Madan et al. (2007).

A review of GIS applications in hydrology and water management has been presented by several research-ers during the early and mid-1990s such as Zhang et al. (1990), De vantier and Feldman (1993), Ross and Tara (1993), Schultz (1993), Deckers and Te Stroet (1996), and Tsihrintzis et al. (1996). These reviews indicate that GIS applications in hydrology and water management are essentially in a modeling dominated context. Longley et al. (1998), on the other hand, while presenting the development of geocomputation, discuss various geo-scientific applications of GIS as well as the role of geo-computation in the development and application of GIS technologies. Although the use of GIS in groundwater modeling studies dates back to 1987, its use for sur-face-water modeling has been more prevalent than for groundwater modeling because the available standard-ized GIS coverage is primarily of the land surface; few standardized coverages of hydrogeological properties are available (Watkins et al., 1996). Watkins et al. (1996) presented an excellent overview of GIS applications in groundwater-flow modeling as well as discussed its usefulness and future directions. On the other hand, Pinder (2002) provides step-by-step procedures for groundwater flow and transport modeling using GIS technology. The application of RS and GIS in different aspects of hydrological management are listed as:

- Exploration and assessment of groundwater resources
- Selection of artificial recharge sites
- GIS-based subsurface flow and pollution mod-eling: model development, applications, and evaluation
- Groundwater-pollution hazard assessment and protection planning
- Estimation of natural recharge distribution

- Aquifer modeling and aquifer character studies
- Hydrogeological data analysis and process monitoring, etc.

The applications of RS technology in groundwater hydrology are very limited compared to other fields of study because of its inherent limitations. Although there is growing interest in exploring this technology, there is a long way to go in order to use RS technol-ogy effectively for the development and management of vital groundwater resources. Based on the present review, the focus of future advancements in RS technol-ogy should be in the following areas of concern:

- A general problem of using remote sens-ing in hydrological studies is that very few remotely sensed data can be directly applied in hydrology. They measure only a part of the electromagnetic spectrum and different hydro-logical parameters are only inferred from them. Therefore, there is an urgent need to improve the accuracy and reliability of remote sensing estimates, which are highly uncertain until now (Beven, 2001). It could be possible by refining analysis techniques as well as developing new and improved sensors and their applications in conjunction with improved field measurements. New RS applications will emerge and mature as new instruments and new types of data become available in the future.

- The major constraint for the utility of RS in hydrogeology is that it can only detect changes at the ground surface or a shallow layer (<1 m deep) of the earth, though the airborne explo-ration of groundwater using electromagnetic prospecting sensors developed for the mineral industry is reported to map aquifers at depths greater than 200 m (Paterson and Bosschart, 1987). However, with the growing pollution and lowering of groundwater worldwide, it is often necessary to explore deep aquifers in which case the usual remote sensing data are of no use, except for the especially acquired data by GPR (ground-penetrating radar), which can penetrate up to about 20 m depth. Although ongoing research activities using GPR, subsur-face methods of groundwater investigations, and tracers are expected to enhance our knowl-edge about complex and hidden subsurface processes, a routine use of any of these tech-niques seems a long way off (Lane et al., 2000; Beven, 2001). Future research should focus on the development of easy-to-use techniques to quantify subsurface water storage and visualize fluid flow and transport processes in the subsurface environment. Furthermore, in our quest to have more accurate and reliable noninvasive techniques for monitoring subsurface processes as well as to combat the heterogeneity and anisotropy of aquifer and vadose-zone systems, the RS technology offers the greatest promise. Future advancements in RS technology in this direction will certainly revolutionize the hydrogeological thinking, theory, and model development.

- More and more RS-based groundwater studies together with field studies should be carried out in order to examine the reliability of RS data. The combined use of multispectral data obtained from different sensor systems is necessary to extract more and better information (Engman and Gurney, 1991). Future research should also be directed toward developing linkages between surface observations and subsurface phenomena. Such studies will not only enhance and refine RS applications in groundwater hydrology, but will also significantly contribute to the sensor development program.

- There is a need to develop an optimal sensor system including both active and passive microwave techniques for more effective soil-moisture monitoring. It will allow a range of applications and the synergism of the two types of measurements to provide more useful and new information (Jackson et al., 1999).

- Recent developments in microwave remote sensing, theory, and sensor availability have resulted in new potential and capabilities. Very limited studies have revealed the potential to extract/detect subsurface parameters and features using these techniques.

- More and more research is required to refine and implement these approaches (Jackson, 2002). The multitemporal and spatial availability of microwave remote sensing data can complement the monitoring and modeling of groundwater recharge. In addition, through the synergistic use of Earth's gravity-field monitoring satellites (e.g., GRACE and CHAMP) data and satellite microwave remote sensing data, it may be possible to monitor seasonal groundwater recharge over large regions in the near future (Jackson, 2002). Future studies should be carried out in this stimulating direction.

- Last, but not the least, there must be strong cooperation between space agencies and soil and water scientists (e.g., soil scientists, hydrologists, hydrogeologists, and environmentalists)

for the planning and development of sensor systems, which will ensure timely implementation of suitable and efficient sensor systems for the effective mapping of land and water resources. Such cooperation will undoubtedly lead to wide-scale research and applications in the fields of hydrology and hydrogeology, which in turn will ensure efficient land and water management by the promising remote sensing technique.

25.3.1 Groundwater Potential Zonation Mapping: A Case Study

RS and GIS are handy tools to determine the quantity of groundwater in an area by doing different analyses. The process starts with preparation of different thematics like rainfall contour, soil, geology, geomorphology, slope, land use and land cover, lineament, drainage, depth to bedrock, depth of water level, etc.

In this study, the study area of Kattankulathur town of 378.53 sq. km has been mapped for groundwater potential. The various thematics like geology, geomorphology, soil, land use and land cover, slope, and drainage have been prepared. These different thematics after giving required weightage and ranking, overlaid to map the groundwater potential of the area. Four different ranks, namely, very poor, poor, moderate, good, and very good have been assigned to the final output.

The different rank and weightage assigned is shown in Table 25.1. The delineate potentiality of the groundwater the various constraint of thematics are considered and depending on the suitability different rank and weightage are assigned. For example, in this current study, the suitability score for water body and river are given different suitability scores. River has been considered a 7 and tanks and other water bodies a 6. Depth of bed rock is classified according to the shallowness of that for a suitable aquifer condition to pertain to the percolated water. Geology as one of the major factors governs the groundwater recharge has been classified into three different divisions. Sand and silt have been given a higher score compared to sandstone and conglomerate and the latter is charnokite, which is of metamorphic origin. Lineament intersection is very much suitable for groundwater recharge and hence it is given the highest rank. Soil as one of the most important factors that govern the percolation of water depending on its porosity and permeability has been classified into different scores. Similarly the different land uses are also given suitability scores, depending on their contribution to groundwater recharge.

The different suitable scores once assigned to various thematics, are further given different weightages.

TABLE 25.1

Rank for Each Category and Weight for Each Theme Assigned for Analysis

S. No.	Layers/Category	Rank	Weight
I	**Drainage**		10
1	River	9	
2	Tank/reservoir	7	
II	**Geomorphology**		16
1	Sedimentary high ground	6	
2	River	9	
3	Buried pediment shallow	5	
4	Buried pediment deep	8	
5	Pediment	4	
6	Structural hill	2	
7	Flood plain	9	
8	Tertiary up land	6	
9	Buried pediment moderate	7	
10	Pediment–Inselberg complex	4	
11	Inselberg	3	
12	Linear/curvilinear ridge	2	
13	Residual hill	1	
III	**Depth to Bed Rock**		9
1	28–34	6	
2	39–45	8	
3	34–39	7	
IV	**Annual Rainfall**		8
1	1140–1340	6	
2	1340–1540	7	
V	**Soil**		12
1	Sandy clay loam	8	
2	Sandy clay	6	
3	Clay	6	
4	Sandy loam	3	
5	Clay loam	3	
6	Loamy sand	2	
VI	**Geology**		6
1	Sand and silt	8	
2	Sandstone and conglomerate	6	
3	Charnokite	4	
VII	**Lineament**		14
1	Lineament (0.5 km buffer)	8	
2	Lineament intersection (1 km buffer)	9	
VIII	**Depth to Weathered Rock**		6
1	19–27	6	
2	11–19	5	
3	3–11	3	
IX	**Depth to Fractured Rock**		5
1	9–17	4	
2	26–35	6	
3	<9	4	

(Continued)

TABLE 25.1 (*Continued*)

Rank for Each Category and Weight for Each Theme Assigned for Analysis

S. No.	Layers/Category	Rank	Weight
4	17–26	6	
X	**Water Level Premonsoon**		5
1	3–5	8	
2	5–7	6	
3	7–9	4	
4	9–11	3	
XI	**Landuse**		9
1	Wet crop	8	
2	Dry crop	6	
3	Barren land	4	
4	Scrub/shrub	5	
5	Built up land	2	
6	Forest	5	
7	Water bodies	9	
8	Hills	2	
	TOTAL		**100**

In the current study as we are looking for groundwater potential mapping, geomorphology has been given the highest weightage. Furthermore, lineament, drainage, and soil have been given higher weightages followed by depth to bed rock and annual rainfall and land uses. Furthermore, the depth to fractured rock, depth to weathered rock, water level in premonsoon and geology have been given the least weightages. Finally, the groundwater potential map (Figure 25.1) has been prepared by overlaying and doing the intersection operation of all these layers with required weightage and ranks.

25.3.2 Water Quality Index Modeling: A Case Study

For the present study, groundwater samples were collected from 35 locations in Ranipet (Figure 25.2), Vellore district; Tamil Nadu was assessed in the monsoon, postmonsoon, and premonsoon season from July 2012 to May 2013. The WHO standards for weights (w_i) and calculated relative weight (W_i) for different parameters are listed in Table 25.2. Water quality assessment was carried out for the parameters like pH, electrical conductivity, total dissolved solids, total alkalinity, total hardness, chloride, sulfate, calcium, magnesium, sodium, potassium, nitrate, chromium, phosphate, and iron. Water quality index and correlation coefficients were determined to identify the highly correlated and interrelated water quality parameters (WQPs) as given in Table 25.3. Regression equations relating these identified and correlated parameters were formulated for highly correlated WQPs for both pre- and postmonsoon periods (Figures 25.3 and 25.4). Comparison of observed and estimated values of the different WQP parameters reveals that

the regression equations developed in the study can be very well used for making water quality monitoring by observing the above-said parameters alone. The analysis reveals that the groundwater of the area needs some degree of treatment before consumption, and it also needs to be protected from the perils of contamination.

The analysis of experimental investigation on the quality of groundwater using 14 physicochemical parameters of the study area indicate that in general the water quality was poor, very poor, and unsuitable for drinking purposes. In this study, the computed WQI values range from **48.69** to **245.24** during monsoon period, from **0** to **351.02** during premonsoon period, and from **0** to **344.62** during postmonsoon period, respectively. The percentage of water quality index shows that maximum in post and premonsoon and minimum in monsoon period.

25.4 Groundwater Contamination Studies

Contamination is defined as the introduction of any undesirable physical, chemical, or microbiological material into a water source. There are two types of contamination sources: point sources and nonpoint sources. Point sources include landfills, leaking gasoline storage tanks, leaking septic tanks, and accidental spills; as the name implies, we can point directly to the source of contamination. A point source is technically defined as any discernible, confined, and discrete conveyance from which pollutants are or may be discharged. Nonpoint sources can be less obvious and can include naturally

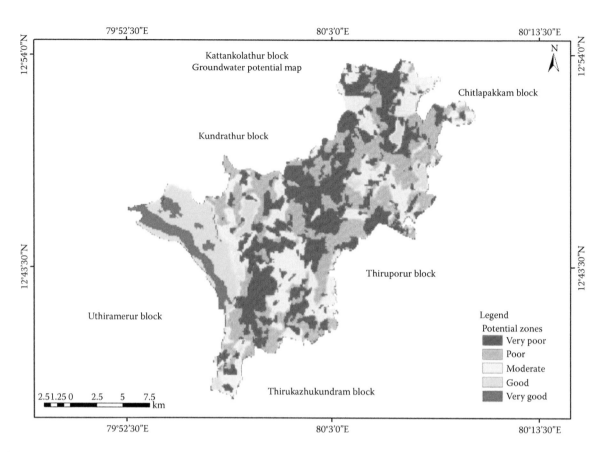

FIGURE 25.1
Groundwater potential mapping.

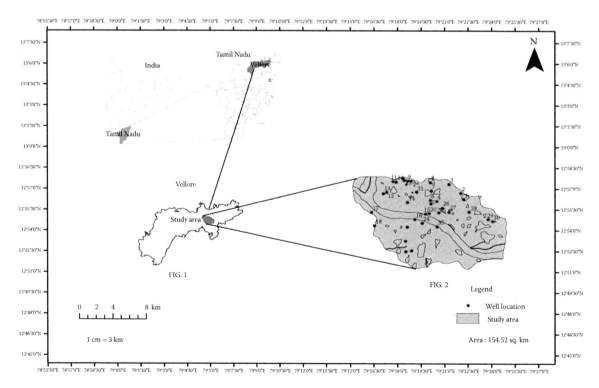

FIGURE 25.2
Well location map.

TABLE 25.2

WHO Standards Weight (w_i) and Calculated Relative Weight (W_i) for Each Parameter

Parameters	Standard Permissible Value (SI) (WHO, 2004)	Weight (w_i)	Relative Weight (W_i)
pH	6.5–8.5	4	0.09
TDS	500	4	0.09
EC	500	4	0.09
Th	200	3	0.06
Ca	75	2	0.04
Mg	50	1	0.02
Nitrate	45	5	0.11
Chloride	250	3	0.06
Fluoride	1–1.5	4	0.09
Sodium	200	2	0.04
Potassium	200	2	0.04
Iron	1	4	0.09
Sulfate	250	4	0.09
Chromium	0.05	5	0.11
Total	**47**	**1.00**	

occurring contaminants, such as iron, arsenic, and radiological runoff from parking lots, and pesticides and fertilizers that infiltrate the soil and make their way into an aquifer (Welch and Remillard, 1991). Some of the point and nonpoint sources are listed here:

Point sources

- On-site septic systems
- Leaky tanks or pipelines containing petroleum products
- Leaks or spills of industrial chemicals at manufacturing facilities
- Underground injection wells (industrial waste)
- Municipal landfills
- Livestock wastes
- Leaky sewer lines
- Chemicals used at wood preservation facilities
- Mill tailings in mining areas
- Fly ash from coal-fired power plants

- Sludge disposal areas at petroleum refineries
- Land spreading of sewage or sewage sludge
- Graveyards
- Road salt storage areas
- Wells for disposal of liquid wastes
- Runoff of salt and other chemicals from roads and highways
- Spills related to highway or railway accidents
- Coal tar at old coal gasification sites
- Asphalt production and equipment cleaning sites

Nonpoint (distributed) sources

- Fertilizers on agricultural land
- Pesticides on agricultural land and forests
- Contaminants in rain, snow, and dry atmospheric fallout

25.4.1 Groundwater Contamination Factors

Many factors can affect how groundwater becomes contaminated. The depth of a well is an obvious factor because a contaminant has to travel farther in deeper wells. The soil and formation, through which the contaminant travel, act as a natural filter. The underlying geologic formations additionally affect the possibility of contamination in two ways. First, the geological formation can be a source of contamination; there are various rock formations that contain minerals such as calcium, magnesium, iron, arsenic, and various radiological. While some of the minerals may not cause any known health effects, naturally occurring arsenic and radiological are known carcinogens. Second, the formation can slow the contaminant or it may have the opposite effect.

25.4.1.1 Contaminants

There are several properties to consider when looking at contaminants. Some of these properties include: persistence, adsorption, solubility, volatility, and molecular size.

TABLE 25.3

Water Quality Classification Based on WQI Values of the Study Area

Water Quality	WQI Values	WQI of Samples for Monsoon	% of Water Samples (Monsoon)	WQI of Samples for Postmonsoon Season	% of Water Samples (Postmonsoon)	WQI of Samples for Premonsoon Season	% of Water Samples (Premonsoon)
Excellent water	<50	48.69	9%	0	0%	0	0%
Good water	50–100	88.12	17%	89.91	11%	0	0%
Poor water	100–200	135.06	26%	136.34	17%	166.17	22%
Very poor water	200–300	245.24	48%	223.45	28%	236.29	32%
Unfit for use	>300	–	–	351.02	44%	344.62	46%

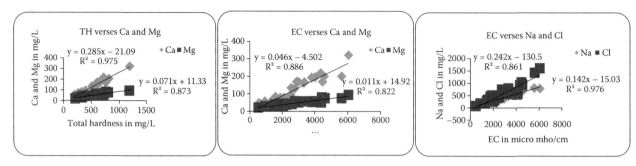

FIGURE 25.3
Linear plot between TH versus Ca and Mg, EC versus Ca and Mg, and EC versus Na and Cl of groundwater in premonsoon season.

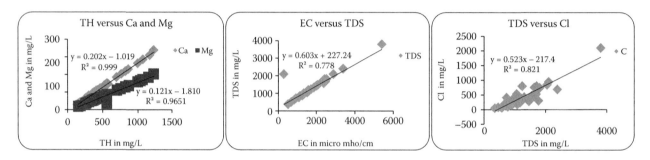

FIGURE 25.4
Linear plot between TH versus Ca and Mg, EC versus TDS and TDS versus Cl of groundwater in the postmonsoon season.

- *Persistence* refers to the staying power of a contaminant. Certain contaminants do not break down easily and will persist in the environment for a long time. PCBs are an example of persistence in the environment.

- *Adsorption* refers to how tightly a compound will attach to soil particles. Compounds that are strongly adsorbed are less likely to leach into the groundwater.

- *Solubility* is the ability of a substance to dissolve in a solvent. Compounds that are highly water-soluble will dissolve in water percolating through the soils down to the water table. Methyl tertiary butyl ether (MTBE) is an example of a contaminant with high water solubility, which is one of the reasons MTBE contamination is so widespread.

- *Volatility* of a substance refers to its tendency to change from a liquid or solid into a gas. One group of contaminants called volatile organic chemicals readily change from liquid or solid to gas when exposed to the atmosphere. The more volatile a substance is, the more likely it will be lost to the atmosphere.

Lastly, the molecular size can play a role. The smaller the molecule, the more likely it will be able to travel between soil particles.

25.4.2 Various Causes of Contamination

25.4.2.1 Groundwater Contamination Due to Industrial Chemical Spill

Manufacturing and service industries have high demands for cooling water, processing water, and water for cleaning purposes. Groundwater pollution occurs when used water is returned to the hydrological cycle. Modern economic activity requires transportation and storage of material used in manufacturing, processing, and construction. Along the way, some of this material can be lost through spillage, leakage, or improper handling. The disposal of wastes associated with the above activities contributes to another source of groundwater contamination. Some businesses, usually without access to sewer systems, rely on shallow underground disposal. They use cesspools or dry holes, or send the wastewater into septic tanks. Any of these forms of disposal can lead to contamination of underground sources of drinking water. Dry holes and cesspools introduce wastes directly into the ground. Septic systems cannot treat industrial wastes. Wastewater disposal practices of certain types of businesses, such as automobile service stations, dry cleaners, electrical component or machine manufacturers, photo processors, and metal platers or fabricators are of particular concern because the waste they generate is likely to contain toxic chemicals. Other industrial sources of contamination include cleaning off holding tanks or spraying equipment on the open ground, disposing of waste in septic systems or dry wells, and storing hazardous materials in uncovered areas or in areas that do not have pads with drains or catchment basins. Underground and aboveground storage tanks holding petroleum products, acids, solvents, and chemicals can develop leaks from corrosion, defects, improper installation, or mechanical failure of the pipes and fittings. Mining of fuel and nonfuel minerals can create many opportunities for groundwater contamination. The problems stem from the mining process itself, disposal of wastes, and processing of the ores and the wastes it creates.

25.4.2.2 Groundwater Contamination Due to Landfill

Municipal solid waste (MSW) landfills accept nonhazardous wastes from a variety of sources, such as households, businesses, restaurants, medical facilities, and schools. Many MSW landfills also can accept contaminated soil from gasoline spills, conditionally exempted hazardous waste from businesses, small quantities of hazardous waste from households, and other toxic wastes. Industrial facilities may utilize their own captive landfill (i.e., a solid waste landfill for their exclusive use) to dispose of nonhazardous waste from their processes, such as sludge from paper mills and wood waste from wood processing facilities. The precipitation that falls into a landfill, coupled with any disposed liquid waste, results in the extraction of the water-soluble compounds and particulate matter of the waste, and the subsequent formation of leachate. The creation of leachate, sometimes deemed "garbage soup," presents a major threat to the current and future quality of groundwater. (Other major threats include underground storage tanks, abandoned hazardous waste sites, agricultural activities, and septic tanks.) Leachate composition varies relative to the amount of precipitation and the quantity and type of wastes disposed. In addition to numerous hazardous constituents, leachate generally contains nonhazardous parameters that are also found in most groundwater systems. These constituents include dissolved metals (e.g., iron and manganese), salts (e.g., sodium and chloride), and an abundance of common anions and cations (e.g., bicarbonate and sulfate). These constituents in leachate typically are found at concentrations that may be an order of magnitude (or more) greater than concentrations present in natural groundwater systems. Leachate from MSW landfills typically has high values for total dissolved solids and chemical oxygen demand, and a slightly low to moderately low pH. MSW leachate contains hazardous constituents, such as VOCs and heavy metals. Wood-waste leachates typically are high in iron, manganese, and tannins and lignin. Leachate from ash landfills is likely to have elevated pH and to contain more salts and metals than other leachates. A release of leachate to the groundwater may present several risks to human health and the environment. The release of hazardous and nonhazardous components of leachate may render an aquifer unusable for drinking-water purposes and other uses. Leachate impacts to groundwater may also present a danger to the environment and to aquatic species if the leachate-contaminated groundwater plume discharges to wetlands or streams. Once leachate is formed and is released to the groundwater environment, it will migrate downward through the unsaturated zone until it eventually reaches the saturated zone. Leachate then will follow the hydraulic gradient of the groundwater system.

25.4.2.3 Groundwater Contamination Due to Chemicals

Chemicals in groundwater can be both naturally occurring or introduced by human interference and can have serious health effects. Fluoride in the groundwater is essential for protection against dental caries and weakening of the bones, but higher levels can have an adverse effect on health. *Arsenic* occurs naturally or is possibly aggravated by overpowering aquifers and by phosphorus from fertilizers. Petrochemicals contaminate the groundwater from underground petroleum storage tanks. *Other heavy metal* contaminants come from mining waste and tailings, landfills, or hazardous waste dumps. Metal and plastic effluents, fabric cleaning, and electronic and aircraft manufacturing are often discharged and contaminate groundwater. Gasoline consists of a mixture of various hydrocarbons (chemicals made up of carbon and hydrogen atoms) that evaporate easily, dissolve to some extent in water, and often are toxic. Benzene, a common component of gasoline, is considered to cause cancer in humans. Underground and aboveground storage tanks are used for storage of chemicals, such as heating oil. If these tanks develop a crack, which often occurs in old and corroded tanks, chemicals seep out and migrate through soil and eventually contaminate groundwater. Abandoned underground tanks pose a threat to groundwater because their locations are not always known. Tracks and trains transporting chemicals pose another potential threat if accidentally the contents of the train are spilled. The oil and gas production industry accounts for a large proportion of the fluids injected into the subsurface. When oil and gas are extracted, large amounts of saltwater (brine) are also brought to the surface. This saltwater can be very damaging if it is discharged into the surface water.

25.4.2.4 Groundwater Contamination Due to Pesticides, Herbicides, and Fertilizers

Pesticides and herbicides are applied to agricultural land to control pests that disrupt crop production.

Pesticide leaching occurs when pesticides mix with water and move through the soil, ultimately contaminating groundwater. The amount of leaching is correlated with particular soil and pesticide characteristics and the degree of rainfall and irrigation.

Leaching is most likely to happen if using a water-soluble pesticide, when the soil tends to be sandy in texture, if excessive watering occurs just after pesticide application, or if the adsorption ability of the pesticide to the soil is low. Leaching may not only originate from treated fields, but also from pesticide mixing areas, pesticide application machinery washing sites, or disposal areas. The nitrogen (N) and phosphorus (P) applied to agricultural land (via synthetic fertilizers, composts, manures, biosolids, etc.) can provide valuable plant nutrients.

Excess N and P can have negative environmental consequences. Excess N supplied by both synthetic fertilizers (as highly soluble nitrate) and organic sources such as manures (whose organic N is mineralized to nitrate by soil microorganisms) can lead to groundwater contamination of nitrate. Nitrate-contaminated drinking water can cause blue baby syndrome. Together with excess P from these same fertilizer sources, eutrophication can occur downstream due to excess nutrient supply, leading to anoxic areas called dead zones.

25.4.2.4.1 Nanoremediation Using Zero-Valent Metals

An emerging abiotic technology for the remediation of recalcitrant pollutants such as chlorinated aliphatic and aromatic compounds is the application of metal systems. Zero-valent metal reduction is a process in which a metal such as iron, platinum, or other zero-valent metals are used to refine polluted waters (Rajan, 2011). These metals are placed in the flow of water, where they begin oxidizing, causing other chain reactions to purify the water. Zero-valent metals such as palladium (Pd0), iron (Fe0), zinc (Zn0), and magnesium (Mg0) have been used for the reductive dechlorination of chlorinated organic compounds. Various compounds that can be removed are linden, DDT, DDE, DDD, chlorobenzenes, chlorinated methane, polychlorinated ethane, herbicide triallate, trichloro ethylene, Zi, Ni, Cr, methyl orange, azo dye, TNT, RDX, etc.

25.5 Innovative and Practical Remediation Techniques

25.5.1 *In Situ* Remedial Measures

25.5.1.1 Air Sparging

The process of blowing air directly into the groundwater is another option for groundwater remediation

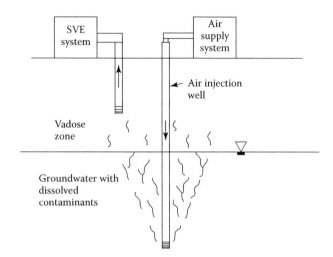

FIGURE 25.5
Schematic diagram of air sparging.

(Figure 25.5) (Reddy and Krishna, 2008). As the bubbles rise, the contaminants are removed from the groundwater by physical contact with the air and are carried up into the unsaturated zone (i.e., soil). As the contaminants move into the soil, a soil vapor extraction system is usually used to remove vapors (Olexsey and Parker, 2006).

25.5.1.2 Bioremediation

Bioremediation or biodegradation is the use of microorganism metabolism to remove pollutants. Technologies can be generally classified as *in situ* or *ex situ*. *In situ* bioremediation involves treating the contaminated material (soil or groundwater) in place, while *ex situ* involves the removal of the contaminated material (soil or groundwater) to be treated. Some examples of specialized bioremediation technologies include phytoremediation, bioventing, bioleaching, land farming, bioreactor, composting, bioaugmentation, rhizofiltration, and biostimulation. Aerobic bioremediation involves microbial reactions that require oxygen to go forward. The bacteria use a carbon substrate as the electron donor and oxygen as the electron acceptor. Anaerobic bioremediation involves microbial reactions occurring in the absence of oxygen and encompasses many processes, including fermentation, methanogenesis, reductive dechlorination, and sulfate- and nitrate-reducing conditions (Suthersan, 1999).

25.5.1.3 In-Well Air Stripping

With in-well air stripping technology, air is injected into a vertical well that has been screened at two depths. The lower screen is set in the groundwater saturated zone, and the upper screen is in the unsaturated zone, often called the vadose zone. Pressurized air is injected into the well below the water table, aerating the water. The aerated water rises in the well and flows out of the system at the upper screen. Contaminated groundwater is drawn into the system at the lower screen. The volatile organic compounds (VOCs) vaporize within the well at the top of the water table, as the air bubbles out of the water. The vapors are drawn off by a soil vapor extraction (SVE) system. The partially treated groundwater is never brought to the surface; it is forced into the unsaturated zone, and the process is repeated as water follows a hydraulic circulation pattern or cell that allows continuous cycling of groundwater. As groundwater circulates through the treatment system *in situ*, contaminant concentrations are gradually reduced. In-well air stripping is a pilot-scale technology (Hamby, 1996).

25.5.1.4 Chemical Oxidation

Oxidation chemically converts hazardous contaminants to nonhazardous or less toxic compounds that are more stable, less mobile, and/or inert. The oxidizing agents most commonly used are ozone, hydrogen peroxide, and hypochlorite, chlorine, and chlorine dioxide. The *in situ* (below ground) chemical oxidation process is designed to destroy organic contaminants either dissolved in groundwater, sorbed to (stuck to) the aquifer material, or present in their free phase (e.g., as gasoline). Oxidants most frequently used in chemical oxidation include hydrogen peroxide (H_2O_2), potassium permanganate ($KMnO_4$), persulfate ($Na_2O_8S_2$), and ozone (O_3). Peroxone, which is a combination of ozone and hydrogen peroxide, is also used. Fenton's reagent, hydrogen peroxide mixed with a metal (commonly iron) catalyst, can also be used. *In situ* chemical oxidation can be accomplished by introducing chemical oxidants into the soil or aquifer at a contaminated site using a variety of injection and mixing apparatuses. Normally, vertical or horizontal injection wells are used to deliver chemical oxidants (Sumathi, 2009).

25.5.1.5 Thermal Treatment

Steam is forced into an aquifer through injection wells to vaporize volatile and semivolatile contaminants. Vaporized components rise to the unsaturated zone where they are removed by vacuum extraction and then treated. Hot water or steam-based techniques include contained recovery of oily waste (CROW), steam injection and vacuum extraction (SIVE), *in situ* steam-enhanced extraction (ISEE), and steam-enhanced recovery process (SERP). Hot water or steam flushing/stripping is a pilot-scale technology. *In situ* biological treatment may follow the displacement and is continued until groundwater contaminant concentrations satisfy statutory requirements. The process can be used to remove large portions of oily waste accumulations and to retard downward and lateral migration of organic contaminants. The process is applicable to shallow and deep contaminated areas and readily available mobile equipment can be used. Hot water/steam injection is typically short to medium duration, lasting a few weeks to several months.

25.5.1.6 Phytoremediation

Phytoremediation is the direct use of living green plants for *in situ*, or in place, removal, degradation, or containment of contaminants in soils, sludge, sediments, surface water, and groundwater. Phytoremediation is a low cost, solar energy driven cleanup technique that is most useful at sites with shallow, low levels of contamination. This method is very effective with, or in some cases, in place of mechanical cleanup methods. There are several ways in which plants are used to clean up, or remediate, contaminated sites. To remove pollutants from soil, sediment, and/or water, plants can break down or degrade organic pollutants or contain and stabilize metal contaminants by acting as filters or traps. The uptake of contaminants in plants occurs primarily through the root system, in which the principal mechanisms for preventing contaminant toxicity are found. In addition, deep-lying contaminated groundwater can be treated by pumping the water out of the ground and using plants to treat the contamination. Plant roots also cause changes at the soil–root interface as they release inorganic and organic compounds (root exudates) in the rhizosphere. These root exudates affect the number and activity of the microorganisms, the aggregation and stability of the soil particles around the root, and the availability of the contaminants. Root exudates, by themselves, can increase (mobilize) or decrease (immobilize) directly or indirectly the availability of the contaminants in the root zone (rhizosphere) of the plant through changes in soil characteristics, release of organic substances, changes in chemical composition, and/or increase in plant-assisted microbial activity (Figure 25.6) (Pivetz, 2001).

25.5.2 *Ex-Situ* Techniques

25.5.2.1 Bioremediation

Bioremediation is the treatment of contaminated soil or water once it has been excavated or pumped out of the location. Slurry phase bioremediation is a process where the contaminated soil is mixed with water and other reagents in a large tank known as a bioreactor. It is mixed in order to keep the microorganisms in contact with the toxins present in the soil. Then oxygen and nutrients are added into the mixing so the microorganisms have an ideal environment to break down the contaminants. Once the process has been completed, the

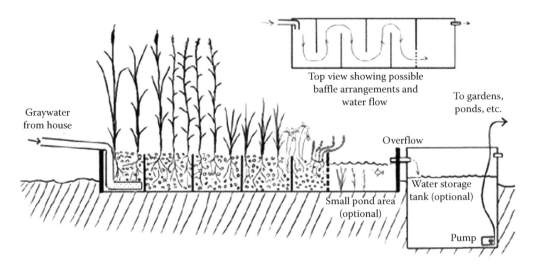

FIGURE 25.6
Phytoremediation of wet land.

water is separated from the soil and the soil is tested and replaced in the environment. This particular process is comparatively fast compared to other bioremediation techniques (Zhang, 2005).

On the other hand, solid-phase bioremediation is a process that treats the contaminated soil in an above-ground treatment center. Conditions inside the treatment areas are controlled to ensure optimum treatment can take place. This type of treatment is easy to maintain, but it requires a lot of space and the process of decontamination will take longer than it would by slurry-phase bioremediation. Solid-phase soil treatments include land farming, soil biopiles, and composting. Land farming

is a more simple process. The soil is spread out over a pad, which has a system that is used to collect any of the residual liquids from the soil which may be toxic. The soil is turned over regularly to allow air to mix in with the excavated soil. In this environment, the microorganisms present in the soil are able to break down the contaminants in the soil. Soil biopiles are a process where piles of soil are placed over the top of a bug vacuum pump. The vacuum pump pulls air through the pile of soil to allow oxygen to get through the soil to the microorganisms. Contaminants that may be turned into gas forms are easily controlled as they are simply sucked up with the air stream through the soil.

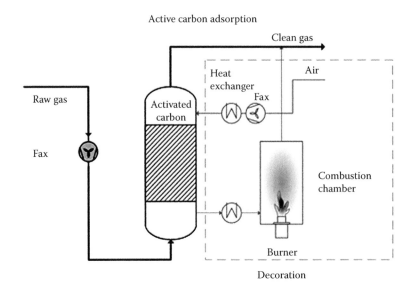

FIGURE 25.7
Active carbon absorption.

25.5.2.2 Oxygen Sparging

In this method of air sparging, the injection of air or oxygen is done through a contaminated aquifer. Injected air traverses horizontally and vertically in channels through the soil column, creating an underground stripper that removes volatile and semivolatile organic contaminants by volatilization. The injected air helps to flush the contaminants into the unsaturated zone. The SVE usually is implemented in conjunction with air sparging to remove the generated vapor-phase contamination from the vadose zone. Oxygen added to the contaminated groundwater and vadose-zone soils also can enhance biodegradation of contaminants below and above the water table.

25.5.2.3 Activated Carbon Adsorption

Adsorption is a process where a solid is used for removing a soluble substance from the water. The most common activated carbon used for remediation is derived from bituminous coal (Figure 25.7).

Activated carbon absorbs volatile organic compounds from groundwater by chemically binding them to the carbon atoms. In this process, active carbon is the solid. Activated carbon is produced specifically to achieve a very big internal surface (between 500 and 1500 m^2/g). This big internal surface makes active carbon ideal for adsorption. Active carbon comes in two variations: powder activated carbon (PAC) and granular activated carbon (GAC). The GAC version is mostly used in treatment. It can absorb the soluble substances, organic and nonpolar substances such as mineral oil, BTEX, poly aromatic hydrocarbons (PACs), and (chloride) phenol (Laumann, 2013).

References

Beven K.J. 2001. *Rainfall-Runoff Modeling: The Primer*. John Wiley & Sons Ltd., Chichester. pp. 297–306.

De Vantier B.A. and Feldman A.D. 1993. Review of GIS applications in hydrologic modeling. *Journal of Water Resources Planning and Management, ASCE* 119(2): 246–261.

Deckers F. and Te Stroet C.B.M. 1996. Use of GIS and database with distributed modeling. In: Abbott M.B. and Refsgaard J.C. (Eds), *Distributed Hydrological Modeling*, Kluwer Academic Publishers, Dordrecht, pp. 215–232.

Engman, E.T. and Gurney, R.J. 1991. *Remote Sensing in Hydrology*. Remote Sensing Applications Series. xiv-t-225 pp. Chapman & Hall, London, New York, Tokyo, Melbourne, Madras. ISBN 0 412 24450 0.

Hamby D.M. 1996. Site remediation techniques supporting environmental restoration activities: A review, *Journal of Science of the Total Environment* 191(3): 203–224.

Jackson T.J. 2002. Remote sensing of soil moisture: Implications for groundwater recharge. *Hydrogeology Journal* 10: 40–51.

Jackson T.J., Engman E.T., and Schmugge T.J. 1999. Microwave observations of soil hydrology. In: Parlange M.B. and Hopmans J.W. (Eds), *Vadose Zone Hydrology: Cutting Across Disciplines*. Oxford University Press, Inc., New York, pp. 317–333.

Jha Madan K., Chowdhury A., Chowdhury V.M., and Peiffer S. 2007. Groundwater management and development by integrated remote sensing and geographic information system: Prospect and constraint, *Journal of Water Resource Management* 21(2): 427–467.

Lane J.W., Buursink M.L., Haeni F.P., and Versteeg R.J. 2000. Evaluation of ground-penetrating radar to detect free-phase hydrocarbons in fractured rocks: Results of numerical modeling and physical experiments. *Ground Water* 38(6): 929–938.

Laumann S. 2013. Assessment of innovative *in situ* techniques for groundwater and soil remediation: Nano remediation and thermal desorption, chemical oxidation for groundwater remediation, TOSC Environmental Briefs for Citizens.

Longley P.A., Brooks S.M., McDonnell R., and Macmillan B. (Eds) 1998. *Geocomputation: A Primer*, John Wiley & Sons Ltd., Chichester, 290 pp.

Meijerink A.M.J. 2000. Groundwater. In: Schultz G.A. and Engman E.T. (Eds), *Remote Sensing in Hydrology and Water Management. Springer*, Berlin, pp. 305–325.

Olexsey R.A. and Parker R.A. 2006. Current and future *in situ* treatment techniques for the remediation of hazardous substances in soil, sediments, and groundwater, *Journal of Soil and Water Pollution Monitoring, Protection and Remediation*, Springer: 3–23.

Paterson N.R. and Bosschart R.A. 1987. Airborne geophysical exploration for groundwater. *Ground Water* 25: 41–50.

Pinder G.F. 2002. *Groundwater Modeling Using Geographical Information Systems*, John Wiley & Sons, New York, 248 pp.

Pivetz B. E. 2001. Phytoremediation of contaminated soil and ground water at hazardous waste sites, *EPA Journal of Ground Water* 1–36.

Rajan C.S. 2011. Nanotechnology in groundwater remediation, *International Journal of Environmental Science and Development*, 2(3): 182–187.

Reddy S. and Krishna R. 2008. Physical and chemical groundwater remediation technologies National risk management research laboratory, in Darnault C.J.G. (Ed.), *Overexploitation and Contamination of Shared Groundwater Resources* Chapter 12, NATO: 257–274. Springer Science+Business Media B.V.

Revzon A.L., Bgatov A.P., Bogdanov A.I., and Bogdanov A.M. 1988. Vsesoiuznyi Nauchno-lssledovatel'skii lnstitut Transportnogo Stroitel'stva, Moscow, USSR. lssledovanie Zemli iz Kosmosa. ISSN 0205-9614. July–August, 1988, pp. 41–48. (In Russian.)

Ross M.A. and Tara P.D. 1993. Integrated hydrologic modeling with geographic information systems. *Journal of Water Resources Planning and Management, ASCE* 119(2): 129–141.

Schultz G.A. 1993. Application of GIS and remote sensing in hydrology. In: Kovar K. and Nachtnebel H.P. (Eds), *Application of Geographic Information Systems in Hydrology and Water Resources Management*, IAHS Pub. No. 211, pp. 127–140.

Singh, P. 2009. *Engineering and General Geology*, 6th ed., S. K. Kataria and Sons, Delhi, India. pp. 502–503.

Sumathi S. 2009. Reductive remediation of pollutants using metals, Centre for environmental science and engineering, Indian institute of technology-Bombay, *The Open Waste Management Journal* Vol. 2: 6–16. Technical Options for the Remediation of Contaminated Groundwater, International Atomic Energy Agency, June 1999.

Suthersan S.S. (Ed.) 1999. *In Situ Air Sparging, Remediation Engineering: Design Concepts*, Boca Raton, FL: CRC Press.

Tsihrintzis V.A., Hamid R., and Fuentes H.R. 1996. Use of geographic information systems (GIS) in water resources: A review. *Water Resources Management* 10: 251–277.

Water dowsing, U.S. Department of the interior/geological survey. 2003. www.st-andrews.ac.uk/~crb/web/gwater1.pdf, 2006.

Watkins D.W. Jr., McKinney D.C., Maidment D.R., and Lin M.-D. 1996. GIS and groundwater modeling. *Journal of Water Resources Planning and Management* 122(2), 88–96.

Welch R. and Remillard M. 1991. Remote sensing/GIS for water resource management applications in the southeast, *Proceedings of the 1991 Georgia Water Resources Conference*, March 19–20, 1991, Athens, Georgia, pp. 282–284.

Zhang H., Haan C.T., and Nofziger D.L. 1990. Hydrologic modeling with GIS: An overview. *Applied Engineering in Agriculture, ASAE* 6(4): 453–458.

Zhang X.H. 2005. Remediation techniques for soil and ground water, *Journal of Point Sources of Pollution, Local Effects and It's Control* II: 350–364.

Zohdy A.A.R., Eaton G.P., and Mabey D.R. 1974. Application of surface geophysics to groundwater investigations, *U.S. Geological Survey, Techniques of Water-Resources Investigations* Book 2, Chapter D1: 1–65.

Section VI

Management of Water Resources and the Impact of Climate Change on Groundwater

26

Managing Groundwater Resources in a Complex Environment: Challenges and Opportunities

J. M. Ndambuki

CONTENTS

26.1 Introduction

In recent years, groundwater has become increasingly important in meeting water needs of industries and agriculture in developing countries. In the same breath, a lot of knowledge has been developed on how to manage groundwater with a view to sustaining those economic activities. With this in mind, the main issue remains as to how the knowledge generated at different parts of the world can be applied to ensure sustainability. While acknowledging the fact that groundwater occurs in large quantities globally, it is not immediately clear how much one can extract sustainably due to the complex nature of the formations in which groundwater exists.

Increasing world population coupled with deterioration of surface water quality due to pollution has resulted in more attention being focused on groundwater storage systems. However, overexploitation of this reserve has adverse impacts, for example, lowered water table and saltwater intrusion. To avoid such undesirable consequences, it is imperative that we understand the behavior of groundwater storage systems when subjected to external stresses. This, coupled with a management scheme will ensure efficient utilization of groundwater resources.

Optimal management of groundwater resources can be done either deterministically or stochastically. In a deterministic approach, it is assumed that all the input data is known without error while the stochastic approach recognizes the fact that uncertainty in input data is real, hence the need to adopt methods that take this fact into account. Furthermore, when the problem of uncertain data is compounded with the issue of different stakeholders' involvement, each having a different objective, the available solution methods of stochastic optimization which presupposes that the stakeholder's preferences are conveniently rendered through a single objective are no longer applicable. Hence, in such situations, the optimization problem should be posed within a multiobjective framework in which the analyst seeks to identify the Pareto set (also known as noninferior set) of solutions.

Pareto set is a collection of optimal solutions such that an improvement of one objective can only be achieved by degrading at least one of the other objectives. Among the works addressing groundwater management in a multiobjective framework include that of Ndambuki (2001), Kaunas and Haimes (1985), Willis and Liu (1984), Shafike et al. (1992), Gharbi and Peralta (1994), and Magnouni and Treichel (1994). Some methods that address the issue of data uncertainty in groundwater management include postoptimality analysis (e.g., Aguado et al., 1977; Gorelick, 1982), chance-constrained programming, and stochastic optimization with recourse method as presented by Ndambuki (2001), Wagner et al. (1992), Ndambuki et al. (2000a), and Mulvey et al. (1995).

26.2 Challenges and Opportunities of Managing Groundwater

From the foregoing, it is apparent that the world today faces many challenges and opportunities. Some of the challenges include: (1) The fact that groundwater aquifers do not respect either geographical boundaries or political demarcations, (2) due to the emerging climatic changes, historical trends can no longer be relied on to predict future occurrences, (3) the parameters that describe groundwater aquifers cannot accurately be determined both in space and in time, and (4) due to individuals' desire to satisfy their needs before consideration of the needs of others, conflicts over water resources are a common occurrence in many parts of the world.

However, the above challenges in themselves provide opportunities for: (a) collaborations among countries sharing the resources with a view to developing an exploitation and management strategy that tends to satisfy all (e.g., Nile Basin Initiative, Lake Victoria Management Programme, Orange-Senqu Initiative), (b) researchers developing management methods which enable communities to cope with the impacts of climate change, (c) scientists and engineers developing new management tools that recognize the uncertainty in groundwater parameter estimations, and (d) stakeholder involvement in the development and management of groundwater resources.

In this chapter, we address challenges (3) and (4), which translates into opportunities (c) and (d) through an optimization procedure that considers groundwater aquifer hydraulic conductivity as uncertain and develops a management strategy within a multiobjective (stakeholders) framework.

26.3 Methodology

Consider a linear optimization problem (LOP) of the following form:

$$\text{minimize } c^T x, \tag{26.1}$$

subject to:

$$a_i^T x \leq b_i, \quad i = 1, \dots, m, \tag{26.2}$$

$$x \geq 0, \tag{26.3}$$

where:

$c, a_i \in R^n$; $b_i \in R$ are the problem parameters while x are the optimization variables. Assuming that all the problem parameters except a_i are accurately known and that a_i are uncertain but lying in ellipsoids ε_i defined as:

$$a_i \in \varepsilon_i = a_i \left\{ \overline{a_i} + \mathbf{P}_i u_i \ \| u_i \| \leq 1 \right\}, \tag{26.4}$$

where:

$\mathbf{P} = \mathbf{P}^T$ are $n \times n$ perturbation matrices; $\overline{a_i}$ are the nominal values and the norm of u_i ensure convexity, then a robust solution of the optimization problem given by Equations 26.1 through 26.3 is as follows:

$$\text{minimize } c^T x, \tag{26.5}$$

subject to:

$$a_i^T x \leq b_i, \quad \forall a_i \in \varepsilon_i, \quad i = 1, \ldots, m, \tag{26.6}$$

$$x \geq 0. \tag{26.7}$$

An optimization problem defined by Equations 26.5 through 26.7, though now deterministic, has infinitely many constraints and a solution to this robust optimization problem is feasible if for all $i = 1, \ldots, m$, the following holds:

$$\left[\overline{a_i}^T x + (P_i u_i)^T x - b_i \right] \leq 0, \quad \forall u_i : u_i^T u_i \leq 1, \tag{26.8}$$

which can equivalently be reformulated by a single constraint as:

$$\overline{a_i}^T x + \|P_i x\| \leq b_i \tag{26.9}$$

Constraints of the form in Equation 26.9 are referred to as second-order cone constraints (otherwise known as Lorentz cone or ice-cream cone constraints). Thus, the optimization problem defined by Equations 26.1 through 26.3 can explicitly be written as a second-order cone optimization (SOCO) problem as follows:

$$\text{minimize } c^T x, \tag{26.10}$$

subject to:

$$\overline{a_i}^T x + \| P_i x \| \leq b_i, \quad i = 1, \ldots, m. \tag{26.11}$$

The norm term is the usual Euclidean norm and can be thought of as a penalty term which introduces some robustness within the optimization problem.

In order to take into consideration stakeholders' objectives, we transform the above single objective into a multiobjective problem. It is worth noting that in multiobjective optimization one is interested in optimizing (minimizing or maximizing) several objectives simultaneously (an approach also known as vector optimization). This approach recognizes the fact that not all the objectives can achieve their optimal values simultaneously unless the objectives are not competing (conflicting). This means that there is no unique solution to such problems. However, one may establish a specific numeric goal (also known as aspiration level) for each of the objectives and then seek a solution that minimizes the sum of deviations of the objective functions from their respective goals. This solution process is known as goal programming.

Consider now the n-dimensional vector space R^n. For any two points $r = (r_1, \ldots, r_n)$ and $\overline{r} = (\overline{r}_1, \ldots, \overline{r}_n)$, we can express the distance $d(r; \overline{r})$ as the norm given by $\|d\| = \|r - \overline{r}\|$. Thus, the distance between the two points can be measured by the L_p–metric (Holder's norm) as:

$$\| d \|_p = \left(\sum_{0=1}^{k} |r_k - \overline{r}_k|^p \right)^{(1-p)} \tag{26.12}$$

Let $\overline{Z} = \{\overline{Z}_1, \ldots, \overline{Z}_K\}^T$ be the aspiration levels (values of objectives which the various stakeholders wish to achieve on the various objectives) and $Z(x) = \{Z_1(x), \ldots, Z_k(x)\}^T$ be the objectives of the stakeholders. The distance between vectors $\overline{Z}$ and $Z(x)$ can be measured by the Holder's norm as:

$$\| \overline{Z} - Z(x) \|_p = \left(\sum_{k=1}^{k} |\overline{Z}_k - Z_k(x)|^p \right)^{(1/p)} \tag{26.13}$$

To avoid biased solution and express stakeholders' preferences toward the considered objectives, weighting, w can be applied to Equation 26.13, which then becomes

$$\| \overline{Z} - Z(x) \|_{w1p} = \left(\sum_{k-1}^{k} w_k |\overline{Z}_k - Z_k(x)|^p \right)^{(1/p)} \tag{26.14}$$

By using the L_2 metric, one would then solve an optimization problem of the following form in order to minimize the distance between $\overline{Z}$ and $Z(x)$ within the feasible set, Ω.

$$\text{minimize } \| \overline{Z} - Z(x) \|_{w,2} \tag{26.15}$$

where x is the optimization variable and the other parameters are as defined before. By introducing a scalar deviational variable δ, Equation 26.15 translates to the following:

$$\text{minimize } \delta \tag{26.16}$$

subject to:

$$\| \overline{Z} - Z(x) \|_{w,2} \leq \delta. \tag{26.17}$$

The above problem is a SOCO problem as long as Ω is defined by linear or second-order cone constraints. Note that inequality 26.17 implies that $\delta \geq 0$.

26.4 Formulation of a Groundwater Management Problem

In this problem, one of our objectives is to minimize the operational cost while the other objective is to maximize the amount of water extracted from the groundwater aquifer (the two objectives represent two stakeholders). This multiobjective SOCO problem is formulated as follows:

$$\text{minimize } [Z_1(x) = c^T x], \qquad (26.18)$$

$$\text{maximize } [Z_2(x) = e^T x], \qquad (26.19)$$

Subject to:

$$\bar{a}_i^T x + \| P_i^T x \| \le b_i, \quad i = 1, \dots, N_c, \qquad (26.20)$$

$$e^T x \ge W_d, \qquad (26.21)$$

$$0 \le x_j \le U_j, \quad j = 1, \dots, N_w, \qquad (26.22)$$

where $c = (c_1, \dots, c_j, \dots, c_{Nw})^T$, $a_i = (a_{i1}, \dots, a_{ij}, \dots, a_{i,Nw})^T$, $e = (1, \dots, 1_j, \dots, 1_{Nw})^T$, $x = (x_i, \dots, x_j, \dots, x_{Nw})^T$ and

- N_w is the number of pumping wells
- N_c is the number of control points
- c_1 is the aggregated daily cost of pumping and transportation in monetary units (MUs) per unit volume at cell j
- x_j is the pumping rate in cell j
- a_{ij} is the response at control point i due to pumping in cell j
- b_l is the constraining value at control point i
- W_d is total water demand
- U_j is the maximum pumping rate allowed in cell j

By introducing a deviational variable, δ, and considering the L_2 metric (Euclidean distance) as the measure of closeness between the aspiration levels $\bar{Z}_1$ and $\bar{Z}_2$ and the feasible objective region, the above multiobjective optimization problem (Equations 26.18 through 26.22) can be reformulated as:

$$\text{minimize } \delta, \qquad (26.23)$$

Subject to:

$$\| \bar{Z} - Z(x) \|_{w,2} \le \delta, \qquad (26.24)$$

$$\bar{a}_i^T x + \| P_i^T x \| \le b_i, \quad i = 1, \dots, N_c, \qquad (26.25)$$

$$e^T x \ge W_d, \qquad (26.26)$$

$$0 \le x_j \le U_j, \quad j = 1, \dots, N_w, \qquad (26.27)$$

where δ is a scalar variable, $\bar{Z} = (\bar{Z}_1, \bar{Z}_2)^T$ are the aspiration levels of objectives $Z(x) = (Z_1(x); Z_2(x))^T$ and objectives $Z(x)$ are as defined by Equations 26.18 and 26.19, respectively. The other parameters and variables are as defined before. To express stakeholders' preferences, preference values can be included in inequality 26.24 when such preferences exist.

SOCO problems can be solved efficiently through the interior-point methods that have been developed (Andersen and Andersen, 1999; Boyd et al., 1994; Sturm, 1999). A few applications of SOCO problems have been reported in the literature. They include antenna array weight design, filter design, grasping force optimization, portfolio optimization, truss design, and equilibrium of systems with piecewise-linear springs design. From the literature, the reported applications are basically in the areas of electrical engineering, mechanical engineering, economics, and structural engineering (Ben-Tal and Nemirovski, 1998; Boyd et al., 1994, 1998; Lobo et al., 1998; Ndambuki et al., 2000a,b).

26.5 Application of the Methodology to an Example

Our goal in this problem is to supply domestic water to a distribution center situated in the middle of a single confined square aquifer of thickness 35 m. The aquifer is an island of dimensions 30 km and with parameters as shown in Figures 26.1 through 26.3.

Objective: The objective is to supply the required amount of water at the lowest possible cost while at the same time satisfying the constraints and ensuring robustness of the optimal solution.

26.5.1 Constraints

- In the specified ecological protect zone, the minimum water level equals 5 m above sea level.
- The hydraulic head in all the nodes except those in contact with the sea are bounded by the bottom of the aquifer.
- To avoid saltwater intrusion, head in cells next to the sea are not allowed to fall below 0.2 m above sea level.

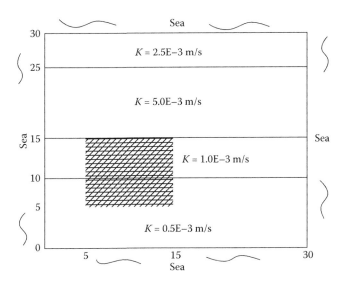

FIGURE 26.1
Ecological protection zone and conductivity zones.

- There is a minimum water demand of 3 m³/s at the distribution center, which should be satisfied.
- Pumping rates from potential wells shown in Figure 26.3 are limited to a maximum yield of 1.5 m³/s.

Unit costs of exploitation measured in monetary units (MUs) defined at each potential well are calculated as a combination of the water pumping costs and water transport costs, which depend on the distance from the well to the distribution center. The costs are assumed to be higher toward the boundaries. The aggregated unit cost coefficients at each cell take on values from 0.1 MU at the center and increase at a rate of 0.1 MU for every 200 m distance. Note that MU can be in any currency units.

The problem is to find the location of wells and the corresponding pumping rates satisfying the constraints. Moreover, the solution should be robust in an environment of uncertain spatial hydraulic conductivity values.

26.5.2 Discussion of the Results

For the analysis of the example we have outlined, 20 realizations of the uncertain hydraulic conductivity, k, were generated using the zonal k values shown in Figure 26.1, a standard deviation (log) of 0.5 and a correlation length of 30,000 m in x-direction and 7500 m in y-direction. These correlation lengths were chosen to replicate the hydraulic conductivity field of the example as used by Wagner et al. (1992) with a view to comparing the results. All the SOCO problems were solved using Sturm's SeDuMi package (Sturm, 1999).

The multiobjective SOCO problem solved (Equations 26.23 through 26.27) resulted in 50 decision variables (49 of them dealing with the location of pumping wells and their strengths and 1 variable measuring the deviation of the optimal solution from the aspiration levels), 101 linear deterministic constraints, 1 deterministic second-order cone constraint to minimize the deviation from the aspiration levels, and 225 second-order cone constraints to capture the robustness (coefficients of these constraints are uncertain, hence stochastic). The targets (aspiration levels) for the two objective functions were computed as $\bar{Z}_1 = 2.07$ MU and $\bar{Z}_2 = 4.47$ m³/s.

We then solved the multiobjective optimization problem defined by Equations 26.23 through 26.27 for various levels of robustness. As in the case of the single objective SOCO problem already discussed, a scaling factor η can be introduced in inequality 26.25 by replacing the perturbation matrices P_i by ηP_i where $\eta \geqslant 0$. If $\eta = 0$, it means that there is no uncertainty (the input parameters are exactly known), $\eta = 1$ means that the uncertainty is given by the matrices $P_i^T P_i$ and $\eta > 1$ it means that the uncertainty is higher than that given by the covariance matrices $P_i^T P_i$.

Table 26.1 shows the levels of robustness and the corresponding number of active wells and values of the operational cost (these results are for a guaranteed volume of water amounting to 3.0 m³/s). The results show that as the level of robustness η is increased from 0 to 0.3, the number of active wells will increase and consequently the operational cost. This is because increasing the level of robustness implies increase in the volume of the ellipsoid and hence one has to search for a solution within a zone of increasingly high uncertainty. The

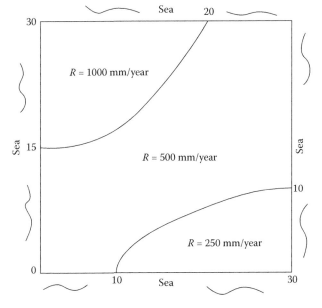

FIGURE 26.2
Recharge zones.

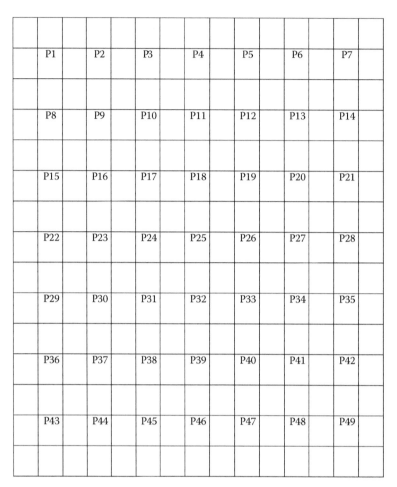

FIGURE 26.3
Location of potential pumping wells.

consequence is that each active well will pump less and since the guaranteed volume of water must be realized, then more pumping wells will have to be mobilized, resulting in higher operational costs.

Table 26.2 and Figure 26.4 show how the deviational variable, δ, varies with the level of robustness (remember that the deviational variable is a measure of the discrepancy between the aspiration level, in this case the ideal vector, and the optimal or compromise solution realized). It is interesting to note that as the level of robustness is increased, the deviational variable increases too. This is because an increase in robustness means an increase in the volume of the ellipsoid which implies that the solution sought must guard against a wider range of uncertainty, hence more conservative. The consequence is that such a solution will lie somewhere in the interior of the feasible set. Thus, as the level of robustness is increased further, the solutions will come from increasingly deeper into the feasible set.

Compromise solutions corresponding to various quantities of water extracted are given in Table 26.3 (these solutions are for 0.2 level of robustness). It

TABLE 26.1

Robustness versus Active Wells and Cost

Level of Robustness	Active Wells	Cost (MU)
0.0	11	2.160
0.1	14	2.176
0.15	14	2.185
0.2	16	2.191
0.25	19	2.193
0.3	19	2.195

TABLE 26.2

Deviational Variable versus Robustness

Level of Robustness	Deviational Variable
0.00	0.086
0.10	0.103
0.15	0.111
0.20	0.119
0.25	0.130
0.30	0.143

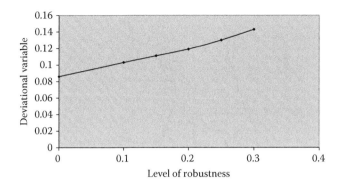

FIGURE 26.4
Robustness versus deviational variable.

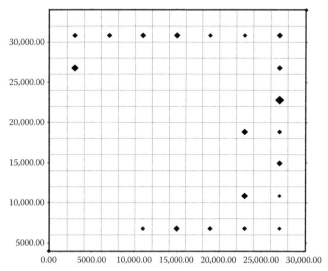

FIGURE 26.6
Optimal solution Number 3.

is apparent that as the quantity of water extracted is increased, the operational cost likewise increases. This depicts some tradeoff between the two conflicting and noncommensurate objectives. The optimal schemes corresponding to the optimal solutions Numbers 1, 3, and 5 are depicted in Figures 26.5 through 26.7. These optimal solutions are chosen to depict how the pumping strategies evolve under different objective tradeoffs.

TABLE 26.3

Volume versus Cost

Optimal Solution Number	Volume (m^3/s)	Cost (MU)
1	3.0	2.191
2	3.15	2.234
3	3.25	2.348
4	3.3	2.419
5	3.45	2.670

The results indicate that as the water demand is increased from a minimum of 3.0 m^3/s to a maximum of 3.45 m^3/s, the operational cost increases from a minimum of 2.191 MU to a maximum of 2.670 MU. The number of active pumping wells range from a minimum of 14 wells to a maximum of 20 wells.

Ndambuki et al. (2000b) compared the optimal solutions for the Monte Carlo approach (which they referred to as stochastic scenario) and the SOCO approach (which they referred to as robust scenario). Their results showed that the robust scenario gives rise to a more expensive optimal strategy. This is because the robust optimal strategy operates more wells (16 wells) compared to the

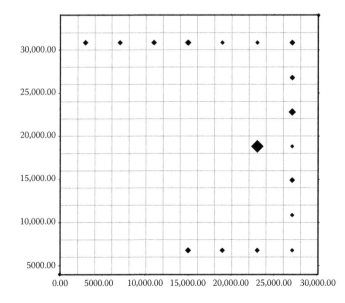

FIGURE 26.5
Optimal solution Number 1.

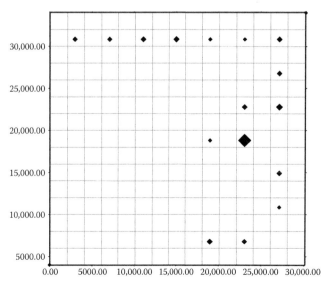

FIGURE 26.7
Optimal solution Number 5.

Monte Carlo optimal strategy, which operates 7 wells. This suggests that Monte Carlo approach gives rise to optimistic solutions compared to the SOCO approach. The authors further performed a postoptimality sensitivity analysis to evaluate how the two optimal solutions would perform in an environment of uncertainty. Their results showed that the SOCO approach gives rise to more stable solutions than the Monte Carlo approach (total constraint violations are less than those of the Monte Carlo approach solution). For a detailed comparison of the performance of the optimal solutions of the two approaches, see Ndambuki (2001).

26.6 Application of a Variant of the Methodology to a Real World Aquifer

This application is based on stochastic programming with recourse (where violations of constraints are penalized in accordance with the magnitude of violations) and differs from those of Wagner et al. (1992), Mulvey et al. (1995), and Ndambuki et al. (2000b) in that the groundwater problem is explicitly solved within a multiobjective framework, that is, we do not use weights (a form of indicator of preference) to synthesize the multiobjective problem into a single objective one.

26.6.1 Study Area

The research was conducted within Ewaso Ng'iro catchment situated in the central part of Kenya. This area extends from about 0° 15′ south of the equator to 1° 30′ north and between longitudes 36° 15′ east and 39° 45′ east (Ministry of Water Development, 1987). It is the domain formed by the area of Upper Ewaso Ng'iro Basin upstream of the confluence of Ewaso Ng'iro and Ewaso Narok Rivers and the area southeast of the confluence. The area shown as Figure 26.8 is wholly underlain by the Precambrian basement rocks and is bounded to the west by the Laikipia Escarpment and to the southeast and southwest by the volcanic rocks of the Aberdares (3999 m) and Mount Kenya (5199 m). The rocks from the higher parts of the study area are mainly volcanic, while in the lower areas, the metamorphic rocks of the basement system predominate.

26.6.2 Boundary Conditions

26.6.2.1 Specified Head Boundary Conditions

This type of boundary condition is also known as Dirichlet condition for which head is given. In any

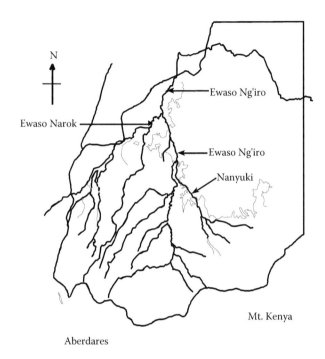

FIGURE 26.8
Study area showing the main rivers.

particular model, this boundary represents points at which the head is known. In real situations, such boundary conditions occur as recharge boundaries or areas beyond the influence of hydraulic stresses and are defined by known equipotential lines. In this study, this boundary condition was specified for the northern side (see Figure 26.9).

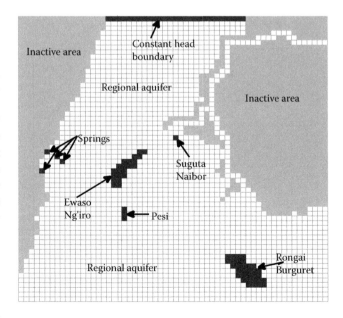

FIGURE 26.9
Modeled area.

26.6.2.2 Specified Flow Boundary Conditions

This type of boundary condition consists of no-flow and constant flux boundaries. In the case of a no-flow boundary condition (also known as Neumann condition), the derivative of the head (flux) across a boundary is given. This boundary condition is set by specifying flux to be zero. This means that flow lines are parallel to such boundaries. In the modeled area considered, this boundary condition was specified for the other three boundaries, that is, south, east, and west (these are areas bounded by Mt. Kenya and the Aberdares ranges).

26.6.3 Initial Conditions

Initial conditions refer to the head distribution everywhere in the system at the start of the simulation. For steady-state simulations, initial condition is not important because the interest here is to determine drawdown in response to imposed stresses (pumping). In such a simulation, relative heads as measured by drawdown are of interest, rather than absolute values of head. In this study, steady-state simulation was considered, so initial condition is not critical.

26.6.4 Discretization of the Model

To delineate the area of interest (regional aquifer system) from the local aquifers, the whole area of study was discretized into 60×60 cells each of 1850×1850 m. The modeled area of interest is shown in Figure 26.9. Three main rivers, that is, Ewaso Ng'iro, Ewaso Narok, and Nanyuki were also modeled within the area of interest (see Figure 26.8). Further, four swamps (Ewaso Narok, Pesi, Rongai Burguret, and Suguta Naibor) and four perennial springs were included (as sinks) in the modeling exercise as well (see Figure 26.9).

26.6.5 Objectives

Three objectives were considered and formulated as follows:

Objective Z_1: minimization of operational cost

$$\text{Minimize } Z_1 = \text{Minimize} \sum_{j=1}^{N_w} \lambda_j r_j x_j \quad (26.28)$$

where:

λ_j is the daily cost of pumping a unit cubic meter of water a height of one meter from pumping well j

r_j is the pumping lift at pumping well j
N_w is the number of potential pumping wells
x_j are the spatially distributed pumping rates (optimization variables)

The cost objective demands that wells are operated in such a manner as to minimize the total cost. This essentially requires the identification of well locations as well as determination of their pumping rates, which, when operated would satisfy the required water demand at the least possible cost. Since the study area is already developed, investment cost was ignored. The general planning problem was, therefore, to determine how the existing well-field should be operated to satisfy the projected water demand. This objective, therefore, entails minimization of operational costs incurred in the process of satisfying the water demand.

The cost of pumping 1 m³ of water a height of 1 m per day, λ_j is proportional to the energy consumed. But energy consumed is proportional to quantity of water pumped and the height of lift, that is, $E \propto x_j r_j$. Here, the proportionality constant is water density (ρ_w) and gravitational acceleration (g_a). Hence, the energy consumed, $E = \rho_w g_a x_j r_j$. Thus, knowing the energy required to lift 1 m³ of water a distance of 1 m, and the cost of 1 kWh (joule) of energy, it is possible to calculate λ_j and hence the minimum operational cost.

Objective Z_2: maximization of total pumping rates

$$\text{Maximize } Z_2 = \text{Maximize} \sum_{j=1}^{Nw} x_j \quad (26.29)$$

Objective Z_3: minimization of penalty cost arising from violation of constraints

As a result of uncertainty in the input parameters (transmissivity), there exists some probability of the designed optimal strategy violating the problem constraints. The cost incurred in the form of penalties due to violation of such constraints is therefore to be minimized. This objective, therefore, addresses uncertainty due to transmissivity. This objective was formulated as:

$$\text{Minimize } Z_3 = \text{Minimize} \sum_{j=1}^{N_w} \bar{\rho}_j x_j \quad (26.30)$$

where $\bar{\rho}_j$ is the aggregated mean penalty cost associated with violation of constraint at all control points for all realizations due to unit pumping rate at cell j.

26.6.6 Constraints

26.6.6.1 Global Water Demand Constraint

The aquifer was considered as the sole source of water. This constraint was considered a hard constraint (not to be violated) to ensure that the minimum water demand was met. It was formulated as follows:

$$\sum_{j=1}^{N_w} x_j \geq D_g \tag{26.31}$$

where D_g is the global (total) water demand within the modeled domain.

26.6.6.2 Local Water Demand Constraints

Because of differences in population distribution and hence differences in water demand within the modeled area, three water demand zone constraints were considered (see Figure 26.3). These constraints were formulated as:

$$\sum_{j=1}^{N_w} x_{j,l} \geq D_l \tag{26.32}$$

where:

D_l are the water demands in demand zones l, $l = 1$, ..., 3

$x_{j,l}$ are the pumping rates in wells j located in local water demand zones l, $l = 1, ..., 3$

26.6.6.3 Pumping Constraint

Pumping rates at potential pumping wells in the water demand zones were constrained to values between some minimum and maximum rates and were of the form:

$$x_j^{\min} \leq x_j \leq x_j^{\max}; \quad j = 1, ..., N_w \tag{26.33}$$

where:

$x_j^{\min}$ and $x_j^{\max}$ are the minimum and maximum permissible pumping rates, respectively, N_w is the total number of potential pumping wells

26.6.6.4 Drawdown Constraints

Drawdown constraints were formulated to avoid mining and were formulated as:

$$\sum_{j=1}^{N_w} a_{i,j} x_j \leq b_i; \quad i = 1, ..., N_c \tag{26.34}$$

where:

$a_{i,j}$ is the drawdown at control point i caused by a unit pumping from pumping well j

b_i is the permissible drawdown at control point i

N_c is the total number of points at which the drawdowns are controlled

Since the transmissivity values are taken as uncertain, the drawdowns $a_{i,j}$ become dependent on the transmissivity field, ω, realized. Hence, inequality Equation 26.34 changes to the following form:

$$\sum_{j=1}^{N_w} a_{\omega,i,j} x_j \leq b_i; \quad i = 1, ..., N_c; \quad \forall \omega \in \Theta \tag{26.35}$$

This set of constraints was formulated as soft constraints, meaning that they could be violated but at some cost commensurate with the degree of violation. These constraints were formulated as follows:

$$\sum_{j=1}^{N_w} a_{\omega,i,j} x_j - b_i \leq q_{\omega,i,j}; \quad i = 1, ..., N_c; \quad \forall \omega \in \Theta \tag{26.36}$$

where:

$q_{\omega,i,j}$ is the amount of violation which is penalized whenever positive

26.6.6.5 Total Recharge Constraint

This constraint was meant to ensure that mining of the resource is avoided and was formulated as follows:

$$\sum_{j=1}^{N_w} x_j \leq R_t \tag{26.37}$$

where:

R_t refers to the total recharge in the well field.

26.7 Statement of the Management Problem

The optimization problem solved, therefore, was as follows:

$$\text{Minimize } Z_1 = \text{Minimize} \sum_{j=1}^{N_w} \lambda_j r_j x_j \tag{26.38}$$

$$\text{Maximize } Z_2 = \text{Maximize} \sum_{j=1}^{N_w} x_j \tag{26.39}$$

$$\text{Minimize } Z_3 = \text{Minimize} \sum_{j=1}^{N_w} \bar{\rho}_j x_j \qquad (26.40)$$

Subject to the following constraints:

$$\sum_{j=1}^{N_w} x_j \geq D_g \qquad (26.41)$$

$$\sum_{j=1}^{N_w} x_{j,l} \geq D_l \qquad (26.42)$$

$$x_j^{\min} \leq x_j \leq x_j^{\max}; \quad j = 1, \ldots, N_w \qquad (26.43)$$

$$\sum_{j=1}^{N_w} a_{\omega,i,j} x_j - b_i \leq q_{\omega,i,j}; \quad i = 1, \ldots, N_c; \quad \forall \omega \in \Theta \qquad (26.44)$$

$$\sum_{j=1}^{N_w} x_j \leq R_i \qquad (26.45)$$

26.7.1 Input Data

The following data were used in this study:

Recharge 24.7×10^3 m^3 d^{-1}

Projected global (year 2000) water demand 9.46×10^3 m^3 d^{-1}

Projected zone 1 water demand 2.07×10^3 m^3 d^{-1}

Projected zone 2 water demand 5.13×10^3 m^3 d^{-1}

Projected zone 3 water demand 2.26×10^3 m^3 d^{-1}

26.7.2 Results

Here, it was assumed that the transmissivity values were precisely known at measurement points and the uncertainty was due to the transmissivity data not being adequate to fully characterize the aquifer. As a result, transmissivity realizations were generated to address the uncertainty arising from inadequacy of the data. Further, the amount of water demand was assumed to correspond to the volume being supplied 9.46×10^3 m^3 d^{-1}. The basic idea was to investigate whether the aquifer could be operated differently (and hopefully in an optimal way) to ensure minimum operational cost while at the same time meeting the requirement of the water demand.

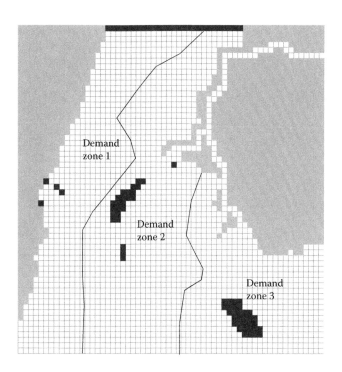

FIGURE 26.10
Local demand zones.

The optimization problem was solved using two standard codes, namely Modflow (McDonald and Harbaugh, 1984) and MINOS (Murtagh and Saunders, 1995). Results show that while two water demand zones (water demand zones 1 and 3, see Figure 26.10) were able to meet their local water demands, water supplied from the other demand zone (water demand zone 2) was not enough to meet its local demand. Comparison of Figure 26.11 (unoptimized scenario) with Figure 26.12 (optimized scenario) show that at water demand zones where the local water demands were met, some pumping wells operate at capacities lower than their maximum. Note that the solid circles in Figures 26.11 and 26.12 indicate that the absolute differences in pumping rates between the unoptimized scenario (Figure 26.11) and optimized scenario (Figure 26.12) are greater than 20%.

26.8 Conclusions

In this chapter, it is shown that when confronted with uncertain multiobjective optimization problems, such problems can be cast as SOCO problems which are efficiently solved by interior-point methods. We have further demonstrated, through an example, how one could apply this novel tool to efficiently manage water

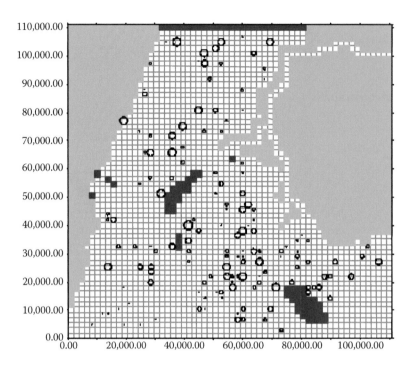

FIGURE 26.11
Existing scenario (unoptimized scenario).

resources in an environment of uncertainty. Using this methodology, one can easily increase or decrease the robustness of the solutions and therefore be able to choose an optimal strategy taking into account the values of the objectives considered and the level of robustness such a solution is able to guarantee to every DM to be presented with such robust solutions. Further, we have shown that through the multiobjective optimization approach, different stakeholder' objectives can be addressed, thus avoiding conflict.

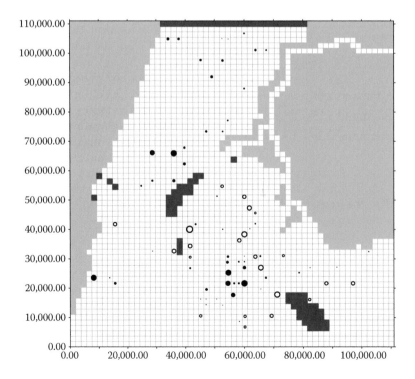

FIGURE 26.12
Optimized scenario.

References

Andersen, E.D. and K.D. Andersen. 1999. Exploiting parallel hardware to solve optimization problems. *SIAM News* 32(4): 1–7.

Aguado, E., Sitar, N., and I. Remson. 1977. Sensitivity analysis in aquifer studies. *Water Resources Research* 13(4): 733–737.

Ben-Tal, A. and A. Nemirovski. 1998. *Convex Optimization in Engineering*. Technion-Israel Institute of Technology, Haifa, Israel.

Boyd, S., Crusius, G., and A. Hansson. 1998. Control applications of non-linear convex programming. *Journal of Process Control* 8(5–6): 313–324.

Boyd, S., Vandenberghe, L., and M. Grant. 1994. Efficient convex optimization for engineering design. In: *Proceedings of the IFAC Symposium on Robust Control Design*. Rio-de-Janeiro, Brazil.

Gharbi, A. and R.C. Peralta. 1994. Integrated embedding optimisation applied to Salt Lake Valley aquifer. *Water Resources Research* 30(3): 817–832.

Gorelick, S.M. 1982. A model for managing sources of groundwater pollution. *Water Resources Research* 18(4): 773–781.

Kaunas, J.R. and Y.Y. Haimes. 1985. Risk management of groundwater contamination in multi-objective framework. *Water Resources Research* 21(11): 1721–1730.

Lobo, M.S., Vandenberghe, L., Boyd, S., and H. Lebret. 1998. Applications of second-order cone programming. *Linear Algebra and its Applications* 284: 193–228.

Magnouni, S.E. and W. Treichel. 1994. A multi-criterion approach to groundwater management. *Water Resources Research* 30(6): 1881–1895.

Mcdonald, M.G. and A.W. Harbaugh. 1984. *A Modular Three-Dimensional Finite Difference Groundwater Flow Model*. Scientific Publishers Co., Colorado.

Ministry of Water Development. 1987. Water resources assessment study of Laikipia District.

Mulvey, J.M., Vanderbei, R.J., and S.A. Zenios. 1995. Robust optimisation of large scale systems. *Operations Research* 43(2): 264–281.

Murtagh, B.A. and M.A. Saunders. 1995. *MINOS Users Guide*. Systems Optimization Laboratory, Stanford University, California.

Ndambuki, J.M. 2001. *Multi-Objective Groundwater Quantity Management: A Stochastic Approach*. Delft University Press, Delft, Netherlands.

Ndambuki, J.M., Otieno, F.A.O., Stroet, C.B.M., and E.J.M. Veling. 2000a. Groundwater management under uncertainty: A multi-objective approach. *Water SA* 26(1): 35–42.

Ndambuki, J.M., Stroet, C.B.M., Veling, E.J.M., and T. Terlaky. 2000b. Robust groundwater management through second order cone optimization approach. In Oliver S. et al. (Ed.) *Groundwater: Past Achievements and Future Challenges* Balkema, Rotterdam, Netherlands, pp. 413–417.

Shafike, N.G., Duckstein, L., and T. Maddock. 1992. Multicriteria analysis of groundwater contamination management. *Water Resources Bulletin* 28(1): 33–43.

Sturm, J.F. 1999. Using SeDuMi 1.02, a Matlab toolbox for optimization over symmetric cones. In: Potra F., Roos C., and Terlaky T. (Eds.) *Optimization Methods and Software* Vol. 11–12: 625–654, Special Issue on Interior Point Methods. Taylor & Francis, UK.

Wagner, J.M., Shamir, U., and H.R. Nemati. 1992. Groundwater quality management under uncertainty: Stochastic programming approaches and the value of information. *Water Resources Research* 28(5): 1233–1246.

Willis, R. and P. Liu. 1984. Optimisation model for groundwater planning. *Journal of Water Resources Planning Management* 110(3): 333–347.

27

Water Availability and Food Security: Implication of People's Movement and Migration in Sub-Saharan Africa (SSA)

Gurudeo Anand Tularam and Omar Moalin Hassan

CONTENTS

27.1 Introduction

The world's population has been increasing over time not only due to natural increase, but also because of the critically important medical advances made over time. In the 1950s, there were around 2.5 billion people (Haub, 2011) but today the population has increased to approximately 7 billion. It is projected to reach about 11 billion by the end of twenty-first century. Although there have been predictions, Figure 27.1 shows three possible scenarios of world population levels based on fixed fertility rates (United Nations Department of Economic and Social Affairs [DESA], 2013a,b). In 1950, about 0.18 billion people lived in Sub-Saharan Africa (SSA) but by 2012 it had reached approximately 0.9 billion. It is predicted that the population of SSA will reach 3.8 billion by the end of twenty-first century (DESA, 2014). According to FAO (Food and Agricultural Organisation—United Nations), the population of SSA will grow faster than the rest of the world given the higher fertility and much improved health care; the death rates among children have dropped significantly (Kohler, 2012). The average birth of the Sub-Saharan African mother is 5.2 children and in some countries the rate is significantly larger such as Niger (7.6), which is the highest in the world today. Overall, the SSA fertility rate is about three times more than that of a European mother (1.6 children per women). Figure 27.2 shows world population in 2013 together with the projected population in 2050 in billions by region.

Urbanization has been growing faster around the globe and especially in the developing countries. In 2000, more than half of the world's population apparently lived in our cities and according to DESA (2014), around 72% of the world population will be living in cities by the year 2050. In SSA, 11% of the population lived in urban areas in 1950 but this figure is projected to reach about 49% by 2025 (Table 27.1) (Bigombe and Khadiagala, 2003). Nsiah-Gyabaah (2003) has argued that SSA has the least population moving to urban areas but the rate of flow is the fastest when compared to the rest of the world. Around 32% of SSA urban population live in only a few large cities such as Lagos, Monrovia, Accra, and Nairobi, among others. The cities attract people mainly because of their higher level of economic activities leading to seemingly better living conditions and employment. United Nations (UN) estimates that half of SSA population will live in cities by 2020 (Nsiah-Gyabaah, 2003). The population increase and rapid urbanization has placed much pressure on water and land but such pressures are noted more in the developing countries, which are more likely to be in water and food stress conditions now and in the future (Asif, 2013; Thornton and Herrero, 2010).

Water is essential for all aspects of life and used for many purposes but is most essential for human consumption (Tularam, 2012; Tularam and Ilahee, 2007a,b, 2008, 2010; Tularam and Krishna, 2009; Tularam and Reza, 2016). The economic activities, agricultural development, and environmental systems could not exist without water

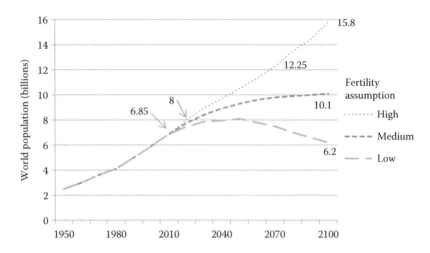

FIGURE 27.1

World population according to three different assumptions about future fertility. (From DESA, 2013a. *Seven Billion and Growing: The Role of Population Policy in Achieving Sustainability.* United Nations Department of Economic and Social Affairs, Population Division. Technical Paper, No. 2011/3.)

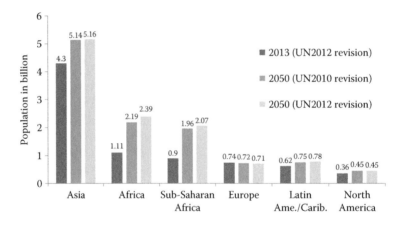

FIGURE 27.2

Current and projected population by region. (From United Nations, 2013. *Making the Most of Africa's Commodities: Industrializing for Growth, Jobs and Economic Transformation.* Economic Commission for Africa, Addis Ababa, Ethiopia.)

TABLE 27.1

Urbanization in Sub-Saharan African Regions (1950–2025)

Region	Percentage of the Population in SSA Urban Areas			
	1950	1975	1996	2025
Sub-Saharan Africa	11	21	32	49
Eastern Africa	5	13	23	39
Middle Africa	14	27	33	50
Southern Africa	38	44	48	62
Western Africa	10	23	37	56

Source: Bigombe, B. and Khadiagala, G.M., *Families in the Process of Development: Major Trends Affecting Families in Sub-Saharan Africa.* United Nations: New York, pp. 155–193, 2003.

(Davidson-Harden et al., 2007; Meeks, 2012; Reza et al., 2013; Roca and Tularam, 2012; Roca et al., 2015; Tularam and Marchisella, 2014; Tularam and Murali, 2014, 2015;

Tularam and Reza, 2016). Nelson Mandela said, "Among the many things that I learnt as president was the centrality of water in the social, political and economic affairs of the country, the continent and the world" (Msangi, 2014, p. 21; Reza et al., 2013; Roca et al., 2015; Steele and Schulz, 2012; Tularam and Reza, 2016). In the developing world including SSA, the lack of water infrastructure seems to be a major problem (Roca and Tularam, 2012; Tularam and Ilahee, 2007a,b). Many households spend hours collecting water for domestic consumption. In rural SSA, it is estimated that an average round trip to collect drinking water is around 36 minutes (Meeks, 2012). The quantity of freshwater per capita in the SSA has been falling—it fell by 30% just in the past two decades (Indraratna et al., 2001b; Jenna et al., 2012; Tularam, 2014).

There are a number of definitions that have been used to describe water scarcity but most of these definitions have not received an unqualified recommendation.

One of the main requirements in the definition is that water scarcity definition must suggest possible ways to conduct both qualitative and quantitative assessments. The World Water Development Report summarizes three of these definitions and defines water scarcity as: "The point at which the aggregate impact of all users impinges on the supply or quality of water under prevailing institutional arrangements to the extent that the demand by all sectors, including the environment, cannot be satisfied fully, a relative concept that can occur at any level of supply or demand. Scarcity may be a social construct (a product of affluence, expectations, and customary behavior) or the consequence of altered supply patterns stemming from climate change. Scarcity has various causes, most of which are capable of being remedied or alleviated…" (FAO, 2012; p. 5).

Water scarcity is said to exist when the freshwater availability per capita per annum is less than 1000 m³— the minimum agreed amount of freshwater for human survival is around 35,318.3 ft³/yr (1000 m³/yr) (Tularam and Marchisella, 2014). Water stress situation exists when freshwater availability per capita per year is higher than 1000 m³/capita/yr and less than 1700 m³/capita/yr (Steinman et al., 2004; Victoria, 2012). On the other hand, water shortage is defined as the ratio of the total freshwater withdrawal to the available rainfall (USDE, 2006). The condition of water scarcity exists when freshwater demand exceeds the water supply available in certain situations and/or places—water requirements of some sectors are not met. Water scarcity can also be expressed in a formula: Water scarcity = an excess of water demand over available supply (FAO, 2012).

The African countries have been playing a crucial role with the international group of action of water reform, which aims for Integrated Water Resources Management (IWRM) (Pietersen et al., 2006). In SSA, the water reform principles are the same as the water reforms in other parts of the world (Barbara, 2003). However, the implementation of these principles has shown several differences. SSA has sufficient water that can satisfy water demands in the region but these water stores have not been utilized in economic terms and this is called economic scarcity of water (Carles, 2009). In the main, this happens when human factors, financial constraints, and poor infrastructure and institutions hinder or in some cases stop people from accessing adequate water (Tularam, 2014; Tularam and Properjohn, 2011). Other factors have further compounded the water security issue such as climate change and rainfall variability, overexploitation of natural resources, and environmental degradation. The increasing population growth rate has itself led to increasing demands and together these factors have had a significant impact on the worsening state of freshwater availability (Pietersen et al., 2006; Tularam and Keeler, 2006; Tularam and Singh, 2009).

While the SSA region appears to have enough water availability throughout the year (UNDP, 2012a,b), the distribution of these waters is uneven between countries. In the central Africa region, the weather is humid and semihumid with sufficient rainfall and there is an abundance of water resources. In contrast, dry and semidry regions experience temperate and semitemperate climate with major differences in the amount of rainfall (Tularam, 2010). The precipitation is of high intensity but within short periods of time often causing floods and rainwater runoff thus washing fertile topsoil downstream (Mekdaschi and Liniger, 2013). The floods of 2000 in Zambia, Angola, and Mozambique have not helped these countries for they have all experienced droughts over the past 3 decades (Boko et al., 2007). The fluctuation, frequency, and amount of rainfall in SSA are equally met by increases in dryness levels of the country or subregion (Climate and Development Knowledge Network, 2012).

In this chapter, the authors will critically review the situation in SSA in terms of water security and food security in the region that may subsequently cause movements of people either within Africa or away from SSA totally. This study will compare and contrast using a country by country analysis, water infrastructure improvements, population dynamics, arable land, and agricultural contributions to the gross domestic product (GDP) of selected countries in SSA. Based on this, the chapter identifies and discusses some pull and push factors that appear to affect the population flow from rural to urban migration in SSA. Finally, some longer term implications of such flows across SSA are also discussed in light of its impact on countries such as Australia. A background section follows that provides an overall view of the SSA region showing the countries within the region, some definitions of water security, food security and environmental refugee, as well as the nature of the tensions that exist within the region in terms of water availability.

27.2 Background

Fourteen African countries have water stress and another 11 countries are estimated to become water stressed by 2025 (Abel Mejía et al., 2012; UN, 2008). Yet Africa has the longest river Nile including others like Congo, Zambezi, and Niger. In addition, Africa has the world's second largest lake; namely, Lake Victoria (Abel Mejía et al., 2012; Ashton, 2002; Ashton and Turton, 2007). Despite this, Africa is the second driest continent in the world after Australia (ADF VII, 2010; Cherry, 2009) and Ashton (2002) reported that around 53% of African land is water abundant sheltering about 61% of the total

TABLE 27.2

Projections of Population and Area and Water Availability According to Three Different Water Availability Conditions

Countries with Water	2000			2025		
	Area %	Population %	Water %	Area %	Population %	Water %
Abundance	52.5	60.8	95.2	34.7	23.9	78.3
Scarcity	26.0	24.3	4.4	39.1	57.3	20.6
Deficit	21.5	14.9	0.4	26.2	18.8	1.1
Total African population		786 million			1428 million	

Source: Ashton, P.J., *Ambio*, 31(3), 236–242, 2002.

population with about 95% of the total renewable water resources. But by 2025, the water abundance area may shrink to 35% and about 24% of the population will live in water abundance areas holding 78% of available renewable water resources. The water deficit and scarcity area, however, will increase from 47% (2000) to 65%, holding only 22% of the total renewable water resources and sheltering 76% of the population (Table 27.2).

SSA represents 18% of the world's landmass (Eberhard et al., 2008) and has an average annual rainfall of 815 mm/yr and rainfall amounts vary greatly in subregions and countries because of climatic differences (Jean-Marc and Guido, 2008). For example, in northern Niger the precipitation reaches as low as 10 mm/yr and less than 100 mm/yr in parts of South Africa and eastern Namibia (Jean-Marc and Guido, 2008). In contrast, rainfall is as high as 2000 mm/yr in countries like Sierra Leone and Liberia in the Gulf of Guinea and Seychelles and Mauritius, which are islands in the Indian Ocean. The rainfall in the African continent is unevenly distributed (Hell et al., 2010) and places where fewer people live often receive large amounts compared to the overpopulated regions. The records show that about 30% of the total available water in the continent flows to areas where only around 10% of Africans live. In SSA countries excepting South Africa, agriculture provides around 27% (2005) of the GDP (Livingston et al., 2011). The agricultural sectors are often self-sustained and rely on rain-fed farming systems but they have the lowest productivity levels when compared to others in the world (Calzadilla et al., 2011; Temesgen, 2012). During 1980–2000, SSA was the only region with a declining food production per capita level. Calzadilla et al. (2011) noted a number of reasons for this; that low investment of infrastructure and irrigation, lack of knowledge and limited access to services in addition to other agro-ecological features. Essentially, this suggests that while the population in the region has been growing faster than the rest of the world, the agricultural industry has not been responding to the growing demand for food and other agricultural products (Biazin et al., 2012).

Food security is a commonly used term in the literature review (Gross et al., 2000; Simon, 2012). Essentially, food security has been used to indicate whether people in an individual country had the chance to get sufficient food with necessary dietary and energy components (Pinstrup-Andersen, 2009). The term self-sufficiency is used to describe a state when a certain country assures and provides the quantity of food that its people demand (Pinstrup-Andersen, 2009). In the African continent and in particular, the SSA region, there has been significant achievements already gained of the millennium goals, which include halving poverty and eradicating hunger by 2015 (UNDP, 2006; UNDPI, 2013). Yet dealing with water availability, climate change, and population growth have had limited the success (Godfray et al., 2010).

According to Jean-Marc and Guido (2008) natural resources are abundant in SSA but there was still a 0.6% fall in the average GDP per capita overall in 2004 compared to 1975 (Biazin et al., 2012; Temesgen, 2012). For the poor people, agriculture is the main source of food and therefore rain-fed agriculture is more likely to be dominating the food production systems. The present study is focused on the SSA region in Africa and it is shown in Figure 27.3. It consists of the whole of the African continent excepting Algeria, Libya, Morocco, Egypt, and Tunisia (UNEP, 2002, 2009). Geographically this region falls below the Sahara Desert, with an estimated total area of about 24.24 million sq. km (Livingston et al., 2011). In fact, the SSA consists of four of Africa's five subregions, namely: East Africa, West Africa, Central Africa, and Southern Africa as shown in Figure 27.4 (Charles and Soumya, 2008). In 2009, about 260 million people (41% of SSA's population) lived in dry land, which is vulnerable to drought (Temesgen, 2012).

Jean-Marc and Guido (2008) reported that SSA receives more than 3880 km^3 of internal renewable water resources yearly. In terms of water resources, the richest countries are Congo (3618 km^3/yr) and Madagascar with 5740 m^3/ha/yr followed by Guinea and Central Africa with 4490 and 3520 m^3/ha/yr, respectively. About 40% of the precipitation in SSA (7500 km^3/yr) falls in Central Africa and this region accounts for only 23% of SSA population. The uneven distribution of rainfall in the SSA dictates the necessity to adapt better systems and policies to maximize the usage of available

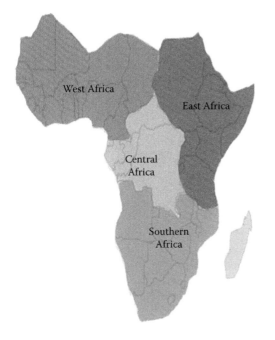

FIGURE 27.3

Four regions of sub-Saharan Africa. (From African Studies Centre. 2014. *Regional Perspectives: Geographic Regions in Africa,* available on: http://exploringafrica.matrix.msu.edu/teachers/curriculum/m20/ activity1.php Michigan State University, retrieved January 20, 2015.)

water (Indraratna et al., 2001a,b). The efficient systems allow the local communities to achieve a sustainable agricultural production, social stability, and economic development (Winterbottom et al., 2013). Indeed, there are several rainwater harvesting and managing systems in SSA; techniques like *in situ* and microcatchment are more popular than rainwater irrigation methods (Temesgen, 2012).

From 1960 to 2005 there has been a significant decline of the per capita share of the internal renewable water resources in SSA (Jean-Marc and Guido, 2008; Rosegrant, 1997). This is due mainly to the population increase. There has been an average fall of 65% (it has dropped from 16,500 m^3 per resident in 1960 to 5500 m^3/resident in 2005). In some cases, in countries like Uganda, Ivory Coast, and Niger, the decline has been more severe. Table 27.3 shows changes in per capita water availabilities by region during 1950–2000. Water availability per capita shows the highest decline in Africa, dropping from 20,000 m^3 of water per person in 1950 to 5100 m^3 of water per person in 2000; that is, Africa lost about three quarters of its share from the available water within 50 years (see Table 27.4).

The arid regions in SSA are more susceptible to food shortage, widespread starvation, frequent hunger, and

FIGURE 27.4

Sub-Saharan Africa. (From Charles, H.T. and A. Soumya, *Reducing Child Malnutrition in Sub-Saharan Africa: Surveys Find Mixed Progress.* Population Reference Bureau, 2008.)

TABLE 27.3

Water and Agriculture in Sub-Saharan Africa

Variable	Unit	SSA	World	SSA as a % of the World
Total area	1000 ha	2,428,795	13,442,788	18.1%
Estimated cultivated area 2007	1000 ha	234,273	1,865,181	12.6%
In % of total area	%	10%	14%	
Per inhabitant	ha	0.34	0.29	
Per economic active person engaged in agriculture	Ha	1.25	1.15	
Estimated total population 2004	1000 inhabitants	689,700	6,389,200	10.8%
Population growth 2003–2004	%/yr	2%	1%	
Population density	inhabitants/km²	28.4	47.5	
Rural population as % of total population	%	62%	51%	
Economically active population engaged in agriculture	%	27%	21%	
Precipitation	km³/yr	19,809	110,000	18.0%
	mm/yr	816	818	
Internal renewable water resources	km³/yr	3880	43,744	9.0%
Per inhabitant	m³/yr	5696	6847	
Total water withdrawal	km³/yr	120.9	3818	3.2%
Agricultural	km³/yr	104.7	2661	3.9%
In % of total water withdrawal	%	86.6%	70%	
Domestic	km³/yr	12.6	380	3.3%
In % of total water withdrawal	%	10.4%	10%	
Industrial	km³/yr	3.6	777	0.5%
In % of total water withdrawal	%	3.0%	20%	
In % of internal renewable water resources	%	3%	9%	
Per inhabitant	m³/yr	171	598	
Irrigation	Ha	7,076,911	277,285,000	2.6%
In % of cultivated area	%	3%	15%	

Source: Jean-Marc, F. and S. Guido (Eds.), 2008. *Water and the Rural Poor Interventions for Improving Livelihoods in Sub-Saharan Africa.* Rome: FAO Land and Water Division, Food and Agriculture Organization of the United Nations.

TABLE 27.4

Water Availability Per Capita by Region, 1950–2000

Region	1950	1960	1970	1980	2000
			(Thousand Cubic Meters)		
Africa	20.0	16.5	12.7	9.4	5.1
Asia (excluding Oceania)	9.6	7.9	6.1	5.1	3.3
Europe (excluding the Soviet Union)	5.9	5.4	4.9	4.6	4.1
North America and Central America	37.2	30.0	25.2	21.3	17.5
South America	105.0	80.2	61.7	48.8	28.3

Source: Rosegrant, M.W. 1997. *Water Resources in the Twenty-First Century: Challenges and Implications for Action.* Food, Agriculture and the Environment Discussion Paper 20. Washington: International Food Policy Research Institute.

poverty than the less arid regions (Hanjra and Qureshi, 2010). The vast majority of SSA countries are net food importers including some that rely on food aid during difficult humanitarian crises (UNDP, 2012a,b). Figure 27.5 shows that the freshwater per capita share will continue to decrease worldwide; and even when the water resources per capita have declined significantly in Northern America, Latin America, and the Caribbean since 1950, the UN predicts that Africa and Asia will have the least water share per person in 2050.

27.3 Challenged Waters in SSA

It is hardly possible to observe a direct relationship between water shortage and conflict among countries, but Ohlsson et al. (1999) categorized water conflicts based on causes and types both within and between the SSA countries. Seventeen major water basins in SSA flow through 35 SSA countries, and each of these countries has certain rights and obligations to use shared

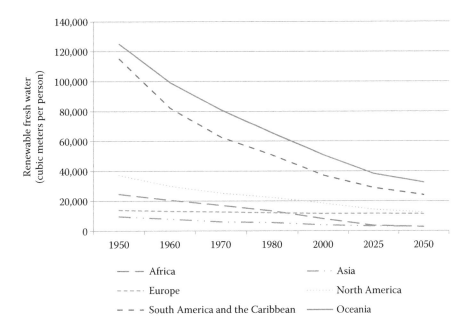

FIGURE 27.5
Water availability per capita in major world areas (1950–2050). (From United Nations. 2006. *World Population Prospects: The 2006 Revision Volume III: Analytical Report*. United Nations Department of Economic and Social Affairs/Population Division.)

waters (Any, 2013). There are a number of difficulties due to this complexity. First, it would be difficult to establish management bodies that can effectively manage natural resources that cross borders like rivers and lakes because of lack of appropriate institutions; and second, the implementation of agreements would require much political will that has been difficult to achieve (Merrey, 2009).

As the amount of water available in the region becomes less, it becomes increasingly difficult to compete for a share of the existing water resources. In this manner, water scarcity generates anxiety and competition among different sectors and consumers of water (Ohlsson et al., 1999). In 2005, a deadly conflict developed between two Kenyan tribes; namely, the Masai and the Kikuyu who lived in the Rift Valley. These two groups formed due to conflicts that originated from some misunderstanding of interest groups in terms of who controls and manages the scarce waters of that valley (Any, 2013; Ministry of Justice, 2011). Further, in the western Sudan region of Darfur, there were armed conflicts among rivals over the last decade; it was noted that the ethnic differences were not the only reason, but rather resource mismanagement appeared to be a factor. There were thousands of people displaced from their villages and/or lands having to resettle in temporary camps in the region causing a rapid desertification and degradation of the livelihood of local people (Abouyoub, 2012). Such disputes do not only occur between people from the same country but also occur between interstates or neighboring countries.

In the case of the Great Lakes in East Africa, there has been border related tensions between Tanzania and Malawi with Tanzania trying to get intervention from the international community to assist in solving the long lasting dispute over its share of Malawi Lake (Mahony et al., 2014). Countries like Niger, Cameroon, Nigeria, and Chad share water from Chad's Lake—the water in this lake has been drained at a significant rate (Sambi, 2010). Even though there was a special body, namely, the Lake Chad Basin Commission, these countries failed to achieve an agreement outlining the quota of each country (SAP, 2008). Moreover, there were signs of new tensions over the River Nile water, which supplies 11 African countries including the two most highly populated; Ethiopia and Egypt (Paisley and Henshaw, 2013). There were two essential agreements to manage water from River Nile, one as old as 1929 that was replaced in 1959; but this appears to favor the downstream countries such as Sudan and Egypt (Paisley and Henshaw, 2013). The upstream states like Ethiopia initiated a challenge to this agreement. The conflict escalated after the independence of the Republic of South Sudan in 2011; during the construction of the Ethiopian Grand Renaissance Dam, which was allowed to keep large amounts of Nile water (about 60,000 million m^3 some of which is essential to the downstream countries, Mulat et al., 2014).

Clearly, there is not enough evidence to claim that the climate change variables have had a direct impact on the origination of conflicts; that is, an increase in the probability of armed conflicts or the severity of conflicts (Fjelde and Uexkull, 2012). But interestingly, Burke et al. (2009) has argued that there is a strong historic relationship between temperature levels in the SSA countries and civil wars in that there is a greater likelihood of observing more wars during warmer years (Adano and Daudi, 2012; Burke et al., 2009). In areas where there is uneven rainfall and limited water or suitable land for farming, the communities may tend to use scarce resources from other areas thus causing tensions and conflicts (Baqe, 2013). It is noted that the consequences are greater in communal tensions when compared to armed conflicts. In any case, such conflicts often result in the movements of larger groups of people away from the areas of conflict (Lecoutere et al., 2010).

27.4 Water Shortage and Social Stability

Even when there is stability among countries in SSA (Bob, 2010), climate change has affected rural communities in their ability to carve out a successful livelihood from their lands (The World Bank, 2009). The experts agree that the phenomenon of environmental refugee that emerged during the twentieth century (Norman, 2001) is defined as the people who are forced to leave their original habitat because of some sort of environmental difficulty (Peters, 2011). The term "environmental refugee" was first coined by Essam El-Hinnawi in 1985 (United Nations Environmental Programme report, 2005). He defined environmental refugees as "…those people who have been forced to leave their traditional habitat, temporarily or permanently, because of a marked environmental disruption (natural and/or triggered by people) that jeopardized their existence and/or seriously affected the quality of their life" (Khan, 2014, p. 127). Since 1985, others have developed similar definitions of the environmental refugee (Ďurková et al., 2012; Myers, 2002). Myers described environmental refugees as those "persons who no longer gain a secure livelihood in their traditional homelands because of environmental factors of unusual scope, notably drought, desertification, deforestation, soil erosion, water shortages, and climate change, also natural disasters as cyclones, storm surges and floods" (Ďurková et al., 2012; Myers, 2002, p. 7). To avoid the word "refuge," the United Nations High Commissioner for Refugees prefers to use "environmentally displaced persons" and refers to the refugees as those "who are displaced from or who feel obliged to leave their usual place of residence, because their lives, livelihoods and welfare have been placed at serious risk as a result of adverse environmental,

ecological or climatic processes and events" (Ďurková et al., 2012, p. 7).

It is noted that the SSA has contributed to the majority of the environmental refugees of the world, and remains the prime locus (Myers, 2002). It is estimated that in 2010 there were around 50 million environmental refugees and this number is projected to reach 200 million by the year 2050 (Myers, 2002; Peters, 2011). In both social and economic development, the reduction of poverty in rural communities is an important issue (Diao et al., 2007). Despite the fall of hungry people worldwide that has occurred mainly in Asia, the SSA poverty levels have not decreased as much, and there has been a limited success toward the reduction of poor people. According to DFID (2004), the number of people living under the poverty line has increased significantly based on the adoption of a new definition for poverty. There were 58 million more poor people in SSA in 1999 compared to the poor people in 1990 (Chikaire and Nnadi, 2012).

This chapter critically reviews the issue of water and food security in SSA. More specifically, the authors analyze water and food related statistics of several SSA countries. Considerations are in terms of rates of population growth, availability of water and food resources, and possible effects of climate change factors with regard to water and food resources and their usage. The population movements and social stability are also analyzed in terms of the movement of people identifying various push and pull factors, identifying some causes of movement mainly from rural to urban regions. Future likelihood of water scarcity or stress is also examined in light of the migration patterns recently observed. Some longer term issues of water and food refugee related migration in the world today are discussed in terms of the numbers presently fleeing to countries such as Italy, Indonesia, and Australia.

In the following, a country by country analysis is conducted and this is followed by discussion of the issues examined. A summary and conclusion completes this chapter. The SSA countries considered are Ethiopia, Kenya, Somalia, and Tanzania from East Africa, Democratic Republic of Congo in the Central Africa region, Nigeria and Niger in West Africa, and the Republic of South Africa in the Southern Africa region. The countries were selected for several reasons including the fact they are in SSA, their high rates of population growth, geographical location, and current and projected water conditions, in addition to the levels of economic development. However, this chapter will only present Somalia and Democratic Republic of Congo (DRC) as examples of our analyses because of the constraints placed on chapter length. Analyses of these and the remaining countries will be published later in another paper titled "A Country by Country Analysis of Water Stress in Sub-Saharan Africa."

27.4.1 Somalia

Somalia is an east African country located in the horn of Africa bordering the Indian Ocean, Djibouti, Ethiopia, and Kenya. The land area of Somalia is 637,657 km^2 with about 13% being arable land. It has the longest coastline in Africa of around 3333 km (Webersik, 2006). The total renewable internal freshwater resources in Somalia are around 6 km^3. There has not been a census since 1983 but UNICEF (2013) estimated the population to be around 9.3 million in 2008. The World Bank (2014) estimates the present population to be around 10.2 million.

Somalia is a hot country with tropical and semitropical, and arid and semi-arid climates. The average temperature is 28°C in coastal areas, but the temperature may reach 47°C and yet it can be as low as 0°C in the highlands. The precipitation in Somalia is highly variable but low most of the time; many parts of the country receive below 500 mm/yr. Coastal areas receive about 50–150 mm/yr while Afgoie receives about 584 mm/yr (Muchiri, 2007). Somalia is presently under water stress conditions and the distribution of water is highly variable. The European Commission Somalia Operations Office estimated the renewable freshwater per capita to be 1685 m^3 but the FAO statistics (2013) showed that the actual renewable freshwater per person was 1500 m^3 in 2012, dropping from 4980 m^3 (1962). Only 30% of the population had access to clean water in 2008, while about 67% and 9% of urban and rural populations had access to improved water sources, respectively (UNICEF, 2013). The country has also performed poorly in terms of sanitation—52% of urban and 6% of rural communities have access to adequate sanitation and overall only 23% of Somali population has access to adequate sanitation (2008) (World Bank, 2014).

According to World Bank (2014), the annual freshwater withdrawal was 54.96% of available water resources. The agriculture water withdrawal was around 99.48%, followed by domestic (0.45%), and then industry with only 0.06%. Figures 27.6 through 27.9 and Tables 27.5 and 27.6 summarize trends in population and water resources in Somalia (1958–2012). Agriculture has been the major contributor of Somali economy, even though there has not been data available since the collapse of the central government in 1991. However, World Bank (2014) shows that agriculture contributed to the national economy (about 50% and 71% in 1958 and 1990, respectively).

Somalia seems to have become vulnerable to droughts and floods. A severe Dabadheer drought (1974–1975) resulted in 20,000 people dying due to famine. Climate change has been blamed for other droughts in 1991–1992, more severe than Dabadheer. It is believed that between 300,000 and 500,000 people lost their lives but indirectly the drought also severely affected another 3 million people (Baumann et al., 2003). A large number

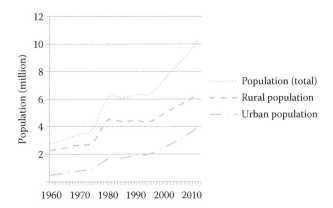

FIGURE 27.6
Population trends in Somalia 1960–2012.

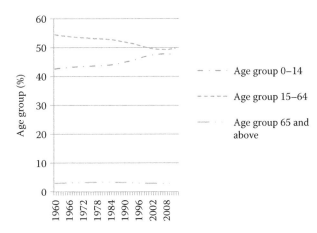

FIGURE 27.7
Age groups in Somalia (1960–2012 % of total population).

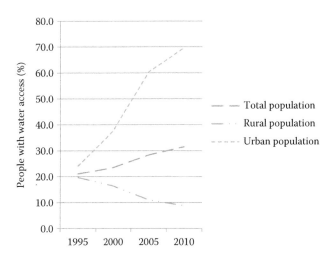

FIGURE 27.8
Improved water sources of Somalia during 1995–2010 (% of population with access).

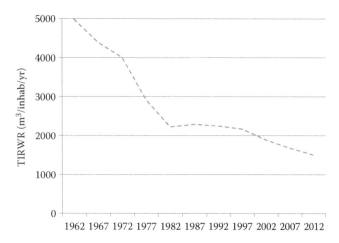

FIGURE 27.9
Somalia: Total IRWR per capita in (m³/inhab/yr) 1962–2012.

of drought-affected persons moved into relief camps and this led to the spread of infectious diseases causing many to die as well. Drought continued to hit some parts of the country and the neighboring countries in 2000 and 2011 (Griffiths, 2003). With similar extremity, the heavy rains caused many floods in 1997 and 2000 destructing many parts of Somalia (Griffiths, 2003). Surprisingly, between 1961 and 2004, 18 floods and 12 droughts were recorded in Somalia (UNEP, 2005). It seems that the rural people are more susceptible to the impact of climate change because of the degradation of ecological systems (Griffiths, 2003). The extreme situation caused many rural people to leave their lands to urban areas (Griffiths, 2003). However, disaster mismanagement, extensive farm production failures, and grazing land overexploitation have also negatively affected the farming lands of Somalia.

27.4.2 Democratic Republic of Congo (DRC)

Democratic Republic of Congo (DRC), often referred to as Zaire, is located in the African Great Lakes region of Central Africa. DRC is a rich country with many natural resources such as oil, diamond, copper, and cobalt. In addition, DRC is also referred to as Africa's water-rich country—DRC possesses 52% and 23% of Africa's surface water reserves and internal renewable water resources, respectively (UNEP, 2011). Despite this, about 60% of the population live under the poverty line, and it is considered one of the world's poorest countries with a ranking of 186 out of 186 countries (also held by Niger in SSA) (UNDP, 2013). DRC is the largest and third most populated in SSA after Nigeria and Ethiopia. The population is growing at a fast rate (see Figure 27.10); it was around 16.2 million in 1962 and was about 69.57 million in 2012 (FAO, 2013). An interesting change in composition has been noted in that in 2012, about 45% of

the population were below the age of 15 years and about 3% of the population were 65 (Figure 27.11).

The population increase has resulted in a significant drop of total renewable water resources per capita—79,291 m³/yr (1962) to 18,441 m³/yr (2011) (Figure 27.12). The annual average precipitation is 1543 m³ (FAO, 2013) with the minimum rainfall in some parts of about 800 m³/yr and maximum around 1800 m³ in other parts. The DRC's total actual renewable water resources are 1283 km³ (FAO, 2013; UNEP, 2011).

The percentage of the population that has access to clean water is low (23%) compared to other SSA countries. As shown in Figure 27.13, Congo has not progressed in terms of water improvements and the people with access to clean water appear to be stable and low—there has been a 10% decline in the number of urban population with access to clean water during the past two decades. Similar to other SSA countries, agriculture is a major contributor to the GDP. In 2006, 47.7% of the GDP came from agriculture, forestry, fishing, and hunting sectors; however, this has fallen to 39.4% in 2011. In the same period, the sectors like mining and quarrying, and finance, real estate, and business services have showed a slight increase overall (AfDB, 2010, 2012).

Table 27.6 shows the arable land in DRC has not increased significantly from 6.4 million ha in 1962 to only 6.8 million ha in 2012. FAO (2013) also noted that other agricultural improvement measures such as percentage of total area cultivated remained almost unchanged during this period. The area of permanent crops was 0.57 million ha in 1962 and peaked at 1.2 million ha in 1992, dropping to 0.755 million ha in 2012. This shows that the country has not responded to the dramatic increase in its population in terms of agricultural production and food security. In contrast to other SSA countries, DRC's agricultural sector consumes the least amount of water when compared to domestic and industry sectors. However, as there are huge amounts of water resources in DRC, the pressure on water is significantly low—only 0.053% in 2007 (FAO, 2013).

27.5 Discussion

It is clear that the SSA is experiencing a number of water and food security challenges. The population increase in SSA is a critical challenge in terms of water and food. The selected countries have shown a higher level of population growth in SSA in general and poverty is widespread with SSA being among the world's poorest regions. The youth are highly disadvantaged particularly in the rural communities. The cities are expanding and SSA is rapidly urbanizing, that is, at a faster rate than others in the

TABLE 27.5

Somalia: Land, Population, and Water Resources (1962–2011)

Somalia	1962		1972		1982		1987		1992		2002		2007		2012	
Land and Population																
Total area (1000 ha)	63,766	E	63,766	E	63,766	E	63,766	E	63,766	E	63,766	E	63,766	E	63,766	E
Arable land (1000 ha)	896	E	946	E	994	E	1018	E	1023	E	1200	E	1000	E	1100	E
Permanent crops (1000 ha)	14	E	14	E	16	E	17	E	20	E	26	E	28	E	29	E
Percentage of total country area cultivated (%)	1.427	E	1.506	E	1.584	E	1.623	E	1.636	E	1.923	E	1.612	E	1.771	E
Total population (1000 inhab)	2952	E	3674	E	6608	E	6427	E	6543	E	7791	E	8733	E	9797	E
Rural population (1000 inhab)	2412	E	2796	E	4808	E	4584	E	4557	E	5142	E	5585	E	6030	E
Urban population (1000 inhab)	540	E	878	E	1800	E	1843	E	1986	E	2649	E	3148	E	3767	E
Population density (inhab/km²)	4.629	E	5.762	E	10.36	E	10.08	E	10.26	E	12.22	E	13.7	E	15.36	E
Rainfall and internal renewable water resources																
Long-term average precipitation in volume (10^9 m³/yr)	179.8	E	179.8	E	179.8	E	179.8	E	179.8	E	179.8	E	179.8	E	179.8	E
Surface water produced internally (10^9 m³/yr)	5.7		5.7		5.7		5.7		5.7		5.7		5.7		5.7	
Groundwater produced internally (10^9 m³/yr)	3.3		3.3		3.3		3.3		3.3		3.3		3.3		3.3	
Overlap between surface water and groundwater (10^9 m³/yr)	3		3		3		3		3		3		3		3	
Total internal renewable water resources (IRWR) (10^9 m³/yr)	6		6		6		6		6		6		6		6	
Total internal renewable water resources per capita (m³/inhab/yr)	4980	K	4001	K	2225	K	2287	K	2247	K	1887	K	1683	K	1500	K
Water withdrawal																
Agricultural water withdrawal (10^9 m³/yr)							0.786				3.28	L	3.281	I		
Industrial water withdrawal (10^9 m³/yr)							0				0	L	0.002	I		
Municipal water withdrawal (10^9 m³/yr)							0.024				0.01	L	0.015	I		
Total water withdrawal (10^9 m³/yr)							0.81				3.29	L	3.298	I		
Agricultural water withdrawal as % of total water withdrawal (%)							97.04				99.7	L	99.48	I		
Industrial water withdrawal as % of total water withdrawal (%)							0				0	L	0.061	I		
Municipal water withdrawal as % of total withdrawal (%)							2.963				0.304	L	0.455	I		
Total water withdrawal per capita (m³/inhab/yr)							126	K			422.3	K	377.6	K		
Freshwater withdrawal as % of total actual renewable water resources (%)							5.51				22.38	I	22.44	I		

Source: FAO. 2005. *Irrigation in Africa in Figures—AQUASTAT Survey 2005.* Ethiopia: FAO Land and Water Division, Food and Agriculture Organization of the United Nations, pp. 2–4.

E—external data, I—AQUASTAT estimate, K—aggregate data, and L—modeled data.

globe. The rural population is decreasing due to migration to cities and elsewhere. Clearly, there are several reasons for this but essentially the rural population has been declining for the past 50 years.

The population of SSA has quadrupled since 1960. Kenya and Niger have increased fivefold, with Tanzania at 4.7 times. South Africa, Nigeria, and Somalia have tripled in the past 50 years, although they are among the lowest in SSA. The rural population has experienced a gradual decline; for example, in Nigeria and Tanzania,

the percentage of rural population has declined about 34% and 22%, respectively. Ethiopia, Niger, and DRC decreases were lower but around 11%–13%. The SSA generally showed a 22% decline in rural population since 1960. Table 27.7 shows the changing population in SSA and selected countries (1960–2012).

Fourteen African countries have water stress, while another 11 countries are predicted to become water stressed by 2025 (Abel Mejía et al., 2012; UN, 2008). Kenya is under water scarcity while water stress exists

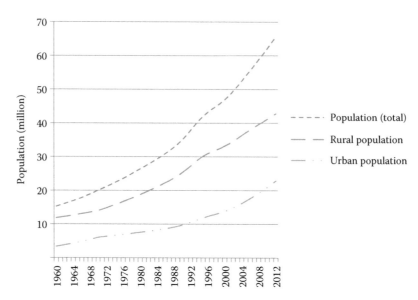

FIGURE 27.10
Population trends in DRC 1960–2012.

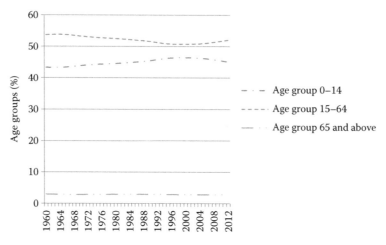

FIGURE 27.11
DRC: Trends of age group percentages in 1960–2012 (% of total population).

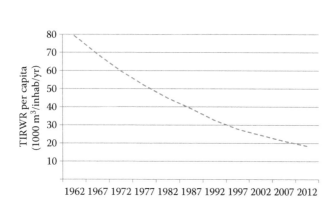

FIGURE 27.12
DRC: Total IRWR per capita in (m³/inhab/yr) 1962–2012.

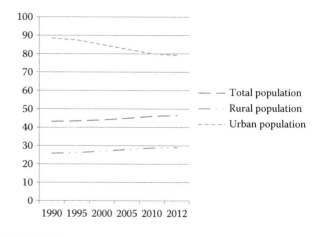

FIGURE 27.13
Improved water sources in DRC 1990–2012 (% of population with access).

TABLE 27.6

Democratic Republic of Congo: Land, Population, and Water Resources 1962–2012

Democratic Republic of Congo	1962		1972		1982		1992		2002		2007		2012	
Land and Population														
Total area (1000 ha)	234,486	E	234,486	E	234,486	E	234,486	E	234,486	E	234,486	E	234,486	E
Arable land (1000 ha)	6400	E	6500	E	6640	E	6700	E	6700	E	6700	E	6800	E
Permanent crops (1000 ha)	570	E	850	E	1030	E	1200	E	750	E	750	E	755	E
Percentage of total country area cultivated (%)	2.972	E	3.135	E	3.271	E	3.369	E	3.177	E	3.177	E	3.222	E
Total population (1000 inhab)	16,181	E	21,431	E	28,551	E	39,444	E	52,491	E	60,772	E	69,575	E
Rural population (1000 inhab)	12,378	E	14,916	E	20,445	E	28,411	E	36,409	E	40,547	E	44,150	E
Urban population (1000 inhab)	3803	E	6515	E	8106	E	11,033	E	16,082	E	20,225	E	25,425	E
Population density (inhab/km²)	6.901	E	9.14	E	12.18	E	16.82	E	22.39	E	25.92	E	29.67	E
Rainfall and internal renewable water resources														
Long-term average precipitation in volume (10^9 m³/yr)	3618	E	3618	E	3618	E	3618	E	3618	E	3618	E	3618	E
Surface water produced internally (10^9 m³/yr)	899		899		899		899		899		899		899	
Groundwater produced internally (10^9 m³/yr)	421		421		421		421		421		421		421	
Overlap between surface water and groundwater (10^9 m³/yr)	420		420		420		420		420		420		420	
Total internal renewable water resources (IRWR) (10^9 m³/yr)	900		900		900		900		900		900		900	
Total internal renewable water resources per capita (m³/inhab/yr)	79,291	K	59,867	K	44,937	K	32,527	K	24,442	K	21,112	K	18,441	K
Water withdrawal														
Agricultural water withdrawal (10^9 m³/yr)									0.072	L	0.072	L		
Industrial water withdrawal (10^9 m³/yr)							0.058		0.123		0.147	L		
Municipal water withdrawal (10^9 m³/yr)							0.216		0.389	L	0.465	L		
Total water withdrawal (10^9 m³/yr)									0.584	L	0.684	L		
Water withdrawal for irrigation (10^9 m³/yr)									0.072	L	0.072	L		
Agricultural water withdrawal as % of total water withdrawal (%)									12.31	L	10.52	L		
Industrial water withdrawal as % of total water withdrawal (%)									21.04	L	21.47	L		
Municipal water withdrawal as % of total withdrawal (%)									66.65	L	68.01	L		
Total water withdrawal per capita (m³/inhab/yr)									11.13	K	11.25	K		
Freshwater withdrawal as % of total actual RWR (%)									0.045	I	0.053	I		

Source: FAO. 2013. *AQUASTAT Database—Food and Agriculture Organization of the United Nations (FAO)*. Website accessed on March 22, 2014.
E—external data, I—AQUASTAT estimate, K—aggregate data, and L—modeled data.

in Somalia, Ethiopia, and South Africa. South Africa has made efforts to rationalize the use of water resources and to ease the pressure posed by water stress. Table 27.8 shows changes in the total renewable water resources per capita (TRWR) in terms of the percentage of water share lost as a result of population increase. It also shows the fresh water withdrawal as percentage of the total actual renewable water resources. Clearly, the TRWR has been decreasing mostly related to the population growth but mismanagement is another factor.

Table 27.8 shows the pressure on water (freshwater withdrawal as a percentage of total actual renewable water resources) varying from 0.05% in DRC, which has the largest storage of water resources in Africa, to 22.44% in Somalia. Others have been withdrawing between 2.92% and 8.91%. So countries are using less than a quarter of the available water. Although the number of people with access to clean water has been increasing, the overall percentage has been falling. For example, in 1991, 79% (25 million) of urban Nigerians

TABLE 27.7

Changes in Total and Rural Populations in SSA

Population by Region/Country	Population (Millions)		Folds Increased	Rural Population (%)		Rural Population Decrease (%)
	1960	2012		1960	2012	
Ethiopia	22	92	4.1	94	83	11
Tanzania	10	48	4.7	95	73	22
South Africa	17	52	3.0	53	38	16
Somalia	3	10	3.7	83	62	21
Kenya	8	43	5.3	93	76	17
DRC	15	66	4.3	78	65	13
Nigeria	45	169	3.7	84	50	34
Niger	3	17	5.1	94	82	12
SSA	228	911	4.0	85	63	22

TABLE 27.8

Total Renewable Water Resources (TRWR) per Capita in SSA

Item	TRWR 1962	TRWR 2012	TRWR Lost 1960–2012	TRWR Lost 1962–2012	Freshwater Withdrawal as % of TARWR[a]	
Country	m³/capita/yr	m³/capita/yr	m³/capita/yr	%	Year	%
Ethiopia	2017[b]	1410	607	30	2002	4.56
Tanzania	9011	2020	6991	78	2002	5.39
South Africa	2805	1013	1792	64	2000	24.28
Somalia	4980	1500	3480	70	2003	22.44
Kenya	3558	718.1	2839.9	80	2003	8.91
DRC	79,291	18,441	60,850	77	2005	0.05
Nigeria	5970	1718	4252	71	2005	4.58
Niger	9771	2022	7749	79	2005	2.92

[a] Freshwater withdrawal as % of total actual renewable water resources (TARWR) (%).
[b] 1997 data.

had access to clean water but 17 years later only 75% have access. Although the country doubled the number of people with access to freshwater, the higher population (55 million) has resulted in the water demand exceeding water supply. This has implications for future food security of residents.

Food production in SSA is another challenge the rapid population growth is not combined with an increase of agricultural production. The arable land has not increased proportionally. Figure 27.14 shows the changes on arable land in SSA since 1962. Vertical integration and growth of agricultural production has become an unreachable target for many SSA countries. The poor land quality and productivity, workforce migration, rainfall variations, water shortage, poor research and development in agriculture, as well as infrastructure are among factors that have negatively affected food security.

The cultivated land in South Africa has been stable at 12 million ha in the past five decades. In DRC, the population has quadrupled but the acreage has not changed to secure the amount of food demanded. The

DRC arable land has been at 6.8 million ha for the past five decades. Tanzania has doubled its arable land while Ethiopia has added another 7 million ha in the past 10 years. Surprisingly, Niger has the largest arable land (15 million ha) after Nigeria, even though it has the second smallest population (17 million) after Somalia (10 million). Nigeria and Somalia have the largest (36 million ha) and smallest (1.1 million ha) arable lands (see Figure 27.15).

SSA is a net food importer and a majority of the population is undernourished. The agriculture is an important sector for GDP and it creates many job opportunities. Agricultural production in SSA is rain-fed and thus vulnerable to rainfall fluctuations except in some limited irrigated areas. Agriculture accounted for about 20% of GDP in 1990 but has fallen to around 15% in 2012. However, agricultural contribution in Niger, Nigeria, DRC, and Ethiopia remains between 40% and 50% in 2012. Kenya and South Africa have showed stability with an average of 19% and 2.5% contribution, respectively. In contrast, agriculture in Tanzania

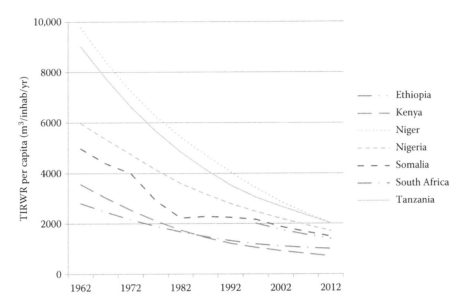

FIGURE 27.14
Total internal renewable water resources per capita (m³/inhab/yr).

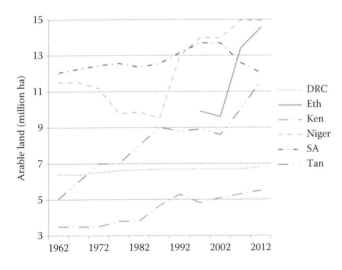

FIGURE 27.15
Changes on arable land of six selected countries (million ha) 1962–2012.

showed a gradual decline from 45% of the GDP in 1990 to 36% in 2012. In Somalia, there are no reliable records of agricultural contribution to GDP since 1991 when agriculture accounted for about 65.5%. Figure 27.16 shows the contribution of agriculture to the GDPs from 1990 to 2012 (World Bank, 2012).

Despite the limitation of agricultural land, other issues like climate change and weather shocks are affecting farm production. Rainfall variations have become more frequent. Crop production has deteriorated and land quality has been declining. Rural life has become tougher causing rural residents to migrate to cities. For

example, Sangare, his wife, and three children migrated from southern parts of Mali hoping to find a sustainable job in the capital city of Bamako. He left his farm because he said, "The fields don't produce anymore" (Min-Harris, 2010, p. 159). The youth migration is constantly occurring and not limited to inside the countries. About 70% of the youth left from their original living areas to cities and many are considering leaving the entire continent in a search for a better life (Min-Harris, 2010). Some SSA youth who recently migrated to Australia have never lived in cities but may use it as a pathway for overseas migration using refugee methods. One SSA young man migrating to Europe through Northern Africa and Mediterranean Sea by boat said, "Better dying in the sea than dying in the valley for thirsty," (Author interview) highlighting the extreme water stress conditions in the semi-desert area around the horn of Africa.

Evidently, rural migration related to environmental factors has been noted often in SSA. The rural areas are more susceptible to floods and droughts such as in Somalia, Kenya, and Ethiopia. Somalia has experienced a lack of rainfall over consecutive seasons, which has caused droughts in 1992 and 2011. In fact, in 1992 there was no agricultural production at all. Lack of pasture led to livestock dying causing much poverty and famine in southern and central Somalia. The rural people migrated to cities specially Mogadishu. Unfortunately, 2 years before the drought the central government collapsed and there were no public institutions to help such migrants. People built camps inside and outside the city including in the Mogadishu stadium, national post, the Independence Square, and many other public areas. In 2011, another drought hit the country,

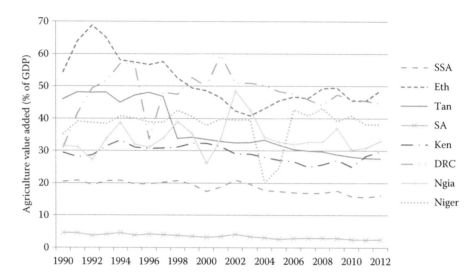

FIGURE 27.16
Agriculture, value added (% of GDP).

causing tens of thousands of rural families to move to Mogadishu adding further pressure whence many crossed the border to Kenya and Ethiopia. However, at this time there were some functioning public institutions that assisted such migrants to resettle but the situation was desperate.

Despite droughts, rural people also experience frequent floods that submerge villages, farmlands, and access roads. The worst flood in 50 years hit the east African region in 2006. This was due to higher rainfall in Somalia causing the Juba River to burst its banks and destroy bridges and wash away roads. UNICEF argues that 0.2 million Somali children under the age of five could die from severe malnutrition by the end of 2014. The unexpected migration of many more people from rural areas to cities has created much pressure on the existing services and facilities. In particular, inadequate sanitation has become a great challenge; nearly 50% of the people in SSA suffer from water-related diseases. The poorest sanitation levels are indeed found in SSA with only 36% having access to basic sanitation (Moe and Rheingans, 2006).

Importantly, those who have moved on are not willing to go back to their farming lands even if the conditions have changed. Some gain some sort of employment enough to send their children to schools. However, the majority of them are unemployed and rely on local and international aid. There have been several initiatives including one from a local charity foundation, which promised to provide the necessary farm machinery to the displaced people to assist them to cultivate their farm lands. As they no longer have any assets, other institutions have offered basic household goods if they voluntarily return to their homelands. In Somalia's case, every rural family has one or

more family members living in the cities and others willing to join that member leaving everything they have behind. There is also encouragement from urbanized family members to be united so they often request their families to move into the cities aiming for a better life for them.

Many SSA rural people (63%) live in their original lands and practice farming and grazing activities. Clearly, not all rural communities in SSA can or are willing to leave their farm lands especially if the agricultural production is good. Others are held back by lack of links or family members in cities, while others do not move because of their larger extended families. Others stay in hope and belief that governments will help and that there are better days to come. Indeed, some communities have thought through their survival plans and have devised possible ways to maintain life on farms. For example, a small farming community of Samsam village in Al Qatarif state (Eastern Sudan) has managed to establish suitable living conditions with minimal support from authorities. These people have built infrastructure for water catchment to harvest rainwater and have made it available for the people and animals throughout the year. They have also built a primary school, health facility, and market. As in many rural communities, people meet at the marketplace to share ideas, socialize, and sell or buy goods, etc. In some cases, the community elders are assisted by the community members who have migrated to cities thus providing a surviving strategy to the people in the village. Such communities have decreased illiteracy and their average time used to collect water thus enhancing their economic and social stability. In their weekly meetings, farmers share useful information about the agricultural season, cost of farm inputs, and profitability of agricultural outputs.

It seems that the lack of confidence of such rural communities in the existing authorities is another pressing factor. This analysis shows some negligence of the rural people by the authorities. The percentage of rural people with access to clean water is as low as 8.8% and 29% in Somalia and Niger, respectively. The lack of education, healthcare, and employment opportunities in rural areas are other pressing factors causing rural youth to migrate to cities. It is true that the youth have a higher probability of leaving their lands than married couples with children. However, the cities in return have become seriously overcrowded and the facilities such as schools, health services, water, sanitation, and transportation are not able to absorb the new arrivals. The security challenges also appear to be escalating as poverty among youth has pushed them to become involved in illegal activities. SSA cities have not delivered what rural youth hoped for such as education, healthcare, and employment in their homelands. In the Sangar emigration case, his employment opportunities in his new urban home were limited to shining shoes and selling sunglasses in the streets of Bamako earning USD 0.22/day.

Despite the environmental factors, rural people migrate as a result of conflict which leads to either internally displaced people or international refugees. In DRC, for example, internal conflict erupted in the country in the 1990s becoming the most deadly conflict worldwide since the two major world wars. The data shows that 63.2% of SSA population lived in rural areas in 2012 and as a result of rural migration it is expected that half of the SSA population will be living in cities by the year 2020. As noted, SSA has a number of water and food security challenges. The analyses have shown a higher level of population growth and poverty being widespread with the rural youth being disadvantaged. The cities are expanding and urbanizing faster than the rest of the world. The rural population has been declining steadily during the past 50 years and the move can be attributed mainly to water and food security.

27.6 Conclusion

This study aimed to review conditions of water availability and stress in SSA region. It also aimed to discuss food security as a consequence of water distress for they are closely related. Another aim was to investigate the phenomenon of rural–urban migration in SSA and consider its links with water availability. This study focused on identifying the nature of the pull and push factors in both rural and urban populations as many in the population determine their future in the water scarce lands. The study also aimed to review the

migration patterns and to observe its long-term impact to the rest of the world particularly in relation to flow toward Australian lands. Two countries (Somalia and DRC) were selected based on specific factors such as the rate of population growth, water resources availability, geographical location and agricultural production as well as economic developments.

During the last 50 years, the SSA region has struggled to reduce the poverty, improve water resources, and provide access to clean water. In the last decade in particular, the region however has achieved some significant outcomes in terms of the millennium development goals. But as the population growth intensifies in the region, the number of people living in poverty has increased remarkably. The study reveals several reasons for the escalating poverty in the region. Some reasons are hardship of clean water access for rural areas, poor infrastructure, and rainfall fluctuations. This study finds that water availability is highly variable in SSA regions, and that there is a gap in terms of water distribution in rural and urban SSA. The region varies in the amount of water available to each country. DRC, which is Africa's water-rich country, possesses 52% and 23% of Africa's surface water reserves and internal renewable water resources, respectively; while Somalia is experiencing a water crisis as the renewable freshwater resources per capita fall below the water stress threshold. DRC has the second lowest percentage of rural communities with access to clean water (29%) while South Africa has the highest percentage of rural people with access to freshwater sources (88%), even though South Africa has limited water resources. Water demand in both rural and urban communities has been increasing and this is found to be closely related to rate of the population growth.

The link between climate change and migration is not clear, but work has been done on this area and the authors are now considering a new model (Perch-Neilsen et al., 2008). The future impact of climate change on availability of water and water quality is also unclear (Calow et al., 2011) but there is an increasing focus that more rainfall fluctuation and variability will be realized (Elasha, 2010). This will directly impact the amount of food produced (Amikuzuno and Donkoh, 2012) and the willingness of investors to invest in agriculture (Wroblewski and Wolff, 2010). Climate change impact on SSA may be high because of the rain-fed agriculture, in addition to the poor adaptive capacity (Cooper et al., 2008; IPCC, 2007; Ringler et al., 2010). A reason for the urban migration from homelands is that the agricultural production has not been improving for a long time mostly because of water shortage (20 years). This has led to a lack of capacity in the rural areas—the low level of land productivity has led to the migration of the rural work force—loss of youth and expertise more generally.

While population has been increasing, the food demand has not been met and indeed SSA is a net food importer. The poor research and development as well as minimal focus and attention received from the public institutions have not aided the situation.

SSA rural migration is predominant mainly among rural youth. In general, rural SSA lacks basic facilities such as health care, education, and employment. All of these have acted as push factors for many rural migrants. As a consequence, there has been an emergence of over-populated cities, placing much pressure on the limited city facilities. There is clear evidence that the move to urban SSA has not fulfilled the expectations of rural migrants. It is also noted that the rural migrants are not willing to move back to their original living areas even if the situation could become better. For this reason, it seems that many have decided to migrate much further to the developed world hoping to get a better life with drastic consequences such as those noted recently in northern Africa. Many of the migrants are living in much hardship in camps often set up by the UN; while others have preferred to take much higher risks by travelling larger distances sometimes by land and sea to more affluent countries such as Italy, Indonesia, and even Australia. Clearly, there is a need to conduct more comprehensive and multidimensional studies to investigate a set of complex interacting factors in rural SSA. There is a need to develop solutions to the problem of SSA urbanization and this means there is a need for more focused studies on water and food stress in SSA that can help recommend ways of creating regional collaboration, better allocation, and indeed maximization of available resources particularly of water and land. There needs to be more research into how to suppress the spread of poverty in SSA and reduce the hardship of the local or rural people in order to help them develop means to create sustainable lives—that is, they may be able to establish a sustainable life in their original lands. This is the only way it seems that the world can deal with the phenomenon of national and international migration of the largest distress and poverty-struck peoples from the most drought-stricken continent on earth in the longer term. It is noted that the consequences are greater in communal tensions caused by water distress when compared to other armed conflicts and this often results in the movements of larger groups of people away from such areas. If such factors are not attended to soon, the many push and pull factors may cause large amounts of people to become mobile and on the move to seek ways to reach major cities around the world and this includes Australia in the longer term as has been noted already. In fact, in more recent times, Australia is seen as the number one destination for those in distress because it is a country where a fair go exists for those starting to make new lives for themselves.

References

Abel Mejía, M.N.H., Enrique, R.S., and D. Miguel. 2012. *Water and Sustainability: A Review of Targets, United Nations Educational, Scientific and Cultural Organization 7*, Place de Fontenoy, Paris. Unescso, France.

Abouyoub, Y. 2012. Darfur—Debunking the myths: The material roots of the Darfur conflict. *IV Online Magazine* IV (449): 43–53.

Adano, W.R. and F. Daudi. 2012. *Links between Climate Change, Conflict and Governance in Africa*. Institute for Security Studies, No. 234.

ADF VII. 2010. Climate change and ecosystem sustainability: Acting on climate change for sustainable development in Africa, seventh African development forum, issues, Paper 9. *ADF* V(2): 10–15.

AfDB. 2010. *Guidelines for User Fees and Cost Recovery for Urban, Networked, Water and Sanitation Delivery. Water Partnership Programme, African Development Bank*. Tunis, Tunisia.

AfDB. 2012. *African Economic Outlook 2012: Democratic Republic of the Congo. African Development Bank*. Tunis, Tunisia. http://www.afdb.org/fileadmin/uploads/afdb/Documents/Project-and-Operations/2011_03%20Guidelines%20for%20User%20Fees%20Cost%20Recovery_Rural.pdf, retrieved January 20, 2015.

African Studies Centre. 2014. *Regional Perspectives: Geographic Regions in Africa*, available on: http://exploringafrica.matrix.msu.edu/teachers/curriculum/m20/activity1.php Michigan State University retrieved January 20, 2015.

Amikuzuno, J. and S.A. Donkoh. 2012. Climate variability and yields of major staple food crops in Northern Ghana. *African Crop Science Journal* 20(2): 349–360.

Any, F. 2013. Brief issue: Water as a stress factor in Sub-Saharan Africa. *European Union Institute for Security Studies* 1: 1–4. http://www.iss.europa.eu/uploads/media/Brief_12.pdf.

Ashton, P.J. 2002. Avoiding conflicts over Africa's water resources, division of water, environment & forestry technology, Pretoria, South Africa. *Ambio* 31(3): 236–242.

Ashton, P.J. and A.R. Turton. 2007. Water and security in Sub-Saharan Africa: Emerging concepts and their implications for effective water resource management in the Southern African Region (55). In: H.G. Brauch, J. Grin, C. Mesjasz, N.C. Behera, B. Chourou, U.O. Spring, P.H. Liotta, and P. Kameri-Mbote (Eds.), *Globalisation and Environmental Challenges*. Berlin: Springer-Verlag.

Asif, M. 2013. *Climatic Change, Irrigation Water Crisis and Food Security in Pakistan*. Uppsala: Department of Earth Sciences, Geotryckeriet, Uppsala University.

Baqe, P.D. 2013. *An Investigation into Conflict Dynamics in Northern Kenya: A Case Study of Marsabit County 1994–2012*. Institute of Diplomacy and International Studies, University of Nairobi, Kenya.

Barbara, V.K. 2003. Water reform in Sub-Saharan Africa: What is the difference? *Physics and Chemistry of the Earth* 28: 1047–1053.

Baumann, R.F., Yates, L.A., and V.F. Washington. 2003. *"My Clan Against the World" US and Coalition Forces in Somalia 1992–1994*. Fort Leavenworth, KS: Combat Studies Institute Press.

Biazin. B., Sterk, G., Temesgen, M., Abdulkedir, A., and Stroosnijder, L. 2012. Rainwater harvesting and management in rain-fed agricultural systems in sub-Saharan Africa–A review. *Physics and Chemistry of the Earth A/B/C* 47: 139–151.

Bigombe, B. and Khadiagala, G.M. 2003. *Families in the Process of Development: Major Trends Affecting Families in Sub-Saharan Africa.* United Nations: New York, pp. 155–193.

Bob, U. 2010. Land-related conflicts in Sub-Saharan Africa. *African Journal on Conflict Resolution* 10(2): 49–64.

Boko, M.I., Niang, A., Nyong, C. et al. 2007. Africa. Climate change 2007: Impacts, adaptation and vulnerability. In: M.L. Parry, O.F. Canziani, J.P. Palutikof, P.J. van der Linden, and C.E. Hanson. (Eds.), *Contribution of Working Group II to the Fourth Assessment Report of the Intergovernmental Panel on Climate Change.* Cambridge, UK: Cambridge University Press, pp. 433–467.

Burke, M., Miguel, E., Satyanath, S., Dykema, J., and D. Lobell. 2009. *Warming increases the risk of civil war in Africa.* PNAS 106(49): 20670–20674.

Calow, R., Bonsor, H., Jones, L., O'Meally, S.O., MacDonald, A., and N. Kaur. 2011. *Climate Change, Water Resources and WASH.* ODI Working Paper 337 ODI, London, UK/British Geological Survey, Nottingham, UK.

Calzadilla, A., Zhu, T., Richard, K., Tol, S.J., and C. Ringler. 2011. *Economy Wide Impacts of Climate Change on Agriculture in Sub-Saharan Africa: How Can African Agriculture Adapt to Climate Change? Insights from Ethiopia and South Africa.* International Food Policy Research Institute and University of Hamburg, Hamburg, Germany.

Carles, A. 2009. *Water Resources in Sub-Saharan Africa, Peace with Water.* Bruxelles: European Parliament.

Charles, H.T. and A. Soumya, 2008. *Reducing Child Malnutrition in Sub-Saharan Africa: Surveys Find Mixed Progress.* Population Reference Bureau, Washington.

Cherry, G.A. 2009. *Review of Australia's Relationship with the Countries of Afric*a. Submission No.41, Joint Standing Committee on Foreign Affairs, Defense and Trade.

Chikaire, J. and F.N. Nnadi. 2012. Agricultural biotechnology: A panacea to rural poverty in Sub-Saharan Africa (review). *Prime Research on Biotechnology (PRB)* 2(1):6–17.

Climate and Development Knowledge Network. 2012. *Managing Climate Extremes and Disasters in Africa: Lessons from the SREX Report.* CDKN.

Cooper, P.J.M., Dimes, J., Rao, K.P.C., Shapiro, B., Shiferaw, B., and S. Twomlow. 2008. Coping better with current climatic variability in the rain-fed farming systems of Sub-Saharan Africa: An essential first step in adapting to future climate change? *Agriculture, Ecosystems & Environment* 126(1–2): 24–35.

Davidson-Harden, A., Naidoo, A., and A. Harden. 2007. The geopolitics of the water justice movement. *Peace Conflict & Development* Issue 11, November. Available at: http://www.peacestudiesjournal.org.uk/dl/PCD%20Issue%2011_Article_ Water%20Justice%2Movement_Davidson%20Naidoo%20Harden.pdf.

DESA, 2013a. *Seven Billion and Growing: The Role of Population Policy in Achieving Sustainability.* United Nations Department of Economic and Social Affairs, Population Division. Technical Paper, No. 2011/3.

DESA, 2013b. *World Population Prospects: The 2012 Revision.* Department of Economic and Social Affairs of United Nations, Population Division. Volume II, Demographic Profiles (ST/ESA/SER.A/345).

DESA, 2014. *World Population Prospects: The 2012 Revision.* United Nations Department of Economic and Social Affairs, London.

DFID. 2004. *Agriculture, Growth and Poverty Reduction.* UK Department for International Development. http://dfid-agriculture-consultation.nri.org/summaries/wp1.pdf, retrieved January 20, 2015.

Diao, X., Hazell, P., Resnick, D., and J. Thurlow. 2007. *The Role of Agriculture in Development Implications for Sub-Saharan Africa.* Washington, D.C.: International Food Policy Research Institute.

Ďurková, P., Gromilova, A., Kiss, B., and M. Plaku. 2012. *Climate Refugees in the 21st Century.* Regional Academy on the United Nations. Online wordpress.com publisher (retreived 8th February, 2016).

Eberhard, A., Foster, V., Cecilia Garmendia, B., Ouedraogo, F., Camos, D., and M. Shkaratan. 2008. *Africa Infrastructure Country Diagnostic: Underpowered: The State of the Power Sector in Sub-Saharan Africa.* Washington, D.C.: The World Bank.

Elasha, B.O. 2010. *Mapping of Climate Change Threats and Human Development Impacts in the Arab Region.* Research Paper Series, Arab Human Development Report, Regional Bureau for Arab States, United Nations Development Programme.

FAO. 2005. *Irrigation in Africa in Figures—AQUASTAT Survey 2005.* Ethiopia: FAO Land and Water Division, Food and Agriculture Organization of the United Nations, pp. 2–4.

FAO. 2012. *Coping with Water Scarcity: An Action Framework for Agriculture and Food Security.* FAO Water Reports 38. Rome: Food and Agriculture Organization of the United Nations, pp. 5–10.

FAO. 2013. *AQUASTAT Database—Food and Agriculture Organization of the United Nations (FAO).* Website accessed on March 22, 2014.

Fjelde, H. and N. Uexkull. 2012. Climate triggers: Rainfall anomalies, vulnerability and communal conflict in Sub-Saharan Africa. Department of Peace and Conflict Research, Uppsala University. *Political Geography* 31: 444–453.

Godfray, H.C.J., Beddington, J.R., and I.R. Crute. 2010. Food security: The challenge of feeding 9 billion people. *Science* 327, 812.

Griffiths, D. 2003. *Forced Migration Online Country Guide: Somalia,* available on: http://www.forcedmigration.org/research-resources/expert-guides/somalia/alldocuments, retrieved January 20, 2015.

Gross, R., Schoeneberger, H., Pfeifer, H., and H.A. Preuss. 2000. The four dimensions of food and nutrition security: Definitions and concepts. Nutrition and food security. *SCN News* 20: 20–25.

Hanjra, M.A. and M.E. Qureshi. 2010. Global water crisis and future food security in an era of climate change. *Food Policy* 35(5): 365–377.

Haub, C. and J. Gribble. 2011. The world at 7 billion. *Population Bulletin* 66(2): 1–13.

Hell, K., Mutegi, C., and P. Fandohan. 2010. Aflatoxin control and prevention strategies in maize for Sub-Saharan Africa. *Tenth International Working Conference on Stored Product Protection.* http://crsps.net/wp-content/downloads/Peanut/Inventoried%208.8/7-2010-4-150.pdf, retrieved January 20, 2015.

Indraratna, B., Tularam, G.A., and B. Blunden. 2001a. Reducing the impact of acid sulphate soils at a site in Shoalhaven Floodplain of New South Wales, Australia. *Quarterly Journal of Engineering Geology and Hydrogeology* 34: 333–346.

Indraratna, B., Glamore, W., Tularam, G.A., Blunden, B.D., and J. Downey. 2001b. Engineering strategies for controlling problems of acid sulphate soils in low-lying coastal areas. *Environmental Geotechnics*, 1(1), 133–146, Newcastle, Australia: Australian Geomechanics Society Inc.

IPCC. 2007. Climate change 2007: The physical science basis. Contribution of working group 1 to the fourth assessment report of the intergovernmental panel on climate change. In: S. Solomon, D. Quin, M. Manning, Z. Chen, M. Marquis, K.B. Averyt, M. Tignor, and H.L. Miller (Eds.), *Regional Climate Projections.* Cambridge, UK: Cambridge University Press, Chapter 11.

Jean-Marc, F. and S. Guido (Eds.) 2008. *Water and the Rural Poor Interventions for Improving Livelihoods in Sub-Saharan Africa.* Rome: FAO Land and Water Division, Food and Agriculture Organization of the United Nations.

Jenna, D., Roz, N., Eran, B., and P. Amy (Eds.) 2012. Water, nutrition, health and poverty in Sub-Saharan Africa. In: *Water Nexus 2012 Connecting the Dots: The Water, Food, Energy and Climate Nexus.* on web publisher; Stanford University. Online available at http://web.stanford.edu/group/tomkat/cgi-bin/docs/ctd/Davis-2012-April-15-Africa-panel-Connecting-the-Dots.pdf (retrieved 4th February, 2016).

Khan, M.R. 2014. *Advanced in Climate Change Research: Toward a Binding Climate Change Adaptation Regime: A Proposed Framework.* New York: Routledge.

Kohler, H.P. 2012. *Copenhagen Consensus 2012: Challenge Paper on "Population Growth."* Revised version: June 4, 2012.

Lecoutere, E., D'Exelle, B., and B. Van Campenhout. 2010. *Who Engages in Water Scarcity Conflicts? A Field Experiment with Irrigators in Semi-arid Africa.* MICROCON Research Working Paper 31, Brighton: MICROCON.

Livingston, G., Schonberger, S., and S. Delaney. 2011. *Sub-Saharan Africa: The State of Smallholders in Agriculture.* Paper presented at the International Fund for Agricultural Development IFAD Conference on New Directions for Smallholder Agriculture, Rome, Italy.

Mahony, C., Clark. H., Bolwell, M., Simcock. T., Potter, R., and J. Meng. 2014. *Where Politics Borders Law: The Malawi-Tanzania Boundary Dispute.* New Zealand Centre for Human Rights Law, Policy and Practice—Working Paper 21.

Meeks, R. 2012. *Water Works: The Economic Impact of Water Infrastructure. Harvard Environmental Economics Program.* Harvard Kennedy School, Discussion Paper: 12–35.

Mekdaschi, S.R. and H. Liniger. 2013. *Water Harvesting: Guidelines to Good Practice.* Bern: Centre for Development and Environment (CDE); Amsterdam: Rainwater Harvesting Implementation Network (RAIN); Rome: MetaMeta, Wageningen; the International Fund for Agricultural Development (IFAD).

Merrey, D.J. 2009. African models for transnational river basin organisations in Africa: An unexplored dimension. *Water Alternatives* 2(2): 183–204.

Min-Harris, C. 2010. *Youth Migration and Poverty in Sub-Saharan Africa: Empowering the Rural Youth.* Topical Review Digest: Human Rights in Sub-Saharan Africa.

Ministry of Justice. 2011. *Active Citizens: Conflict Mapping: An Insider's Perspective (Action Research).* Report on National Conflict Mapping for the Active Citizens Programme. Ministry of Justice, National Cohesion and Constitutional Affairs.

Moe, C.L. and R.D. Rheingans. 2006. Global challenges in water, sanitation and health. *Journal of Water and Health,* 04(Suppl.).

Msangi, J.P. 2014. *Managing Water Scarcity in Southern Africa: Policy and Strategies: Combating Water Scarcity in Southern Africa.* Springer Briefs in Environmental Science, DOI: 10.1007/978-94-007-7097-3_2. p. 21

Muchiri, P.W. 2007. *Climate of Somalia.* Technical Report No W-01. Nairobi, Kenya: FAO-SWALIM.

Mulat, A.G., Moges, S.A., and Y. Ibrahim. 2014. Impact and benefit study of Grand Ethiopian Renaissance Dam (GERD) during impounding and operation phases on downstream structures in the Eastern Nile. In: A.M. Melesse, W. Abtew, G. Shimelis, and S.G. Setegn (Eds.), *Nile River Basin,* Springer, New York. Chapter 27, pp. 543–564.

Myers, N. 2002. Environmental refugees: A growing phenomenon of the 21st century. *Philosophical Transactions of the Royal Society of London B,* 357: 609–613. DOI 10.1098/rstb.2001.0953.

Norman, M. 2001. *Environmental Refugees: A Growing Phenomenon of the 21st Century.* UK: The Royal Society.

Nsiah-Gyabaah, K. 2003. Urbanization, environmental degradation and food security. In: *Poster Presentation at the Open Meeting of the Global Environmental Change Research Community,* Montreal, Canada.

Ohlsson, L., Spillmann, K.R., Krause, J., Muller, D., and C. Nicolet. 1999. International security challenges in a changing world: Water scarcity and conflict. *Studies in Contemporary History and Security Policy* 3: 211–234.

Paisley, K.R. and T.W. Henshaw. 2013. Transboundary governance of the Nile River Basin: Past, present and future. *Environmental Development* 7, 59–71. http://dx.doi.org/10.1016/j.envdev.2013.05.003, retrieved January 20, 2015.

Perch-Neilsen, S.L., Battig, M.B., and D. Imboden. 2008. Exploring the link between climate change and migration. *Climate Change* 91: 375–393.

Peters, K.L. 2011. *Environmental Refugees.* Social Sciences Department, College of Liberal Arts, California Polytechnic State University, San Luis Obispo, CA.

Pietersen, K., Beekman, H., Abdelkader, A. et al. 2006. Africa environment outlook 2: Our environment, our wealth: Section 2; Environmental state-and-trends: 20-Year retrospective. In: *Chapter 4: Freshwater.* United Nations Environment Programme, Nairobi, Kenya, pp. 119–154.

Pinstrup-Andersen, P. 2009. Food security: Definition and measurement. Springer Science and Business Media B.V. & International Society for plant pathology. *Food Security* 1: 5–7.

Reza, S.R., Roca, E., and G.A. Tularam. 2013. *Fundamental Signals of Investment Profitability in the Global Water Industry.* Available at SSRN: http://ssrn.com/abstract=2371115 or http://dx.doi.org/10.2139/ssrn.2371115, retrieved January 20, 2015.

Ringler, C., Zhu, T., Cai, X., Koo, J., and D. Wang. 2010. *Climate Change Impacts on Food Security in Sub-Saharan Africa: Insights from Comprehensive Climate Change Scenarios.* Environment and Production Technology Division, International Food Policy Research Institute, IFPRI Discussion Paper 01042.

Roca, E.D. and G.A. Tularam. 2012. Which way does water flow? An econometric analysis of the global price integration of water stocks. *Applied Economics* 44(23): 2935–2944.

Roca, E., Tularam, G.A., and R. Reza. 2015. Fundamental signals of investment profitability in the global water industry. *International Journal of Water Resources Development,* 9(4): 395–424.

Rosegrant, M.W. 1997. *Water Resources in the Twenty-First Century: Challenges and implications for Action.* Food, Agriculture and the Environment Discussion Paper 20. Washington: International Food Policy Research Institute.

Sambi, N. 2010. An assessment of lake extent changes using four sets of satellite imagery from the Terralook database a case study of Lake Chad, Africa (Master's thesis). Institutionen fur naturgeografioc hkvartär geologi, Stockholms Universitet.

SAP. 2008. *Reversal of Land and Water Degradation Trends in the Lake Chad Basin Ecosystem Strategic Action Programme for the Lake Chad Basin.* Agreed by the LCBC Member States of Cameroon. Chad, Niger, and Nigeria: Central African Republic.

Simon, G. 2012. *Food Security: Definition, Four Dimensions, History. Basic Readings as an Introduction to Food Security for Students from the IPAD.* Master, SupAgro, Montpellier attending a joint training programme in Rome from March 19–24, 2012. Master in Human Development and Food Security, Faculty of Economics, University of Roma Tre.

Steele, M. and N. Schulz. 2012. *Water Hungry Coal Burning South Africa's Water to Produce Electricity.* Greenpeace Report 2012. Johannesburg Greenpeace.

Steinman, A.D., Luttenton, M., and K.E. Havens. 2004. Sustainability of surface and subsurface water resources: Case studies from Florida and Michigan, U.S.A. Universities council on water resources. *Water Resources Update* 127: 100–107.

Temesgen, B.B. 2012. Rainwater harvesting for dryland agriculture in the Rift Valley of Ethiopia. Thesis, Wageningen University, Wageningen, NL.

Thornton, P.K. and M. Herrero, 2010. *The Inter-Linkages between Rapid Growth in Livestock Production, Climate Change, and the Impacts on Water Resources, Land Use, and Deforestation.* Policy Research Working Paper 5178. Development Economics, The World Bank.

Tularam, G.A. 2010. Relationship between El Niño Southern Oscillation Index and rainfall (Queensland, Australia). *International Journal of Sustainable Development and Planning* 5(4): 378–391.

Tularam, G.A. 2012. Water security Issues of Asia and their implications for Australia. *Fifth International Groundwater Conference,* Artworks, India.

Tularam, G.A. 2014. Review of investing in water for a green economy. *Australian Planner* 51(4): 372–373. DOI:10.1080/07293682.2014.948987.

Tularam, G.A. and M. Ilahee. 2007a. Initial loss estimates for tropical catchments of Australia. *Environmental Impact Assessment Review* 27(6): 493–504.

Tularam, G.A. and M. Ilahee. 2007b. Environmental concerns of desalinating seawater using reverse osmosis. *Journal of Environmental Monitoring* 9(8): 805–813, Cambridge, UK: Royal Society of Chemistry. High impact: Included in South Australian Parliamentary report on Desalination.

Tularam, G.A. and M. Ilahee. 2008. Exponential smoothing method of base flow separation and its impact on continuous loss estimates. *American Journal of Environmental Sciences* 4 (2): 136–144.

Tularam, G.A. and M. Ilahee. 2010. Time series analysis of rainfall and temperature interactions in coastal catchments. *Journal of Mathematics and Statistics* 6(3): 372–380.

Tularam, G. A. and Keeler, H.P. 2006. The study of coastal groundwater depth and salinity variation using time-series analysis. *Environmental Impact Assessment Review* 26(7): 633–642.

Tularam, G.A. and M. Krishna. 2009. Long term consequences of groundwater pumping in Australia: A review of impacts around the globe. *Journal of Applied Sciences in Environmental Sanitation* 4(2): 151–166.

Tularam, G.A. and P. Marchisella. 2014. Water scarcity in Asia and its long term water and border security implications for Australia. In: R.M. Clark and S. Hakim (Eds.), *International Practices for Protecting Water and Wastewater Infrastructure.* New York: Springer Publication.

Tularam, G.A. and K.K. Murali. 2014. *Water Security Problems in Asia and Longer Term Implications for Australia.* Brisbane, Australia: Environmental Futures Centre, Faculty of Science Environment Engineering and Technology, Griffith University, 1p.

Tularam, G.A. and K.K. Murali. 2015. Water security problems in Asia and longer term implications for Australia. In: W.L. Filho and V. Sumer (Eds.), *Sustainable Water Use and Management Green Energy and Technology.* New York: Springer Publications, pp. 119–149.

Tularam, G.A. and M. Properjohn. 2011. An investigation into water distribution network security: Risk and implications. *Security Journal* 4: 1057–1066.

Tularam, G.A. and R. Reza. 2016. Water exchange traded funds: A study on idiosyncratic risk using Markov switching analysis. *Cogent Economics & Finance* 4(1): 1–12. http://dx.doi.org/10.1080/23322039.2016.1139437.

Tularam, G.A. and R. Singh. 2009. Estuary, river and surrounding groundwater quality deterioration associated with tidal intrusion. *Journal of Applied Sciences in Environmental Sanitation* 4(2): 141–150.

UNDP. 2006. *Getting Africa on Track to Meet the MDGs on Water Supply and Sanitation: A Status Overview of Sixteen African Countries.* United Nations Development Programme, New York, NY, pp. 19–25.

UNDP. 2012a. Africa Human Development Report 2012: Towards a food secure future. In: *Chapter 4: Sustainable Agricultural Productivity for Food, Income and Employment.* Regional Bureau for Africa, United Nations Development Programme, New York, NY, pp. 63–82.

UNDP. 2012b. Africa Human Development Report 2012: Towards a food secure future. In: *Chapter 2: How Food Insecurity Persists amid Abundant Resources.* Regional Bureau for Africa, United Nations Development Programme, New York, NY, pp. 27–44.

UNDP. 2013. *Human Development Report 2013: The Rise of the South: Human Progress in a Diverse World.* New York: United Nations Development Programme, pp. 139–197.

UNDPI. 2013. *Sub-Saharan Africa Continues Steady Progress on Millennium Development Goals.* The Millennium Development Goals Report 2013. Nairobi, Kenya: United Nations Department of Public Information, pp. 1–2.

UNEP. 2002. *Regionally Based Assessment of Persistent Toxic Substances, Chemicals: Sub-Saharan Africa Regional report.* United Nations Environment Programme, Nairobi, Kenya.

UNEP. 2005. *The State of the Environment in Somalia: A Desk Study.* United Nations Environment Programme, Nairobi, Kenya.

UNEP. 2009. Water sustainability of agribusiness activities in South Africa. United Nations Environmental Programme. *Chief Liquidity Series* 1: 42–46.

UNEP. 2011. *Water Issues in the Democratic Republic of the Congo: Challenges and Opportunities.* Nairobi, Kenya: United Nations Environment Programme.

UNICEF. 2013. *Somalia: Fast Facts.* Unicef Somalia Support Centre, The United Nations Children's Fund, New York.

United Nations. 2006. *World Population Prospects: The 2006 Revision Volume III: Analytical Report.* United Nations Department of Economic and Social Affairs/Population Division.

United Nations. 2008. *Trends in Sustainable Development: Africa Report.* Division for Sustainable Development, Department of Economic and Social Affairs, United Nations.

United Nations. 2013. *Making the Most of Africa's Commodities: Industrializing for Growth, Jobs and Economic Transformation.* Economic Commission for Africa, Addis Ababa, Ethiopia.

USDE. 2006. *Energy Demands on Water Resources: Report to Congress on the Interdependency of Energy and Water United States.* Department of Energy.

Victoria, L. 2012. Review of water scarcity in Northern China: Why is the water scarce, and what can be done about it? *ENVR* 200: 1–14.

Webersik, C. 2006. Fighting for the plenty—The banana trade in southern Somalia. *Paper presented to the Conference on Multinational Corporations, Development and Conflict,* Queen Elizabeth House, Oxford.

Winterbottom, R., Reij, C., Garrity, D. et al. 2013. *Improving Land and Water Management.* World Resources Institute. Available at http://www.wri.org/sites/default/files/improving_land_and_water_management 0.pdf.

World Bank. 2009. South Asia: Shared views on development and climate change. In: *Chapter 10: The Social Dimensions of Climate Change.* South Asia Region Sustainable Development Department, World Bank, 131–137. Online reports and documents: published on websites, Washington. http://siteresources.worldbank.org/SOUTHASIAEXT/Resources/Publications/448813-1231439344179/5726136-1259944769176/SAR_Climate_Change_Full_Report_November_2009.pdf.

World Bank. 2012. The future of water in African cities: Why waste water? Chapter 1. In: M. Jacobsen, M. Webster, K. Vairavamoorthy (Eds.), *Africa's Emerging Urban Water Challenges.* Washington, D.C.: The World Bank, Washington.

World Bank. 2014. *World Development Indicators 1960–2013.* The World Bank, Washington. Online reports and documents: published on websites. http://data.worldbank.org/sites/default/files/wdi-2014-book.pdf.

Wroblewski, J. and H. Wolff. 2010. *Risks to Agribusiness Investment in Sub-Saharan Africa.* Evans School of Public Affairs, University of Washington, Washington; available on http://faculty.washington.edu/hgwolff/Evans%20UW_Request%2057_Agribusiness%20and%20Investment%20Risks_02-12-2010.pdf.

28

Utilization of Groundwater for Agriculture with Reference to Indian Scenario

S. Ayyappan, A. K. Sikka, A. Islam, and A. Arunachalam

CONTENTS

28.1 Introduction

Though irrigation has contributed significantly in boosting India's food production and creating grain surpluses, there are vast irrigated areas where agricultural productivity continue to remain low. Many of these low productivity areas may be classified as economically water-scarce areas, where water is not a limiting factor but lack of financial means restricts development of the available resources. Irrigation development is suggested as a key strategy to enhance agricultural productivity. Several studies suggest that the irrigation needs to play a larger role toward the goal of achieving a higher agricultural productivity and national food security (Persaud and Rosen 2003; Kumar 1998).

Irrigated agriculture plays a vital role in the securing of global food production. Globally, irrigation accounts for more than 70% of total water withdrawals and more than 90% of total consumptive water use (Siebert et al., 2010). Earth's fresh groundwater resources are estimated at about 10 million km^3, which is more than 200 times the global annual renewable water resource provided by rain. As shown in Table 28.1, groundwater shares 1.7% of the total global water resources and 30% of the total freshwater resources of the earth (Shiklomanov

and Rodda, 2003). Global aquifers hold an enormous water reserve that is several times greater than surface water resources. About 35% of the area of the continents is underlain by relatively homogeneous aquifers, 18% is endowed with groundwater, some of which is extensive, in geologically complex regions. Table 28.2 summarizes the global survey of groundwater irrigation data. Out of the total 301 M ha area equipped for irrigation globally, 38% is equipped for irrigation with groundwater (Siebert et al., 2010). The countries with the largest extent of areas equipped for irrigation with groundwater are India (39 million ha), China (19 million ha), and the United States (17 million ha).

28.2 Water Resources of India

India supports 17% of the human and 15% of the livestock population of the world with only 2.4% of the land and 4% of the water resources. Water resources potential of the country has been estimated as 1869 km^3, which includes replenishable groundwater resources (Central Water Commission of India [CWC], 2013). The

TABLE 28.1

Distribution of Earth's Water Resources

Water Source	Total Water (1000 km³)	Fresh Water (1000 km³)	Percent of Total Water	Percent of Fresh Water
Ocean, seas, and bays	1,338,000	–	96.5	–
Glaciers and permanent snow covers	24,064	24,064	1.74	68.70
Groundwater	23,400	10,530	1.7	30.06
Soil moisture	16.5	16.5	0.001	0.05
Ground ice of permafrost zone	300	300	0.022	0.86
Lakes	176.4	91	0.013	0.26
Atmosphere	12.9	12.9	0.001	0.04
Swamp water	11.5	11.5	0.0008	0.03
River stream	2.12	2.12	0.0002	0.006
Biological water	1.12	1.12	0.0001	0.003
Total	1,385,984.54	35,029.14	100	100

Source: Shiklomanov, I.A. and J.C. Rodda. 2003. *World Water Resources at the Beginning of the Twenty-First Century.* Cambridge University Press, Cambridge, UK.

TABLE 28.2

Global Survey of Groundwater Irrigation

Region	Groundwater Irrigation (Mha)	Groundwater Volume Used (km³)
Global total	112.9 (38%)	545 (43%)
Africa	2.5 (19%)	18 (18%)
America	21.5 (44.1%)	107 (48%)
Asia	80.6 (38%)	399 (45%)
South Asia	48.3 (57%)	262 (56%)
East Asia	19.3 (29%)	57 (34%)
Europe	7.3 (32.4%)	18 (38%)
Oceania	0.9 (23.9%)	3 (21%)

Source: Siebert, S. et al. *Hydrology and Earth System Sciences* 14, 1863–1880, 2010.

Note: Values within parentheses refers proportionate of total.

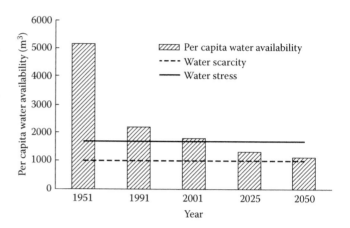

FIGURE 28.1

Per capita water availability. (Adapted from CWC. 2013. Water and related statistics. Water Resources Information System Directorate, Information System Organisation, Water Planning & Project Wing, Central Water Commission, New Delhi.)

domestic and industrial needs may further worsen if anticipated impact of climate change on water resources is also taken into account. Unless timely and properly managed, water scarcity may lead to adverse situations. The challenge is thus to produce more from less water by efficient use of utilizable water resources in irrigated areas, enhance productivity of challenged ecosystems, that is, rainfed and waterlogged areas, and utilize a part of gray water for agriculture production in a sustainable manner.

28.2.1 Groundwater Irrigation

Groundwater has rapidly emerged to occupy a dominant place in India's agriculture and food security in the recent years. It plays a vital role in agricultural development by providing assured irrigation to the farmers and by enhancing the productivity of other inputs. In the agriculture sector, the importance of groundwater has been increasing manifold due to factors such as technological breakthrough in water extraction technology, soft loans for installation of groundwater extraction mechanism, and remunerative relative price ratio in favor of water intensive, commercial, and horticultural crops. Groundwater offers reliability and control of water in irrigation, which is very important. The aquifers that host groundwater are the primary buffers against drought for both human requirements and crop production. It can be drawn on demand and also relatively more risk-free from pollution than surface water resources, making it more attractive to many usages and many groups of users. Its "on-demand" availability in controlled conditions helps in adoption of modern and diversified agriculture and thus achieving higher agricultural and water productivity and improved livelihoods. Experimental results indicate

utilizable water resources of the country have been assessed as 1121 km³, of which surface and groundwater water contribute 690 and 431 km³, respectively (CWC, 2013). As depicted in Figure 28.1, India is facing water stress conditions with per capita availability of water declining sharply from 5177 m³ in 1951 to 1544 m³ in 2011 (CWC, 2013). It is projected to reduce further to 1465 m³ and 1235 m³ by the years 2025 and 2050, respectively, under high population growth scenarios (Kumar et al., 2005). It has been projected that population and income growth will boost the water demand in the future to meet food production, and domestic and industrial requirements. The water scarcity situation for various uses such as agriculture, drinking water, and that water control alone can bridge the gap between potential and actual yields by about 20% (Herdt and Wickham, 1978). Studies conducted at five sites in the Indo-Gangetic basin under the Challenge Program on Water & Food (CPWF) Groundwater Governance in Asia (GGA) project revealed that groundwater use produced higher benefits as compared to canal irrigation and conjunctive water use (Sharma et al., 2008).

The net annual groundwater availability for the country is estimated as 398 BCM whereas the annual groundwater draft for irrigation, domestic, and industrial sector is 245 BCM. The stage of groundwater development for the country as a whole is 62%. Out of the 245 BCM, the irrigation sector accounts for 222 BCM (91% of total groundwater draft) (Central Groundwater Board [CGWB], 2014). However, there is a wide inter- as well as intraregional disparity as far as groundwater development is concerned. The stage of groundwater development in the states of Punjab, Haryana, Rajasthan, and Delhi is more than 100%, whereas in the states of Himachal Pradesh, Tamil Nadu and Uttar Pradesh, and UTs of Daman and Diu, and Puducherry stage of groundwater development is 70% and above. On the other hand, in the Eastern part of the country, the level of groundwater development is very low. Out of 6607 assessment units (Blocks/Mandals/Talukas) in the country, 1071 units (16%) in various states have been categorized as "overexploited," that is, the annual groundwater extraction exceeds the net annual groundwater availability and significant decline in long-term groundwater level trend has been observed either in premonsoon, postmonsoon, or both. In addition, 217 units (3%) are "critical," that is, the stage of groundwater development is above 90% and within 100% of net annual groundwater availability and significant decline is observed in the long-term water level trend in both premonsoon and postmonsoon periods. There are 697 semicritical units (11%), where the stage of groundwater development is between 70% and 100%, and significant decline in long-term water level trend has been recorded in either premonsoon or postmonsoon season. Apart from this, there are 92 assessment units (~1%) completely underlain by saline groundwater. The remaining 68% units are safe. Thus, the major groundwater development challenge in India is that there is extreme overexploitation of the resource in some parts of the country coexisting with relatively low levels of extraction in other parts of the country. Groundwater pollution is another major cause of concern. Instances of high fluoride have been reported in 13 states, arsenic in West Bengal, and iron in the northeastern states, Orissa, and other parts of the country. In West Bengal, arsenic toxicity has been reported as a result of overdraft, with excessive groundwater withdrawal during lean periods for irrigating summer paddy. In the canal-irrigated

lands of Haryana, Punjab, Delhi, Rajasthan, Gujarat, Uttar Pradesh, Karnataka, and Tamil Nadu, groundwater is affected due to high salinity, the affected area being over 193,000 km^2 (Thatte et al., 2009). Long-term decline in groundwater levels, deterioration of groundwater quality, problems of salinity, drying up of wetlands and low-flows in streams and rivers during summer months are all highlighted as environmental ill-effects of unplanned intensification of groundwater use in agriculture (Shah, 1993; Seckler et al., 1999; Postel, 1999; Burke and Moench, 2000). The major concern at the national level is the sustainable agricultural growth through equitable development of groundwater resources. This requires region-specific policy interventions in various aspects of groundwater development for irrigation purposes.

Groundwater extraction primarily depends on availability of mechanical devices (pumps and tubewells), and reliable and economically accessible and efficient sources of energy. According to Minor Irrigation Census (2006–2007), the number of wells and tube wells has increased from 11.4 million in 1986–87 to 19.7 million in 2006–07. Tube wells have become the largest single source of irrigation water in India (Shankar et al., 2011). Groundwater can be accessed only by expending some form of energy: electricity in developed regions and diesel in remote and underdeveloped regions. As such, availability of energy for groundwater extraction and pricing policies has great impact on the use of groundwater and are often termed "water-energy nexus." In the eastern states in India with progressive rural deelectrification since the mid-1980s (Shah, 2001), groundwater irrigation has become dependent on the more costly and dirtier diesel fuel. Further, because of its diesel-dependence, groundwater-rich eastern India is unable to take full advantage of this abundant groundwater resource.

Most of the groundwater development has taken place through private investment. Tubewell and pump technology, which are widely used for groundwater extraction, together with government policies on subsidizing credit and rural energy supplies, have led to the phenomenal growth in groundwater development in India. However, reduced farm profitability due to increasing pumping cost with declining water table, deceleration in productivity of irrigation water, and equity issues in groundwater distribution are the major causes of concern. Groundwater pollution is another emerging threat to the sustainability of water resources. Therefore, an understanding of the interrelationship of hydrogeological, agroclimatic, socioeconomic, and policy factors for sustainable development and equitable distribution of these precious natural resources needs urgent attention.

Indian Council of Agricultural Research (ICAR) initiated water management research through river valley projects in the late 1960s and early 1970s, and later All India Coordinated Research Project (AICRP) was sanctioned in 1970, which became operational as AICRP on Optimization of Groundwater Utilization through Wells and Pumps at Water Technology Centre (WTC), Indian Agricultural Research Institute (IARI), New Delhi with four cooperating centers. The name of the scheme was changed from "AICRP on Optimization of Groundwater Utilization through Wells and Pumps" to "AICRP on Groundwater Utilization (GWU)" in 2007 with broadened mandate to take care of new thrust areas of research. At present, the scheme has been mandated to ensure optimum utilization of groundwater for sustainable agriculture through its proper assessment, modeling different use patterns, work out strategies for its efficient utilization and augmentation, develop efficient hardware, and study groundwater pollution problems in varying agricultural management situations. In the XIIth plan, this has been merged with AICRP on water management for integrated irrigation management.

28.3 Enhancing Groundwater Productivity through Efficient Water Management

As pumping of groundwater for irrigation requires motive power either using diesel or electricity operated pumps, it becomes too costly for farmers to irrigate their fields. Efficient water management therefore becomes an important issue for groundwater users. Broadly, the important water management measures for enhancing productivity of groundwater involve (a) proper irrigation scheduling, (b) improved agronomic practices, (c) optimization of rice transplanting date, (d) conjunctive use of rain, canal, and groundwater, (e) acceleration of resource conserving technologies, for example, zero tillage, conservation tillage, raised bed, etc., (f) use of water and energy efficient irrigation methods, for example, sprinkler irrigation, microirrigation, low energy water application (LEWA) devices, and (g) multiple use of pumped water. By adopting an optimum schedule of irrigation, a substantial amount of water can be saved and the yields can be increased. Though it is a common perception that rice needs continuous submergence, several studies have shown that the intermittent irrigation with ponding depth of 5 ± 2 cm of water in the rice fields 3 days after disappearance (DAD) of previously ponded water is the most optimum irrigation schedule for the entire period of rice growth. Such practices could result in 23%–65% saving of water as compared to traditional methods of growing rice under continuous submerged conditions.

Water application methods play an important role in controlling the water losses through deep percolation, surface runoff, and direct evaporation from soil surface. Pressurized irrigation systems, namely, rainguns, sprinklers, microsprinklers, and microirrigation systems, namely, drip and subsurface drip, are efficient irrigation methods with application efficiency ranging between 80% and 95%. The water saving varies in the order of 25%–40% with various sprinklers. Similarly, 40%–60% savings in irrigation water can be achieved in the drip method of irrigation as compared to gravity methods, in addition to improvement in quantity and quality of produce. The low-cost star microtube drip irrigation method has been found beneficial for banana and vegetables in farmers' fields of south Bihar with benefit cost ratio for the system varying in the range of 1.18–1.24 depending on the variety and crop spacing used (Sikka and Bhatnagar, 2006). Analysis of energy requirement for pumping water from varying groundwater depths (3, 7, 11, and 15 m) for surface (gravity) and drip irrigation showed that the energy requirement gets reduced substantially with drip irrigation as compared to a gravity irrigation system (Srivastava and Upadhyaya, 1998). ICAR Research Complex for Eastern Region, Patna has also developed and demonstrated the effectiveness of LEWA in not only saving water but also in energy saving, and its practical utility for small holders (Singh and Islam, 2007). Adoption of improved irrigation schedules and improving irrigation methods with higher application efficiencies will not only result in increase in water productivity, but will also result in reduction in irrigation water withdrawals from groundwater with no reduction in yields (Karimi et al., 2012). Such reduction in groundwater withdrawal can result in decline in energy consumption and subsequently carbon emission from groundwater pumping, which is a key climate change mitigation strategy.

It is well recognized that promoting multiple uses of water results in enhanced water productivity and efficient use of available water resources. Different forms of multiple use systems, that is, multiple use of harvested rainwater, canal water after outlet and in the network itself, groundwater where pumped water from tubewells used for other productive purposes before it is delivered to the field for irrigating field crops, etc. have been investigated at different locations (Sikka, 2009). For example, in a tubewell-based irrigation system at ICAR Research Complex for Eastern Region (ICAR-RCER), Patna, multiple use of water was demonstrated by routing pumped irrigation water through a secondary reservoir, where pumped water was stored up to a desired capacity for aquaculture and then released in an appropriate stream size for irrigation purposes. The routed water, containing a good amount of nutrients, provided opportunity for applying water to the fields in correct amounts and at an appropriate time and enhances yield and quality of agricultural produce.

28.3.1 Conjunctive Use of Surface and Groundwater Resources

The National Water Policy calls for the conjunctive use of all the available water resources, namely, rainwater, river and surface water sources, groundwater, sea water, and recycled wastewater. The conjunctive use of ground and surface water can help to improve cropping intensity by using surface and rainwater during *kharif* and groundwater during *rabi* and summer seasons. Integrated water resources management is vital for maximizing the benefits of the available irrigation water.

Conjunctive use of surface and groundwater consists of harmoniously combining the use of both surface and groundwater resources in order to minimize the undesirable physical, environmental, and economic effects of each solution and to optimize the water demand/supply balance. Usually conjunctive use of surface and groundwater is considered within a river basin management program, that is, both the river and the aquifer belong to the same basin. There are several approaches for conjunctive use of surface and groundwater resources, namely, cyclic mode and blending mode. In the cyclic mode, canal and groundwater are used separately whereas in the bending mode these waters are used simultaneously. Cyclic use is generally adopted to accommodate fluctuations in the canal supplies due to the rotational system. This approach also helps to maintain a favorable salt balance in the root zone. Simultaneous use or blending strategy is practiced where canal supplies are usually inadequate, and mixing of groundwater with the canal water is necessary to get the required flow rate for proper irrigation. Further, mixing saline groundwater with good quality canal water helps to decrease the salinity of the irrigation water in order to reduce the risk of soil salinization.

The Haroti region of southeastern Rajasthan has poor quality groundwater. This region has canal water but the supply of canal water is not sufficient especially to the tail ender. In this area, there is the possibility to use the marginally saline groundwater for crop production. Experimental results showed decrease in grain yield with increasing number of groundwater irrigation as compared to irrigation with canal water alone. As shown in Table 28.3, with the application of two irrigations through canal water followed by one irrigation through groundwater in cyclic mode, one can achieve as good production as in canal water irrigation on a sustainable basis (AICRP on GWU, 2012). Experimental results have shown that the salt buildup in postharvest soil was also nonsignificant. Thus, there is 33% saving (two irrigations out of six) of good quality water without any economic yield reduction and soil health deterioration. By adopting this particular technology (conjunctive use of poor quality groundwater and good quality canal water in cyclic mode) in the Bundi district, the area of wheat, mustard, chickpea, or coriander can be doubled with the existing water resources. Conjunctive water use of harvested/canal water and saline groundwater for wheat and mustard crop developed by Udaipur center of AICRP on GWU was recommended in Rajasthan.

Similarly, conjunctive water use planning for wheat crop in Junagadh region has been found beneficial to farmers and state. Conjunctive water use of harvested water and groundwater for wheat crop has the potential of saving 110 mm of groundwater and has applicability in the Saurashtra region of Gujarat. Large numbers of water harvesting structures (check dams) were constructed in Gujarat and are a source of groundwater recharge only after monsoon. These structures lose a large volume of their storage through evaporation and seepage. Although seepage can recharge aquifers, one has to pump it for irrigation which consumes more power as compared to pumping from surface storage directly. Hence, it is advisable to use stored water from check dams as early as possible after a monsoon. It has been found economical when at least two irrigations are given from these water harvesting structures (surface source). Under conjunctive water use planning for

TABLE 28.3

Effect of Conjunctive Use of Water on Grain and Straw Yield of Wheat

Irrigation Water	Grain Yield (q ha⁻¹)			Change in Grain Yield (%)		
	2008–2009	2009–2010	2010–2011	2008–2009	2009–2010	2010–2011
Canal water (CW)	36.88	38.07	37.05	–	–	–
Groundwater (GW)	27.44	28.34	28.53	−25.6	−25.6	−23.0
1CW + 1GW	33.75	34.55	35.02	−8.5	−9.2	−5.5
2CW + 1GW	36.51	37.69	38.23	−1.0	−1.0	3.2
1CW + 2GW	32.26	33.31	32.59	−12.5	−12.5	−12.0

Source: AICRP on GWU. 2012. Annual Report 2011–12, AICRP on Groundwater Utilization, Directorate of Water Management, Bhubaneswar, Odisha, India, p. 126.

wheat crop in Junagadh region, 533.94 m³ of groundwater draft (7.72%) per hectare can be reduced per irrigation and 123.8 kilowatt hour power (4.9%) per hectare can be saved per irrigation. Thus, conjunctive use planning provides ample opportunity to avoid deep pumping, reduce groundwater draft, and achieve higher economy by utilizing spillover water before it escapes from water harvesting structures.

28.3.2 Improving Groundwater Productivity through Improved Pumping Efficiency

A typical tubewell irrigation system consists of tubewell, pumpset, and conveyance network. Faulty design of tubewell, that is, filter, gravel pack, and inadequate development of the well after construction often causes excessive resistance resulting in higher energy requirement as well as low pump discharge. Efficiency of pumping units plays an important role in efficient utilization of groundwater resources as low efficiency of pumping units result in excessive energy consumption, reduced irrigation efficiency, and low productivity. A wide variety of pumps are available in the market. In areas with shallow water tables in India most of the centrifugal pumps are operated with 2.5–10 hp engines. In a field study conducted in northwest Bengal, Bom and Steenbergen (1997) observed that 5 hp pumpset used for lifting water from shallow tubewells were generally oversized, and that there was excessive friction loss in the suction pipe due to installation of poorly designed check valves. Recently, there has been increase in the use of small (3.5 hp) portable diesel as well as kerosene operated pumpsets due to ease in transportation from one location to other location. Studies conducted under the aegis of AICRP on "Optimization of Ground Water Utilization through Wells and Pumps" has revealed the reasons for lower efficiencies of the pumps at different locations in India as: (1) mismatch of selected pumping units with the well conditions, (2) mismatch of the drive units with the pump requirement, (3) excessive length of delivery pipe and excessive suction lift, (4) use of inefficient foot valves (offer very high head loss), (5) use of reducer at the delivery side (use of 10 × 8 cm nipple at the outlet of delivery pipe of 10-cm diameter pump), (6) poor quality of pipe fittings with unnecessary short radius bends, (7) loose foundation causing excessive vibrations during operation, and (8) lack of technical service and awareness to the farmers on selection, installation, and operation of pumps.

Rectification measures by removal of nipple at the delivery side and replacement of short radius bends with long radius bends in the Tarai region of Pant Nagar resulted in 4.83% discharge rate increase and saving in the fuel consumption of about 190 L/year. Similarly, adoption of improved foot valves in Tarai area of

FIGURE 28.2
Propeller pump.

Pantnagar resulted in energy savings of about 11% (Sikka and Bhatnagar, 2006). The Pantnagar center under the AICRP on GWU has also developed and tested different sizes of propeller pumps (Figure 28.2) ranging in sizes from 15 to 30 cm to deliver 20 to 100 L/s for 1- to 3-m head. The performance test was carried out at different speeds of rotation of the pump and lift ranging from about 1 m to about 2 m, and revealed that the pump performed best at a speed of 1440 rpm. The efficiency of the pump at this speed varied from 65%–40% at a static head of 1.0–2.5 m and discharge and horse power ranging from 65–30 L/s and 2.3–3.2 HP, respectively. A low-cost high-efficiency propeller pump (2900 rpm and 1440 rpm inclined as well as vertical models) for lifting water from irrigation channels, ponds, drains, etc. has also been developed by the PAU Ludhiana and is being widely used in southwest Punjab, where fields are at a higher level than the distributaries, for lifting water from distributaries. A tractor-driven propeller pump (1440 rpm and 35 hp) with a capacity of 97.42 L/s against a total head of 2.56 m has also been designed at Ludhiana center.

A considerable amount of pumped water is lost when conveyed through unlined earthen channels. These losses can be attributed to seepage from the porous earthen channels, "start-up" losses for wetting of the dry perimeter of the earthen channel, and overtopping of the undersized channels. The combined losses resulting from seepage and start-up losses has been reported to be at least 15% per 100 m (Bom and Steenbergen, 1997), which will be much higher in cases of undersized channels. As remedial measures, farmers have adopted polythene tubes as a means of water conveyance in various parts of Bihar and West Bengal. This is particularly more suited to scattered and highly fragmented land holdings in Bihar, West Bengal, and Eastern UP. Unlike field channels, these polythene pipes also allow irrigation of uneven lands/higher elevation areas and also no cultivable land is lost. As shown in Figure 28.3, introduction

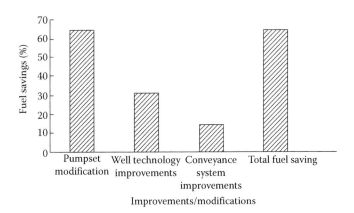

FIGURE 28.3
Fuel efficiency improvements for groundwater pumping in North Bengal. (Adapted from Steenbergen, F. and G.J. Bom. 2000. *GRID IPTRID Network Magazine* 16: 8–9.)

of different modification/improvements in well, pump, and conveyance systems in northwest Bengal resulted in fuel savings ranging from 15% to 50% with a combined savings translating into 70% (Steenbergen and Bom, 2000).

28.3.3 Groundwater Markets

Emergence of low-cost pump technology and tubewell technology has led to spontaneous growth of groundwater markets for irrigation in many parts of the country, and more particularly in Gangetic plains of India. It involves localized informal sale and purchase of pumped water mostly through privately owned shallow tubewells and low-lift pumps. Groundwater markets have helped small and marginal farmers with small and scattered land holdings in providing access to groundwater for irrigation and enhancing productivity in the eastern region of UP, North Bihar, and West Bengal. As the ownership of mechanical water extraction/lifting devices remains out of reach of the marginal farmers in most parts of eastern Gangetic plains, the small farmers with land holdings up to 0.4 ha in eastern UP are the biggest beneficiaries of the groundwater market (Pant, 2004). Farmers owning wells generally achieve higher yields while those purchasing water from well owners achieve yields higher than those dependent on canal irrigation alone but not as high as the yields achieved by well owners. Table 28.4 presents the comparison of crop productivity and groundwater productivity for pump owner and nonpump owner based on the study conducted in the three villages of Vaishali district, Bihar. Higher crop productivity and groundwater productivity (yield per unit of groundwater used) was observed for pump owner as compared to the nonpump owner (Islam and Gautam, 2009).

Though the informal groundwater markets operating without government interventions are able to increase access to water for some of the poorest farmers, but this hinders the reallocation of water for more productive use with water trading remaining informal. There is no clear policy or legal measures regarding water markets in India. The basic issue, therefore, is that of evaluating a legally and institutionally enforceable system, which will ensure sustainability and provide the parameters within which water markets could operate. The introduction of more formal water markets, wherever feasible, could provide opportunity for efficient reallocation using market mechanisms (World Bank, 1999).

TABLE 28.4

Productivity of Major Crops in Vaishali District of Bihar, India

	Paddy		Wheat	
	Productivity (kg/ha)	Water Productivity (kg/m³)	Productivity (kg/ha)	Water Productivity (kg/m³)
Pump owner	2375	3.20	2680	2.20
Nonpump owner	1740	2.70	2240	1.80
All	2115	3.00	2480	2.10

Source: Islam, A. and R.S. Gautam, 2009. *Groundwater Governance in the Indo-Gangetic and Yellow River Basins: Realities and Challenges (Selected Papers on Hydrogeology, v.15).* A. Mukherji, K.G. Villholth, B.R. Sharma, and J. Wang, Eds., CRC Press, Taylor & Francis Group, Boca Raton, FL, Chapter 6, pp. 105–118; Sharma, B.R., A. Mukherjee, R. Chandra, A. Islam, B. Dass, and M.D. Ahmed. 2008. *Proceedings of the CGIAR Challenge Program on Water and Food, 2nd International Forum on Water and Food,* Addis Ababa, Ethiopia, November 10–14, 2008. The CGIAR Challenge Program on Water and Food, Colombo, pp. 73–76.

28.3.4 Enhancing Groundwater Productivity through Group Tubewells

The concept and functioning of group tubewells owned and operated by small and marginal farmers in regions of eastern UP (Deoria) and north Bihar (Vaishali) has been demonstrated by the Vaishali Area Small Farmers Association (VASFA). VASFA, a registered NGO established in 1979, facilitated group borings and tubewells in the Vaishali district. They installed 35 group tubewells with 6″ or 4″ delivery (7.5–10 hp pumps) with command area of 20 to 40 acres. Each group tubewell has Small Farmers Association (SFA) with 20–40 farmers having nominal membership fee as a service charge, each consisting of a subcommittee of three members (a president, secretary, and treasurer). Water distribution, operation, and maintenance are managed by SFA with the help of VASFA. Cost of water is realized based on operating charges plus service charges as decided by the committee. Associations also helped the small farmers in meeting their requirements of quality inputs, technology, and sale of output at reasonable prices besides managing valuable water resources. In the absence of electric operated pumps, the community/group tubewells with lower water charges are an option for bringing more area under irrigation and increasing the production and productivity. When integrated and supplemented with improved resource conservation technology and on-farm water management practices, this could further enhance water productivity of groundwater (Sikka and Bhatnagar, 2006). It also demonstrates a good example of participatory irrigation management in groundwater management. Though these group tubewells are favored due to higher discharge and lower water charges compared to private tubewells, there needs to be a proper conflict resolution mechanism, and effective and transparent governance for its success (Islam and Gautam, 2009).

28.3.5 Participatory Groundwater Management

The check dam movement in the Saurashtra region of Gujarat for improving the groundwater availability is a good example of participatory groundwater management. This involved the formation of village level local institutions in hundreds of villages to undertake the planning, finance, and construction of a system of check dams in and around the village to collect and hold rainwater to recharge the underground aquifers and thereby recharge the dug wells. These check dams also improved surface water availability for conjunctive water use and improved groundwater quality in poor quality coastal areas and enhanced groundwater recharge in high elevated areas of the region. Junagadh center under AICRP on GWU, on the basis of groundwater potential estimation and groundwater quality mapping for southwest Saurashtra region have recommended policy guidelines for groundwater management and crop planning.

28.3.6 Solar-Powered Groundwater Irrigation

With rapidly falling prices of photovoltaic cells, solar photovoltaic (PV) and wind turbine pumps are now rapidly becoming more attractive than the traditional power sources. These technologies, powered by renewable energy sources (solar and wind), are especially useful in remote locations where a steady fuel supply is problematic and skilled personnel are scarce. Groundwater pumping with renewable energy sources has significant local environmental and socioeconomic benefits as they directly displace greenhouse gas emissions while contributing to sustainable rural development. The solar pumps are gaining popularity for groundwater pumping and such solar-powered pumping systems are already operational in Nalanda district of Bihar (Shah and Kishore, 2012). In Eastern India, where groundwater is available at shallow depths, solar pumps offer an ideal technology for utilization of available groundwater resources to boost agricultural production in the region. Subsidizing capital investment in solar pumps and piped distribution systems may be a suitable option for encouraging groundwater use in Eastern India, which otherwise remains underutilized. But it may not be a very attractive option in areas of high pumping depth of groundwater because of the capital cost and land taken up by solar panels. In Western and Southern India, the solar pumping system with near-zero marginal cost of groundwater pumping can increase the stress on already depleted groundwater resources. With an appropriate policy, promotional strategy, and incentives, solar pumps can solve the nexus between energy and groundwater irrigation.

28.3.7 Groundwater and Disaster Risk Management

Groundwater is intimately connected to global change phenomena and it can play an important role in disaster risk management for both flooding and drought. As the response of aquifers to climatic variability and climatic fluctuations is much slower than that of surface water storages, aquifers act as a more resilient buffer during dry spells, especially when they have large storages. The buffering capacities of aquifers are being used in techniques such as managed aquifer recharge for purposeful and actively managed recharge of aquifers for subsequent recovery and use. The managed aquifer recharge can improve the sustainability of groundwater resources and mitigate disaster risk by augmenting the quantity or improving the quality of groundwater resources; reduce the effects of seasonal and climate variability; and manage excessive or unpredictable surface water flows.

28.4 Strategies for Sustainable and Efficient Utilization of Groundwater

1. *Areas facing the problem of overexploitation of groundwater*: Mostly northern, southern, central, and western parts of India under arid and semi-arid climatic zones characterized by low to medium rainfall are facing the problem of overexploitation of groundwater. It has been observed that due to free access to exploit the groundwater through electric tubewells, farmers withdraw excess groundwater for irrigation purposes. The following strategies may be followed for these areas:

   a. Rational planning for raising high water demanding crops in overexploited areas and/or shifting to less water demanding crops. Encouraging use of energy efficient pumps through taxation and energy pricing policy as well as adoption of microirrigation systems.

   b. Incentivization for water conservation.

   c. Recharge structures to tap high rainfall events.

2. *Areas facing the problem of underutilization of groundwater*: In eastern India, where much of the topography is flat, groundwater is relatively abundant and rainfall and recharge are high, but groundwater use is low. Hence, a few strategies for these areas could be:

   a. Investments in rural electrification to intensify groundwater use which could boost agricultural productivity.

   b. Provision for subsidy and credits for pumping devices.

   c. Availability of low-cost and low-hp pump sets for small and marginal farmers.

   d. Proper design criteria of wells and tubewells should be strictly followed in the development of groundwater resources.

   e. Change in groundwater law which will make it easier for small and marginal farmers to invest in wells and tubewells.

   f. Encouraging community/group tubewell with provision of piped distribution network, and effective and transparent community-based institutional system should be developed for community tubewells.

   g. Incentive and government subsidy for solar operated pumping systems, along with provision of microirrigation systems.

3. *Groundwater management in coastal areas*: Based on the aquifer properties of the coastal areas, pump size should be specified and optimal pumping rate and schedules should be estimated for exploiting groundwater resources in the coastal areas in order to check upcoming of saline water intrusion. This should be strictly followed while providing subsidy through government organizations.

28.5 Summary

India has a large groundwater resource but its availability and utilization varies substantially from state to state and area to area. While the benefits of groundwater irrigation are widely recognized, there is growing concern about groundwater depletion and quality deterioration due to unregulated withdrawal. For long-term sustainability, groundwater development should be consistent with the availability of the resource in different regions. Some institutional provisions need to be made to check overexploitation of groundwater. Integrated and coordinated development of surface and groundwater resources and their conjunctive use should be encouraged for efficient utilization of available water resources. There is a need to evolve and demonstrate innovative and cost-effective technologies aimed at value addition through multiple uses and increasing water productivity for efficient utilization of pumps for groundwater utilization. Increasing contamination of groundwater with special reference to arsenic and fluoride is another major area of concern, and needs immediate attention and strategies to combat it. Challenges for the research and development include development of energy efficient pumping units, low-cost small capacity diesel pump sets, promoting participatory on-farm water management practices, multiple water use interventions, cooperative movement in groundwater utilization and management, and research on institutional, legal and policy issues relating to groundwater development and water markets. There is a need to formulate a suitable energy policy for the holistic and sustainable development of this precious natural resource, which, in turn, will emerge as a precursor for the development of agriculture as a whole. Appropriate policy interventions for promoting solar pumps will not only solve the nexus between energy and groundwater irrigation, but also pave the way for a second green revolution in the otherwise "High Potential Low Productive" Eastern region of India. There is a strong need to develop policy initiative (1) to enhance equity in groundwater development as presently it is skewed in favor of large farmers; (2) to create a balance among different users of agriculture, domestic, industry, power, and others regarding their rights, prices, and cross-subsidies, and (3) to develop comprehensive environmental guidelines for groundwater usage and contamination, and a legal framework to implement it.

References

AICRP on GWU. 2012. Annual Report 2011–12, AICRP on Groundwater Utilization, Directorate of Water Management, Bhubaneswar, Odisha, India, p. 126.

Bom, G.J. and F. Steenbergen. 1997. Fuel efficiency and inefficiency in private tubewell development. *Energy for Sustainable Development*, 5(III): 46–50.

Burke, J. and M. Moench. 2000. Groundwater and society: Resources, tensions and opportunities. Themes in groundwater management for the twenty-first century. United Nations Publications, Sales No. E.99.ll.A.1.

CGWB. 2014. Dynamic groundwater resources of India (as on March 31, 2011). Central Ground Water Board, Ministry of Water Resources, River Development and Ganga Rejuvenation, Government of India, Faridabad.

CWC. 2013. Water and related statistics. Water Resources Information System Directorate, Information System Organisation, Water Planning & Project Wing, Central Water Commission, New Delhi.

Herdt, R.W. and T. Wickham. 1978. *Exploring the Gap between Potential and Actual Rice Yields: The Philippine Case. Economic Consequences of the New Rice Technology.* Los Banos, Philippines, International Rice Research Institute.

Islam, A. and R.S. Gautam, 2009. Groundwater resource conditions, socio-economic impacts and policy-institutional options: A case study of Vaishali district of Bihar, India. In *Groundwater Governance in the Indo-Gangetic and Yellow River Basins: Realities and Challenges (Selected papers on Hydrogeology, v.15).* A. Mukherji, K. G. Villholth, B. R. Sharma, and J. Wang, Eds., CRC Press, Taylor & Francis Group, Boca Raton, FL, Chapter 6, pp. 105–118.

Karimi, P., A.S. Qureshi, R. Bahramloo, and D. Molden. 2012. Reducing carbon emissions through improved irrigation and groundwater management: A case study from Iran. *Agricultural Water Management*, 108:52–60.

Kumar, P. 1998. *Food Demand and Supply Projections for India.* Agricultural Economics Policy Paper 98-IARI, New Delhi.

Kumar, R., R.D. Singh, and K.D. Sharma. 2005. Water resources of India. *Current Science*, 89(5): 794–811.

Pant, N. 2004. Trends in Groundwater Irrigation in Eastern and Western UP. *Economic and Political Weekly*, XXXIX(31): 3463–3468.

Persaud, S. and S. Rosen. 2003. India's consumer and producer price policies: Implications for food security. *Economics Research Service, USDA, Food Security Assessment*, GFA-14, 32–39.

Postel, S. 1999. *Pillar of Sand: Can the Irrigation Miracle Last?* Norton, New York.

Seckler, D., R. Barker, and U.A. Amarasinghe. 1999. Water scarcity in the twenty-first century. *International Journal for Water Research & Development*, 15(1/2): 29–42.

Shah, T. 1993. *Water Markets and Irrigation Development: Political Economy and Practical Policy.* Oxford University Press, Bombay.

Shah, T. 2001. Wells and welfare in the Ganga Basin: Public policy and private initiative in eastern Uttar Pradesh, India. IWMI research report 54, Colombo, Sri Lanka.

Shah, T. and A. Kishore. 2012. Solar-Powered Pump Irrigation and India's Groundwater Economy: A Preliminary Discussion of Opportunities and Threats. Water Policy Research Highlight-26. IWMI-Tata Water Policy Program (ITP), www.iwmi.org/iwmi-tata/apm2012.

Shankar, P.S.V., H. Kulkarni, and S. Krishnan. 2011. India's groundwater challenge and the way forward. *Economic & Political Weekly*, xlvi 38(2): 37–45.

Sharma, B.R., A. Mukherjee, R. Chandra, A. Islam, B. Dass, and M.D. Ahmed. 2008. Groundwater Governance in the Indo-Gangetic Basin: An Interplay of Hydrology and Socio-Ecology. Fighting Poverty through Sustainable Water Use: Volumes I. Proceedings of the CGIAR Challenge Program on Water and Food, 2nd International Forum on Water and Food, Addis Ababa, Ethiopia, November 10–14, 2008. The CGIAR Challenge Program on Water and Food, Colombo. 73–76.

Shiklomanov, I.A. and J.C. Rodda. 2003. *World Water Resources at the Beginning of the Twenty-First Century.* Cambridge University Press, Cambridge, UK.

Siebert, S., J. Burke, J.M. Faures, K. Frenken, J. Hoogeveen, P. Döll, and F.T. Portmann. 2010. Groundwater use for irrigation—A global inventory. *Hydrology and Earth System Sciences*, 14: 1863–1880.

Sikka, A.K. 2009. Water Productivity of different agricultural systems. In strategic analyses of the National River Linking Project (NRLP) of India, series 4. *Water Productivity Improvements in Indian Agriculture: Potentials, Constraints and Prospects.* Kumar, M.D. and U.A. Amarasinghe, Eds. International Water Management Institute, Colombo, Sri Lanka, pp. 73–84.

Sikka, A.K. and P.R. Bhatnagar. 2006. Realizing the potential: Using pumps to enhance productivity in the Eastern Indo-Gangetic plains. In *Groundwater Research and Management: Integrating Science into Management Decisions.* B.R. Sharma, K.G. Villholth, and K.D. Sharma, Eds. Proceedings of IWMI-ITP-NIH International Workshop on "Creating Synergy between Groundwater Research and Management in South and Southeast Asia," Roorkee, India, pp. 200–212. International Water Management Institute, Colombo, Sri Lanka.

Singh, A.K. and A. Islam. 2007. Development of low pressure and low energy sprinkling nozzles. *Journal of Agricultural Engineering*, 44(1): 26–32.

Srivastava, R.C. and A. Upadhyaya. 1998. Study of feasibility of drip irrigation in shallow ground water zones of eastern India. *Agricultural Water Management*, 36: 71–83.

Steenbergen, F. and G.J. Bom. 2000. Recent efficiency improvements in affordable, low-lift tubewell irrigation in India. *GRID IPTRID Network Magazine*, 16: 8–9.

Thatte, C.D., A.C. Gupta, and M.L. Baweja. 2009. Water resources development in India. Indian National Committee on Irrigation and Drainage, New Delhi.

World Bank. 1999. *Groundwater Regulation and Management, World Bank.* Washington DC/Allied Publishers, New Delhi.

29

Groundwater Resource Management Using GIS Tools

K. B. V. N. Phanindra

CONTENTS

29.1 Introduction

Groundwater is the largest source of readily available freshwater in the world (Subramanya, 2005). Proper accounting and management of this precious resource is vital for human survival and economic development of any nation. Population growth assisted by urbanization, climate change, disasters, and improper management has resulted in overexploitation and degradation of this resource in almost all parts of the world. Water resource management problems are primarily dominated by regional topography, geology, hydrology, economy, climate, regulations on water withdrawals/use, and implementation strategies. Hence, multiple solution strategies depending on the region of interest are to be evaluated to solve a given water resource management problem.

Conventional data collection methodologies have the greatest disadvantage of handling large volumes of spatial datasets with acceptable level of accuracy. Hence, any modeling and/or management strategy derived from this data will lead to an error-prone analysis and results. This is particularly true at regional to global watershed scales that spread over a large area and demand huge spatial datasets. This limitation is completely eliminated with the development of remote sensing (RS) and geographic information

system (GIS) techniques. The ability of GIS to store, retrieve, and analyze spatial and a-spatial data at high speed is unmatched by conventional methods (Engel et al., 2007). Most of the hydrologic cycle components can be estimated, analyzed, and displayed in relation to their position on Earth's surface, thus making full use of GIS capabilities.

Groundwater has an important role in meeting the water requirements of agriculture, industrial, and domestic sectors of India. About 85% of India's rural domestic water requirements, 50% of urban water requirements, and more than 50% of irrigation requirements are being met from groundwater resources. A proper auditing, accounting, and management of groundwater databases on a real-time basis in a GIS environment are essential for use with hydrogeologists, groundwater modelers, and water managers in developing sustainable solution strategies.

This chapter presents the role of GIS tools in various application domains of groundwater including: (1) hydrogeology and data management; (2) groundwater potential zone identification; (3) groundwater flow and transport modeling; (4) vulnerability and quality mapping; and (5) groundwater management. The list of GIS tools (and their functionalities) that can be applied in each application domain was briefly supported with recently published research findings.

29.2 Remote Sensing in Groundwater Hydrology

Remote sensing (RS) is the process of inferring about surface to near-surface features using the observations made from remotely located, noncontact sensor systems (Campbell, 2011). RS techniques are broadly classified into *active* (uses naturally available energy source) and *passive* (uses own energy source for illumination) depending on the nature of the energy detected by the sensor (Lillesand and Kiefer, 2000). Basic components of an RS system include *energy source* (sun–passive, radar–active), *transmission path* (atmosphere, near surface), *target* (observing features on Earth or ocean surface), and *sensor/satellite* (for energy detection) (Jha et al., 2007). RS predominantly uses the visual and infrared bands of the electromagnetic (EM) spectrum for generating images of the Earth and ocean surfaces. Microwave RS techniques are preferred for measurements during cloudy/night times (soil moisture, water content estimation). Satellite data obtained from RS sensors require correction to remove geometric and radiometric errors resulting from the interaction of the two moving bodies (satellite and Earth), sensor drift, and changes in atmospheric/cloud conditions. Radiometric (spectral reflectance) errors resulting from the sun's azimuth, elevation, and atmospheric conditions can be corrected using readily available image processing software. Geometric correction requires a number of ground control points (GCPs) within the image to place in a correct spherical coordinate location on the Earth's surface. The final corrected image has to be processed (through a series of operations including image enhancement, image interpretation, and image classification) for use with various applications.

Based on the purpose served, satellite systems used in RS are broadly classified into *Earth/ocean observing* and *meteorological/environmental* satellites. Earth-observing satellites have less revisit interval, high spatial resolution, and are used for terrestrial applications. Environmental satellites have high revisit interval, low spatial resolution, and are used for weather/climate applications. Sensors used in RS can be broadly classified into four categories, namely, coarse resolution sensors, very high resolution sensors, multispectral sensors, and hyperspectral sensors. A detailed list of sensors used in RS (for groundwater applications) along with brief specifications is provided in Table 29.1. Coarse (low) resolution sensors have a spatial resolution of more than about 200 m, with high revisit interval and are primarily used to make inferences at regional to global scales. Multispectral (MS) sensors can capture the image across different bands of the EM spectrum

separated by filters. MS images require a prior knowledge on spectral signatures of various observing features for interpretation. Very high resolution sensors can sense the Earth's features at a spatial resolution of less than about 5 m and are mainly used for decision making at local/urban scale. Hyperspectral sensors cover the features of Earth in a wide number of spectral bands, and are mainly used in the fields of geology and mining.

Advances in spatial data collection methods using RS have significantly improved the representation and analysis of terrain, hydrologic, geologic, and geomorphic features. A wide range of satellite sensors (LANDSAT, AVHRR, MODIS, MERIS, CARTOSAT) can be used to retrieve hydrological parameters of terrain, vegetation, and soil at different scales. Application of RS in the fields of hydrogeologic mapping, exploration of well sites, groundwater management, and aquifer vulnerability is well documented (Jha et al., 2007; Meijerink et al., 2007). Groundwater flow and storage characteristics of a region are mainly controlled by hydrogeology, topography, land use, land cover, surface water features, surface geology, and drainage pattern. Processed images derived from RS sensors can be best utilized to represent the terrain features (including slope, aspect, drainage pattern, land use, and land cover), hydrologic features (including precipitation, stream network, and watershed), soil characteristics (including moisture, and texture), and geologic characteristics (including lineaments, faults, and lithology) both in spatial and temporal domains. Satellite image processing techniques will help in the identification of various derived features (for use with groundwater applications) using the interpretation tasks such as classification, enumeration, measurement, texture, tone, shape, size, pattern, and association.

29.3 About GIS

GIS is designed to input, store, edit, retrieve, analyze, and output geographically referenced data (Demers, 2013). The rapid growth of GIS and its applications in various fields (geography, biology, ecology, forest science, archeology, business, economics, agriculture, water resources, etc.) has resulted in multiple definitions for GIS. In short, GIS is the collection of computer hardware and software systems to manage spatially referenced data. GIS has gained popularity as a visualization tool to represent spatially referenced maps and their associate data, though applications can be extended to develop standalone or coupled models in various domains. As a database, GIS can store, query,

TABLE 29.1

List of Major Satellite Sensors and Their Specifications Used in Groundwater Applications

Satellite Sensor (Country of Origin)	Spectral Bands/ Wavelength	Spatial Resolution	Temporal Resolution	Major Applications
Coarse/Low Resolution Sensors				
AVHRR (NOAA-USA)	Red (1 band); infrared (4 bands)	1090 m	1 day	Discern clouds, land water bodies, snow/ice extent; inception of snow/ice melting, vegetation change detection
MODIS (NASA–USA)	36 (0.4 to 14.4 μm)	250–1000 m	1 day	Monitor changes in cloud cover, radiation budget, processes occurring in the oceans, on land, and in the lower atmosphere
METEOSAT (Europe)	12 (0.6 to 14.4 μm)	2500–5000 m	30 min	Short-term forecasting; numerical weather prediction and climatological studies
Multispectral (MS) Sensors				
LANDSAT-7 ETM+ (USA)	7 (0.45 to 2.35 μm)	15 m (Pan); 30 m (MS)	16 days	Change detection of earth features (agricultural development, deforestation, urbanization, development, and degradation of water resources) over periods of months to decades
ASTER (Japan)	14 (0.52 to 11.65 μm)	15 m/30 m/90 m	16 days	Digital terrain mapping and modeling; monitoring vegetation and ecosystem dynamics; land surface climatology
SRTM (USA)	C-band, X-band	90 m	1 month	Digital terrain modeling; regional weather forecasting
RESOURCESAT-2 (India)	3 (0.52 to 1.7 μm)	5.8 m (LISS 4)	24 days	Land and water related mapping; monitoring water, crop, and soil parameters bodies
SPOT-7 (France)	0.45 to 0.85 μm	1.5 m (Pan); 6 m (MS)	1 day	3D terrain modeling; urban and rural planning; natural disaster management; oil and gas exploration
High Resolution Commercial Sensors				
IKONOS (GeoEye – USA)	RGB, NIR (0.445 to 0.853 μm)	0.8 m (Pan); 4 m (MS)	3–5 days	Environmental monitoring; geological studies; rapid image collection; disaster response
Worldview-3 (Digital Globe, US)	8 MS, 8 SWIR, 12 CAVIS bands	0.31 m (Pan); 1.24 m (MS)	<1 day	Land classification; disaster response; feature extraction and change detection; environment modeling
CARTOSAT-2 (India)	0.5 to 0.85 μm	<1 m	4 days	Cartography in India, urban and rural infrastructure development, management and mapping
QUICKBIRD (Digital Globe, US)	4 (0.45 to 0.9 μm)	0.6 m (Pan); 2.44 m (MS)	4 days	Telecommunications; land use and infrastructure planning; environmental assessment; oil and gas exploration
Hyper Spectral Sensors				
HYPERION (USGS-USA)	220 (0.4 to 2.5 μm)	30 m	16 days	Land cover classification; soil and crop mapping; crop health monitoring and yield prediction; contaminant mapping

manipulate, and analyze the attribute data associated with spatial features. The uniqueness of GIS lies in linking the spatial features with a-spatial attributes, and performing numerous operations (including overlay, measurement, classification, analysis, and modeling) on the spatial maps.

Spatial objects in the real world can exist as point, line, polygon, or surface features. GIS is capable of representing the first three features by their respective symbols, whereas surface features are represented by adding elevation to point, line, and polygon features. Point features have no spatial dimension, linear objects require at least two coordinate pairs to represent their spatial location, and polygon features are represented as two-dimensional objects. Data collection sources for use with GIS include: ground survey data (topo sheets, geological maps, GPS data), point remote sensing data (weather station, sensor networks), and areal remote sensing data (satellite images, radar). Quality of the input datasets used in GIS is evaluated using four indicators, namely, accuracy, time, scale, and completeness. Data representation in GIS is done in either vector or raster forms. Vector data model is an object-based approach, and is efficient in representing real world discrete objects and associated fields. Vector data structure (land marks, roads, streams, watershed, and buildings) allows the representation of geographic features in a visually intuitive and more reminiscent form (Demers, 2013). In the vector system, surfaces are represented by connecting points of known elevation into triangulated flat surfaces. Raster data model is a field-based approach, and is efficient in representing geographic phenomena that are continuous over an area. Raster structure (land use, surface geology, DEM, precipitation) allows the representations of a particular theme/attribute in a two-dimensional array of grid cells. The choice between vector and/or raster GIS is chiefly driven by the application.

GIS databases are subjected to *entity* and/or *attribute* errors. Identification and rectification of such errors before performing the analysis is essential, as it leads to error-prone analysis results. Vector models are primarily subjected to *entity (positional) errors* that can occur through (1) missing entities, (2) incorrectly placed entities, and (3) disordered entities. Attribute errors (resulting from wrong code, placing, and misspelling) can occur in both vector and raster models (Demers, 2013). Satellite images derived from different sensors (and representing different parameters) are to be standardized in GIS (bring down to one common Cartesian coordinate system based on reference globe) for performing spatial analysis tasks (such as overlay and intersection). Other editing techniques on the spatial maps include *edge matching* (physically link to adjacent maps for analysis of larger area), *conflation* (combine multiple datasets with an acceptable level of accuracy), and *templating* (rectify graphical discrepancies arising from multiple maps of same theme).

GIS software is highly diverse in functionality, database structure, and hardware requirement. Desktop or standalone GIS software can be grouped into two categories:

1. Open source (public domain) software, with the source code made available with license. These include: GRASS GIS, ILWIS, MapWindow GIS, QGIS, SAGA GIS

2. Commercial (proprietary) software such as: ArcMap and ArcGIS (ESRI), ERDAS IMAGINE, GeoMedia (Intergraph), MapInfo

29.4 Groundwater Scenario in India

Groundwater in the Indian subcontinent occurs in a wide range of geological formations ranging from unconsolidated alluvial sediments of river basins (specific yield: 0.04–0.20) to consolidated hard rock terrains (specific yield: 0.002–0.04). Groundwater yield potential ranges from less than 1 lps (in hilly regions) to more than 25 lps in river alluvium. Groundwater resource estimation across the country is done on an annual basis by the Central Groundwater Board (CGWB) using water level fluctuation or rainfall infiltration factor methods. Annual replenishable groundwater resources are obtained by adding monsoon and nonmonsoon recharge. Net annual groundwater availability is estimated after deducting natural discharge during nonmonsoon periods from the annual replenishable source. Groundwater draft is the quantity of water being extracted from the aquifer for irrigation, domestic, and industrial needs. Stage of groundwater development is the ratio of annual groundwater draft and net annual groundwater availability, and is expressed as a percentage. CGWB classified all the administrative units (blocks/mandals/districts) of the country based on the stage of groundwater development (Table 29.2).

The annual replenishable groundwater resource for the entire country is estimated to be 431 billion cubic meters (bcm). The share of rainfall and other sources (recharge from recycled irrigation water and water conservation structures, seepage from canals) and annual replenishable groundwater source are, respectively, 67% and 33%. The stage of groundwater development for the entire country is 61%. Out of the total

TABLE 29.2

Criteria for Categorization of Assessment Units

Stage of Groundwater Development	Significant Long-Term Decline		Category
	Premonsoon	Postmonsoon	
≤70%	No	No	Safe
>70% and ≤90%	No	No	Safe
	Yes/No	No/Yes	Semicritical
>90% and ≤100%	Yes/No	No/Yes	Semicritical
	Yes	Yes	Critical
>100%	Yes/No	No/Yes	Overexploited
	Yes	Yes	Overexploited

Source: CGWB, 2009. *Dynamic Groundwater Resources of India (as on 31 March 2009).* Central Groundwater Board, Ministry of Water Resources, Government of India, Faridabad.

5842 administrative units, 802 are overexploited (lying mostly in Delhi, Punjab, Haryana, and Rajasthan), 169 units are critical, 523 units are semicritical, 4277 units are safe, and 71 units are saline (CGWB, 2009).

India is the largest consumer of groundwater for agricultural needs across the globe. Tube wells are the largest single source (about 40% of the irrigated area) of irrigation water in India (Shah, 2009). Groundwater dependency of urban and rural India is highly varied and is dominant by individual needs and socioeconomic status. Irrigation (80%) and domestic (7%) sectors are the major consumers of groundwater across the country (Vijayshankar et al., 2011). In general, groundwater dependency of urban India decreases with increase in size and population of the city (Patel and Krishnan, 2008). The peninsular and hard rock cities of the country show a low dependence of groundwater when compared to alluvial aquifer cities. Nearly 90% of the rural water supply (7% for drinking and 80% for irrigation) is sourced from groundwater. Intensive groundwater abstraction has led to localized aquifer depletion (especially in semiconfined urban aquifer systems). Region specific guidelines (and regulations) on groundwater withdrawals are nearly absent or not being implemented across the country. Developing modeling strategies for scientific understanding of the aquifer behavior under projected climate, land use, demographic, and anthropogenic changes can help in sustainable management of groundwater resources of the country.

29.5 GIS Applications in Groundwater Hydrology

Hydrologic models demand large volumes of spatial and associated input datasets (topography, meteorological, land cover, land use, water levels, geologic, geomorphologic, and water quality) from a number of sources. GIS is a powerful tool that can integrate data from multiple sources (satellite imagery, hard maps, boreholes, physical observations, and empirical sets) simultaneously for use with groundwater applications. Representation of all features in a common and least erroneous projection system is vital before performing spatial analysis with GIS. Advances in GIS and information technology have resulted in sharing of spatial datasets across the global users. In addition to providing spatial information on the Internet, these web-based GIS services in conjunction with sensor network tools help in developing decision support systems to solve real world, complex, groundwater problems on a real time basis.

A number of analysis and modeling tools in GIS are being used (or can be developed) to analyze spatially

referred maps and associated data that suits a given groundwater application. Examples of these GIS tools include:

1. Measurement tools that work on point, line, and polygon datasets to measure density, distance, sinuosity, shape, and area

2. Classification tools that mainly work on raster datasets to reclassify the images using neighborhood functions and filters

3. Terrain analysis tools that work on elevation datasets to generate slope, view shed, contour maps, drainage network, watershed maps, cut, and fill

4. Topology tools that perform connectivity, contiguity, and proximate functions on both vector and raster datasets

5. Statistical tools that work on interval or ratio datasets to generate spatial statistics, prediction surfaces, and error maps

6. Geo-processing tools that perform spatial analysis (such as search, extract, proximity, editing, tracking, and network) on geo-referenced maps

7. Spatial arrangement tools that deal with placement, ordering, concentration, connectedness, and dispersion of multiple objects within the spatial datasets

8. Overlay tools that mainly work on vector datasets (using Boolean search strategies like union, intersection, and complement) and raster datasets (using map algebra principles) to generate new thematic layers by combining the individual datasets

9. Retrieval tools that query and retrieve the spatial and attributed data stored in GIS

10. Modeling and simulation tools that operate on geographic and attribute datasets using a combination of commands/tools to answer about spatial phenomena

Availability of high resolution digital elevation models (DEM) led to the development of microwatershed atlas of India, which in combination with soil and land database is vital for use with groundwater applications. Datasets used in groundwater applications come from a variety of sources and occur in various forms. GIS is an efficient tool in managing, coordinating, analyzing, and representing these datasets for direct and/or indirect use with groundwater studies. Geo-statistical tools in GIS can help in generating continuous surfaces of various hydrological parameters from the measurements at random locations. Spatial analysis (from various sources) in GIS can help in mapping hydrologic features in relation to interaction of surface and groundwater, mapping recharge patterns, and potential aquifer sites. GIS aids in generating groundwater flow field from hydrogeologic datasets. Geo-processing and modeling tools in GIS can be integrated with numerical groundwater models for use in decision making. Prioritization of groundwater vulnerable zones and their spatial relation to various contributing factors can be done using overlay tools in GIS. Review of GIS applications in groundwater hydrology has been presented by many researchers (Meijerink et al., 2007; Jha et al., 2007).

GIS applications to groundwater can be broadly categorized into the following five domains. The following sections describe the role of GIS in each application domain supported by recent published literature (primarily after 2007).

1. Hydrogeology and data management
2. Groundwater potential zone identification
3. Groundwater flow and transport modeling
4. Vulnerability and quality mapping
5. Groundwater management

29.6 Hydrogeology and Data Management

GIS is emerging as a standard practice for managing and analyzing spatial groundwater datasets (Strassberg et al., 2007). Development of standardized GIS datasets for various classes of geospatial phenomena is essential for any country to optimally manage its resources, and implement disaster mitigation programs. India is making its efforts to bring all the data sources (topography, hydrogeology, surface geology, groundwater, meteorological, and hydrology) onto a common GIS platform for use with various agencies and researchers. A variety of data models and standards have been developed for representing groundwater information. These include standards focused on describing site information and standards for representing data in groundwater simulation models (Strassberg et al., 2007). The main advantage of using a GIS database system is that the information from various sources can be quickly assimilated, viewed, analyzed, and displayed based on spatial relationships.

Data design in a GIS environment uses relational database management systems (RDBMS) such as Microsoft Access, Oracle, and Microsoft SQL Server. Hence, GIS is capable of performing any relational operations on geospatial features. Geodatabase objects used in groundwater include features, feature classes, feature data sets, relationships, rasters, and raster catalogs. Features are spatial vector objects (points, lines, polygons, and multipatches) with attributes to describe their properties.

Features are instances within a feature class, which is a collection of features with the same geometry type, attributes, and relationships. Relationships are objects that define associations between other object classes based on key fields. Raster data sets represent imaged, sampled, or interpolated data on a uniform rectangular grid, and raster catalogs are used for storing, indexing, and attributing raster data sets.

Data requirement for groundwater studies broadly falls under two categories, namely, physical (topography, geology, aquifer characteristics, and lithology) and hydrological (water levels, recharge, evapotranspiration, stream-aquifer flows, and pumping) framework. Most of the datasets are acquired from local agencies/departments, and comes in varieties of formats and quality, that requires GIS processing (using conversion tools, aggregation tools, and editing tools). Secondary data obtained from field observations and experiments requires the application of geo-statistical tools in GIS to represent the variability on a continuous spatial domain.

Gogu et al. (2001) developed a hydrogeologic and GIS database for the Walloon region, Belgium. A total of 14 layers of primary hydrogeological database were created in GIS environment. Data obtained from various sources (organizations) have dissimilarities in terms of data type/format, quantity, quality, and storage media. A structured spatio-temporal database methodology was developed for use with hydrogeologists, modelers, managers, and decision makers. The constraints that define final HYGES database include: (1) maximum information, (2) minimum data redundancy, (3) reduction of storage capacity, and (4) optimum data retrievability for the analysis.

Strassberg et al. (2007) demonstrated GIS capabilities to represent the 3D nature of the subsurface, and described hydrogeologic information in the form of 3D geospatial objects. A groundwater data model that includes 2D and 3D object classes for representing aquifers, wells, and borehole data was developed to support mapping and analysis procedures. Relations were established to provide connectivity between the features to aid in performing logical operations. Temporal objects (water levels, water quality) were embedded into the data model through time series (TS) and TSType tables (linking through common spatial feature).

Kambhammettu et al. (2011) applied geo-statistical tools in GIS to interpolate water table elevations from the observations at 38 well locations for the years 1996 and 2003 for the Carlsbad alluvial aquifer in the United States. Low order polynomials were used to model the trend in water table elevations. The GIS analysis was loosely coupled to MATLAB® code for generating omni and directional semi-variograms. Contour maps of water table and estimation variance surfaces were developed using raster-based GIS tools (Figure 29.1).

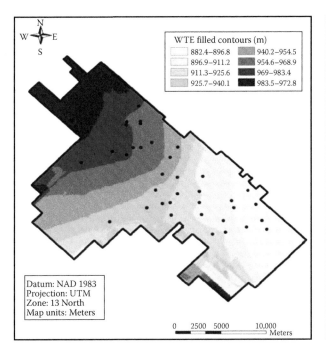
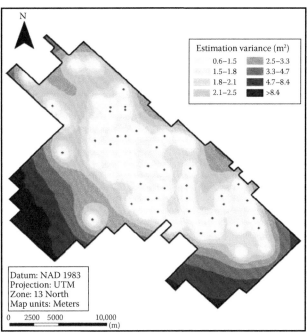

FIGURE 29.1
Maps of water table surface (top) and estimation variance (bottom) developed using geo-statistical tool in GIS, for the Carlsbad alluvial aquifer in Southern New Mexico.

Kambhammettu et al. (2010) developed a new tool in GIS (named "modified aggregation tool") to generate the surface elevation at any desired resolution, from the grid available at much finer resolution. Partially occupied cells were given due weightage based on the area in estimating average elevation for the aggregated cell. The procedure followed in aggregating the base digital elevation model (DEM) to the user-specified model resolution is illustrated using GIS-assisted model builder tool (Figure 29.2). The developed tool can be used in lieu of resampling techniques to minimize the spatial information loss. The developed tool was applied to the Rincon Valley DEM data in the United States, and observed to be more efficient and robust for direct use with hydrologic models.

29.7 Groundwater Potential Zone Identification

GIS tools are widely applied in identification and mapping of groundwater potential zones in alluvial, hard rock, mountainous, and coastal aquifers. Most of the literature in this domain was limited to performing overlay and map algebraic functions (on raster datasets) using weight- and rank-based approach (to various thematic map units). However, GIS tools in conjunction

with evolutionary, probability, statistical, and belief function algorithms were proven to identify the groundwater prospect zones in a more logical sense, and aid in validating the developed models. Table 29.3 lists major data requirements in delineating groundwater potential zones using GIS.

Batelaan and Smedt (2007) developed a spatially distributed water balance model (WetSpass) to simulate long-term average recharge considering land cover, soil texture, topography and hydrometeorological parameters. The water balance model was further coupled to a regional groundwater model and applied to several catchments in the Netherlands. The two models were coupled using raster datasets at the same resolution, to investigate the contribution of various factors on spatial distribution of recharge.

Ganapuram et al. (2009) used RS datasets and GIS tools in mapping groundwater prospect zones for the Musi Basin in Telangana, India. SRTM DEM of the region at 90 m spatial resolution was used to derive various terrain features using GIS tools. Hydrogeomorphic maps were prepared by integrating lithologic, geomorphologic, and hydrologic units. The final hydrogeomorphic units were classified into various groundwater potential zones using intersection tools in GIS.

Chenini et al. (2010) and Chenini and Mammou (2010) used GIS techniques in conjunction with a numerical groundwater flow model for demarcating suitable sites for artificial recharge of groundwater in the Maknassy Basin

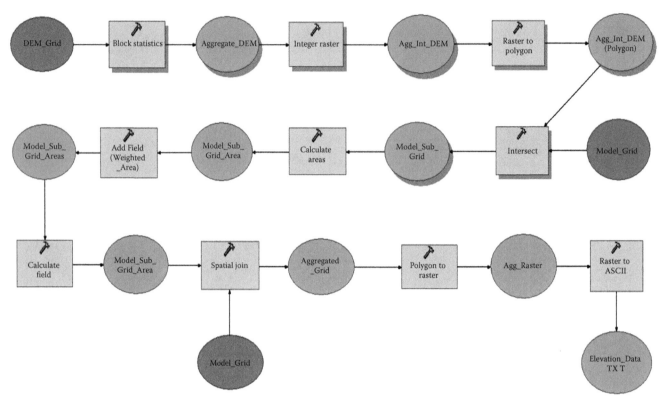

FIGURE 29.2
Workflow of various GIS operations in aggregating the base DEM (using Model Builder).

TABLE 29.3

Spatial Parameters Used in Delineating Groundwater
Prospect Zones

Contributing Feature	Derived Parameters	Data Sources
Topography	Drainage map, slope map, watershed limits, drainage density	SoITopo sheets, satellite imageries, DEM, field acquired data
Geology	Lithology, fractured outcrops, lineament density, geomorphic units	Surface geological maps, aerial photographs, satellite imageries,
Hydrology	Rainfall pattern, boundary conditions, piezometric levels	Satellite imageries, field observations,
Soil and vegetation	Texture, land cover, land use, permeability	Satellite imageries, field observations, experimental data

in Central Tunisia. Topographic, geologic, hydrogeologic, and groundwater datasets were spatially integrated and indexed for generating the recharge zone maps. Effect of water recharge on the piezometric behavior of a hydrologic system was evaluated using a steady-state groundwater model, MODFLOW, for one hydrologic year.

Joo Oh et al. (2011) used GIS tools and a probability model to map the regional groundwater potential zones of Pohang City in Korea. The relation between groundwater specific capacity and various contributing factors are established using the frequency ratio model and sensitivity analysis. They observed that soil texture and ground elevation have respectively the highest and lowest contributing effects on groundwater potential.

Sukumar and Sankar (2010) used GIS tools for digitization, registration, and generation of thematic maps of geology, drainage intensity, soil texture, permeability, effective soil depth, and water holding capacity for the Vaigai River Basin in Tamil Nadu, India. Each contributing parameter was assigned a weight based on its influence on groundwater recharge. Potential sites for artificial recharge were identified using overlay and intersect tools and ranked using ordinal levels.

Ozdemir (2011) produced the groundwater spring potential map (GSPM) of Sultan Mountains in Turkey using logistic regression tools in GIS environment. A total of 17 explanatory variables were further reclassified into 118 classes to create input datasets for a logistic regression model that is used to estimate model coefficients. The interpretations of the potential map showed that stream power index, relative permeability of lithologies, geology, elevation, aspect, wetness index, plan curvature, and drainage density play a major role in spring occurrence and distribution.

Nampak et al. (2014) applied an evidential belief function (EBF) model for spatial prediction of groundwater productivity in the Langat Basin in Malaysia. Groundwater wells of the study area were classified based on yield and the spatial correlation between potential zones and well yield was evaluated using a number of contributing factors. The groundwater potential map developed from the EBF model was compared to the logistic regression method and found to be efficient.

Quantification of groundwater recharge for the Wu River watershed in Taiwan was done by Yeh et al. (2014). A total of five contributing parameters (lithology, land cover/land use, lineaments, drainage, and slope) were weighed, digitized, processed, and overlaid in a GIS environment to generate the groundwater potential map. Further, they used stable base flow (SBF) technique to estimate the groundwater recharge using base flow separation principles.

29.8 Groundwater Flow and Transport Modeling

Groundwater models are useful in predicting the behavior of an aquifer system in response to projected climate, anthropogenic, and operational changes. Groundwater models are classified into flow (solves for groundwater heads) and transport models (solves for solute concentration using advection, dispersion, and reaction). Numerical groundwater models are classified into gridded (discretized) and nongridded (mesh-free) methods. The gridded method (finite element, finite difference) solves the flow/transport equation at each element/grid cell and then integrates using mass conservation principles across the boundaries. Examples of such tools include MODFLOW, PLASM, FEFLOW, SUTRA, MT3D, and SEAWAT. Mesh-free methods (analytic element, boundary element) are discretized at boundaries or along flow elements (Ashraf and Ahmad, 2012). Datasets required for groundwater modeling are broadly categorized into (1) physical (topography, geology, elevation contours, isopach maps, stream, and lake sediments) and (2) hydrogeologic (water table and potentiometric maps, aquifer storage properties, hydraulic conductivity/transmissivity, groundwater fluxes, source, and sink components) (Anderson and Woessner, 2002). Every country is making efforts to develop/update spatial datasets (in GIS format) required for groundwater models. A graphical user interface (GUI) to a groundwater numerical model (such as GMS, Groundwater Vistas, PMWIN, Visual-MODFLOW) has the ability to create and analyze spatial input datasets; calibrate and validate the model; and process the output using a GIS environment.

GIS is an efficient tool in importing, creating, editing, and converting "geographic map layers" into "numerical grid layers" for use with groundwater simulation. GIS is effective in managing time invariant spatial datasets including hydrostratigraphic boundaries, hydraulic heads, and aquifer parameters (hydraulic conductivity, transmissivity, specific yield, storage coefficient, and effective porosity) for use with modeling. Time variant parameters including groundwater fluxes and variable aquifer storage demand are needed for spatio-temporal data sets in GIS. Link between GIS and groundwater model can happen at three levels: (1) *loose coupling*, where GIS is used as a data processor to groundwater model, (2) *tight coupling*, where data transfer between the two models is automated, and (3) *embedded coupling*, where the groundwater model is created completely using GIS programming/code (Gogu et al., 2001). The process-based models used in groundwater include simulation of steady and transient groundwater flow, advection, hydrodynamic dispersion, adsorption, desorption, retardation, and multicomponent chemical reaction. However, the role of GIS is mostly limited to pre/postprocessing of the spatial datasets for use with groundwater modeling.

Hernandez and Gaskin (2006) developed a groundwater modeling module in GRASS (GMTG), an open source GIS system. Applicability of the developed module was evaluated for two cases, one having a 1-layered unconfined aquifer, and the other with a 3-layer alluvial aquifer system. GMTG simulation results were observed to be in good agreement with those obtained from PMWIN.

Dams et al. (2008) assessed the impact of land-use changes over 2000–2020 on the hydrological balance and groundwater quantity for the KleineNete Basin, Belgium. WetSpass model in conjunction with steady-state MODFLOW was used to predict the groundwater recharge and levels under different climate projection scenarios. GIS tools were used in preparing spatial data sets and linking the models.

Wang et al. (2008) have used GIS functionalities for groundwater modeling in the Northern China Plain, as most of the input datasets are available in MAPGIS format. The GIS-based MODFLOW model was further integrated under the Internet environment using server-based GIS tools. The estimated water budget component shows a good agreement with the observations.

Impact of irrigation water-saving practices and projected groundwater abstractions on the groundwater dynamics in the upper Yellow River Basin, China was assessed by Xu et al. (2009, 2011). GIS (time invariant parameters) was loosely coupled to MODFLOW. Simulation results conclude that water-saving practices (canal lining), upgraded hydraulic structures, and upgraded farm irrigation technology has resulted in sustainable development of agriculture.

Ajami et al. (2012) developed a GIS enabled application, RIPGIS-NET, using .NET framework for use with riparian evapotranspiration package (RIP-ET) in MODFLOW. RIPGIS-NET is designed to derive all spatial input data sets for each plant function subgroup and provide visualization tools of groundwater heads and depth to water table maps in the GIS environment.

Elfeki et al. (2011) developed concentration envelop maps of a wastewater lake in Jeddah by quantifying the uncertainty using GIS tools and Monte-Carlo simulations. The scarcity of published and satellite imagery data lead to the consideration of various flow scenarios to match the aquifer response. Random walk particle tracking method was used to solve the transport equation. The envelope maps aid in decision making processes under uncertainty for management of the resource.

Singhal and Goyal (2011) developed a GIS-based methodology to estimate average flux through the model boundary from the readily available groundwater levels for part of the Pali district in Rajasthan, India. GIS tools were used to determine flow velocity and direction, and generate flow field using Darcy's law along the boundary cells. Deterministic interpolation techniques were applied to generate the groundwater profiles from the measurements at the 108 well locations.

Spatial and temporal identification of waterlogged areas and associated groundwater behavior for the Indus Basin in Pakistan was done by Ashraf and Ahmad (2012). GIS is integrated with finite element groundwater flow model (Feflow), simulated under transient-state condition to analyze and monitor resource status and land conditions of waterlogged areas.

Kambhammettu et al. (2012, 2014) used GIS tools to develop MODFLOW datasets for investigating the effect of grid size on evapotranspiration simulations for the lower Rio Grande Basin in the United States. GIS tools were used in delineating the model boundary and fluxes at various spatial resolutions (Figure 29.3) that can help the numerical model in performing multiple simulation runs simultaneously. Raster-based DEM aggregation techniques have resulted in a quick and efficient way of developing MODFLOW datasets (of ground surface elevation and hydraulic heads) for use with simulation.

Vahid (2013) applied statistical and GIS tools in modeling groundwater levels for the Mazandaran province in Iran. Mean groundwater levels are assumed to be dependent on four independent variables, and solved using multivitiate regression analysis. GIS was used to provide groundwater level map using two raster layers, namely, transmissivity of the aquifer and average precipitation over catchment.

29.9 Groundwater Vulnerability and Quality Mapping

GIS is efficient in spatial representation of regions vulnerable to groundwater contamination for use with groundwater management and policy strategies. Such maps are helpful in evaluating the potential hazard of existing aquifers to various contaminants. Groundwater vulnerability mapping can be classified into sensitivity mapping (based on hydrogeologic setting) and vulnerability modeling (groundwater sensitivity to human activity and climate change) (Engel et al., 2007). GIS-assisted DRASTIC and SEEPAGE models are highly utilized in developing regional maps of groundwater vulnerability to point (domestic and industrial effluents) and nonpoint (agricultural runoffs) pollutions. DRASTIC is a regional scale groundwater quality assessment model, with index estimated by accumulating the weight and rank factors of seven contributing parameters. Raster-based GIS tools are highly efficient in managing and analyzing datasets used with vulnerability and quality modeling (Figure 29.4). Most of the GIS-assisted vulnerability models do not account for temporal variability in agronomic management, land use, climate, and other contributing factors. Also, most of the published literature is limited to classification of groundwater vulnerable zones using DRASTIC index approach, with little attention given to spatially corelating with surficial waste deposit sites and quality parameters. Modifying the DRASTIC parameters (and weights) to suit a given application and region can help in better understanding and characterization of aquifer groundwater vulnerability to contamination. Table 29.4 details the list of DRASTIC parameters, their data sources, and the basis for ranking the parameter.

Dixon (2005) developed a methodology to generate contamination potential maps of Woodruff County in the United States using land use, pesticide, and soil structure information in conjunction with some selected parameters from the DRASTIC model. The DRASTIC model was modified to reflect the hydrogeologic conditions of the intensely irrigated aquifer. RS, GIS, GPS, and fuzzy rule-based tools were applied to generate groundwater sensitivity maps. Three different vulnerability modeling approaches were compared with modified DRASTIC index and field water quality data.

Almasri and Kaluarachchi (2007) developed a framework for modeling the impacts of land use changes and protection alternatives on nitrate pollution of groundwater for the Sumas–Blaine aquifer, United States. GIS tools were used to develop input datasets, spatial distribution of on-ground nitrogen sources, and corresponding loadings. The developed framework was used to

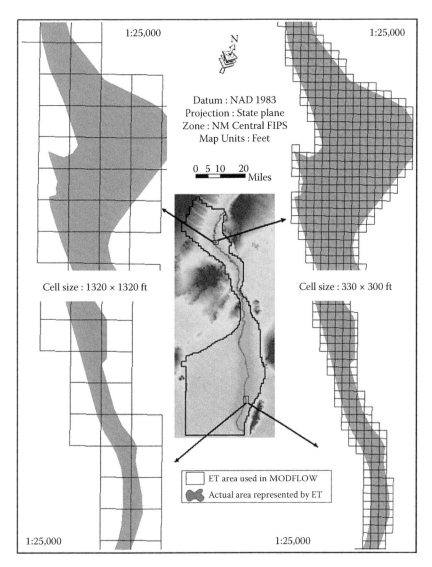

FIGURE 29.3
Representation of model cells at different resolution for use with groundwater modeling.

estimate nitrates leached to groundwater using a soil nitrogen dynamic model. Simulated results were well agreed with nitrate concentrations observed at the monitoring well site.

Asadi et al. (2007) evaluated groundwater quality in relation to land use/land cover for a part of Hyderabad City in India. Groundwater quality index is estimated in GIS environment using a combination of physical and chemical contributing parameters that is similar to DRASTIC index. Water quality index map is spatially correlated with land use/change images in raster format.

Berkhoff (2008) developed an evaluation scheme to assess groundwater vulnerability for the Hase River catchment in Germany to suit a water framework directive (WFD). Vulnerability was considered to be a function of exposure, sensitivity, and adaptive capacity. Nitrogen load caused by land use was evaluated using STOFFBILANZ nutrient model. Sensitivity was computed using the DRASTIC index of natural groundwater potential and adaptive capacity in the catchment was observed to be very low. Groundwater protection areas and remedial measures were suggested based on vulnerability indices.

Klug (2009) assessed the risk of groundwater contamination in Houston County, United States using modified DRASTIC model (to suit local hydrogeologic conditions). DRASTIC parameter weights were modified to account for the leaching of pesticides. GIS tools were used to collect, process, and display the spatial data sets.

Bai et al. (2012) evaluated the DRASTIC model for the Baotou City watershed in China. GIS tools were used to

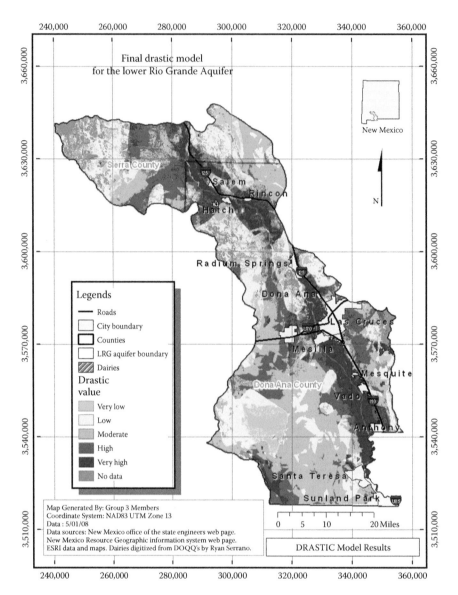

FIGURE 29.4
Groundwater vulnerability (DRASTIC) model for the Lower Rio Grande Aquifer.

TABLE 29.4

DRASTIC Parameters and Their Data Sources Used in Groundwater Vulnerability

DRASTIC Parameter	Data Format	Data Source	Weight	Basis for Rank
Depth to groundwater (D)	Point data (vector)	Monitoring wells (need to apply kriging techniques)	5	Depth to water table (1–10)
Net recharge (R)	Point data/grid data	Average rainfall distribution, land use	4	Land use characteristics (1,3,8)
Aquifer media (A)	Imagery/grid	Surface geology, hydrogeologic datasets	3	Primary aquifer material (1–10)
Soil media (S)	Raster/tabular	National soil and landuse board, soil database	2	Texture (1–10)
Topography (T)	Raster/grid	Soltopo sheets, DEM	1	Slope (1–10)
Impact of vadose zone (I)	Raster/point	Geologic map, field experiments	5	Material depth (1–10)
Hydraulic conductivity (C)	Point/aerial data	Pumping test data, *in situ*/laboratory experiments	3	Hydraulic conductivity (1–10)

prepare the raster datasets. The DRASTIC model parameters, weights, and grade classification were modified based on extension theory and analytic hierarchy process. The performance of the modified DRASTIC was tested using groundwater quality parameters.

Wang et al. (2012) developed a groundwater contamination risk assessment methodology by integrating hazards, intrinsic vulnerability, and groundwater value, and applied it to the Beijing plain in China. DRASTIC parameter ratings were modified in accordance with China geological survey to estimate intrinsic vulnerability. Spatial overlay tools in GIS were used to generate the groundwater contamination risk map and the results were validated using nitrate concentration distribution in the shallow aquifer.

Khan et al. (2014) aimed at understanding the aquifer vulnerability characteristics of the lower Kali watershed in western Uttar Pradesh, India. The model was named DRASIC, as the topography (T) component is neglected in the analysis due to negligible slopes. Vulnerability maps were prepared at the administrative block level, and were spatially correlated with major ion concentrations in the groundwater, during pre- and postmonsoon seasons.

Majumdar (2014) applied ArcSWAT model (a GIS interface to SWAT) to simulate nitrate leaching phenomena, in the unsaturated zone for the Sngur-Manjeera watershed in Telangana, India. The source code was modified to account for nitrate leaching dynamics past the root zone in an agriculture intense watershed. Simulated nitrate concentrations were in congruence with the measurements made at random borehole locations. Spatial distribution of nitrate leaching for the watershed was prepared in a GIS environment, for further use with agricultural managers.

29.10 Groundwater Management

Advances in wireless sensor and communication technologies have led to the development of real-time decision support systems for use with resource management. Desktop and server-based GIS applications can help in performing spatial analysis, coupling with modeling and programming tools, and displaying the results in a real-time user interactive environment. Defining regulations on groundwater withdrawals using spatial (GIS) tools (by delineating wellhead protection areas using buffer tools, identifying vulnerable zones, etc.) can help water managers in properly planning the resources without compromising on the degradation.

Scott et al. (2003) developed a GIS-based management tool to estimate groundwater demand from riparian vegetation along the San Pedro River in the United States. The tool estimates the groundwater use by multiplying (cell statistics in raster dataset) phreatic vegetation and seasonal groundwater per unit of vegetation. The user has the ability to choose vegetation type and amount using the defined polygon. Evapotranspiration (ET) from plant subgroups was estimated using a combination of micrometeorological (eddy covariance) and eco-psychological measurements for use with calibration.

Dawoud et al. (2005) developed a GIS-based multilayered aquifer modeling system of the western Nile delta in Egypt using FEM-based TRIWACO software. The calibrated model was used to estimate water balance and the groundwater aquifer potentiality. Effect of potential water management strategies for the river aquifer system was evaluated using two management scenarios (reducing surface water inflow and increasing annual abstraction; construction of a new canal).

Kolm et al. (2007) developed a GIS-based groundwater evaluation model for the Capitol and Snowmass Creek (CSC) in Colorado. Groundwater resource availability, economical exploitability, long-term resource sustainability, and vulnerability to contamination were evaluated. The GIS layers used in the analysis included: (1) general geographic information containing landmarks, (2) hydrologic information containing precipitation, water bodies, and irrigated regions, (3) hydrogeologic information containing aquifers, wells, and stratigraphic units, and (4) topographic information containing DEM and its contours.

El-Kadi et al. (2014) used numerical groundwater models (MODFLOW-SEAWAT with GMS as the interface) to assess groundwater sustainability on Jeju island in South Korea. Effects of projected climate, changes in land use, and increasing pumping conditions were evaluated and the results are evaluated by assessing negative effects on groundwater sustainability indicators. The model was calibrated for hydraulic heads, spring flows, and salinity. Sustainability of groundwater resources was assessed for various recharge scenarios, each with varying pumping–sustainable yield ratios.

Pardo and Connelly (2014) developed a groundwater management tool for the East Mendip Hills, England using GIS interface. GIS was applied into four stages, namely, (1) identification of the themes of interest and data sources, (2) data collection, (3) data collation and entry, and (4) grouping the datasets based on category. The developed GIS tool can assist the local planning authority in decision making processes regarding future mineral development and management of limestone dominant aquifers. Zones of high and low sensitivity with regard to potential impact of mineral extraction on groundwater resources can facilitate decision making in water management.

References

Ajami, H., Maddock III, T., Meixner, T., Hogan, J. F., and D. P. Guertin. 2012. RIPGIS-NET: A GIS tool for riparian groundwater evapotranspiration in MODFLOW. *Groundwater Journal*. 50(1), 154–158.

Almasri, M. N. and D. P. J. J. Kaluarachchi. 2007. Modeling nitrate contamination of groundwater in agricultural watersheds. *Journal of Hydrology*. 343, 211–229.

Anderson, M. P. and W. W. Woessner. 2002. *Applied Groundwater Modeling: Simulation of Flow and Advective Transport*. Academic Press, London.

Asadi, S. S., Vuppala, P., and M. A. Reddy. 2007. Remote sensing and GIS techniques for evaluation of groundwater quality in Municipal Corporation of Hyderabad (Zone-V), India. *International Journal of Environmental Research and Public Health*. 4(1), 45–52.

Ashraf, A. and Z. Ahmad. 2012. Integration of groundwater flow modeling and GIS, in *Water Resources Management and Modeling*, Purna Nayak (Ed.), InTech Publishers. pp. 239–262.

Bai, L., Wang, Y., and. F. Meng. 2012. Application of DRASTIC and extension theory in the groundwater vulnerability evaluation. *Water and Environment Journal*. 26, 381–391.

Batelaan, O. and F. D. Smedt. 2007. GIS-based recharge estimation by coupling surface–subsurface water balances. *Journal of Hydrology*. 337, 337–355.

Berkhoff, K. 2008. Spatially explicit groundwater vulnerability assessment to support the implementation of the Water Framework Directive—A practical approach with stakeholders. *Hydrology Earth System Science (HESS)*. 12, 111–122.

Campbell, J. B. and R. H. Wynne. 2011. *Introduction to Remote Sensing*. 5th ed. The Guilford Press, New York.

CGWB. 2009. *Dynamic Groundwater Resources of India (as on 31 March 2009)*. Central Groundwater Board, Ministry of Water Resources, Government of India, Faridabad.

Chenini, I. and A. B. Mammou. 2010. Groundwater recharge study in arid region: An approach using GIS techniques and numerical modeling. *Computers & Geosciences*. 36, 801–817.

Chenini, I., Mammou, A. B., and M. E. May. 2010. Groundwater recharge zone mapping using GIS-based multi-criteria analysis: A case study in Central Tunisia (Maknassy Basin). *Water Resources Management*. 24, 921–939.

Dams, J., Woldeamlak, S. T., and O. Batelaan. 2008. Predicting land-use change and its impact on the groundwater system of the KleineNete catchment, Belgium. *Hydrology Earth System Science (HESS)*. 12, 1369–1385.

Dawoud, M. A., Darwish, M. M., and M. M. El-Kadi. 2005. GIS-based groundwater management model for western Nile Delta. *Water Resources Management*. 19, 585–604.

Demers, M. N. 2013. *Fundamentals of Geographic Information Systems*. Fourth Edition. John Wiley & Sons, Inc., New York.

Dixon, B. 2005. Groundwater vulnerability mapping: A GIS and fuzzy rule based integrated tool. *Applied Geography Journal*. 25, 327–347.

Elfeki, A., Ewea, H., and N. Al-Amri. 2011. Linking groundwater flow and transport models, GIS Technology, satellite images and uncertainty quantification for decision making: Buraiman Lake Case Study Jeddah, Saudi Arabia. *International Journal of Water Resources and Arid Environments*. 1(4), 295–303.

El-Kadi, A. I., Tillery, S., Whittier, R. B., Hagedorn, B., Mair, A., Ha, K., and G. W. Koh. 2014. Assessing sustainability of groundwater resources on Jeju Island, South Korea, under climate change, drought, and increased usage. *Hydrogeology Journal*. 22, 625–642.

Engel, B., Storm, D., White, M., Arnold, J., and M. Arabi. 2007. A hydrologic/water quality model application protocol. *Journal of American Water Resources Association*. 43(5), 1223–1236.

Ganapuram, S., Vijayakumar, G. T., Muralikrishna, I. V., Kahya, E., and M. C. Demirel, 2009. Mapping of groundwater potential zones in the Musi Basin using remote sensing data and GIS. *Advances in Engineering Software*. 40, 506–518.

Gogu, R. C., Carabin, G., Hallet, V., Peters, V., and A. Dassargues. 2001. GIS-based hydrogeological databases and groundwater modeling. *Hydrogeology Journal*. 9, 555–569.

Hernandez, J. J. C. and S. J. Gaskin. 2006. The groundwater modeling tool for GRASS (GMTG): Open source groundwater flow modeling. *Computers & Geosciences*. 32, 339–351.

Jha, M. K., Chowdhury, A., Chowdary, V. M., and S. Peiffer. 2007. Groundwater management and development by integrated remote sensing and geographic information systems: Prospects and constraints. *Water Resources Management*. 21, 427–467.

Joo Oh, H., Kim, Y. S., Choi, J. K., Park, E., and S. Lee. 2011. GIS mapping of regional probabilistic groundwater potential in the area of Pohang City, Korea. *Journal of Hydrology*. 399, 158–172.

Kambhammettu, B. P., Allena, P., and J. P. King. 2011. Application and evaluation of universal kriging for optimal contouring of groundwater levels. *Journal of Earth System Science*. 120(3), 413–422.

Kambhammettu, B. P., Chandramouli, S., and J. P. King. 2010. An improved aggregation tool in GIS for hydrologic models with uniform resolution. *International Journal of Geomatics and Geosciences*. 1(4), 962–970.

Kambhammettu, B. P., King, J. P., and W. Schimid. 2014. Grid-Size dependency of evapotranspiration simulations in shallow aquifers—An optimal approach. *Journal of Hydrologic Engineering, ASCE*. 19(10), 014018-1–014018-9. .

Kambhammettu, B. P., Schmid, W., and J. P. King. 2012. Effect of elevation resolution on evapotranspiration simulations using MODFLOW. *Groundwater Journal*. 50(3), 367–375.

Khan, A., Khan, H. H. Umar, R., and M. H. Khan. 2014. An integrated approach for aquifer vulnerability mapping using GIS and rough sets: Study from an alluvial aquifer in North India. *Hydrogeology Journal*. 22, 1561–1572.

Klug, J. 2009. *Modeling the Risk of Groundwater Contamination Using DRASTIC and Geographic Information Systems in Houston County, Minnesota. Papers in Resource Analysis*. Saint Mary's University of Minnesota, University Central Services Press, Vol 11.

Kolm, K. E., Heijde, P. K., and M. Dechesne. 2007. *GIS based Groundwater Resource Evaluation of the Capitol and Snowmass Creek (CSC) Study Areas.* Pitkin County, CO.

Lillesand, T. M. and R. W. Kiefer. 2000. *Remote Sensing and Image Interpretation.* 4th ed., John Wiley & Sons, Inc., New York.

Majumdar, S. 2014. Investigation of Nutrient Transport in A Semi-arid, Tropical, Agricultural Watershed Using A Process Based Model. M.Tech. Thesis Report, Indian Institute of Technology Hyderabad.

Meijerink, A. M. J., Bannert, D., Batelaan, O., Lubczynski, M. W., and T. Pointet. 2007. Remote Sensing Applications to Groundwater. IHP-VI, Series on Groundwater No. 16, United Nations Educational (UNESCO).

Nampak, H., Pradhan, B., and M. A. Manap. 2014. Application of GIS based data driven evidential belief function model to predict groundwater potential zonation. *Journal of Hydrology.* 513, 283–300.

Ozdemir, A. 2011. Using a binary logistic regression method and GIS for evaluating and mapping the groundwater spring potential in the Sultan Mountains (Aksehir, Turkey). *Journal of Hydrology.* 405, 123–136.

Patel, A. and S. Krishnan. 2008. Groundwater situation in urban India: Overview, opportunities and challenges. In U. A. Amarasinghe, T. Shah and R. P. S. Malik (Eds.), *Strategic Analysis of the National River-Linking Project of India, Series 1: India's Water Future: Scenarios and Issues.* International Water Management Institute, Colombo, pp. 367–380.

Prado, M. and R. Connelly. 2014. A GIS based groundwater management tool for long term mineral planning. Downloaded from: https://water.tallyfox.com/documents/gis-based-groundwater-management-tool.

Scott, R. L., Goodrich, D. C., and L. R. Levick. 2003. A GIS-based management Tool to quantify riparian vegetation groundwater use. Proc. 1st Interagency Conference on Research in the Watersheds, Benson, AZ. pp. 222–227.

Shah, T. 2009. *Taming the Anarchy: Groundwater Governance in South Asia.* Resources for the Future, Washington DC and International Water Management Institute, Colombo.

Singhal, V. and R. Goyal. 2011. GIS based methodology for groundwater flow estimation across the boundary of the study area in groundwater flow modeling. *Journal of Water Resource and Protection.* 3, 824–831.

Strassberg, G., Maidment, D. R., and N. L. Jones. 2007. A geographic data model for representing ground water systems. *Groundwater Journal.* 45(4), 515–518.

Subramanya, K. 2005. *Engineering Hydrology.* 2nd ed. Tata McGraw-Hill Publishing Company Limited, New Delhi.

Sukumar, S. and K. Sankar, 2010. Delineation of potential zones for artificial recharge using GIS in Theni district, Tamilnadu, India. *International Journal of Geomatics and Geosciences,* 1(3), 639–648.

Vahid, G., 2013. Modeling of ground water level using statistical method and GIS. A case study: Amol-Babol Plain, Iran. *International Journal of Water Resources and Environmental Sciences.* 2(3), 53–59.

Vijayshankar, P. S., Kulkarni, H., and S. Krishnan. 2011. India's groundwater challenge and the way forward. *Economical & Political Weekly.* XLVI(2), 37–45.

Wang, J., He, J., and H. Chen. 2012. Assessment of groundwater contamination risk using hazard quantification, a modified DRASTIC model and groundwater value, Beijing Plain, China. *Science of the Total Environment.* 432, 216–226.

Wang, S., Shao, J., Song, X., Zhang, Y., Huo Z., and X. Zhou, 2008. Application of MODFLOW and geographic information system to groundwater flow simulation in North China Plain, China. *Environmental Geology.* 55, 1449–1462.

Xu, X., Huang, G., and Qu, Z. 2009. Integrating MODFLOW and GIS technologies for assessing impacts of irrigation management and groundwater use in the Hetao Irrigation District, Yellow River basin. *Science in China.* 52: 3257.

Xu, X., Huang, G., Qu, Z., and L. S. Pereira. 2011. Using MODFLOW and GIS to assess changes in groundwater dynamics in response to water saving measures in irrigation districts of the Upper Yellow River Basin. *Water Resources Management.* 25(8), 2035–2059.

Yeh, H. F., Lin, H. I., Lee, S. T., Chang, M. H., Hsu, K. C., and C. H. Lee. 2014. GIS and SBF for estimating groundwater recharge of a mountainous Basin in the Wu River watershed, Taiwan. *Journal of Earth System Science.* 123(3), 503–516.

30

Application of Natural and Artificial Isotopes in Groundwater Recharge Estimation

A. Shahul Hameed and T. R. Resmi

CONTENTS

30.1 Introduction

Agriculture, domestic, and industries are the three major consumers of water. The ever-increasing demand for freshwater for all these sectors has made its availability dwindle day by day. Unscientific management of the available freshwater resources further aggravates the freshwater availability. With surface water having been put to use and reached its peak of utilization, attention has already been turned toward abstracting groundwater sources and in many regions of the world, groundwater has been exploited to full availability resulting in depletion in its level. Thus, groundwater recharge estimation has assumed global importance, especially in the arid and semi-arid regions as part of the sustainable management of groundwater sources. Groundwater is recharged naturally by rain and to a lesser extent by surface water bodies like rivers and streams. For sustainable groundwater resource management, quantification of groundwater recharge is pivotal and a prerequisite. Hydrologists and hydrogeologists over the years have developed many

methods and techniques for the estimation of groundwater recharge. Several approaches are employed to quantify groundwater recharge from precipitation and from other sources. The most commonly used methods are direct measurements water balance including the hydrograph method and the tracer methods. Each method has its own advantages as well as limitations.

With the discovery of isotopes and isotopes as tracers, they perform a key role for solving many hydrological problems including investigations pertaining to groundwater. Isotopes, especially of water forming elements—hydrogen and oxygen—are widely used to address many hydrology related issues including recharge and subsurface flow. The isotopes employed for the purpose are both radioactive and stable in nature, the latter taking a major role compared to the radioactive isotopes owing to their inherent limitations to employ them in open field experiments. The tracer technique based on environmental and radioactive isotopes of water is extensively used to get the point value of recharge in a particular location.

This chapter mainly concentrates on the use of stable and radioactive isotopes for groundwater recharge investigations; as a prelude, a brief discussion on the conventional methods of groundwater recharge estimation is also presented herein.

30.2 Water Table Fluctuation (WTF) Method

Among the various methods used for groundwater recharge studies, the most widely used technique is based on groundwater level changes. This is most likely due to the simplicity of applying the method to estimate the recharge from temporal water level data and the abundance of water level data from observation bores. The water table fluctuation (WTF) method provides an estimate of groundwater recharge by analysis of water level fluctuations in observation wells and is based on the assumption that a rise in water table elevation measured in observation wells is caused by the addition of recharge across the water table, lateral inflows or interlayer leakage. As the water level measured in an observation well is representative of an area of at least several tens of square meters, the WTF method can be viewed as an integrated approach and less of a point measurement than those methods that are based on data in the unsaturated zone. The WTF method links the change in groundwater storage over time with resulting groundwater fluctuation from observed monitoring data, through the specific yield of the unconfined aquifer which is represented by the following equation:

$$\Delta S = A \times \Delta h \times Sy \qquad (30.1)$$

where
ΔS is the change in groundwater storage in a defined time interval (e.g., t_0 to t in m^3)
A is the surface area of the aquifer (m^2)
Δh is water level rise in observation wells at a defined time interval (e.g., t_0 to t in m)
S_y is the specific yield of the aquifer

For applying the WTF method, the wet and dry period is first defined and the water level change (Δh) is measured. The specific yield (in the defined dry period) defined as the quantity of water that a unit volume of saturated permeable rock or soil will yield when drained by gravity is then calculated using the following equation:

$$Sy = \frac{\Delta S}{A\Delta h}$$

where Sy is specific yield, Δh is the water level change, ΔS is change in storage, and A is the surface area of the aquifer. The main limitations of the WTF technique are the method requires a specific yield, which is representative of the whole aquifer to be calculated; the observation wells should be located such that the monitored water levels are representative of the area as a whole; and the method cannot account for a steady rate of recharge, where the rate of recharge is constant and equal to the rate of drainage away from the water table.

30.3 Direct Measurement of Groundwater Recharge—Lysimeter

Direct measurement of groundwater recharge is made in lysimeters installed at depth to limit the influence of surface processes including interception and surface runoff. A lysimeter is an instrument that measures the exchange of soil water (or groundwater) between a sample monolith of soil enclosed in an open-topped container and the unenclosed soil that surrounds it. Provided that the exposed surface and vegetation of the sample are typical of the immediate surroundings, the lysimeter can be used to infer the water exchange of the whole surrounding area of vegetation and soil. Usually, the water exchange that the instrument measures is a discharge because the input of precipitation to the soil surface exceeds the evapotranspiration from it; but this need not necessarily be so and the assumption of discharge is invalid in soils supplied with groundwater by adjacent springs and the like (Ingram et al., 2011). The discharge D is calculated from Equation 30.2:

$$P - E - D - \Delta W - \eta = 0 \qquad (30.2)$$

where P = precipitation, D = ground and soil water discharge, ΔW = change in storage (increase reckoned positive), η = error, and E = evapotranspiration, which can be inferred by difference provided η is known to be small; and where all terms are flux densities, that is, volumes per unit area per unit time expressed as equivalent depths. The advantages are that the method directly measures balance components and can cover a wide range of time steps from brief events to seasonal variations.

30.4 Water Budget (Balance) Method

The water budget (or balance) of a groundwater system can be determined by calculating the inputs and outputs of water, and the storage changes of the groundwater system. The water balance approaches measure recharge as a residue of other fluxes that are easier to measure or estimate such as rainfall, evaporation, and discharge. The major inputs of water are from rainfall, irrigation return flow, and seepage from rivers. The major outputs from a groundwater system are evapotranspiration, base flow to rivers, and groundwater pumping. Changes in groundwater storage can be attributed to the difference between inflows and outflows of water over a defined time interval. The storage change can be expressed as:

$$\text{Change in storage } (\Delta S) = \text{inflows} - \text{outflows}$$

Expanding this, the water budget equation with all the relevant inflow and outflow components can be mathematically expressed as follows:

$$R_{rain} + RF + Q_{on} + Q_{River} = ET + PG + Q_{off} + Q_{bf} + \Delta S \quad (30.3)$$

where
   R_{rain} is direct recharge from rainfall (m^3)
   RF is irrigation return flow (m^3)
   Q_{on} and Q_{off} are lateral groundwater flows into and out of the groundwater system (m^3)
   Q_{River} is river seepage recharge (m^3)
   ET is evapotranspiration (m^3)
   PG is extraction of groundwater by pumping (m^3)
   Q_{bf} is the base flow (groundwater discharge to streams or springs in m^3)
   ΔS is the change in groundwater storage (m^3)

A positive ΔS indicates an increase in groundwater storage, while a negative ΔS represents a decrease in groundwater storage.

The limitation of this method is that all the components of the water balance equation other than the rainfall recharge are estimated using the relevant hydrological and meteorological information. The rainfall recharge is then calculated by substituting these estimates in the water balance equation. This approach is valid for the areas where the year can be divided into monsoon and nonmonsoon periods and the water balance is carried out separately. Also, the elements of the water balance equation are computed using independent methods wherever possible. Owing to the shortcomings in the techniques used, computations of water balance elements always involve errors. To apply the above equation correctly, it is thus essential that both the area and the period for which the balance is assessed be carefully chosen.

30.5 Chloride Balance Method

This method is based on the availability of environmental chloride on land by atmospheric deposition processes (rainfall + dry fallout). It is assumed that all the chloride present in the unsaturated zone has atmospheric deposition as its source and there exists no other source or sink in the unsaturated zone for the chloride ions (Allison and Hughes, 1978; Edmunds and Gaye, 1994). Under steady state conditions assuming piston flow, it is possible to obtain a chloride mass balance for the chloride flux entering and leaving the root zone, by the following equation:

$$R_e = \frac{Cp}{Cz} \times P \quad (30.4)$$

where
   R_e = recharge rate leaving the root zone (mm/yr)
   C_p = chloride ion concentration in rainfall (mg/L)
   C_z = mean chloride ion concentration in soil water (mg/L)
   P = precipitation (mm/yr)

The limitation of this method is the assumption that all the chloride has the atmospheric deposition as its source, whereas chloride can also originate from soils or from any anthropogenic sources.

30.6 Isotopes: Environmental and Artificial

Isotopes are atoms of the same element having the same atomic number but different mass numbers as they differ in the number of neutrons in the nucleus. In general, the isotopes are divided into two groups—stable and radioactive isotopes. Further, based on their origin, they can be classified as environmental (or natural) and artificial isotopes. Environmental isotopes may be defined as those isotopes, both stable as well as radioactive, which occur in the environment in varying amounts and over which the investigator has no control. The environmental isotopes most commonly used as tracers in hydrology are deuterium, oxygen-18, and tritium (isotopes of the water molecule), carbon-13, carbon-14, nitrogen-15, and sulfur-34 (isotopes of dissolved solutes in water). Of these, deuterium, oxygen-18, carbon-13, nitrogen-15, and sulfur-34 are stable and the remaining are radioactive.

The artificial isotopes are those produced in the reactor or laboratory as per their requirement. These isotopes are introduced into the system by the investigator to trace the movement of water and its components. The common artificial isotopes in use in hydrological investigations are ^3H, ^{14}C, ^{82}Br, ^{60}Co, ^{51}Cr, ^{131}I, ^{22}Na, and ^{198}Au.

In groundwater studies, isotope techniques are broadly employed in the following fields.

- Occurrence and recharge mechanism
- Origin, identification of recharge sources and areas
- Soil moisture movement in unsaturated zone
- Interconnection between groundwater bodies
- Pollution source and mechanism
- Age determination

30.7 Groundwater Recharge Estimation Using Environmental Radioactive Isotopes

Environmental isotopes help in studying the problems concerning groundwater recharge qualitatively to understand whether groundwater recharge occurs, that is, whether a particular water body is a renewable resource. Besides, it also helps to determine quantitatively the annual recharge of precipitation to the aquifer in a particular region.

30.7.1 Qualitative Evidence of Recharge Using Tritium

Nuclear bomb tests, which began in 1952 in Europe and North America, have resulted in a large amount of artificial tritium in the atmosphere. Since 1961, tritium concentrations in monthly samples have been measured at a worldwide network of stations. Mean monthly concentrations in excess of 5000 TU were recorded at several northern hemisphere stations during the early 1960s (Rao, 2006). The international treaty stopped surface nuclear bomb tests in 1963 and tritium concentrations in precipitation decreased steadily afterward. If a particular groundwater body contains tritium significantly above 5 TU (1 TU is 3.2 pCi/l or 0.12 Bq/l), then it has been recharged in the postthermonuclear era (i.e., after 1952). Tritium can hence be used to detect recharge that occurred in the past three decades. The following are the semiquantitative ages as indicated by their tritium content.

<0.8 TU indicates submodern water (prior to the 1950s)

0.8–4 TU indicates a mix of submodern and modern water

5–15 TU indicates modern water (<5–10 years)

15–30 TU indicates some bomb tritium

>30 TU recharge occurred in the 1960s to 1970s

Fifty years have elapsed since the maximum tritium peak of 1963 owing to the thermonuclear tests. This accounts for four half-lives (1963–2013) of tritium and subsequently the tritium concentrations have been reduced by a factor of 16. With no further anthropogenic additions, tritium will continue to drop to near natural background levels. However, in arid areas due to scarcity of rainfall, high potential evaporation and low infiltration through the unsaturated zone, the groundwater may not receive any modern recharge. This is indicated by low tritium content in groundwater. Environmental carbon-14 could be used to date very old waters up to an age of 35,000 years (Rao, 2006).

30.7.2 Quantitative Estimation of Recharge Using Tritium

The distribution of tritium in precipitation has strongly varied with time, particularly during 1963–1970 due to thermonuclear tests. Hence, determination of depth at which the tritium peak of 1963 is present in the unsaturated zone soil water could be used for estimating groundwater recharge. The total water content above the depth at which the tritium peak of 1963 was found is the recharge over the intervening years. The method cannot be used if the 1963 peak has already reached the water table. This is mostly the case, except in arid regions. Studies carried out by Unnikrishnan Warrier et al. (2010) have shown that tritium in precipitation in Kozhikode and Hyderabad stations in southern India has reached the levels prevailed in the prebomb era resulting in the usage of environmental tritium for age dating/groundwater recharge has almost become invalid. Most of the studies carried out on groundwater recharge has concluded that the estimation of recharge is best carried out as an iterative process as the data is always limited and circumstances vary both in space and time. Studies carried out by Lerner et al. (1990) state that estimates based on the summation of shorter time steps are better than those based on longer time steps for the same duration.

30.7.3 Recharge Estimation Using Chloride (as ^{36}Cl)

Chloride ion is a well known chemical tracer, as it is practically conservative in all environments. The application of its isotope in hydrology is limited thus far because there is no suitable isotope of chlorine for convenient application in isotope hydrology. However, ^{36}Cl—the radioactive isotope of chlorine—has recently paved the way in hydrological studies with the introduction of modern analytical techniques. Over the past 2–3 decades, the interest in ^{36}Cl applications has broadened owing to its possible use for the investigations on groundwater recharge (Andrews and Fontes, 1992), groundwater infiltration rates (Walker, et al., 1992), and rates of erosion. The ^{36}Cl decays to ^{36}Ar by beta emission ($E_{max} \sim 0.7$ MeV). It has a half life of 301,000 years. The conservative nature of chloride ion and the long half life of ^{36}Cl have made the isotope particularly attractive for dating of extremely old waters. Radioactive ^{36}Cl is produced in the upper atmosphere by reactions of protons with argon at extremely small quantities of about 26 atoms m^2/S. It is also produced near the Earth's surface by the interaction of cosmic ray neutrons with the most abundant ^{35}Cl, which is about 75%. It is also produced to a certain extent by the terrestrial neutrons in deeper layers. The latter will depend on the concentration of uranium and thorium in the aquifer matrix. The contribution from the thermonuclear weapon tests too has added considerably to the existing natural concentration of ^{36}Cl.

Groundwater dating using ^{36}Cl is based on two fundamental methods, namely, (1) the decay of cosmogenic and epigenic ^{36}Cl over long periods of time in the subsurface or (2) in-growth of hypogenic ^{36}Cl produced radiogenically in the subsurface. In both cases, it is essential to have an understanding of the initial concentration of ^{36}Cl in the recharging groundwater (Clark and Fritz, 1997). The initial $^{36}Cl/^{35}Cl$ ratio is very small on the order of 5 to 30×10^{-15}. The small concentrations are now amenable to measurement with the introduction of high precision accelerator mass spectrometry (AMS) techniques even though these measurements are very expensive (Rao, 2006). The specific concentration of ^{36}Cl is measured on the total chloride extracted from water or mineral samples. Hence, its concentration is expressed as atoms of ^{36}Cl per total chloride (moles of chloride × Avagadro's number)

$$R^{36}Cl = \frac{atoms^{36}Cl}{Cl}$$

By convention, the very small concentration of ^{36}Cl is multiplied by 10^{15} for ease of expression as it gives a number between 0 and several thousand (Clark and Fritz, 1997). In groundwater, the concentration of ^{36}Cl per liter, represented as $A^{36}Cl$, is an important expression as it is independent of the Cl^- content. $A^{36}Cl$ is calculated as:

$$A^{36}Cl = atoms \ ^{36}Cl/litre = (^{36}Cl/Cl) \times MCl^- \times 6.022 \times 10^{23}$$

where $MCl^- =$ moles of chloride per liter $\{mg/L \times 10^{-3}/35.5\}$ and $^{36}Cl/Cl = R^{36}Cl$. $A^{36}Cl$ is usually multiplied with 10^{-7} to give values between 0 and 100. These two parameters provide insights into the origin and behavior of ^{36}Cl in groundwater. While $R^{36}Cl$ (or simply ^{36}R) will not change during concentration by evaporation, the $A^{36}Cl$ will increase or in another words the $A^{36}Cl$ will not necessarily change with increases in Cl^- in groundwater by leaching or evaporate dissolution ($^{36}Cl^-$ free additions of Cl^-), whereas $R^{36}Cl$ will decrease proportionally (Clark and Fritz, 1997).

^{36}Cl techniques were utilized for the dating of groundwater in the Milky River aquifer in Canada (Philips et al., 1986). Another application of this technique was the confirmation of hydraulically estimated groundwater ages of up to a million years in the Great Artesian basin in Australia (Herczeg et al., 1999).

30.8 Direct Recharge through Precipitation: Case Study 1

Artificial tracer methods have been extensively used in India and in other countries to estimate rates of infiltration in the unsaturated zone for determining the direct rainfall recharge to the groundwater. The method involves tagging a horizontal layer at a certain depth below the root zone with a suitable tracer followed by monitoring the tracer profile at regular intervals.

The vertically downward movement of moisture is very slow and the moisture flows in such a way that if any freshwater is added to the top surface of the soil, the infiltrated layer of the water pushes the older layer downward in the soil system until the last layer of the moisture reaches the saturated zone. The moisture movement through the soil in this way is termed piston flow movement, the added water replacing the older water further downward but in the same amount thus keeping the affective moisture content constant.

Radiotracers used for soil moisture movement studies should have specific characteristics like (1) tracer should follow the movement of water as closely as possible, (2) it should be easily detectable, (3) it should not change the hydraulic characteristics of an aquifer, (4) it should not be absorbed/adsorbed by the medium through which it will be transmitted along with water, (5) its natural concentration must be much lower than the intended dose, and (6) its half life should be optimum and long enough for the duration of the experiment, but short enough so that radioactive contamination is minimum (Athavale, 1983).

Tritium as tritiated water is the most commonly used tracer, although other tracers like Cobalt-60, a gamma emitting isotope, as $K_3Co(CN)_6$, have also been used to take advantage of *in situ* detectability in the field.

Among the radioactive tracers available for the investigation of soil moisture movement, tritium—a radioactive isotope of hydrogen—is considered to be an ideal tracer. It has all the desirable qualities mentioned in the proceeding section. Besides, it has the additional advantages such as: it behaves similar to ordinary water, it is a beta emitter of low energy, the energy level being 18.6 kev, and belongs to the lowest radiotoxicity class, it can be measured with high detection sensitivity, and it has a comparatively long half life (12.43 yr) and hence is useful for soil moisture movement studies. However, tritium as a tracer is not entirely free from limitation. Being a soft beta emitter, it cannot be measured in the field *in situ*. The application of artificially injected tritium will interfere with the use of naturally occurring tritium in hydrological investigations, if present in significant levels.

30.8.1 Experimental Sites

Field experiments were conducted at two locations, namely, Mooliyar and Cheemeni in Kasargode District of Kerala, India (Figure 30.1). These sites fall under the Chandragiri and Kavvayi River basins, respectively. Both sites are located equidistant from the Arabian Sea and have similar hydrometeorological characteristics typical of humid tropics. The area receives an uneven distribution of rainfall in different months. The average annual rainfall in Mooliyar is 3800 mm and that in Cheemeni is 3890 mm. The number of rainy days during the year of the investigation in the selected areas is 110 days with a monthly rainfall of the order of 1000 to 1500 mm in the months of July–August and practically no rain in the months from January–April. The maximum temperature recorded during the year of the study was 39°C and the minimum was 19°C. The annual potential evaporation was 1332 mm.

Both study areas form part of the midland region with an elevation of 7.6–76 m above mean sea level in the northern part of Kerala. The locations have a general slope toward the west. The morphological unit of the area is residual flat topped hills with gentle slopes and concave depositional valleys. The laterite form as the capping over the mounds and in low lying areas they occur as detrial formations. The depth of laterite overburden in the topography is about 10 m in both areas. The area is represented with granulitic rocks of Precambrian age. The area is devoid of much vegetation except for the presence of grass and trees like cashew, which are scattered. The laterites are hard in nature, which occurs to a depth of up to 11 m. However, the loose soil occurs only to a depth less than a meter. Below the laterite zone, a lithomarge zone containing clays is observed. The kaolin is underlain by weathered basement rocks.

30.8.1.1 Tracer Injection

A total of seven sites (four in Mooliyar and three in Cheemeni) in Kasargode district were chosen for the tracer application. Tritium activity equal to 100 μCi was prepared for each site and taken to the sites. At each site, 5–10 holes equidistant along the perimeter of a circle of radius 5 cm were driven to a depth of 50–75 cm using an iron rod, ensuring the absence of vegetative roots. In each hole, tritiated water with tritium of specific activity of 20 μCi/ml was introduced using a syringe through small copper tubes. The openings in the hole were closed by filling with the loose soil to prevent the loss of the tracer due to evaporation. Iron nails were plugged at the site of injection to identify the site of injection. Necessary identification marks available in the location were also recorded. After the monsoon season, the soil samples were collected at an interval of 5–10 cm depths beyond the depth of the injection point to ascertain the downward movement of the tracer. Necessary precautions were taken while collecting and transporting the soil samples to the laboratory for analysis. The soil samples were collected in polythene bags or in airtight polythene containers to avoid loss of moisture during transit. The tritiated water was extracted from the soil by vacuum distillation method. The moisture content of the soil before the injection of the tracer as well as at the time of collection of soil samples was also determined. Moisture determination using an infrared moisture meter is always preferred to that by the gravimetric method.

The tritium activity in the extracted moisture was determined by counting on a liquid scintillation counting system (LSC-WALLAC-1410). Each sample was counted for a minimum period of 2 mins depending on the total number of counts to minimize error. The tritium counts per minute (cpm) obtained are plotted against depth and the center of gravity of the tritium peak was calculated (Figure 30.2). By subtracting the depth of injection from the center of gravity of the tritium peak, the shift in the tritium peak could be known. Then, the percentage of recharge to groundwater was calculated.

The recharge to groundwater was determined by finding the tritium peak shift and multiplying it by average field capacity in the tritium peak shift region. This recharge is the recharge to the groundwater during the time interval of tritium injection and sampling. Mathematically, the percentage of recharge of precipitation to groundwater can be found using the following relationship.

$$R = \frac{A_v.d}{P} \times 100$$

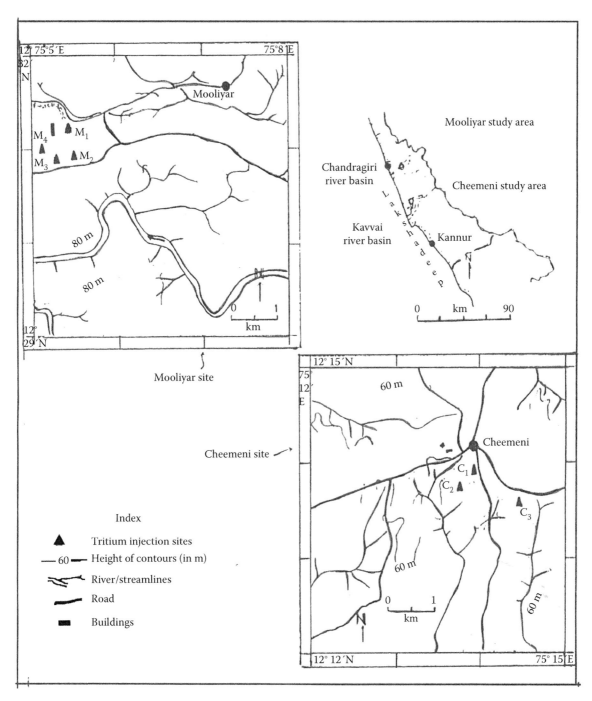

FIGURE 30.1
Location map of the recharge experiment sites at Kasargode, Kerala.

where

R = Percentage of recharge to groundwater

A_v = Average volumetric moisture content in the tritium peak region

d = Shift in tritium peak (cm)

P = Precipitation (cm)

The shift in the tracer was of the order of 2.5 cm indicating highly impervious layers below the soil zone, which was shallow in depth. The percentage recharge was estimated to be below 1% of the precipitation during the study period in the region. The tracer movement through the soil at the site of injection indicated that there was very little vertical movement of tritium even after the second monsoon was over. The downward movement of moisture through the laterite is generally low as the hydraulic conductivity in the vertical direction is considered to be less compared to that in the horizontal direction.

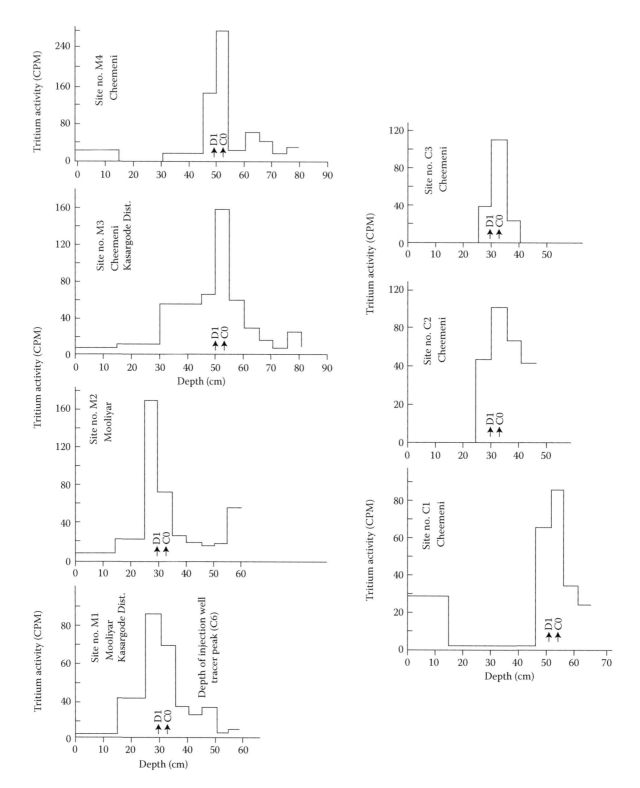

FIGURE 30.2
Depth-wise distribution of measured tritium activity at the test sites.

30.9 Recharge Estimation Using Environmental Stable Isotopes

The use of stable isotopes in hydrology depends on variations in natural waters. The variations result from isotope fractionation, which occurs during some physical and chemical processes. Examples of physical processes that could lead to isotopic fractionation are evaporation of water or condensation of vapor. During evaporation, the residual liquid is enriched in the heavier isotope molecule because the lighter molecule moves more rapidly and hence has a greater tendency to escape from the liquid phase, that is, there is a difference in the volatility between the two molecular species. Chemical fractionation effects occur because a chemical bond involving a heavy isotope will have a lower vibrational frequency than an equivalent bond with a lighter isotope. The bond with the heavy isotope will thus be stronger than that with the light isotope. Fractionation may occur during both equilibrium and nonequilibrium chemical reactions. During nonequilibrium or irreversible reactions, kinetic fractionation leads to the enrichment of the lighter isotope in the reaction product because of the ease with which the bond between lighter isotopes could be broken.

Stable environmental isotopes are measured as the ratio between the least abundant and the most abundant isotopes of an element under consideration compared to a standard. Fractionation processes modify this ratio slightly for any given compound. In hydrology, only variations in stable isotope concentrations are of interest, rather than actual abundance. Hence, isotopic concentrations are expressed using δ notation as the difference between the measured ratios of the sample and reference over the measured ratio of the reference. As fractionation processes do not impart huge variations in isotope concentrations, δ values are expressed as parts per thousand or per mill (‰) difference from the reference. Thus, ^{18}O concentration in a sample is expressed as

$$\delta^{18}O\,Sample = \left\{ \frac{\delta^{18}O/\delta^{16}O\,sample}{\delta^{18}O/\delta^{16}O\,reference} - 1 \right\} X\,1000\text{‰}\,VSMOW$$

Large variations in the isotopic composition of natural waters occur in the atmospheric part of the water cycle and the surface waters which are exposed to the atmosphere. Soil and subsurface waters inherit the isotopic composition of the atmospheric and surface water inputs and further changes take place negligibly except as a result of mixing of waters of different isotopic composition. Determining the meteoric end member is the first necessary step in recharge estimation of a groundwater body. Being an integral part of the water

molecule, meteoric water in a particular environment exhibits a characteristic isotopic signature that can be used to trace groundwater recharge (Clark and Fritz, 1997, Gat, 1971; Lee et al., 1999; Price and Swart, 2006). Stable environmental isotopes (^{18}O and ^2H) applied naturally over the entire drainage catchments are not subject to chemical reactions during contact with mineral matter at temperatures encountered at or near Earth's surface (Drever, 1988). However, they undergo fractionation during evaporation and condensation. Variation in isotopic content of precipitation may be dampened as water passes through the vadoze zone to the water table, with the result that groundwater δ values can approach uniformity in time and space, and are changed only by mixing with waters with different isotopic contents (Sklash, 1990). This greater temporal variability in the isotopic content of precipitation relative to that of soil and groundwater means that there is frequently a difference between the δ of water input to the catchment's surface and water stored in the catchment prior to the event (Buttle, 1998). It is this difference that permits use of the mass balance equation to separate the rainwater component of the groundwater after the heavy storm season.

30.9.1 Case Study 2

Groundwater resources are the primary drinking water sources in Kerala. The present study is aimed to estimate the rain water component in shallow aquifers of a tropical river basin, Chaliyar. The Chaliyar River basin lies between 11°08′ and 11°38′ N and 75°45′ and 76°35′ E and spreads over parts of Malapuram, Kozhikode, and Wayanad districts of Kerala and Nilgiri district of Tamil Nadu. Chaliyar is the third largest river in Kerala with a total catchment area of 2535 km^2 and originates from the Elambaleri Hills in the Wayanad plateau at an elevation of 2066 m. The Chaliyar River joins the Arabian Sea south of Calicut after flowing for a distance of 169 km. The major tributaries are Chalipuzha, Punnapuzha, Pandipuzha, Karimpuzha, Cherupuzha, Kanjirapuzha, Iruvahnipuzha, Kurumbanpuzha, and Vadapurampuzha, which constitute the Chaliyar River drainage system. Most of these rivers have their origin in the Nilgiri Hills in the east and Wayanad Hills in the north. The drainage map of the Chaliyar River with the sampling location is depicted in Figure 30.3. The drainage area is dominated by rain forests of medium to high productivity.

The basin enjoys a tropical humid climate with sweltering summer and high monsoon rainfall. Generally March and April are the hottest and December and January are the coolest months. The maximum temperature ranges from 22°C to 32.9°C and the minimum temperature from 22°C to 25.8°C. The average annual

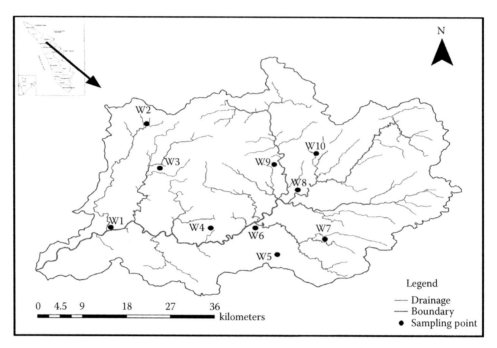

FIGURE 30.3
Drainage map of the Chaliyar River with sampling locations.

maximum temperature is 30.9°C and minimum is 23.7°C. The temperature starts rising from January reaching a peak in April. It decreases during the monsoon months. On average about 3000 mm of rainfall occurs annually in the basin of which 75% is obtained during the southwest monsoon and the remaining in the northeast and premonsoon seasons. The premonsoon rainfall is highly variable and is characterized by thunderstorms.

30.9.1.1 Geology, Hydrogeology, and Geomorphology

The basin includes parts of highly rugged Nilgiri hills, the Nilambur valley with moderate relief, and the more or less plain land between Nilambur and River valley. The lower reaches of the Chaliyar River is blessed with fertile alluvial soil and is densely populated and cultivated by the farming community. The annual average stream flow is 5902 Mm³. It is observed that the Chaliyar River is a truly monsoonal river, sediment discharge approaching 0 during nonmonsoon months. Geomorphologically the Chaliyar drainage basin includes parts of distinct provinces like the Wayanad plateau and the Nilgiri hills at higher altitude, the Nilambur valley forming the slopes of the foot hills and low lands adjoining the main trunk of the river. The lower reaches of the Chaliyar main channel show a sudden change in the geomorphology beyond 110 km from the source in the downstream direction. The channel takes a sharp bend at 110 km and beyond this the river is meandering at consistent intervals. The main stream Chaliyar is a 7th order stream

and the drainage network analysis shows that the pattern is dendritic combined with rectangular. The latter is more characteristic for the area close to the confluence of Punnapuzha with the Chaliyar River. The overprint of the rectangular drainage pattern over the dendritic pattern as seen to the north of Nilambur is due to the presence of a right angle fault system. During its course, the river cuts through a number of lithologies like gneisses, charnockite, metapelites, schists, and quartz reefs of Precambrian age (Ambili, 2010; Hariharan, 2001). Laterites, older and younger gravels, sediments along river terraces, alluvium and soil representing subrecent to recent deposits are seen in the basin. It is worth noting that the headwater tributaries flow exclusively above laterites developed over gneissic country rocks. However, the downstream tributaries that join the lower reaches of the Chaliyar main stream is underlain essentially by charnockite (Figure 30.4).

Groundwater samples from 10 wells from different subbasins of the Chaliyar River basin were collected during the premonsoon and postmonsoon periods in 2006. Rainwater samples were collected as monthly composites for the whole rainy period. The samples were subjected to oxygen and hydrogen isotope measurements and the data are presented in Table 30.1. Water samples for oxygen isotope analysis were prepared by H_2O–CO_2 equilibration. About 200 µL of each sample was equilibrated with laboratory standard CO_2 gas at 32°C. Hydrogen isotope was analyzed by the equilibration technique aided by the Pt catalyst coated on hydrophobic polymer. Both the isotopes were measured on

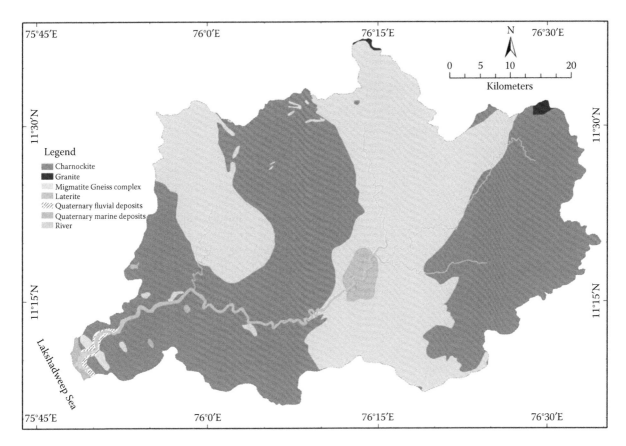

FIGURE 30.4
Geology map of the Chaliyar River basin. (District Resource Maps, GSI.)

Continuous Flow Finnigan Delta Plus[XP] Isotope Ratio Mass Spectrometer with reference to the standard, V-SMOW. The analytical reproducibility is ±0.18‰ for $\delta^{18}O$ and ±1.8‰ for δD.

The isotopic composition of rainwater for the whole rainy period of the year was taken for the construction of local meteoric water line (LMWL). The δ values showed wide variation for both the isotopes in accordance with the climatic regime. The southwest monsoon rains (June–September) were having more enriched values (–2.53‰ to –29.6‰ for δD and –1.61‰ to –4.83‰ for $\delta^{18}O$) than for northeast monsoon rains (October–November; –38.9‰ to –42.9‰ for δD and –6.72‰ to –6.85‰ for $\delta^{18}O$). The summer rains (April and May) were also enriched. The summer rains are caused by the localized disturbances in the atmosphere and due to the prevailing high temperature, evaporative enrichment in the falling raindrops driven by non-equilibrium fractionation. Thus, the rains that reach the ground will be enriched in heavier isotopes than at the cloud base. In the case of northeast monsoon rains, the vapor source is mainly continental and traverses vast areas of northern India, and as a consequence, the heavier isotope will be preferentially removed from the air mass during its journey. In addition to it, during a

monsoon cycle heavier isotopes will be greater in the early rains and as the season advances, the rains will be progressively depleted. Thus, during the withdrawal phase (northeast) of the monsoon currents from the mainland, the heavier isotope content in the rains will be much less.

The deuterium excess was also quite different for the two rains. Deuterium excess is defined as the excess deuterium that cannot be accounted by equilibrium fractionation of isotopes between water and vapor (Clark and Fritz, 1997) and is calculated as

$$\text{Deuterium excess (‰)} = \delta D - 8\delta^{18}O$$

Although wind speed and temperature can influence the kinetic fractionation, relative humidity is the most important driving force. The deuterium excess values of regional precipitation deviate from that of global meteoric water line (GMWL). If deuterium excess values are greater than 10, then evaporation under low humidity conditions takes place at the source area. Significant re-evaporation of local surface waters under low humidity creates vapor mass with high deuterium excess. If such vapor mixes with atmospheric reservoir and recondenses, the resultant precipitation will have

TABLE 30.1

Oxygen and Hydrogen Isotopic Composition and Deuterium Excess Values of Rainwater and Groundwater in the Pre- and Postmonsoon Periods

	$\delta^{18}O$ (‰)	δD (‰)	Deuterium Excess (‰)
a) Rainwater			
Apr	−2.61	−15.3	5.60
May	−3.12	−9.01	16.0
Jun	−2.56	−6.78	13.7
Jul	−1.61	−2.53	10.4
Aug	−2.10	−6.62	10.2
Sep	−4.83	−29.6	9.00
Oct	−6.85	−42.9	11.8
Nov	-6.72	−38.9	14.9
b) Groundwater—Premonsoon			
W1	−3.09	−14.2	10.5
W2	−3.02	−5.48	18.7
W3	−3.12	−9.57	15.4
W4	−4.34	−15.2	19.5
W5	−3.25	−10.4	15.6
W6	−2.75	−9.44	12.6
W7	−2.75	−8.15	13.9
W8	−3.09	−10.4	14.3
W9	−2.76	−8.69	13.4
W10	−2.29	−5.02	13.3
Groundwater—Postmonsoon			
W1	−3.29	−19.1	7.30
W2	−3.37	−19.6	7.40
W3	−3.48	−18.5	9.30
W4	−4.22	−23.8	9.90
W5	−3.52	−17.5	10.7
W6	−3.32	−13.5	13.0
W7	−3.32	−19.1	7.40
W8	−3.41	−20.6	6.70
W9	−3.31	−15.2	11.3
W10	−3.23	−18.5	7.40

high deuterium excess (Clark and Fritz, 1997). The deuterium excess value thus serves as a tool to differentiate air masses causing rainfall in a region. In this study, for the southwest monsoon rain, the value was close to that of the global meteoric line (+10), whereas greater values were obtained for the northeast monsoon rains. Thus, two types of air masses of different origin can be inferred. The southwest monsoon rains are caused by the marine vapors, carrying moisture from the Indian Ocean and Arabian Sea. The relative humidity at the vapor source region is around 85%, and correspondingly the deuterium excess value is around 10‰. In the case of northeast monsoon rains, the winds are in the reverse direction, that is, oriented toward the southern hemisphere. The vapors reach the peninsular region after traversing the vast northern Indian plains. During

its journey, the air mass acquires re-evaporated vapor from the land and the relative humidity of the vapor source region is also considerably less (<70%). The combined effect of the two processes produces rainfall with high deuterium excess value. The two different vapor source origins were reported in this region earlier (Deshpande et al., 2003; Gupta et al., 2004; Resmi, 2011).

The regression plot of δD and $\delta^{18}O$ of the rainwater and groundwater samples was constructed (Figure 30.5). The rainwater samples (LMWL) were found to plot close to the global meteoric waterline. The regression line obtained for LMWL is $\delta D = 7.5 \, \delta^{18}O + 11$. The slope is slightly less than the GMWL (8 indicating that evaporative enrichment of heavier isotopes in the regional precipitation). The d-intercept value is also close +10, that of GMWL. The groundwater samples of the premonsoon season were found to plot very close to the LMWL, but slightly above the line. The slope and intercept values were also considerably less; 4.7 and 4.6, respectively (Table 30.2). The stable hydrogen and oxygen isotopes of groundwater are generally considered to be transported conservatively in the absence of high temperature water–rock interaction and significant evaporation. However, the regression parameters obtained indicated evaporative enrichment. This may be due to mixing of the water with different isotopic composition. Evaporation from the unsaturated zone may be very high in the summer season and the water that is retained in the pore spaces will be enriched very much. The rain water of the summer period itself is enriched due to kinetic fractionation of isotopes and as this water infiltrates the unsaturated zone to reach the water table, it may get mixed with the highly enriched pore water and as a consequence, the isotopic composition of the groundwater in this season is showing signals of evaporative enrichment.

Groundwater samples in the postmonsoon season were found to be clustered with very little variation in both the isotopes. The slope of the regression line of this season is higher than that of premonsoon but still less than the LMWL. This increase in 's' denotes that the groundwater is substantially replenished with the monsoon rains. However, indication of evaporative enrichment is also seen. The mixing of incoming infiltrated rain water and the relatively enriched groundwater of the premonsoon season has produced groundwater with totally different isotopic composition in the postmonsoon. Thus, from the regression plots, two conclusions can be reached: (1) the groundwater in the basin is recharged mainly by the meteoric water and (2) the isotopic signals are different for the rain and groundwater of the two seasons facilitating estimation of relative contribution of the recharging source.

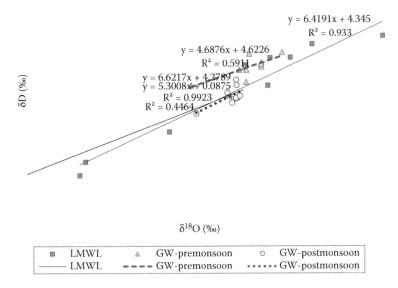

FIGURE 30.5

δD—δ¹⁸O plot of rainwater (LMWL) and groundwater in the two seasons.

TABLE 30.2

Slope and y-Intercept Values of Regression Lines Obtained for Rainwater and Groundwater in the Two Seasons

Water Type	Slope (s)	y-Intercept (d)
LMWL	7.5	11
Groundwater (premonsoon)	4.7	4.6
Groundwater (postmonsoon)	6.6	4.3

30.10 Estimation of Recharge of the Groundwater by Rain

The isotopic mass balance method has been widely used for estimating the recharge characteristics of groundwater and hydrograph separation (Deshpande et al., 2003; Gupta et al., 2004, Jeelani et al., 2010; Langhoff et al., 2005; Liu et al., 2008; Maurya et al., 2011; Paternoster et al., 2008; Saravanakumar et al., 2010; Sharda, et al., 2006). The prerequisite for such estimations is that there should be at least two isotopically different end members. The mass balance equation used in this study is given next.

$$M_1R_1 + M_2R_2 = R_{AM}$$

where

M_1 is the fraction of groundwater in the premonsoon season

M_2 is the fraction of rainwater

R_1 and R_2 are the corresponding isotopic compositions

R_{AM} is the isotopic composition of the admixture, that is, groundwater in the postmonsoon season

In this isotope mass balance method, it is assumed that the groundwater present after the postmonsoon season is a mixture of infiltrated rainwater of the monsoon storms and the groundwater present in the aquifer prior to the monsoon season, that is, premonsoon groundwater. Thus, the point recharge in different parts of the Chaliyar basin was estimated, the data of which are given in Table 30.3. The rainwater fraction ranged between 23% and 61% in the groundwater of the river basin.

In certain wells, more than 50% recharge by rainwater was observed and in certain others the rainwater fraction was only up to 30%. The groundwater level fluctuation was measured bimonthly as given in Figure 30.6. The water level fluctuations can be divided into two

TABLE 30.3

Estimated Fraction of Rainwater in Groundwater of Chaliyar River Basin in the Postmonsoon Season and the Data Selected for Estimation

Code	Premonsoon (GW) δ¹⁸O‰ (R1)	Rain δ¹⁸O‰ (R2)	Postmonsoon (GW) δ¹⁸O‰ (RAM)	Percentage
W1	−3.09	−3.82	−3.29	27
W2	−3.02	−3.82	−3.37	44
W3	−3.12	−3.82	−3.48	51
W4	−4.34	−3.82	−4.22	23
W5	−3.25	−3.82	−3.52	47
W6	−2.75	−3.82	−3.32	53
W7	−2.75	−3.82	−3.32	53
W8	−3.09	−3.82	−3.41	30
W9	−2.76	−3.82	3.31	44
W10	−2.29	−3.82	−3.23	61

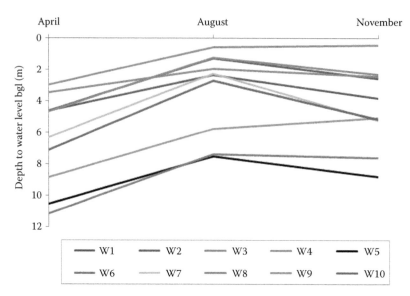

FIGURE 30.6
Groundwater level fluctuations in the three seasons.

groups; one with fluctuation <2.5 m and the other with >2.5 m. In wells that showed higher water level fluctuation, the rainwater fraction was also found to be more (>50%). Similarly, wells with <2.5 m seasonal difference in water levels showed minimum rain water fraction in them (<30%). Thus, the isotopic estimation of rain water fraction in groundwater could explain the meteoric recharge to the groundwater system.

Rainwater samples showed two characteristic isotope signatures; the southwest monsoon samples were enriched in heavier isotopes with low deuterium excess and the northeast monsoon samples were depleted with high deuterium excess. This peculiar pattern revealed that the area is affected by two types of precipitation with different vapor source origins. The groundwater samples in both seasons were found to plot close to the local meteoric waterline in the δD versus $\delta^{18}O$ regression plot indicating that the groundwater is recharged by precipitation. The isotope mass balance method applied to find out the rainwater fraction revealed that the rainwater fraction varied from 23%–61% in groundwater in different locations of the river basin. Thus, the results obtained give useful and necessary information on the hydrological features of the area.

References

Allison, G.B. and M.W. Hughes. 1978. The use of environmental chloride and tritium to estimate total recharge in an unconfined aquifer. *Australian Journal of Soil Research*, 16: 181–195.

Ambili, V. 2010. Evolution of Chaliyar River Drainage Basin: Insights from Tectonic Geomorphology. PhD thesis, Cochin University of Science and Technology, Cochin, India.

Andrews, J.N. and J.-C. Fontes. 1992. Importance of the in situ production of 36Cl, 36Ar and 14C in hydrology and hydrogeochemistry. In *Isotope Techniques in Water Resources Development 1991*, IAEA Symposium 319, March 1991, Vienna, 245–269.

Athavale, R.N. 1983. Injected tracers in studies on the unsaturated zone. *Proceedings of the Workshop on Hydrology, BARC*, December 13–14, 1983: 179–197.

Athavale, R.N., Murty, C.S., and R. Chand. 1980. Estimation of recharge to the phreatic aquifers of the Lower Manar Basin, India by using the tritium injection method. *Journal of Hydrology*, 45:185.

Buttle, J.M. 1998. Fundamentals of small catchment hydrology. In: Kendall C. and McDonnell, J. (Eds.), *Isotope Tracers in Catchment Hydrology*. Elsevier, Amsterdam, the Netherlands, 40.

Clark, I.D. and P. Fritz. 1997. *Environmental Isotopes in Hydrogeology*. Lewis Publishers, Boca Raton, FL, p. 328.

CWRDM. 1998. *Application of Nuclear Techniques to Hydrological Problems of Kerala*. Final report to Board of Research in Nuclear Sciences (BRNS), Department of Atomic Energy, BARC, Mumbai.

Datta, P.S., Gupta, S.K., and A. Jayasurya. 1978. Soil moisture movement through vadose zone in alluvial plains of Sabarmati Basin. Report No. HYD-78-0333, Physical Research Laboratory, Ahmedabad, India.

Deshpande, R.D., Bhattacharya, S.K., Jani, R.A., and S.K. Gupta. 2003. Distribution of oxygen and hydrogen isotopes in shallow groundwater from Southern India: Influence of a dual monsoon system. *Journal of Hydrology*, 271, 226–239.

Drever, J.I. 1988. *The Geochemistry of Natural Waters*. Prentice-Hall, Englewood Cliffs, NJ, p. 437.

Edmunds, M.W. and C.B. Gaye. 1994. Estimating the spatial variability of groundwater recharge in the Sahel using chloride. *Journal of Hydrology*, 156: 47–59.

Gat, J.R. 1971. Comments on the stable isotope method in regional groundwater investigations. *Water Resources Research*, 7, 980–993.

Goel, P.S., Datta, P.S., Rama, Sangal, S.P., Kumar, H., Bahadur, P., Sabherwal, R.K., and B.S. Tanwar. 1975. Tritium tracer studies on groundwater recharge in the alluvial deposits of Indo-Gangetic plains of Western U.P., Punjab and Haryana. In: Athavale, R.N., Srivastava, V.B. (Eds.), *Approaches and Methodologies for Development of Groundwater Resources*: Indo-German Workshop, Hyderabad, India, Proceedings, National Geophysical Research Institute, pp. 309–322.

Gupta, S.K., Deshpande, R.D., Bhattacharya, S.K., and R.A. Jani 2004. Groundwater $\delta^{18}O$ and δD from central Indian Peninsula: Influence of Arabian Sea and Bay of Bengal branches of the summer monsoon. *Journal of Hydrology*, 128, 223–236.

Hariharan, G.N. 2001. Geochemistry and sedimentology of Chaliyar River sediments with special reference to the occurrence of placer gold. PhD thesis, Cochin University of Science and Technology, Cochin, India.

Herczeg, A.L., Love, A.J., Sampson, L., Cresswell, R.G., and L.K. Fifield. 1999. Flow velocities estimated from Cl-36 in South-West Great Artesian Basin, Australia. *International Symposium on Isotope Techniques in Water Resources Development and Management*, IAEA, Vienna, SM 361, p. 34.

Ingram, H.A.P., Coupar, A.M., and O.M. Bragg. 2011. Theory and practice of hydrostatic lysimeters for direct measurement of net seepage in a patterned mire in north Scotland. *Hydrology and Earth System Sciences*, 5(4), 693–709.

Jeelani, G., Bhat, N.A., and K. Shivanna. 2010. Use of ^{18}O tracer to identify stream and spring origins of a mountainous catchment: A case study from Liddar watershed, Western Himalaya, India. *Journal of Hydrology*, 393, 257–264.

Langhoff, J.H., Rasmussen, K.R., and S. Christensen. 2005. Quantification and regionalization of groundwater-surface water interaction along an alluvial stream. *Journal of Hydrology*, 320, 342–358.

Lee, K.S., Wenner, D.B., and I. Lee. 1999. Using H- and O-isotopic data for estimating the relative contributions of rainy and dry season precipitation to groundwater: Example from Cheju Island, Korea. *Journal of Hydrology*, 222, 65–74.

Lerner, D.N., Issar, A., and I. Simmers. 1990. A guide to understanding and estimating natural recharge. *International Contribution to Hydrogeology*, I.A.H. Publication, 8, Verlag Heinz Heisse, 345pp.

Liu, Y., Fan, N., and S. An, et al. 2008. Characteristics of water isotopes and hydrograph separation during the wet season in the Heishui River, China. *Journal of Hydrology*, 353, 314–321.

Maurya, A.S., Shah, M., Deshpande, R.D., Bhardwaj, R.M., Prasad, A., and S.K. Gupta. 2011. Hydrograph separation and precipitation source identification using stable water isotopes and conductivity: River Ganga at Himalayan foothills. *Hydrology Processes*, 25, 1521–1530.

Nair, A.R., Pendharkar, A.S., Navada, S.V., and S.M. Rao. 1979. Groundwater recharge studies in Maharashtra. Development of isotopes techniques and field experiences. *Proceedings of the Symposium on Isotope Hydrology, Vienna*, 803–826.

Paternoster, M., Liotta, M., and R. Favara. 2008. Stable isotope ratios in meteoric recharge and groundwater at Mt. Vulture Volcano, Southern Italy. *Journal of Hydrology*, 348, 87–97.

Philips, F.M., Bentley, H.W., Davis, S.N., Elmore, D., and G.B. Swanick. 1986. Chlorine-36 dating of very old groundwater, Milky River aquifer, Alberta, Canada. *Water Resources Research*, 22(13), 2003–2016.

Price, R.M. and P.K. Swart. 2006. Geochemical indicators of groundwater recharge in the surficial aquifer system, Everglades National Park, Florida, USA. *Perspectives on Karst Geomorphology, Hydrology and Geochemistry—A Tribute Volume to Derek C. Ford and William B. White*, Harmon, R.S., Wicks, C. Eds., Geological Society of America, 251–266.

Rao, S.M. 2006. *Practical Isotope Hydrology*. New India Publishing Agency, New Delhi, p. 201.

Resmi, T.R. 2011. Finger printing of waters of a tropical River basin using isotope systematics. Abstract volume, *23rd Kerala Science Congress*, January 29–31, 2011, p. 103.

Saravanakumar, U., Kumar, B., Rai, S.P., and S. Sharma. 2010. Stable isotope ratios in precipitation and their relationship with meteorological conditions in the Kumaon Himalayas, India. *Journal of Hydrology*, 391, 1–8.

Shahul Hameed, A., Reshmi, T.R., and S. Suraj. 2012. Estimation of groundwater recharge by rain in a tropical river basin using stable isotope technique. *Proceedings of the International Groundwater Conference (IGWC)*. Dec. 18–21, Association of Geologists and Hydrogeologists (GEOFORUM), Maharashtra, India, pp. 135–145.

Sharda, V.N., Kurothe, R.S., Sena, D.R., Pande, V.C., and S.P. Tiwari. 2006. Estimation of groundwater recharge from water storage structures in semi-arid climate of India. *Journal of Hydrology*, 329, 224–243.

Sklash, M.G. 1990. Environmental isotope studies of storm and snowmelt runoff generation. *Process Studies in Hill Top Hydrology*. In: Anderson, M.G. and Burt, T.P. Eds., John Wiley & Sons, Chichester, UK, pp. 401–435.

Unnikrishnan Warrier, C., Praveen Babu, M., Manjula, P., Velayudhan, K.T., Shahul Hameed, A., and K. Vasu. 2010. Isotopic characterization of dual monsoon precipitation—Evidence from Kerala, India. *Current Science*, 98(11), 1487–1495.

Walker, R.G., Jolly, I.D., and M.H. Stadter et al. 1992. Evaluation of the use of ^{36}Cl in recharge studies. In *Isotope Techniques in Water Resources Development, IAEA Symposium* 319, March 1991 Vienna, pp. 19–32.

Zimmerman, U., Munich, K.O., Roether, W., Kreutz, W., Schubech, K., and O. Siegel. 1966. Tracers determine movement of soil moisture and evapotranspiration. *Science*, 152, 346–347. doi: 10.1126/science.152.3720.346.

31

Impact of Climate Change on Groundwater

Avdhesh Tyagi, Nicholas Johnson, and William Logan Dyer

CONTENTS

31.1 Introduction

Although approximately 70% of the Earth's surface is covered by water, freshwater makes up only 3% of the total water on the planet. The majority of freshwater is stored as ice in glaciers and polar ice sheets. Although humans rely heavily on freshwater from rivers and lakes, this surface water amounts to only 0.02% of all water on Earth. Most liquid freshwater is stored in aquifers as groundwater. Still, groundwater makes up only 1% of all water on the planet. Groundwater can be viewed as a product of climate. This is because the groundwater available for use is deposited by atmospheric precipitation. Changes in climate will then inevitably affect groundwater, its resources, and quality. Publications discussing this problem are numerous, dissimilar, and contradictory; however, the effects of climate changes on groundwater have only been discussed in a limited manner. Geological science has demonstrated continuous climate change throughout the history of Earth. Changes developed both slowly and relatively quickly in the geological time scale. Past climatic changes have been caused by changes in solar activity, meteorite showers, variations in Earth's axis position, volcanic activity, and a wide array of other natural activities, which caused changes in the Earth's albedo and the greenhouse effect of the atmosphere. Figure 31.1 presents a schematic flowchart showing a relationship between climate change and loss of fresh groundwater in coastal aquifers.

31.2 Climate Change

Paleo-climates of the past allow the development of an analog of the probable future climate. An example of how these relationships can be made can be seen by comparing temperatures today with the recorded temperatures found in ice cores such as the Vostok Ice Core temperature graph in Figure 31.2. Global warming by 1°C can be the climate of the Holocene Optimum; by 2°C the climate of the Mikulian Interglacial Period; and warming by 3–4°C, the Pliocene Optimum (Kovalevskii, 2007). These time periods can be used to characterize the likely future climate. These estimates of potential global warming are based on an extrapolated relationship between the air temperature and chemical content of the atmosphere (Tucker, 2008). Current predictions are commonly referred to as wide time intervals in the future. The global warming by 1°C is most often believed to occur in the first quarter of the twenty-first century; 2°C in the mid-twenty-first century; and 3°C at the beginning of the next century (Kovalevskii, 2007). This determines possible hydrogeological forecasts.

Based on the forecasts by Kovalevskii (2007), in "Effect of climate changes on goundwater," there will be a regular and gradual growth of the air temperature increments from the south to the north. Some temperature changes have already been observed and can be seen in Figures 31.3 and 31.4, showing NOAA average sea surface temperatures in 1985 and 2006. Figures 31.2 and 31.3 can be compared with the Annual Mean Temperature in

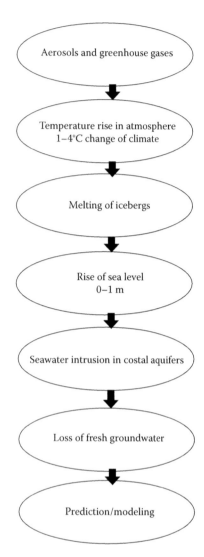

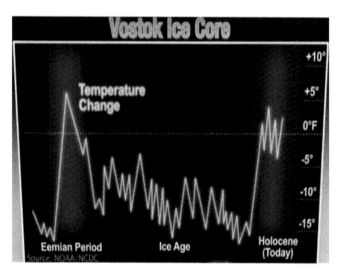

FIGURE 31.2
Average temperature over different ages.

FIGURE 31.1
Relationship between climate change and fresh groundwater in coastal aquifers.

Figures 31.4 and 31.5. Predicted precipitation increases in the middle latitudes are many times smaller than those in the low and high latitudes. Model forecasts show even a likely decrease in precipitation in the middle latitudes (Joigneaux, 2011). Precipitation decrease is shown to spread from the western boundaries of Russia to the Urals, the primary area of concern for Kovalevskii's research, including the central and southern regions of Russia. Around the world, the anticipated changes in climatic conditions will entail changes in the entire complex of hydrogeological conditions; in the water, heat, and salt balances of groundwater, as well as in the environment interconnected with groundwater. Taking into account the highest importance of hydrodynamic forecasts, it is practical to consider, first of all, the potential changes in groundwater resources (Kovalevskii, 2007).

Significant climate change is expected to alter India's hydroclimate regime over the course of the twenty-first century. Wide agreement has been reached that the Indo-Gangetic basin is likely to experience increased water availability from increasing snow-melt up until around 2030 but face gradual reductions thereafter. Most parts of the Indo-Gangetic basin will probably also receive less rain than in the past; however, all the rest of India is likely to benefit from greater precipitation. According to the Intergovernmental Panel on Climate Change, most Indian landmass south of the Ganges Plain is likely to experience a 0.5–1°C rise in average temperature by 2029 and 3.5–4.5°C rise by 2099. Many parts of peninsular India, especially the Western Ghats, are likely to experience a 5%–10% increase in total precipitation; however, this increase is likely to be accompanied by a greater variance in temperature (Shah, 2009). Throughout the subcontinent, it is expected that very wet days are likely to contribute more and more total precipitation, suggesting that most of India's precipitation may be received in fewer than 100 h of thunderstorms.

This will generate more flooding events, and may reduce total infiltration as a matter of more concentrated runoff. The higher precipitation intensity and larger number of dry days in a year will also increase evapotranspiration. Increased frequency of extremely wet rainy seasons is also likely to mean increased runoff. In Shah's "Climate change and groundwater," (2009) a comparison of the 1900–1970 period and 2041–2060, most of India is likely to experience 5%–20% increase in annual runoff. India can expect to receive more of its water via rain than via snow. Snow-melt will occur faster and earlier. Less soil moisture in summer and higher crop evapotranspirative demand can also be expected as a consequence. As climate change results in spatial and temporal changes in precipitation, it will significantly influence natural recharge. Moreover, as much of natural aquifer recharge occurs in areas with

FIGURE 31.3
NOAA average sea surface temperature in 1985.

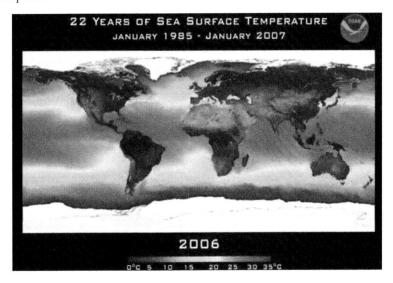

FIGURE 31.4
NOAA average sea surface temperature in 2006.

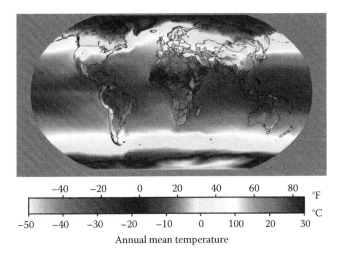

FIGURE 31.5
Annual mean temperature.

vegetative cover, such as forests, changing evapotranspiration rates resulting from rising temperatures may reduce infiltration rates form natural precipitation and therefore reduce recharge. Recharge clearly has a strong response to the temporal pattern on precipitation as well as soil cover and soil properties. In the African context, Shah cites arguments that replacing natural vegetation by crops can increase natural recharge by nearly a factor of 10. If climate change results in changes in natural vegetation in forests or savanna, these too may influence natural recharge; however, the direction of the net effect will depend on the pattern of changes in the vegetative cover (McCallum, 2010).

Simulations developed by Australian scientists have shown that changes in temperatures and rainfall may influence the growth rates and the leaf size of plants that have an effect on groundwater recharge. The direction of change is contextually sensitive. In some places, the vegetation response to climate change might cause the average recharge to decrease, but in other areas, groundwater recharge is likely to more than double (McCallum, 2010). We have an inadequate understanding of how exactly rainfall patterns will change, but increased variability seems almost guaranteed. This will lead to intense and large rainfall events in brief monsoons followed by longer dry spells. While evidence suggests that groundwater recharge through natural infiltration occurs only beyond a certain threshold level of precipitation, it also demonstrates that the runoff coefficient increases with increased rainfall intensity. Increased variability in precipitation will negatively impact natural recharge in general. The Indo-Gangetic aquifer system has been getting a significant portion of its natural recharge from Himalayan snow-melt (Shah, 2009). As snow-melt-based runoff continues to increase during the coming decades, their contribution to potential recharge will likely increase; however, a great deal of this may end up as a form of "rejected recharge," enhancing river flows and intensifying the flood proneness of eastern India and Bangladesh. As the snow-melt-based runoff begins declining, one should expect a decline in runoff as well as groundwater recharge in that vast basin.

31.3 Melting Glaciers

Glaciers are an important part of the current global ecosystem. They are found in the lower, mid, and upper latitudes. These glaciers generally have a melt and replenish cycle that coincides with the local seasons. However, most of the regularly observed glaciers have been receding over the past years. In Greenland, portions of the country have gone from completely covered by glaciers to rocky and without a continuous ice sheet (see Figures 31.6 through 13.11).

In Alaska, coastal glaciers have been melting and shedding icebergs at an increasing rate. Figures 31.6

FIGURE 31.7
Greenland melting.

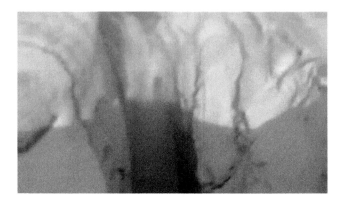

FIGURE 31.8
Greenland melting.

FIGURE 31.9
Melting of icebergs.

FIGURE 31.6
Greenland melting.

through 31.11 show a glacier going through a melt/erosion cycle with a dramatic collapse into the ocean.

The Gangotri Glacier in India is the main source for the Ganges river system. This glacier has been responsible for providing freshwater to a main river across southeastern Asia and is receding at continually increasing

FIGURE 31.10
Melting of icebergs.

FIGURE 31.11
Melting of icebergs.

rates (see Figure 31.12). The reduction of this glacier will greatly impact the flow of the Ganges and the ecosystem it supplies.

31.4 Seawater Rise and Intrusion

Climate change and groundwater will show some of their most drastic interrelations in coastal areas. Data from coastal tidal gauges in the north Indian Ocean are readily available for more than the last 40 years; in Shah's "Climate change and groundwater: India's opportunities for mitigation and adaptation," estimates are presented for a sea level rise between 1.06 and 1.75 mm per year. This is consistent with a 1–2 mm per year global sea level rise which has been estimated by the IPCC. Rising sea levels will of course present a threat to coastal aquifers. Many of India's coastal aquifers are already increasing in saline intrusion. The problem is especially acute in the Saurashtra Coast in Gujarat and the Minjur Aquifer in Tamil Nadu. In coastal West Bengal, mangrove forests are threatened by saline intrusion overland (Shah, 2009). This will affect the aquifers supplying these ecosystems.

The sea-level rise that accompanies climate change will reduce the freshwater supply in many coastal communities by infiltrating groundwater and rendering it brackish and undrinkable without excessive treatment (McCallum, 2010). In this publication, Ibaraki describes that most people are probably aware of the damage that rising sea levels can do aboveground, but not underground, which is where the freshwater is. According to Ibaraki, coastlines are made of many different layers

FIGURE 31.12
Gangotri (Himalayas) glacier recession.

and kinds of sand. Coarse sands let water through to aquifers and can lead to contaminated, brackish water. Ibaraki plans to create a world salinity hazard map showing areas that have the potential for the most groundwater loss due to sea-level rise. An example of the extensive and severe problems of water sufficiency and quality, Florida has the largest concentration of desalination plants in the United States. Ninety-three percent of Florida's 16 million residents rely on groundwater as their drinking water supply (Meyland, 2008).

The saline/freshwater interface location and behavior can be approximated by several model types. The first is a U-Tube manometer. In the manometer, the hydrostatic balance between fresh and saline water can be seen. The freshwater is less dense than the saline water and will therefore float on one side of the manometer. This shows that in an aquifer there will be an interface with freshwater on top and denser saline water intruding to the bottom of the aquifer (Todd and Mays, 2005). Another model is the Glover model. This is a conceptual model that relies on some basic simplifying assumptions about the aquifer involved, but still gives good approximations of saline/freshwater interface.

31.5 Effect on Groundwater

Scientists have suggested that climate change may alter the physical characteristics of aquifers. They argue that higher CO_2 concentrations in the atmosphere are influencing carbonate dissolution and promote the formation of karstified soils, which in turn may have a negative effect on the infiltration properties of top soils. Others have argued the opposite. From experimental data, some scientists have claimed that elevated atmospheric CO_2 levels may affect plants, vadose zone, and groundwater in ways that may hasten infiltration from precipitation by up to 119% in a Mediterranean climate to up to 500% in a subtropical climate (Shah, 2009).

Diffusive groundwater recharge is the most important process in the restoration of groundwater resources. Changes to any of the variables that have an effect on diffuse recharge may have an impact on the amounts of water entering aquifers (Shah, 2009). Some efforts have been made to model changes predicted in diffuse aquifer recharge. To determine the impacts of climate change on the Edwards Aquifer in central Texas, a doubled atmospheric concentration of carbon dioxide was modeled for precipitation adjustments (McCallum, 2010). Changes to rainfall and stream flow were scaled based on this model, and by using a water-balance technique, the impact on recharge was determined. McCallum's review in "Impacts of climate change on

groundwater in Australia" observed that changes to rainfall and stream flow under such scenarios would yield reduced groundwater levels in the aquifer even if groundwater extraction was not increased. The reduction in groundwater levels might allow for additional seawater intrusion, impacting groundwater quality. This is inferred from the simple relationships between recharge and climate change.

Saltwater intrusion is not the only issue changing climates can create in groundwater systems. Certain hydrological conditions allow for spring flow in karst systems to be reversed. The resulting back flooding represents a significant threat to groundwater quality. The surface water could be contaminated and carry unsafe compounds back into the aquifer system (Joigneaux, 2011). Joigneaux and his team examined the possible impacts of future climate change on the frequency and occurrences of back flooding in a specific karst system in their article "Impact of climate change on groundwater point discharge." They first established the occurrence of such events in the study area over the past 40 years. Preliminary investigations showed that back flooding in this Loiret, France karst has become more frequent since the 1980s. Adopting a downscaled algorithm relating large-scale atmospheric circulation to local precipitation special patterns, they viewed large-scale atmospheric circulation as a set of quasi-stationary and recurrent states, called weather types, and its variability as the transition between them (Joigneaux, 2011). Based on a set of climate model projections, simulated changes in weather type occurrence for the end of the century suggests that back flooding events can be expected to increase until 2075, at which point the event frequency will decrease.

As Joigneaux explains, alluvial systems and karst hydrogeological systems are very sensitive to small changes in hydrological components. Stream back flooding and the subsequent appearance of sink holes can occur because of relative changes between surface and underground drainage, which are controlled by both precipitation and discharge (Joigneaux, 2011). Consequently, this type of system is sensitive to small climate variations, even at temperate mid-latitudes. Dry-weather stream flow is closely related to the rise and fall of groundwater tables. Since the 1980s, stream flow has deleted rapidly, owing to limited precipitation during the dry period and immoderate groundwater pumping for agricultural, domestic, and industrial uses. Ecologic and environmental disasters such as decreased number of species and population sizes, water quality deterioration, and interference with navigable waterways, have resulted from these changes. Kil Seong Lee Chung (2007), in "Hydrological effects of climate change, groundwater withdrawal, and land use in a small korean watershed," analyzes the influences on total runoff during the dry periods and simulates its variability. Understanding these factors is very important for the watershed level planning and management of water resources, especially in tropical climate areas. Chung particularly investigated how changing dry-weather climate would affect the use and withdrawal of water from stream and groundwater systems. By using surface waters as a set of boundary conditions, models like Chung's help demonstrate the effects of climate change on groundwater resources.

31.6 Loss of Freshwater

The use of freshwater supplies will have a growing impact on a variety of issues. Desalination might be used to ensure supplies of drinkable water, but it is an energy-intensive process. "Our energy use now could reduce the availability of freshwater and groundwater through the climate change process," Ibaraki says. "These resources are decreasing due to human activities and population increase." Another approach to protecting water supplies is to transfer water from regions that have it in abundance to regions that face water shortage. Unfortunately, both approaches require much energy (Tucker, 2008).

In the United States, much of the agricultural land depends on irrigating crops using water from aquifers. However, these aquifers are being "mined" for agriculture at rates that exceed the recharge rate, depleting them. The Ogallala Aquifer stretches across the U.S Midwest, running from South Dakota, down to New Mexico and Texas; it is being pumped faster than the natural replacement rate, leading to a significant drop in the water table, possibly by hundreds of feet. When fossil aquifers like the Ogallala and the North China Plain are depleted, pumping will become impossible (Meyland, 2008). This will make the existing agricultural system unfeasible.

Groundwater is harder to manage and protect than surface water because it is difficult to monitor and model. Large efforts are needed to put groundwater systems under the management and protection of agencies dedicated to the job. Managing authorities could equitably administer intrastate, interstate, and international aquifer basins using scientific research and management plans, implemented by educated professionals. The management agencies can conduct studies, prepare management strategies, quantify the resources,

determine equitable distributions of the water, and establish safety margins for allocations, anticipating climate swings such as severe drought. Groundwater will only become more important as a resource in the future. Effective management and protection of groundwater sources will become critical as the United States and the rest of the world work toward sustainable use of the Earth's water resources.

31.7 Summary

There is no doubt that climate change is taking place around the globe. The temperature rise may range between 1°C and 4°C. This is going to result in melting of icebergs, no matter how slow or fast. Such an action will raise the seawater level less than 1 m (or 3 ft). This rise will result in pushing the seawater interface into the coastland aquifers in each continent, leading to loss of freshwater in coastal areas.

References

Chung, Kil Seong Lee, and Eun-Sung. 2007. Hydrological effects of climate change, groundwater withdrawal, and land use in a small Korean watershed. *Hydrological Processes*, 3046–3056.

Joigneaux, E. 2011. Impact of climate change on groundwater point discharge: Backflooding of karstic springs (Loiret, France). *Hydrology and Earth Systems Sciences*, 8, 2459–2470.

Kovalevskii, V.S. 2007. Effect of climate changes on groundwater. *Water Resources*, 140–152.

McCallum, J.L. 2010. Impacts of climate change on groundwater in Australia: A sensitivity analysis of recharge. *Hydrogeology Journal*, 1625–1638.

Meyland, S.J. 2008. Rethinking groundwater supplies in light of climate change: How can groundwater be sustainably managed while preparing for water shortages, increased demand, and resource depletion? *Forum on Public Policy*. Oxford: Oxford Round Table, 1–14.

Shah, T. 11th August 2009. Climate change and groundwater: India's opportunities for mitigation and adaptation. *Environmental Research Letters*, 1–13.

Todd, D., K. Mays, and W. Larry. 2005. *Groundwater Hydrology*, 3rd ed., Hoboken, NJ: John Wiley & Sons Inc., 590.

Tucker, P. 2008. Climate change imperils groundwater sources. *The Futurist*, 42(2), 10.

32

Multimodeling Approach to Assess the Impact of Climate Change on Water Availability and Rice Productivity: A Case Study in Cauvery River Basin, Tamil Nadu, India

V. Geethalakshmi, K. Bhuvaneswari, A. Lakshmanan, Nagothu Udaya Sekhar, Sonali Mcdermid, A. P. Ramaraj, R. Gowtham, and K. Senthilraja

CONTENTS

32.1 Introduction

Understanding the interactions among different components of the Earth's physical system and the influence of human activities on the environment has progressed immensely; however, the climatic impacts of increasing greenhouse gases (GHGs) are still quite uncertain (Jones, 2000). The most detailed atmosphere-ocean general circulation models (AOGCMs) differ in the predicted climate change for a given increase in GHGs (Murphy et al., 2004; Stainforth et al., 2005). This large uncertainty introduced in projecting the future climate is further confounded by uncertainty in the impact assessment using impact models such as hydrological or crop simulation models (Wilby, 2005; Challinor et al., 2009; Hawkins and Sutton, 2009; Osborne et al., 2013).

A comprehensive understanding of uncertainties in projected impacts is a key element of making vigorous assessments (Collins, 2007; Challinor et al., 2013; Katz et al., 2013). Uncertainty has two implications for adaptation practices. First, adaptation procedures need to be developed that do not rely on precise projections of changes in river discharge, groundwater, etc. In the future warmer climate, even with small perturbations in precipitation frequency and/or quantity, distribution and availability of water resources would be severely affected as a result of impact on hydrological parameters such as mean annual discharge, runoff, evapotranspiration, etc. (Whitfield and Cannon, 2000; Muzik, 2001; Srikanthan and McMahon, 2001; Xu and Singh, 2004). Second, based on the studies done thus far, it is difficult to assess water-related consequences of climate policies and emission pathways with high credibility and accuracy.

Climate change will affect current water management practices and the existing operational rules for supply of water (Kundzewicz et al., 2007). It is also widely recognized that improved incorporation of current climate variability into water-related management would make adaptation to future climate change easier. Traditionally, it has been conveniently assumed that the natural water resource base is constant, and hydrological design rules have been based on the assumption of stationary hydrology, tantamount to the principle that the past is the key to the future. However, the validity of this principle is limited and, therefore, the current procedures for designing water-related infrastructure must be revised.

For better management of the available water resources and to develop location-specific adaptation strategies in the context of climate change, it is important to understand the hydrological changes at a river basin scale. In the present chapter, an attempt was made to understand the hydrology and rice productivity of Cauvery River Basin (CRB) in India by integrating climate model outputs in impact models through a multimodel approach.

32.2 Future Climate Projections

Climate models are used to generate projections of future climate on large spatial and temporal scales. Climate models are mathematical representations of the climate, which divide the earth, ocean, and atmosphere into grids and calculate the weather variables, such as surface pressure, wind, temperature, humidity, and rainfall at each grid point over time for the current and future condition.

Global climate models (GCMs) have evolved from the atmospheric general circulation models (AGCMs) widely used for daily weather prediction. GCMs have been used for a range of applications, including investigating interactions among processes of the climate system, simulating evolution of the climate system, and providing projections of future climate states under scenarios that might alter the evolution of the climate system. The most widely recognized application is the projection of future climate states under various scenarios of increasing atmospheric carbon dioxide (CO_2). The main problem in GCMs includes the too coarse scale and that the GCMs calculate precipitation with large errors. Even as GCM grid sizes tend toward one or two degrees, there is still a significant mismatch with the scale at which many hydrological and water resource studies are conducted (Varis et al., 2004).

Regional climate models (RCMs) have been developed using dynamic downscaling techniques to attain horizontal resolution on the order of tens of kilometers,

over selected areas of interest. This nested regional climate modeling technique consists of using initial conditions, time-dependent lateral meteorological conditions derived from GCMs (or analyses of observations), and surface boundary conditions to drive high-resolution RCMs (e.g., Cocke and LaRow, 2000; Von Storch et al., 2000). The basic strategy is, thus, to use a global model to simulate the response of the global circulation to large-scale forcings and an RCM to account for sub-GCM grid-scale forcing (e.g., complex topographical features and land-cover in homogeneity) in a physically based way; and enhance the simulation of atmospheric circulations and climatic variables at fine spatial scales (up to 10 to 20 km or less). More recently, significant improvements have been achieved in the area of nested RCMs (Christensen et al., 2001; Varis et al., 2004). The ability of RCMs to reproduce the present day climate has substantially improved. New RCM systems have been introduced including multiple nesting and atmospheric RCMs coupled with, for example, lake and hydrology models (e.g., Rummukainen et al., 2001, 2004; Hay et al. 2002; Samuelsson et al., 2003). The effects of domain size, resolution, boundary forcing, and internal model variability in RCMs are now better understood. The main theoretical limitations of this technique that remain to be improved include (Hay et al., 2002; Varis et al., 2004): (1) the inheritance of systematic errors in the driving fields provided by global models. For example, boundary conditions from a GCM might themselves be so biased that they impact the quality of the regional simulation, complicating the evaluation of the regional model itself (e.g., Hay et al., 2002), (2) lack of two-way interactions between regional and global climate, and (3) the algorithmic limitations of the lateral boundary interface.

Climate projections are done based on scenarios given by Intergovernmental Panel on Climate Change (IPCC). Initially, a Special Report on Emission Scenarios (SRES) given by IPCC in the year 2000 was used for studies under Coupled Model Intercomparison Project 3 (CMIP3) that used a simplistic assumption of GHG emissions. A climate scenario is a plausible representation of future climate that has been constructed for explicit use in investigating the potential impacts of anthropogenic climate change. Climate scenarios often make use of climate projections (descriptions of the modeled response of the climate system to scenarios of GHG and aerosol concentrations), by manipulating model outputs and combining them with observed climate data. SRES scenarios described four major storylines, namely, A1 (a future world of very rapid economic growth, global population that peaks in mid-century and declines thereafter, and rapid introduction of new and more efficient technologies), A2 (a very heterogeneous world with continuously increasing global population and regionally oriented economic growth that is more fragmented and slower than in other storylines),

B1 (a convergent world with the same global population as in the A1 storyline but with rapid changes in economic structures toward a service and information economy, with reductions in material intensity, and the introduction of clean and resource-efficient technologies), and B2 (a world in which the emphasis is on local solutions to economic, social, and environmental sustainability, with continuously increasing population but lower than A2 and intermediate economic development).

In the recent days, RCP scenarios (representative concentration pathways) are used. RCP2.6 scenario describes the radiative forcing level reaches a value around 3.1 W/m mid-century, returning to 2.6 W/m^2 by 2100. Under this scenario, GHG emissions and emissions of air pollutants are reduced substantially over time. RCP4.5 is a stabilization scenario where total radiative forcing is stabilized before 2100 by employing a range of technologies and strategies for reducing GHG emissions. RCP6.0 is a stabilization scenario where total radiative forcing is stabilized after 2100 without overshoot by employing a range of technologies and strategies for reducing GHG emissions. RCP8.5 is characterized by increasing GHG emissions over time representative of scenarios in the literature leading to high GHG concentration levels.

Different regionalization techniques can give different local projections, even when based on the same GCM. With the same technique, different RCMs will give different regional projections, even when based on the same GCM output (Jones et al., 2004). Uncertainty in future climate change derives from three main sources: forcing, model response, and internal variability (Tebaldi and Knutti, 2007; Hawkins and Sutton, 2009). Forcing uncertainty arises from incomplete knowledge of external factors influencing the climate system, including future trajectories of anthropogenic emissions of GHGs, stratospheric ozone concentrations, land use change, etc. Model uncertainty, also termed response uncertainty, occurs because different models may yield different responses to the same external forcing as a result of differences in, for example, physical and numerical formulations. Internal variability is the natural variability of the climate system that occurs in the absence of external forcing, and includes processes intrinsic to the atmosphere, the ocean, and the coupled ocean-atmosphere system. To reduce the uncertainty in climate projection, multimodel ensembles are used in recent days.

32.3 Impact Assessment

The potential effects of climate change on the hydrology and crops' productivity are assessed using hydrological models and dynamic crop simulation models.

32.3.1 Hydrological Model

Improved hydrological models and methodologies are continuously emerging from the research community. Computer-based hydrological models have been developed and applied at an ever-increasing rate during the past four decades. Fleming (1975) and Singh and Valiron (1995) have reviewed the status and development trends in catchment scale hydrological modeling. The concept of using regional hydrologic models for assessing the impacts of climatic change has several attractive characteristics (Gleick, 1986; Schulze, 1997), namely, (1) the models are tested for different climatic/physiographic conditions and the models structured for use at various spatial scales, which permits flexibility in identifying and choosing the most appropriate approach to evaluate any specific region; (2) hydrologic models can be tailored to fit the characteristics of available data, including GCM-derived climate perturbations (at different levels of downscaling) as model input and a variety of responses to climate change scenarios can hence be modeled; (3) regional-scale hydrologic models are considerably easier to manipulate than general circulation models; (4) regional models can be used to evaluate the sensitivity of specific watersheds to both hypothetical changes in climate and to changes predicted by large-scale GCMs, and (5) methods that can incorporate both detailed regional hydrologic characteristics and output from large-scale GCMs will be well situated to take advantage of continuing improvements in the resolution, regional geography, and hydrology of global climate models. The choice of a model for a particular case study depends on many factors (Gleick, 1986), while the study purpose, model, and data availability have been the dominant ones (Ng and Marsalek, 1992; Xu, 1999). For example, for assessing water resources management on a regional scale, monthly rainfall-runoff (water balance) models were found useful for identifying hydrologic consequences of changes in temperature, precipitation, and other climatic variables (e.g., Gleick, 1986; Schaake and Liu, 1989; Mimikou et al., 1991; Arnell, 1992; Xu and Halldin, 1997; Xu and Singh, 1998). For detailed assessments of surface flow, conceptual lumped-parameter models are used.

The soil and water assessment tool (SWAT) is a conceptual watershed scale simulation model that was developed as a decision support tool to assess the impact of land management practices on the water resources of the United States by Arnold et al. (1998) and Srinivasan et al. (1998). Currently, SWAT is applied worldwide and considered a versatile model that can be used to integrate multiple environmental processes, enabling more effective watershed management and better-informed policy decision. SWAT can be used to study the long-term effects due to natural variations in climate as well as the manufactured interventions on water yields in ungauged catchments. The model operates on a daily time step and can be used for the assessment of existing and anticipated water uses and water shortages (Gosain et al., 2005).

Inputs required for running SWAT models are digital elevation map, land use, soil particulars, and weather parameters. It has the capability of simulating a high level of spatial details by subdividing the watershed into a large number of subwatersheds, which are further subdivided into hydrologic response units (HRUs) based on unique soil/land use and topographic characteristics. It also uses the management practices such as planting and harvesting time of the crops grown, irrigation and fertilizer application details, and crop rotation details to simulate the hydrology and crops productivity.

The hydrology part of the SWAT model includes snowmelt, surface runoff, evapotranspiration, groundwater percolation, lateral flow, and groundwater flow (or return flow). The soil profile is subdivided into multiple layers that support soil water processes including infiltration, evaporation, plant uptake, lateral flow, and percolation to lower layers. The soil percolation component of SWAT uses a storage routing technique to simulate flow through each soil layer in the root zone. Downward flow occurs when field capacity of a soil layer is exceeded and the layer below is not saturated. Groundwater flow contribution to total stream flow is simulated by routing a shallow aquifer storage component to the stream (Arnold et al., 1993).

Much attention has been paid to uncertainty issues in hydrological modeling due to their effects on prediction and further on decision making (Van et al., 2008; Sudheer et al., 2011). Uncertainty in hydrological modeling is from model structures, input data, and parameters (Lindenschmidt et al., 2007). In general, structural uncertainty could be improved by comparing and modifying the diverse model components (Hejberg and Refsguard, 2005). The uncertainty of model input occurs because of changes in natural conditions, limitations in measurement, and lack of data (Beck, 1987).

Gosain et al. (2006) used the SWAT model for simulating the impacts of 2041–2060 climate change scenarios on the stream flows of 12 major river basins in India, ranging in size from 1668 to 87,180 km². Surface runoff was found to generally decrease, and the severity of both floods and droughts increased, in response to the climate change projection. The SWAT model was applied to simulate the existing hydrological situation and to evaluate the impact of future climate change of the Bhavani River Basin in Tamil Nadu (Lakshmanan et al., 2011). Model prediction showed an increase in surface flow from the current level during the near future (2011–2040), slight decline in the middle time (2041–2070), and again increasing trend toward the end of the century (2071–2100).

32.3.2 Crop Simulation Model

Crop simulation model is one of the important tools to assess the impact of climate change on crop growth and production. Several impact assessment studies in the developed countries have employed simulation models as a tool to evaluate the impact of global climatic changes on agriculture. Few modeling studies have been done in India to assess the impact of climate change on crop production (Agarwal and Sinha, 1993; Rao and Sinha, 1994; Hundal and Kaur, 1996; Saseendran et al., 2000). Dynamic crop model systems, as decision supporting tools, have extensively been utilized by agricultural scientists to evaluate possible agricultural consequences from interannual climate variability and/or climate change (Paz et al., 2007; Semenov and Doblas-Reyes 2007; Challinor and Wheeler, 2008). Using the CERES-rice model, Lal et al. (1998) predicted 20% decline in rice yields in northwestern India due to elevated CO_2 and temperature. Saseendran et al. (1998) conducted sensitivity experiments of the CERES-rice model to CO_2 concentration changes and reported that for a positive change in temperature up to 5°C, there is a continuous decline in the yield and for every 1° increment, the decline in rice yield was about 6%. In another experiment, it was noticed that the physiological effect of ambient CO_2 at 2°C increase in temperature was compensated for the yield losses at 425 ppm CO_2 concentration (Saseendran et al., 1998). Using the CERES-rice and CERES-maize model, Lansigan and Salvacion (2007) assessed the effect of climate change on rice and corn yields in selected areas over the Philippines and reported that rice and corn respond differently to climate change because of a difference in photosynthetic pathways, which determines carbon assimilation and under potential climate change, corn performed better than rice. Challinor et al. (2008) assessed the vulnerability of crop productivity to climate change under A2 emission scenario derived from the PRECIS model using a crop simulation model and reported the negative impact of high temperature thresholds at the time of flowering on yield of crops.

The SWAT model has the plant growth components as a simplified version of EPIC (environment policy integrated climate) plant growth model. The plant growth component of SWAT utilizes routines for phenological plant development based on plant-specific input parameters, namely, energy and biomass conversion, precipitation and temperature constraints, canopy height and root depth, and shape of the growth curve. These parameters have been developed (and provided in a crop database of the model) for plant species like agricultural crops, forests, grassland, and rangeland. Conversion of intercepted light into biomass is simulated assuming a plant's species-specific radiation use efficiency (RUE). The RUE quantifies the efficiency of a plant in converting light energy into biomass and is assumed to be independent of the plant's growth stage. The effects of increased CO_2 are directly accounted for in the model by changes in plant growth, biomass production, and evapotranspiration rates (Arnold et al., 1998).

The SWAT model predicted a reduction of 35% in wheat crop yield by 2020 in Anjeni watershed (Yakob Mohammed, 2011). Lakshmanan et al. (2011) showed that yields of the Kharif season rice with change in temperature, rainfall, and CO_2 is decreasing with advancement of time from current to end of the century, while the Rabi rice yield levels were not affected considerably over the time.

32.4 Integration of Climate and Impact Models

It is questionable that the principal limitation of assessing climate change impacts is that the uncertainty involved in the model estimates due to either being unquantified or so large that meaningful conclusions cannot be drawn from them. This is particularly true for estimates of resilience and adaptive capacity for farming systems because weather is a primary driver and estimates of alternative future climates derived from GCMs have significant uncertainty, particularly for precipitation (Murphy et al., 2004). The uncertainties and systematic biases in input data are propagated by and possibly magnified by uncertainties in biophysical processes models and compounded in socioeconomic analyses by uncertainties in policy and macroeconomic systems. There is therefore the need to devote significant effort toward "integrated assessments." According to IPCC (2001) and Harremoes and Turner (2001), the definition for integrated assessment is "an interdisciplinary process that combines, interprets, and communicates knowledge from diverse scientific disciplines from the natural and social sciences to investigate and understand causal relationships within and between complicated systems" that it should provide information suitable for decision making. The framework used in the present investigation for integrating climate information in the impact models (hydrological models and crop simulation models) is presented in Figure 32.1.

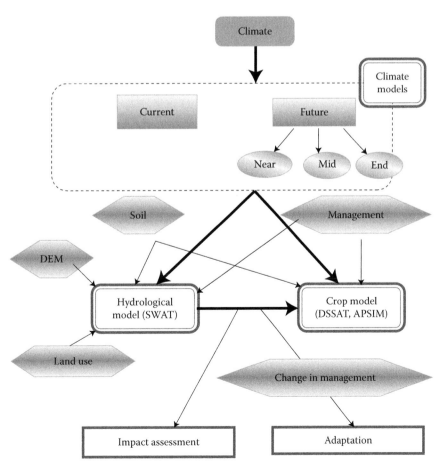

FIGURE 32.1
Framework for integration of climate information in the impact models.

32.5 A Case Study in Cauvery River Basin

Cauvery is the fourth largest river in the Indian peninsula running across two states, namely, Karnataka and Tamil Nadu. CRB covers an area of 81,155 km², of which 44,016 km² lies in Tamil Nadu state between 10.00 to 11.30°N latitudes and 78.15 to 79.45°E longitudes and the rest of the area lies in Karnataka state (Figure 32.2).

The basin receives an annual average rainfall of 1129 mm, of which, 50% is received from the southwest monsoon (June–September), 33% during the northeast monsoon (October–January) period, and the rest during the summer months (February–March). Most of the upstream part and catchment areas of the CRB receive rainfall during the southwest monsoon season filing up the Mettur reservoir on the River Cauvery that supplies water to the Tamil Nadu part. However, Cauvery delta area of Tamil Nadu receives a major share of its rainfall during the northeast monsoon season. Since this river basin receives rainfall from both the monsoons, rice crop is cultivated in both kharif (southwest monsoon)

and Rabi (northeast Monsoon) seasons. The river supports irrigation in about 5.8 million ha of cultivable area spread over the Tamil Nadu and Karnataka states.

Rice is cultivated in more than 3.1 million ha that contributes to 40% of the food grain production of Tamil Nadu. Due to the region's fertility and contribution of food grains to the state, Cauvery delta has been known as the "Granary of South India." As rice is grown mainly under flooded conditions and consumes about 80% of all the irrigation water diverted to agriculture in Tamil Nadu, any change in the climate that leads to reduction in water availability impacts rice production and productivity to a greater extent. Climate change induced variation in precipitation pattern (onset and withdrawal, amount, intensity, and distribution) and frequent extreme weather events (flood and drought) in Cauvery basin are major threats to rice production and small holders' livelihoods who constitute a majority of the farming community in the area. Hence, researchers are looking at the adaptation strategies that could help the farmers to sustain the rice production and their income in the context of changing climate and increasing water scarcity.

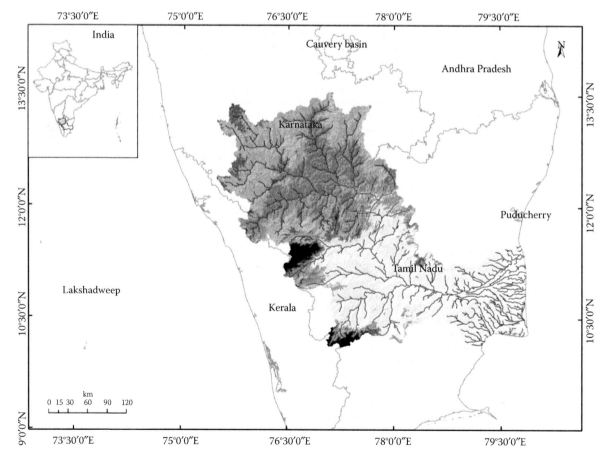

FIGURE 32.2
Location of Cauvery river basin in India.

32.5.1 Climate Analysis

32.5.1.1 Current Climate Variability

Climate variability refers to the climatic parameter of a region varying from its long-term mean. Every year in a specific time period, the climate of a location is different. Some years have below average rainfall, whereas others have average or above average rainfall. Annamalai et al. (2013) analyzed the historical rainfall data of South Asia (Figure 32.3).

Annamalai et al. (2013) reported decreasing rainfall tendency in both southwest and northeast monsoon seasons in most parts of central and northern India. In contrast, the peninsular parts of India, particularly over the region 9–16°N encompassing the rice growing areas showed an increasing rainfall tendency. This increase was particularly strong during the northeast monsoon season.

To analyse the current climate variability of the CRB, 60 years (1950–2010) of rainfall and temperature data were analyzed (Udaya Sekhar Nagothu, 2010). Observation over the year indicates that the temperature was increasing over the period, especially from the 1980s, most of the years showed positive anomaly. Rainfall in recent years showed a decreasing trend and the maximum decrease was observed during 2002–2003 (Figure 32.4).

32.5.1.2 Future Climate Projections

Climate models at different spatial scales and levels of complexity provide the major source of information for constructing scenarios. Global climate models (GCMs) produce information at the global scale. Future climate scenarios derived from 16 GCMs output downscaled using bias-correction/spatial downscaling method (Wood et al., 2004) to a 0.5-degree grid are available for the entire globe (www.climatewizard.org). To reduce the uncertainty in future climate projections, an ensemble of 16 different GCM outputs for A1B scenario (medium emission scenario) with 60% probability was extracted for the CRB for mid- (2040–2069) and end-(2070–2099) century time scale.

32.5.1.2.1 Rainfall

The spatial distribution of annual rainfall as well as rainfall expected in different seasons (premonsoon, monsoon, and postmonsoon) during baseline, mid-, and end-century is depicted in Figure 32.5.

Annual rainfall is expected to be more in the mid-century compared to the baseline (Figure 32.5). The rate of increase is expected to be higher in the upper Cauvery region, that is, Karnataka part of Cauvery, where the rainfall is expected to be 21% higher than the baseline, while in the middle Cauvery and delta region, there is a possibility for 11% and 7% increase than the baseline, respectively. The end-century shows significant increase in precipitation compared to mid-century and baseline. In the upper Cauvery region, the precipitation would be higher by 33% than the current quantity. In the middle Cauvery region and in the delta region, the annual rainfall is expected to increase by 15% and 10%, respectively.

In general, during the summer (premonsoon) season, not much rainfall is received in the Cauvery basin and the quantity of rainfall ranges from 90 to 150 mm. The model predictions indicated no definite changes in premonsoon rainfall during mid- and end-century

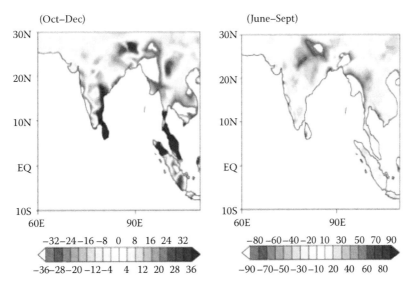

FIGURE 32.3
Observed changes in long-term rainfall (1950–2010).

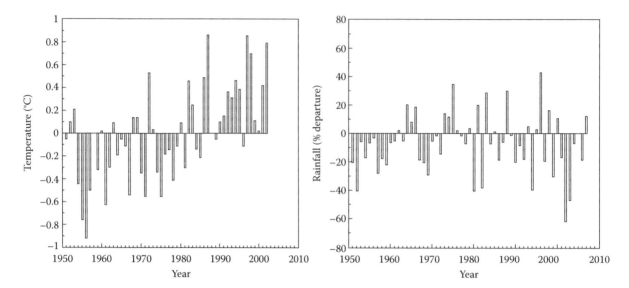

FIGURE 32.4
Observed temperature and rainfall anomaly over Cauvery river basin.

(Figure 32.5). During the monsoon season (June–September), under current climate, most of the regions in Cauvery basin receive about 350–500 mm of rainfall. In the mid-century, the southwest monsoon rainfall quantity in the upper Cauvery would increase by 29% than baseline, while in middle Cauvery and in delta region the increase would be 11% and 7%, respectively. Rainfall would further increase toward the end-century and in upper, middle, and delta Cauvery region, rainfall is expected to increase by 41%, 16%, and 13%, respectively.

In contrast to the SWM rainfall, which is expected to become stronger in the upper Cauvery basin and weaker over the lower Cauvery basins, the postmonsoon season is predicted to become stronger over the middle and the lower Cauvery basins during mid- and end-century (Figure 32.5). In the mid-century, the postmonsoon rainfall in the upper Cauvery is not expected to change and would remain as in baseline scenario whereas across all the other basins of Cauvery the postmonsoon rainfall is predicted to increase up to 8% than the baseline. In the end-century, the upper Cauvery basin is expected to receive 5% more rainfall, while the other parts of Cauvery basin would get 10%–14% more during the postmonsoon period compared to baseline.

32.5.1.2.2 Temperature

The spatial distribution of expected maximum temperature during baseline, mid-, and end-century is depicted in Figure 32.6.

Annual maximum temperature is expected to increase compared to baseline. In mid-century, annual maximum temperature is expected to increase in the magnitude of 1–1.9°C in the Cauvery basin. The increase in expected temperature ranges from 2.5 to 3.7°C during end-century over the Cauvery basin.

The spatial distribution of expected minimum temperature during baseline, mid-, and end-century is depicted in Figure 32.7.

It is observed that the baseline temperature is 17°C, 22.5°C, and 25°C in upper, mid-Cauvery, and delta Cauvery regions, respectively. Annual temperature is expected to be more in the mid-century compared to the baseline. The rate of increase is expected to be higher in the Cauvery delta region where the temperature is expected to increase by 1.8°C than the baseline, while in the middle and upper Cauvery, the increase would be 1.3°C and 1.1°C, respectively. The end-century projections show a significant increase in temperature compared to mid-century and baseline. In end-century, the increase is expected to be 3.3°C.

In general, during the summer season, temperature is higher than SWM and postmonsoon season in the Cauvery basin. Model predictions clearly indicate that in all the seasons, minimum temperature is expected to increase with the advancement in time. In premonsoon, monsoon, and postmonsoon seasons, the temperature increase would be 2.1°C, 1.7°C, and 1.8°C in the mid-century and 3.3°C, 2.5°C, and 2.3°C during end-century, respectively, in the Cauvery basin.

32.5.2 Impact of Climate Change on Hydrology of Cauvery River Basin

For assessing the hydrology of the Cauvery basin in the current and future climate scenarios, the SWAT

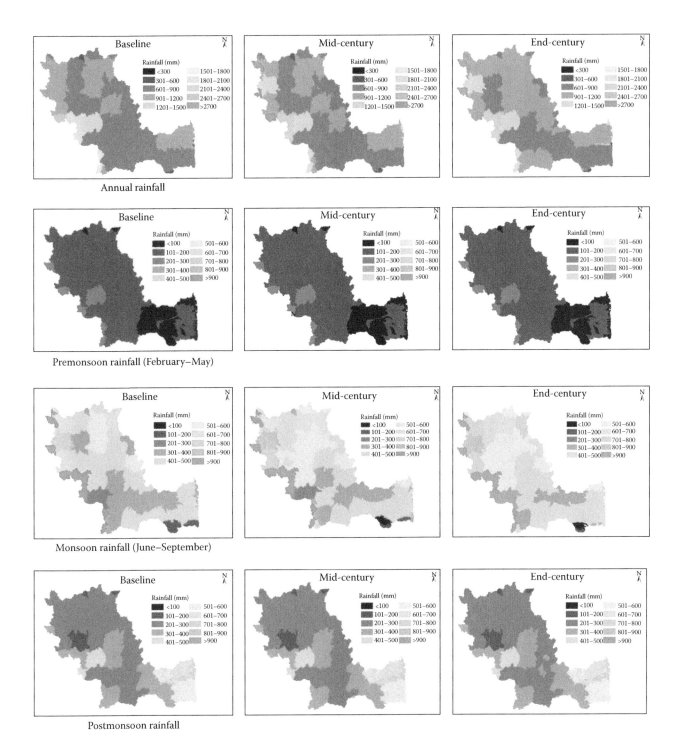

FIGURE 32.5
Spatial distribution of rainfall in Cauvery river basin.

model was employed. The daily-observed gridded data of precipitation at 0.50×0.50 resolutions and maximum and minimum temperature at $1° \times 1°$ resolutions obtained from the India Meteorological Department (IMD) was used for deriving the baseline (1971–2005) climate. Weather data on solar radiation, wind speed, and relative humidity were generated using long-term statistics through the weather generator built in the SWAT model. The future climate scenarios for A1B scenario with respect to two timelines, namely, mid-(2040–2069) and end-century (2070–2099) were used in the model as input.

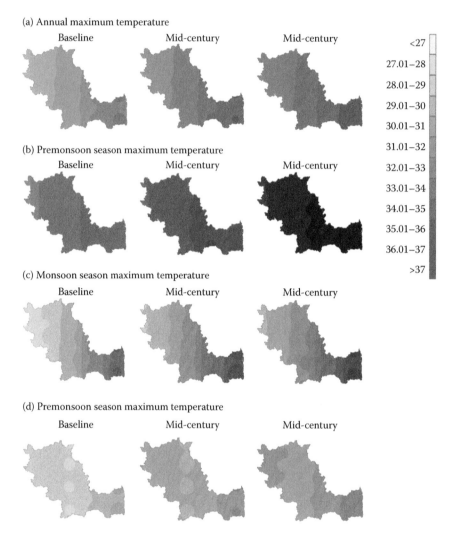

FIGURE 32.6
Spatial distribution of maximum temperature in Cauvery river basin.

32.5.2.1 Hydrological Parameters

Average annual water yield, PET, and soil water storage for baseline, mid-century, and end-century time scale are presented in Figure 32.8.

As per model prediction, the annual water yield (sum of surface flow, lateral flow, and ground water flow contribution to stream flow) is expected to increase by 21%, 15%, and 14% in the upper Cauvery, middle Cauvery, and delta regions, respectively, in the mid-century compared to current levels. In the end-century, it is expected to increase by 27%, 20%, and 21%, respectively, for upper Cauvery, middle Cauvery, and delta region compared to current conditions (Figure 32.8). Increase in precipitation would lead to increased water yield as a result of increase in the surface flow, lateral flow, and ground water flow.

In case of annual PET, the increase is expected to be higher during end-century, whereas a slight increase is expected in the mid-century (Figure 32.8). In the upper Cauvery, the annual PET is expected to increase by 4.3% and 8.4% in the mid- and end-century, respectively, compared to baseline. The predicted change ranges from 3% to 3.7% during mid-century and from 5% to 8.7% in the end-century over the mid Cauvery. The increase is expected at the delta region by 4.5% and 9.3% in mid- and end-century, respectively. The magnitude of changes in PET in the mid-century is lesser than the end-century, which is obviously due to an increase in temperature. However, PET is not only affected by the temperature but also by the CO_2 concentration and it is a known fact that any increase in CO_2 would tend to increase the water use efficiency due to decrease in stomatal (Morison, 1987; Rosenberg et al., 1988) conductance of the leaves, which would reduce the rate of transpiration. During the mid-century, the temperatures are expected to increase by around 1.8°C and the impact of this might get nullified due to the simultaneous increase in CO_2 concentration in the atmosphere.

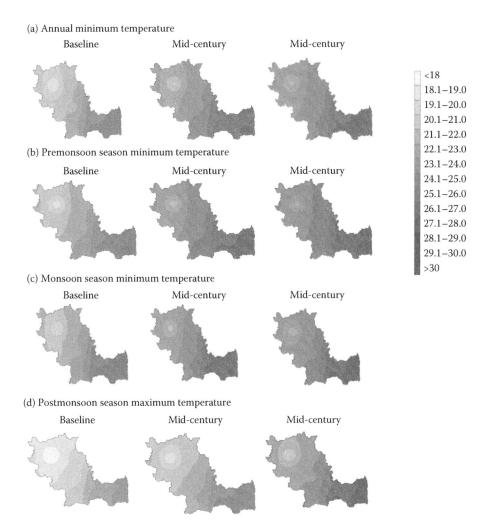

FIGURE 32.7
Spatial distribution of minimum temperature in Cauvery river basin.

The annual soil water storage is also predicted to increase in mid- and end-century (Figure 32.8). In the mid-century, the annual soil water storage would increase by 14%, 7%, and 5% in the upper Cauvery, mid Cauvery, and delta region, respectively. In end-century, it is expected to increase by 18%, 10%, and 7% in upper Cauvery, mid Cauvery, and delta region, respectively.

32.5.2.2 Water Availability at Major Reservoirs of Cauvery Basin

The spatial and temporal distribution of water availability at some of the major reservoirs, namely, Krishnarajasagar, Mettur, Bhavanisagar, Kodivery, Kalingrayan, Amaravathi, Upper anicut, Grand anicut (barrage), and Lower anicut (barrage) (Figure 32.9) distributed across the subdivisions of Cauvery was studied for baseline, mid-, and end-century.

The water availability was assessed in terms of average flow duration curves (Figure 32.10).

The flow duration curve indicates that more inflows could be expected with mid- and end-century climate scenarios when compared to the baseline. The increase in the daily flows during end-century is higher than the mid-century (Figure 32.10). The flow duration curve also indicates more high flows in the future climate at all reservoir locations. A slight decrease in low flows is observed at all the reservoirs with the exception of Krishnarajasagar, Bhavani, and Kodivery under mid-century scenario and the low flows would further decrease under end-century scenarios at all the reservoirs excluding Krishnarajasagar where it is more than mid-century and baseline. The high flows will be higher in the end-century followed by the mid-century than the baseline. Hence, the inflows in Cauvery basin are expected to be intensified in mid-century, which would increase further by the end-century.

The dependability of flows for major water resources project is usually assessed at 75% dependability level. Hence, the water flows at major reservoirs in future climatic conditions were analyzed at 75% dependability. In

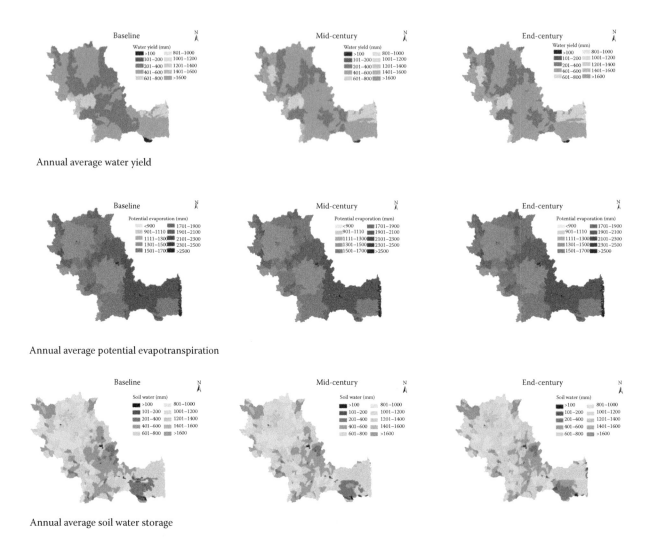

FIGURE 32.8
Spatial and temporal variation of hydrological parameters in Cauvery river basin.

addition to that, the flow was also compared under baseline, mid-, and end-century at 25% and 50% dependability levels. With the stream flow for baseline, mid-century, and end-century scenarios at different dependability levels (25%, 50%, and 75%) for major reservoir sites of Cauvery river basin (Table 32.1), it could be inferred that the flow for all dependability levels at all the nine major reservoir locations would increase in the future climate scenario.

32.5.2.3 Irrigation Water Requirement over Cauvery Delta Region

For assessing the water demand, it is necessary to have accurate information on the spatial extent of paddy cultivation, the season during which it is grown at a particular place and the length of the growing season, the cropping sequence/rotation, the type of irrigation system, and the irrigation source (surface or ground water) as inputs. Hence, impact of climate change on irrigation

water requirement was done only for the Cauvery delta region. For this purpose, district maps were superimposed on the Cauvery basin and identified the districts falling under each subbasin. Area under each crop during Kharif and Rabi seasons at district level were calculated from the statistical data available with the Department of Economics and Statistics, Tamil Nadu (www.tn.gov.in/crop/districttables.htm). The irrigation control module of the SWAT model allows irrigating the plant under manual irrigation system besides auto irrigation systems. SWAT model was executed by keeping all the input parameters constant except climate variables, which were changed according to mid- and end-century period of simulation. The level of CO_2 maintained in the model was 350 ppm for baseline, 550 ppm for mid-century, and 650 ppm for end-century. The effect of climate change on the potential evapotranspiration and irrigation water requirement for rice was assessed in the Cauvery delta region.

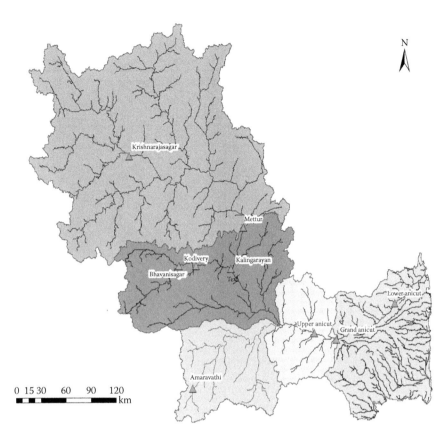

FIGURE 32.9
Major reservoirs of Cauvery river basin.

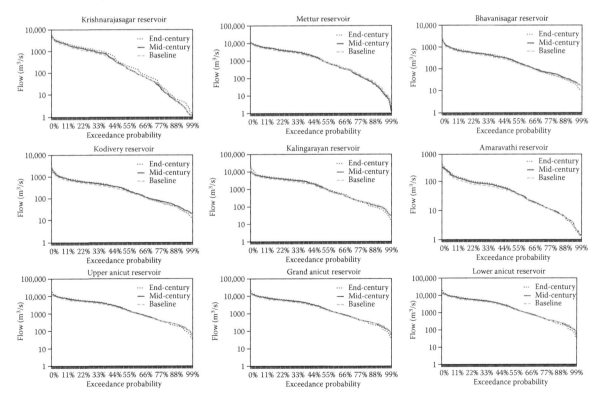

FIGURE 32.10
Flow duration curve under baseline, mid-, and end-century for major reservoirs sites.

TABLE 32.1

Comparison of Daily Natural Dependable Flows (m³/s) at 25%, 50%, and 75% Level Under Baseline, Mid-, and End-Century over Major Reservoir Locations of the Cauvery Basin

Reservoirs	25% Dependability Flow (m³/s)			50% Dependability Flow (m³/s)			75% Dependability Flow (m³/s)		
	Baseline	Mid-Century	End-Century	Baseline	Mid-Century	End-Century	Baseline	Mid-Century	End-Century
Krishnarajasagar	1108	1303	1504	246	258	303	39	38	51
Mettur	3077	3498	3776	1059	1062	1276	197	194	198
Bhavanisagar	472	534	566	228	274	275	62	72	72
Kodivery	527	587	620	258	324	326	75	88	89
Kalingarayan	3053	3601	3859	1146	1392	1523	227	235	237
Amaravathy	85	99	106	48	54	55	16	17	17
Upper anicut	4805	5518	5776	1879	2209	2281	458	476	479
Grand anicut	4863	5651	5909	1928	2297	2332	474	482	510
Lower anicut	4880	5656	5915	1940	2306	2344	476	490	516

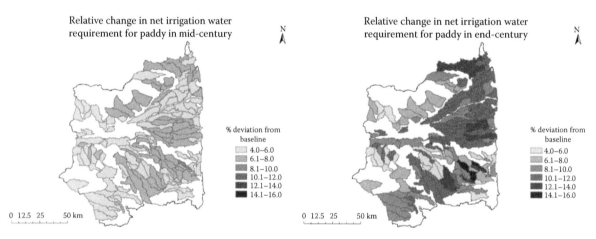

FIGURE 32.11
Changes (%) in irrigation water requirement for paddy during mid- and end-century compared to the baseline.

The potential irrigation water requirement for paddy was simulated in SWAT by assuming that an unlimited supply of water is available to meet the crop consumptive use requirements. The assumption made in this simulation run is that the extent of irrigated area remains the same in current and future climate. In the mid-century, the irrigation water requirement is projected to increase from 4% to 8.1% and at the end-century by 4.5% to 14.7% at various locations in the Cauvery delta region (Figure 32.11). The increasing trend in potential evapotranspiration is the main cause for the increase in water demand.

32.6 Impact of Climate Change on Rice Productivity

In the current study, to understand the impact of climate change on rice crop productivity of the Cauvery delta region, the SWAT model was applied and the EPIC model that is built in with SWAT has predicted the changes in crops productivity. Additional inputs used for simulating rice yield include management practices, namely, details on tillage operations, time and method of sowing; quantity, method, and time of fertilizer/irrigation application; and harvesting details adopted at different districts of the Cauvery Basin that were collected from the Department of Agriculture and from Crop Production Guide of Tamil Nadu state.

SWAT model simulated rice yields for baseline, mid-, and end-century time scales are presented in Figure 32.12.

The results of the study indicated that future climate impacted negatively the rice yield over the Cauvery delta region. SWAT model predicted reduction in rice yield for mid- (2040–2069) and end-(2070–2099) century from the current (1971–2005) yield level of 4163 kg ha⁻¹. However, the reduction is expected to be higher toward end-century compared to mid-century. The reduction ranged between 0.1% and 8.9% during mid-century and 10.0%

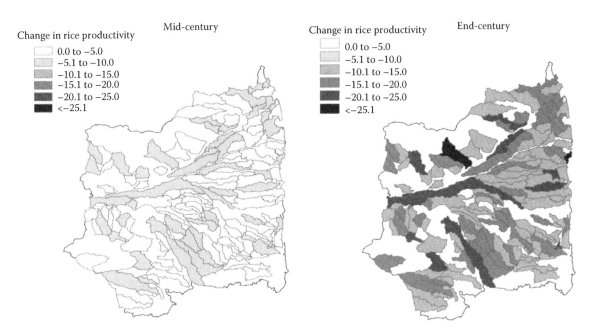

FIGURE 32.12
Impact of climate change on rice productivity over Cauvery river delta region.

and 25.3% in end-century (Figure 32.12). This reduction in yield with the advancement of time might be due to temperature increments over the Cauvery Delta region as predicted by 16 ensemble climate models.

Cauvery delta region, the irrigation water requirement of rice is expected to shoot up under future warmer climatic conditions at the rate of 4%–8.1% and 4.5%–14.7% in the mid- and end-century, respectively. Rice productivity of the Cauvery delta region is also expected to be negatively affected by climate change, which would have marked implications on regional food security.

32.7 Summary of the Results in Cauvery River Basin

From the case study, it is well demonstrated that climate model projections could be integrated into impact models (hydrological models and crop simulation models) for assessing the impact of changing climate. In Cauvery basin, temperature is expected to increase in mid- and end-century time period. Annual rainfall, water yield, potential evapotranspiration (PET), and soil water storage are expected to increase with the advancement of time. Annual rainfall, water yield, and soil water storage would increase at a higher rate in the upper Cauvery region compared to the mid-Cauvery and Delta region. However, the expected increase in PET is higher at the Delta region than at the upper Cauvery. The inflow at major reservoir locations (Krishnarajasagar, Mettur, Bhavanisagar, Kodivery, Kalingrayan, Amaravathi, Upper anicut, Grand anicut, and Lower anicut) of the Cauvery basin is expected to be higher under mid- and end-century time slices and the prediction for 75% dependable flow, which is commonly used for planning purposes, is also expected to be high with the maximum increase of 30.8% in the end-century. In the

32.8 Future Directions

Among the urgent research needs are those that may lead to reducing uncertainty, both to better understand how climate change might affect freshwater and to assist water managers who need to adapt to climate change. However, this is an old plea, and easier said than done. After a call to reduce uncertainty was issued in the IPCC First Assessment Report in 1990, major funds have been spent worldwide on reducing uncertainties in understanding, observations, and projections of climate change, its impacts and vulnerabilities. Meanwhile, uncertainties in projections of future changes remain high, even if characterization of uncertainty has been improved recently (IPCC, 2007a).

Precipitation, the principal input signal to freshwater systems, is not adequately simulated in present climate models. Consequently, quantitative projections of changes in river flow at the basin scale, relevant to water management, remain largely uncertain (Milly et al., 2005; Nohara et al., 2006). In high latitudes and parts of the tropics, climate models are consistent in projecting future precipitation increase, while in some subtropical and lower mid-latitude regions, they are consistent in projecting precipitation decrease. Between these areas of robust increase and decrease in model projections, there are areas with high uncertainty, where the current generation of climate models do not agree on the sign of precipitation and runoff changes (Milly et al., 2005; Nohara et al., 2006; IPCC, 2007a,b). Hence, impact assessments based on only one or a few model scenarios may yield contrasting river flow projections, so that a new framework for handling uncertainty is needed to support the process of decision making. Wilby and Harris (2006) show how components of uncertainty can be weighted, leading to conditional probabilities for future impact assessments.

This suggests that a useful parallel pathway to identification of research needs would be to focus on providing a better basis for decisions that must invariably be made under high uncertainty. For example, improved characterization of uncertainty (joint analysis of ensembles of climate models, cf. IPCC, 2007a) could help water managers in their efforts to adapt to uncertain future hydrological changes. In addition, incorporating that type of climate change information in a risk management approach to water resource planning would be useful.

It is necessary to evaluate social and economic costs and benefits (in the sense of avoided damage) of adaptation, at several time scales.

Estimation, in quantitative terms, of future climate change impacts on freshwater resources and their management should be improved. Progress in understanding is conditioned by adequate availability of observation data, which calls for enhancement of monitoring endeavours worldwide, addressing the challenges posed by projected climate change to freshwater resources and reversing the tendency of shrinking observation networks. The lack of information is notorious, and critical, in developing countries. A recent example of data-related difficulties is the continental runoff study by Gedney et al. (2006) and related discussion (Peel and McMahon, 2006) challenging the representativeness of the data set and the practice of runoff reconstruction. Adequate data are crucial to understanding observed changes and to improve models, which can be used for future projections.

If only short hydrometric records are available, the full extent of natural variability can be understated and detection studies confounded. Data on water use, water quality, groundwater, sediment transport, and water-related systems (e.g., aquatic ecosystems) are even less available.

Climate change impacts on groundwater, water quality, and aquatic ecosystems (not only via temperature,

but also altered flow regimes, water level, and ice cover) are not adequately understood.

On the modeling side, climate change modeling and impact modeling have to be better integrated and this requires solving a range of difficult problems related to scale mismatch and uncertainty. The effects of CO_2 enrichment on evapotranspiration and, hence, runoff and recharge, remain uncertain, with different representations producing different estimates of impact.

Acknowledgment

The authors thank the Ministry of Foreign Affairs, Norway and The Royal Norwegian Embassy, New Delhi for the financial support to carry out this research work through the Project ClimaAdapt. The authors also thank the Agricultural Model Intercomparison and Improvement Project (www.agmip.org) for the use of AgMIP Protocols for Regional Integrated Assessments and support.

References

Agarwal, P. K. and S. K. Sinha. 1993. Effect of probable increase in carbon dioxide and temperature on productivity of wheat in India. *Journal of Agricultural Meteorology*, 48:811–814.

Annamalai, H., J. Hafner, K.P. Sooraj, and P. Pillai. 2013. Global warming shifts the monsoon circulation, drying South Asia. *Journal of Climate*, 26(9):2701–2718.

Arnell, N. W. 1992. Factors controlling the effects of climate change on river flow regimes in a humid temperate environment. *Journal of Hydrology*, 132:321–342.

Arnold, J. G., P. M. Allen, and G. Bernhardt. 1993. A comprehensive surface-groundwater flow model. *Journal of Hydrology*, 142:47–69.

Arnold, J. G., R. Srinivasan, R. S. Muttiah, and J. R. Williams. 1998. Large area hydrologic modeling and assessment. Part I. Model Development. *Journal of American Water Resources Association*, 34(1):73–89.

Beck, M. B. 1987. Water quality modeling: A review of the analysis of uncertainty. *Water Resources Management*, 23:1393–1442.

Challinor, A. J., F. Ewert, S. Arnold, E. Simelton, and E. Fraser. 2009. Crops and climate change: Progress, trends, and challenges in simulating impacts and informing adaptation. *Journal of Experimental Botany*, 60:2775–2789.

Challinor, A. J., P. Thornton, and M. S. Smith. 2013. Use of agro-climate ensembles for quantifying uncertainty and informing adaptation. *Agricultural and Forest Meteorology*, 170:2–7.

Challinor, A. J. and T. R. Wheeler. 2008. Crop yield reduction in the tropics under climate change: Processes and uncertainties. *Agricultural and Forest Meteorology*, 148:343–356.

Challinor, A. J., T. R. Wheeler, P. Q. Craufurd, C. A. T. Ferro, and D. B. Stephenson. 2008. Adaptation of crops to climate change through genotypic responses to mean and extreme temperatures. *Agriculture, Ecosystems & Environment*, 119(1–2):190–204.

Christensen, J., M. Hulme, H. Von Storch, P. Whetton, R. Jones, L. Mearns, and C. Fu. 2001. Regional climate information—Evaluation and projections. In: *Climate Change 2001: The Scientific Basis. Contribution of Working Group I to the Third Assessment Report of the Intergovernmental Panel on Climate Change*. J. T. Houghton et al., Eds. Cambridge University Press, 583–638.

Cocke, S. D. and T. E. LaRow. 2000. Seasonal prediction using a regional spectral model embedded within a coupled ocean-atmosphere model. *Monthly Weather Review*, 128:689–708.

Collins, M. 2007. Ensembles and probabilities: A new era in the prediction of climate change. *Philosophical Transaction of the Royal Society*, 365:1957–1970.

Fleming, G. 1975. Computer Simulation Techniques in Hydrology. Elsevier, New York. Singh, V.P. (Ed) 1995. *Computer Models of Watershed Hydrology*. Water Resources Publications, Highlands Ranch, CO.

Gedney, N., P. M. Cox, R. A. Betts, O. Boucher, C. Huntingford, and P. A. Stott. 2006. Detection of a direct carbon dioxide effect in continental river runoff records. *Nature*, 439:835–838.

Gleick, P. H. 1986. Methods for evaluating the regional hydrologic impacts of global climatic changes. *Journal of Hydrology*, 88:97–116.

Gosain, A. K., S. Rao, and D. Basuray. 2006. Climate change impact assessment on hydrology of Indian River basins. *Current Science*, 90(3):346–353.

Gosain, A. K., S. Rao, R. Srinivasan, and N. Gopal Reddy. 2005. Return-flow assessment for irrigation command in the Palleru river basin using SWAT model. *Hydrological Processes*, 19:673–682.

Harremoes, P. and R. K. Turner. 2001. Methods for integrated assessment. *Regional Environment Change*, 2:57–65.

Hawkins, E. and R. Sutton. 2009. The potential to narrow uncertainty in regional climate predictions. *Bulletin of the American Meteorological Society*, 90:1095–1107.

Hay, L. E., M. P. Clark, R. L. Wilby et al. 2002. Use of regional climate model output for hydrologic simulations. *Journal of Hydrometeorology*, 3(5):571–590.

Hejberg, A. L. and J. C. Refsgaard. 2005. Model uncertainty parameteruncertainty versus conceptual models. *Water Science Technology*, 52:177–186.

Hundal, S. S. and P. Kaur. 1996. Climate change and its impact on crop productivity in Punjab, India. In: *Climate Variability and Agriculture*. Y. P. Abrol, Ed. Narosa Publishing House, Northeast Delhi, India, 377–393.

IPCC. 2001. The Scientific Basis. In: *IPCC Third Assessment Report: Climate Change*. Cambridge University Press, 881.

IPCC. 2007a. Summary for policymakers. In: *Climate Change 2007: The Physical Science Basis. Contribution of Working Group I to the Fourth Assessment Report of the Intergovernmental Panel on Climate Change*. S. Solomon, D. Qin, M. Manning, Z. Chen, M. Marquis, K. B. Averyt, M. Tignor, and H. L. Miller, Eds. Cambridge University Press, UK, 1–18. http://www.ipcc.ch/pdf/assessment report/ar4/wg1/ar4-wg1-spm.pdf.

IPCC. 2007b. Summary for policymakers. In: *Climate Change 2007: Impacts, Adaptation and Vulnerability. Contribution of Working Group II to the Fourth Assessment Report of the Intergovernmental Panel on Climate Change*. M. L. Parry, O. F. Canziani, J. P. Palutikof, P. J. van der Linden, and C. E. Hanson. Cambridge University Press, UK, 7–22. http://www.ipcc.ch/pdf/assessment-report/ar4/wg1/ar4-wg1-spm.pdf.

Jones, R. G., M. Noguer, D. C. Hassell et al. 2004. *Generating high resolution climate change scenarios using PRECIS*. Met Office Hadley Centre, Exeter, UK, 40.

Jones, R. N. 2000. Managing uncertainty in climate change projections: Issues for impact assessment. *Climatic Change*, 45:403–419.

Katz, R. W., P. F. Craigmile, P. Guttorp, M. Haran, B. Sansó, and M. L. Stein. 2013. Uncertainty analysis in climate change assessments. *National Climate Change*, 3:769–771.

Kundzewicz, Z. W., L. J. Mata, N. Arnell et al. 2007. Freshwater resources and their management. In: *Climate Change 2007: Impacts, Adaptation and Vulnerability. Contribution of Working Group II to the Fourth Assessment Report of the Intergovernmental Panel on Climate Change*. M. L. Parry, O. F. Canziani, J. P. Palutikof, P. J. van der Linden, and C. E. Hanson, Eds. Cambridge University Press, UK, 173–210. http://www.ipcc.ch/pdf/assessment-report/ar4/wg2/ar4-wg2-chapter3.pdf.

Lakshmanan, A., V. Geethalakshmi, R. Srinivasan, N. U. Sekhar, and H. Annamalai. 2011. Climate change adaptation in Bhavani Basin using SWAT model. *Applied Engineering in Agriculture*, 27(6):887–893.

Lal, M., P. H. Whettori, A. B. Pittodi, and B. Chakraborty. 1998. The greenhouse gas induced climate change over the Indian Sub-continent as projected by GCM model experiments. *Terrestrial, Atmospheric and Oceanic Sciences, TAO*, 9(4):663–669.

Lansigan, F. P. and A. R. Salvacion. 2007. Assessing the Effect of Climate Change on Rice and Corn Yields in Selected Provinces in the Philippines. A paper presented in 10th National Convention on Statistics (NCS). Manila, Philippines, October 1–2, 2007.

Lindenschmidt, K. E., K. Fleischbein, and M. Baborowski. 2007. Structural uncertainty in a river water quality modelling system. *Ecological Modelling*, 204:289–300.

Milly, P. C. D., K. A. Dunne, and A. V. Vecchia. 2005. Global pattern of trends in streamflow and water availability in a changing climate. *Nature*, 438:347–350.

Mimikou, M., Y. Kouvopoulos, G. Cavadias, and N. Vayianos. 1991. Regional hydrological effects of climate change. *Journal of Hydrology*, 123:119–146.

Morison, J. I. L. 1987. Intercellular CO_2 Concentration and Stomatal Response to CO_2. In: *Stomatal Function*. Stanford University Press, Stanford, CA, 242–243.

Murphy, J. M., D. M. H. Sexton, D. N. Barnett et al. 2004. Quantifying uncertainties in climate change from a large ensemble of general circulation model predictions. *Nature*, 430:768–772.

Muzik, I. 2001. Sensitivity of hydrologic systems to climate change. *Canadian Water Resources Journal*, 26(2):233–253.

Ng, H. Y. F. and J. Marsalek. 1992. Sensitivity of streamflow simulation to changes in climatic inputs. *Nordic Hydrology*, 23:257–272.

Nohara, D., A. Kitoh, M. Hosaka, and T. Oki. 2006. Impact of climate change on river runoff. *Journal of Hydrometeorology*, 7:1076–1089.

Osborne, T., G. Rose, and T. Wheeler. 2013. Variation in the global-scale impacts of climate change on crop productivity due to climate model uncertainty and adaptation. *Agricultural and Forest Meteorology*, 170:183–194.

Paz, J. O., C. W. Fraisse, L. U. Hatch et al. 2007. Development of an ENSO-based irrigation decision support tool for peanut production in the southeastern US. *Computers and Electronics in Agriculture*, 55:28–35.

Peel, M. C. and T. A. McMahon. 2006. A quality-controlled global runoff data set. *Nature*, 444:E14. doi:10.1038/nature05480

Rao, D. and S. K. Sinha. 1994. Impact of climate change on simulated wheat production in India. In: *Implications of Climate Change for International Agriculture: Crop Modelling Study*. Rosenzweig, C. and A. Iglesias, Eds. US Environmental Protection Agency, Washington, DC. EPA 230-B-94-003.

Rosenberg, M. J., B. A. Kimball, P. Martin, and C. Cooper. 1988. Climate change, CO_2 enrichment and evapotranspiration. In: *Climate and Water: Climate Change, Climatic Variability, and the Planning and Management of U.S. Water Resources*. P. E. Waggoner, Ed. John Wiley & Sons, New York, 131–147.

Rummukainen, M., S. Bergstrom, G. Persson, J. Rodhe, and M. Tjernstrom. 2004. The Swedish regional climate modeling programme, SWECLIM: A review. *Ambio*, 33:176–182.

Rummukainen, M., J. Raisanen, B. Bringfelt et al. 2001. A regional climate model for northern Europe: Model description and results from the downscaling of two GCM control simulations. *Climate Dynamics*, 17:339–359.

Samuelsson, P., B. Bringfelt, and L. P. Graham. 2003. The role of aerodynamic roughness for runoff and snow evaporation in land-surface schemes—Comparison of uncoupled and coupled simulations. *Global and Planetary Change*, 38:93–99.

Saseendran, S. A., K. Singh, L. S. K. Rathore et al. 1998. Evaluation of the CERES-Rice v 3 for the climatic conditions of the Kerala State, India. *Journal of Applied Meteorology and Climatology*, 5:385–392.

Saseendran, S. A., K. K. Singh, L. S. Rathore, S. V. Singh, and S. K. Sinha. 2000. Effects of climate change on rice production in the tropical humid climate of Kerala, India. *Climatic Change*, 44(4):459–514.

Schaake, J. C. and C. Liu. 1989. Development and applications of simple water balance models to understand the relationship between climate and water resources, New Directions for Surface Water Modelling. *Proceedings of the Baltimore Symposium*, May 1989, IAHS Publ. 181:343–352.

Schulze, R. E. 1997. Impacts of global climate change in a hydrologically vulnerable region: Challenges to South African hydrologists. *Progress in Physical Geography*, 21:113–136.

Semenov, M. A. and F. J. Doblas-Reyes. 2007. Utility of dynamical seasonal forecasts in predicting crop yield. *Climate Research*, 34:71–81.

Singh, V. P. and Valiron, F. 1995. *Computer Models of Watershed Hydrology*. Water Resources Publications, Highlands Ranch, Colorado, USA.

Srikanthan, R. and T. A. McMahon. 2001. Stochastic generation of annual, monthly and daily climate data, a review. *Hydrology and Earth Systems Sciences*, 5(4):653–670.

Srinivasan, R., T. S. Ramanarayanan, J. G. Arnold, and S. T. Bednarz. 1998. Large area hydrologic modeling and assessment part II: Model application. *Journal of the American Water Resources Association*, 34:91–102.

Stainforth, D. A., T. Aina, C. Christensen et al. 2005. Uncertainty in predictions of the climate response to rising levels of greenhouse gases. *Nature*, 433:403–406.

Sudheer, K. P., G. Lakshmi, and I. Chaubey. 2011. Application of a pseudo simulator to evaluate the sensitivity of parameters in complex watershed models. *Environmental Modelling & Software*, 26:135–143.

Tebaldi, C. and R. Knutti. 2007. The use of the multi-model ensemble in probabilistic climate projections. *Philosophical Transactions of the Royal Society*, 365:2053–2075.

Udaya Sekhar Nagothu, H. Annamalai, and V. Geetha Lakshmi. 2010. Summary, policy recommendations, future course. p. 11 and 51. In: *Sustainable Rice Production in a Warmer Planet: Linking Science, Stakeholder and Policy*. Macmillan Publishers India Ltd., New Delhi, India, 1–224.

Van, G. A., T. Meixner, R. Srinivasan, and S. Grunwals. 2008. Fit for purpose analysis of uncertainty using split-sampling evaluations. *Hydrological Sciences Journal*, 53:1090–1103.

Varis, O., T. Kajander, and R. Lemmela. 2004. Climate and water: From climate models to water resources management and vice versa. *Climatic Change*, 66:321–344.

Von Storch, H., H. Langenberg, and F. Feser. 2000. A spectral nudging technique for dynamical downscaling purposes. *Monthly Weather Review*, 128:3664–3673.

Whitfield, H. P. and Cannon, J. A. 2000. Recent variation in climate and hydrology in Canada. *Canadian Water Resources Journal*, 25(1):19–65.

Wilby, R. L. 2005. Uncertainty in water resource model parameters used for climate change impact assessment. *Hydrological Processes*, 19:3201–3219.

Wilby, R. L. and Harris, I. 2006. A framework for assessing uncertainties in climate change impacts: Low-flow scenarios for the River Thames, UK. *Water Resources Research*, 42: W02419. doi:10.1029/2005WR004065.

Wood, A. W., L. R. Leung, V. Sridhar, and D. P. Lettenmaier. 2004. Hydrologic implications of dynamical and statistical approaches to downscaling climate model outputs. *Climatic Change*, 62:189–216.

Xu, C. Y. 1999. From GCM's to river flow: A review of downscaling methods and hydrologic modelling approaches. *Progress in Physical Geography*, 23(2):229–249. 1–224

Xu, C. Y. and S. Halldin. 1997. The effect of climate change on river flow and snow cover in the NOPEX area simulated by a simple water balance model. *Nordic Hydrology,* 28:273–282.

Xu, C. Y. and V. P. Singh. 1998. A review on monthly water balance models for water resources investigations. *Water Resources Management,* 12:31–50.

Xu, C.-Y. and V. P. Singh. 2004. Review on regional water resources assessment models under stationary and changing climate. *Water Resources Management,* 18:591–612.

Yakob Mohammed. 2011. Climate change impact assessment on soil water availability and crop yield in Anjeni watershed Blue Nile basin. PhD Thesis.

Index

Printed and bound by CPI Group (UK) Ltd, Croydon, CR0 4YY

24/10/2024

01778285-0015